Honoring America

For Americans, the flag has always had a special meaning. It is a symbol of our nation's freedom and democracy.

Flag Etiquette

Over the years, Americans have developed rules and customs concerning the use and display of the flag. One of the most important things every American should remember is to treat the flag with respect.

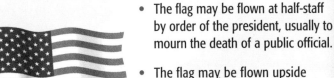

- The flag should be raised and lowered by hand and displayed only from sunrise to sunset. On special occasions, the flag may be displayed at night, but it should be illuminated.

- The flag may be displayed on all days, weather permitting, particularly on national and state holidays and on historic and special occasions.

- No flag may be flown above the American flag or to the right of it at the same height.

- The flag should never touch the ground or floor beneath it.

- The flag may be flown at half-staff by order of the president, usually to mourn the death of a public official.

- The flag may be flown upside down only to signal distress.

- The flag should never be carried flat or horizontally, but always carried aloft and free.

- When the flag becomes old and tattered, it should be destroyed by burning. According to an approved custom, the Union (stars on blue field) is first cut from the flag; then the two pieces, which no longer form a flag, are burned.

★ ★ ★ ★ ★ ★ ★ ★

The American's Creed

I believe in the United States of America as a Government of the people, by the people, for the people, whose just powers are derived from the consent of the governed; a democracy in a republic; a sovereign Nation of many sovereign States; a perfect union, one and inseparable; established upon those principles of freedom, equality, justice, and humanity for which American patriots sacrificed their lives and fortunes.

I therefore believe it is my duty to my Country to love it; to support its Constitution; to obey its laws; to respect its flag, and to defend it against all enemies.

The Pledge of Allegiance

I pledge allegiance to the Flag of the United States of America and to the Republic for which it stands, one Nation under God, indivisible, with liberty and justice for all.

The Star-Spangled Banner

O! say, can you see, by the dawn's early light,
What so proudly we hail'd at the twilight's last gleaming?
Whose broad stripes and bright stars, thro' the perilous fight,
O'er the ramparts we watched were so gallantly streaming?
And the rockets' red glare, the bombs bursting in air,
Gave proof thro' the night, that our flag was still there.
O! say, does that Star-Spangled Banner yet wave
O'er the land of the free and the home of the brave?

On the shore, dimly seen thro' the mist of the deep,
Where the foe's haughty host in dread silence reposes,
What is that which the breeze, o'er the towering steep,
As it fitfully blows, half conceals, half discloses?
Now it catches the gleam of the morning's first beam,
In full glory reflected now shines on the stream.
'Tis the Star-Spangled Banner. O long may it wave
O'er the land of the free and the home of the brave.

And where is that band who so vauntingly swore,
That the havoc of war and the battle's confusion
A home and a country should leave us no more?
Their blood has wash'd out their foul footstep's pollution.
No refuge could save the hireling and slave
From the terror of flight or the gloom of the grave,
And the Star-Spangled Banner in triumph doth wave
O'er the land of the free and the home of the brave.

O thus be it e'er when free men shall stand
Between their lov'd home and war's desolation,
Blest with vict'ry and peace, may the Heav'n-rescued land
Praise the pow'r that hath made and preserv'd us a nation.
Then conquer we must, when our cause it is just,
And this be our motto, "In God is our Trust."
And the Star-Spangled Banner in triumph shall wave
O'er the land of the free and the home of the brave.

Glencoe

OUR WORLD TODAY

People, Places, and Issues

Senior Author
Richard G. Boehm, Ph.D.

David G. Armstrong, Ph.D.

Francis P. Hunkins, Ph.D.

Dennis Reinhartz, Ph.D.

Merry Lobrecht

NATIONAL GEOGRAPHIC

 **Glencoe
McGraw-Hill**

New York, New York   Columbus, Ohio   Chicago, Illinois   Peoria, Illinois   Woodland Hills, California

ABOUT THE AUTHORS

NATIONAL GEOGRAPHIC

The National Geographic Society, founded in 1888 for the increase and diffusion of geographic knowledge, is the world's largest nonprofit scientific and educational organization. The Society uses sophisticated communication technologies to convey geographic knowledge to a worldwide membership. The School Publishing Division supports the Society's mission by developing innovative educational programs—ranging from traditional print materials to multimedia programs including CD-ROMS, videos, and software.

David G. Armstrong

David G. Armstrong, Ph.D., is Dean of the School of Education at the University of North Carolina at Greensboro. A social studies education specialist with additional advanced training in geography, Dr. Armstrong was educated at Stanford University, University of Montana, and University of Washington. He taught at the secondary level in the state of Washington before beginning a career in higher education. Dr. Armstrong has written books for students and teachers.

Merry Lobrecht

Merry Lobrecht is the Social Studies Curriculum Coordinator for the Humble ISD. She was the recipient of both the 2001 National Council for Geographic Education Distinguished Teacher Achievement Award and the Texas Council of Social Studies Texas Alliance for Geographic Distinguished Teacher Award for 2001.

SENIOR AUTHOR
Richard G. Boehm

Richard G. Boehm, Ph.D., was one of seven authors of *Geography for Life,* national standards in geography, prepared under Goals 2000: Educate America Act. In 1991 he received the George J. Miller award from the National Council for Geographic Education (NCGE) for distinguished service to geographic education. He has twice won the *Journal of Geography* award for best article. He presently holds the Jesse H. Jones Distinguished Chair in Geographic Education at Southwest Texas State University in San Marcos, Texas.

Francis P. Hunkins

Francis P. Hunkins, Ph.D., is Professor of Education at the University of Washington. He began his career as a teacher in Massachusetts. He received his master's degree in education from Boston University and his doctorate from Kent State University with a major in general curriculum and a minor in geography. Dr. Hunkins has written numerous books and articles.

Dennis Reinhartz

Dennis Reinhartz, Ph.D., is Professor of History and Russian at the University of Texas at Arlington. A specialist in Russian and East European history, as well as in the history of cartography and historical geography, Dr. Reinhartz has written numerous books in these fields. He is a consultant to the U.S. State and Justice Departments and to the U.S. Holocaust Memorial Museum in Washington, D.C.

Glencoe/McGraw-Hill

A Division of The McGraw·Hill Companies

TIME Reports © Time Inc. Prepared by TIME School Publishing in collaboration with Glencoe/McGraw-Hill.

Printed in the United States of America.
Send all inquiries to:
Glencoe/McGraw-Hill
8787 Orion Place
Columbus, Ohio 43240-4027

ISBN 0-07-827382-X (Student Edition)   ISBN 0-07-827383-8 (Teacher Wraparound Edition)
3 4 5 6 7 8 9   027/043   05 04 03

CONSULTANTS

TEACHER REVIEWERS

Contents

CONTENTS

Features

NATIONAL GEOGRAPHIC EYE on the Environment

NATIONAL GEOGRAPHIC GEOGRAPHY & HISTORY

▲ Poison arrow frog

Skills

Social Studies Skills

Critical Thinking Skills

Technology Skills

Study and Writing Skills

Making Connections

Culture

Government

People

Technology

▲ Inuit greet with a nose rub

Exploring Culture

CONTENTS

Exploring GOVERNMENT

teen Scene

Teen from Senegal ▶

Believe It or Not!

Primary Source

Literature

▼ Teens washing cars to earn money

BUILDING CITIZENSHIP

Exploring Economics

Heroes at work ▶

TIME REPORTS — FOCUS ON WORLD ISSUES

Maps

Unit 5   Russia and the Eurasian Republics

Unit 6   Africa South of the Sahara

Unit 7   North America and Middle America

Unit 8   South America

Unit 9   Australia, Oceania, and Antarctica

Charts and Graphs

Unit 6 Africa South of the Sahara

Unit 7 North America and Middle America

Unit 8 South America

Unit 9 Australia, Oceania, and Antarctica

REFERENCE ATLAS

ATLAS KEY

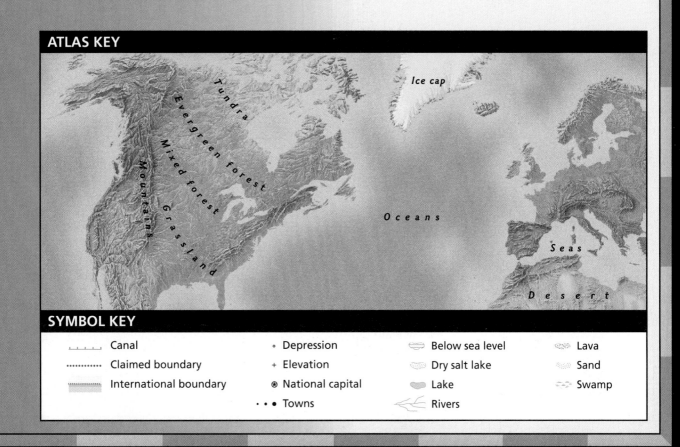

SYMBOL KEY

⌐⌐⌐⌐	Canal	∘	Depression	⬭	Below sea level	🌊	Lava
··········	Claimed boundary	+	Elevation	🌊	Dry salt lake	🌊	Sand
▓▓▓▓	International boundary	⊛	National capital	⬮	Lake	🌊	Swamp
		· · ●	Towns	⤙⤛	Rivers		

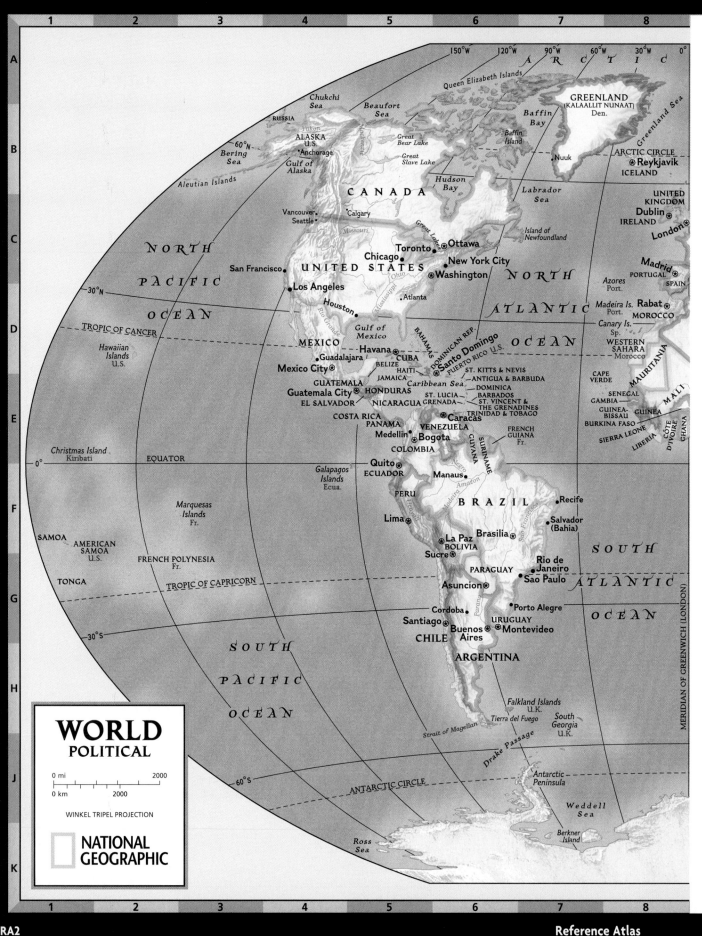

WORLD
POLITICAL

0 mi   2000
0 km   2000

WINKEL TRIPEL PROJECTION

NATIONAL
GEOGRAPHIC

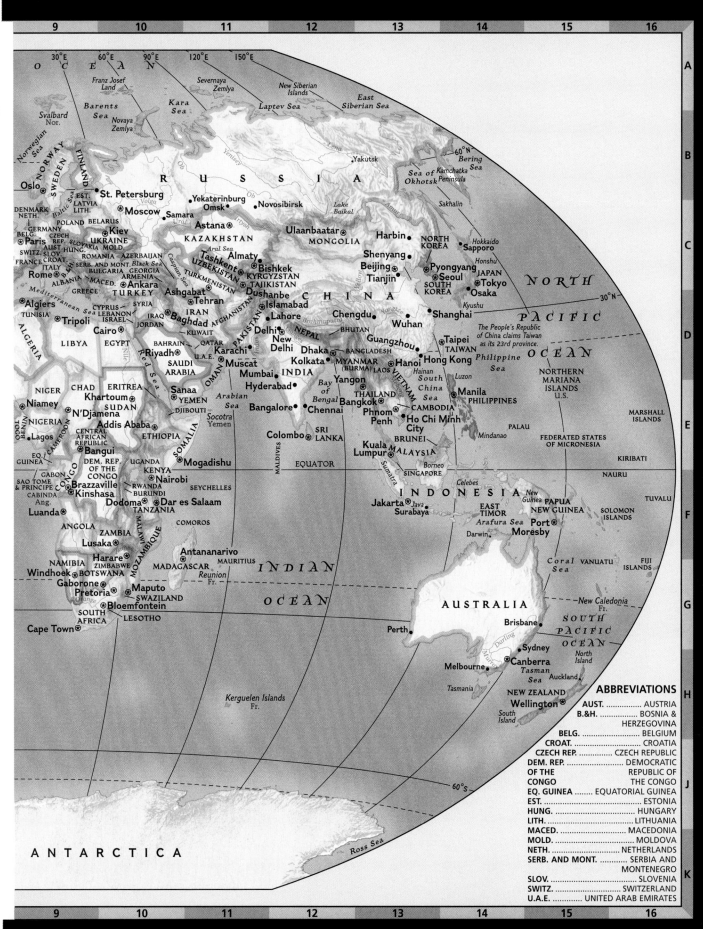

O   C   E   A   N
30°E   60°E   90°E   120°E   150°E
Franz Josef
Land
Severnaya
Zemlya
New Siberian
Islands
Barents
Sea
Kara
Sea
Laptev Sea
East
Siberian Sea
Svalbard
Nor.
Novaya
Zemlya
Norwegian
Sea
60°N
Bering
Sea
Yakutsk
Kamchatka
Peninsula
Sea of
Okhotsk
Oslo
NORWAY
SWEDEN
FINLAND
DENMARK
NETH.
GERMANY
BELG.
Paris
CZECH
SWITZ. SLOV.
FRANCE CROAT.
ITALY
Rome
ALBANIA
GREECE
POLAND BELARUS
EST.
LATVIA
LITH.
St. Petersburg
Yekaterinburg
Omsk
Novosibirsk
R   U   S   S   I   A
Moscow
Samara
Ob
Volga
Ural
Yenisey
Lena
Amur
Sakhalin
Hokkaido
Sapporo
Honshu
JAPAN
Tokyo
Osaka
Kyushu
Kiev
UKRAINE
SLOVAKIA
AUST. HUNG.
SERB. AND MONT.
BULGARIA
MACED.
B.&H.
Astana
KAZAKHSTAN
Aral Sea
Almaty
Ulaanbaatar
MONGOLIA
Harbin
NORTH
KOREA
Shenyang
Beijing
Tianjin
Pyongyang
Seoul
SOUTH
KOREA
ROMANIA AZERBAIJAN
Black Sea
Tashkent
Bishkek
KYRGYZSTAN
Caspian Sea
UZBEKISTAN
TURKMENISTAN
GEORGIA
ARMENIA
TURKEY
Ashgabat
Dushanbe
TAJIKISTAN
CHINA
Chengdu
Wuhan
Shanghai
Algiers
TUNISIA
Mediterranean Sea
CYPRUS
SYRIA
LEBANON
ISRAEL
Tripoli
Cairo
JORDAN
IRAQ
Baghdad
IRAN
Tehran
Islamabad
AFGHANISTAN
PAKISTAN
Lahore
Delhi
New
Delhi
NEPAL
BHUTAN
Brahmaputra
Indus
Yellow
30°N
NORTH
PACIFIC
OCEAN
Taipei
TAIWAN
Hong Kong
Guangzhou
The People's
Republic
of China claims Taiwan
as its 23rd province.
Philippine
Sea
NORTHERN
MARIANA
ISLANDS
U.S.
Luzon
ALGERIA
LIBYA
EGYPT
Nile
Red Sea
Riyadh
SAUDI
ARABIA
KUWAIT
QATAR
BAHRAIN
U.A.E.
OMAN
Muscat
Karachi
Dhaka
BANGLADESH
Kolkata
MYANMAR
(BURMA)
LAOS
Hanoi
South
China
Sea
Hainan
Manila
PHILIPPINES
MARSHALL
ISLANDS
NIGER
CHAD
ERITREA
Khartoum
SUDAN
YEMEN
Sanaa
DJIBOUTI
Arabian
Sea
Mumbai
INDIA
Hyderabad
Bay
of
Bengal
Yangon
THAILAND
Bangkok
VIETNAM
CAMBODIA
Phnom
Penh
Ho Chi Minh
City
BRUNEI
Mindanao
PALAU
FEDERATED STATES
OF MICRONESIA
Niamey
NIGERIA
N'Djamena
CENTRAL
AFRICAN
REPUBLIC
Addis Ababa
ETHIOPIA
SOMALIA
Socotra
Yemen
Bangalore
Chennai
SRI
LANKA
Colombo
MALDIVES
EQUATOR
Kuala
Lumpur
MALAYSIA
SINGAPORE
Sumatra
Borneo
KIRIBATI
NAURU
TOGO
BENIN
Lagos
CAMEROON
EQ.
GUINEA
GABON
SAO TOME
& PRINCIPE
CABINDA
Ang.
Congo
DEM. REP.
OF THE
CONGO
Bangui
UGANDA
KENYA
Nairobi
RWANDA
BURUNDI
Mogadishu
SEYCHELLES
I   N   D   O   N   E   S   I   A
New
Guinea
PAPUA
NEW GUINEA
SOLOMON
ISLANDS
TUVALU
Brazzaville
Kinshasa
Dodoma
Dar es Salaam
TANZANIA
Jakarta
Java
Surabaya
EAST
TIMOR
Arafura Sea
Darwin
Port
Moresby
Luanda
ANGOLA
ZAMBIA
Lusaka
COMOROS
MALAWI
MOZAMBIQUE
FIJI
ISLANDS
VANUATU
Coral
Sea
Antananarivo
MAURITIUS
MADAGASCAR
Reunion
Fr.
I   N   D   I   A   N
Harare
ZIMBABWE
NAMIBIA
Windhoek
BOTSWANA
Gaborone
Pretoria
Maputo
SWAZILAND
Orange
O   C   E   A   N
New Caledonia
Fr.
AUSTRALIA
SOUTH
PACIFIC
OCEAN
Bloemfontein
SOUTH
AFRICA
LESOTHO
Cape Town
Perth
Brisbane
Darling
Sydney
Canberra
Murray
Melbourne
Tasman
Sea
North
Island
Auckland
Kerguelen Islands
Fr.
Tasmania
NEW ZEALAND
Wellington
South
Island
60°S
A   N   T   A   R   C   T   I   C   A
Ross Sea

ABBREVIATIONS

AUST.	AUSTRIA
B.&H.	BOSNIA &
	HERZEGOVINA
BELG.	BELGIUM
CROAT.	CROATIA
CZECH REP.	CZECH REPUBLIC
DEM. REP.	DEMOCRATIC
OF THE	REPUBLIC OF
CONGO	THE CONGO
EQ. GUINEA	EQUATORIAL GUINEA
EST.	ESTONIA
HUNG.	HUNGARY
LITH.	LITHUANIA
MACED.	MACEDONIA
MOLD.	MOLDOVA
NETH.	NETHERLANDS
SERB. AND MONT.	SERBIA AND
	MONTENEGRO
SLOV.	SLOVENIA
SWITZ.	SWITZERLAND
U.A.E.	UNITED ARAB EMIRATES

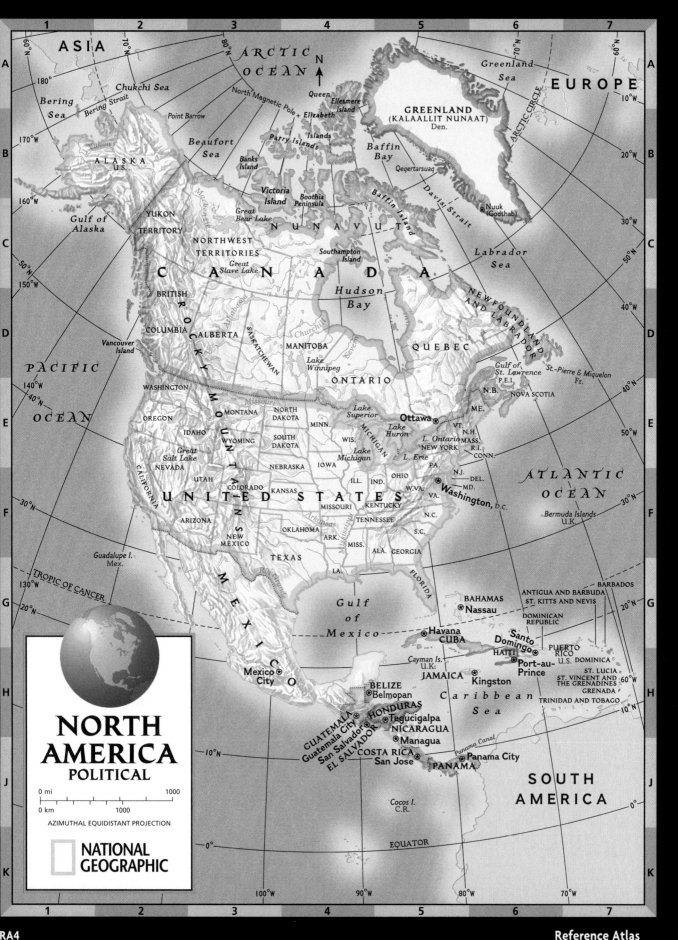

NORTH AMERICA
POLITICAL

0 mi 1000
0 km 1000
AZIMUTHAL EQUIDISTANT PROJECTION

NATIONAL GEOGRAPHIC

ASIA

ARCTIC OCEAN

180°
170°W
160°W
150°W
140°W
130°W

Bering Sea
Chukchi Sea
Bering Strait
Point Barrow
Beaufort Sea

ALASKA U.S.
Yukon
Gulf of Alaska

North Magnetic Pole
Queen Elizabeth Islands
Ellesmere Island
Parry Islands
Banks Island
Victoria Island
Boothia Peninsula

GREENLAND
(KALAALLIT NUNAAT)
Den.

Greenland Sea

EUROPE

10°W
20°W
30°W
40°W
50°W
60°W

Baffin Bay
Qeqertarsuaq
Nuuk (Godthab)
ARCTIC CIRCLE

YUKON TERRITORY
NORTHWEST TERRITORIES
Great Bear Lake
Great Slave Lake
Mackenzie

NUNAVUT

Baffin Island
Davis Strait

CANADA

Labrador Sea

Southampton Island
Hudson Bay

BRITISH COLUMBIA
Vancouver Island
ALBERTA
SASKATCHEWAN
Athabasca
Churchill
MANITOBA
Lake Winnipeg
ONTARIO
Severn
QUEBEC

NEWFOUNDLAND AND LABRADOR

St.-Pierre & Miquelon Fr.
Gulf of St. Lawrence
P.E.I.
N.B.
NOVA SCOTIA

PACIFIC OCEAN

WASHINGTON
OREGON
IDAHO
MONTANA
NORTH DAKOTA
SOUTH DAKOTA
WYOMING
Great Salt Lake
NEVADA
UTAH
CALIFORNIA
COLORADO
ARIZONA
NEW MEXICO

MINN.
WIS.
Lake Superior
MICHIGAN
Lake Huron
Lake Michigan
IOWA
NEBRASKA
KANSAS
MISSOURI
OKLAHOMA
ARK.
TEXAS

UNITED STATES

ROCKY MOUNTAINS

Missouri
Arkansas
Mississippi
Rio Grande

ME.
VT.
N.H.
N.Y.
NEW YORK
MASS.
R.I.
CONN.
PA.
N.J.
DEL.
MD.
Ottawa
L. Ontario
L. Erie
OHIO
IND.
ILL.
W.VA.
VA.
Washington, D.C.
KENTUCKY
TENNESSEE
N.C.
S.C.
MISS.
ALA.
GEORGIA
LA.
FLORIDA

ATLANTIC OCEAN

Bermuda Islands U.K.

Guadalupe I. Mex.
TROPIC OF CANCER

MEXICO

Mexico City

Gulf of Mexico

Havana
CUBA
Cayman Is. U.K.
JAMAICA
Kingston

BAHAMAS
Nassau

ANTIGUA AND BARBUDA
ST. KITTS AND NEVIS
BARBADOS

DOMINICAN REPUBLIC
Santo Domingo
HAITI
Port-au-Prince
PUERTO RICO U.S.
DOMINICA
ST. LUCIA
ST. VINCENT AND THE GRENADINES
GRENADA
TRINIDAD AND TOBAGO

BELIZE
Belmopan
GUATEMALA
Guatemala City
San Salvador
EL SALVADOR
HONDURAS
Tegucigalpa
NICARAGUA
Managua
COSTA RICA
San Jose

Caribbean Sea

Panama Canal
Panama City
PANAMA

Cocos I. C.R.

SOUTH AMERICA

EQUATOR

100°W
90°W
80°W
70°W

50°N
40°N
30°N
20°N
10°N
0°

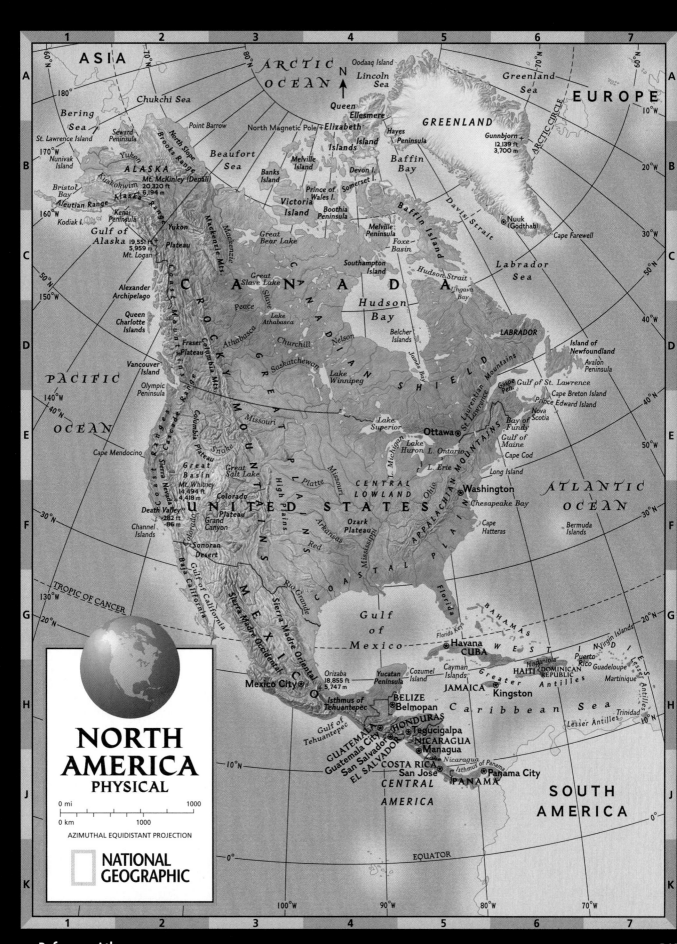

NORTH AMERICA
PHYSICAL

0 mi   1000

0 km   1000

AZIMUTHAL EQUIDISTANT PROJECTION

NATIONAL GEOGRAPHIC

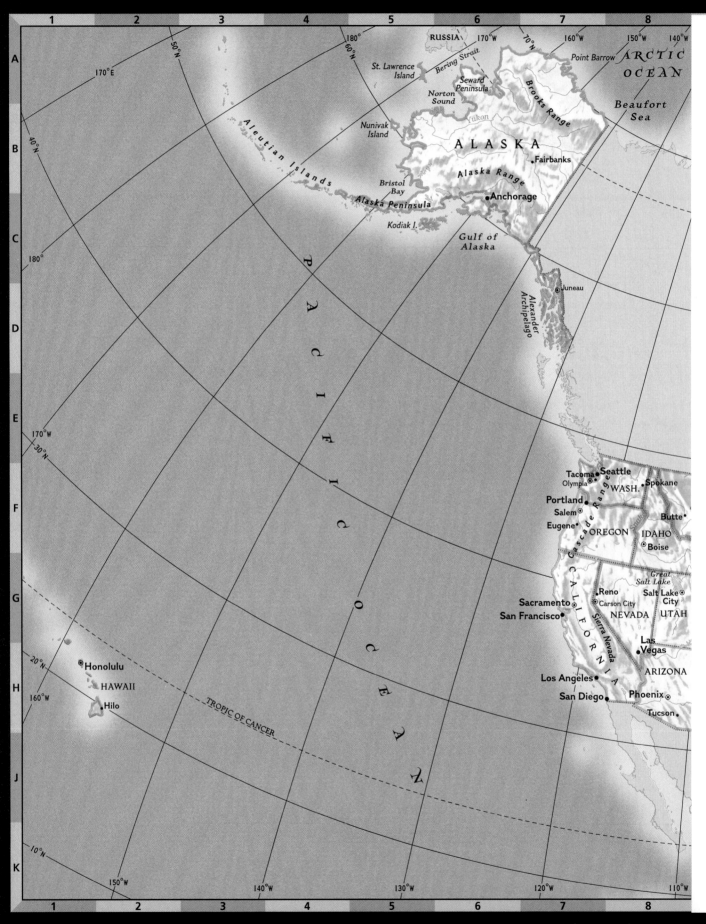

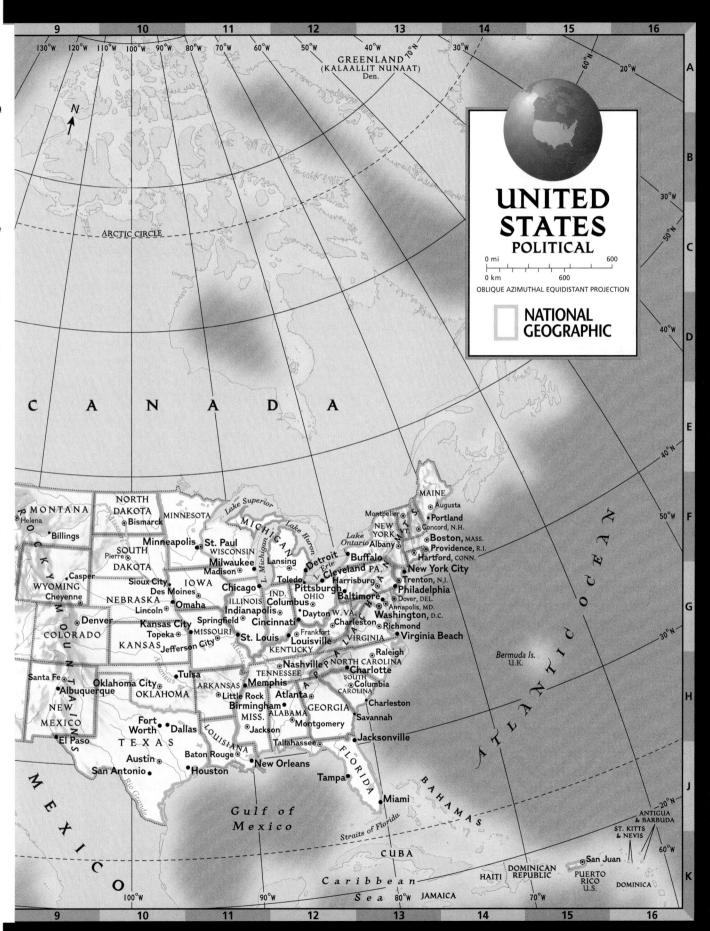

UNITED
STATES
POLITICAL

0 mi   600
0 km   600

OBLIQUE AZIMUTHAL EQUIDISTANT PROJECTION

NATIONAL
GEOGRAPHIC

GREENLAND
(KALAALLIT NUNAAT)
Den.

ARCTIC CIRCLE

C A N A D A

MONTANA
Helena
Billings

NORTH
DAKOTA
Bismarck

MINNESOTA

Lake Superior

MICHIGAN

Lake Huron

MAINE
Augusta
Montpelier
Portland
Concord, N.H.
NEW
YORK
Albany
Boston, MASS.
Providence, R.I.
Hartford, CONN.

SOUTH
DAKOTA
Pierre

Minneapolis St. Paul
WISCONSIN
Milwaukee
Madison

Lansing
Detroit
L. Erie
Cleveland PA.
Harrisburg

Buffalo

L. Michigan

Lake
Ontario

New York City

WYOMING
Casper
Cheyenne

Sioux City
IOWA
Des Moines
Chicago
ILLINOIS
Omaha

Toledo
Pittsburgh
OHIO
IND.
Columbus
Indianapolis

Trenton, N.J.
Philadelphia
Dover, DEL.
Baltimore
Annapolis, MD.
Washington, D.C.
Dayton W. VA.

Denver
COLORADO

NEBRASKA
Lincoln

Kansas City
Topeka
MISSOURI

Springfield
Cincinnati

Frankfort
St. Louis
KANSAS
Jefferson City
Louisville
KENTUCKY

Charleston
Richmond
VIRGINIA

Virginia Beach

ROCKY MOUNTAINS

Santa Fe
Albuquerque

NEW
MEXICO
El Paso

Oklahoma City
OKLAHOMA

Tulsa

Arkansas

ARKANSAS
Little Rock

TENNESSEE
Nashville
Memphis

Raleigh
NORTH CAROLINA
Charlotte
SOUTH
Columbia
CAROLINA

Charleston

Birmingham
ALABAMA
GEORGIA
Atlanta

Savannah

Fort
Worth
Dallas
TEXAS

Austin
San Antonio

LOUISIANA
Jackson
MISS.
Montgomery

Baton Rouge
Houston
New Orleans

Tallahassee

Jacksonville

FLORIDA

Tampa

Miami

Rio Grande

M E X I C O

Gulf of
Mexico

Straits of Florida

BAHAMAS

CUBA

Caribbean
Sea
JAMAICA

HAITI

DOMINICAN
REPUBLIC

PUERTO
RICO
U.S.

San Juan

ANTIGUA
& BARBUDA

ST. KITTS
& NEVIS

DOMINICA

ATLANTIC OCEAN

Bermuda Is.
U.K.

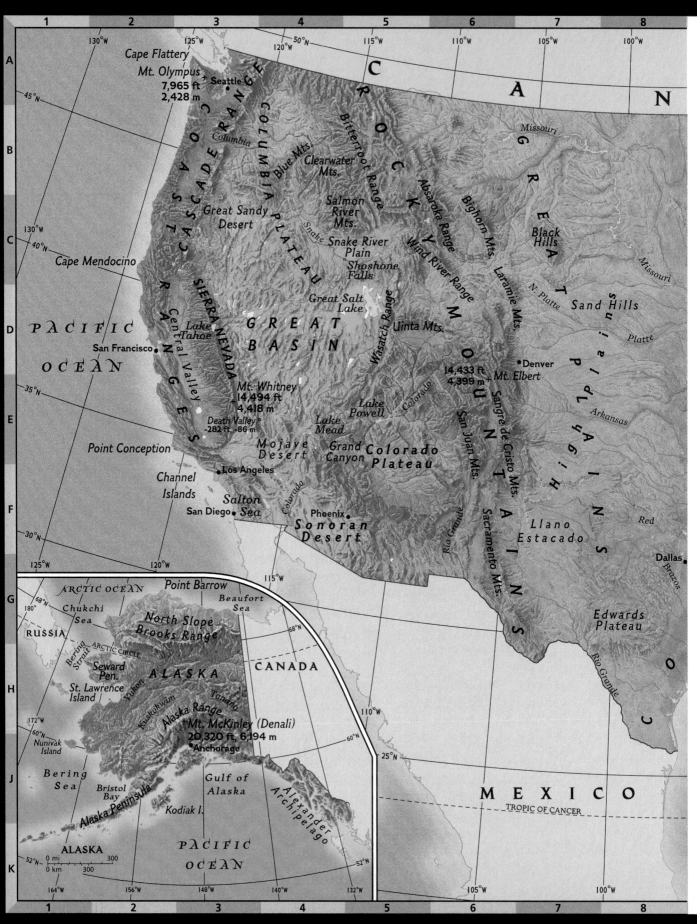

130°W   125°W   50°N   120°W   115°W   110°W   105°W   100°W

A

Cape Flattery
Mt. Olympus
7,965 ft
2,428 m
Seattle

C A N A D A

45°N

B

Columbia

Blue Mts.

Clearwater
Mts.

Bitterroot Range

Missouri

G R E A T

Black
Hills

130°W
40°N

C

Cape Mendocino

Great Sandy
Desert

Salmon
River
Mts.

Snake

Snake River
Plain

Shoshone
Falls

Absaroka Range

Bighorn Mts.

Laramie Mts.

Wind River Range

N. Platte

Missouri

Sand Hills

Platte

D

PACIFIC

OCEAN

San Francisco

Lake
Tahoe

Central Valley

Great Salt
Lake

Wasatch Range

GREAT
BASIN

Uinta Mts.

M O U N T A I N S

Denver

Mt. Elbert

14,433 ft
4,399 m

H i g h   P l a i n s

35°N

E

Point Conception

Mt. Whitney
14,494 ft
4,418 m

Death Valley
-282 ft, -86 m

Lake
Mead

Lake
Powell

Colorado

Grand
Canyon

Colorado
Plateau

San Juan Mts.

Sangre de Cristo Mts.

14,433 ft
4,399 m

Arkansas

F

Los Angeles

Mojave
Desert

Channel
Islands

Salton
Sea

San Diego

Colorado

Phoenix

Sonoran
Desert

Rio Grande

Sacramento Mts.

Llano
Estacado

Red

Dallas

Brazos

30°N

125°W   120°W   115°W

G

Point Barrow

Beaufort
Sea

Edwards
Plateau

68°N
180°

Chukchi
Sea

North Slope
Brooks Range

68°N

RUSSIA

Bering Strait

ARCTIC CIRCLE

Seward
Pen.

ALASKA

CANADA

H

St. Lawrence
Island

Yukon

Kuskokwim

Tanana

Alaska Range

Mt. McKinley (Denali)
20,320 ft, 6,194 m

110°W

172°W
60°N

Nunivak
Island

Anchorage

60°N

25°N

J

Bering
Sea

Bristol
Bay

Gulf of
Alaska

Alexander Archipelago

M E X I C O

Alaska Peninsula

Kodiak I.

K

52°N

0 mi   300
0 km   300

ALASKA

PACIFIC

OCEAN

52°N

TROPIC OF CANCER

C O

164°W   156°W   148°W   140°W   132°W   105°W   100°W

SIERRA NEVADA

CASCADE RANGE

COAST RANGES

COLUMBIA PLATEAU

R O C K Y

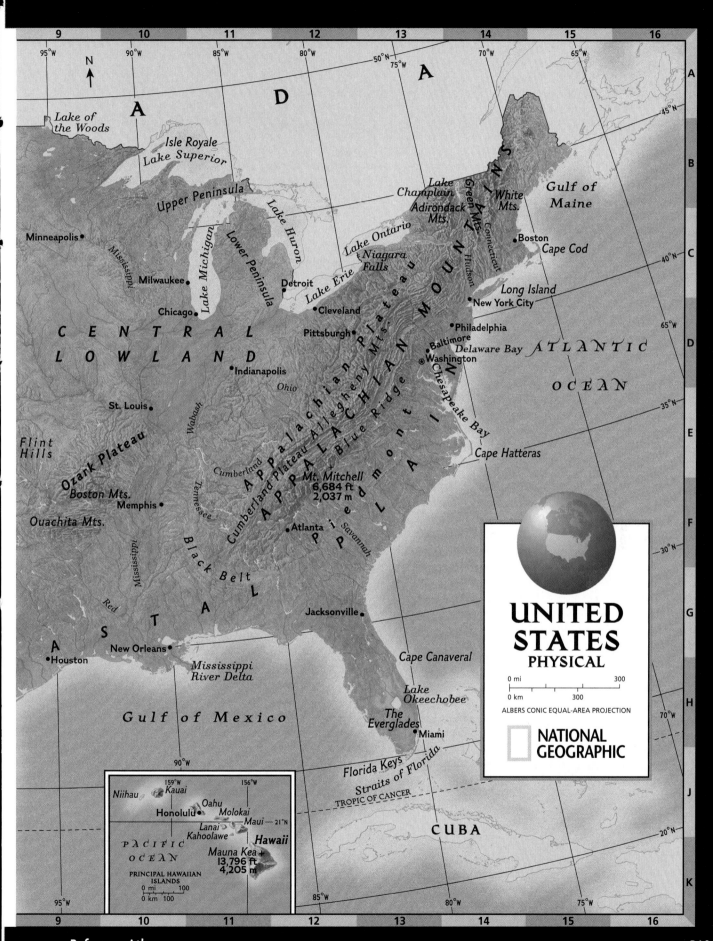

UNITED STATES PHYSICAL

0 mi 300
0 km 300

ALBERS CONIC EQUAL-AREA PROJECTION

NATIONAL GEOGRAPHIC

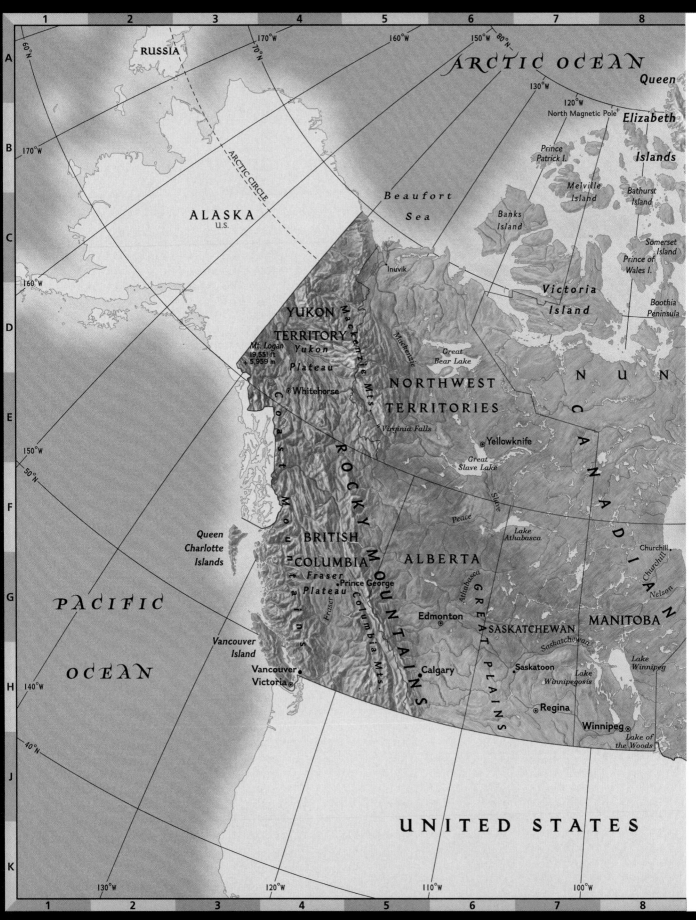

ARCTIC OCEAN

RUSSIA

170°W
70°N

ARCTIC CIRCLE

ALASKA
U.S.

60°N
170°W
160°W

150°W
50°N

40°N

PACIFIC

OCEAN

140°W

130°W
120°W
110°W
100°W

Queen

North Magnetic Pole

Elizabeth

Prince
Patrick I.

Melville
Island

Bathurst
Island

Islands

Banks
Island

Somerset
Island

Prince of
Wales I.

Beaufort

Sea

Inuvik

Victoria

Island

Boothia
Peninsula

YUKON

TERRITORY

Mt. Logan
19,551 ft
+ 5,959 m

Yukon

Plateau

Whitehorse

Mackenzie Mts.

Mackenzie

Great
Bear Lake

NORTHWEST

TERRITORIES

Virginia Falls

Yellowknife

Great
Slave Lake

Slave

Peace

Lake
Athabasca

N U N

C
A
N
A
D
I
A
N

ROCKY

Queen
Charlotte
Islands

BRITISH

COLUMBIA

Fraser
Plateau

Fraser

Prince George

Coast
Mountains

Columbia Mts.

MOUNTAINS

ALBERTA

Athabasca

Edmonton

Calgary

G
R
E
A
T

P
L
A
I
N
S

Saskatchewan

SASKATCHEWAN

Saskatoon

Regina

Lake
Winnipegosis

Churchill

Nelson

MANITOBA

Lake
Winnipeg

Vancouver
Island

Vancouver

Victoria

Winnipeg

Lake of
the Woods

UNITED STATES

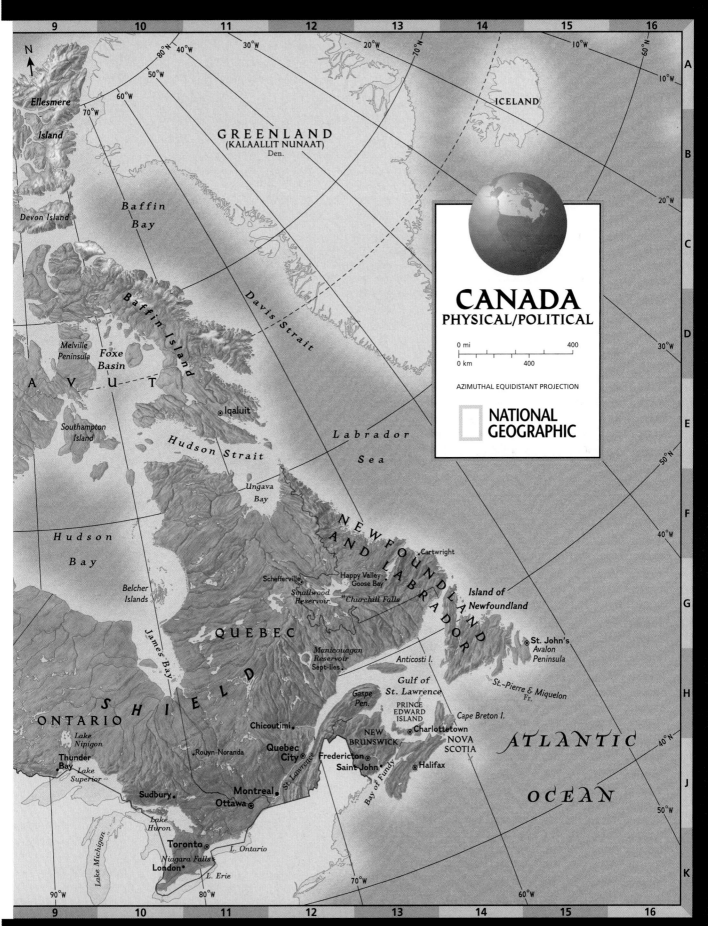

N

Ellesmere
Island

Devon Island

*Baffin
Bay*

Melville
Peninsula

*Foxe
Basin*

Southampton
Island

N U N A V U T

⊙ Iqaluit

Baffin Island

GREENLAND
(KALAALLIT NUNAAT)
Den.

ICELAND

Davis Strait

Hudson Strait

*Labrador
Sea*

*Ungava
Bay*

CANADA
PHYSICAL/POLITICAL

0 mi   400
0 km   400

AZIMUTHAL EQUIDISTANT PROJECTION

**NATIONAL
GEOGRAPHIC**

Hudson

Bay

*Belcher
Islands*

James Bay

QUEBEC

S H I E L D

ONTARIO

Lake
Nipigon

Thunder
Bay

Lake
Superior

Sudbury.

Rouyn-Noranda.

Chicoutimi.

Quebec
City ⊙

Montreal.

Ottawa ⊛

Lake
Huron

Toronto ⊙

Niagara Falls
London•

L. Ontario

L. Erie

Lake Michigan

Scheffervile.

*Smallwood
Reservoir*

Happy Valley-
Goose Bay

"Churchill Falls

N E W F O U N D L A N D

A N D   L A B R A D O R

Cartwright

*Island of
Newfoundland*

⊙ St. John's
*Avalon
Peninsula*

St.-Pierre & Miquelon
Fr.

*Manicouagan
Reservoir*
Sept-Iles.

Anticosti I.

*Gulf of
St. Lawrence*

Gaspe
Pen.

PRINCE
EDWARD
ISLAND

Cape Breton I.

NEW
BRUNSWICK

⊙ Charlottetown

NOVA
SCOTIA

Fredericton ⊙
Saint John•

⊙ Halifax

St. Lawrence

Bay of Fundy

ATLANTIC

OCEAN

A

UNITED STATES

Tijuana

Mexicali

Sonoran Desert

B

30°N

Baja California

Gulf of California

Ciudad Juarez

Chihuahua

C

Nuevo Laredo

Monterrey

La Paz

Matamoros

Gulf of Mexico

D

Sierra Madre Occidental

False Cape

Mazatlan

M   E   X   I   C   O

Sierra Madre Oriental

20°N

Ciudad Madero
Tampico

San Luis Potosi

Leon

E

Guadalajara

Revillagigedo Islands
Mex.

Merida

Cozumel Island

Bay of Campeche

Yucatan Peninsula

Mexico City

Orizaba
18,855 ft
+5,747 m

Popocatepetl
17,802 ft
5,426 m

Veracruz

F

Sierra Madre del Sur

Acapulco

Isthmus of Tehuantepec

Sierra Madre

Belmopan

Belize City

BELIZE

Gulf of Honduras

Gulf of Tehuantepec

HON

GUATEMALA

Guatemala City

Tegucigalpa

G

10°N

EL SALVADOR
San Salvador

Leon

CENTRAL

MIDDLE AMERICA
PHYSICAL/POLITICAL

0 mi   400

0 km   400

AZIMUTHAL EQUIDISTANT PROJECTION

NATIONAL GEOGRAPHIC

AMERICA

H

PΛCIFIC

J

OCEΛN

Cocos Island
C.R.

K

0°

ATLANTIC OCEAN

TROPIC OF CANCER

BAHAMAS

•Freeport

⊛Nassau

Straits of Florida

Andros Island

Turks & Caicos Islands U.K.

W **E** **S** **T** **I** **N** **D** **I** **E** **S**

⊛Havana

CUBA

•Camaguey

•Holguin

•Santiago de Cuba

Isle of Youth

Cayman Islands U.K.

G *r* *e* *a* *t* *e* *r*

Montego Bay•

JAMAICA ⊛Kingston

Hispaniola

HAITI
Port-au- ⊛
Prince

•Santiago

Santo •
Domingo

DOMINICAN REPUBLIC

A *n* *t* *i* *l* *l* *e* *s*

San Juan ⊛

Puerto Rico U.S.

Virgin Islands U.S. & U.K.

ST. KITTS & NEVIS

ANTIGUA & BARBUDA

Guadeloupe Fr.

DOMINICA

Bird I. Venez.

Martinique Fr.

ST. LUCIA

ST. VINCENT & THE GRENADINES

BARBADOS

GRENADA

L *e* *s* *s* *e* *r*

A *n* *t* *i* *l* *l* *e* *s*

C *a* *r* *i* *b* *b* *e* *a* *n* *S* *e* *a*

Aruba Neth.

Curacao Neth. Bonaire

Lesser Antilles

TRINIDAD & TOBAGO
Port-of-Spain ⊛

Tobago

Trinidad

DURAS

Coco

NICARAGUA

⊛Managua

Lake Nicaragua

Mosquito Coast

COSTA

Puerto Limon•

San Jose ⊛ **RICA**

Gulf of Mosquitos

Isthmus of Panama

PANAMA

⊛ Panama City

•David

Gulf of Panama

SOUTH

AMERICA

EQUATOR

80°W

70°W

60°W

30°N

20°N

10°N

0°

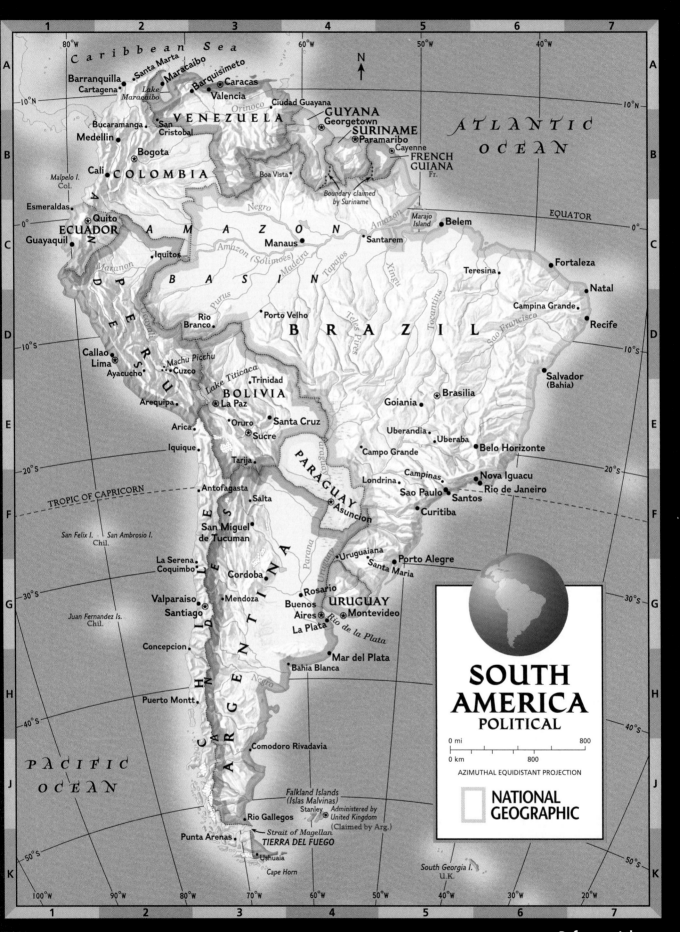

SOUTH AMERICA
POLITICAL

0 mi   800

0 km   800

AZIMUTHAL EQUIDISTANT PROJECTION

NATIONAL GEOGRAPHIC

SOUTH AMERICA
PHYSICAL

0 mi   800

0 km   800

AZIMUTHAL EQUIDISTANT PROJECTION

NATIONAL GEOGRAPHIC

Map labels:

Caribbean Sea

Caracas
Lake Maracaibo
VENEZUELA
Orinoco
LLANOS
Bogota
COLOMBIA
Angel Falls Total drop 3,212 ft 979 m
GUIANA HIGHLANDS
GUYANA
Georgetown
SURINAME
Paramaribo
Cayenne
FRENCH GUIANA
ATLANTIC OCEAN
Boundary claimed by Suriname
Malpelo I.
Quito
ECUADOR
Negro
Amazon
Marajo Island
EQUATOR
AMAZON
Selvas
BASIN
Amazon
Ucayali
Purus
Madeira
Tapajos
Xingu
Teles Pires
Tocantins
Sao Francisco
BRAZIL
PERU
Lima
Machu Picchu
Lake Titicaca
BOLIVIA
La Paz
MATO GROSSO
PLATEAU
BRAZILIAN HIGHLANDS
Brasilia
Sucre
Altiplano
Salar de Uyuni
GRAN CHACO
PARAGUAY
Asuncion
Iguazu Falls
TROPIC OF CAPRICORN
San Felix I.
San Ambrosio I.
Paraguay
Parana
Uruguay
Uruguay
ANDES
CORDILLERA
PAMPAS
ARGENTINA
Juan Fernandez Is.
Aconcagua 22,834 ft 6,960 m
Santiago
Buenos Aires
URUGUAY
Montevideo
Rio de la Plata
Negro
PATAGONIA
Chiloe Island
-131 ft -40 m
Valdes Peninsula
PACIFIC OCEAN
Taitao Peninsula
Gulf of San Jorge
Wellington I.
Falkland Islands (Islas Malvinas)
Stanley
South Georgia I.
Strait of Magellan
Tierra del Fuego
Cape Horn

EUROPE
POLITICAL

0 mi 400
0 km 400

AZIMUTHAL EQUIDISTANT PROJECTION

NATIONAL GEOGRAPHIC

ATLANTIC OCEAN

Akureyri
Reykjavik
ICELAND

ARCTIC CIRCLE

Faroe Islands
Den.
Torshavn

Rockall
U.K.

Shetland Islands
Lerwick

Isle of Lewis
Orkney Islands

Inverness

UNITED
SCOTLAND
Glasgow
Edinburgh
Aberdeen

NORTHERN IRELAND
Belfast

IRELAND
Irish Sea
Dublin
Liverpool
Cork
Manchester
KINGDOM
WALES
Birmingham
Cardiff
ENGLAND

Celtic Sea

Land's End
London
Southampton

English Channel
Brest
Le Havre
Brussels
Rennes
Paris

FRANCE

Nantes

La Rochelle

Bay of Biscay
Limoges
Bordeaux

A Coruna
Vigo

Porto
Coimbra
Valladolid

Bilbao
Donostia-San Sebastian
Pyrenees
Toulouse

PORTUGAL
SPAIN
Madrid
ANDORRA
Zaragoza
Barcelona

Lisbon

Valencia

Cape St. Vincent
Cordoba
Seville
Murcia
Palma
Balearic Islands Sp.

Cadiz
Malaga
Cartagena
GIBRALTAR
U.K.
Strait of Gibraltar

AFRICA

Tromso

Norwegian Sea

MERIDIAN OF GREENWICH (LONDON)

N O R W A Y
S W E D E N

Trondheim
Are
Alesund
Sundsvall

Bergen
Oslo
Uppsala
Stavanger
Stockholm

Skagerrak
Goteborg
Gotland

Arhus
DENMARK
Copenhagen
Malmo
Baltic

Kiel
Gdansk
Hamburg

NETH.
The Hague
Amsterdam
Berlin
Bydgoszcz

Brussels
BELGIUM
Bonn
GERMANY
POLAND
LUX.
Frankfurt
Wroclaw
Lodz

Strasbourg
Prague
CZECH REP.
Bratislava
SLOVAKIA

Munich
Vienna
Zurich
LIECH.
Bern
AUSTRIA
Geneva
SWITZERLAND
Budapest
Lyon
ALPS
SLOVENIA
HUNGARY
Milan
Ljubljana
Zagreb
Turin
Venice
CROATIA

MONACO
Genoa
BOSNIA & HERZEGOVINA
Nice
SAN MARINO
Sarajevo
Marseille

Corsica Fr.
ITALY

VATICAN CITY
Rome
Adriatic Sea

Tirana
ALBANIA

Naples

Sardinia It.

Cagliari
Tyrrhenian Sea
Ionian Sea

M e d i t e r r a n e a n

Palermo
Sicily
Messina
Catania

Valletta
MALTA

North Sea

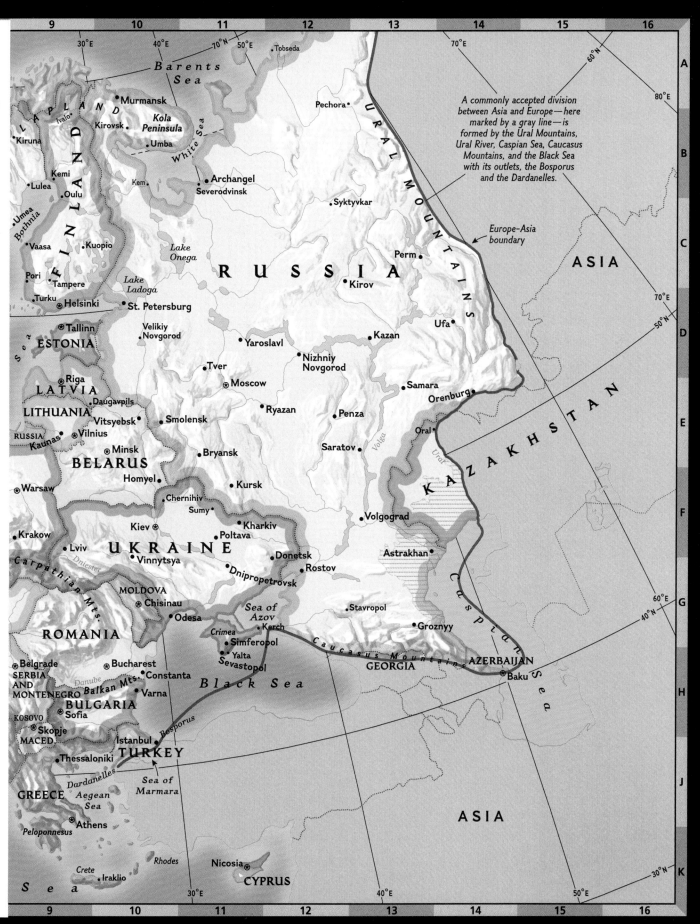

A commonly accepted division between Asia and Europe—here marked by a gray line—is formed by the Ural Mountains, Ural River, Caspian Sea, Caucasus Mountains, and the Black Sea with its outlets, the Bosporus and the Dardanelles.

Europe-Asia boundary

ASIA

RUSSIA

KAZAKHSTAN

Barents Sea

Tobseda
Pechora

Kola Peninsula
Murmansk
Kirovsk
Umba
White Sea

LAPLAND
Ivalo
Kiruna
Kemi
Lulea
Oulu
Umea
Vaasa
Kuopio
Pori
Tampere
Turku
Helsinki

FINLAND

Kem
Archangel
Severodvinsk

Syktyvkar

Lake Onega
Lake Ladoga

St. Petersburg
Tallinn
ESTONIA
Riga
LATVIA
Daugavpils
LITHUANIA
Vitsyebsk
Vilnius
RUSSIA
Kaunas
Minsk

BELARUS
Homyel
Warsaw
Smolensk
Bryansk

Velikiy Novgorod
Yaroslavl
Tver
Moscow
Ryazan

Nizhniy Novgorod

Penza
Saratov

Kirov
Kazan
Ufa
Volga
Samara
Orenburg
Oral

Perm

URAL MOUNTAINS

Chernihiv
Sumy
Kiev
Kursk
Kharkiv
Poltava

UKRAINE
Vinnytsya
Donetsk
Dnipropetrovsk
Rostov

Volgograd
Astrakhan

Carpathian Mts.
Dniester
Lviv
Krakow

MOLDOVA
Chisinau
Odesa

Sea of Azov
Kerch
Crimea
Simferopol
Yalta
Sevastopol

Stavropol
Groznyy

Ural
Caspian Sea

ROMANIA
Bucharest
Belgrade
SERBIA AND MONTENEGRO
Constanta
Danube
Balkan Mts.
Varna

Black Sea

Caucasus Mountains
GEORGIA
AZERBAIJAN
Baku

KOSOVO
Skopje
MACED.
BULGARIA
Sofia
Thessaloniki
GREECE

Istanbul
Bosporus
TURKEY
Dardanelles
Sea of Marmara
Aegean Sea
Athens
Peloponnesus

Crete
Iraklio
Rhodes
Nicosia
CYPRUS

ASIA

Sea

30°E · 40°E · 70°N · 50°E · 70°E · 60°N · 80°E · 70°E · 50°N · 60°E · 40°N · 50°E · 30°N

EUROPE
PHYSICAL

0 mi   400
0 km   400

AZIMUTHAL EQUIDISTANT PROJECTION

NATIONAL GEOGRAPHIC

Reykjavik
ICELAND

ARCTIC CIRCLE

Faroe
Islands

Shetland
Islands

Orkney
Islands

Outer Hebrides

*British
Isles*

Highlands

Edinburgh

Belfast
UNITED

IRELAND
Dublin

*Irish
Sea*

*Great
Britain*

KINGDOM

Cardiff

London

*North
Sea*

Norwegian Sea

MERIDIAN OF GREENWICH (LONDON)

N

SCANDINAVIA

SWEDEN

NORWAY

Oslo

Stockholm

Gulf of

Baltic

Jutland

DENMARK
Copenhagen

Zealand

Amsterdam
NETH.

BELGIUM
Brussels
Lux.

Berlin

GERMANY

Rhine

POLAND

N O R

Oder

English Channel

Brittany

Seine
Paris

Loire

FRANCE

Danube

Elbe
Prague
CZECH REP.

Bratislava
Vienna

SLOVAKIA

*ATLANTIC
OCEAN*

*Bay of
Biscay*

Mont Blanc
15,771 ft.
4,807 m

Bern
SWITZ.

LIECH.

AUSTRIA

Budapest
HUNGARY

Massif
Central

Rhone

A L P S

SLOVENIA

Ljubljana

Drava

Zagreb
CROATIA

Danube

Sava

Cantabrian Mountains

Pyrenees

MONACO

Po

Riviera

Adriatic Sea

BOSNIA &
Sarajevo
HERZEGOVINA

Douro

IBERIAN

Ebro

ANDORRA

Corsica

SAN MARINO

Apennines

ITALY

VATICAN
CITY
Rome

PORTUGAL

Madrid

SPAIN

Tagus

Lisbon

PENINSULA

Tirana
ALBANIA

GIBRALTAR
Strait of Gibraltar

Baetic Mountains

Balearic Islands

Sardinia

*Tyrrhenian
Sea*

*Ionian
Sea*

M e d i t e r r a n e a n

AFRICA

Sicily
Etna
10,902 ft
3,323 m

Valletta
MALTA

30°N

10°W

0°

10°E

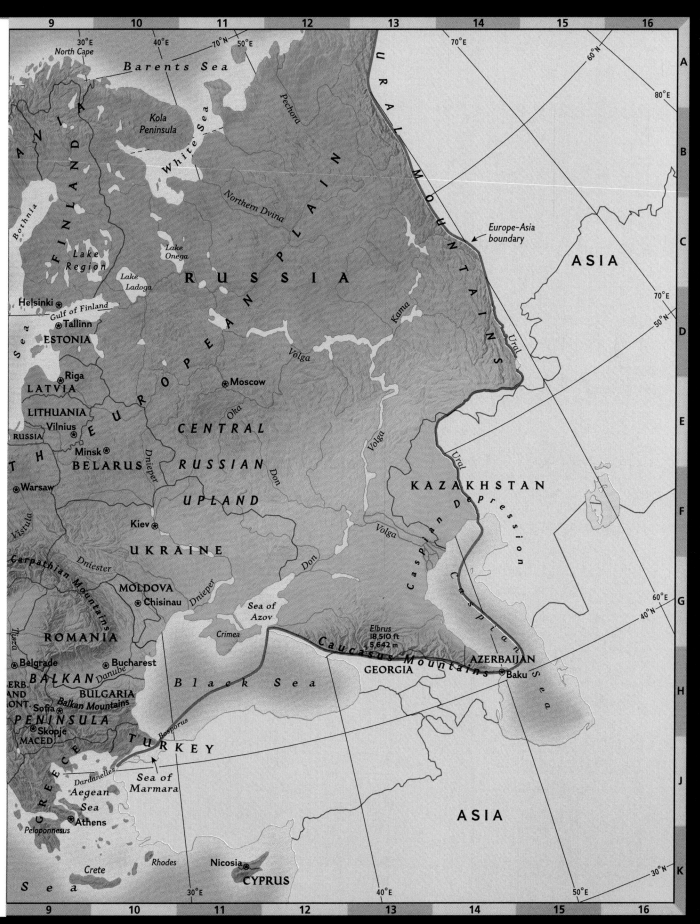

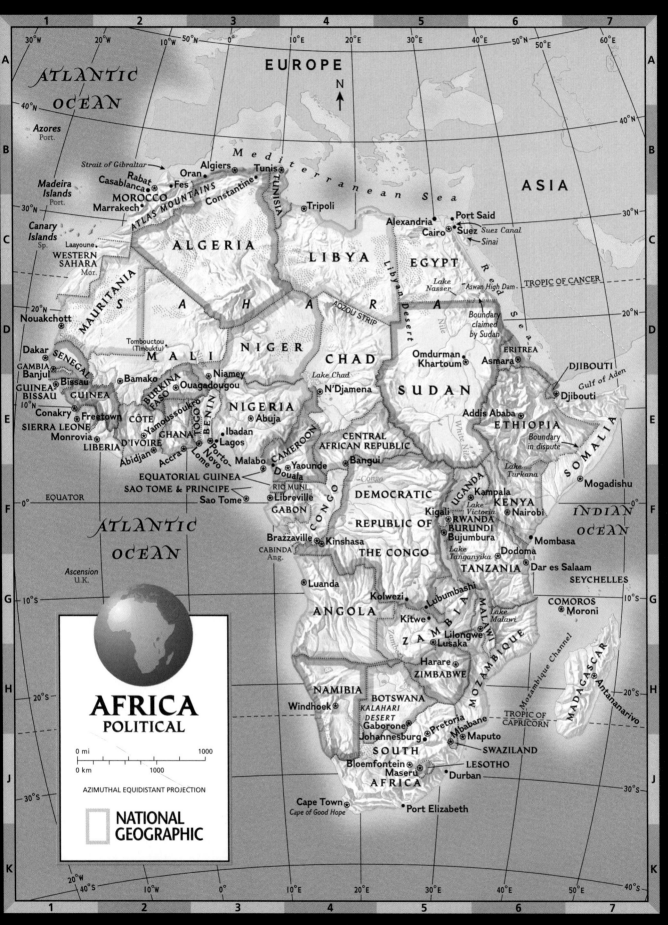

AFRICA
POLITICAL

0 mi   1000

0 km   1000

AZIMUTHAL EQUIDISTANT PROJECTION

NATIONAL GEOGRAPHIC

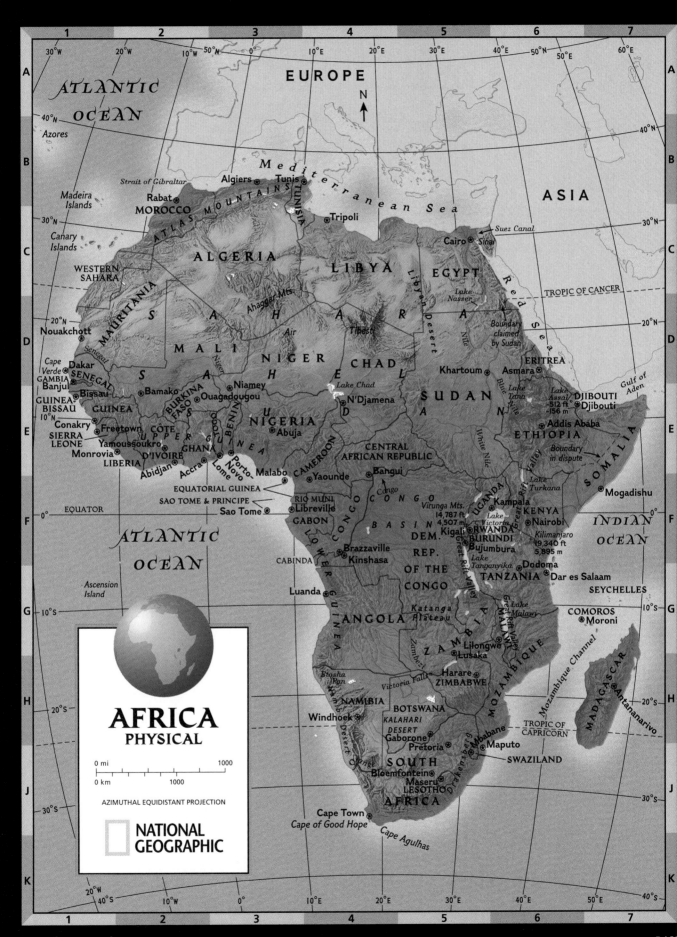

AFRICA
PHYSICAL

0 mi 1000
0 km 1000

AZIMUTHAL EQUIDISTANT PROJECTION

NATIONAL GEOGRAPHIC

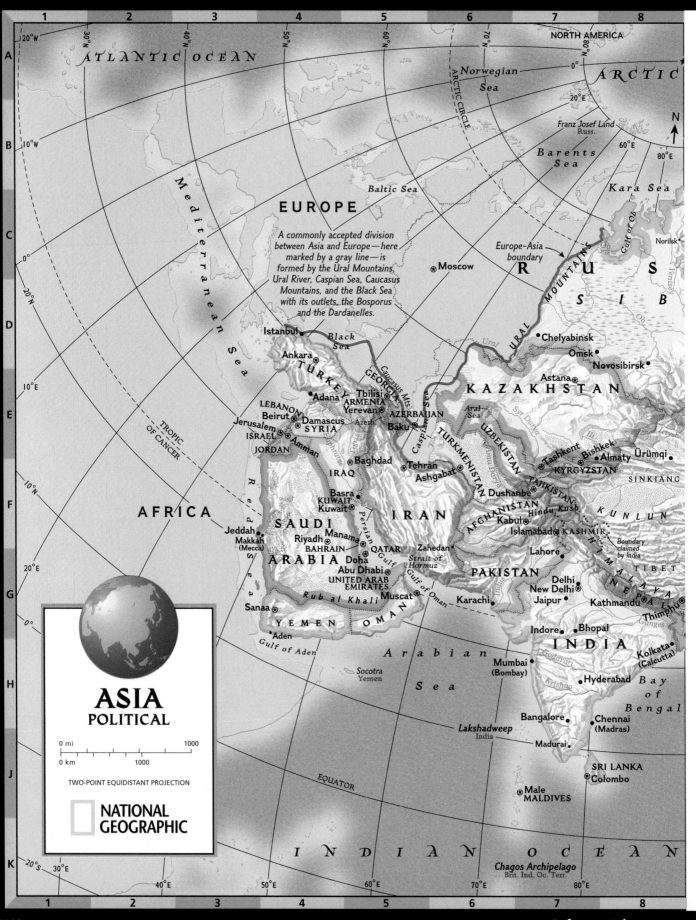

ATLANTIC OCEAN

Mediterranean Sea

EUROPE

A commonly accepted division between Asia and Europe—here marked by a gray line—is formed by the Ural Mountains, Ural River, Caspian Sea, Caucasus Mountains, and the Black Sea with its outlets, the Bosporus and the Dardanelles.

Moscow

Europe-Asia boundary

Norwegian Sea

ARCTIC CIRCLE

NORTH AMERICA

ARCTIC

Franz Josef Land Russ.

Barents Sea

Kara Sea

Baltic Sea

Gulf of Ob

Ob

Yenisey

Norilsk

R U S S I B

URAL MOUNTAINS

Chelyabinsk
Omsk
Novosibirsk

Astana

KAZAKHSTAN

Black Sea

Istanbul
Ankara
Adana
TURKEY
LEBANON
Beirut
Jerusalem
ISRAEL
JORDAN
Damascus
SYRIA
Amman
Euphrates
Tigris

Caucasus Mts.
Tbilisi
GEORGIA
ARMENIA
Yerevan
AZERBAIJAN
Azerb.
Baku

Aral Sea

Syr Darya

Amu Darya

UZBEKISTAN

Tashkent
Bishkek
KYRGYZSTAN
Almaty
Ürümqi

SINKIANG

Ili
Irtysh

TROPIC OF CANCER

AFRICA

Red Sea

Jeddah
Makkah
(Mecca)
Riyadh

SAUDI
ARABIA

Rub al Khali

Sanaa

YEMEN
Aden
Gulf of Aden

Socotra
Yemen

Baghdad
IRAQ
Basra
KUWAIT
Kuwait

Manama
BAHRAIN
QATAR
Doha
Abu Dhabi
UNITED ARAB
EMIRATES

Persian Gulf

Caspian Sea

Tehran
Ashgabat

TURKMENISTAN

Dushanbe
TAJIKISTAN

IRAN

Zahedan

Strait of Hormuz

Gulf of Oman

Muscat

OMAN

Karachi

Arabian
Sea

AFGHANISTAN
Kabul
Islamabad
KASHMIR

Hindu Kush

KUNLUN

Boundary claimed by India

TIBET

Lahore

PAKISTAN

Indus

Delhi
New Delhi
Jaipur

HIMALAYA

Kathmandu
NEPAL
Thimphu

Boundary claimed by India

Indore
Bhopal

INDIA

Godavari

Mumbai
(Bombay)

Krishna

Hyderabad

Kolkata
(Calcutta)

Bay of Bengal

Bangalore

Chennai
(Madras)

Madurai

Lakshadweep
India

SRI LANKA
Colombo

Male
MALDIVES

EQUATOR

I N D I A N   O C E A N

Chagos Archipelago
Brit. Ind. Oc. Terr.

N

60°E
80°E

ASIA
POLITICAL

0 mi ————— 1000
0 km ————— 1000

TWO-POINT EQUIDISTANT PROJECTION

NATIONAL GEOGRAPHIC

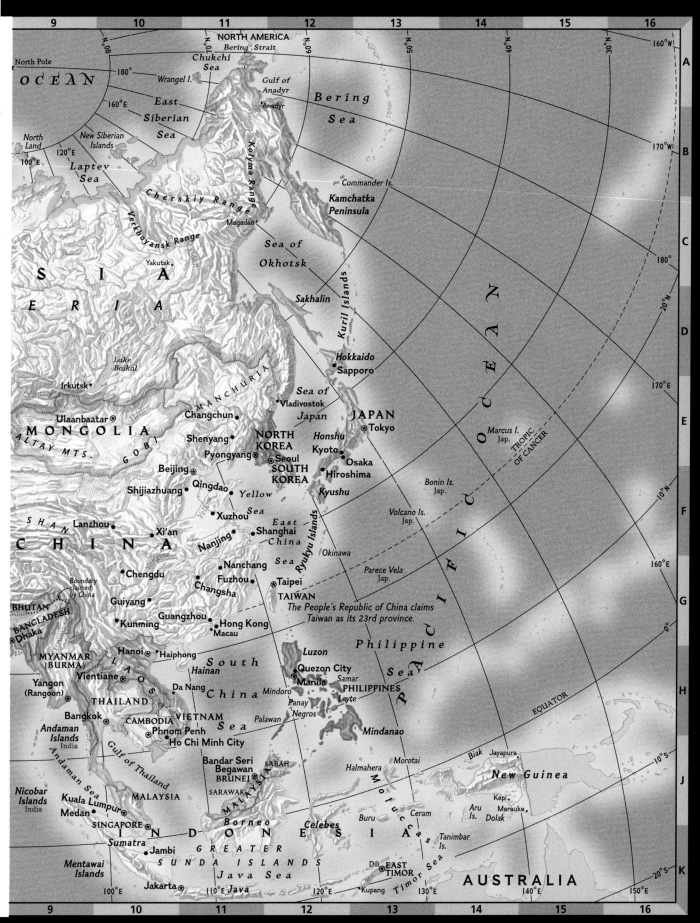

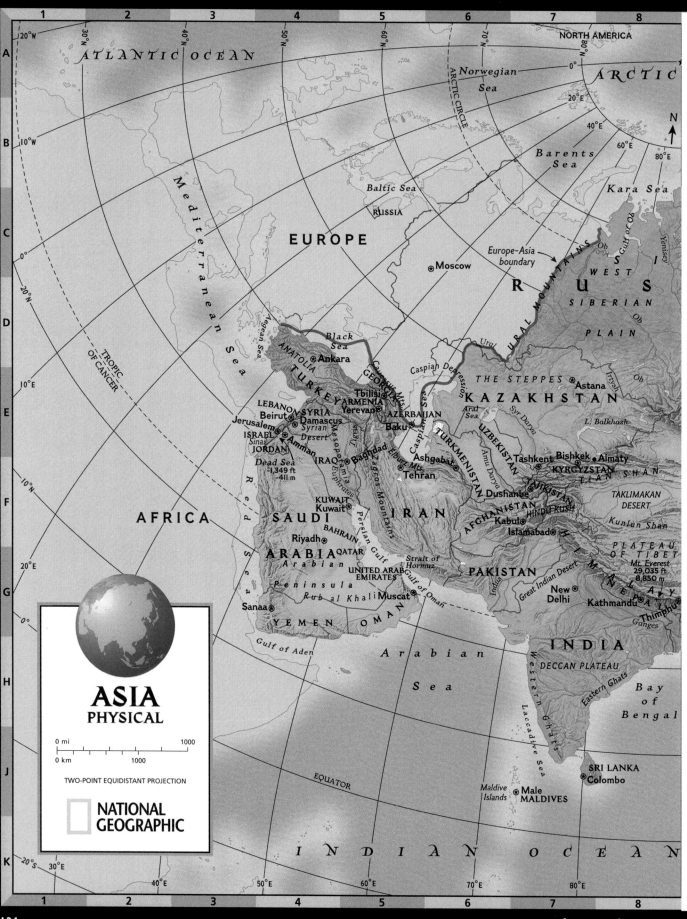

ASIA
PHYSICAL

0 mi   1000
0 km   1000

TWO-POINT EQUIDISTANT PROJECTION

NATIONAL GEOGRAPHIC

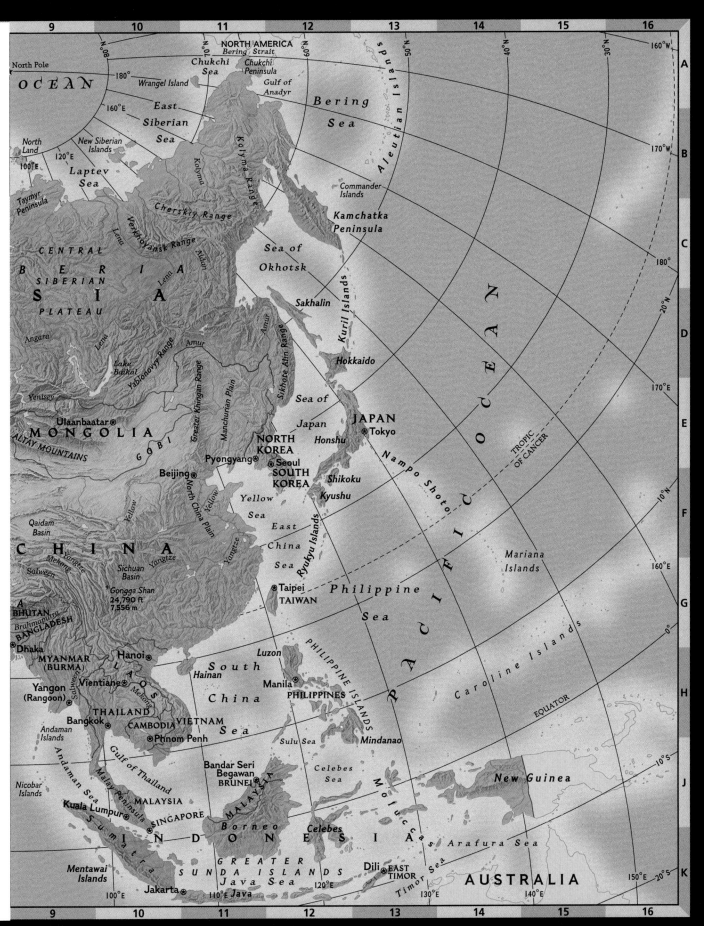

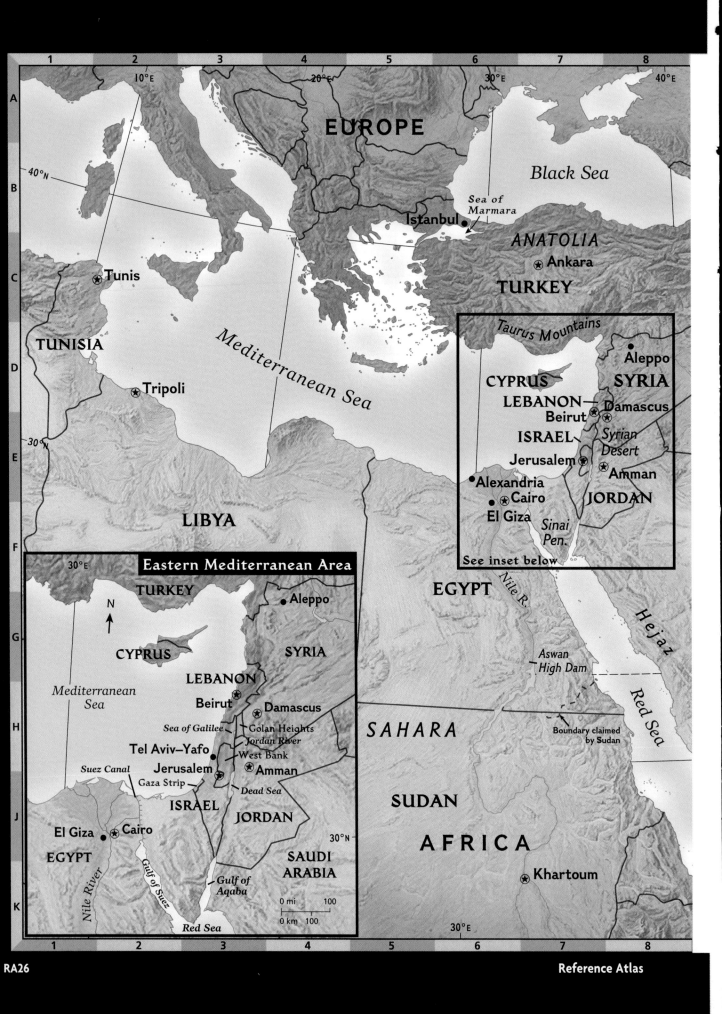

1   **2**   **3**   **4**   **5**   **6**   **7**   **8**

A

40°N

B

EUROPE

Black Sea

Sea of Marmara

Istanbul

C

Tunis

ANATOLIA

Ankara

TURKEY

TUNISIA

Mediterranean Sea

Tripoli

Taurus Mountains

Aleppo

D

CYPRUS   **SYRIA**

LEBANON   Damascus

Beirut

ISRAEL   *Syrian Desert*

Jerusalem   Amman

30°N

E

Alexandria

Cairo   **JORDAN**

El Giza

Sinai Pen.

LIBYA

See inset below

F

Eastern Mediterranean Area

30°E

TURKEY

N

EGYPT

Nile R.

Aleppo

Hejaz

G

Mediterranean Sea

CYPRUS

SYRIA

Aswan High Dam

LEBANON

Beirut

Damascus

Red Sea

H

Sea of Galilee   Golan Heights

Jordan River

Tel Aviv–Yafo   West Bank

SAHARA

Boundary claimed by Sudan

Suez Canal

Jerusalem   Amman

Gaza Strip   *Dead Sea*

J

El Giza   Cairo

ISRAEL   **JORDAN**

30°N

SUDAN

EGYPT

Nile River

Gulf of Suez

SAUDI ARABIA

A F R I C A

K

Gulf of Aqaba

0 mi   100

0 km   100

Red Sea

Khartoum

30°E

1   **2**   **3**   **4**   **5**   **6**   **7**   **8**

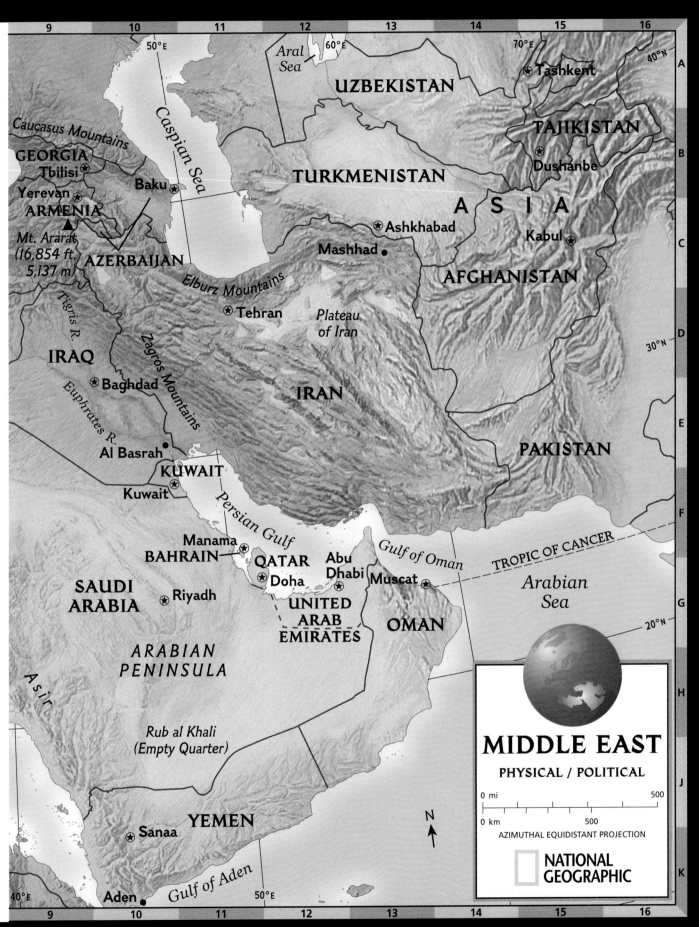

Aral Sea

UZBEKISTAN

TAJIKISTAN

⊛ Tashkent

Caucasus Mountains

Caspian Sea

TURKMENISTAN

Dushanbe

GEORGIA
Tbilisi ⊛

Baku ⊛

A S I A

Yerevan ⊛
ARMENIA

⊛ Ashkhabad

Kabul ⊛

▲ Mt. Ararat
(16,854 ft.
5,137 m)

AZERBAIJAN

Elburz Mountains

Mashhad •

AFGHANISTAN

Tigris R.

⊛ Tehran

Plateau
of Iran

IRAQ

Zagros Mountains

IRAN

⊛ Baghdad

Euphrates R.

PAKISTAN

Al Basrah •

KUWAIT

Kuwait •

Persian Gulf

Manama
BAHRAIN ⊛

Gulf of Oman

TROPIC OF CANCER

QATAR
Doha ⊛

Abu
Dhabi
⊛

Muscat ⊛

Arabian
Sea

SAUDI
ARABIA

⊛ Riyadh

UNITED
ARAB
EMIRATES

OMAN

ARABIAN
PENINSULA

Asir

Rub al Khali
(Empty Quarter)

N

MIDDLE EAST

PHYSICAL / POLITICAL

0 mi   500

0 km   500

AZIMUTHAL EQUIDISTANT PROJECTION

YEMEN

Sanaa •

NATIONAL
GEOGRAPHIC

Aden •

Gulf of Aden

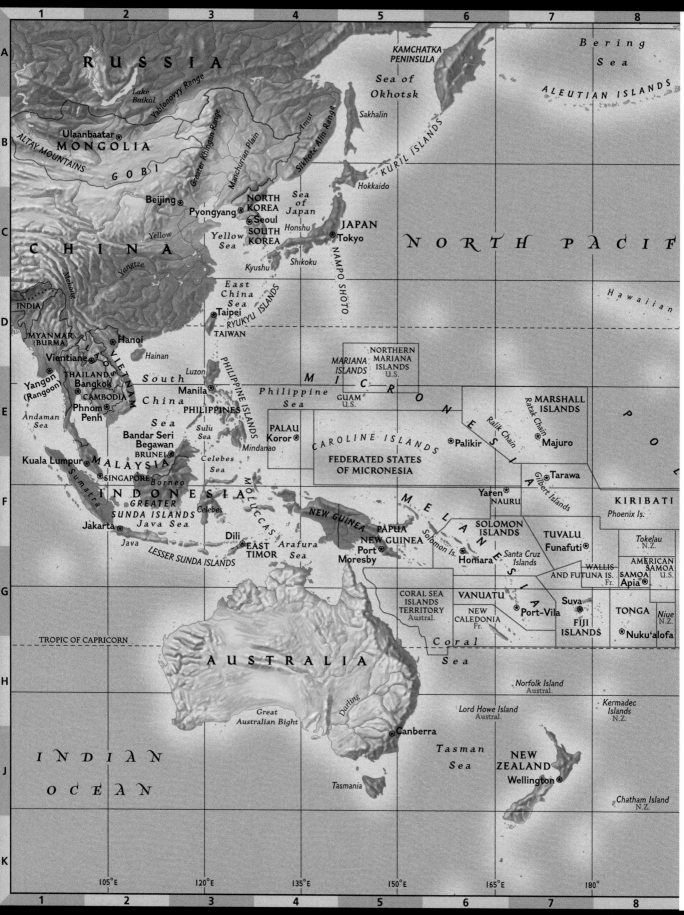

9 **10** **11** **12** **13** **14** **15** **16**

Gulf of
Alaska

Kodiak I.

Alexander
Archipelago

Coast Mountains

*Queen
Charlotte
Islands*

R O C K Y   M O U N T A I N S

G R E A T

C A N A D I A N   S H I E L D

*Hudson
Bay*

C A N A D A

A

*Vancouver
Island*

Cascade Range

G R E A T   P L A I N S

Missouri

Great Lakes

APPALACHIAN MTS.

Ottawa ⊛

B

45°N

I C   O C E A N

C E N T R A L
L O W L A N D

Mississippi

Washington ⊛

C

30°N

U N I T E D   S T A T E S

C O A S T A L   P L A I N

ATLANTIC
OCEAN

Baja California

Gulf of California

Sierra Madre Occidental

Sierra Madre Oriental

TROPIC OF CANCER

Islands

HAWAII
U.S.

M E X I C O

Gulf of
Mexico

Nassau ⊛

BAHAMAS

D

Havana ⊛

CUBA

DOMINICAN
REPUBLIC

15°N

Mexico
City ⊛

JAMAICA HAITI

Santo
Domingo

GUATEMALA
Guatemala City ⊛
San Salvador ⊛
EL SALVADOR

BELIZE
HONDURAS
⊛ Tegucigalpa
NICARAGUA
Managua ⊛

C a r i b b e a n
S e a

Caracas
⊛

E

San Jose ⊛
COSTA RICA

Panama City ⊛
PANAMA

VENEZ.

Kiritimati

Line Islands

0°

EQUATOR

*Galapagos
Islands
Ecua.*

Quito ⊛

LLANOS

Bogota
⊛

COLOMBIA

F

ECUADOR

AMAZON
BASIN

Y

E

Marquesas Is.

PERU

BRAZIL

G

COOK
ISLANDS
N.Z.

Society Is.

Tuamotu Archipelago

15°S

Lima ⊛

S

I

A

FRENCH POLYNESIA
Fr.

Austral Is.

Pitcairn Island U.K.

La Paz ⊛

BOLIVIA

H

30°S

S O U T H

Santiago ⊛

ARGENTINA

ANDES

CHILE

J

P A C I F I C

*Chiloe
Island*

PATAGONIA

O C E A N

45°S

K

165°W **150°W** **135°W** **120°W** **105°W** **90°W** **75°W**

9 **10** **11** **12** **13** **14** **15** **16**

PACIFIC
RIM

PHYSICAL/POLITICAL

0 mi   1500

0 km   1500

MILLER CYLINDRICAL PROJECTION

**NATIONAL
GEOGRAPHIC**

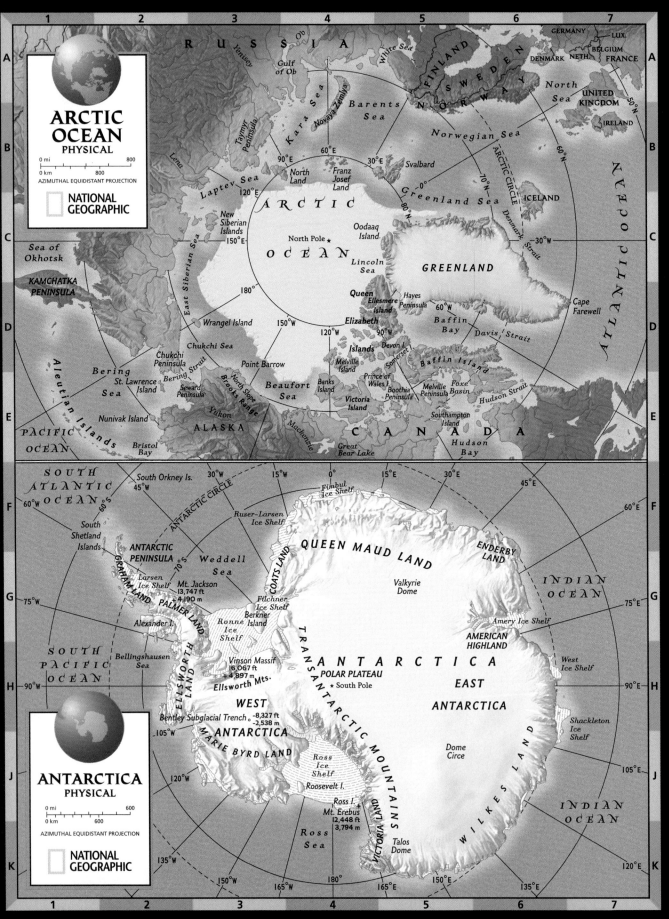

RA30

NATIONAL GEOGRAPHIC

GEOGRAPHY HANDBOOK

A geographer is a person who studies the earth and its people. Have you ever wondered if you could be a geographer? One way to learn how is to use this Geography Handbook. It will show you that geography is more than studying facts and figures. It also means doing some of your own exploring of the earth. By learning how to use geographic tools, such as globes, maps, and graphs, you will get to know and appreciate the wonders of our planet.

Geologists studying a volcano ▲

Table of Contents

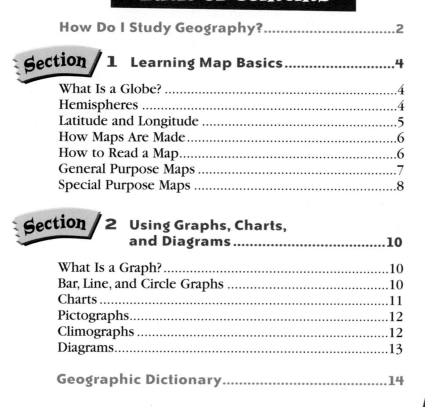

Polar ice cap ▼

◄ Lava flow

How Do I Study

Everything you see, touch, use, and even hear is related to geography—the study of the world's people, places, and environments. How can you possibly study such a huge amount of information in your geography class? Where do you start?

Geographers—people who study geography—ask themselves this question, too. To understand how our world is connected, some geographers have broken down the study of geography into five themes. The Five Themes of Geography are (1) location, (2) place, (3) human/environment interaction, (4) movement, and (5) regions. These themes are highlighted in blue throughout this textbook.

Most recently, as suggested in the *Geography Standards for Life,* geographers have begun to look at geography a different way. They break down the study of geography into Six Essential Elements, which are explained for you below. Being aware of these elements will help you sort out what you are learning about geography.

ELEMENT 2 Places and Regions

Geographers also look at places and regions. Place includes those features and characteristics that give an area its own identity or personality. These can be physical characteristics—such as landforms, climate, plants, and animals—or human characteristics—such as language, religion, architecture, music, politics, and way of life.

To make sense of all the complex things in the world, geographers often group places or areas into regions. Regions are united by one or more common characteristics.

Des Moines, Iowa, is a place ▲ characterized by farms. It is also part of a region known as the Corn Belt.

ELEMENT 1 The World in Spatial Terms

Geographers first take a look at where a place is located. Location serves as a starting point by asking "Where is it?" Knowing the location of places helps you to orient yourself in space and to develop an awareness of the world around you.

◀ This street sign is located in Paris, France.

ELEMENT 3 Physical Systems

Why do some places have mountains and other places have flat deserts? When studying places and regions, geographers analyze how physical systems—such as volcanoes, glaciers, and hurricanes—interact and shape the earth's surface. They also look at ecosystems, or communities of plants and animals that are dependent upon one another and their particular surroundings for survival.

◀ A glacier carved this deep valley in New Zealand.

Geography?

ELEMENT 4 Human Systems

Geographers also examine human systems, or how people have shaped our world. They look at how boundary lines are determined and analyze why people settle in certain places and not in others. An ongoing theme in geography is the continual movement of people, ideas, and goods.

People, vehicles, and goods move quickly through Asmara, the capital of Eritrea. ▶

ELEMENT 5 Environment and Society

The study of geography includes looking at human/environment interaction, or how and why people change their surroundings. Throughout history, people have cut forests and dammed rivers to build farms and cities. Some activities have led to air and water pollution. The physical environment affects human activities as well. The type of soil and amount of water in a place determines if crops can be grown. Earthquakes and floods also affect human life.

◀ Romanian farmers work in a field near a nuclear power plant.

ELEMENT 6 The Uses of Geography

Understanding geography, and knowing how to use the tools and technology available to study it, prepares you for life in our modern society. Individuals, businesses, and governments use geography and maps of all kinds on a daily basis. Computer programs, such as geographic information systems (GIS), allow us to make informed decisions about how to make the best use of our place and region.

◀ A cartographer uses GIS to make a map.

3

Learning Map Basics

Guide To Reading

Main Idea

Globes and maps provide different ways of showing features of the earth.

Terms to Know

- hemisphere
- latitude
- longitude
- scale bar
- scale
- relief
- elevation
- contour line

What Is a Globe?

A globe is a model of the earth that shows the earth's shape, lands, distances, and directions as they truly relate to one another. A world globe can help you find your way around the earth. By using one, you can locate places and determine distances.

Hemispheres

To locate places on the earth, geographers use a system of imaginary lines that crisscross the globe. One of these lines, the Equator, circles the middle of the earth like a belt. It

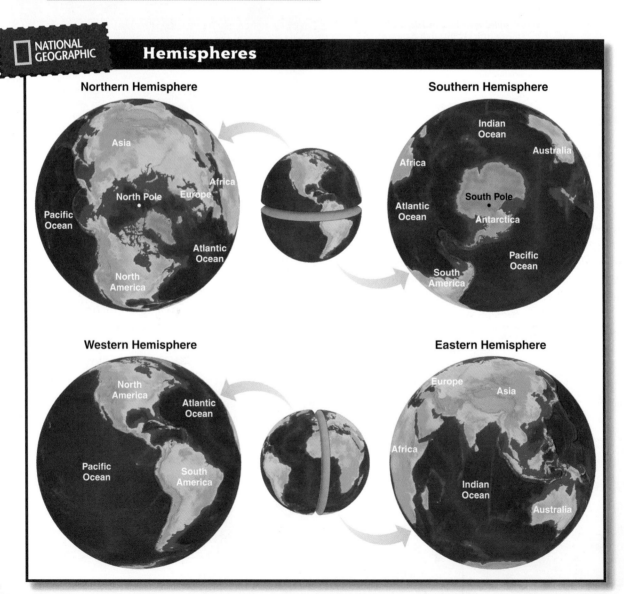

NATIONAL GEOGRAPHIC

Hemispheres

Northern Hemisphere

Asia
Africa
Europe
North Pole
Pacific Ocean
Atlantic Ocean
North America

Southern Hemisphere

Indian Ocean
Australia
Africa
Atlantic Ocean
South Pole
Antarctica
Pacific Ocean
South America

Western Hemisphere

North America
Atlantic Ocean
Pacific Ocean
South America

Eastern Hemisphere

Europe
Asia
Africa
Indian Ocean
Australia

divides the earth into "half spheres," or **hemispheres.** Everything north of the Equator is in the Northern Hemisphere. Everything south of the Equator is in the Southern Hemisphere.

Another imaginary line runs from north to south. It helps divide the earth into half spheres in the other direction. Find this line—called the Prime Meridian or the Meridian of Greenwich—on a globe. Everything east of the Prime Meridian for 180 degrees is in the Eastern Hemisphere. Everything west of the Prime Meridian for 180 degrees is in the Western Hemisphere. In which hemispheres is North America located? It is found in both the Northern Hemisphere and the Western Hemisphere.

Latitude and Longitude

The Equator and the Prime Meridian are the starting points for two sets of lines used to find any location. *Parallels* circle the earth like stacked rings and show **latitude,** or distance measured in degrees north and south of the Equator. The letter *N* or *S* following the degree symbol tells you if the location is north or south of the Equator. The North Pole, for example, is at 90°N (North) latitude, and the South Pole is at 90°S (South) latitude.

Two important parallels in between the poles are the Tropic of Cancer at 23½°N latitude and the Tropic of Capricorn at 23½°S latitude. You can also find the Arctic Circle at

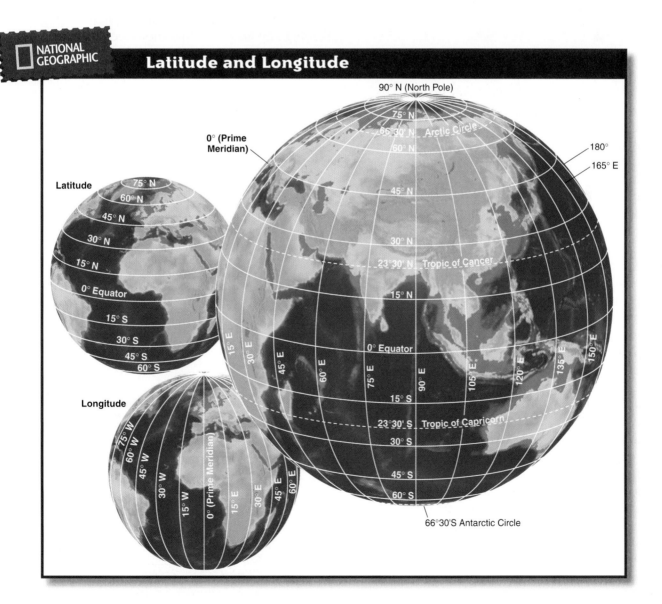

Latitude and Longitude

66½°N latitude and the Antarctic Circle at 66½°S latitude.

Meridians run from pole to pole and crisscross parallels. Meridians signify **longitude,** or distance measured in degrees east *(E)* or west *(W)* of the Prime Meridian. The Prime Meridian, or 0° longitude, runs through Greenwich, England. On the opposite side of the earth is the 180° meridian, also called the International Date Line.

Lines of latitude and longitude cross each other in the form of a grid system. You can find a place's absolute location by naming the latitude and longitude lines that cross exactly at that place. For example, the city of Tokyo, Japan, is located at about 36°N latitude and about 140°E longitude.

🌐 How Maps Are Made

For more than 4,000 years, people have made maps to organize their knowledge of the world. The reason for producing maps has not changed over the centuries, but the tools of mapmaking have. Today satellites located thousands of miles in space gather data about the earth below. The data are then sent back to the earth, where computers change the data into images of the earth's surface. Mapmakers analyze and use these images to produce maps.

For modern mapmakers, computers have replaced pen and paper. Most mapmakers use computers with software programs called *geographic information systems (GIS)*. With GIS, each kind of information on a map is kept as a separate electronic "layer" in the map's computer files. Because of this modern technology, mapmakers are able to make maps—and change them—more quickly and easily than before.

🌐 How to Read a Map

Maps can direct you down the street, across the country, or around the world. An ordinary map holds all kinds of information. Learn the map's code, and you can read it like a book.

Map Key   The map key explains the lines, symbols, and colors used on a map. Look at the map of Spain below. Its key shows that dots mark major cities. A circled star indicates the national capital—in Spain's case, the city of Madrid. Some keys tell which lines stand for national boundaries, roads, or railroads. Other map symbols may represent human-made or natural features, such as canals, forests, or natural gas deposits.

Compass Rose   An important step in reading any map is to find the direction marker. A map has a symbol that tells you where the *cardinal directions*—north, south, east, and west—are positioned. Sometimes all of these directions are shown with a compass rose. An *intermediate direction,* such as southeast, may also be on the compass rose. Intermediate directions fall between the cardinal directions.

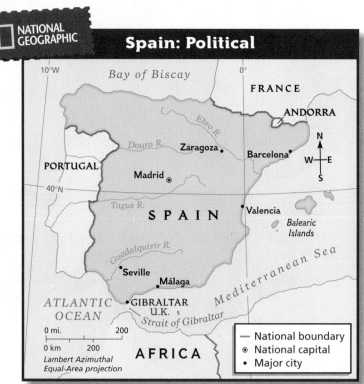

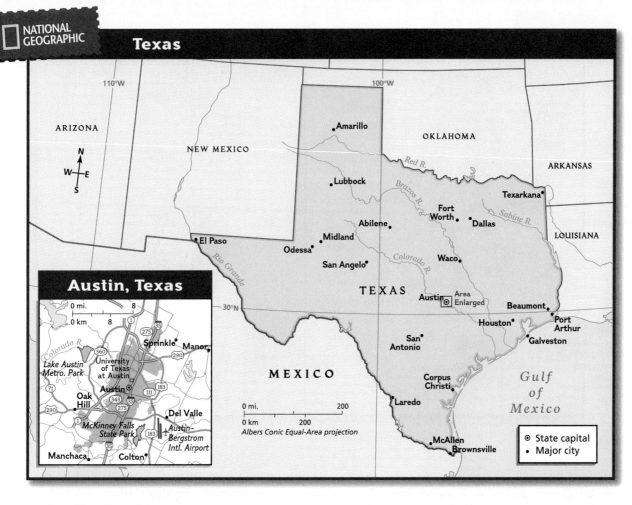

Texas

Map labels:

ARIZONA

NEW MEXICO

OKLAHOMA

ARKANSAS

LOUISIANA

110°W   100°W

N W E S

Amarillo

Lubbock

Red R.

Texarkana

Brazos R.

Fort Worth

Dallas

Sabine R.

Abilene

El Paso

Midland

Odessa

Waco

Colorado R.

San Angelo

Rio Grande

30°N

TEXAS

Austin Area Enlarged

Beaumont

Houston

Port Arthur

Galveston

San Antonio

MEXICO

Corpus Christi

Gulf of Mexico

Laredo

0 mi. 200
0 km 200
Albers Conic Equal-Area projection

McAllen
Brownsville

⊙ State capital
• Major city

Austin, Texas

0 mi. 8
0 km 8

Colorado R.

Sprinkle   Manor

Lake Austin Metro. Park

University of Texas at Austin

Oak Hill

Austin

Del Valle

McKinney Falls State Park

Austin-Bergstrom Intl. Airport

Manchaca   Colton

Latitude and Longitude Lines

Like globes, maps have lines of latitude and longitude that form a grid. Every place on the earth has a unique position or "address" on this grid. Knowing this address makes it easier for you to locate cities and other places on a map. For example, what is the grid address of Madrid, Spain? The map on page 6 shows you that the address is about 41°N latitude and about 4°W longitude.

Scale   A measuring line, often called a scale bar, helps you determine distance on a map. The map's scale tells you what distance on the earth is represented by the measurement on the scale bar. For example, 200 miles on the earth may be represented by 1 inch on the map. Knowing the scale allows you to see how large an area is. Map scale is usually given in both miles and kilometers.

Each map has its own scale. What scale a mapmaker uses depends on the size of the area shown on the map. If you were drawing a map of your backyard, you might use a scale of 1 inch equals 5 feet. In contrast, the scale bar on the inset map above of Austin, Texas, shows that about ⅝ inch represents 8 miles. Scale is important when you are trying to compare the size of one area to another.

General Purpose Maps

Maps are amazingly useful tools. You can use them to preserve information, to display data, and to make connections between seemingly unrelated things. Geographers use many different types of maps. Maps that show a wide range of general information about an area are called *general purpose maps.* Two of the most common general purpose maps are political and physical maps.

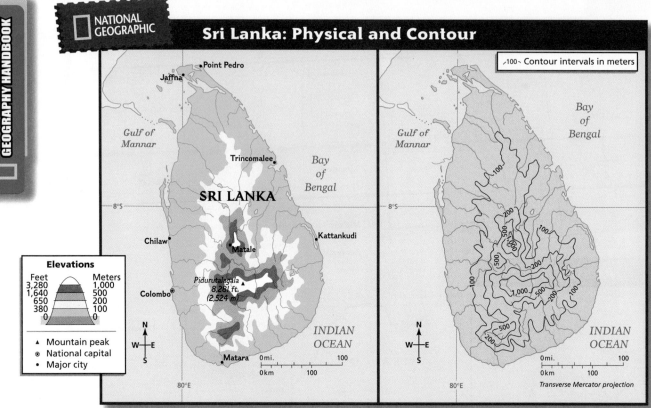

Sri Lanka: Physical and Contour

Political Maps   *Political maps* show the names and boundaries of countries, the location of cities and other human-made features of a place, and often identify major physical features. The political map of Spain on page 6, for example, shows the boundaries between Spain and other countries. It also shows cities and rivers within Spain and bodies of water surrounding Spain.

Physical Maps   *Physical maps* call out landforms and water features. The physical map of Sri Lanka above shows rivers and mountains. The colors used on physical maps include brown or green for land, and blue for water. These colors and shadings may show relief—or how flat or rugged the land surface is. In addition, physical maps may use colors to show elevation—the height of an area above sea level. A key explains what each color and symbol stands for.

Contour Maps   One kind of physical map, called a *contour map,* also shows elevation. A

contour map has contour lines—one line for each major level of elevation. All the land at the same elevation is connected by a line. These lines usually form circles or ovals—one inside the other. If contour lines come very close together, the surface is steep. If the lines are spread apart, the land is flat or rises very gradually. Compare the contour map of Sri Lanka above to its physical map.

Special Purpose Maps

Some maps are made to present specific kinds of information. These are called *thematic* or *special purpose maps.* They usually show themes or patterns, often emphasizing one subject or theme. Special purpose maps may present climate, natural resources, and population density. They may also display historical information, such as battles or territorial changes. The map's title tells what kind of special information it shows. Colors and symbols in the map key are especially important on these types of maps.

One type of special purpose map uses colors to show population density, or the average number of people living in a square mile or square kilometer. As with other maps, it is important to first read the title and the key. The population density map of Egypt to the right gives a striking picture of differences in population density. The Nile River valley and delta are very densely populated. In contrast, the desert areas east and west of the river are home to few people.

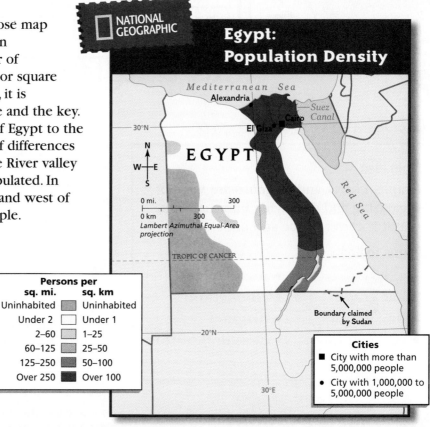

Egypt: Population Density

Persons per		
sq. mi.		**sq. km**
Uninhabited		Uninhabited
Under 2		Under 1
2–60		1–25
60–125		25–50
125–250		50–100
Over 250		Over 100

Boundary claimed by Sudan

Cities
■ City with more than 5,000,000 people
● City with 1,000,000 to 5,000,000 people

Section 1 Assessment

Defining Terms
1. Define hemisphere, latitude, longitude, scale bar, scale, relief, elevation, contour line.

Recalling Facts
2. Why do people make maps?

3. What are the four cardinal directions?

4. What are two of the most common types of general purpose maps?

Critical Thinking
5. Comparing and Contrasting Describe the similarities and differences between physical maps and contour maps.

6. Synthesizing Information What imaginary line divides the earth into the Eastern and Western Hemispheres?

Graphic Organizer
7. Organizing Information Create a diagram like the one below. In each of the outer ovals, write an example of a feature that you would find on a typical physical map.

Map Features

Applying Social Studies Skills

8. Analyzing Maps Look at the map of Egypt above. At what latitude and longitude is Alexandria located? Use the key to describe the population density of Alexandria and its surrounding area.

Using Graphs, Charts, and Diagrams

Guide To Reading

Main Idea

Graphs, charts, and diagrams are ways of organizing and displaying information so that it is easier to see and understand.

Terms to Know

- axis
- bar graph
- line graph
- circle graph
- chart
- pictograph
- climograph
- diagram
- elevation profile

What Is a Graph?

A graph is a way of summarizing and presenting information visually. Each part of a graph gives useful information. First read the graph's title to find out its subject. Then read the labels along the graph's axes—the vertical line along the left side of the graph and the horizontal line along the bottom of the graph. One axis will tell you what is being measured. The other axis tells what units of measurement are being used.

Bar, Line, and Circle Graphs

Bar Graphs Graphs that use bars or wide lines to compare data visually are called bar graphs. Look carefully at the bar graph below, which compares world languages. The vertical axis lists the languages. The horizontal axis gives speakers of the language in millions. By comparing the lengths of the bars, you can quickly tell which language is spoken by the most people. Bar graphs are especially useful for comparing quantities, and they may show the bars rising up from the bottom of the graph or extending out from the vertical axis.

Line Graphs A line graph is a useful tool for showing changes over a period of time. The amounts being measured are plotted on the grid above each year, and then are connected by a line. Line graphs sometimes have two or more lines plotted on them. The line graph on page 11 shows that the number of farms in the United States has decreased since 1940. The vertical axis lists the number of farms in millions. The horizontal axis shows the passage of time in ten-year periods from 1940 to 1998.

Comparing World Languages

Language	Number of Native Speakers (in millions)
Chinese (Mandarin)	885
English	322
Spanish	266
Bengali	189
Hindi	182
Portuguese	170
Russian	170
Japanese	125
German	98
Chinese (Wu)	77

Source: *National Geographic Atlas of the World*, 1999.

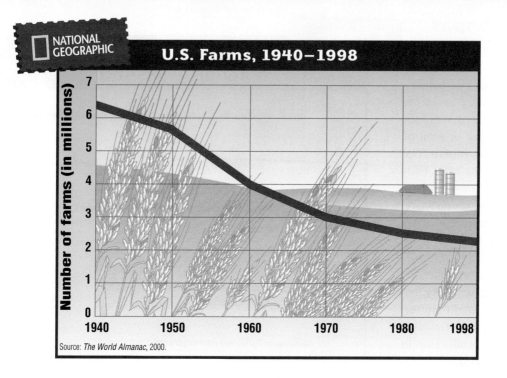

NATIONAL GEOGRAPHIC

U.S. Farms, 1940–1998

Source: *The World Almanac*, 2000.

Circle Graphs You can use circle graphs when you want to show how the *whole* of something is divided into its *parts.* Because of their shape, circle graphs are often called pie graphs. Each "slice" represents a part or percentage of the wholc "pie." On the circle graph below, the whole circle (100 percent) represents the world's population in 2000. The slices show how this population is divided among the world's five largest continents.

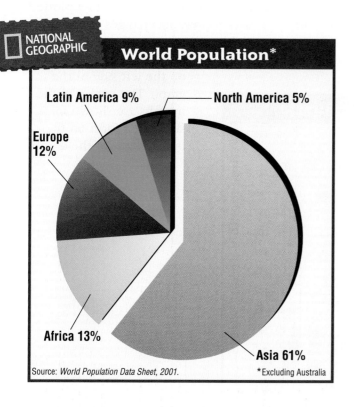

NATIONAL GEOGRAPHIC

World Population*

Latin America 9%   North America 5%

Europe 12%

Africa 13%

Asia 61%

Source: *World Population Data Sheet, 2001.*   *Excluding Australia

Charts

Charts present related facts and numbers in an organized way. They arrange data, especially numbers, in rows and columns for easy reference. Look at the chart on page 36. To interpret the chart, first read the title. It tells you what information the chart contains. Next, read the labels at the top of each column and on the left side of the chart. They explain what the numbers or data on the chart are measuring. One kind of chart, a *flowchart,* joins certain elements of a chart and a diagram. It can show the order of how things happen or how they are related to each other. The flowchart on page 520 presents the branches of the United States government. Notice how the chart shows the relationship among the tasks and the offices or bodies of each branch.

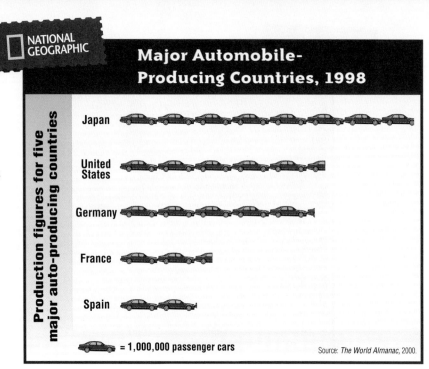

Pictographs

Like bar and circle graphs, pictographs are good for making comparisons. **Pictographs** use rows of small pictures or symbols, with each picture or symbol representing an amount. The pictograph on the right shows the number of automobiles produced in the world's five major automobile-producing countries. The key tells you that one car symbol stands for 1 million automobiles. Pictographs are read like a bar graph. The total number of car symbols in a row adds up to the auto production in each selected country.

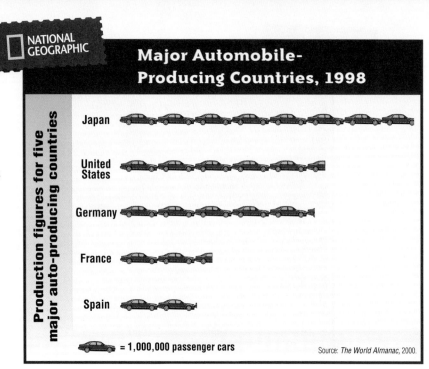

Major Automobile-Producing Countries, 1998

Production figures for five major auto-producing countries

Japan
United States
Germany
France
Spain

🚗 = 1,000,000 passenger cars

Source: *The World Almanac*, 2000.

Climographs

A **climograph,** or climate graph, combines a line graph and a bar graph. It gives an overall picture of the climate—the long-term weather patterns—in a specific place. Because climographs include several kinds of information, you need to read them carefully. Note that the vertical bars on the climograph below represent average amounts of precipitation (rain, snow, or sleet) in each month of the year. These bars are measured against the axis on the right side of the graph. The line plotted above the bars represents changes in the average monthly temperature. You measure this line against the axis on the left side of the graph. The names of the months are shown in shortened form on the bottom axis of the graph.

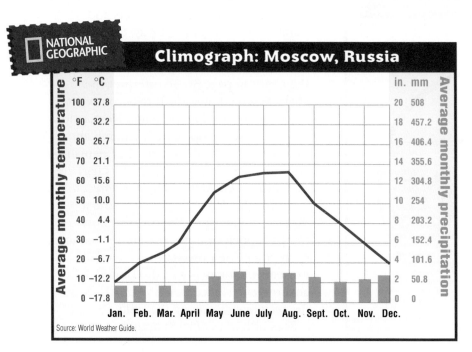

Climograph: Moscow, Russia

Average monthly temperature — Average monthly precipitation

Jan. Feb. Mar. April May June July Aug. Sept. Oct. Nov. Dec.

Source: World Weather Guide.

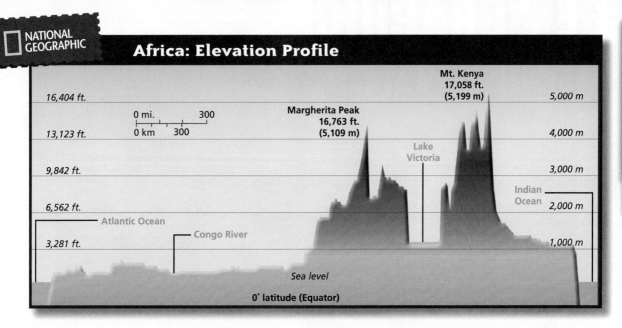

NATIONAL GEOGRAPHIC

Africa: Elevation Profile

16,404 ft.

13,123 ft.

9,842 ft.

6,562 ft.

3,281 ft.

0 mi. 300
0 km 300

Mt. Kenya
17,058 ft.
(5,199 m)

Margherita Peak
16,763 ft.
(5,109 m)

Lake
Victoria

Atlantic Ocean

Congo River

Indian
Ocean

Sea level

0° latitude (Equator)

5,000 m

4,000 m

3,000 m

2,000 m

1,000 m

Diagrams

Diagrams are drawings that show steps in a process, point out the parts of an object, or explain how something works. You can use a diagram to assemble a stereo. The diagram on page 562 shows how locks enable ships to move through a canal. An **elevation profile** is a type of diagram that can be helpful when comparing the elevations of an area. It shows an exaggerated profile, or side view, of the land as if it were sliced and you were viewing it from the side. The elevation profile of Africa above clearly shows low areas and mountains. The line of latitude at the bottom tells you where this profile was "sliced."

Section 2 Assessment

Defining Terms
1. Define axis, bar graph, line graph, circle graph, chart, pictograph, climograph, diagram, elevation profile.

Recalling Facts
2. How does a bar graph differ from a line graph?
3. What percentage does the whole circle in a circle graph always represent?
4. What two features does a climograph show?

Critical Thinking
5. Synthesizing Information Draw and label a flowchart showing the steps in some simple process—for example, making a sandwich or doing laundry.

Graphic Organizer
6. Organizing Information Create a chart like the one below. In the left column, list the types of graphs that are discussed in this section. In the right column, list what each type of graph is most useful for showing.

Types of Graphs	Useful for showing . . .

Applying Social Studies Skills

7. Analyzing Graphs Look at the bar graph on page 10. Which language is the most widely spoken? About how many people speak it?

GEOGRAPHIC DICTIONARY

NATIONAL GEOGRAPHIC

Volcano · **Mountain peak** · **Strait** · **Sound** · **Valley** · **Ocean** · **Cape** · **Island** · **Cliff** · **Isthmus** · **Bay** · **Harbor** · **Peninsula** · **Gulf** · **Delta** · **Seacoast**

As you read about the world's geography, you will encounter the terms listed below. Many of the terms are pictured in the diagram.

absolute location exact location of a place on the earth described by global coordinates

basin area of land drained by a given river and its branches; area of land surrounded by lands of higher elevations

bay part of a large body of water that extends into a shoreline, generally smaller than a gulf

canyon deep and narrow valley with steep walls

cape point of land that extends into a river, lake, or ocean

channel wide strait or waterway between two landmasses that lie close to each other; deep part of a river or other waterway

cliff steep, high wall of rock, earth, or ice

continent one of the seven large landmasses on the earth

delta flat, low-lying land built up from soil carried downstream by a river and deposited at its mouth

divide stretch of high land that separates river systems

downstream direction in which a river or stream flows from its source to its mouth

elevation height of land above sea level

Equator imaginary line that runs around the earth halfway between the North and South Poles; used as the starting point to measure degrees of north and south latitude

glacier large, thick body of slowly moving ice

gulf part of a large body of water that extends into a shoreline, generally larger and more deeply indented than a bay

harbor a sheltered place along a shoreline where ships can anchor safely

highland elevated land area such as a hill, mountain, or plateau

hill elevated land with sloping sides and rounded summit; generally smaller than a mountain

island land area, smaller than a continent, completely surrounded by water

isthmus narrow stretch of land connecting two larger land areas

lake a sizable inland body of water

latitude distance north or south of the Equator, measured in degrees

longitude distance east or west of the Prime Meridian, measured in degrees

lowland land, usually level, at a low elevation

map drawing of the earth shown on a flat surface

meridian one of many lines on the global grid running from the North Pole to the South Pole; used to measure degrees of longitude

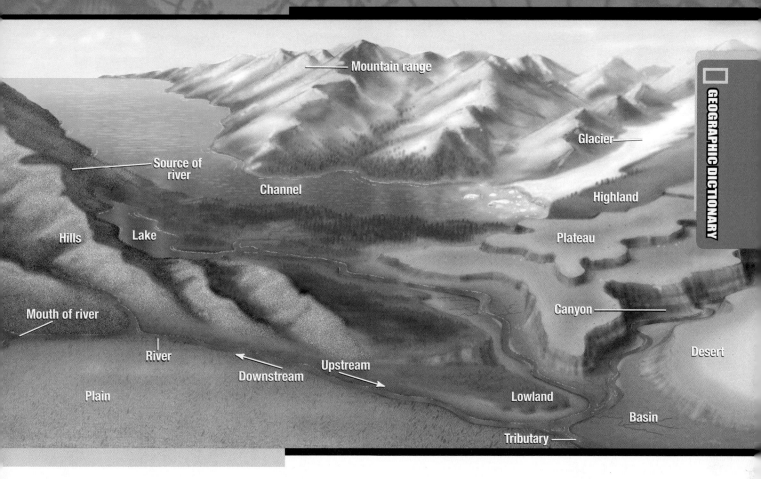

mesa broad, flat-topped landform with steep sides; smaller than a plateau

mountain land with steep sides that rises sharply (1,000 feet or more) from surrounding land; generally larger and more rugged than a hill

mountain peak pointed top of a mountain

mountain range a series of connected mountains

mouth (of a river) place where a stream or river flows into a larger body of water

ocean one of the four major bodies of salt water that surround the continents

ocean current stream of either cold or warm water that moves in a definite direction through an ocean

parallel one of many lines on the global grid that circles the earth north or south of the Equator; used to measure degrees of latitude

peninsula body of land jutting into a lake or ocean, surrounded on three sides by water

physical feature characteristic of a place occurring naturally, such as a landform, body of water, climate pattern, or resource

plain area of level land, usually at low elevation and often covered with grasses

plateau area of flat or rolling land at a high elevation, about 300–3,000 feet high

Prime Meridian line of the global grid running from the North Pole to the South Pole through Greenwich, England; starting point for measuring degrees of east and west longitude

relief changes in elevation over a given area of land

river large natural stream of water that runs through the land

sea large body of water completely or partly surrounded by land

seacoast land lying next to a sea or an ocean

sound broad inland body of water, often between a coastline and one or more islands off the coast

source (of a river) place where a river or stream begins, often in highlands

strait narrow stretch of water joining two larger bodies of water

tributary small river or stream that flows into a large river or stream; a branch of the river

upstream direction opposite the flow of a river; toward the source of a river or stream

valley area of low land between hills or mountains

volcano mountain created as liquid rock and ash erupt from inside the earth

Be an Active Reader!

How Should I Read My Textbook? Reading your social studies book is different from other reading you might do. Your textbook has a great amount of information in it. It is an example of nonfiction writing—it describes real-life events, people, ideas, and places.

Here are some reading strategies that will help you become an active textbook reader. Choose the strategies that work best for you. If you have trouble as you read your textbook, look back at these strategies for help.

✔ Before You Read

Set a Purpose
- Why are you reading the textbook?
- How might you be able to use what you learn in your own life?

Preview
- Read the chapter title to find out what the topic will be.
- Read the subtitles to see what you will learn about the topic.
- Skim the photos, charts, graphs, or maps.
- Look for vocabulary words that are boldfaced or in color. How are they defined?

Draw From Your Own Background
- What do you already know about the topic?
- How is the new information different from what you already know?

If You Don't Know What A Word Means...
- think about the setting, or *context*, in which the word is used.
- check if prefixes such as *un, non,* or *pre* can help you break down the word.
- look up the word's definition in a dictionary or glossary.

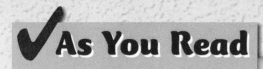

✔As You Read

Question

- What is the main idea?
- How well do the details support the main idea?
- How do the photos, charts, graphs, and maps support the main idea?

Connect

- Think about people, places, and events in your own life. Are there any similarities with those in your textbook?

Predict

- Predict events or outcomes by using clues and information that you already know.
- Change your predictions as you read and gather new information.

Visualize

- Use your imagination to picture the settings, actions, and people that are described.
- Create graphic organizers to help you see relationships found in the information.

Reading Do's

Do . . .

- ✔ establish a purpose for reading.
- ✔ think about how your own experiences relate to the topic.
- ✔ try different reading strategies.

Reading Don'ts

Don't . . .

- ⊘ ignore how the textbook is organized.
- ⊘ allow yourself to be easily distracted.
- ⊘ hurry to finish the material.

✔After You Read

Summarize

- Describe the main idea and how the details support it.
- Use your own words to explain what you have read.

Assess

- What was the main idea?
- Did the text clearly support the main idea?
- Did you learn anything new from the material?
- Can you use this new information in other school subjects or at home?

Young boy from the island of New Guinea

City of Hong Kong, China

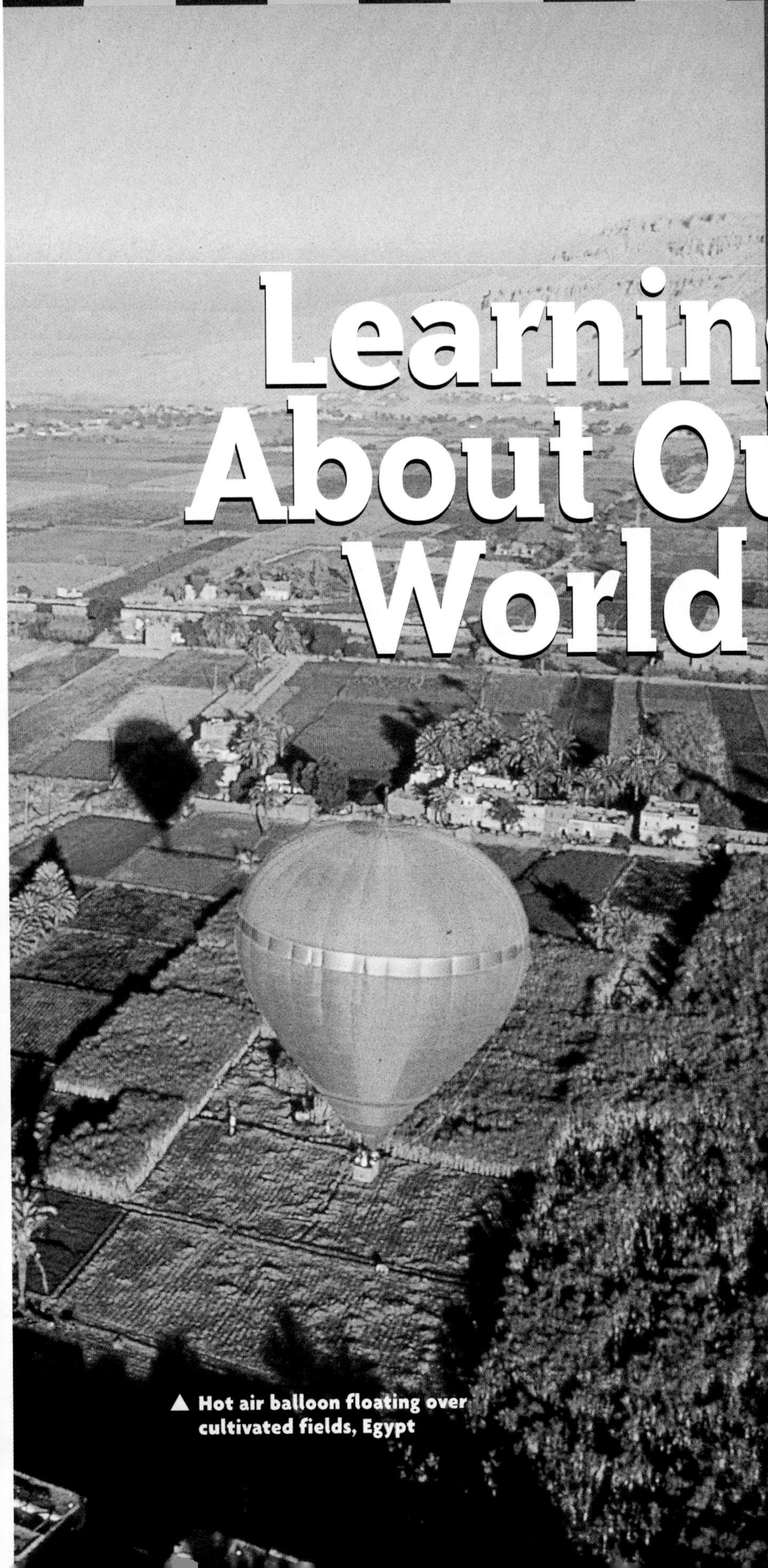

NATIONAL GEOGRAPHIC

Learning About Our World

You are about to journey to dense rain forests, bleak deserts, bustling cities and marketplaces, and remote villages. You are entering the many worlds of culture. In your study of the earth you will learn about different places and different peoples. Imagine that you could visit any place in the world. Where would you want to go? What would you want to see?

▲ Hot air balloon floating over cultivated fields, Egypt

NGS ONLINE
www.nationalgeographic.com/education

Our Social World

The World and Its People

NATIONAL GEOGRAPHIC

To learn more about the world's culture regions, view **The World and Its People Chapter 3** video.

Our World Today online

Chapter Overview Visit the **Our World Today: People, Places, and Issues** Web site at owt.glencoe.com and click on **Chapter 1–Chapter Overviews** to preview information about the world's people.

FOLDABLES™
Study Organizer

Categorizing Information Study Foldable Make this foldable to help you organize what you learn about our world, its people, and their cultures.

Step 1 Fold a sheet of paper into thirds from top to bottom.

This forms three sections.

Step 2 Open the paper and refold it into fourths from side to side.

Fold it in half, then in half again. → *This forms four sections.*

Step 3 Unfold the paper and draw lines along the folds.

Step 4 Label your table foldable as shown.

	Near	Far
World		
People		
Cultures		

Reading and Writing As you read the chapter, write key words and phrases in your table foldable to help you remember main ideas.

◄ **Painted elephants are part of the Dussehra festival in India.**

Why It Matters

Discovering Other Cultures

A while ago it was common for people to spend most of their lives in the same town or place in which they were born. Today, your neighbor may be someone from another state, another country, or another continent. How do people in the rest of the world live? How do we get along with them? This book will help you learn about other people and places and what issues are important to them.

People Far and Near

Guide to Reading

Main Idea

Modern technology has helped to bring the world's diverse peoples closer together.

Terms to Know

- ethnic group
- custom
- minority group
- majority group

Reading Strategy

Create a diagram like this one. On the spokes list reasons the world may be getting smaller.

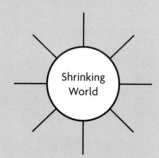

Shrinking World

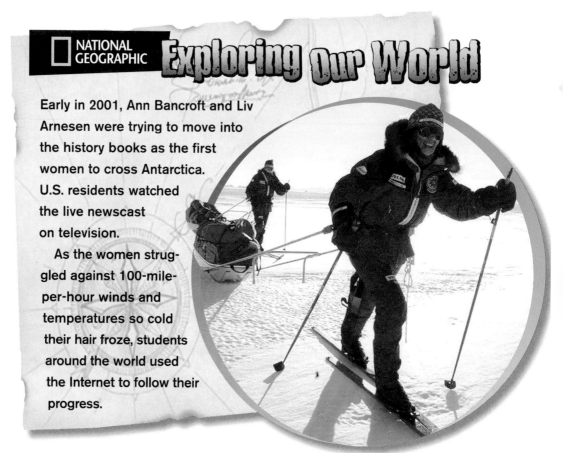

NATIONAL GEOGRAPHIC

Exploring Our World

Early in 2001, Ann Bancroft and Liv Arnesen were trying to move into the history books as the first women to cross Antarctica. U.S. residents watched the live newscast on television.

As the women struggled against 100-mile-per-hour winds and temperatures so cold their hair froze, students around the world used the Internet to follow their progress.

Today, people can talk across an ocean as easily as across a backyard fence. This is what is meant when you hear people say that the world today is shrinking. *Our World Today: People, Places, and Issues* is a book about our shrinking world. It is about the world's peoples, who they are, where they live, and how past experiences helped shape the world they live in today.

Technology Shrinks the World

With modern technology, the world's people have been brought closer together. Because the world is getting smaller, the chance that you will meet people from other cultures is increasing. By studying other people and countries, you will become able to see connections between the United States and the world around us. Learning to understand and respect what makes each culture unique, and recognizing common experiences that link all people will help you become an informed member of the global village.

Inventions Change the World

When the first telephone cable was laid along the bottom of the Atlantic Ocean in 1956, it could carry only 36 calls between Europe and North America at one time. Nearly forty years later, glass cables as fine as hairs were carrying 300,000 long-distance calls at once. Inventions are changing the world and it is changing faster every year.

Jet planes can cross oceans in brief hours, carrying people from one continent to another. Bullet trains speed workers from city to city while subways shuttle them across towns.

Communication satellites receive radio, television, and other signals in outer space. News can be broadcast live to the entire world so that more people than ever can watch. The result is a smaller world.

Internet Technology Millions of people today can use the **Internet,** a global network of computers, because of improved telephone cables and satellites. But other inventions made the Internet possible in the first place. The most important of these is the computer. Today's personal computers have more processing power than the large computers of the 1960s that helped put an American on the moon! Today, millions of people use the Internet to exchange mail, shop, do research, play games with friends in other countries, and much more. Again, the world seems to have grown a little smaller.

✓ Reading Check | **Name two ways in which technology makes the world seem smaller.**

The World Next Door

Sometimes the world really has become "smaller." A woman in a Houston suburb describes her neighborhood this way: "I was born in Chicago, Illinois, and have lived in the Houston area for 10 years. Inge, the woman next door, is from Denmark. Shiv, her husband,

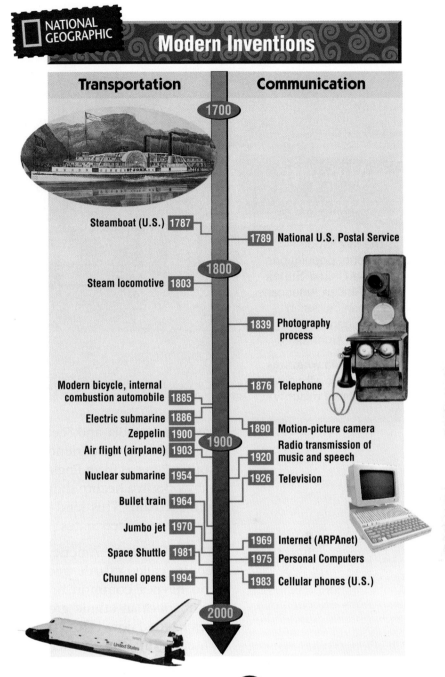

NATIONAL GEOGRAPHIC

Modern Inventions

Transportation | **Communication**

1700

Steamboat (U.S.) 1787

1789 National U.S. Postal Service

1800

Steam locomotive 1803

1839 Photography process

Modern bicycle, internal combustion automobile 1885

1876 Telephone

Electric submarine 1886
Zeppelin 1900
Air flight (airplane) 1903

1890 Motion-picture camera

1900

Radio transmission of music and speech 1920

Nuclear submarine 1954

1926 Television

Bullet train 1964

Jumbo jet 1970

1969 Internet (ARPAnet)

Space Shuttle 1981

1975 Personal Computers

Chunnel opens 1994

1983 Cellular phones (U.S.)

2000

Analyzing the Time Line

In the last century, communication and transportation technologies have evolved at an amazing rate.

Technology Which 19th century inventions are still used today?

Our Social World

23

Minority Groups

The largest ethnic minority populations in the United States are African Americans, Hispanics, Asian Americans, and Native Americans.

Issues In what way are people with disabilities also a minority?

was born in India, and Seeta and Rajiv, their kids, are Canadian citizens. My daughter's best friend, Ellen, was born in South Africa, and her mother, Janet, is from England. My best friend, Irma, came from Mexico. She works in a beauty shop owned by Van and her husband—a couple from Vietnam." This kind of neighborhood is rapidly becoming common in the United States.

Ethnic Groups American communities include various groups of peoples called ethnic groups. An ethnic group is a group of people who have a common origin and share a language and a history. Members of an ethnic group often follow the same customs—practices handed down from the past. How many ethnic groups are represented in the Houston neighborhood described above?

Minority Groups Often ethnic groups are also minorities. A minority group is a group of people whose culture, race, or ethnic origin is different from that of most of the people in the region. Sometimes the minority group is treated differently from the majority group. Majority is normally defined as a number greater than half of a total. When studying societies however, the majority is the group in society that controls most of the wealth and power. However, the majority group is not always the largest group in numbers. For example, South Africa at one time had a government that favored its white citizens and passed laws that restricted the black African population. The blacks were considered a minority even though they made up a much larger percentage of the population. In many cases women are thought of as a minority group because they have less influence than men have in business and politics.

It is important to remember that a minority group is not always the same as an ethnic group. For example, Asian Americans are a minority group in the United States, but "Asian" is not a specific ethnic term. All Asian Americans do not share the same origin, language, or history. Asian Americans may be Vietnamese, Chinese, or Japanese, for example. Still others may come from many other countries in Asia.

Building Bridges   Schools are the place where most young people first meet people from other ethnic groups. Public schools reflect the values in their neighborhoods. As one student says: "Making friends just depends on what you like to do, and who likes to do those things." Curiously, about the time students enter middle school and high school, things begin to change. **Sociologists,** the scientists who study organized groups of people, have found that student friendships start forming along racial lines. Skin color is not the only reason for division between ethnic groups, however. Sometimes groups divide over religious beliefs and values. Building bridges between the "different worlds" in your community and school is possible. You can start by learning about the beliefs and values of other people in the world.

✓ Reading Check   **Why is "Asian" not a specific ethnic term?**

Assessment

Defining Terms
1. **Define** ethnic group, custom, minority group, majority group.

Recalling Facts
2. **Culture** Explain the difference between a minority group and an ethnic group.
3. **Place** In what country was the group that controlled the wealth and power not the largest ethnic group?
4. **Technology** In what way is technology shrinking the world? Give examples.

Critical Thinking
5. **Understanding Cause and Effect** Why do you think minority groups are sometimes treated differently from the majority?
6. **Making Comparisons** What do you think it means that some groups are divided by religious beliefs and values?

Graphic Organizer
7. **Organizing Information** Create a diagram like this one that describes features of your culture. On the lines write the types of food, clothing, language, music, and so on found in your culture.

```
        Your
       Culture
```

Applying Social Studies Skills

8. **Interpreting Time Lines** Look at the time line on page 23. During which century were most modern communication and modern transportation invented?

Counting Heads

How do we know there are more than 280 million people in the United States? Who counts the people? Every 10 years since 1790, the United States Census Bureau has counted heads in this country. Why and how do they do this?

The First Census

After the American colonies fought the Revolutionary War and won their independence, the new government required a census. By knowing how many people were in each state, the government could divide the war expenses fairly. The census would also determine the number of people that each state could send to Congress.

The census began in August 1790, about a year after George Washington became president. The census law defined who would be counted, and it required that every household be visited by census takers. These workers walked or rode on horseback to gather their data. By the time it was completed, the census counted 3.9 million people.

The first census asked for little more than one's name and address. Over time, the census added questions to gather more than just population data. By 1820, there were questions about a person's job. Soon after, questions about crime, education, and wages appeared. In 1950 the census used its first computer to process data. Census takers go door-to-door to gather information from those who do not return their census forms in the mail. Now census data are released over the Internet.

▲ The Electric Tabulating Machine processed the 1890 census in 2½ years, a job that would have taken nearly 10 years to complete by hand.

The Census and Race

Sometimes the census itself can cause debate. Many people are uncomfortable with answering questions they believe are their own business. Others worry about how the information might be used or misused. The 2000 census gave rise to concerns about how people are counted and classified into races.

Before 2000, the census form directed individuals to mark only one box from a list of different races. This meant people of mixed races were forced to claim membership in one race. Tiger Woods, for example, might have had to choose between identifying himself as African American or Asian American.

Now, for the first time, the census had allowed people to mark one or more categories for race. Figures show that nearly 7 million people have taken advantage of this new way to be counted.

Making the Connection

1. In what two ways were population data from the first census used?

2. How has technology changed the way census data are collected and processed?

3. **Drawing Conclusions** Why do you think the national and state governments want information about people's education and jobs?

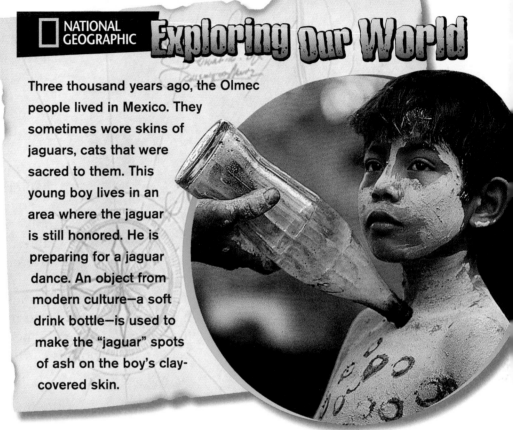

NATIONAL GEOGRAPHIC Exploring Our World

Three thousand years ago, the Olmec people lived in Mexico. They sometimes wore skins of jaguars, cats that were sacred to them. This young boy lives in an area where the jaguar is still honored. He is preparing for a jaguar dance. An object from modern culture—a soft drink bottle—is used to make the "jaguar" spots of ash on the boy's clay-covered skin.

Thanks to technology, many of us have the power to "tune in" to the world. By simply pushing a button or clicking a mouse, you can find out why polite Egyptians do not cross their legs in public, who introduced rabbits to Australia, who has made the most points in an NBA basketball game, when Russians celebrate Women's Day, and where people dress in *kangas*. But just knowing facts is not enough. It is far better to learn how to organize information so that it is meaningful and so that it helps you better understand the world.

The Social Sciences

Organizing information to help people understand the world around them is the role of social scientists. Social scientists study the interaction of people and society. There are many important disciplines, or fields of study, in the social sciences. Four types of social scientists deal directly with society. **Anthropologists** study people and

Celebrations

In most cultures, people often bring out their most beautiful clothes for events like weddings. This wedding guest is from Morocco (upper right), and this bride is from Mauritius (above).

Place What aspects of daily life besides clothing reflect culture?

societies. **Sociologists** study human behavior as it relates to groups of people. **Historians** study how societies came to be what they are today. **Human geographers** look at people and their environments.

Geographers organize facts about Earth's surface and people. They do this with maps, graphs, and other tools. These tools help geographers to find patterns in the earth's organization. For example, geographers examine why people live where they do and how different industries change the environment. In fact, the work geographers do is so important that you will read more about it in the next chapter.

Social scientists also give us tools to examine new beliefs and values. They look at what makes different people and places special. As one scientist warns: "There are two ways to make a person feel homeless—one is to destroy his home and the other is to make his home look and feel like everybody else's home." In this quotation, the home being talked about is *culture*.

✓ Reading Check What is the role of social scientists?

What Is Culture?

Waking up to rock music, putting on denim jeans, and celebrating the Fourth of July are part of the culture of the United States. Culture is the way of life of people who share similar beliefs and customs. These people may speak the same language, follow the same religion, and dress in a certain way. The culture of a people also includes their government, food, music, literature, and the ways they make a living. In the United States, people of many cultures live together. But people who live in the United States and who call themselves Americans believe in certain political values, such as freedom of speech, free public education, and the right to practice a religion of their choosing. Americans believe that hard work should be rewarded. Americans pride themselves on getting things done quickly and in a practical way. These beliefs, among others, are part of what defines us as Americans. What other beliefs can you name that define American culture?

Once people learn their own culture, it is sometimes hard for them to imagine any other way of life. They may want to judge people in terms of their own culture and their own standards. This practice is called ethnocentrism. (Remember the term *ethnic group* from Section 1?) Ethnocentrism means your values are "centered" or based on those of your own particular ethnic group. It is very common for people to prefer their own cultures. In fact, it is very hard not to. Many positive qualities, such as patriotism and taking pride in your nation's history, are ethnocentric feelings. Not all ethnocentric expressions are positive, however. Jokes about certain races or religions are really just ethnocentric statements about a different group of people. By making fun of the ways people are different, we are really saying that our way is better.

Cultural Borrowing Although all people have a culture, a large percentage is "borrowed" from other cultures. Cultural borrowing is the adoption of one group's culture traits by another group. A culture trait is a normal practice in a specific culture. The **Maoris** of New Zealand traditionally press their noses and foreheads together when they meet. When the Maoris shake hands instead, they are using a borrowed culture trait. Today, baseball is a popular sport in Japan. The Japanese borrowed this game from the United States. Possibly, Americans first borrowed the idea of baseball from a similar game played in Britain.

Cultural Diffusion Cultural diffusion is how a culture spreads its knowledge and skills from one area to another. Merchants and traders used to be the major agents, or causes, of cultural diffusion. They spread cultures when they bought and sold goods.

Today, cultural diffusion occurs through radio, television, telephones, computers, and the Internet. For instance, as many as one-third of the world's people have learned to speak English, often because of

EXPLORING CULTURE

Culture

Cultural differences can be interesting but they can also cause misunderstandings. These can be avoided if we learn how cultures differ from our own. In the photos, Inuit greet with a nose rub, Japanese businessmen bow in meeting (top right), and the French women kiss each other on the cheek (bottom right).

Looking Closer How do members of your community greet one another?

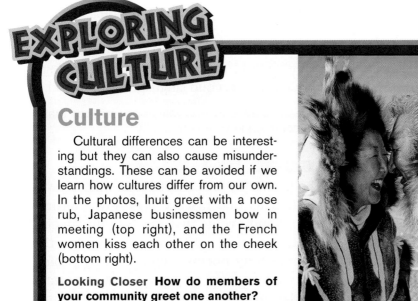

World Culture Regions

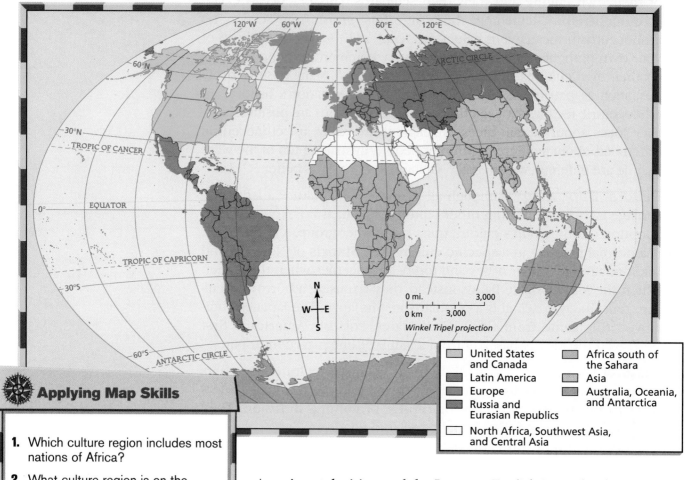

120°W 60°W 0° 60°E 120°E

60°N

ARCTIC CIRCLE

30°N

TROPIC OF CANCER

0° EQUATOR

TROPIC OF CAPRICORN

30°S

N
W—E
S

0 mi.   3,000
0 km   3,000
Winkel Tripel projection

60°S ANTARCTIC CIRCLE

United States and Canada
Latin America
Europe
Russia and Eurasian Republics
North Africa, Southwest Asia, and Central Asia
Africa south of the Sahara
Asia
Australia, Oceania, and Antarctica

Applying Map Skills

1. Which culture region includes most nations of Africa?

2. What culture region is on the continents of both Africa and Asia?

Find NGS online map resources @ www.nationalgeographic.com/maps

American television and the Internet. English is used in between 80 and 90 percent of Web sites worldwide. Airline pilots use English to communicate when they fly across national borders. Many experts predict that businesspeople worldwide will have to learn English—the language of most computer software—to keep up with world trade. Because of technology, English is the most widely used second language in the world.

✓Reading Check   **Name two ethnocentric values or practices that are common in your community.**

Important Lessons in History

History is the story of the past. For thousands of years, history was passed on by word of mouth. This oral history was how cultures passed on the record of people and events that made them special. Famous epics like *The Iliad*, the story of a 10-year war between the Greeks and the Trojans, began as long poems memorized by professional storytellers. These storytellers were respected and important members of their cultures. Today, **Aborigines** of Australia still memorize much of their history and repeat it in the form of stories to the young people in their clan, or family groups.

Eventually, writing became the most important way to keep records of the past. Historians today search through legal documents, diaries, newspapers, and many other written sources for information about the past. They also use artifacts, such as tools and household goods, to try to recreate what happened long ago.

History also tells how past conflicts influence the present. For example, why are millions of people in **Sudan** starving when this African country has large areas of rich farmland? A historian will know that the conflicts of today have their roots in religious wars dating back hundreds of years. Acres of farmland are left unplanted because soldiers burn the land and kill the farmers who try to grow crops in disputed areas.

Studying history can also teach us important lessons and can guide our behavior in the present. A well-known saying is "Those who cannot remember the past are condemned to repeat it." This means that if we do not study our own history, we may end up making the same mistakes of earlier generations.

✓ Reading Check **How was the history of a culture first passed on?**

▲ An African griot, or storyteller, shares cultural stories through song.

Government Is Necessary

Most families have rules that guide how the family members behave. These rules might be about mealtimes, chores, homework, or allowances. Without rules, family members would not know how to behave or what to expect from one another. People need rules in order to live together without conflict. In countries, rules are created by governments. In a limited government even the people who make the laws must obey them. Constitutions, statements of rights, or other laws set limits on how much power government officials have so that they cannot take advantage of the people.

Democracy is a form of limited government. In a direct democracy, the people govern themselves by voting individually on issues. Direct democracy is still practiced in some small New England towns and in parts of Switzerland. In a representative democracy, people elect representatives. Then the representatives make and enforce laws. The United States is a representative democracy.

In an unlimited government power belongs to the ruler or rulers. No rules or laws exist to limit what the ruler can or cannot do. Unlimited governments include dictatorships and absolute monarchies. An example of a dictatorship is **Saddam Hussein's** rule in Iraq. In an absolute monarchy kings or queens are born into ruling families. Their power is inherited and unlimited. King Fahd ibn Abdul Aziz of Saudi Arabia is an absolute monarch.

Not all monarchies are examples of unlimited government. **Queen Elizabeth II** of England is a monarch, but not an absolute ruler. Britain has a parliament and laws that limit the power of the British kings and queens. Great Britain is a constitutional monarchy and is a type of limited government.

✓ Reading Check **Describe the characteristics of limited and unlimited governments.**

Analyzing the Chart

Service industries embrace a wide range of areas, including banking and finance, education, health care, communication, and many others.

Economics What classification would describe most industry where you live?

Classification	Description	Example Product	
Primary	Takes natural resources from the earth – mining, forestry, fishing, and agriculture are included here.	Fishing	
Secondary	Makes products using the natural resources – construction, factories, and processing plants are in this classification.		Canning plant processes the fish
Tertiary	Provides a service – such as restaurants, supermarkets, hospitals, education, and emergency services.	Supermarket sells the fish	
Quaternary	Gathers information– industries in this classification research, gather, and provide information.	2 5 7 1 8 5	Bar codes tell the market when to reorder the fish

Balancing Our Wants and Needs

The different ways people and nations go about meeting their daily needs are known as *economic systems.* All economic systems are concerned with producing goods.

Traditional Economies In a traditional economy, people meet their needs on the basis of their customs. These have been handed down over many years. In some parts of Africa and South America, for example, children learn their trades from their parents who learned from their parents. In this way the same family does the same work generation after generation.

Command Economies Under a command economy, government makes all the decisions. Individuals have little or no say about basic economic questions such as what and how much to produce and what to charge. North Korea is a country with a command economy. Communism is an example of a command economy.

Market Economies In a market economy, individuals determine for themselves what to produce, who will want it **(demand),** how much to produce **(supply),** and how much to charge **(price).** No country has a pure market economy because governments regulate, or control, some parts of businesses. This system is sometimes called a "free enterprise system."

Mixed Economies Most nations have a mixed economy. The Chinese, for example, have mostly a command economy but are working toward a market system by allowing some private businesses. The United States prides itself on its market economy. However, the government may regulate prices or set rules as in the airline industry and companies that provide gas and electricity.

Differences in Development

Countries differ in how much manufacturing and industry they have. **Industrialized countries** hold 97 percent of all patents ownership (rights to inventions). Many countries in Europe and North America, as well as Australia, Japan, and South Korea, among others, are industrialized countries. Other countries have only a few manufacturing centers. Many people in these countries grow only enough food for their own families. Countries that are working toward industrialization are called **developing countries.** Most developing countries are found in Africa, Asia, and Latin America.

Many corporations from industrialized countries are now building plants in developing countries. They have found a valuable "resource" in these places—people. The spread of industry has created growing economies in places like Mexico and China.

Issues faced by developing and industrialized nations can be very different. Increasing population, not enough jobs, poor schools, and lack of social and health services are problems in many developing countries. In more industrialized countries, leaders are looking at ways to clean up pollution, fight crime and drugs, and protect their economies. However, no country is unaffected by the problems of its neighbors.

✓Reading Check **What are two issues facing most developing countries?**

Assessment

Defining Terms

1. Define social scientist, culture, ethnocentrism, cultural borrowing, cultural diffusion, limited government, democracy, unlimited government, dictatorship, absolute monarchy, constitutional monarchy.

Recalling Facts

2. Culture What four groups of scientists study society?

3. Government People in the United States who call themselves Americans believe in certain political values. What are they?

4. Culture What are the main agents of cultural diffusion today?

Critical Thinking

5. Understanding Cause and Effect How does history shape a culture?

6. Making Comparisons Analyze two kinds of economic systems.

Graphic Organizer

7. Organizing Information Create a diagram like this one. In the outer ovals list practices that are characteristic of your government.

Government

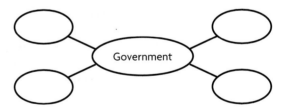

Applying Social Studies Skills

8. Analyzing Charts Look at the Types of Industries chart on page 32. Choose your own example product and show how it would go through the different processes.

Patterns in Today's World

Guide to Reading

Main Idea

All over the world people are moving from one place to another looking for freedom, jobs, and a better life.

Terms to Know

- migrate
- urbanization
- refugee
- rights
- responsibilities
- interdependence
- globalization

Reading Strategy

Create a chart like this one. List two regions under each that are experiencing conflict or cooperation at this time.

Conflict	Cooperation

NATIONAL GEOGRAPHIC Exploring Our World

Imagine that you and your friends are in Berlin, Germany. Can you hear the music? Every summer, hundreds of thousands of young people gather here for a music festival. Although most of these young people are here only to visit, many thousands of others come to find jobs and new lives. Germany faces challenges in finding room for its newcomers.

People have been moving from place to place since the dawn of time. Some places "pull" people with opportunities for freedom and a better life. Other places, where poverty is widespread or where governments are unstable, tend to "push" people away. The push and pull of migration is just one of the social forces that shape today's world.

Human Migration

Throughout the world, people migrate, or move, in great numbers. More and more people leave villages and farms and move to cities. This movement is called urbanization. Nearly half the world's people live in cities—a far higher percentage than ever before.

People move to cities for many reasons. The most common reason is to find jobs. This movement is putting a tremendous strain on the ability of cities to provide basic services such as clean water, sewage removal, housing, and health care.

When movement is from country to country, problems can increase dramatically. Refugees are people who are forced to leave their homelands because of wars or unjust governments. Refugees often do not speak the language or know the customs of the people in their adopted countries. They usually must work at the lowest-paying jobs and often without benefits earned by other workers.

Reading Check What is the most common reason people move to cities from rural areas?

Population Growth

How fast has the earth's population grown? The chart on page 36 shows the rapid increase in world population in recent years. Rapid population growth presents many challenges. An increase in the number of people means that more food is needed. Fortunately, since 1950 world food production has increased faster than population on all continents except Africa. Because so many people there need food, bad weather or war can ruin crops and bring disaster. Millions may suffer from a lack of food.

Also, populations that grow rapidly may use resources more quickly than populations that do not grow as fast. Some countries face shortages of water and housing.

Reading Check Why is rapid population growth a problem for many countries?

Our World Today Online

Web Activity Visit the *Our World Today: People, Places, and Issues* Web site at owt.glencoe.com and click on **Chapter 1— Student Web Activities** to learn more about the world population "clock."

NATIONAL GEOGRAPHIC On Location

Kosovo

In 1999 a civil war exploded in Kosovo, a province of Serbia. Thousands of people were forced from their homes.

Movement What causes people to become refugees?

World Population

Analyzing the Graph and Chart

The world's population is expected to reach about 9 billion by 2050.

Place Which country has the second-largest number of people?

Visit owt.glencoe.com and click on **Chapter 1— Textbook Updates.**

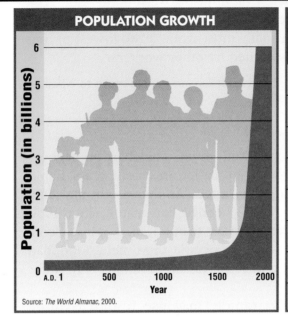

POPULATION GROWTH

Source: *The World Almanac*, 2000.

MOST POPULOUS COUNTRIES

Country	Millions of People
China	1,273.3
India	1,033.0
United States	284.5
Indonesia	206.1
Brazil	171.8
Pakistan	145.0
Russia	144.4

Source: *World Population Data Sheet*, 2001.

Conflict and Cooperation

Very few, if any, countries have been free of conflict in their history. As technology brings us closer together, however, nations are beginning to understand the importance of cooperation. Disagreement over land is a common reason for conflict between groups and nations. This is especially true when one culture has been displaced, or moved, by another culture. The **Israelis** and **Palestinians,** for example, have been fighting over land for more than 50 years. In eastern Europe, wars have broken out over who should control the land in countries once ruled by the former Soviet Union. These conflicts are also about religion, race, and politics. When groups and nations cooperate, they work together to find peaceful solutions to problems. Or, in the case of the **European Community,** they work together to prevent problems and to benefit from their combined strength. An example of economic cooperation can be found close to home. The **North American Free Trade Agreement (NAFTA)** is an agreement among Canada, the United States, and Mexico to improve trade relations among these countries.

Reading Check What regions of the world have seen economic cooperation among nations?

Civic Participation

Civic participation is being concerned with the public affairs of a community, state, nation, or the world. It is being an involved citizen. Some forms of government demand more involvement from their citizens than other types. In a democracy, for example, citizens need to be aware of their rights and responsibilities. **Rights** are benefits and protections guaranteed to you by law. In the United States, for example, you have the right to own property. Responsibilities are duties that you owe to your fellow citizens to make sure that the government continues. A major responsibility of democratic citizenship is voting. You also have a responsibility to respect the property and privacy of others. In totalitarian governments, the people have no rights. Their responsibilities are to obey the laws of the land.

Reading Check How are rights different from responsibilities?

Globalization

Think of the many ways you use products from other countries. The fruit you put on your breakfast cereal might have come from Mexico or South America. Your running shoes were likely made in China or Taiwan. Your book bag might have been made in India. Interdependence exists when countries depend on one another for goods, raw materials to make goods, and markets in which to sell goods. You might hear the world referred to as a "global village." In a village, people depend upon one another to provide what they need to live.

Many people are working to preserve the cultures of peoples such as the Masai in Kenya, Africa. They point out that globalization—the

Primary Source

GLOBALIZATION

Kofi Annan, secretary-general of the United Nations, spoke to the General Assembly about globalization.

"*If one word [describes] the changes we are living through, it is 'globalization.' . . . What are [the] global issues? I have grouped them under three headings, each of which I relate to a fundamental human freedom . . . First, freedom from want. How can we call human beings free and equal in dignity when over a billion of them are struggling to survive on less than one dollar a day? . . . The second . . . is freedom from fear. . . . We must do more to prevent conflicts from happening at all. . . . The third [is] the freedom of future generations to sustain their lives on this planet. . . . We need to remember the old African wisdom which I learned as a child—that the earth is not ours. It is a treasure we hold in trust for our descendents.*"

Millennium Report, April 3, 2000.

Analyzing Primary Sources

Do you think these are the only global issues? Do these issues affect you in your daily life? If they do, how? If they don't, do you think you should have to worry about them?

development of a world culture and interdependent economy—might erase traditions and customs of smaller groups. An important issue in the world today is how small countries can use products and services of developing nations and still preserve local cultures and values. A saying that has become popular in recent years is that we should "think globally and act locally." What does that expression mean to you?

✓ Reading Check What is meant by the words "the world is becoming a global village"?

Technology and World Issues

Technology is a tool. Like any tool, it can be used both wisely and foolishly. The Internet, if used wisely, can help develop better citizens. Citizens can stay better informed. They can organize more easily. They can also communicate with leaders and representatives quickly and directly by e-mail.

Beyond the problem of how technology is used, is the problem of how technology can be shared. Developing countries complain that they do not have access to the information that technology provides. What responsibility, if any, do industrialized countries have to share technology? Because progress, in many ways, is determined by technology, this may be the most important issue of all.

✓ Reading Check In what way is technology a tool? Give two examples.

Section 3 Assessment

Defining Terms
1. **Define** migrate, urbanization, refugee, rights, responsibilities, interdependence, globalization.

Recalling Facts
2. **Place** About how much of the world's population lives in cities?
3. **Movement** What is the most common reason people move to cities?
4. **Government** What responsibilities do people in democracies have?

Critical Thinking
5. **Synthesizing Information** What products found in your classroom were made in other countries?
6. **Understanding Cause and Effect** How can conflict affect human migration?

Graphic Organizer
7. **Organizing Information** Create a diagram like this one and list three results of human migration.

Human Migration

Applying Social Studies Skills
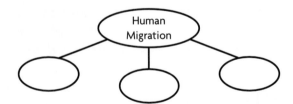

8. **Understanding Citizenship** Describe the civic participation expected of citizens of the United States.

TIME
REPORTS

ciudada

Our Shrinking World

Indians in Peru use the Internet to line up buyers for their farm goods.

THOMAS MULLER

The Global Economy and Your Future

Two forms of globalization: In Cuba, a student and her professor develop medicines to sell abroad. A woman in China makes goods for export.

How Trade Changes Lives

For Nora Lydia Urias Perez, life has never been easy. A single mother, she lived with her five-year-old daughter in the Mexican state of Veracruz. The only work she could find there was on a farm, earning $5 a day. That just wasn't enough.

In 2000 she moved to Nogales, a city just south of the New Mexico border. She got a job in a stapler factory that had moved to Nogales from New York City. Ms. Urias's job paid her $10 a day. To her, it was a fortune.

Thanks to the North American Free Trade Agreement (NAFTA), hundreds of thousands of Mexicans work in factories like Ms. Urias's. Companies in the United States, Japan and Europe own their factories. Their workers assemble products with parts that come from the United States. They send the finished goods—everything from dresses to TVs—back to the United States and Canada.

Global Relationships

This relationship is an example of **globalization**, the linking together of the world's nations through trade. What's driving globalization today is the search for cheap labor. Cheap labor helps manufacturers keep costs low. Low costs can mean lower prices for many things you buy.

A Sweatshirt's Global Journey

This map follows the route cotton has actually taken to a popular store near you.

1. Uzbekistan: Workers harvest cotton.

2. Iran: A freight train moves bales of raw cotton to the Arabian Sea.

3. Indian and Pacific Oceans: A ship carries the cotton 4,000 miles to South Korea.

4. South Korea: Workers spin cotton into thread and weave it into cloth.

5. Sea of Japan: A ship carries finished cloth to Russia's Far East.

6. Russia's Far East: Workers cut and stitch the cloth into sweatshirts.

7. Pacific Ocean: A ship takes the finished sweatshirts to California.

8. The United States: Trucks haul the sweatshirts to stores.

Source: The Nation

INTERPRETING MAPS

Making Inferences How do you think the price of the sweatshirt might be affected if the sweatshirt were made entirely in the United States?

Some fear global companies may neglect the environment.

Police block a march by globalization's foes in 1999.

A U.S. resident made this Taiwanese movie in China.

Globalization is changing far more than prices. More people, money, and goods are crossing national borders than ever before.

Pop Goes Global

Popular entertainment is no exception. A movie popular in the United States is likely to be a favorite elsewhere. Asians love basketball as much as Americans do. Kids everywhere listen to Latin pop music and wear jeans and sneakers to school.

That doesn't mean all kids think and act the same way. "It is important to see individual differences from one country to the next," advises a woman who has studied teens in 44 nations.

Culture Clash

Companies that forget that advice can get into trouble. A U.S. company opened a theme park outside Paris, France. But the French stayed away. They hated the fast food the park sold. They didn't even like the park's name. It contained the word "Euro," short for "European." The French see themselves as French first, Europeans second.

When the park's owners figured all this out, they made the park more French. They offered food and drinks that suited French tastes, for example. They even put the word "Paris" in the park's name. Today that theme park is one of the most popular in the world.

Good for Everyone?

As the park's owners learned, globalization isn't **"Americanizing"** the world. Local cultures are too strong for that.

But globalization hasn't been good for everyone. The poorest countries have seen little or no increase in trade. Many Americans' jobs have moved to countries where wages are low. And so far the lives of people like Ms. Urias haven't improved much. It costs more to live in Nogales than in Veracruz. So Ms. Urias is still poor.

Gaining Skills

Experts say these problems are only temporary. In recent years trade has created millions of jobs. It has enabled people in poorer countries like Mexico to pick up new skills. The more skilled workers are, the more they get paid.

Ms. Urias looks forward to better times. "I am not saying it will be easy to start life [in Nogales]," she told a reporter. "But at last there is a chance that things for me will get better. There was no chance of that in Veracruz. I had no hope."

EXPLORING THE ISSUE

1. **Cause and Effect** How might the health of the U.S. economy shape Ms. Urias's life?

2. **Making Inferences** Why do you think that the poorest countries have seen few gains from globalization?

41

Globalization's New Face

The Phoenicians were great sailors. They lived in Southwest Asia, on the coast of Lebanon. They set up **trade routes** all around the Mediterranean Sea. Some experts think they may have sailed to England to bring back tin. They did all this as far back as 1200 B.C.

High-speed cargo ships crisscross the oceans, carrying goods from nation to nation.

The Internet

The Internet has changed the way we swap goods, too. Twenty-five years ago, an American importer might have used "snail mail" to order a shipment of French bikes. Today she can check out the manufacturer's stock on his Web page. Then, in seconds, she can e-mail her order halfway around the world.

The deals she makes aren't much different from those the Phoenicians made. They traded timber for horses. She trades money for bicycles.

What's different is that she makes her trades in a flash, and without leaving her seat. She can do more business in the same time, and she can do business anywhere. The Phoenicians could do business only where they could sail.

▲ A Yagua tribesman (right) takes part in an Internet poll in the rain forests of Peru.

As the Phoenicians showed, globalization is not new. People have traded in faraway lands, moved around, and mixed cultures for thousands of years.

What is new is the speed at which these exchanges take place. Technology is shrinking the world. Telephones zip our voices around the world. Jet planes carry us great distances in a few hours.

EXPLORING THE ISSUE

1. **Making Inferences** How might trade help people from different cultures understand one another?

2. **Analyzing Information** How does the Internet make growing up different for you than it was for your parents?

Sharing Globalization's Gains

A little more than 6 billion people live on Earth. About half of them get by on less than $2 a day. What does globalization mean to them? So far, not much.

Overall the impact of increased trade has been amazing. The ability of people to make and spend money has grown almost everywhere.

Yet the fruits of globalization haven't been spread evenly. **Industrialized countries** have more to trade than **developing countries**. Foreign companies prefer to build more factories in rich countries than in poor ones.

The result is that countries like Kenya tend to create new jobs slowly. Places like Canada tend to create them more quickly. Some countries in Asia and Africa are barely able to create any new jobs at all.

A Wider Gap

Those differences worry a lot of people. If the trend continues, experts say that the gap between rich and poor countries can only get wider.

What can be done to narrow that gap? There are no easy answers. International businesses are certainly part of it. During the 1990s, private companies spent more than $1 trillion to build factories in developing countries.

Rich nations are also part of the answer. They are already helping poorer countries pay for new roads, phone lines, seaports, and airports. And they are encouraging poor nations to

MARIE DORIGNY/TIMEPIX

▲ **Nowhere is the gap between rich and poor clearer than in Pakistan. Here a child laborer makes soccer balls for sale around the world.**

produce things that people elsewhere want to buy.

China figured out how to do that years ago. Thanks to trade, the ability of the Chinese to earn and spend money now doubles every 10 years. Finding ways to help about 200 other nations equal that success is one of today's biggest challenges. ▪

EXPLORING THE ISSUE

1. **Making Inferences** Why do you think experts worry about the widening gap between rich and poor countries?

2. **Problem Solving** What would you do to help spread the fruits of globalization more evenly around the globe?

Preparing for a Smaller World: What Can One Person Do?

Every day in 2000, half a million airline passengers, 1.4 billion e-mail messages, and $1.5 trillion crossed national borders. All that shifting about of people, ideas, and money would have been unthinkable 10 years earlier. The Internet was a toddler. The World Wide Web had just been born.

What will the world look like 10 years from now? No one can say. But two things are sure. Inventions that create faster ways to communicate will make the world seem a lot smaller than it is today. And more and more Americans will have jobs that require them to deal with people from other nations.

DAVID STRICK/CORBIS OUTLINE

▲ Which of Pepperdine University's nine teammates was born in the U.S.? It's Anh Nguyen, fourth from left.

Learning About Other Cultures

You will be able to do that well if you have taken the time to learn about other countries. To really get to know people from other cultures, you need to understand what makes them tick. You can do that best by speaking to them in their own language.

You won't have to leave the United States to need that knowledge. Globalization has enabled more and more people to cross borders to find work. Employers will want to hire people who can work well with people born in other countries.

They will also want to know if you are committed to a **lifetime of learning**. As technology changes, your job will, too. Your need to learn new things won't stop when you leave high school or college.

Globalization is shaping tomorrow's job market. Only you can prepare yourself to thrive in it. And there's no time like today to start. ▪

EXPLORING THE ISSUE

1. **Determining Cause and Effect** How does the Internet make the world seem smaller?

2. **Analyzing Information** Modern companies require employees at every level to solve problems they face on the job. Why are lifetime learners better equipped than others to solve problems?

REVIEW AND ASSESS

UNDERSTANDING THE ISSUE

1. Defining Key Terms Write definitions for the following terms: *globalization, communication, trade route, Americanizing, culture, developing country, lifetime learner*.

2. Writing to Inform Write a short article about how globalization shapes the way people live and what they do. Use as many words as you can from the above list.

3. Writing to Persuade Overall, is globalization good or bad for the world? Defend your answer in a letter to an imaginary friend who lives in a developing country in Africa.

INTERNET RESEARCH ACTIVITY

4. With your teacher's help, use Internet resources to contact two classrooms—one in an industrialized country and one in a developing country. Exchange lists on what imported goods kids in your country and theirs own or use. Compare the lists, and discuss what they say about the importance of trade.

5. Use the Internet to find information on the history of the Internet. Write an essay telling how the Internet sped up communication, noting the key milestones described on the time line.

BEYOND THE CLASSROOM

6. Look through your local newspaper for a week. Find articles on topics related to globalization. For example, look for stories about the Internet, imports and exports, immigration, and even crimes like drug-smuggling. In an oral report, tell how the articles suggest that globalization is making the world smaller.

▲ **More and more Americans are crossing the borders for fun.**

PHOTODISC

7. Take an inventory of your room at home. Write down the name of each item made in another country. Count the items imported from the same country. Then make a bar graph to show how many imported items you own. Have each bar stand for one category—clothing, CDs, or sports equipment, for example. Write a caption explaining what the graph says about how important trade is to you.

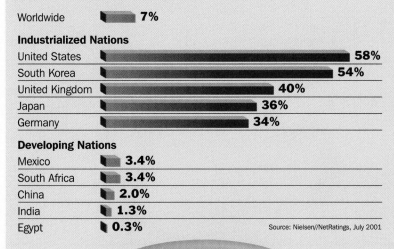

The Digital Divide
(Individuals with home access to the Internet in 2001)

Worldwide	7%
Industrialized Nations	
United States	58%
South Korea	54%
United Kingdom	40%
Japan	36%
Germany	34%
Developing Nations	
Mexico	3.4%
South Africa	3.4%
China	2.0%
India	1.3%
Egypt	0.3%

Source: Nielsen//NetRatings, July 2001

Where in the world are people wired to the Internet at home? Almost everywhere. But industrialized nations have a big lead. People with home access make up a big chunk of the populations of these richer nations. It's just the opposite with developing nations. People with home access make up a tiny part of the populations of these poorer nations. Experts call this gap the "digital divide," and it worries them. The Internet is a tool. Nations must use it to participate fully in world trade.

BUILDING GRAPH READING SKILLS

1. Comparing Compare the amount of Internet use in industrialized and developing nations.

2. Determining Cause and Effect What does a nation need besides Internet access to succeed in world trade?

FOR UPDATES ON WORLD ISSUES GO TO
www.timeclassroom.com/glencoe

Social Studies Skill

Reading Thematic Maps

Thematic (special purpose) maps focus on a single theme. This theme may be to show the battles of a particular war or locations of endangered species, for example.

Learning the Skill

To read a special purpose map, follow these steps:

- Read the map title. It tells what kind of special information the map shows.
- Find the map's scale to determine the general size of the area.
- Read the key. Colors and symbols in the map key are especially important on this type of map.
- Analyze the areas on the map that are highlighted in the key. Look for patterns.

Practicing the Skill

Look at the map below to answer the following questions.

1. What is the title of the map?
2. Read the key. What four civilizations are shown on this map?
3. Which civilization was farthest west? East?
4. What do the locations of each of these civilizations have in common?

Applying the Skill

Find a special purpose map in a newspaper or magazine. Pose three questions about the map's purpose, then have a classmate answer the questions.

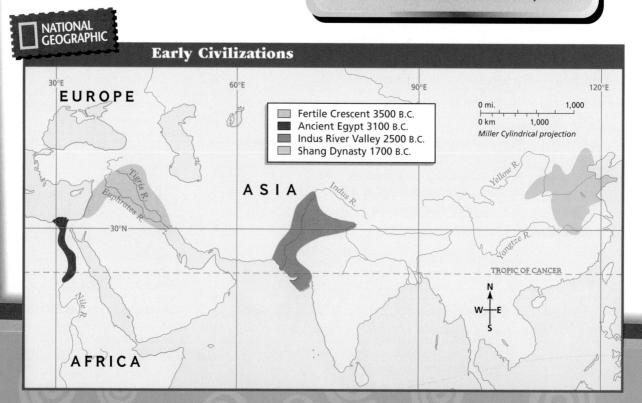

GO TO Practice key skills with **Glencoe Skillbuilder Interactive Workbook, Level 1.**

NATIONAL GEOGRAPHIC

Early Civilizations

Key:
- Fertile Crescent 3500 B.C.
- Ancient Egypt 3100 B.C.
- Indus River Valley 2500 B.C.
- Shang Dynasty 1700 B.C.

0 mi. 1,000
0 km 1,000
Miller Cylindrical projection

EUROPE

ASIA

AFRICA

Tigris R.
Euphrates R.
Nile R.
Indus R.
Yellow R.
Yangtze R.

30°E   60°E   90°E   120°E
30°N
TROPIC OF CANCER

Reading Review

Section 1 | People Far and Near

Terms to Know
ethnic group
custom
minority group
majority group

Main Idea
Modern technology has helped to bring the world's diverse peoples closer together.

✓ Culture Our shrinking world makes it more likely that the people around us will represent other ethnic groups.

✓ Culture It is important to understand what makes people similar and different so that we can get along in a world that is changing.

✓ Place Schools are good places to build bridges between different peoples living in the same region.

Section 2 | Understanding Culture

Terms to Know
social scientist
culture
ethnocentrism
cultural borrowing
cultural diffusion
limited government
democracy
unlimited government
dictatorship
absolute monarchy
constitutional
   monarchy

Main Idea
People all over the world usually live close to others who follow similar beliefs and like the same foods, music, and clothing.

✓ Culture Anthropologists, sociologists, historians, and human geographers are all social scientists who study the interaction of people.

✓ Culture Culture is the way of life of people who share similar beliefs and customs.

✓ Culture Most people have pride in their ethnic group and prefer their own culture.

✓ Culture Culture is continually spreading around the world. It spreads in two main ways, through cultural diffusion and cultural borrowing.

✓ History The story of a people's past helps us to understand its present and possibly its future.

✓ Government People need rules in order to live together.

✓ Economics People all over the world use natural resources to fill their wants and needs.

Section 3 | Patterns in Today's World

Terms to Know
migrate
urbanization
refugee
rights
responsibilities
interdependence
globalization

Main Idea
All over the world people are moving from one place to another looking for freedom, jobs, and a better life.

✓ History Throughout the world people are continually moving toward a better life and away from conflict and poverty.

✓ History As the world is getting smaller, people are living closer together and more conflicts are arising. People must learn to cooperate with each other.

✓ Region Americans living in the United States enjoy freedom and democracy. They have a civic responsibility to their government to make sure it continues.

✓ Interdependence The movement of goods and services is helping to bring our world closer together.

Chapter 1 Assessment and Activities

Using Key Terms

Match the terms in Part A with their definitions in Part B.

A.

1. ethnic group
2. minority group
3. majority group
4. culture
5. ethnocentrism
6. cultural diffusion
7. rights
8. responsibilities
9. urbanization
10. interdependence

B.

a. countries depending on one another
b. people moving from the country to the cities
c. duties that you owe to your government
d. how a country spreads its knowledge and skills
e. people believe their way of life is best
f. way of life of people who share similar beliefs and customs
g. people who have a common origin
h. group of people whose culture, race, or ethnic origin is different from most of the people in the region
i. group in society that controls most of the wealth and power
j. benefits and protections guaranteed by law

Reviewing the Main Idea

Section 1 People Far and Near

11. **Culture** In what way is the world shrinking?
12. **Culture** Why are schools good places to "build bridges" between ethnic groups?
13. **Culture** List three traits that would be common to an ethnic group.

Section 2 Understanding Culture

14. **Culture** If you wanted to study people, what type of scientist might you want to be?
15. **Culture** Give one example of cultural borrowing and one example of cultural diffusion.
16. **Government** Why do countries need governments?
17. **Economics** What is the difference between an industrialized country and a developing country?

Section 3 Patterns in Today's World

18. **Place** Why are so many people moving to cities?
19. **Region** Name two places where there is conflict going on in the world today.
20. **Government** Name one of the rights we have as citizens of the United States.

 **World Culture Regions**

Place Location Activity

On a separate sheet of paper, using chapter or unit maps, match the letters on the map with the numbered places listed below.

1. Latin America
2. North Africa, Southwest Asia, and Central Asia
3. Europe
4. Russia
5. East Asia
6. United States and Canada
7. Australia, Oceania, and Antarctica
8. Africa south of the Sahara

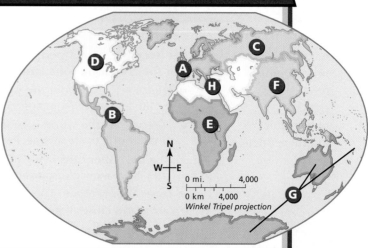

Self-Check Quiz Visit the *Our World Today: People, Places, and Issues* Web site at owt.glencoe.com and click on **Chapter 1—Self-Check Quizzes** to prepare for the Chapter Test.

Critical Thinking

21. **Making Predictions** In what ways do you think a company investing in a developing country could help the people there? How could that same company harm the culture?

22. **Analyzing Information** Imagine that you are a sociologist visiting your school. What characteristics of the local culture would you observe? What conclusions might you develop about your campus culture?

23. **Organizing Information** Create an outline that includes information from each section in this chapter. On your outline, list the title of each section and then list at least three important facts under each title.

Current Events Journal

24. **Analyzing Quotations** Read the following quote and analyze it using information you learned in this chapter. "Own only what you can always carry with you. Know languages, know countries, know people. Let your memory be your travel bag."

Mental Mapping Activity

25. **Focusing on the Region** Create a simple outline map of North America and South America. On the map label the following culture regions.
    • United States and Canada
    • Latin America

Technology Skills Activity

26. **Developing Multimedia Presentations** You are on assignment for your school newspaper. Walk around the halls, attend sports events and fine arts productions, and take photos of many faces around your campus. Scan the photos into presentation software on your computer. Develop a presentation that represents the many cultures that make up your school climate.

Standardized Test Practice

Directions: Study the graph, and then answer the following question.

Exports by Culture Region

Africa South of the Sahara $84.4
Asia $1,389.3
Europe and Russia $2,290.0
Latin America $311.5
United States and Canada $1,005.9
Oceania $88.0
Southwest Asia, North Africa, and Central Asia $263.3

Billions of Dollars

Source: *Britannica Book of the Year*, 1999.

1. **According to the graph, what is the combined value of the goods exported by the United States and Canada?**
   A $1,005,900,000,000
   B $1,005,900,000
   C $1,005,900
   D $1,005

Test-Taking Tip: In order to understand any type of graph, look carefully around the graph for keys that show how it is organized. On this bar graph, the numbers along the left side represent billions of dollars. Therefore, you need to multiply the number on the graph by 1,000,000,000 to get your answer.

Chapter 2 Earth Patterns

The World and Its People NATIONAL GEOGRAPHIC

To learn more about Earth's structure and landforms, view *The World and Its People* Chapters 1 and 2 videos.

Our World Today online

Chapter Overview Visit the *Our World Today: People, Places, and Issues* Web site at owt.glencoe.com and click on **Chapter 2–Chapter Overviews** to preview information about Earth.

Why It Matters

Spaceship Earth

A famous inventor once compared the planet Earth to a large spaceship hurtling through the galaxy. The spaceship-planet carries all the resources needed for its journey. As passengers on this ship, it is important to know something about how the "ship" works to avoid costly repairs and breakdowns.

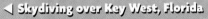

◄ **Skydiving over Key West, Florida**

FOLDABLES™
Study Organizer

Summarizing Vocabulary Study Foldable To fully understand what you read you must be able to identify and explain key vocabulary terms. Use this foldable to identify, define, and use important terms in Chapter 2.

Step 1 Fold a sheet of notebook paper in half from side to side.

Step 2 On one side, cut along every third line.

Tabs will form as you cut.

Step 3 Label your foldable as you read the chapter. The first vocabulary term is labeled on the model below.

Usually forms 10 tabs.

Reading and Writing As you read the chapter, select and write key vocabulary terms on the front tabs of your foldable. Then write the definition of each term under the tabs. After each definition, write a sentence using each vocabulary term correctly.

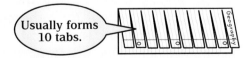

Thinking Like a Geographer

Guide to Reading

Main Idea

Geographers use various tools to understand the world.

Terms to Know

- geography
- landform
- environment
- region
- Global Positioning System (GPS)
- geographic information systems (GIS)

Reading Strategy

Create a chart like this one and write three details or examples for each heading.

How Geographers View the World
1.
2.
3.

Tools of Geography
1.
2.
3.

Uses of Geography
1.
2.
3.

NATIONAL GEOGRAPHIC Exploring Our World

How would *you* go about making an accurate map of the world? Scientists decided the best way to map the earth was to see it from space. In February 2000, the space shuttle *Endeavour* used a special camera called a radar camera to take pictures of the land below. By using radar, the camera was not hampered by clouds or darkness.

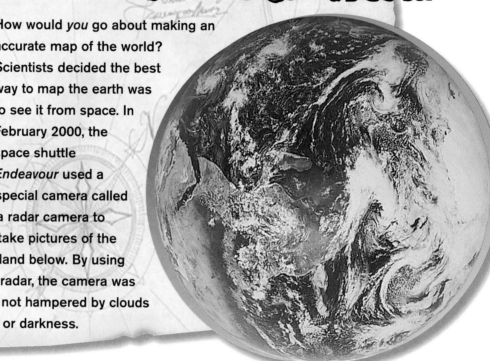

Why do geographers want to know exactly what the earth looks like? Think about the following: In China, the spring flooding of the **Yangtze** (YANG•SEE) **River** threatens people and crops every year. In 1998 the floods killed more than 4,000 people. Using information gathered from land and climate studies, the Chinese government built dams to hold some of the floodwaters back.

This is just one example of how people around the world use geographic knowledge collected from various sources. When you study geography, you learn about the earth's land, water, plants, and animals. You learn about how the continents were formed and what causes erosion. This is physical geography. Geographers also study people—where they live, how they live, how they change and are influenced by their environment, and how different groups compare to one another. This is human geography.

A Geographer's View of Place

Geographers look at major issues—like the flooding of the Yangtze—that affect millions of people. They also look at local issues—such as where is the best place for a company to build a new store in town. Whether an issue is major or local, geographers try to understand both its physical characteristics and human systems.

Physical Characteristics   Geographers study places. They look at *where* something is located on the earth. They also try to understand what the place is *like*. They ask: What features make a place similar to or different from other places?

To answer this question, geographers identify the landforms of a place. Landforms are individual features of the land, such as mountains and valleys. Geographers also look at water. Is the place near an ocean or on a river? Does it have plentiful or very little freshwater? They consider whether the soil will produce crops. They see how much rain the place usually receives and how hot or cold the area is. They find out whether the place has minerals, trees, or other resources.

Human Systems   Geographers also observe the social characteristics of the people living in the place. Do many or only a few people live there? Do they live close together or far apart? Why? What kind of government do they have? What religions do they follow? What kinds of work do they do? What languages do they speak?

NATIONAL GEOGRAPHIC   On Location

Varied Landforms

This mountain valley in France and this desert in Africa have very different physical characteristics.

Place List physical features shown in the photographs.

Earth Patterns

GIS Day

Teens, like Jenna, are putting away their computer video games and booting up GIS software. Jenna joined one of thousands of events taking place on the first International Geographic Information Systems Day in Washington, D.C. Here, the easy-to-use software was demonstrated by the National Geographic Society. GIS is software that helps users visualize geographic situations and problems by mapping and analyzing large amounts of geographic data.

People and Places Geographers are especially interested in how people interact with their environment, or natural surroundings. People can have a major impact on the environment. Remember how the Chinese built dams along the Yangtze? When they did so, they changed the way the river behaved in flood season.

Regions Geographers also look at the big picture, or how individual places relate to other places. In other words, geographers look at regions, or areas that share common characteristics. Regions can be relatively small—like your state or town. They can also be huge—like all of the western United States. Some regions may even include several countries.

✓ Reading Check **What do geographers study to determine the social characteristics of a place?**

The Tools of Geography

Geographers need tools to study people and places. Maps and globes are two tools they use to organize information about places.

Collecting Data for Mapping Earth How do geographers gather information so that they can make accurate maps? One way is to take photographs from high above the earth. Some of these are called LANDSAT images and show details such as the shape of the land, what plants cover an area, and how land is being used. Special cameras can even reveal information hidden by ice and snow.

How do geographers accurately label the exact locations of places on a map? Believe it or not, the best way to find a location is from outer space. Satellites traveling around the earth make up the Global Positioning System (GPS). A GPS receiver is a special device that receives signals from these satellites. When the receiver is put at a location, the GPS satellite can tell the exact latitude and longitude of that place. As a result, a mapmaker can know where exactly on the earth the particular area is located. GPS devices are even installed in vehicles to help drivers find their way.

Geographic Information Systems Today geographers use another powerful tool in their work—computers. Special computer software called geographic information systems (GIS) helps geographers gather many different kinds of information about the same place. After typing in all the data they collect, geographers use the software to combine and overlap the information on special maps.

In the 1990s, a logging company in California wanted to cut down parts of a forest. Environmental groups said that doing so would destroy the nesting areas of some rare birds. Geographers using GIS software created one map that showed the forest and another map that showed the nesting sites—and then overlapped these two maps. People could then see which areas had to be protected and which could be cut.

✓ Reading Check **What is the difference between GPS and GIS?**

Uses of Geography

Geographic information is used in planning. Government leaders use geographic information to plan new services in their communities. They might plan how to handle disasters or how much new housing to allow in an area. Companies can see where people are moving in a region in order to make plans for expanding.

Some businesses offer geographic information to their customers. Suppose you were looking for an apartment to rent. Some real estate agents have computer programs that can identify all the apartments of a certain size and price in an area. They can create a map so that you can see exactly where each apartment is located.

Finally, geographic information helps people manage resources. Many natural resources, such as oil or coal, are available only in limited supply and must be managed carefully so that they do not run out. Geographic information can help locate more natural resources.

✓ Reading Check   Why do people have to manage resources carefully?

Our World Today Online

Web Activity Visit the ***Our World Today: People, Places, and Issues*** Web site at owt.glencoe.com and click on **Chapter 2—Student Web Activities** to learn more about geographic information systems.

Section 1 Assessment

Defining Terms

1. Define geography, landform, environment, region, Global Positioning System (GPS), geographic information systems (GIS).

Recalling Facts

2. Place What two kinds of characteristics of a place do geographers study?

3. Technology What are the main tools of geography?

4. Human/Environment Interaction What are three uses for geography?

Critical Thinking

5. Understanding Cause and Effect Identify three physical characteristics of your region. How have these characteristics affected the way people live there?

6. Categorizing Information Give five examples of regions. Begin with an area near you that shares common characteristics, then look for larger and larger regions.

Graphic Organizer

7. Organizing Information Create a diagram like this one. In the center, write the name of a place you would like to visit. In the outer ovals, identify the types of geographic information you would like to learn about this place.

Applying Social Studies Skills

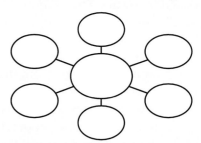

8. Analyzing Maps Find Egypt on the map on page RA21 of the Reference Atlas. Along what physical feature do you think most Egyptians live? Why? Turn to the population map of Egypt on page 9 of the Geography Handbook to see if you are correct.

Making Connections

CULTURE   GOVERNMENT   PEOPLE   TECHNOLOGY

Geographic Information Systems

What if a farmer could save money by applying fertilizer only to the crops that needed it? Today, thanks to computer technology called geographic information systems (GIS), farmers can do just that.

The Technology

Geographic information systems (GIS) use computer software to combine and display a wide range of information about an area. It starts with a map showing a specific location on the earth. This map is then linked with other information about that same place, such as satellite photos, amounts of rainfall, or housing locations.

Think of geographic information systems as a stack of transparencies. Each transparency shows the same general background but highlights different information. The first transparency may show a base map of an area. Only the borders may appear. The second transparency may show only rivers and highways. The third may highlight mountains and other physical features, buildings, or cities.

In a similar way, GIS technology places layers of information onto a base map. It can then switch each layer of information on or off, allowing data to be viewed in many different ways. In the case of the farmer mentioned above, GIS software combines information about soil type, plant needs, and last year's crop to pinpoint exact areas that need fertilizer.

How It Is Used

GIS technology allows users to quickly pull together data from many different sources and construct maps tailored to specific needs. This helps people analyze past events, predict future scenarios, and make sound decisions.

A person who is deciding where to build a new store can use GIS technology to help select the best location. The process might begin with a list of possible sites. The store owner gathers information about the areas surrounding each place. This could include shoppers' ages, incomes, and educations; where shoppers live; traffic patterns; and information about other stores in the area. The GIS software then builds a computerized map composed of these layers of information. The store owner can use the information to decide on a new store location.

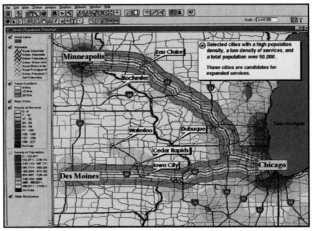

Graphic image created using ArcView® GIS software, and provided courtesy of Environmental Systems Research Institute, Inc.

Making the Connection

1. What is GIS technology?

2. How does GIS software analyze data in a variety of ways?

3. **Drawing Conclusions** Create a list of questions that you would want answered to locate the best place to add a new school to your district.

Guide to Reading

Main Idea

Landforms in all their variety affect how people live.

Terms to Know

- plate tectonics
- fault
- plain
- plateau
- canyon
- aquifer
- climate
- tropics
- greenhouse effect

Reading Strategy

Create a diagram like this one. In each of the surrounding circles, write the name of a landform and a fact about it.

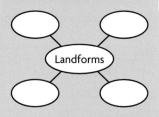

Landforms

NATIONAL GEOGRAPHIC Exploring Our World

Forces beneath the earth's surface shape the land and the lives of the people who live on it. Here in the Azores, a volcano makes cooking easy. People wrap meat and vegetables in a cloth and bury the bundle in a hole where heat from deep inside the earth rises to the surface. The temperature reaches 200°F (93°C), which is hot enough to steam the food.

It is amazing to think that the earth thousands of miles beneath your feet is so hot that it has turned metal into liquid. Although you may not feel these forces, what lies inside the earth affects what lies on top. Mountains, deserts, and other landscapes were formed over time by forces acting below the earth's surface—and they are still changing today. Some forces work slowly and show no results for thousands of years. Others appear suddenly and have dramatic, and sometimes very destructive, effects.

Forces Beneath the Earth's Crust

You have probably watched science shows about earthquakes and volcanoes. You have probably also seen news on television discussing the destruction caused by earthquakes. These disasters result from forces at work inside the earth.

Plate Movements   Scientists have developed a theory about the earth's structure called **plate tectonics.** This theory states that the crust is not an unbroken shell but consists of plates, or huge slabs of rock, that

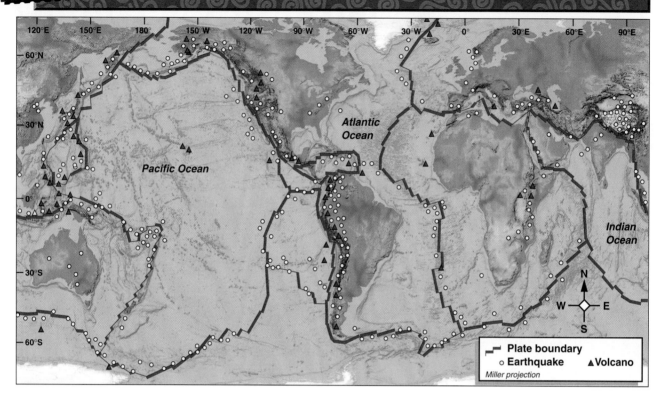

Plate boundary
o **Earthquake** ▲**Volcano**
Miller projection

Analyzing the Diagram

Most of North America sits on one plate.

Region What pattern do you see among plate boundaries, earthquakes, and volcanoes?

move. The plates float on top of liquid rock just below the earth's crust. They move—but often in different directions. Oceans and continents sit on these gigantic plates, as the diagram above shows.

When Plates Meet Sometimes, these plates push against each other. When this happens, one of three events occurs, depending on what kinds of plates are involved.

If two continental plates smash into each other, the collision produces high mountain ranges. This kind of collision produced the **Himalaya** mountain range in South Asia.

If a continental plate and an ocean plate move against each other, the thicker continental plate slides over the thinner ocean plate. The downward force of the lower plate causes molten rock to build up. Then, as magma, it erupts to form volcanic mountains.

Sometimes two plates do not meet head-on but move alongside each other. To picture this, put your two hands together and then move them in opposite directions. When this action occurs in the earth, the two plates move apart. This movement creates faults, or cracks in the earth's crust. Violent earthquakes can happen near these faults. One of the most famous faults in the United States is the **San Andreas Fault** in California. The earth's movement along this fault caused a severe earthquake in San Francisco in 1906 and another, less serious quake there in 1989.

✓ Reading Check What happens when two continental plates collide?

Types of Landforms

Look at the illustration on pages 14 and 15 of the Geography Handbook. Notice the many different forms that the land may take. Which ones are familiar to you? Which ones are new to you?

On Land   Mountains are huge towers of rock formed by the collision of the earth's tectonic plates or by volcanoes. Some mountains may be a few thousand feet high. Others can soar higher than 20,000 feet (6,096 m). The world's tallest mountain is Mount Everest, located in South Asia's Himalaya mountain ranges. It towers at 29,035 feet (8,850 m)— nearly 5.5 miles (8.9 km) high.

In contrast, plains and plateaus are mostly flat. What makes them different from each other is their elevation, or height above sea level. Plains are low-lying stretches of flat or gently rolling land. Plateaus are also flat but have higher elevations.

Between mountains or hills lie valleys. A valley is a long stretch of land lower than the land on either side. You often find rivers at the bottom of valleys. Canyons are steep-sided lowlands that rivers have cut through plateaus. One of the most famous canyons is the **Grand Canyon** in Arizona. Over time the **Colorado River** flowed over a plateau and carved through rock to form the Grand Canyon.

Reading Check How are plains and plateaus different?

Bodies of Water

About 70 percent of the earth's surface is water. Most of that water is salt water, which people and animals cannot drink. Only a small percentage is freshwater, which is drinkable. People and most animals need freshwater to live. Many other creatures, however, make their homes in the earth's saltwater oceans and seas.

Valleys vs. Canyons

The Great Rift Valley in Africa is surrounded by mountains (left). Canyons, like the Grand Canyon in Arizona (below), are carved from plateaus.

Place   How are valleys and canyons similar?

NATIONAL GEOGRAPHIC On Location

Believe It or Not!

Mt. Pinatubo

Mt. Pinatubo (PEE•nah•TOO•boh) is a volcanic mountain in the Philippine Islands. Its eruption in the early 1990s impacted the world's climate. The powerful explosion shot ash and sulfur dioxide into the earth's atmosphere, which blocked some of the sun's rays from reaching the earth. The world's climate was cooler for two years after the volcano's blast.

Salt Water All the oceans on the earth are part of a huge, continuous body of salt water—almost 98 percent of the planet's water. The four major oceans are the Pacific Ocean, the Atlantic Ocean, the Indian Ocean, and the Arctic Ocean.

Freshwater Only about 2 percent of the water on the earth is freshwater. Eighty percent of that freshwater is frozen in glaciers, or giant sheets of ice. Only a tiny fraction of the world's freshwater—not even four-hundredths of a percent—is found in lakes and rivers.

When you think of freshwater, you probably think of mighty rivers and huge lakes. People can get freshwater from another source, though. Groundwater is water that fills tiny cracks and holes in the rock layers below the surface of the earth. This is a vital source of water because there is 10 times more groundwater than there is water in rivers and lakes. Groundwater can be tapped by wells. Some areas have aquifers, or underground rock layers that water flows through. In regions with little rainfall, both farmers and city dwellers sometimes have to depend on aquifers and other groundwater for most of their water supply.

✓Reading Check **Which is more plentiful, salt water or freshwater?**

Climate

Climate is the usual, predictable pattern of weather in an area over a long period of time. It is one reason people decide to live in a particular place. One important influence on climate is the angle at which the sun's rays hit the earth. In general, temperatures are higher in places where the rays strike the earth's surface more directly than in places where they strike the surface at an angle. This means that places in low latitudes—regions near the Equator—usually have warmer climates than places at higher latitudes. Generally, warmer climates are found in the tropics—areas near the Equator that lie between the **Tropic of Cancer** (23½°N latitude) and the **Tropic of Capricorn** (23½°S latitude).

The combined effects of water, wind, and land also influence the climate at a given place. In general, water warms and cools more slowly than land. In land areas where strong winds blow in from over the ocean, temperatures tend to be less varied than in places at the same latitude that do not benefit from these winds.

Finally, altitude, or elevation, of a place affects its climate. As altitude increases, air temperatures decrease. This means that places at high elevations that are quite close to the Equator can have cool climates.

The Impact of People on Climate You may have noticed that temperatures in large cities are generally higher than those in nearby rural areas. Why is that? City streets and buildings absorb more of the sun's rays than do the plants and trees of rural areas.

Cities are warmer even in winter. People burn fuels to warm houses, power industry, and move cars and buses along the streets. This burning raises the temperature in the city. The burning also releases a cloud of chemicals into the air. These chemicals blanket the city and hold in more of the sun's heat, creating a so-called heat island. The burning of fuels is creating a worldwide problem. In the past 200 years, people have burned coal, oil, and natural gas as sources of energy. Burning these fuels releases certain gases into the air.

Some scientists believe that the buildup of these gases presents dangers. It creates a greenhouse effect—like a greenhouse, the gases prevent the warm air from rising and escaping into the atmosphere. As a result, the overall temperature of the earth will increase.

In some countries, people are clearing large areas of rain forests. They want to sell the lumber from the trees. They also want to use the land to grow crops or as pasture for cattle. People often clear the forests by burning down trees in an area. This burning releases gases into the air, just like burning oil or natural gas does.

Another danger of clearing the rain forests is related to rainfall. Water on the earth's surface evaporates and then falls as rain. In the rain forests, much of this water evaporates from the leaves of trees. If the trees are cut, less water will evaporate. As a result, less rain will fall. Scientists worry that, over time, the area that now holds rain forests will actually become dry and unable to grow anything.

Not all scientists agree about the greenhouse effect. Some argue that the world is not warming. Others say that even if it is, predictions of disaster are extreme. Many scientists are studying world temperature trends closely. They hope to be able to discover whether the greenhouse effect is a real threat.

✓ Reading Check **What is the greenhouse effect?**

Assessment

Define Terms
1. Define plate tectonics, fault, plain, plateau, canyon, aquifer, climate, tropics, greenhouse effect.

Recalling Facts
2. **Region** Name three types of landforms.
3. **Movement** What is one reason people decide to settle in a particular area?
4. **Place** What are the world's four oceans?
5. **Region** Why do the areas near the Equator have warmer climates than other areas?

Critical Thinking
6. **Drawing Conclusions** Why do you think it is important to keep groundwater free of dangerous chemicals?
7. **Summarizing** How does clearing of the rain forests affect climate?

Graphic Organizer
8. **Organizing Information** Create a diagram like this one. List at least four sources of freshwater and salt water on the lines under each heading.

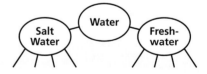

Applying Social Studies Skills

9. **Analyzing Diagrams** Look at the diagram of tectonic plate boundaries on page 58. Why might it be a problem that most of the world's population lives along the western edge of the Pacific Ocean?

Social Studies Skill

Using a Map Key

To understand what a map is showing, you must read the **map key,** or legend. The map key explains the meaning of special colors, symbols, and lines on the map.

Learning the Skill

Colors in the map key may represent different elevations or heights of land, climate areas, or languages. Lines may stand for rivers, streets, or boundaries.

Maps also have a compass rose showing directions. The cardinal directions are north, south, east, and west. North and south are the directions of the North and South Poles. If you stand facing north, east is the direction to your right. West is the direction to your left. The compass rose might also show intermediate directions, or those that fall between the cardinal directions. For example, the intermediate direction northeast falls between north and east. To use a map key, follow these steps:

- Read the map title.
- Read the map key to find out what special information it gives.
- Find examples of each map key color, line, or symbol on the map.
- Use the compass rose to identify the four cardinal directions.

Practicing the Skill

Look at the map of Washington, D.C., below, to answer the following questions.

1. What does a red square represent?
2. What does a blue square represent?
3. Does the Washington Monument lie east or west of the Lincoln Memorial?
4. From the White House, in what direction would you go to get to the Capitol?

Applying the Skill

Find a map in a newspaper or magazine. Use the map key to explain three things the map is showing.

GO TO

Practice key skills with **Glencoe Skillbuilder Interactive Workbook, Level 1.**

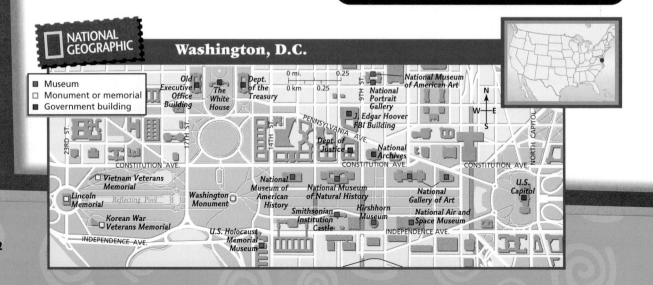

NATIONAL GEOGRAPHIC

Washington, D.C.

Museum
Monument or memorial
Government building

People and the Environment

Guide to Reading

Main Idea

The actions people take can have serious effects on the environment.

Terms to Know

- conservation
- pesticide
- ecosystem
- crop rotation
- erosion
- deforestation
- acid rain

Reading Strategy

Create a diagram like this one. On the lines, write at least two problems that are caused by human use of water, land, and air.

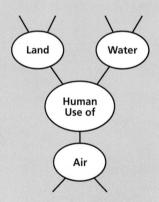

NATIONAL GEOGRAPHIC

Exploring Our World

What happens when we harm the environment? It is possible that some plants and animals might be gone forever. This scientist hopes to prevent that. She works at a seed bank. Behind her, stored at −4°F (−20°C), are jars of plant seeds from around the world. Should any of these 4,000 types of plants become extinct, these seeds can start growing them again.

The rapidly growing number of people threatens the delicate balance of life in the world. More and more people use more water. They need more land to live on and to grow more food. Spreading industry fouls the air. Humans must act carefully to be sure not to destroy the earth that sustains life.

Water Use

People, plants, and animals need freshwater to live. People need clean water to drink. They also need water for their crops and their animals. In fact, as much as 70 percent of the water used is for farming. You read earlier that 70 percent of the earth's surface is water. However, only a small fraction of the world's water is freshwater. Because the earth's supply of freshwater is limited, people must manage this precious resource carefully.

Water Management   Some regions receive heavy rainfall in some months of the year and little, or none, in other months. They can manage their water supply by building storage areas to hold the heavy rains for later use.

Managing water supplies involves two main steps. The first step is conservation, or the careful use of resources so that they are not wasted. Did you know that 6 or 7 gallons (23 to 27 liters) of water go down the drain every minute that you shower? Taking short showers instead of baths is an easy way to prevent wasting water.

The second approach to managing the water supply is to avoid polluting water. Some manufacturing processes use water. Sometimes those processes result in dangerous chemicals or dirt entering the water supply. Many farmers overuse fertilizers to help their crops grow. Many also use pesticides, or powerful chemicals that kill crop-destroying insects. These products are needed to increase food production, but they can seep into the water supply and cause harm.

✓Reading Check How can manufacturing and farming harm the water supply?

Land Use

As humans expand their communities, they invade ecosystems. These are places where the plants and animals are dependent upon one another—and their particular surroundings—for survival. For example, some people may want to drain a wet, marshy area to get rid of disease-carrying mosquitoes and make the soil useful for farming. When the area

NATIONAL GEOGRAPHIC   On Location

Madeira Islands

This hillside has been terraced to allow the owner to build a house and plant crops.

Human/Environment Interaction How would terraced fields help prevent erosion?

is drained, however, the ecosystem is destroyed. In managing land, humans must recognize that some soils are not suited for growing certain crops. Benefits must be weighed against losses.

Soil   To grow food, soil needs to have certain minerals. Farmers add fertilizers to the soil to supply some of these minerals. Some also practice crop rotation, or changing what they plant in a field to avoid using up all the minerals in the soil. Some crops—like beans—actually restore valuable minerals to the soil. Many farmers now plant bean crops every three years to build the soil back up.

If people do not carefully manage the soil, it can erode away. In erosion, wind or water carries soil away, leaving the land less fertile than before. Have you ever seen a group of trees alongside a farmer's field? The farmer may have planted those trees to block the wind and prevent erosion. In the tropics, erosion presents a problem—especially if farmers plant their crops on sloping land. When heavy rains come, the soil may simply wash down the hillside.

Forests   Some areas of the world have thick forests of tall trees. Growing populations in these countries often turn to these lush forests as a source of land to grow food. Yet deforestation, or cutting down forests, is a problem—especially in the tropics. Rains in these areas are extremely heavy. When the tree roots are no longer there to provide nutrients or to hold the soil, the water can wash the nutrients and the soil away. Then farmers have to move and cut trees in another section of forest. As a result more of the forest is lost.

✔ Reading Check   **What problem can result from erosion?**

NATIONAL GEOGRAPHIC **On Location**

Taiwan

Burning fossil fuels adds harmful chemicals to the air.

Human/Environment Interaction What are some effects of air pollution?

Air Pollution

Industries and vehicles that burn fossil fuels are the main sources of air pollution. Throughout the world, fumes from cars and other vehicles pollute the air. When sulfur oxides from coal-burning power plants and nitrogen oxides from cars combine with moisture in the air, they form acids. When acidic moisture falls to Earth as rain or snow, it is called acid rain. Acid rain can corrode structures, damage forests, and harm organisms.

Some scientists believe that increasing amounts of pollutants in the atmosphere will cause the earth to warm. While not all experts agree, some scientists say that the increase in temperature can have disastrous effects. Glaciers and ice caps may melt, raising the level of the world's seas. Higher seas could flood coastal cities. Warmer temperatures can also affect the land, making it no longer able to produce food.

✓ Reading Check **What causes highly acidic rain?**

Balancing People and Resources

Water, land, and air are among people's most precious resources. We need water and air to live. We need land to grow food. Only by caring for these resources can we be sure that we will still have them to use in the future.

Sometimes, though, protecting the environment for the future seems to clash with feeding people in the present. For example, farmers destroy the rain forests not because they want to but because they need to feed their families. They dislike being told by people in other countries that they should save the rain forests. Before they stop cutting down the rain forests, these farmers will need to find new ways to meet their needs.

✓ Reading Check **How does saving the rain forests clash with current human needs?**

Assessment

Defining Terms

1. Define conservation, pesticide, ecosystem, crop rotation, erosion, deforestation, acid rain.

Recalling Facts

2. Human/Environment Interaction How do human activities affect ecosystems?

3. Human/Environment Interaction What are two ways of managing water?

4. Economics Why do farmers practice crop rotation?

Critical Thinking

5. Making Comparisons Which resource—water, soil, or air—do you think is most precious to people? Why?

6. Analyzing Information What ecosystems were affected by the growth of your community?

Graphic Organizer

7. Organizing Information Create a diagram like this one and list three potential results of global warming.

Global warming

Applying Social Studies Skills

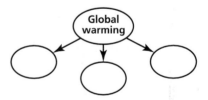

8. Analyzing Photographs Look at the On Location photo of Taiwan on page 65. List ways the environment has been changed and ways it is still being changed.

Reading Review

Section 1 | Thinking Like a Geographer

Terms to Know

geography
landform
environment
region
Global Positioning System (GPS)
geographic information systems (GIS)

Main Idea

Geographers use various tools to understand the world.

✓ Place Geographers study the physical and social characteristics of places.

✓ Human/Environment Interaction Geographers are especially interested in how people interact with their environment.

✓ Technology To study the earth, geographers use maps, globes, photographs, the Global Positioning System, and geographic information systems.

✓ Economics People can use information from geography to plan, make decisions, and manage resources.

Section 2 | Physical Geography

Terms to Know

plate tectonics
fault
plain
plateau
canyon
aquifer
climate
tropics
greenhouse effect

Main Idea

Landforms in all their variety affect how people live.

✓ Movement Earthquakes and volcanoes can reshape the land.

✓ Location Mountains, plateaus, valleys, and other landforms are found on the earth.

✓ Science Humans and most animals need freshwater to live. Only a small fraction of the world's freshwater is found in rivers and lakes.

✓ Region The tropics, near the Equator, receive more of the sun's direct rays than other regions.

✓ Culture Human actions such as building cities, burning fuels, and clearing the rain forests can affect climate.

Section 3 | People and the Environment

Terms to Know

conservation
pesticide
ecosystem
crop rotation
erosion
deforestation
acid rain

Main Idea

The actions people take can have serious effects on the environment.

✓ Human/Environment Interaction People need to manage water resources because the earth's freshwater supply is limited.

✓ Human/Environment Interaction Air pollution has damaging effects on the land and on people's health.

This pool formed by an underground spring is an unusual sight in Mexico's Chihuahua Desert. ▶

Using Key Terms

Match the terms in Part A with their definitions in Part B.

A.

1. environment
2. landform
3. region
4. plate tectonics
5. ecosystem
6. geographic information systems
7. erosion
8. canyon
9. fault
10. tropics

B.

a. area that shares common characteristics
b. wearing away of the earth's surface
c. theory that the earth's crust consists of huge slabs of rock that move
d. system of plant, animal, and human interdependence
e. special software that helps geographers gather and use information
f. natural surroundings
g. areas near the Equator
h. crack in the earth's crust
i. steep-sided lowland that a river has cut through a plateau
j. particular feature of the land

Reviewing the Main Ideas

Section 1 Thinking Like a Geographer

11. **Place** Give three examples of the physical characteristics of a place.
12. **Region** How is a region different from a place?
13. **Human/Environment Interaction** Give an example of how people use geographic knowledge.

Section 2 Physical Geography

14. **Movement** In what ways do the plates in the earth's crust move?
15. **Place** Which have a higher elevation—plains or plateaus?
16. **Human/Environment Interaction** Why are cities warmer than nearby rural areas?
17. **Human/Environment Interaction** What problem has been caused by the burning of fuels?

Section 3 People and the Environment

18. **Human/Environment Interaction** What two steps are involved in water management?
19. **Human/Environment Interaction** How can farmers restore the minerals in soil?
20. **Human/Environment Interaction** What two problems can result from air pollution?

 **The World**

Place Location Activity

On a separate sheet of paper, match the letters on the map with the numbered places listed below.

1. North America
2. Pacific Ocean
3. Africa
4. South America
5. Antarctica
6. Australia
7. Atlantic Ocean
8. Asia

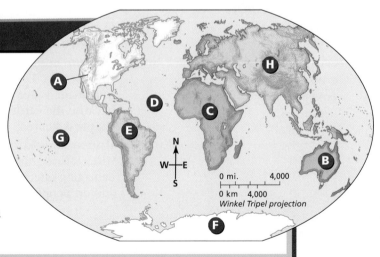

Self-Check Quiz Visit the *Our World Today: People, Places, and Issues* Web site at owt.glencoe.com and click on **Chapter 2—Self-Check Quizzes** to prepare for the Chapter Test.

Critical Thinking

21. **Analyzing Information** From where does the freshwater in your community come? How can you find out?

22. **Sequencing Information** Make a chart like the one below, and list the ways you use electricity from the moment you wake up until you go to sleep. In the second column, write how you would perform the same activity if you had no electricity to rely on.

Activities With Electricity	Without Electricity

Current Events Journal

23. **Writing a Paragraph** Global Positioning System units are now available for people to use for such activities as driving, hiking, and boating. Write a paragraph describing how the use of these units could save lives.

Mental Mapping Activity

24. **Focusing on the Region** Create a map of the world's oceans and continents. Label the following items:

- Equator
- North America
- high latitude climate regions
- tropical climate regions
- Pacific Ocean
- Africa

Technology Skills Activity

25. **Developing Multimedia Presentations** Research how your state's climate influences its culture, including tourist attractions, types of clothing, and the economy. Use your research to develop a television or radio commercial promoting your state.

Standardized Test Practice

Directions: Study the maps below, and then answer the question that follows.

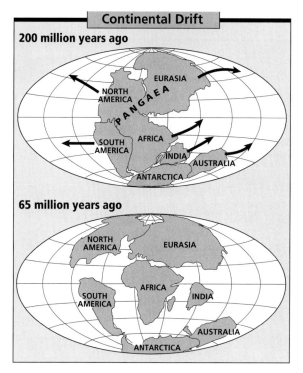

Continental Drift
200 million years ago
65 million years ago

1. What "supercontinent" do many scientists believe existed 200 million years ago?

   A Eurasia

   B Pangaea

   C Gondwana

   D Antarctica

Test-Taking Tip: Use information on the maps to answer this question. Read the title above the maps and then the two subtitles. If you reread the question, you see it is asking about a certain time period. Make sure you use the correct map above to answer the question.

Endangered Spaces

Shrinking Habitats When you think of Africa, what images come to mind? Roaring lions? Sprinting cheetahs? Lumbering elephants? Unless conditions change, some wild African animals may soon be only memories. Many are endangered, primarily because their habitats—their grassland and forest homes—are being destroyed in many ways.

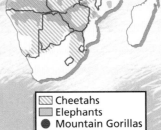

 Population growth — Africa south of the Sahara has the world's highest population growth rate. Farmers and ranchers turn wild lands into fields and pastures to raise food. Urban sprawl also takes over habitats.

 Logging — Logging companies build roads and cut valuable trees, destroying forest habitats.

Mining — Open pit mines scar the land, pollute waters, and destroy trees.

☐ Cheetahs
☐ Elephants
● Mountain Gorillas

As habitats shrink, so do populations of African animals.

 Cheetahs live in Africa's grasslands. As people move into the cheetahs' home, the big cats struggle to survive. Only about 12,000 cheetahs are left in the wild.

 Mountain gorillas live in the misty mountain forests of Central and East Africa. Logging and mining are destroying these forests. Only about 650 mountain gorillas remain.

These and other endangered African animals will survive only if their habitats are saved.

Loggers destroy a forest in the Democratic Republic of the Congo.

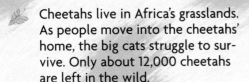

Cheetahs are running out of room in Africa.

Making a Difference

The Cheetah Conservation Fund Cheetahs in Africa are getting a helping hand from the Cheetah Conservation Fund (CCF). This organization is based in Namibia, which is home to about 2,500 cheetahs. Namibian ranchers often trap and shoot cheetahs to protect their livestock. The CCF has donated about 80 special herding dogs to ranchers. The dogs protect the livestock and keep cheetahs out of harm's way at the same time. The CCF also teaches villagers and schoolchildren about cheetahs and about why it is important to save these big cats and their habitats.

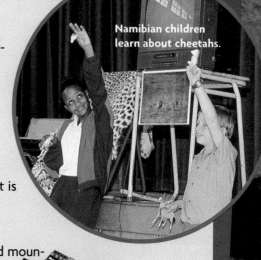

Namibian children learn about cheetahs.

Protecting Gorillas For nearly 20 years, Dian Fossey studied mountain gorillas in Rwanda. Through her book, *Gorillas in the Mist*, which was made into a movie, Fossey told others about mountain gorillas and how their survival was threatened by habitat destruction and poaching. Fossey established the Karisoke Research Center and an international fund to support gorilla conservation.

Dian Fossey fought fiercely to end gorilla poaching. Although Fossey was murdered at Karisoke in 1985, the Dian Fossey Gorilla Fund International continues its work protecting mountain gorillas and their habitat.

What Can You Do?

Adopt a Cheetah
You and your classmates can help save cheetahs in the wild by adopting one. To learn more, contact the Cheetah Conservation Fund at www.cheetah.org

Find Out More
What animal habitats are endangered where you live? Work with a partner to investigate endangered spaces in your area. Summarize your findings in a report to the class.

A mountain gorilla

Unit 2

Business district at dusk, Dubai, United Arab Emirates

Man gazing out across ▶ the vast expanses of the Sahara

North Africa and Southwest Asia

Ancient Egyptian pyramids overlook industrial smokestacks. Three-thousand-year-old stone temples tower over sparkling new oil derricks. Remote mountain villages and endless desert seas of sand and gravel contrast with modern beaches overrun by tourists. All of these extremes can be found within the culture region of North Africa and Southwest Asia.

▲ Shepherd tending sheep, Atlas Mountains, Morocco

NGS ONLINE
www.nationalgeographic.com/education

73

North Africa and Southwest Asia

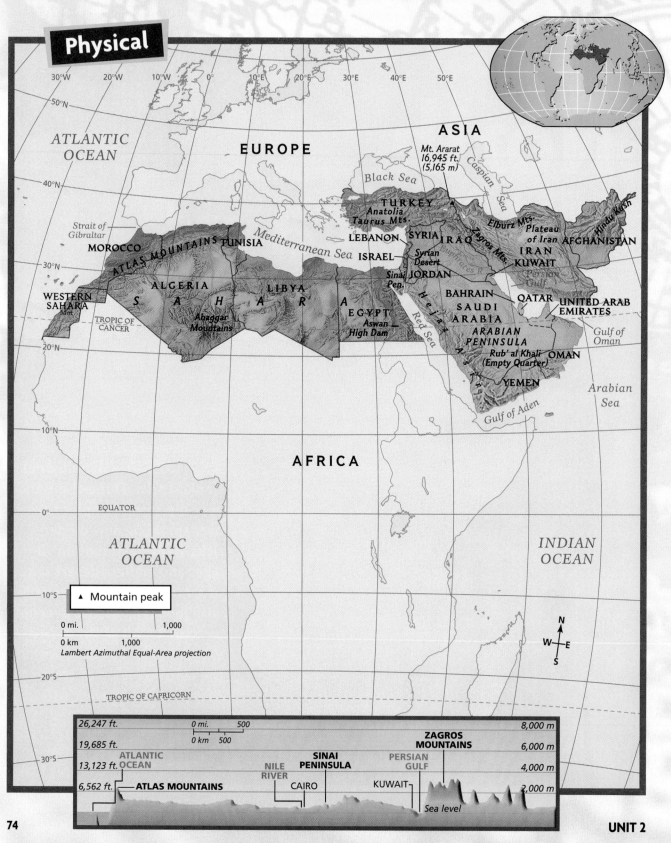

Physical

ATLANTIC OCEAN

EUROPE

ASIA

30°W 20°W 10°W 0° 10°E 20°E 30°E 40°E 50°E

50°N

40°N

Mt. Ararat
16,945 ft.
(5,165 m)

Black Sea

Caspian Sea

TURKEY
Anatolia
Taurus Mts.

Elburz Mts.
Plateau
of Iran

Hindu Kush

Strait of
Gibraltar

MOROCCO

Mediterranean Sea

LEBANON

SYRIA

IRAQ

Zagros Mts.

AFGHANISTAN

ATLAS MOUNTAINS TUNISIA

ISRAEL

Syrian
Desert

IRAN

KUWAIT

30°N

Sinai
Pen.

JORDAN

Euphrates R.

Persian
Gulf

WESTERN
SAHARA

ALGERIA

LIBYA

S A H A R A

EGYPT

Aswan
High Dam

Hejaz

BAHRAIN

SAUDI
ARABIA

QATAR

UNITED ARAB
EMIRATES

Mt.

TROPIC OF
CANCER

Ahaggar
Mountains

Red Sea

ARABIAN
PENINSULA

Gulf of
Oman

20°N

Rub' al Khali
(Empty Quarter)

OMAN

YEMEN

Arabian
Sea

Gulf of Aden

10°N

AFRICA

0° EQUATOR

ATLANTIC
OCEAN

INDIAN
OCEAN

10°S

▲ Mountain peak

0 mi.   1,000

0 km   1,000
Lambert Azimuthal Equal-Area projection

N
W E
S

20°S

TROPIC OF CAPRICORN

30°S

26,247 ft.			8,000 m
19,685 ft.	ZAGROS MOUNTAINS		6,000 m
13,123 ft.	ATLANTIC OCEAN	SINAI PENINSULA	PERSIAN GULF		4,000 m
6,562 ft.	ATLAS MOUNTAINS	NILE RIVER	CAIRO	KUWAIT		2,000 m

0 mi. 500
0 km 500

Sea level

Political

EUROPE

ASIA

ATLANTIC OCEAN

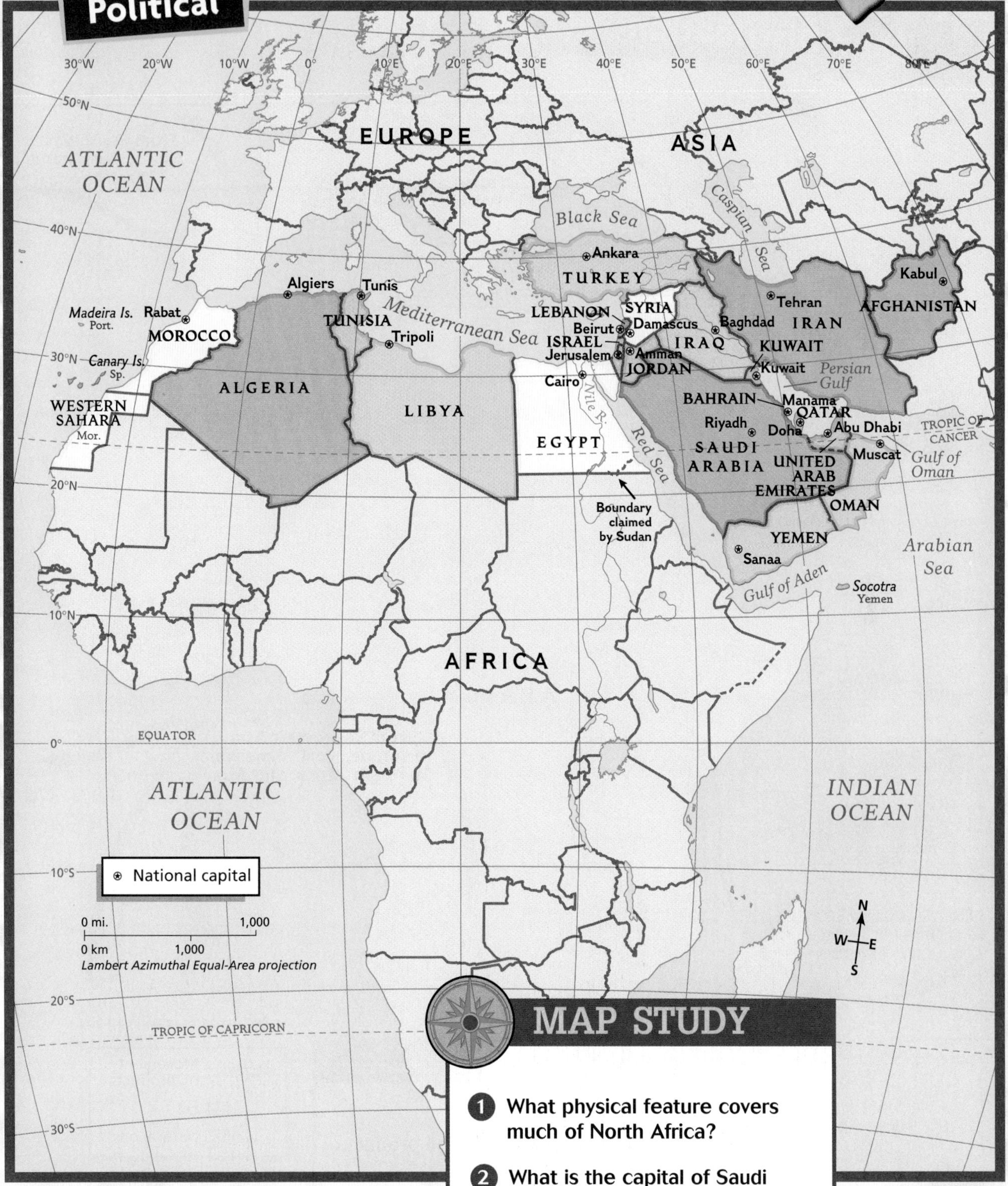

30°W 20°W 10°W 0° 10°E 20°E 30°E 40°E 50°E 60°E 70°E 80°E

50°N

40°N

Black Sea

Caspian Sea

⊛ Ankara

TURKEY

Kabul ⊛

Algiers ⊛ ⊛ Tunis

AFGHANISTAN

Madeira Is. Rabat ⊛
Port.
MOROCCO

TUNISIA

Mediterranean Sea

LEBANON **SYRIA**
Beirut ⊛ ⊛ Damascus

⊛ Tehran

IRAN

⊛ Baghdad

Tripoli ⊛

ISRAEL
Jerusalem ⊛ ⊛ Amman

IRAQ

KUWAIT

Canary Is.
Sp.

30°N

Cairo ⊛

JORDAN

⊛ Kuwait

Persian Gulf

WESTERN SAHARA
Mor.

ALGERIA

LIBYA

Nile R.

BAHRAIN ⊛ Manama

⊛ Riyadh Doha ⊛ **QATAR** ⊛ Abu Dhabi TROPIC OF CANCER

EGYPT

20°N

Red Sea

SAUDI ARABIA

UNITED ARAB EMIRATES

⊛ Muscat Gulf of Oman

OMAN

Boundary claimed by Sudan

YEMEN

Arabian Sea

⊛ Sanaa

10°N

Gulf of Aden

Socotra
Yemen

AFRICA

0° EQUATOR

ATLANTIC OCEAN

INDIAN OCEAN

10°S

⊛ National capital

0 mi. 1,000
0 km 1,000
Lambert Azimuthal Equal-Area projection

N
W E
S

20°S

TROPIC OF CAPRICORN

MAP STUDY

30°S

1 What physical feature covers much of North Africa?

2 What is the capital of Saudi Arabia?

North Africa and Southwest Asia

North Africa and Southwest Asia

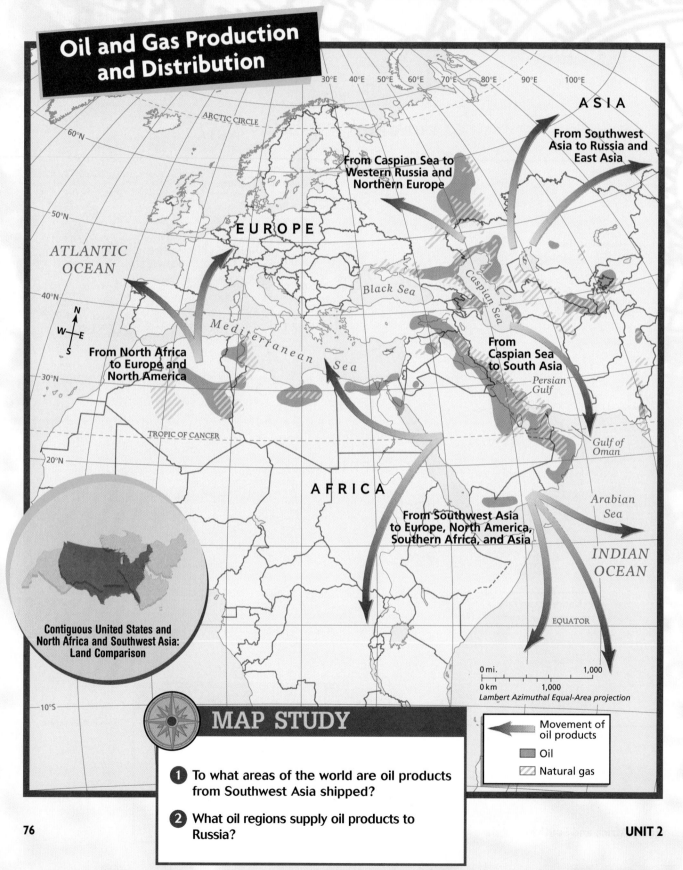

Oil and Gas Production and Distribution

From Caspian Sea to Western Russia and Northern Europe

From Southwest Asia to Russia and East Asia

From North Africa to Europe and North America

From Caspian Sea to South Asia

From Southwest Asia to Europe, North America, Southern Africa, and Asia

ARCTIC CIRCLE

60°N

50°N

40°N

30°N

TROPIC OF CANCER

20°N

EQUATOR

10°S

30°E 40°E 50°E 60°E 70°E 80°E 90°E 100°E

ASIA

EUROPE

ATLANTIC OCEAN

Black Sea

Caspian Sea

Mediterranean Sea

Persian Gulf

Gulf of Oman

Arabian Sea

INDIAN OCEAN

AFRICA

N
W E
S

0 mi. 1,000
0 km 1,000
Lambert Azimuthal Equal-Area projection

Contiguous United States and North Africa and Southwest Asia: Land Comparison

MAP STUDY

Movement of oil products

Oil

Natural gas

1 To what areas of the world are oil products from Southwest Asia shipped?

2 What oil regions supply oil products to Russia?

Fast Facts

COMPARING POPULATION:
United States and Selected Countries of North Africa and Southwest Asia

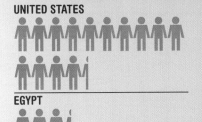

UNITED STATES

EGYPT

ALGERIA

SAUDI ARABIA

= 20,000,000

ISRAEL

Source: *World Population Data Sheet, 2001.*

URBAN POPULATIONS:
Selected Cities of North Africa, Southwest Asia

CAIRO, EGYPT

TEHRAN, IRAN

ALEXANDRIA, EGYPT

ANKARA, TURKEY

= 500,000

ISTANBUL, TURKEY

Source: *National Geographic Atlas of the World,* 7th Edition, 1999.

Data Bits

Country	Automobiles per 1,000 people	Telephones per 1,000 people
Kuwait	317	227
Lebanon	299	179
Libya	159	68
Morocco	38	50
Yemen	15	13

Population: Urban vs. Rural

Country	Urban	Rural
Kuwait	97%	3%
Lebanon	87%	13%
Libya	86%	14%
Morocco	48%	52%
Yemen	34%	66%

Source: *World Desk Reference, 2000.*

GRAPHIC STUDY

1 What generalization can you make about the rural and urban population split and the number of telephones and automobiles for each country?

2 How does the population of Cairo compare to that of Tehran? How does the population of Ankara compare to that of Cairo?

REGIONAL ATLAS

Country Profiles

AFGHANISTAN

POPULATION:
26,800,000
106 per sq. mi.
41 per sq. km

LANGUAGES:
Pashto, Dari

MAJOR EXPORTS:
Fruits and Nuts

MAJOR IMPORT:
Foods

CAPITAL:
Kabul

LANDMASS:
251,772 sq. mi.
652,090 sq. km

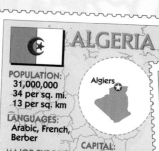

ALGERIA

POPULATION:
31,000,000
34 per sq. mi.
13 per sq. km

LANGUAGES:
Arabic, French,
Berber

MAJOR EXPORT:
Petroleum

MAJOR IMPORT:
Machinery

CAPITAL:
Algiers

LANDMASS:
919,591 sq. mi.
2,381,741 sq. km

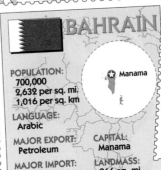

BAHRAIN

POPULATION:
700,000
2,632 per sq. mi.
1,016 per sq. km

LANGUAGE:
Arabic

MAJOR EXPORT:
Petroleum

MAJOR IMPORT:
Machinery

CAPITAL:
Manama

LANDMASS:
266 sq. mi.
698 sq. km

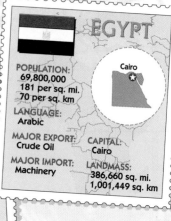

EGYPT

POPULATION:
69,800,000
181 per sq. mi.
70 per sq. km

LANGUAGE:
Arabic

MAJOR EXPORT:
Crude Oil

MAJOR IMPORT:
Machinery

CAPITAL:
Cairo

LANDMASS:
386,660 sq. mi.
1,001,449 sq. km

IRAN

POPULATION:
66,100,000
105 per sq. mi.
41 per sq. km

LANGUAGES:
Persian, Kurdish

MAJOR EXPORT:
Petroleum

MAJOR IMPORT:
Machinery

CAPITAL:
Tehran

LANDMASS:
630,575 sq. mi.
1,633,189 sq. km

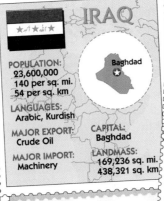

IRAQ

POPULATION:
23,600,000
140 per sq. mi.
54 per sq. km

LANGUAGES:
Arabic, Kurdish

MAJOR EXPORT:
Crude Oil

MAJOR IMPORT:
Machinery

CAPITAL:
Baghdad

LANDMASS:
169,236 sq. mi.
438,321 sq. km

ISRAEL

POPULATION:
6,400,000
787 per sq. mi.
304 per sq. km

LANGUAGES:
Hebrew, Arabic

MAJOR EXPORT:
Polished
Diamonds

MAJOR IMPORT:
Chemicals

CAPITAL:
Jerusalem *

LANDMASS:
8,131 sq. mi.
21,059 sq. km

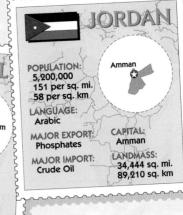

JORDAN

POPULATION:
5,200,000
151 per sq. mi.
58 per sq. km

LANGUAGE:
Arabic

MAJOR EXPORT:
Phosphates

MAJOR IMPORT:
Crude Oil

CAPITAL:
Amman

LANDMASS:
34,444 sq. mi.
89,210 sq. km

* Israel has proclaimed Jerusalem
as its capital, but many countries'
embassies are located in Tel Aviv.

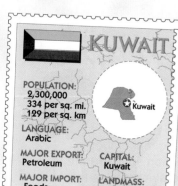

KUWAIT

POPULATION:
2,300,000
334 per sq. mi.
129 per sq. km

LANGUAGE:
Arabic

MAJOR EXPORT:
Petroleum

MAJOR IMPORT:
Foods

CAPITAL:
Kuwait

LANDMASS:
6,880 sq. mi.
17,819 sq. km

LEBANON

POPULATION:
4,300,000
1,071 per sq. mi.
414 per sq. km

LANGUAGES:
Arabic, French

MAJOR EXPORT:
Paper

MAJOR IMPORT:
Machinery

CAPITAL:
Beirut

LANDMASS:
4,015 sq. mi.
10,399 sq. km

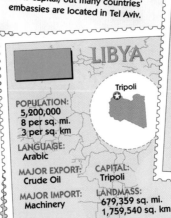

LIBYA

POPULATION:
5,200,000
8 per sq. mi.
3 per sq. km

LANGUAGE:
Arabic

MAJOR EXPORT:
Crude Oil

MAJOR IMPORT:
Machinery

CAPITAL:
Tripoli

LANDMASS:
679,359 sq. mi.
1,759,540 sq. km

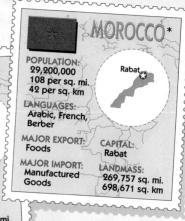

MOROCCO *

POPULATION:
29,200,000
108 per sq. mi.
42 per sq. km

LANGUAGES:
Arabic, French,
Berber

MAJOR EXPORT:
Foods

MAJOR IMPORT:
Manufactured
Goods

CAPITAL:
Rabat

LANDMASS:
269,757 sq. mi.
698,671 sq. km

* Morocco claims the Western
Sahara area, but other countries
do not accept this claim.

Countries and flags not drawn to scale

For more information on countries in this region, refer to the Nations of the World Data Bank on pages 690–699.

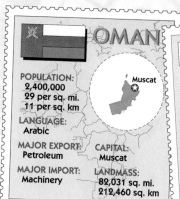

OMAN

POPULATION:
2,400,000
29 per sq. mi.
11 per sq. km

LANGUAGE:
Arabic

MAJOR EXPORT:
Petroleum

MAJOR IMPORT:
Machinery

CAPITAL:
Muscat

LANDMASS:
82,031 sq. mi.
212,460 sq. km

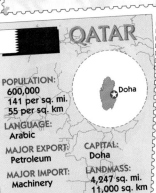

QATAR

POPULATION:
600,000
141 per sq. mi.
55 per sq. km

LANGUAGE:
Arabic

MAJOR EXPORT:
Petroleum

MAJOR IMPORT:
Machinery

CAPITAL:
Doha

LANDMASS:
4,247 sq. mi.
11,000 sq. km

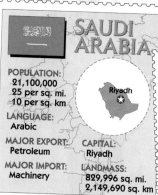

SAUDI ARABIA

POPULATION:
21,100,000
25 per sq. mi.
10 per sq. km

LANGUAGE:
Arabic

MAJOR EXPORT:
Petroleum

MAJOR IMPORT:
Machinery

CAPITAL:
Riyadh

LANDMASS:
829,996 sq. mi.
2,149,690 sq. km

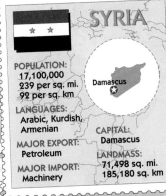

SYRIA

POPULATION:
17,100,000
239 per sq. mi.
92 per sq. km

LANGUAGES:
Arabic, Kurdish,
Armenian

MAJOR EXPORT:
Petroleum

MAJOR IMPORT:
Machinery

CAPITAL:
Damascus

LANDMASS:
71,498 sq. mi.
185,180 sq. km

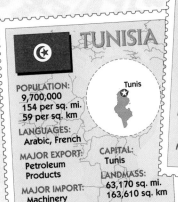

TUNISIA

POPULATION:
9,700,000
154 per sq. mi.
59 per sq. km

LANGUAGES:
Arabic, French

MAJOR EXPORT:
Petroleum
Products

MAJOR IMPORT:
Machinery

CAPITAL:
Tunis

LANDMASS:
63,170 sq. mi.
163,610 sq. km

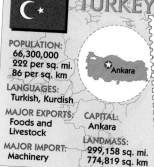

TURKEY

POPULATION:
66,300,000
222 per sq. mi.
86 per sq. km

LANGUAGES:
Turkish, Kurdish

MAJOR EXPORTS:
Foods and
Livestock

MAJOR IMPORT:
Machinery

CAPITAL:
Ankara

LANDMASS:
299,158 sq. mi.
774,819 sq. km

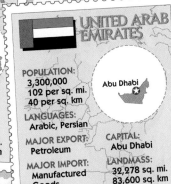

UNITED ARAB EMIRATES

POPULATION:
3,300,000
102 per sq. mi.
40 per sq. km

LANGUAGES:
Arabic, Persian

MAJOR EXPORT:
Petroleum

MAJOR IMPORT:
Manufactured
Goods

CAPITAL:
Abu Dhabi

LANDMASS:
32,278 sq. mi.
83,600 sq. km

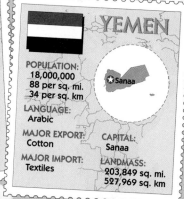

YEMEN

POPULATION:
18,000,000
88 per sq. mi.
34 per sq. km

LANGUAGE:
Arabic

MAJOR EXPORT:
Cotton

MAJOR IMPORT:
Textiles

CAPITAL:
Sanaa

LANDMASS:
203,849 sq. mi.
527,969 sq. km

BUILDING CITIZENSHIP

Religious Tolerance In Southwest Asia there are holy places of many religions, including temples, shrines, tombs, and mosques. Because Muslims are forbidden to worship statues or images, in some Islamic countries officials have destroyed ancient shrines and statues revered by Hindus or Buddhists.

1. Who owns religious properties in the United States?
2. Do you think government officials have a responsibility to protect valuable and sacred objects of all religions?

Write a short script that could be read by a television news broadcaster. The script should report on the destruction of a holy site by members of another religion. Present both points of view.

▲ Destroyed Buddhist statue

North Africa and Southwest Asia–Early Cultures

Our World Today online

Chapter Overview Visit the **Our World Today: People, Places, and Issues** Web site at owt.glencoe.com and click on **Chapter 3—Chapter Overviews** to preview information about North Africa and Southwest Asia.

Compare-Contrast Study Foldable Make and use this foldable to help you determine how Mesopotamia and ancient Egypt were similar and different.

Step 1 Fold a sheet of paper from side to side, leaving a 2-inch tab uncovered along the side.

Fold it so the left edge lays 2 inches from the right edge.

Step 2 Turn the paper and fold it into thirds.

Step 3 Unfold and cut along the two inside fold lines.

Cut along the two folds on the front flap to make 3 tabs.

Step 4 Label your foldable as shown.

Ancient Civilizations

Mesopotamia | Both | Ancient Egypt

Reading and Writing As you read the chapter, write what you learn about these ancient civilizations under the tabs. Be sure to list similarities and differences under the appropriate tabs.

Why It Matters

Lessons of History

Knowing how people and cultures developed in the past will help us make better decisions today about our own future. Learning about other people and cultures will help us understand our community and ourselves. Studying history helps us to discover new things about our relationship to the rest of the world.

◄ **Avenue of the Sphinxes, Luxor, Egypt**

Mesopotamia and Ancient Egypt

Guide to Reading

Main Idea

Learning about how past cultures lived helps us better understand our own cultures.

Terms to Know

- civilization
- irrigation system
- city-state
- theocracy
- cuneiform
- empire
- delta
- pharaoh
- polytheistic
- embalm
- pyramid
- hieroglyphics
- papyrus
- scribe

Reading Strategy

Create a chart like the one below. Write facts about Mesopotamia in the M column. Write facts about Egypt in the E column.

	M	E
Religion		
Writing		
Economy		

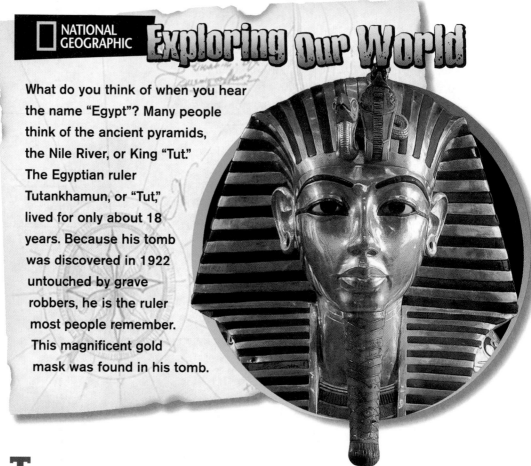

NATIONAL GEOGRAPHIC Exploring Our World

What do you think of when you hear the name "Egypt"? Many people think of the ancient pyramids, the Nile River, or King "Tut." The Egyptian ruler Tutankhamun, or "Tut," lived for only about 18 years. Because his tomb was discovered in 1922 untouched by grave robbers, he is the ruler most people remember. This magnificent gold mask was found in his tomb.

Thanks to television and films, Egypt is probably the most familiar of the ancient civilizations. Movies like *The Mummy* have helped spread knowledge of Egyptian culture, however inaccurate, through the world. **Egypt,** in North Africa, and **Mesopotamia,** in Southwest Asia, were the earliest known human civilizations.

Civilization is a term historians use to describe a culture that has reached a certain level of development. This development includes a system of writing, building cities, and specialized workers, such as farmers, blacksmiths, builders, and priests. The earliest civilizations date back only about 5,000 years.

Mesopotamia

One of the first civilizations grew in the fertile crescent of land between the **Tigris** and **Euphrates Rivers.** The map on page 83 shows you this area. The region then was called Mesopotamia, which means "between the rivers." This area of rich farmland—in what is today Syria and Iraq—was the site of the first permanent human settlements. The term *Fertile Crescent* has been used to describe this region. The era of Mesopotamia and the other early civilizations is known as the **Bronze Age** because these cultures often made wide use of the metal bronze, which is a mixture of copper and tin.

Farming   Around 4500 B.C. wandering peoples who hunted animals and gathered plants for food settled in large numbers along the banks of the Tigris and Euphrates Rivers. They saw the rich, fertile soil left by the waters from yearly floods and knew it would be a productive area to farm. Over the next 500 years these people dug ditches and built an irrigation system to control the flooding and to better water more land. A 12-month calendar, based on the phases of the moon, was developed to better predict the coming of the flood waters. Here, crops like the grain barley and possibly wheat were grown for the first time. The plow was used here for the first time, as well.

Government   Some of the villages and towns grew into cities with populations of up to 40,000 people. Each city was considered a small state, or nation. The city-state of this time was made up of the city and farmland around it. The city-state was a theocracy—it was ruled by an individual who was both the religious leader and the king. These priest-kings had almost absolute power.

Sumer   The earliest of these city-states rose in an area called **Sumer,** located near the **Persian Gulf,** where the rivers flow closest to each other. The Sumerians created a form of writing known as cuneiform. It was written with wooden triangular-shaped sticks in the form of hundreds of different wedge-shaped markings on moist clay tablets. Cuneiform records became permanent when the clay tablets were allowed to harden in the sun.

Akkad and Babylon   Eventually, around 2300 B.C. the warlike kingdom of **Akkad** conquered Sumer and several other city-states to create the first empire, or group of states under one ruler. The Akkadian Empire gradually lost control of its empire and gave way to **Babylon** in about 1800 B.C. Babylon's greatest king was **Hammurabi,** who pushed the boundaries of his empire to the Mediterranean Sea. To better rule, Hammurabi wrote a set, or code, of laws. Some of his laws seem cruel by today's standards. However, at that time laws changed often and not all people had to follow the laws. The punishment for the same offense could be a fine or could

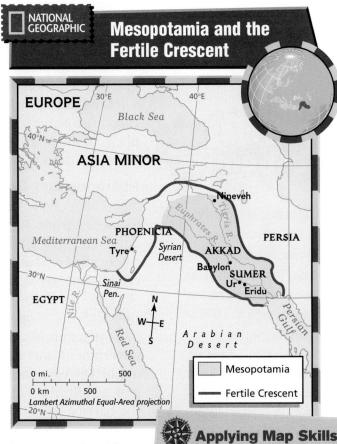

Mesopotamia and the Fertile Crescent

NATIONAL GEOGRAPHIC

Applying Map Skills

1. What feature were most early cities located near?

2. Compare this map to the map on page 75. In what modern-day countries would these cities be located?

Find NGS online map resources@ www.nationalgeographic.com/maps

be death. **Hammurabi's Code** was an attempt to bring some justice and fairness to the idea of law. Hammurabi borrowed cuneiform writing from Sumer and many of their laws as well to create his code.

The Babylonians contributed to the field of mathematics by developing a number system based on 60. From them, we have borrowed the 60-minute hour, 60-second minute, and 360-degree circle. They also used a clock controlled by drops of water to tell time.

The Phoenicians As the peoples of the area warred with neighboring states such as Egypt, they also traded, spreading ideas and cultures. Among the most important traders were the **Phoenicians,** who were located mainly in what is today **Lebanon.** By about 1200 B.C. the Phoenicians had sailed as far as southern Europe and around the southern tip of Africa. They used the sun and the stars to navigate. The Phoenicians also developed an alphabet that gave rise to the Hebrew, Greek, and Latin alphabets still in use today.

✓ Reading Check **What were three important city-states of the Fertile Crescent?**

Ancient Egypt

Like Mesopotamia, Egypt also grew out of a river valley—the **Nile.** The Nile is the longest river in the world. It runs north from its source for 4,160 miles (6,693 km). It passes through the mountains of East Africa, called the Mountains of the Moon by later Greek and Roman geographers, to the **Mediterranean Sea.** At the Mediterranean it forms a great delta, a fan-shaped or triangular piece of richly fertile land.

Pharaohs and Gods Hunter-gatherers first began to farm in the Nile Valley at its delta in about 5000 B.C. Eventually, two kingdoms formed, **Upper Egypt** to the south and **Lower Egypt** around the delta. The two kingdoms were united about 3100 B.C. under a great ruler called a pharaoh. Pharaoh means "great house," which probably refers to his lavish palace.

The pharaohs were both kings and gods in the Egyptian polytheistic religion. Polytheistic means "many gods," which describes most religions of ancient times. The pharaoh ruled as a god, so Egypt's government was a theocracy like those in Mesopotamia. The most important Egyptian gods were Horus, the god of light, Hapi, the river god, and Re, the sun god. Another important god was Osiris, the god of the harvest and eternal life. The Nile brought the Egyptians water and fertile soil, while the sun helped grow their crops.

The Egyptians believed in a form of life after death. To preserve the body for the next life, it was embalmed, or preserved immediately after death. Embalming was necessary, for Egyptians believed that the soul could not exist without the body. The embalmed body, or mummy, was wrapped in long strips of linen, a cloth made from the flax plant, which grew along the Nile.

The mummies of poor people were often buried in caves or the desert sand, but rich people's mummies were placed in coffins in often

Exploring Economics

Centers of Trade

Why did some cities develop into rich centers of trade? To succeed at trade, there must be a demand for products and a way to move goods. The peoples of Mesopotamia produced extra food, but they lacked trees for construction and mineral resources to make metals. They were able to trade extra food for other raw materials. Since water is the easiest way to transport goods, cities located nearest rivers and seas became important trading centers.

Ancient Egypt

This mummy (right) was uncovered in the Valley of the Kings, Egypt. The coffin, or sarcophagus (left), made to hold a mummy, was created around 1000 B.C. The hieroglyphics (top) are from the Kom Ombo Temple in Egypt.

History What were the biggest tombs called?

very elaborate tombs. Many of these tombs contained fabulous treasures. The biggest tombs belonged to the pharaohs and were called **pyramids.** The largest of these is the **Great Pyramid of Khufu** (Cheops) at El Giza on the west bank of the Nile. It is over 4,500 years old and originally stood 482 feet (147 m) high on 13 acres. That it still stands is proof of ancient Egyptian engineering and building skills. Egyptian sculpture, especially the statues of great pharaohs such as Ramses II, were often monuments, built to glorify the rulers. It is in large part because of these tombs and statues that we know as much as we do about the ancient Egyptian civilization.

Egyptian Writing   The ancient Egyptian system of writing is called **hieroglyphics.** Hieroglyphics is a form of picture writing with about 800 signs. The signs were cut into stone or painted on walls and **papyrus** (peh•PIE•rus). Papyrus, like flax, is a plant that grows along the Nile. It was used to make a form of paper, and gives us our root word for paper. Pharaohs kept hundreds of royal officers known as **scribes** to record the government's business.

Language experts only learned to read hieroglyphics in the early 1800s. During Napoleon Bonaparte's invasion of Egypt, a French soldier found what seemed to be a very old stone tablet near the town of Rosetta by the Nile. This stone tablet provided the key that cracked the code for hieroglyphics. The **Rosetta Stone** had the same message

Web Activity Visit the *Our World Today: People, Places, and Issues* Web site at owt.glencoe.com and click on **Chapter 3– Student Web Activities** to learn more about the Phoenicians.

written three times—in hieroglyphics, a more modern Egyptian language called Demotic, and Greek. Since both the Demotic and Greek could be read, scientists in 1821 were finally able to put meanings to the symbols used in hieroglyphics.

Trade and Conquest Egypt conquered many lands during its long history. It expanded as far as Mesopotamia in the north and southward to the present-day **Sudan.** Egypt also suffered defeats. In the 1700s B.C. invaders from Asia, the **Hyksos** or "rulers of foreign lands," conquered Egypt. They ruled for about 150 years until they were overthrown. They left behind various technologies, including horse-drawn chariots. Even then, people and nations that could use technology were able to dominate, or control, other cultures.

Egypt's trade routes stretched far and wide. Egypt sought more gold and gems in the African kingdom of Kush, also near present-day Sudan. The first female pharaoh, **Hatsheptsut,** sent a trading expedition even farther south to Punt (near present-day **Somalia**) to trade for hard woods, incense, ivory, and other products. To the north, Egypt traded across the eastern Mediterranean with the Phoenicians and the Greeks. Along with trade goods, the Egyptians spread ideas and accomplishments. Egypt, like all great empires of the past, grew weak and was eventually conquered by even greater empires—the Greeks and the Romans.

✓ Reading Check   Why was trade important to the ancient Egyptians?

Assessment

Defining Terms

1. Define civilization, irrigation system, city-state, theocracy, cuneiform, empire, delta, pharaoh, polytheistic, embalm, pyramid, hieroglyphics, papyrus, scribe.

Recalling Facts

2. History Where was the first civilization, and what was the region called then and now?

3. History What were the two early forms of writing, and where did they develop?

4. Economics Why were the Phoenicians so important to trade and spreading culture?

Critical Thinking

5. Understanding Cause and Effect How did the advancements in early farming methods in the Fertile Crescent lead to increased population growth?

Graphic Organizer

6. Organizing Information On a separate sheet of paper, create a diagram like this one. In the outer ovals write the three languages or writing systems that are found on the Rosetta Stone. Next, write what you know about each.

Rosetta Stone

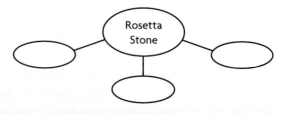

Applying Social Studies Skills

7. Analyzing Maps Look at the map on page 83. What do the locations of the towns have in common?

Making Connections

The Egyptian Pyramids

The ancient Egyptians viewed the pharaoh, or king, as the most important person on the earth. They believed he was a god who would continue to guide them after his death. A pyramid served as a tomb for the pharaoh and provided a place where the body would safely pass into the afterlife. Rooms inside the pyramid held food, clothing, weapons, furniture, jewels, and everything else the pharaoh might need in the afterlife.

The Great Pyramid at El Giza

The largest of Egypt's pyramids is the Great Pyramid of Khufu, built nearly 4,500 years ago. When the pyramid was new, it stood 482 feet (147 m) high. The square base of the pyramid covers 13 acres (about 5 ha). More than 2 million limestone and granite blocks were used in building it. These are no ordinary-sized blocks, however. The huge stones weigh an average of 2.5 tons (2.3 t) each.

Construction

For thousands of years, people have wondered how the Egyptians built the pyramids without modern tools or machinery. In the fifth century B.C., a Greek historian thought it took 100,000 people to build the Great Pyramid. Today archaeologists believe a workforce of about 20,000 did the job in about 20 years. Barges carried supplies and building materials for the pyramid down the Nile River. Nearby quarries supplied most of the stone. Skilled stonecutters carved the stones into the precise size and shape so that no mortar, or cementing material, was needed to hold the stones together.

Engineers think that workers built ramps and used papyrus twine to drag the huge stones to the pyramid. They formed ramps up all four sides of the pyramid and made the ramps higher and longer as the pyramid rose. They then dragged the stones up the ramps. Once finished, the ramps were cleared away. Then stonemasons smoothed and polished the stone, and the finished pyramid towered over the surrounding desert.

▶ Making the Connection

1. Why did the Egyptians build the pyramids?
2. How many workers did ancient historians and modern archaeologists each think it took to build the Great Pyramid?
3. **Sequencing Information** Describe the process experts think Egyptians used to build the pyramids.

◀ The Great Pyramid at El Giza, Egypt

Religions of the Middle East

NATIONAL GEOGRAPHIC Exploring Our World

The term *Middle East* is often used to refer to countries in Southwest Asia. The term *Middle East* is Eurocentric, meaning that it is based on the European perspective of "east." Southwest Asia was called the Middle East, since it was about halfway between Europe and China. Historians and policy makers, as well as news reporters refer to "conflict in the Middle East."

Guide to Reading

Main Idea

Three of the world's monotheistic religions—Judaism, Christianity, and Islam—all developed in Southwest Asia.

Terms to Know

- monotheism
- Diaspora
- scapegoat
- messiah
- disciple
- minister
- Crusades
- five pillars of faith
- hajj

Reading Strategy

Create and complete a chart like this one. List important beliefs of each religion.

Religion	Beliefs
Judaism	
Christianity	
Islam	

Today, Judaism, Christianity, and Islam have become major world faiths. All three religions are examples of monotheism, or the belief in one supreme god. All three look to the ancient city of **Jerusalem** as a holy site.

Judaism

Judaism is the oldest of these three world religions. It was first practiced by a small group of people in the Middle East called the Israelites. The followers of Judaism today are known as Jews. We know about the early history of the Jewish people and their religion from their holy book—the **Torah.**

According to Jewish belief, the Jews are descended from Abraham and Sarah, who first worshipped the one God, or Yahweh. The Jews believe that they are God's chosen people and will remain so for as long as they follow God's laws. The most well-known of these laws were revealed by the prophet **Moses** and are known as the **Ten Commandments.** The five books of Moses, along with the books of laws and teachings, make up the complete Torah.

Israel became an important and prosperous state under its first three kings—Saul, David, and Solomon. David made the city of Jerusalem Israel's capital in about 1000 B.C. After Solomon's death, the kingdom split in two. Thereafter, the Jews would be conquered and exiled many times.

Eventually, the Jewish people spread to countries in many parts of the world. This scattering of the Jews outside of the Holy Land was called the Diaspora. For many hundreds of years, Jews have lived and worked as citizens of different countries. Jewish scholars, writers, artists, and scientists have greatly increased the world's knowledge.

In some areas Jews have been treated with tolerance and understanding. In other areas they have been viewed with suspicion and hatred. Some governments have used Jewish communities as a scapegoat, or someone to blame for their troubles. Property belonging to Jewish people has been seized and their lives threatened. More than 6 million Jews were murdered in Europe during the **Holocaust** in the 1940s.

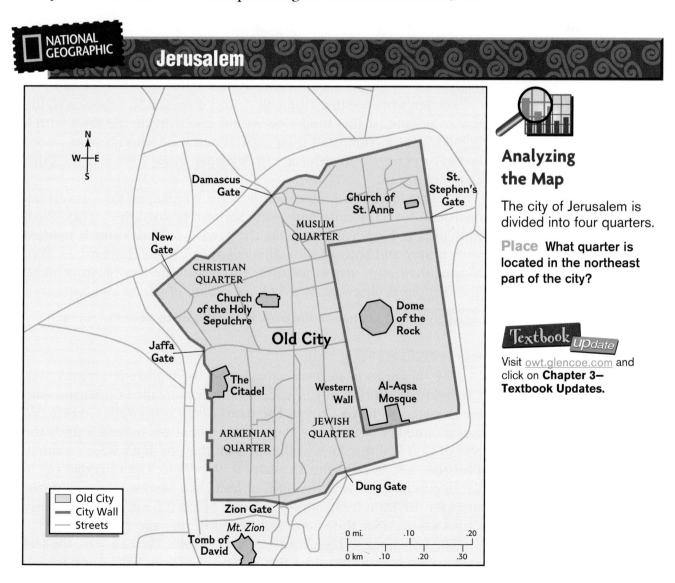

NATIONAL GEOGRAPHIC
Jerusalem

Analyzing the Map

The city of Jerusalem is divided into four quarters.

Place What quarter is located in the northeast part of the city?

Textbook *Update*

Visit owt.glencoe.com and click on **Chapter 3— Textbook Updates.**

NATIONAL GEOGRAPHIC On Location

Three Religions

Jewish scholars examine Torah scrolls (left). A Christian missionary priest ministers to a Native American family (center). Muslims study the Quran (right).

Religion What do all three religions have in common?

The Jews believe that God will deliver a messiah, or savior, to the Jewish people. At that time, God would also provide the Jews with a homeland, or a country of their own. When the United Nations voted in 1947 to create a Jewish state in Palestine, many Jews accepted this as an act of God.

Judaism has several important holy days, including Rosh Hashanah (RAHSH huh·SHAH·nuh) and Yom Kippur (YAHM kih·POOR). Rosh Hashanah is New Year's Day on the Jewish calendar and is marked with prayer and solemn thoughts. Following Rosh Hashanah is Yom Kippur, the holiest day in Judaism. Also called the Day of Atonement, Yom Kippur is observed by a 24-hour period of prayer and fasting.

✓ Reading Check What are the basic laws revealed to Moses known as?

Christianity

The traditions of Judaism gave rise to the world's largest religion, **Christianity.** Christianity is made up of people, called Christians, who are followers of **Jesus Christ.** The word *Christ* is from the Greek word for "anointed," which means chosen one. Christians believe Jesus is the Son of God and that he was the Messiah that the Jews were awaiting. Christians see Jesus as the Messiah for all people. The Christian calendar begins in A.D. 1 with the birth of Jesus. "B.C." and "A.D." are abbreviations for the Latin terms meaning "years before Christ" and "years after Christ's birth." (See the Social Studies Skill on page 240.)

The holy book of the Christians is the **Bible.** The first part, the Old Testament, is composed of the books of Moses and other Jewish writers. It contains the history and traditions of Judaism that led up to the birth of Jesus. The second part, the New Testament, deals with the birth, the life, and the teachings of Jesus as recorded by his followers, called **disciples.** Christians believe that after Jesus was killed, he rose from the dead, thereby proving the existence of an afterlife with God for all who truly believe.

The disciples spread Jesus' teachings across the Roman world and beyond. However, until about A.D. 300, Christians were persecuted in the Roman Empire. Then the emperor **Constantine the Great** proclaimed that Christianity was to be a lawful religion. At this point, Christians were no longer persecuted.

The spread of Christianity was achieved primarily through the work of individuals and missions. The teachings of Jesus were carried to far parts of the globe by missionaries who built churches, schools, and hospitals to **minister** to, or take care of, new Christians. Europe—especially Rome and Constantinople—became the center of Christianity. For hundreds of years, the Christian church shared power with the rulers of many of the nations of Europe. The most famous universities of Europe

Primary Source

COMPARING SCRIPTURE

Although there are many differences between the world's major religions, there are also many similarities. These quotes from Judaism, Christianity, and Islam illustrate the belief in good deeds.

When the holy one loves a man, He sends him a present in the shape of a poor man, so that he should perform some good deed to him, through the merit of which he may draw a cord of grace. **The Torah; Genesis 104a**

He who has two coats, let him share with him who has none: and he who has food, let him do likewise. **The Bible; Luke 3:11**

Every person's every joint must perform a charity every day the sun comes up: to act justly between two people is a charity. . . . a good word is a charity; every step you take in prayers is a charity; and removing a harmful thing from the road is a charity. **Saying of the Prophet Muhammad**

Analyzing Primary Sources

All three religions share the message of helping others. Why do you suppose there has been such conflict among them?

Spread of Islam

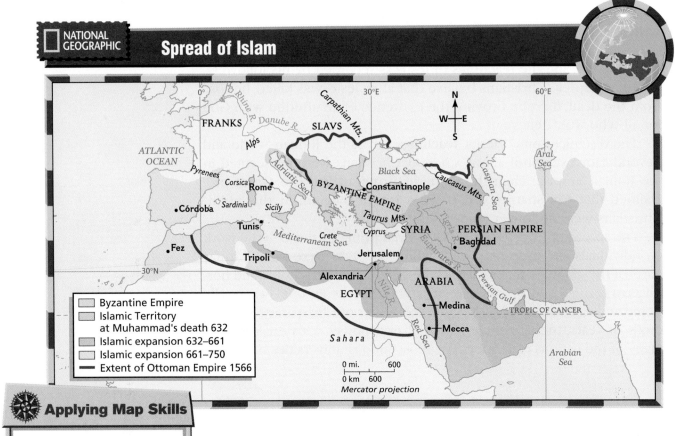

Byzantine Empire

Islamic Territory at Muhammad's death 632

Islamic expansion 632–661

Islamic expansion 661–750

Extent of Ottoman Empire 1566

0 mi.   600

0 km   600

Mercator projection

Applying Map Skills

1. Which Islamic expansion included Egypt?

2. Did Islamic territory completely encircle the Mediterranean Sea?

Find NGS online resources @ www.nationalgeographic.com/maps

were begun by Christian scholars. Catholic popes and kings organized military campaigns, called the Crusades, to capture the city of Jerusalem. Today, Christians of many denominations look to Jerusalem as a holy city where important churches and shrines are located.

Christians around the world mark important events in the life of Jesus. Christmas celebrates Jesus' birth. Most Christians celebrate Palm Sunday, the occasion when Jesus entered Jerusalem in triumph, and Good Friday, the day of Jesus' crucifixion, or death on a cross. Easter, believed to be the day that Jesus rose from the dead, is the most important day of the Christian calendar.

√ Reading Check Why was Constantine important to spreading Christianity?

Islam

Some Jewish and Christian traditions are honored by the world's second-largest religion, **Islam.** In the Arabic language, the word *Islam* means "surrender" to the will of God, or Allah. The followers of Islam are called Muslims.

Muslims believe that **Muhammad** is the last and greatest prophet of Allah. The Muslim holy book, the **Quran** (Koran), contains revelations from God to Muhammad. There are three holy cities in Islam. **Makkah** (Mecca) and Madinah are in Saudi Arabia. Jerusalem is important in Palestine because Muslims believe Muhammad ascended to heaven from there in A.D. 632.

The Quran describes the five pillars of faith, or the five obligations all Muslims must fulfill. The first duty is to state your belief in the faith: "There is no god but God, and Muhammad is his messenger." Second, Muslims must pray five times a day, facing the holy city of

Makkah. The third obligation is to give to charity. The fourth is to fast, which means not eating or drinking anything from dawn to sunset, during the lunar holy month of Ramadan. Only the young, sick people, pregnant women, and travelers do not have to fast. The last pillar of faith is a pilgrimage, called the hajj. Once in each Muslim's life, he or she must, if able, journey to Makkah to pray. The reward for fulfilling all these religious duties is paradise.

The Muslim calendar begins in A.D. 622, the year of the *Hijrah,* when Muhammad was forced to flee for safety from Makkah to Madinah. Muslim dates are written with an "A.H." to signify the years after the *Hijrah.*

Since the time of Muhammad, the Islamic faith has spread widely. In the A.D. 700s, victorious Muslim armies expanded the Muslim empire. Soldiers believed that to die for the faith was to go to Heaven. Merchants and scholars also spread the Muslim faith. Muslims exchanged ideas with those they conquered. Muslim knowledge in art and architecture, mathematics, medicine, astronomy, geography, history, and other fields was greatly increased.

As mentioned before, Ramadan is a very important holiday on the Muslim calendar. This is the month, according to Muslim beliefs, in which God began to reveal the Quran to Muhammad. Muslims observe Ramadan by fasting from dawn to sunset and refraining from any acts that take their attention away from God.

Reading Check What is the Islamic pilgrimage known as?

Section 2 Assessment

Defining Terms
1. Define monotheism, Diaspora, scapegoat, messiah, disciple, minister, Crusades, five pillars of faith, hajj.

Recalling Facts
2. Religion What are the world's three largest monotheistic religions?

3. History Why do historians refer to this area as the Middle East instead of Southwest Asia?

4. History Which city in Israel do the three religions look to as a holy site?

Critical Thinking
5. Making Inferences Look at the map on page 92. What might be one explanation for Islam not spreading from Spain to France?

Graphic Organizer
6. Organizing Information Create a chart like the one below and write key terms to summarize the quotes in the **Primary Source** on page 91.

Key Terms

Applying Social Studies Skills

7. Analyzing Diagrams Examine the map of Jerusalem on page 89. Explain how the city has been divided.

Social Studies Skill

Using Latitude and Longitude

Learning the Skill

To find an exact location, geographers use a set of imaginary lines. One set of lines—latitude lines—circles the earth's surface east to west. The starting point for numbering latitude lines is the Equator, which is 0° latitude. Every other line of latitude is numbered from 1° to 90° and is followed by an N or S to show whether it is north or south of the Equator. Latitude lines are also called parallels.

A second set of lines—longitude lines—runs vertically from the North Pole to the South Pole. Each of these lines is also called a meridian. The starting point—0° longitude—is called the Prime Meridian (or Meridian of Greenwich). Longitude lines are numbered from 1° to 180° followed by an E or W—to show whether they are east or west of the Prime Meridian.

To find latitude and longitude, choose a place on a map. Identify the nearest parallel, or line of latitude. Is it located north or south of the Equator? Now identify the nearest meridian, or line of longitude. Is it located east or west of the Prime Meridian?

Practicing the Skill

1. On the map below, what is the exact location of Washington, D.C.?
2. What cities on the map lie south of 0° latitude?
3. What city is located near 30°N, 30°E?

Applying the Skill

Turn to page RA2 of the **Reference Atlas.** Determine the latitude and longitude for one city. Ask a classmate to use the information to find and name the city.

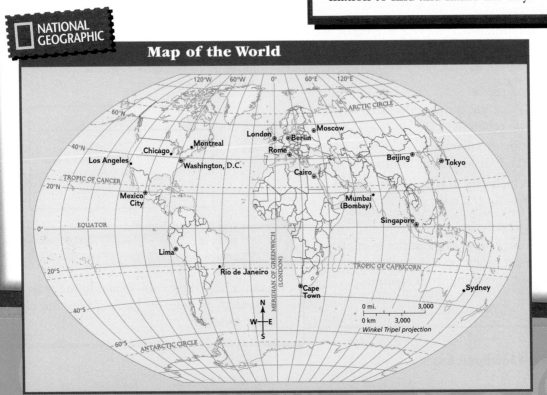

Map of the World

NATIONAL GEOGRAPHIC

Reading Review

Section 1 | Mesopotamia and Ancient Egypt

Terms to Know

civilization
irrigation
   system
city-state
theocracy
cuneiform
empire
delta

pharaoh
polytheistic
embalm
pyramid
hieroglyphics
papyrus
scribe

Main Idea

Learning about how past cultures lived helps us better understand our own cultures.

✓ History One of the first civilizations developed in the Fertile Crescent.

✓ History Early advancements in Mesopotamia, Sumer, and Phoenicia were in farming, writing, and government.

✓ History Ancient Egypt is known for pharaohs, pyramids, hieroglyphics, and mummies.

Section 2 | Religions of the Middle East

Terms to Know

monotheism
Diaspora
scapegoat
messiah
disciple
minister

Crusades
five pillars
   of faith
hajj

Main Idea

Three of the world's monotheistic religions—Judaism, Christianity, and Islam—all developed in Southwest Asia.

✓ History Judaism is the world's oldest monotheistic religion. The Jews believe they are God's chosen people.

✓ Religion Christians believe Jesus is the Messiah and the Son of God.

✓ Religion Muslims are followers of Islam. Muslims believe Allah is the one God and Muhammad is the messenger.

◄ Desert areas begin where the fertile Nile River Valley ends.

Assessment and Activities

Using Key Terms

Match the terms in Part A with their definitions in Part B.

A.

1. civilization
2. theocracy
3. cuneiform
4. pharaoh
5. polytheistic
6. Diaspora
7. monotheism
8. hajj
9. disciple
10. city-state

B.

a. holy journey in Islam
b. early believer in Jesus Christ
c. culture that has reached level of development where people can specialize their skills
d. god-king of ancient Egypt
e. believing in many gods
f. belief in one God
g. ruled by religious leader who is also a king
h. ancient form of writing in Sumer
i. city and its surrounding countryside
j. scattering of the Jewish people

Reviewing Main Ideas

Section 1 Mesopotamia and Ancient Egypt

11. **History** What was one of the early advancements in ancient Mesopotamia?
12. **History** Where did the earliest city-states arise?
13. **Economics** The Phoenicians were the most important early traders of the region. How did they navigate? Where and how did they travel?
14. **History** Who were the Asians that invaded ancient Egypt?
15. **History** Why did the Egyptians embalm their dead?

Section 2 Religions of the Middle East

16. **Religion** What is the similarity between Yahweh and Allah?
17. **Religion** What is the role of the Messiah in Jewish and Christian religious belief?
18. **Religion** Judaism, Christianity, and Islam are similar in their belief of one supreme god and in viewing Jerusalem in Israel as a holy site. List some of the differences among the three religions.
19. **Religion** What are some of the important holidays in Judaism, Christianity, and Islam?

 **NATIONAL GEOGRAPHIC** **Southwest Asia**

Place Location Activity

On a separate sheet of paper use unit or chapter maps to match the letters on the map with the numbered places listed below.

1. Persian Gulf
2. Zagros Mountains
3. Euphrates River
4. Turkey
5. Iran
6. Israel
7. Iraq
8. Saudi Arabia
9. Makkah
10. Jerusalem

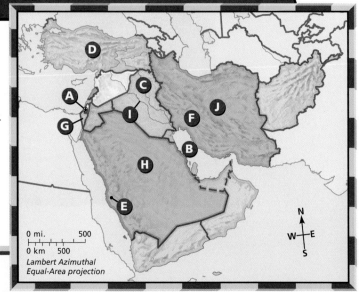

0 mi.   500
0 km   500
Lambert Azimuthal
Equal-Area projection

Self-Check Quiz Visit the *Our World Today: People, Places, and Issues* Web site at owt.glencoe.com and click on **Chapter 3—Self-Check Quizzes** to prepare for the Chapter Test.

Critical Thinking

20. **Analyzing Information** Hammurabi wrote a code of laws to help him rule better. How would laws help a king rule?

21. **Categorizing Information** Create a chart like the one below. Complete the name of the holy book for each of the religions listed.

Religion	Holy Book
Judaism	
Christianity	
Islam	

Current Events Journal

22. **Summarizing Information** Look through one of the weekly newsmagazines in your library or at home for information on the Middle East. Summarize the article for your journal.

Mental Mapping Activity

23. **Focusing on the Region** Create a simple outline map of Egypt and the Middle East. Draw in the Nile, Tigris, and Euphrates Rivers. Shade the areas where the early civilizations of ancient Egypt and Mesopotamia were located. Locate the cities of Tyre, Ninevah, Babylon, and Ur on your map.

Technology Skills Activity

24. **Using the Internet** Search the Internet and find several newspapers that publish online. Use at least three different sources to research recent discoveries about any ancient cultures in this region. Use the computer to create a report on this topic. You may want to include visual materials for display.

Standardized Test Practice

Directions: Study the map below, and then answer the question that follows.

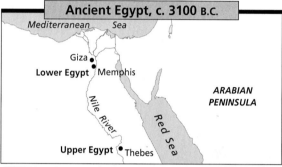

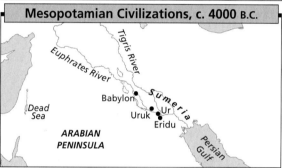

1. **What characteristic did the first Egyptian and Mesopotamian civilizations share?**

   A They were established in the same year.

   B Both civilizations began in North Africa.

   C They both developed on the banks of rivers.

   D People in both civilizations relied on hunting to obtain food.

Test-Taking Tip: When you answer a map question, do *not* rely on your memory of the map. Instead, check each answer choice against the information on the map and get rid of answer choices that are incorrect. Eliminating even one wrong choice will help you locate the correct answer.

North Africa and Southwest Asia Today

The World and Its People ▪ NATIONAL GEOGRAPHIC

To learn more about the people and places of North Africa and Southwest Asia, view **The World and Its People Chapters 16** and **17** videos.

Our World Today online

Chapter Overview Visit the **Our World Today: People, Places, and Issues** Web site at owt.glencoe.com and click on **Chapter 4—Chapter Overviews** to preview information about North Africa and Southwest Asia.

FOLDABLES™
Study Organizer

Categorizing Information Study Foldable Asking yourself questions while reading material helps you to focus on what you are reading. Make this foldable to help you ask and answer questions about the people and places in North Africa and Southwest Asia.

Step 1 Fold a sheet of paper in half from side to side, leaving a ½ inch tab along the side.

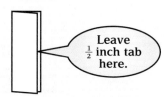

Leave ½ inch tab here.

Step 2 Turn the paper and fold it into fourths.

Fold in half, then fold in half again.

Step 3 Unfold and cut up along the three fold lines.

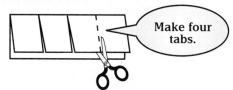

Make four tabs.

Step 4 Label as shown.

Egypt, Libya, the Maghreb | Turkey, Israel | Syria, Lebanon, Jordan, Arabia | Iraq, Iran, Afghanistan

Reading and Writing As you read, ask yourself questions about these countries. Write your questions and answers under each appropriate tab.

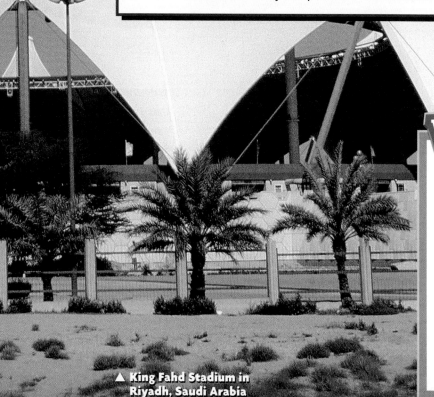

▲ **King Fahd Stadium in Riyadh, Saudi Arabia**

Why It Matters

Crossroads

Because of their location near Europe and Asia, North Africa and Southwest Asia remain even today at the "crossroads of the world." The world depends upon the oil and gas resources found here. Because of this situation, many nations have more reasons than usual for watching closely the events in these oil-rich countries. Achieving peace in this region is of global importance.

North Africa

Main Idea

North Africa's desert landscape has shaped the people and culture for many centuries, as has the Islamic religion.

Terms to Know

- delta
- silt
- oasis
- aquifer
- dictator
- terrorism
- secular
- constitutional monarchy

Reading Strategy

Create a diagram like this one. In the outer ovals, list facts about five of the countries in this section. In the center oval, write three facts that all five countries have in common.

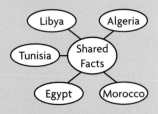

NATIONAL GEOGRAPHIC Exploring Our World

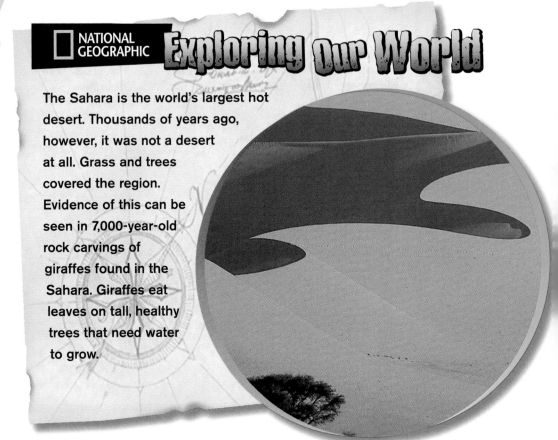

The Sahara is the world's largest hot desert. Thousands of years ago, however, it was not a desert at all. Grass and trees covered the region. Evidence of this can be seen in 7,000-year-old rock carvings of giraffes found in the Sahara. Giraffes eat leaves on tall, healthy trees that need water to grow.

The people of North Africa and Southwest Asia have many things in common. Islam is the major religion and Arabic is the most common language. Much of North Africa and Southwest Asia is desert, and the cultures have adapted to the desert way of life. This section will look at some of the countries in North Africa.

Egypt

Egypt is a large country about the same size as Texas and New Mexico together. Yet most of it is desert. Therefore, Egypt's people crowd into about 5 percent of the land, or an area about twice the size of Maryland. The lifeline of Egypt is the Nile River, which supplies 85 percent of the country's water. From its sources in eastern Africa, the Nile flows 4,160 miles (6,693 km) north to the Mediterranean Sea, making it the world's longest river. Where the river empties into the Mediterranean Sea, you find the Nile's delta. A **delta** is the area formed from soil deposited by a river at its mouth.

For centuries, the Nile's waters would rise in the spring. The swollen river carried **silt,** or small particles of rich soil. When it reached

Egypt, the Nile flooded its banks. As the floodwaters dried, the silt was left behind, enriching the soil and making the land especially good for farming. Dams and channels control the river's flow and its use for farming and generating electric power.

The triangle-shaped **Sinai** (SY•ny) **Peninsula** lies southeast of the Nile delta. This area is a major crossroads between Africa and Southwest Asia. The **Suez Canal** separates the Sinai Peninsula from the rest of Egypt. Egyptians and Europeans built the canal in the mid-1860s. Today the Suez Canal is still one of the world's most important waterways. Ships use the canal to pass from the Mediterranean Sea to the Red Sea. In making this journey, they avoid traveling all the way around Africa.

Egypt's Economy   About 29 percent of Egypt's people work in agriculture, though only about 2 percent of Egypt's land is used for farming. Important crops include sugarcane, grains, vegetables, fruits, cotton, and cattle. Look at the physical map on page 74. Where do you think the best farmland in Egypt is located?

Farmers rely on dams to control the water needed for their fields. Egypt's largest dam and a major source of electric power is the **Aswan High Dam.** The dams give people control over the Nile's floodwaters. They can store the water for months behind the dams, then release it several times during the year, rather than having just the spring floods. This control allows farmers to harvest two or three crops a year.

The dams bring challenges as well as benefits. They block the flow of silt, making the farmland less fertile. Farmers now rely more heavily on chemical fertilizers to grow crops. In addition, the dams prevent less freshwater from reaching the delta. So, salt water from the Mediterranean Sea flows deeper into the delta, making the land less fertile.

Egyptian factories make food products and textiles. Egypt's main energy resource, however, is oil, found in and around the Red Sea. Petroleum products make up almost half the value of Egypt's exports. Another important industry is tourism. Visitors come to see the magnificent ruins of ancient Egypt.

✓Reading Check **Why is the Aswan High Dam important?**

Libya

Except for the coastal lowlands, **Libya** is a desert with only a few oases. An oasis is a green area in a desert fed by underground water. In fact, the Sahara covers more than 90 percent of Libya. During the spring and fall, fierce dust-heavy winds blow from the desert, creating temperatures in coastal areas as high as 115°F (46°C).

Libya has no permanent rivers, but aquifers lie beneath the vast desert. Aquifers are underground rock layers that store large amounts of water. Aquifers are also found in the United States, in drier regions such as northern and western Texas. In the 1990s, the Libyan government built pipelines to carry underground water from the desert to coastal areas.

Politics in Egypt
Egypt is a republic, a government headed by a president. A legislature makes the laws, but the Egyptian president has broad powers in running the country. In the 1990s, some Islamic political and religious groups opposed the government. Some of these groups used violence to reach their political goals. However, by the early 2000s, the government had stopped these attacks.

Poor soil and a hot climate mean that Libya has to import about three-fourths of its food. The discovery of oil in Libya in 1959 brought the country great wealth. Libya's government uses oil money to buy food, build schools and hospitals, and maintain a strong military.

Libya's People Almost all of Libya's 5.2 million people have mixed Arab and Berber heritage. The **Berbers** were the first people known to live in North Africa. During the A.D. 600s, the Arabs brought Islam and the Arabic language to North Africa. Since then, Libya has been a Muslim country, and most of its people speak Arabic. More than two-thirds of Libyans live along the Mediterranean coast. Most live in two modern cities—**Tripoli,** the capital, and **Benghazi** (behn•GAH•zee).

Libya's Government In 1969 a military officer named **Muammar al-Qaddhafi** (kuh•DAH•fee) gained power and overthrew the king. Qaddhafi is a dictator, or an all-powerful leader. For many years, the United States and other democratic nations have accused Qaddhafi of encouraging terrorism against the United States and its citizens. Terrorism is the use of violent acts against civilians to achieve certain goals. The U.S. government worked many years through international courts to prove Libya's involvement in the 1988 terrorist bombing of Pan Am Flight 103 over Lockerbie, Scotland. One Libyan was eventually found guilty in an international court.

✓ Reading Check **Who has governed Libya since 1969?**

The Maghreb

Tunisia, Algeria, and **Morocco** form a region known as the **Maghreb.** *Maghreb* means "the land farthest west" in Arabic. These three countries make up the westernmost part of the Arabic-speaking Muslim world.

Tunisia In terms of land area, Tunisia is North Africa's smallest country. Along the fertile eastern coast, farmers grow wheat, olives, fruits, and vegetables. Fishing is also an important industry. Tunisian factories produce food products, textiles, and oil products. In addition, tourism is a growing industry.

In ancient times, Phoenician sailors founded the city of **Carthage** in what is now the northern part of Tunisia. Carthage became a powerful trading center and challenged Rome for control of the Mediterranean. Carthage was defeated by Rome, the city completely destroyed, and its citizens sold into slavery. Tunisia's largest city today is **Tunis,** the capital city of about 1 million people.

Algeria More than one and a half times the size of Alaska, Algeria is the largest country in North Africa. Like neighboring Libya, Algeria must import much of its food, which it pays for by selling oil and natural gas. These sales have helped Algeria's industrial growth, but widespread poverty still exists. Many Algerians have moved to France and other European countries to find work.

About 31 million people live in Algeria and are—as in Libya and Tunisia—of mixed Arab and Berber heritage. If you visited Algeria, you would discover centuries-old Muslim traditions blending with those of France. Why? Algeria became a French possession in 1834. Many people in Algeria's cities speak French as well as Arabic. Also, French dishes are popular, along with local foods.

In 1954, Algerian Arabs rose up against the French and eventually gained their freedom in 1962. Today Algeria is a republic, with a strong president and a legislature. In the early 1990s, members of Muslim political parties opposed many of the government's secular, or non-religious, policies. The Muslims gained enough support to win a national election. The government, however, rejected the election results and imprisoned many Muslim opponents, and a civil war began that has taken many lives.

Morocco   Morocco has an economy based on agriculture and industry. Farmers grow sugar beets, grains, fruits, and vegetables for sale to Europe during the winter. Morocco leads the world in the export of phosphate rock and is a leading producer of phosphates, used in fertilizers. An important service industry in Morocco is tourism.

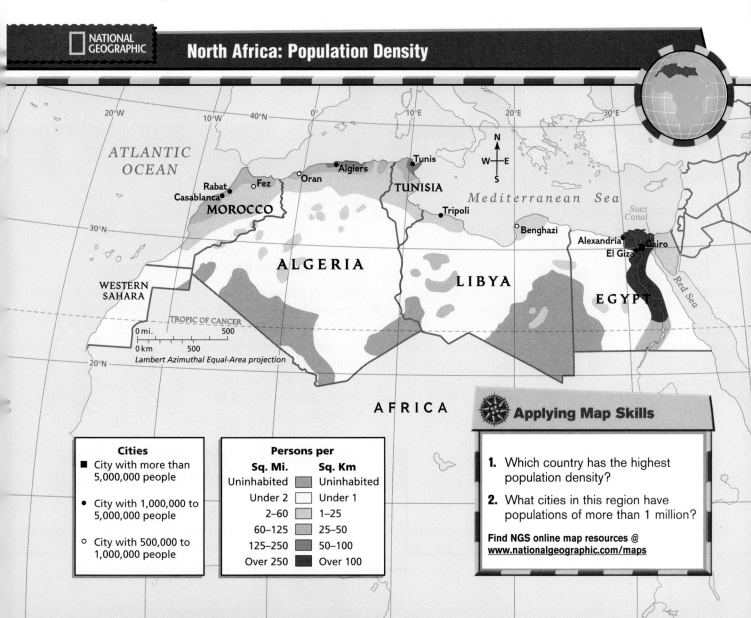

NATIONAL GEOGRAPHIC

North Africa: Population Density

Cities
- ■ City with more than 5,000,000 people
- ● City with 1,000,000 to 5,000,000 people
- ○ City with 500,000 to 1,000,000 people

Persons per

Sq. Mi.	Sq. Km
Uninhabited	Uninhabited
Under 2	Under 1
2–60	1–25
60–125	25–50
125–250	50–100
Over 250	Over 100

Applying Map Skills

1. Which country has the highest population density?

2. What cities in this region have populations of more than 1 million?

Find NGS online map resources @ www.nationalgeographic.com/maps

The map on page 75 shows you that Morocco's northern tip almost touches Europe. Here you will find the **Strait of Gibraltar.** It separates Africa and Europe—or Morocco and Spain—by only 8 miles (13 km). Like Libya, Morocco was first settled by the Berbers thousands of years ago. During the A.D. 600s, Arab invaders swept into Morocco. A century later, the Arabs and Berbers together crossed the Strait of Gibraltar and conquered Spain. Their descendants, called **Moors,** ruled parts of Spain and developed an advanced civilization until Christian Spanish rulers drove them out in the late 1400s.

Morocco's traditional culture is based on Arab, Berber, and African traditions. Moroccan music today blends the rhythms of these groups, sometimes with a dash of European pop and rock. Morocco also is known for its skilled artisans who make a variety of goods, such as carpets, pottery, jewelry, brassware, and woodwork.

Today Morocco's government is a constitutional monarchy. In this form of government, a king or queen is head of state, but elected officials run the government. In Morocco, the monarch still holds many powers, however. Beginning in the 1970s, the Moroccan king claimed the desert region of **Western Sahara,** formerly under Spanish control. The discovery of minerals there sparked a costly war between Morocco and a rebel group wanting Western Sahara to be independent. A referendum sponsored by the United Nations to allow the people of Western Sahara to decide their own future has been repeatedly postponed.

Reading Check **Who were the Moors?**

Assessment

Defining Terms
1. Define delta, silt, oasis, aquifer, dictator, terrorism, secular, constitutional monarchy.

Recalling Facts
2. Human/Environment Interaction Why is the Nile River so important to Egypt?
3. Region What is a main source of water in Libya?
4. Culture Why do many Algerians speak French?

Critical Thinking
5. Making Predictions Centuries ago, Moroccan invaders crossed the Strait of Gibraltar into Spain. What effects of this invasion might you expect to see today?

6. Making Generalizations Human alterations of earth's features can have positive benefits and also negative consequences. How has the Aswan High Dam helped and hurt Egypt?

Graphic Organizer
7. Organizing Information Create a diagram like the one here. Choose one country from this section and fill in each outer part of the diagram with a fact about the country.

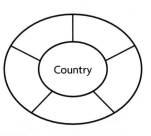

Country

Applying Social Studies Skills

8. Analyzing Maps Study the North Africa population density map on page 103. Why is the interior of Algeria so lightly populated?

Southwest Asia: Turkey and Israel

Guide to Reading

Main Idea

Turkey and Israel both have strong ties to Europe and the United States.

Terms to Know

- mosque
- kibbutz
- moshav

Reading Strategy

Create a chart like this one, filling in at least two key facts about Turkey and Israel in each category.

Turkey	Fact #1	Fact #2
Land		
Economy		
People		
Israel		
Land		
Economy		
People		

NATIONAL GEOGRAPHIC Exploring Our World

There is only one city in the world that lies on two continents. The Bosporus (BAHS•puhr•uhs), a strait in Turkey, divides this city—Istanbul. The Bosporus also separates Europe from Asia. It is an important seaway that links the Black Sea to the Sea of Marmara and, eventually, to the Mediterranean Sea.

Both **Turkey** and **Israel** have strong cultural ties to Europe. Turkey has a unique location—it bridges the continents of Asia and Europe. The large Asian part of Turkey occupies the peninsula known as **Asia Minor.** The much smaller European part lies on Europe's Balkan Peninsula. Israel is completely in Southwest Asia, but its people include many immigrants from around the world, especially from Europe, Central Asia, and the United States.

Turkey

Three important waterways—the **Bosporus,** the **Sea of Marmara** (MAHR•muh•ruh), and the **Dardanelles** (DAHRD•uhn•EHLZ) separate the Asian and European parts of Turkey. Together, these waterways are called the Turkish Straits. Find these bodies of water on page RA19 of the **Reference Atlas.**

Festival Time

Kudret Özal lives in Söğüt, Turkey. Every year, she and her family attend a festival that honors a warrior ancestor. She says, "At night everyone gathers to sing, dance, and tell jokes and stories." According to custom, Kudret wears clothing that covers her head, arms, shoulders, and legs.

Turkey's climate varies throughout the country. The large, central **Anatolian Plateau** experiences hot, dry summers and cold, snowy winters. People living in the coastal areas enjoy hot, dry summers and mild, rainy winters. Many of Turkey's people are farmers, growing cotton and fruits on the coast and raising livestock. Wheat is grown in the inland areas. Turkey also has many mineral resources such as coal, copper, and iron ore. Tourism is a growing industry, thanks to the country's beautiful beaches and historic sites.

Turkey's People   About 97 percent of Turkey's more than 66 million people are Muslims. Turkish is the official language, but Arabic and Kurdish are also spoken. Kurdish is the language of the **Kurds,** an ethnic group who make up about 20 percent of Turkey's people. The Turkish government has tried to turn the Kurds away from Kurdish culture and language. Unwilling to abandon their identity, the Kurds have demanded their own independent state. Tensions between the two groups have resulted in violent clashes. Ultimately, the Kurds of Turkey are seeking to unite with other Kurds from Iraq, Iran, and Syria to form an independent homeland called "Kurdistan" in the heart of Southwest Asia. None of the countries involved is willing to see this happen.

About 75 percent of Turkey's people live in cities or towns. **Istanbul** is Turkey's largest city with more than 9 million people. It is the only city in the world located on two continents. Istanbul is known for its beautiful palaces, museums, and mosques. Mosques are places of worship for followers of Islam. Thanks to its location at the entrance to the Black Sea, Istanbul is a major trading center.

History and Government   Istanbul began as a Greek port called **Byzantium** more than 2,500 years ago. Later it was renamed **Constantinople** after the Roman emperor Constantine the Great. For almost a thousand years, the city was the glittering capital of the Byzantine Empire.

Many of Turkey's people today are descendants of an Asian people called Turks. These people migrated to Anatolia during the A.D. 900s. One group of Turks—the **Ottomans**—conquered Constantinople in the 1400s. They, too, renamed the city, calling it Istanbul. The city served as the brilliant capital of a powerful Muslim empire called the **Ottoman Empire.** At its height, this empire ruled much of southeastern Europe, North Africa, and Southwest Asia.

World War I led to the breakup of the Ottoman Empire. During most of the 1920s and 1930s, **Kemal Atatürk,** a military hero, served as Turkey's first president. Atatürk introduced many political and social changes to modernize the country. Turkey soon began to consider itself European as well as Asian. Most Turkish people, however, continued to practice the Muslim faith. During the 1990s, secular, or nonreligious, political groups struggled for control of Turkey's government. Turkey has been a member of the North Atlantic Treaty Organization (NATO) since 1952 and is seeking to join the European Union. (See Chapter 10.)

Reading Check **What is unusual about Turkey's largest city?**

Israel

Israel is slightly larger than the state of New Jersey. The mountains of Galilee lie in Israel's north. East of these mountains is a plateau called the **Golan Heights.** South of the Golan Heights, between Israel and Jordan, is the **Dead Sea.** At 1,349 feet (411 m) below sea level, the shores of the Dead Sea are the lowest place on the earth's surface. The Dead Sea is also the earth's saltiest body of water—about nine times saltier than ocean water. It is an important source of potash, a type of mineral salt.

In southern Israel, a desert called the **Negev** (NEH•gehv) covers almost half the country. A fertile plain no more than 20 miles (32 km) wide lies along the country's Mediterranean coast. To the east, the **Jordan River** cuts through the floor of a long, narrow valley before flowing into the Dead Sea.

Israel's Economy Israel's best farmland stretches along the Mediterranean coastal plain. For centuries, farmers here have grown citrus fruits, such as oranges, grapefruits, and lemons. Citrus fruits are still Israel's major agricultural export. Farther inland, you find that the desert actually blooms. This is possible because farmers add fertilizers to the soil and carefully use scarce water resources.

In very dry areas, crops are grown with drip irrigation. This method uses computers to release specific amounts of water from underground tubes to the roots of plants. Israeli farmers plant fruits and vegetables that do not absorb salts, such as the Negev tomatoes. As a result of technology, Israel's farmers are able to export some food to other countries.

NATIONAL GEOGRAPHIC On Location

Israel

Cartloads of apples are transported to market from a farm settlement called a kibbutz in Metulla, Israel.

Place How do Israeli farmers make the desert bloom?

About 6 percent of Israelis live and work on farm settlements. Some Israelis live in a settlement called a kibbutz (kih•BUTS), where property is shared. Another kind of settlement is called a moshav (moh•SHAHV). People in a moshav share in farming, production, and selling, but each person may own private property as well.

Israel is the most industrialized country in Southwest Asia. It has been helped by large amounts of aid from Europe and the United States. Israel's skilled workforce produces electronic products, clothing, chemicals, food products, and machinery. Diamond cutting and polishing is also a major industry. **Tel Aviv-Yafo** is the largest manufacturing center. Jerusalem was made the capital of Israel in 1950.

The Israeli People The area that today is Israel has been home to different groups of people over the centuries. The ancient traditions of these groups have led to conflict among their descendants today. About 80 percent of Israel's more than 6 million people are Jews. They have moved to Israel from many countries. The other 20 percent belong to an Arab people called **Palestinians.** Most Palestinians are Muslims, but some are Christians. Time Reports: Focus on World Issues (see pages 115–121) looks at the history of the people in this region and explains some of the reasons for the ongoing conflicts between the Israelis and the Arab world and how they are working toward a peaceful settlement of these conflicts.

✓Reading Check **What city is the largest manufacturing center in Israel?**

Assessment

Defining Terms
1. **Define** mosque, kibbutz, moshav.

Recalling Facts
2. **Place** What bodies of water form the Turkish Straits?
3. **History** What other names has the city of Istanbul had?
4. **Human/Environment Interaction** What kind of technology allows Israeli farmers to grow crops in very dry areas?

Critical Thinking
5. **Analyzing Information** How has Istanbul's location made it a trading center?
6. **Understanding Cause and Effect** Why have violent clashes occurred between the Kurds and the Turkish government?

Graphic Organizer
7. **Organizing Information** On a diagram like this one, label an example of Turkey's culture at the end of each line.

Turkey's Culture

Applying Social Studies Skills

8. **Analyzing Maps** Study the political map on page 75. What country borders Israel to the north? To the northeast? To the east?

Carpet Weaving

For thousands of years, people have been making the hand-knotted floor coverings sometimes called Persian or Turkish rugs. Valued for their rich color and intricate design, these handmade rugs are unique works of art.

History

Most experts think that the nomadic peoples of Asia were among the first to make hand-knotted carpets. They used their carpets as wall coverings, curtains, and saddlebags, as well as covering for the bare ground in their tents. The soft, thick rugs blocked out the cold and could also be used as a bed or blanket.

As the nomads moved from place to place, they spread the art of carpet making to new lands and peoples. Throughout the years, the greatest carpet-producing areas have included Turkey, the republics of the Caucasus, Persia (Iran), and Turkmenistan. People in other countries, including Afghanistan, Pakistan, Nepal, India, and China, also became skilled carpet weavers.

Weaving and Knotting

Early nomads wove their carpets from sheep's wool on simple wooden looms that could be rolled up for traveling. Each carpet was woven with two sets of threads. The *warp* threads run from top to bottom, and the *weft* threads are woven from side to side. Hand-tied knots form the carpet's colorful pattern. A skillful weaver can tie about 15 knots a minute. The best carpets, however, can have more than 500 knots per square inch!

Color and Design

The beauty of woven carpets comes from the endless combination of colors and designs. Over the years, various regions developed their own

▲ Turkish carpet weavers

carpet patterns. These were passed down from generation to generation. Often the images hold special meanings. For instance, the palm and coconut often symbolize happiness and blessings.

The very first rugs were colored gray, white, brown, or black—the natural color of the wool. Then people learned to make dyes from plants and animals. The root of the madder plant, as well as certain insects, provided red and pink dye. Turmeric root and saffron supplied shades of yellow, while the indigo plant provided blue.

Making the Connection

1. How did the art of carpet weaving spread from one place to another?

2. What creates the pattern in a Turkish carpet?

3. **Drawing Conclusions** In what way do hand-knotted carpets combine art with usefulness?

Syria, Lebanon, Jordan, and Arabia

Guide to Reading

Main Idea

Syria, Lebanon, Jordan, and Saudi Arabia have Arab populations but different economies and forms of government.

Terms to Know

- Bedouin
- wadi
- desalinization
- hajj

Reading Strategy

Create a chart like this one, listing three key economic activities for each country.

Country	Economic Activities
Syria	
Lebanon	
Jordan	
Saudi Arabia	

NATIONAL GEOGRAPHIC Exploring Our World

Wherever you go in Syria, you will probably hear the word *tafaddal,* meaning "welcome." It is common for families, in particular the desert dwellers known as Bedouins, to welcome strangers into their homes. This practice developed from the harshness of life in the desert. Without food, water, and shelter freely offered, many desert travelers would die.

People in desert countries greatly value hospitality. Large parts of the countries you will read about in this section are made up of desert. Saudi Arabia makes up most of the large Arabian Peninsula. Yemen, Oman, Kuwait, Qatar, and the United Arab Emirates make up the rest. Offshore, in the Persian Gulf, lies the island country of Bahrain.

Syria

Syria's land includes fertile coastal plains and valleys along the Mediterranean Sea. The vast Syrian Desert covers the eastern region. Agriculture is Syria's main economic activity. Farmers raise mostly cotton, wheat, and fruits. The Syrian government has built dams on the **Euphrates River,** which provide water for irrigation as well as hydroelectric power for cities and industries. Future conflict with both Turkey and Iraq over Euphrates water is a possibility.

Syria's People Almost half of Syria's 17.1 million people live in rural areas. A few are Bedouins—nomadic desert peoples who follow a traditional way of life. Most other Syrians live in cities. **Damascus,** the capital, is one of the oldest continuously inhabited cities in the world. It was founded as a trading center more than 4,000 years ago.

Islam has deeply influenced Syria's traditional arts and buildings. In many Syrian cities, you can see spectacular mosques and palaces. As in other Arab countries, hospitality is a major part of life in Syria. Group meals are a popular way of strengthening family ties and friendships. The most favorite foods are lamb, flat bread, and bean dishes flavored with garlic and lemon.

Syria's Government In 1946 Syria became an independent country. Since the 1960s, one political party has controlled Syria's government. It does not allow many political freedoms. As of May 2002, Syria was one of several nations named by the U.S. government as being "state sponsors" of terrorism. This means that the United States believes that these countries help organize terrorist attacks by providing money or a base of operations. Other countries suspected of supporting terrorists include Libya, Iran, Iraq, Sudan, North Korea, and Cuba.

✓ Reading Check **On what river has Syria built dams?**

Lebanon

Lebanon is about half the size of New Jersey. Because the country is so small, you can swim in the warm Mediterranean Sea, then play in the snow in the mountains, both on the same day.

EXPLORING CULTURE

Music

The most common stringed instrument of Southwest Asia is the oud. Often pear-shaped, its neck bends sharply backward. Music from this region uses semitones that are not heard in Western music. Semitones are the "invisible" notes that lie between the black and white keys of a piano. Legend says that the oud owes its special tones to the birdsongs absorbed by the wood from which the oud was made.

Looking Closer What instrument in our culture do you think came from the oud?

GO TO

World Music: A Cultural Legacy
Hear music of this region on Disc 1, Track 25.

Cedar trees once covered Lebanon. King Solomon used them to build the Jewish Temple. Cedars were used in the Bible as symbols of beauty, strength, endurance, and pride. Now only a few groves survive in a protected area. Still, Lebanon is one of the most densely wooded Southwest Asian countries.

The Lebanese People   About 88 percent of Lebanon's nearly 4.3 million people live in coastal urban areas. **Beirut** (bay•ROOT), the capital and largest city, was once a major banking and business center. European tourists called Beirut "the Paris of the East" because of its elegant shops and sidewalk cafés. Today, however, Beirut is still rebuilding after a civil war that lasted from 1975 to 1990.

Lebanon's civil war arose among groups of Muslims and Christians. About 70 percent of the Lebanese are Arab Muslims and most of the rest are Arab Christians. Many lives were lost in the war, many people fled as refugees, and Lebanon's economy was almost destroyed. Israel invaded Lebanon during the civil war.

Arabic is the most widely spoken language in Lebanon. French is also an official language. Why? France ruled Lebanon before the country became independent in the 1940s. Local foods reflect a blend of Arab, Turkish, and French influences.

✓ Reading Check   **Why is Beirut in the process of rebuilding?**

Jordan

A land of contrasts, Jordan stretches from the fertile Jordan River valley in the west to dry, rugged country in the east. Jordan lacks water resources. However, small amounts of irrigated farmland lie in the Jordan River valley. Here farmers grow wheat, fruits, and vegetables. Jordan's desert is home to tent-dwelling Bedouins, who raise livestock.

Jordan also lacks energy resources. The majority of its people work in service and manufacturing industries. The leading manufactured goods are phosphate, potash, pottery, chemicals, and processed foods.

People and Government   Most of Jordan's 5.2 million people are Arab Muslims. They include about 1.3 million Palestinian refugees. **Amman** is the capital and largest city. On a site occupied since prehistoric times, Amman is sprinkled with Roman and other ancient ruins.

During the early 1900s, the Ottoman Empire ruled this area. After the Ottoman defeat in World War I, the British set up a territory that became known as Jordan, which became an independent country in 1946.

Jordan has a constitutional monarchy. Elected leaders govern, but a king or queen is the official head of state. From 1952 to 1999, **King Hussein I** (hoo•SAYN) ruled Jordan. He worked to blend the country's traditions with modern ways of life. In 1994, Hussein signed a peace treaty with neighboring Israel. He was helped in his work by his American-born wife, **Queen Noor.** Since 1978, Queen Noor has played a major role in promoting Arab-Western relations. The present ruler of Jordan is Hussein's son, **King Abdullah II** (uhb•dul•LAH).

✓ Reading Check   **What are Jordan's leading manufactured goods?**

Petra

Petra was built during the 300s B.C. The city's temples and monuments were carved out of cliffs in the Valley of Moses in Jordan. It was a major center of the caravan trade that reached as far as China, Egypt, Greece, and India. Archaeologists have found dams, rock-carved channels, and ceramic pipes that brought water to the 30,000 people who once lived here.

World Oil Reserves

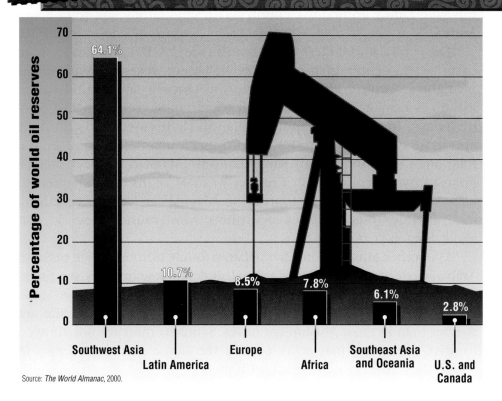

Percentage of world oil reserves

- Southwest Asia: 64.1%
- Latin America: 10.7%
- Europe: 8.5%
- Africa: 7.8%
- Southeast Asia and Oceania: 6.1%
- U.S. and Canada: 2.8%

Source: *The World Almanac*, 2000.

Analyzing the Graph

Southwest Asia has more oil than all other regions of the world combined.

Region What percentage of the world's oil reserves does Southwest Asia hold?

Visit owt.glencoe.com and click on **Chapter 4— Textbook Updates.**

Saudi Arabia

Saudi Arabia, the largest country in Southwest Asia, is about the size of the eastern half of the United States. Vast deserts cover this region. The largest and harshest desert is the Rub' al-Khali, or Empty Quarter, in the southeast.

Saudi Arabia has no rivers or permanent bodies of water. Water for farming sometimes comes from seasonal wadis, or dry riverbeds filled by rainwater from rare downpours. The desert also holds oases.

An Oil-Based Economy Saudi Arabia holds about 25 percent of the world's oil. This entire region of North Africa and Southwest Asia is by far the world's leading producer of oil. The graph above compares the amount of oil reserves in Southwest Asia with those of other regions. Saudi Arabia belongs to the **Organization of Petroleum Exporting Countries (OPEC).** Today the 11 OPEC countries supply more than 40 percent of the world's oil. By increasing or reducing supply, they are able to influence world oil prices.

Oil has helped Saudi Arabia boost its standard of living. Money earned by selling oil has built schools, hospitals, roads, and airports. Aware that someday its oil will run out, Saudi Arabia's government has been trying to broaden its economy. In recent years, it has given more emphasis to industry and agriculture. To get more water and grow more food, the government of Saudi Arabia has spent much money on irrigation and another process called desalinization, which takes salt out of seawater.

History and People The people of Saudi Arabia were once divided into many different family groups. In 1932 a monarchy led by the Saud family unified the country. The Saud family still rules Saudi Arabia.

The capital and largest city, **Riyadh** (ree•YAHD), sits amid a large oasis in the center of the country. In recent years, oil wealth has brought sweeping changes to Riyadh. Once a small rural town, Riyadh is now a modern city with towering skyscrapers and busy highways. In western Saudi Arabia, Makkah (Mecca) is another important city. In the A.D. 600s, the prophet Muhammad preached the religion of Islam in Makkah. Since that time, Makkah has been Islam's holiest city. As you read in Chapter 3, all Muslims are expected to make a hajj, or religious journey, to Makkah at least once during their lifetime. Today, millions of Muslims from around the world visit Makkah each year.

As in other Muslim countries, Islam strongly influences life in Saudi Arabia. Government, business, school, and home schedules are timed to Islam's five daily prayers and two major yearly celebrations. Saudi customs concerning the roles of men and women in public life are stricter than in most Muslim countries. Saudi women are not allowed to drive cars. They may work outside the home, but only in jobs in which they avoid close contact with men.

✓ Reading Check What influences almost every part of Saudi Arabian culture?

Section 3 Assessment

Defining Terms
1. Define Bedouin, wadi, desalinization, hajj.

Recalling Facts
2. **Government** Why are many people in the United States suspicious of the Syrian government?
3. **Culture** Why was Beirut called "the Paris of the East"?
4. **Culture** What is the importance of the city of Makkah?

Critical Thinking
5. **Understanding Cause and Effect** How could a dam on the Euphrates River cause conflict between Turkey, Syria, and Iraq?
6. **Drawing Conclusions** How could the nations belonging to OPEC affect your life?

Graphic Organizer
7. **Organizing Information** Create a diagram like this one. Inside the large oval, list characteristics that Syria, Lebanon, and Jordan share.

Syria → ○ ← Jordan, Lebanon ↑

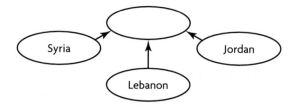

Applying Social Studies Skills

8. **Analyzing Graphs** Examine the World Oil Reserves graph on page 113. Which region of the world has the third largest reserves of oil?

TIME
REPORTS

The Arab-Israeli Conflict

Will words replace violence in the Holy Land's quarrel?

AMIT SHABI—REUTERS

TIME REPORTS

FOCUS ON WORLD ISSUES

Sacred symbols: the ancient Temple of the Jews, the birth of Jesus, and Islam's Dome of the Rock.

United in Grief

In April 2001 two couples shook hands in Amman, the capital of Jordan. One couple, Ayelet and Tzvika Shahak, were Israeli. They are citizens of Israel, a nation that Jews consider their **homeland**.

The other couple, Amal and Jamal al-Durra, were **Palestinian**. Most Palestinian Arabs are Muslims, and about 1 million of them are Israeli citizens. But like the al-Durras, most have no country to call their own. They believe that the land Israel occupies is their homeland.

Since Israel's founding in 1948, the conflict between the Israelis and the Palestinians has taken thousands of lives. Those of the Shahaks' daughter and the al-Durras' son were among them.

A Borderless Battleground

A peace group brought the Shahaks and the al-Durras together in Amman. Their common grief united them. Yet in all other ways they stood miles apart. The al-Durras want the 3.7 million Palestinians now outside Israel to be able to live on Israeli soil. The Shahaks, like most Israelis, oppose that idea. They fear that Palestinians would soon take over their nation.

The meeting gave Mr. Shahak some hope for the future, however. "Just the fact of Jamal's willingness to look for a

Historic moment: Israeli Prime Minister Yitzhak Rabin and Palestinian leader Yasser Arafat shake hands at the White House in 1993.

solution means to me that perhaps eventually there will be someone to talk to," he said. "But it seems to me that there's still a long way to go."

Most Israelis and Palestinians would agree. Israelis and Arabs have fought four major wars since 1948. The violence has never completely stopped. And it has boiled over into the nations of North Africa and Southwest Asia.

One Small Nation

Israel is a tiny country, a bit larger than New Jersey. Its population of more than 6 million, mostly made up of Jews, is smaller than New York City's.

Outside of Israel, about 395 million people live in 19 North African and Southwest Asian nations. Almost all of them are Muslims, followers of the Islamic faith. Together, those 19 nations

CYNTHIA JOHNSON

ART RESOURCE—PHOTO RESEARCHERS; SUPERSTOCK

Palestinian militant: student leader Qais Adwan

ALFRED YAGOBZADEH—SIPA FOR TIME

ALBERT FACELLY—SIPA FOR TIME

Israeli military officer: Colonel Noam Tibon

cover more land than the United States and Mexico combined.

Outsized Impact

Despite its small size, Israel has had a major impact on its neighbors. This is especially true of Muslim nations populated by Arabs, or people who speak Arabic. (People in Turkey, Iran, and Afghanistan speak other languages.)

In the early part of the 20th century, most Arabs lived in small villages. They felt a strong loyalty to their **clan**, or group of related families. Those Arabs who moved to cities began to see themselves as citizens of a nation. Few Arabs, however, thought of themselves as members of a community of nations with a common purpose.

Israel's birth in 1948 changed all that, creating the region's only true democracy. Despite support for Israel from

EXPLORING THE ISSUE

1. Analyzing Information What might make people who share a language and a religion feel loyal to one another, wherever they live?

2. Making Inferences In the year 2000, 600,000 Palestinians still lived in 59 refugee camps set up by the United Nations. How might growing up in a camp make young people angry?

3. Cause and Effect Why might Israel's democracy and economic success make the region's kings and self-appointed leaders uneasy?

the United Nations, Arabs questioned Israel's right to the land. Suddenly the Muslim world shared a goal—driving the Israelis out.

Israel survived war with its Arab neighbors in 1948, 1956, 1967, and 1973. In 1979 Israel signed a peace treaty with Egypt. Israel also signed a treaty with Jordan in 1994.

Yet some Muslim nations still consider themselves at war with Israel. Among them are a few oil-rich nations, such as Iran, Iraq, Libya, and Syria. Those nations send money—and sometimes weapons—to the Palestine Liberation Organization (PLO), headed by Yasser Arafat. The PLO's main goal is to set up an independent Palestinian nation. People like the Shahaks and the al-Durras know that. And they hope their leaders find a fair way to get both sides what they want. ▪

GARO NALBANDIAN

Inside view of the Dome of the Rock, a Muslim holy site in Jerusalem

Conflict on Holy Ground

A 35-acre plot of land sits atop a hill in the heart of Jerusalem, Israel's capital city. The hilltop is a beautiful place, filled with fountains, gardens, buildings, and domes. Many key events in the early history of Christianity took place here. The hill also has a special hold on Jews and Muslims.

Muslims call the hilltop Haram al-Sharif, or the Noble Sanctuary. A house of worship called the al-Aqsa Mosque sits there, as does a glittering, gilded dome. The dome covers a sacred rock. Muslims believe that Muhammad, the last and greatest prophet of Islam, rose to heaven from that rock.

Jews call the hill the Temple Mount. It is the most sacred site of Judaism, the Jewish religion. Two great Jewish temples once stood there. On the western side of the hill is a wall that held back the earth below the second of the temples. Jews come from all over the world to pray at the Western Wall.

From Insult to Injury

Generally only Muslims are allowed to enter the Noble Sanctuary. In September 2000, some heavily guarded Israeli politicians entered the site to challenge that rule. Rioting broke out. It spread to the Gaza Strip and to an area called the **West Bank**, where 900,000 Arabs live.

Palestinians named the daily, bloody rioting that followed the **al-Aqsa Intifadeh** (uprising). Fighting stopped for a while in June 2001. But the mutual trust that had been growing since 1993, when serious peace talks began, was dead. ▪

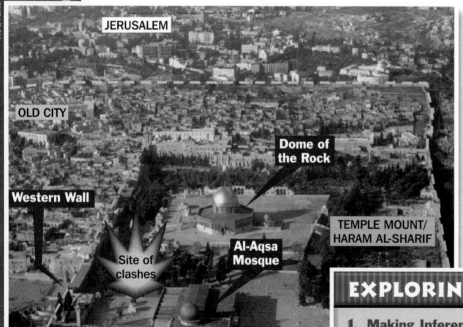

JERUSALEM

OLD CITY

Dome of the Rock

Western Wall

Al-Aqsa Mosque

Site of clashes

TEMPLE MOUNT/ HARAM AL-SHARIF

RUETERS

▲ Temple Mount/Haram al-Sharif was originally the site of King Solomon's Temple, built around 960 B.C.

EXPLORING THE ISSUE

1. **Making Inferences** Why might the Israeli politicians have thought their tour of the mount would help them win the support of Israeli voters?

2. **Drawing Conclusions** Do you think the al-Aqsa Intifadeh created sympathy for the Palestinians? Why or why not?

The Search for Peace

Kiyan Khaled al-Sayfi is a Palestinian teenager. She lives in a refugee camp in Bethlehem. Bethlehem is a city in the West Bank, an area between Israel and Jordan. "I expect to die at any moment by a stray bullet from an Israeli soldier," she said one month after the al-Aqsa Intifadeh began. "It is a terrible feeling."

About 15 miles away, near the town of Hebron, Colonel Noam Tibon, 38, was in charge of 2,000 Israeli soldiers. "My 70-year-old parents," he said, "are very worried about the situation in Israel. They went through all the wars here. But they don't stop dreaming of peace for my children, their grandchildren."

Missing Deadlines

Will that dream ever come true? The answer depends on what Israelis and Palestinians are willing to give up. Mainly, the Palestinians want land they can call their own. The Israelis want to be able to live in peace in their own country. In 1993, both groups agreed on a plan to reach those goals. If all went well, by 2000 the Palestinians would have their own nation in the West Bank and the Gaza Strip.

Yet all didn't go well. By the summer of 2000, several problems still had to be solved. For one thing, Israel wanted all of Jerusalem to be its capital. The Palestinians wanted to make the eastern half of Jerusalem their capital.

Control of the Temple Mount was another sticking point. Another obstacle was Israel's firm refusal to give

Teenager Kiyan Khaled al-Sayfi in Bethlehem

ALBERT FACELLY—SIPA FOR TIME

Palestinians everywhere the right to move to Israel. Still another was the nearly 300 Jewish settlements in the West Bank, the Gaza Strip, and East Jerusalem. The Palestinians wanted them destroyed.

The truce that halted the al-Aqsa Intifadeh in June 2001 wasn't a real peace. But when one comes, teens like Kiyan won't have to fear for their lives. And Colonel Tibon's children—or their children—will see his parents' dreams come true. ▪

EXPLORING THE ISSUE

1. **Comparing** What hopes and fears might Kiyan and Colonel Tibon have in common?

2. **Explaining** Why would Israel not want to give all Palestinians the right to move to Israel?

Promoting Peace: What Can One Person Do?

What can an individual do to help reduce tensions in Southwest Asia? Israelis Ayelet and Tzvika Shahak asked themselves that question after a Palestinian's bomb killed their daughter. They decided to promote understanding. Their group brought them together with Palestinians Amal and Jamal al-Durra.

John Wallach, an American, set up another group, Seeds of Peace. Every summer Seeds of Peace brings about 150 Muslim and Jewish teenagers to a camp in Maine. "The whole point," Wallach says, "is to let young people see that the differences are wide, that they are deep, but that it's up to them to find a way to resolve them."

▲ The camp broke down barriers for Israeli Dana Gadalyahu and Palestinian Ayah el-Rozi, both 15.

Learning from Friends

Campers one recent summer came from the nations of Israel,

John Wallach (left) with camp counselor Jared Fishman

Egypt, Jordan, Tunisia, Morocco, and Qatar. Palestinian Sara Jabari, 15, came from the West Bank. She and Dana Naor, 13, an Israeli from a Tel Aviv suburb, became fast friends. Their friendship helped them understand each other's point of view. "She is so nice," Sara said of Dana. "I really got a new idea of the Israelis from Dana."

Do students in your school understand the Arab and Israeli points of view? If not, you could help your fellow students see how complex the issue is. Set up a discussion panel. Bring together people on each side and have them discuss their views. Ask your teachers if they can suggest other ideas. ◼

EXPLORING THE ISSUE

1. **Analyzing Information** Why is Seeds of Peace an appropriate name for John Wallach's group?

2. **Making Inferences** Do you think it was harder for the Shahaks and al-Durras to meet than it was for the campers?

REVIEW AND ASSESS

▲ Trust-building exercise at Seeds of Peace.

UNDERSTANDING THE ISSUE

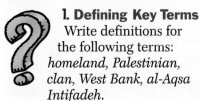

1. Defining Key Terms Write definitions for the following terms: *homeland, Palestinian, clan, West Bank, al-Aqsa Intifadeh.*

2. Writing to Inform In a 300-word article, describe the roadblocks that kept Israel and the Palestinians from ending their conflict in 2000. Use the terms listed above.

3. Writing to Persuade Israel once proposed letting Palestinians control the top of the Temple Mount, or Haram al-Sharif. Israelis would control the bottom portion, where they believe the remains of their ancient temples are buried. Write a short essay on why this plan might be a good idea, a bad one, or a little of each.

INTERNET RESEARCH ACTIVITIES

4. With your teacher's help, find resources on the Internet that include information on relations between the Palestinians and Israel today. What does each side stress about its history? What does each side leave out? Report your findings to your class.

5. Christians trace their religion's roots to the Holy Land. With your teacher's help, use Internet resources to learn about how some of the many Christian groups and their missionaries in the region are working to help end the violence in Israel. Find out what goals these groups have.

BEYOND THE CLASSROOM

6. Research another ethnic or religious conflict—the 1994 war in Rwanda, for example, or the 1999 conflict in Kosovo. How is the conflict you chose like the Arab-Israeli conflict? How is it different? Summarize your conclusions.

7. With your teacher's help, find information on efforts that people in your community are making to promote international understanding and cooperation. Contact these people to learn more about their work.

Years of Tears

Jerusalem Syria
Gaza Strip — West Bank
Israel
Egypt Jordan
Sinai Peninsula
60 mi.
60 km

Timeline of Major Arab-Israeli Conflicts, 1948-2001

1967 Third Arab-Israeli War. Israel wins control of the Gaza Strip and the Sinai Peninsula from Egypt; the Golan Heights from Syria; and the West Bank, including East Jerusalem, from Jordan.

1987-1993 First Palestinian Intifadeh (uprising). Ends when both sides agree to a timetable for creating a Palestinian state.

1948-49 First Arab-Israeli War. Israel wins control of Western Jerusalem and most of the former Palestine.

1956 Second Arab-Israeli War. The UN ends it; no side gains.

1973-74 Fourth Arab-Israeli War. Israel defeats Arab invaders.

2000-2001 Second Palestinian Intifadeh takes about 600 lives.

1940 1950 1960 1970 1980 1990 2000

BUILDING SKILLS FOR READING TIMELINES

1. Analyzing the Data Timelines help you understand the sequence of events. What was the longest period of time between wars?

2. Making Inferences Make a list of the land Israel won after each war. What does that list suggest about the role wars play in setting boundaries between nations?

FOR UPDATES ON WORLD ISSUES GO TO www.timeclassroom.com/glencoe

JAMES LUKOWSKI

Iraq, Iran, and Afghanistan

Guide to Reading

Main Idea

Iraq, Iran, and Afghanistan have recently fought wars and undergone sweeping political changes.

Terms to Know

- alluvial plain
- embargo
- shah
- Islamic republic

Reading Strategy

Create a chart like this one and list one fact about the people in each country.

Country	People
Iraq	
Iran	
Afghanistan	

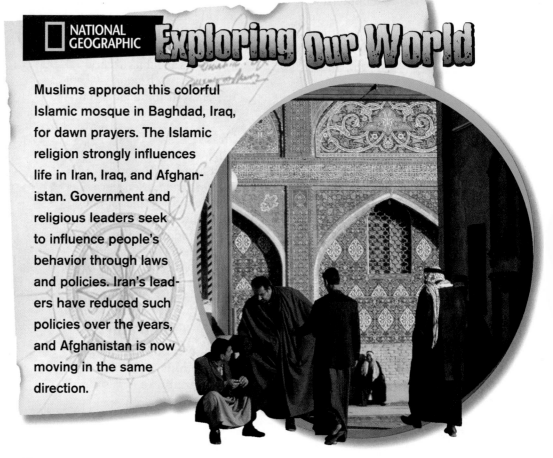

NATIONAL GEOGRAPHIC Exploring Our World

Muslims approach this colorful Islamic mosque in Baghdad, Iraq, for dawn prayers. The Islamic religion strongly influences life in Iran, Iraq, and Afghanistan. Government and religious leaders seek to influence people's behavior through laws and policies. Iran's leaders have reduced such policies over the years, and Afghanistan is now moving in the same direction.

Iraq, Iran, and Afghanistan are located in a region where some of the world's oldest civilizations developed. This region has experienced turmoil throughout history and even today.

Iraq

As you read in Chapter 3, the world's first known cities arose between the Tigris and Euphrates Rivers. These rivers are the major geographic features of **Iraq.** Between the two rivers is an alluvial plain—an area that is built up by rich fertile soil left by river floods. Most farming takes place here—growing wheat, barley, dates, cotton, and rice.

Oil is the country's major export. Iraq's factories process foods and make textiles, chemicals, and construction materials.

People and Government   About 70 percent of Iraq's 23.6 million people live in urban areas. **Baghdad,** the capital, is the largest city.

From the A.D. 700s to 1200s, Baghdad was the center of a large Muslim empire that made many advances in the arts and sciences. Muslim Arabs make up the largest group in Iraq's population. The second-largest group consists of another Muslim people, the Kurds, who want to form their own country.

Modern Iraq gained its independence as a kingdom in 1932. In 1958 the last king was overthrown in a revolt. Since then, military leaders have governed Iraq as a dictatorship. The current leader, **Saddam Hussein** (sah•DAHM hoo•SAYN), rules with an iron hand.

In the 1980s, Iraq, with aid from Western and Arab countries, fought a bloody war with its neighbor Iran. The fighting cost thousands of lives and billions of dollars in damage to cities and oil-shipping ports in the Persian Gulf. In 1990, partly because of a dispute over oil, Iraq invaded neighboring **Kuwait.** By April of 1991, at the end of the **Persian Gulf War,** a United Nations force led by the United States pushed Iraqi troops out of Kuwait. This operation was known as "Desert Storm."

After the Persian Gulf War, Saddam Hussein refused to cooperate with the demands of the United Nations. In response, the United States and other nations put an embargo on trade with Iraq. An embargo is an order that restricts trade with another country. Since then, Iraq has not exported as much oil as before and cannot import certain goods. This has severely damaged Iraq's economy.

✓Reading Check What two rivers have influenced the history of Iraq?

NATIONAL GEOGRAPHIC On Location

Baghdad, Iraq

Iraqis shop for vegetables at an outdoor market in Baghdad.

Place What kind of outdoor market have you seen?

Iran and Afghanistan

An Iranian family picnics in the hills above Tehran (left). Under the Taliban, women in Afghanistan were rarely allowed in public (right).

Government What form of government does Iran have? What group led Afghanistan in the 1990s?

Iran

Once known as Persia, **Iran** is slightly larger than Alaska. Iran is an oil-rich nation and is where the first oil wells in Southwest Asia were drilled in 1908. Like Saudi Arabia, Iran is trying to promote other industries in order to become less dependent on oil earnings. With limited supplies of water, less than 12 percent of Iran's land can be farmed. Some farmers use ancient underground channels to bring water to their fields. They grow wheat, rice, sugar beets, and cotton. Iran is also the world's largest producer of pistachio nuts.

The Iranian People   Iran's 66.1 million people differ from those of other Southwest Asian countries. More than one-half are Persians, not Arabs or Turks. The Persians' ancestors migrated from Central Asia centuries ago. They speak **Farsi,** or Persian, the official language of Iran. Other languages include Kurdish, Arabic, and Turkish. About 60 percent of Iranians live in urban areas. **Tehran,** located in northern Iran, is the largest city and the capital. Iran is also home to about 2 million people from Iraq and Afghanistan who have fled recent wars. Nearly 98 percent of Iran's people practice some form of Islam.

Iran's Government   About 2,000 years ago, Iran was the center of the powerful Persian Empire ruled by kings known as shahs. In 1979 Muslim religious leaders overthrew the last monarchy. Iran now has an Islamic republic, a government run by Muslim religious leaders. The government has introduced laws based on its understanding of the Quran. Followers of religions other than Islam have been persecuted. Many Western customs seen as a threat to Islam are now forbidden. Like Syria, Iran has been accused by many Western governments of supporting terrorists.

✓ Reading Check   How do Iranians differ in ethnic background from most other Southwest Asians?

Afghanistan

A landlocked nation, **Afghanistan** (af•GA•nuh•STAN) is mostly covered with the rugged peaks of the **Hindu Kush** mountain range. The Khyber (KY•buhr) Pass cuts through the mountains and for centuries has been a major trade route linking Southwest Asia with other parts of Asia. The capital city, **Kabul** (KAH•buhl), lies in a valley. The country's almost 27 million people are divided into about 20 different ethnic groups.

A Country at War During the 1980s, the Afghan people fought against Soviet troops who had invaded their country. When the Soviets pulled out in 1989, the Afghan people faced poverty, food shortages, and rising crime. The country collapsed into civil war. Many people turned to the **Taliban,** a group of fighters educated at strict Islamic schools in Pakistan. By 1996, the Taliban had taken control of the capital, Kabul, and about 80 percent of the country. They set up very strict laws based on their view of the religion of Islam. For example, men had to wear beards, and women had to cover themselves in public and could not hold jobs or go to school. In the north, a group known as the **Northern Alliance** continued to battle with the Taliban but with little success. In October 2001, after the attack on the World Trade Center, the United States accused the Taliban of supporting terrorists and began bombing Taliban forces. The United States also sent aid to the Northern Alliance. By mid-November, the Taliban government had collapsed and the Northern Alliance had captured Kabul. The United Nations then began working with local leaders to create a new government for Afghanistan.

✓ Reading Check What trade route cuts through the Hindu Kush?

Section 4 Assessment

Defining Terms

1. Define alluvial plain, embargo, shah, Islamic republic.

Recalling Facts

2. Economics What is Iraq's major export?

3. Culture What group had gained control of most of Afghanistan by 1996?

4. Government What type of government does Iran have? Why do so many Western nations consider it a threat?

Critical Thinking

5. Understanding Cause and Effect Why have Iraq and the United States continued to treat each other with hostility?

6. Drawing Conclusions Why do you think the Afghan people turned to the Taliban for leadership after the Soviets pulled out?

Graphic Organizer

7. Organizing Information Create a chart like this one for each of the following countries: Iraq, Iran, and Afghanistan. Then write one fact about the country under each heading.

Country		
Capital	Landforms	Agriculture
People	Religion	Government

Applying Social Studies Skills

8. Analyzing Maps Study the physical map on page 74. Between what two bodies of water is Iran located?

Technology Skill

Using the Internet

To learn more about almost any topic imaginable, use the **Internet**—a global network of computers. Many features, such as e-mail, interactive educational classes, and shopping services are offered on the Net.

▲ **The National Geographic Society's Web site**

Learning the Skill

To get on the Internet, you need three things: (a) a personal computer, (b) a device that connects your computer to the Internet, and (c) an account with an Internet service provider (ISP). An ISP is a company that enables you to log on to the Internet, usually for a fee.

After you are connected, the easiest way to access Internet sites is to use a "Web browser," a program that lets you view and explore information on the World Wide Web.

The Web consists of many documents called "Web sites," each of which has its own address, or Uniform Resource Locator (URL). Many URLs start with the keystrokes *http://*

If you do not know the exact URL of a site, commercial "search engines" such as Google or Yahoo! can help you find information. Type a subject or name into the "search" box, and then press Enter. The search engine lists available sites that may have the information you are looking for.

Practicing the Skill

Follow these steps to learn how the Internet can help you find information about Iran.

1. Log on to the Internet and access a search engine.
2. Search by typing "Iran" in the search box.
3. Scroll the list of Web sites that appears when the search is complete. Select a site to bring up and read or print.
4. If you get "lost" on the Internet, click on the back arrow key at the top of the screen until you find a familiar site.
5. Continue selecting sites until you have enough information to write a short report on natural resources found in Iran.

Applying the Skill

Follow the above steps to locate information about *oil and natural gas in Iraq.* Use the information you gather to create a chart or graph showing the amount of oil and natural gas Iraq has exported in the past five years.

Section 1 — North Africa

Terms to Know
delta
silt
oasis
aquifer
dictator
terrorism
secular
constitutional monarchy

Main Idea
North Africa's desert landscape has shaped the people and culture for many centuries, as has the Islamic religion.
✓ Location Most people of Egypt live along the Nile River or in its delta.
✓ Economics Forty percent of Egypt's workers live by farming, but industry has grown in recent years.
✓ Region North Africa includes Egypt, Libya and the three countries called the Maghreb—Tunisia, Algeria, and Morocco.
✓ Economics Oil, natural gas, and phosphates are among the important resources in these countries.
✓ Culture Most of the people in these countries are Muslims and speak Arabic.

Section 2 — Southwest Asia: Turkey and Israel

Terms to Know
mosque
kibbutz
moshav

Main Idea
Turkey and Israel both have strong ties to Europe and the United States.
✓ Location Turkey lies in both Europe and Asia.
✓ Economics Tourism is a growing industry in Turkey, thanks to beautiful beaches and historic sites.
✓ Culture About 80 percent of Israel's population are Jews. They have moved to Israel from many countries.
✓ History Israel and its Arab neighbors continue to work toward a peaceful settlement of issues that divide them.

Section 3 — Syria, Lebanon, Jordan, and Arabia

Terms to Know
Bedouin
wadi
desalinization
hajj

Main Idea
Syria, Lebanon, Jordan, and Saudi Arabia have Arab populations but different economies and forms of government.
✓ Economics Farming is the main economic activity in Syria.
✓ History Lebanon is rebuilding and recovering after a civil war.
✓ Economics Saudi Arabia is the world's leading oil producer.

Section 4 — Iraq, Iran, and Afghanistan

Terms to Know
alluvial plain
embargo
shah
Islamic republic

Main Idea
Iraq, Iran, and Afghanistan have recently fought wars and undergone sweeping political changes.
✓ Economics Iraq is suffering because of an international trade embargo.
✓ Culture Oil-rich Iran is ruled by Muslim religious leaders.
✓ Economics Afghanistan is mountainous and relatively undeveloped.

Assessment and Activities

Using Key Terms

Match the terms in Part A with their definitions in Part B.

A.

1. oasis
2. secular
3. delta
4. shah
5. hajj
6. kibbutz
7. silt
8. terrorism
9. aquifer
10. embargo

B.

a. Israeli community
b. underground rock layer that stores water
c. Iran's former monarch
d. religious pilgrimage
e. area formed by soil at a river's mouth
f. water and vegetation surrounded by desert
g. nonreligious
h. small particles of rich soil
i. violence against civilians
j. restricts trade with another country

Reviewing the Main Ideas

Section 1 North Africa

11. **Movement** What two bodies of water does the Suez Canal connect?
12. **Government** What type of government does Morocco have today?
13. **Human/Environment Interaction** Why must Libya depend on aquifers for water?
14. **Region** What does *Maghreb* mean?
15. **Economics** What energy resource is important to almost all of North Africa's countries?

Section 2 Southwest Asia: Turkey and Israel

16. **Place** What is Turkey's largest city?
17. **History** In what year was Jerusalem made the capital of Israel?

Section 3 Syria, Lebanon, Jordan, and Arabia

18. **History** What makes Damascus an important city?
19. **Economics** What is OPEC?

Section 4 Iraq, Iran, and Afghanistan

20. **Culture** How do the people of Iran differ from other Southwest Asian peoples?

NATIONAL GEOGRAPHIC North Africa

Place Location Activity

On a separate sheet of paper, match the letters on the map with the numbered places listed below.

1. Red Sea
2. Morocco
3. Libya
4. Algeria
5. Atlas Mountains
6. Nile River
7. Cairo
8. Tunisia
9. Tripoli
10. Sinai Peninsula

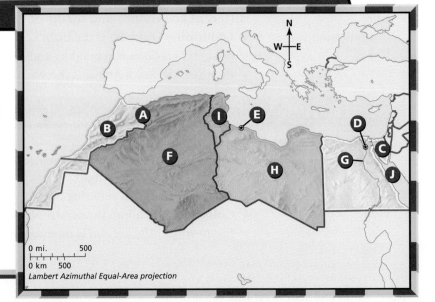

Our World Today online

Self-Check Quiz Visit the *Our World Today: People, Places, and Issues* Web site at <u>owt.glencoe.com</u> and click on **Chapter 4—Self-Check Quizzes** to prepare for the Chapter Test.

Critical Thinking

21. **Understanding Cause and Effect** Why are the most densely populated areas of North Africa along the Mediterranean Sea and the Nile River?

22. **Analyzing Information** On a chart like this, list a reason for the importance of oil and water to Southwest Asia and one result of their abundance or scarcity.

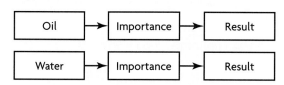

| Oil | → | Importance | → | Result |

| Water | → | Importance | → | Result |

Current Events Journal

23. **Writing a Report** Jerusalem is a holy city to three of the world's major religions. Research these religions and identify at least three holy sites you might visit to learn more about them.

Mental Mapping Activity

24. **Focusing on the Region** Draw a simple outline map of North Africa, and then label the following:

- Mediterranean Sea
- Egypt
- Red Sea
- Libya
- Atlantic Ocean
- Morocco
- Nile River
- Tunisia
- Atlas Mountains
- Algeria

Technology Skills Activity

25. **Using the Internet** Use the Internet to research life in the desert. Besides the Sahara, what other large deserts are there in the world? What kinds of life do deserts support? How do humans adapt to life in the desert? Are deserts changing in size and shape? Why? Use your research to create a bulletin board display on "Desert Cultures."

The Princeton Review

Standardized Test Practice

Directions: Study the graph, and then answer the question that follows.

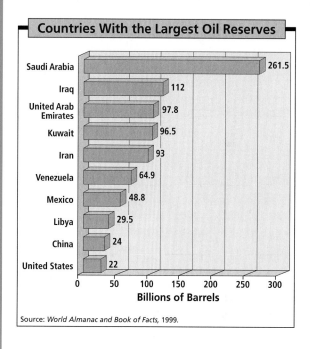

Countries With the Largest Oil Reserves

Saudi Arabia — 261.5
Iraq — 112
United Arab Emirates — 97.8
Kuwait — 96.5
Iran — 93
Venezuela — 64.9
Mexico — 48.8
Libya — 29.5
China — 24
United States — 22

Billions of Barrels
0 50 100 150 200 250 300

Source: *World Almanac and Book of Facts*, 1999.

1. **How many of the 10 countries with the largest oil reserves are located in Southwest Asia?**

   A one

   B three

   C five

   D seven

Test-Taking Tip: You need to rely on your memory as well as analyze the graph to answer this question. Look at each country, and then think back to the countries you studied in Chapter 4. Which of those listed on the graph did you just learn about?

EYE on the Environment

A Water Crisis

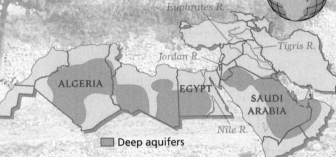

Draining the Rivers The ball game is over. You are hot, sweaty, and thirsty. You press the button on the drinking fountain, but no water comes out. A crisis? Consider this: Many people in Southwest Asia and North Africa never have enough water to meet their needs.

☐ Deep aquifers

Most of the usable water in this region comes from aquifers—underground areas that store large amounts of water—and from the Jordan, Tigris, Euphrates, and Nile Rivers. Despite these great rivers, water is scarce. The rivers flow through several countries. As each country takes its share of water, less remains for those downstream. A few countries have desalinization plants that turn seawater into freshwater. Desalinization is expensive, though. Water resources are further strained by many factors.

- Population growth — By 2025, about 570 million people will inhabit the region. That is too many people for the existing water supplies.

- Irrigation — About 90 percent of water supplies in Southwest Asia are used to irrigate crops.

- Pollution — River water in many places is polluted by salt, sewage, and chemicals.

Finding Solutions Faced with growing demand and decreasing supplies, countries in this region are looking for creative solutions to the water crisis.

- Some countries are recycling wastewater to use on crops.

- Advances in technology are making desalinization more affordable.

- Countries are building dams to regulate water. They also are constructing pipelines to carry water to where it is most needed.

Camels crossing Egypt's desert drink water piped from the Nile River, 300 miles (483 km) away.

Making a Difference

Wise Water Ways Scientist Sandra Postel is trying to educate others on ways to use water more wisely. In her book, *Last Oasis: Facing Water Scarcity*, Postel argues that we can no longer meet rising demands for water by building larger dams and drilling deeper wells. Instead of reaching out for more water, Postel argues, everyone needs to do more with less water. People need to conserve and recycle water and to use it more efficiently. Through her research, Postel has found that farmers, industries, and cities could cut their water use by as much as 50 percent. Water could be saved by practicing water conservation methods such as drip irrigation and water recycling. Postel hopes that governments around the world will work together to protect one of the earth's most precious resources.

Author Sandra Postel

Meeting Demand A group in Southwest Asia and North Africa is studying ways to ease water shortages in the region. The Water Demand Management Research Network (WDMRN) is made up of scientists and government representatives who are studying ways to meet the growing needs for water. The WDMRN shares information with other water researchers and holds meetings to encourage cooperation between all countries in the region.

Worker takes a drink of water at a desalinization plant, Kuwait.

What Can You Do?

💧 **Conserve Water**
Saving water is as easy as turning off a faucet. Practice water conservation by taking shorter showers and by turning off the water while brushing your teeth. What other ways can you conserve water at home or at school?

💧 **Find Out More**
Investigate the pathway drinking water takes in your community. Collaborate with classmates to create a bulletin board display showing how water gets from its source to a drinking fountain in your school.

Taj Mahal,
Agra, India

Macaques in a hot
spring, Japan

NATIONAL GEOGRAPHIC

Asia

For many people in the Western Hemisphere, the region of Asia—in the Eastern Hemisphere—brings to mind exotic images. Ancient temples stand in dense rain forests. Farmers work in flooded rice fields. Pandas nibble bamboo shoots. Yet bustling cities, gleaming skyscrapers, and high-technology industries also can be found here. Turn the page to learn more about this region and its 3 billion people.

▲ **Monks wrapping statue of Buddha in yellow cloth, Thailand**

NGS ONLINE
www.nationalgeographic.com/education

REGIONAL ATLAS

Asia

Physical

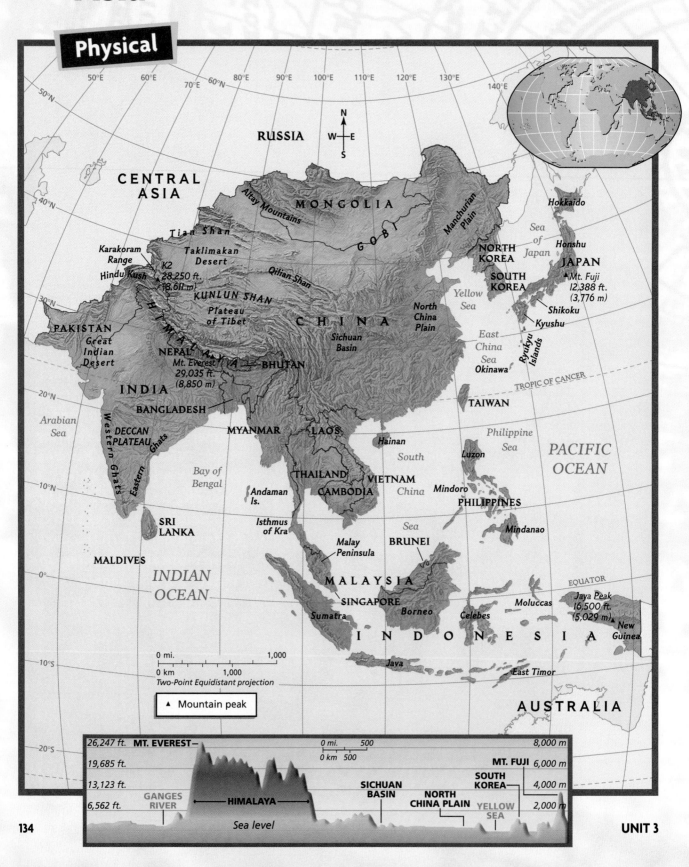

RUSSIA

CENTRAL ASIA

Altay Mountains

MONGOLIA

G O B I

Manchurian Plain

Hokkaido

NORTH KOREA

Sea of Japan

Honshu

JAPAN

▲Mt. Fuji 12,388 ft. (3,776 m)

SOUTH KOREA

Tian Shan

Karakoram Range

Taklimakan Desert

K2 28,250 ft. (8,611 m)

Qilian Shan

Hindu Kush

KUNLUN SHAN

Plateau of Tibet

C H I N A

Shikoku

Kyushu

Yellow Sea

North China Plain

East China Sea

Ryukyu Islands

Okinawa

PAKISTAN

Great Indian Desert

NEPAL

Mt. Everest 29,035 ft. (8,850 m)

BHUTAN

Sichuan Basin

TROPIC OF CANCER

H I M A L A Y A

I N D I A

BANGLADESH

Arabian Sea

DECCAN PLATEAU

Western Ghats

Eastern Ghats

MYANMAR

LAOS

Hainan

South China

TAIWAN

Philippine Sea

Luzon

PACIFIC OCEAN

Bay of Bengal

Andaman Is.

THAILAND

CAMBODIA

VIETNAM

Mindoro

PHILIPPINES

SRI LANKA

Isthmus of Kra

Sea

BRUNEI

Mindanao

MALDIVES

Malay Peninsula

INDIAN OCEAN

MALAYSIA

SINGAPORE

Borneo

Celebes

Moluccas

EQUATOR

Jaya Peak 16,500 ft. (5,029 m)▲

New Guinea

Sumatra

I N D O N E S I A

Java

East Timor

0 mi. 1,000

0 km 1,000

Two-Point Equidistant projection

▲ Mountain peak

AUSTRALIA

50°E 60°E 70°E 80°E 90°E 100°E 110°E 120°E 130°E 140°E

50°N 40°N 30°N 20°N 10°N 0° 10°S 20°S

60°N

N W E S

26,247 ft.	**MT. EVEREST**	8,000 m
19,685 ft.		6,000 m
13,123 ft.	SICHUAN BASIN	**MT. FUJI**
		SOUTH KOREA 4,000 m
6,562 ft.	GANGES RIVER **HIMALAYA**	NORTH CHINA PLAIN YELLOW SEA 2,000 m
	Sea level	

0 mi. 500

0 km 500

Political

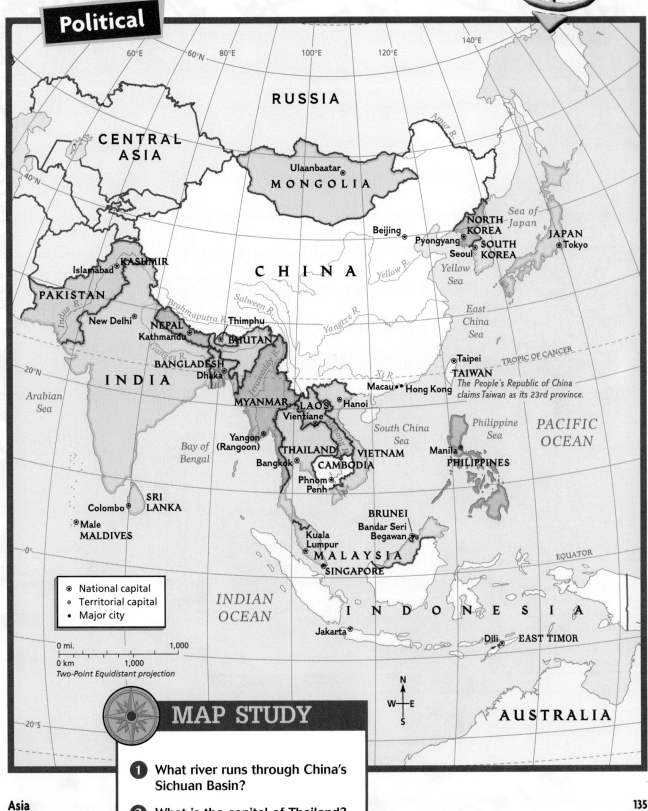

RUSSIA

CENTRAL ASIA

MONGOLIA
⊛ Ulaanbaatar

CHINA

NORTH KOREA
⊛ Beijing
⊛ Pyongyang
SOUTH KOREA
⊙ Seoul

Sea of Japan

JAPAN
⊛ Tokyo

Yellow R.

Yellow Sea

KASHMIR
⊛ Islamabad

PAKISTAN

Indus R.

New Delhi ⊛

Salween R.

Brahmaputra R.

NEPAL
⊛ Thimphu
⊛ Kathmandu
BHUTAN

Ganges R.

BANGLADESH
⊛ Dhaka

INDIA

Yangtze R.

East China Sea

⊛ Taipei
TAIWAN
TROPIC OF CANCER

The People's Republic of China claims Taiwan as its 23rd province.

Xi R.
Macau • • Hong Kong

Irrawaddy R.

MYANMAR
LAOS
⊙ Vientiane
• Hanoi

South China Sea

Philippine Sea

PACIFIC OCEAN

Arabian Sea

Bay of Bengal

Yangon (Rangoon) ⊛

THAILAND
⊛ Bangkok

VIETNAM

CAMBODIA
⊛ Phnom Penh

Mekong R.

⊛ Manila
PHILIPPINES

SRI LANKA
⊛ Colombo

⊛ Male
MALDIVES

BRUNEI
⊛ Bandar Seri Begawan

Kuala Lumpur ⊛

MALAYSIA

⊛ SINGAPORE

- ⊛ National capital
- ⊙ Territorial capital
- • Major city

INDIAN OCEAN

I N D O N E S I A

EQUATOR

0 mi.   1,000
0 km   1,000
Two-Point Equidistant projection

Jakarta ⊛

Dili ⊙ EAST TIMOR

N
W E
S

MAP STUDY

AUSTRALIA

1 What river runs through China's Sichuan Basin?

2 What is the capital of Thailand?

Asia

Monsoons

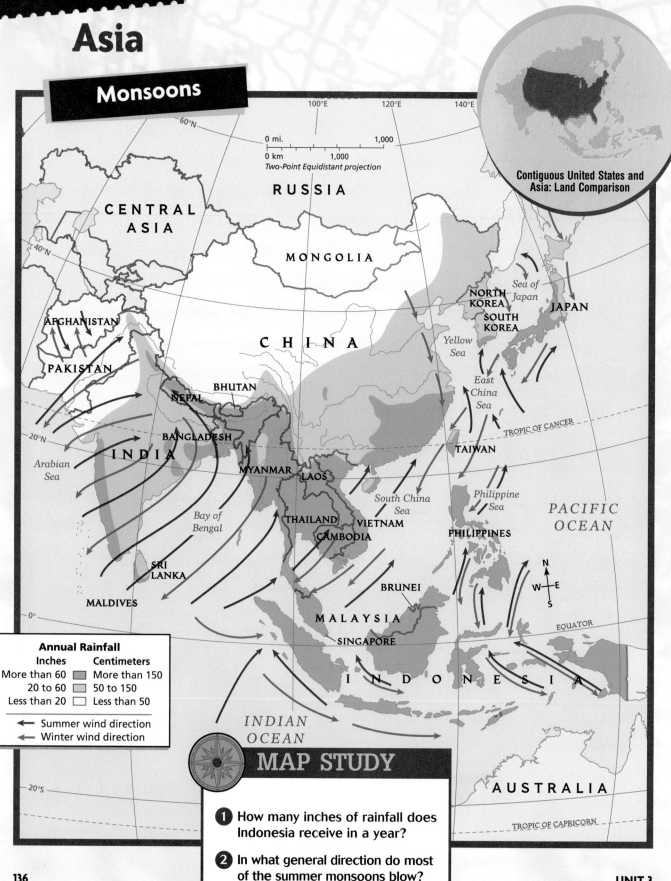

Contiguous United States and Asia: Land Comparison

RUSSIA

CENTRAL ASIA

MONGOLIA

CHINA

AFGHANISTAN

PAKISTAN

NEPAL

BHUTAN

BANGLADESH

INDIA

MYANMAR

LAOS

Arabian Sea

Bay of Bengal

THAILAND

CAMBODIA

VIETNAM

SRI LANKA

MALDIVES

MALAYSIA

SINGAPORE

BRUNEI

INDONESIA

NORTH KOREA

SOUTH KOREA

Sea of Japan

JAPAN

Yellow Sea

East China Sea

TAIWAN

South China Sea

Philippine Sea

PHILIPPINES

PACIFIC OCEAN

TROPIC OF CANCER

EQUATOR

INDIAN OCEAN

AUSTRALIA

TROPIC OF CAPRICORN

100°E   120°E   140°E

60°N

40°N

20°N

0°

20°S

0 mi.   1,000
0 km   1,000
Two-Point Equidistant projection

Annual Rainfall

Inches		Centimeters
More than 60		More than 150
20 to 60		50 to 150
Less than 20		Less than 50

← Summer wind direction
← Winter wind direction

MAP STUDY

1 How many inches of rainfall does Indonesia receive in a year?

2 In what general direction do most of the summer monsoons blow?

Fast Facts

COMPARING POPULATION:
United States and Selected Countries of Asia

UNITED STATES

CHINA

INDIA

INDONESIA

= 70,000,000

JAPAN

Source: *Population Reference Bureau,* 2000.

WORLD POPULATION:
Asia's Share of the World's People

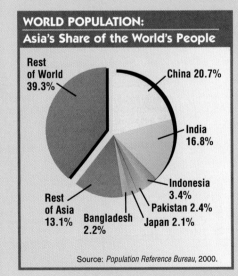

Rest of World 39.3%

China 20.7%

India 16.8%

Indonesia 3.4%

Pakistan 2.4%

Japan 2.1%

Bangladesh 2.2%

Rest of Asia 13.1%

Source: *Population Reference Bureau,* 2000.

China

Data Bits

🚗	Automobiles per 1,000 people	3
📱	Telephones per 1,000 people	56
VOTE	Democratic elections	No

Ethnic Makeup

Other 6%

Zhaung 1%

Hui 1%

Han Chinese 92%

World Ranking

GNP per capita in US $ — 125 $860

Life expectancy — 80 70 years

Literacy — 110 84%

1st
50th
100th
150th

Population: Urban vs. Rural

30%   70%

Source: *World Desk Reference,* 2000.

GRAPHIC STUDY

1. Which Asian country's population is nearest the United States?

2. What percentage of the world's population lives in Asia?

REGIONAL ATLAS

Country Profiles

BANGLADESH

POPULATION:
133,500,000
2,401 per sq. mi.
927 per sq. km

LANGUAGE:
Bengali

MAJOR EXPORT:
Clothing

MAJOR IMPORT:
Machinery

CAPITAL:
Dhaka

LANDMASS:
55,598 sq. mi.
143,999 sq. km

Dhaka

BHUTAN

POPULATION:
900,000
50 per sq. mi.
19 per sq. km

LANGUAGES:
Dzonkha, Local
Languages

MAJOR EXPORT:
Cardamom

MAJOR IMPORT:
Fuels

CAPITAL:
Thimphu

LANDMASS:
18,147 sq. mi.
47,001 sq. km

Thimphu

BRUNEI

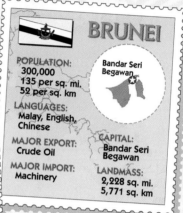

POPULATION:
300,000
135 per sq. mi.
52 per sq. km

LANGUAGES:
Malay, English,
Chinese

MAJOR EXPORT:
Crude Oil

MAJOR IMPORT:
Machinery

CAPITAL:
Bandar Seri
Begawan

LANDMASS:
2,228 sq. mi.
5,771 sq. km

Bandar Seri
Begawan

CAMBODIA

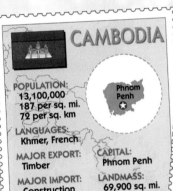

POPULATION:
13,100,000
187 per sq. mi.
72 per sq. km

LANGUAGES:
Khmer, French

MAJOR EXPORT:
Timber

MAJOR IMPORT:
Construction
Materials

CAPITAL:
Phnom Penh

LANDMASS:
69,900 sq. mi.
181,041 sq. km

Phnom
Penh

CHINA

POPULATION:
1,273,300,000
345 per sq. mi.
133 per sq. km

LANGUAGE:
Mandarin Chinese

MAJOR EXPORT:
Machinery

MAJOR IMPORT:
Machinery

CAPITAL:
Beijing

LANDMASS:
3,696,100 sq. mi.
9,572,899 sq. km

Beijing

EAST TIMOR

POPULATION:
800,000
139 per sq. mi.
54 per sq. km

LANGUAGES:
Tetun, Javanese,
Portuguese

MAJOR EXPORT:
Coconut Products

MAJOR IMPORT:
Manufactured
Goods

CAPITAL:
Dili

LANDMASS:
5,741 sq. mi.
14,869 sq. km

Dili

INDIA

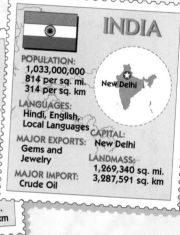

POPULATION:
1,033,000,000
814 per sq. mi.
314 per sq. km

LANGUAGES:
Hindi, English,
Local Languages

MAJOR EXPORTS:
Gems and
Jewelry

MAJOR IMPORT:
Crude Oil

CAPITAL:
New Delhi

LANDMASS:
1,269,340 sq. mi.
3,287,591 sq. km

New Delhi

INDONESIA

POPULATION:
206,100,000
280 per sq. mi.
108 per sq. km

LANGUAGES:
Bahasa Indonesia,
Javanese

MAJOR EXPORT:
Crude Oil

MAJOR IMPORT:
Manufactured
Goods

CAPITAL:
Jakarta

LANDMASS:
735,355 sq. mi.
1,904,569 sq. km

Jakarta

JAPAN

POPULATION:
127,100,000
871 per sq. mi.
336 per sq. km

LANGUAGE:
Japanese

MAJOR EXPORT:
Machinery

MAJOR IMPORT:
Manufactured
Goods

CAPITAL:
Tokyo

LANDMASS:
145,869 sq. mi.
377,801 sq. km

Tokyo

LAOS

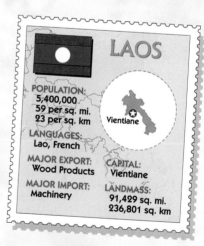

POPULATION:
5,400,000
59 per sq. mi.
23 per sq. km

LANGUAGES:
Lao, French

MAJOR EXPORT:
Wood Products

MAJOR IMPORT:
Machinery

CAPITAL:
Vientiane

LANDMASS:
91,429 sq. mi.
236,801 sq. km

Vientiane

Countries and flags not drawn to scale

For more information on countries in this region, refer to the Nations of the World Data Bank on pages 690–699.

MALAYSIA

POPULATION:
22,710,000
178 per sq. mi.
69 per sq. km

LANGUAGES:
Malay, English,
Chinese

MAJOR EXPORT:
Electronic
Equipment

MAJOR IMPORT:
Machinery

CAPITAL:
Kuala Lumpur

LANDMASS:
127,317 sq. mi.
329,751 sq. km

Kuala Lumpur

MALDIVES

POPULATION:
300,000
2,586 per sq. mi.
1,000 per sq. km

LANGUAGES:
Maldivian Divehi,
English

MAJOR EXPORT:
Fish

MAJOR IMPORT:
Machinery

CAPITAL:
Male

LANDMASS:
116 sq. mi.
300 sq. km

Male

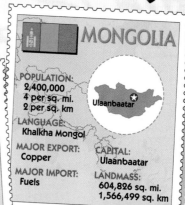

MONGOLIA

POPULATION:
2,400,000
4 per sq. mi.
2 per sq. km

LANGUAGE:
Khalkha Mongol

MAJOR EXPORT:
Copper

MAJOR IMPORT:
Fuels

CAPITAL:
Ulaanbaatar

LANDMASS:
604,826 sq. mi.
1,566,499 sq. km

Ulaanbaatar

MYANMAR

POPULATION:
47,800,000
183 per sq. mi.
71 per sq. km

LANGUAGES:
Burmese,
Local Languages

MAJOR EXPORT:
Beans

MAJOR IMPORT:
Machinery

CAPITAL:
Yangon
(Rangoon)

LANDMASS:
261,228 sq. mi.
676,581 sq. km

Yangon
(Rangoon)

NEPAL

POPULATION:
23,500,000
414 per sq. mi.
160 per sq. km

LANGUAGE:
Nepali

MAJOR EXPORT:
Clothing

MAJOR IMPORT:
Petroleum
Products

CAPITAL:
Kathmandu

LANDMASS:
56,826 sq. mi.
147,179 sq. km

Kathmandu

NORTH KOREA

POPULATION:
22,000,000
473 per sq. mi.
183 per sq. km

LANGUAGE:
Korean

MAJOR EXPORT:
Minerals

MAJOR IMPORT:
Petroleum

CAPITAL:
Pyongyang

LANDMASS:
46,541 sq. mi.
120,541 sq. km

Pyongyang

PAKISTAN

POPULATION:
145,000,000
472 per sq. mi.
182 per sq. km

LANGUAGES:
Urdu, English,
Punjabi, Sindhi

MAJOR EXPORT:
Cotton

MAJOR IMPORT:
Petroleum

CAPITAL:
Islamabad

LANDMASS:
307,375 sq. mi.
796,101 sq. km

Islamabad

PHILIPPINES

POPULATION:
77,200,000
667 per sq. mi.
257 per sq. km

LANGUAGES:
Tagalog, English

MAJOR EXPORT:
Electronic
Equipment

MAJOR IMPORT:
Raw Materials

CAPITAL:
Manila

LANDMASS:
115,831 sq. mi.
300,000 sq. km

Manila

SINGAPORE

POPULATION:
4,100,000
17,155 per sq. mi.
6,624 per sq. km

LANGUAGES:
Chinese, Malay,
Tamil, English

MAJOR EXPORT:
Computer Equipment

MAJOR IMPORT:
Aircraft

CAPITAL:
Singapore

LANDMASS:
239 sq. mi.
619 sq. km

Singapore

REGIONAL ATLAS

Country Profiles

SOUTH KOREA
POPULATION:
48,800,000
1,273 per sq. mi.
492 per sq. km
LANGUAGE:
Korean
MAJOR EXPORT:
Electronic
Equipment
MAJOR IMPORT:
Machinery
CAPITAL:
Seoul
LANDMASS:
38,324 sq. mi.
99,259 sq. km
Seoul

SRI LANKA
POPULATION:
19,500,000
770 per sq. mi.
297 per sq. km
LANGUAGES:
Sinhalese, Tamil,
English
MAJOR EXPORT:
Textiles
MAJOR IMPORT:
Machinery
CAPITAL:
Colombo
LANDMASS:
25,332 sq. mi.
65,610 sq. km
Colombo

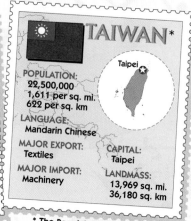

TAIWAN*
POPULATION:
22,500,000
1,611 per sq. mi.
622 per sq. km
LANGUAGE:
Mandarin Chinese
MAJOR EXPORT:
Textiles
MAJOR IMPORT:
Machinery
CAPITAL:
Taipei
LANDMASS:
13,969 sq. mi.
36,180 sq. km
Taipei

* The People's Republic of China claims Taiwan as its 23rd province.

THAILAND
POPULATION:
62,400,000
315 per sq. mi.
122 per sq. km
LANGUAGES:
Thai, Local Languages
MAJOR EXPORT:
Manufactured
Goods
MAJOR IMPORT:
Machinery
CAPITAL:
Bangkok
LANDMASS:
198,116 sq. mi.
513,120 sq. km
Bangkok

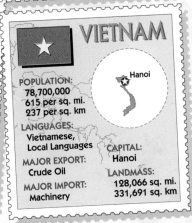

VIETNAM
POPULATION:
78,700,000
615 per sq. mi.
237 per sq. km
LANGUAGES:
Vietnamese,
Local Languages
MAJOR EXPORT:
Crude Oil
MAJOR IMPORT:
Machinery
CAPITAL:
Hanoi
LANDMASS:
128,066 sq. mi.
331,691 sq. km
Hanoi

Countries and flags not drawn to scale

BUILDING CITIZENSHIP

Women's Rights Not all countries have the same laws for men and women. In some countries, women are not allowed to own property, vote, go to school, or work. Part of the reason for this is that women's contributions to society in the area of raising children and running a household are not valued as much as men's contributions.

Why is it important in the United States that men and women have equal rights and that those rights are protected by the law?

Imagine that you are a sixth grade exchange student from an Asian country. Write a letter to your sister at home describing some activities that girls in your American school take part in on an equal basis with boys.

Vietnamese mother and baby ▶

Three generations of a
Chinese family

5 South Asia

The World and Its People
NATIONAL GEOGRAPHIC

To learn more about the people and places of South Asia, view **The World and Its People Chapter 23** video.

Our World Today
ONLINE

Chapter Overview Visit the **Our World Today: People, Places, and Issues** Web site at owt.glencoe.com and click on **Chapter 5–Chapter Overviews** to preview information about South Asia.

Categorizing Information Study Foldable Make this foldable to organize information from the chapter to help you learn more about the land, economy, government, history, and religions of six South Asian countries.

Step 1 Collect four sheets of paper and place them about ¹/₂ inch apart.

Keep the edges straight.

Step 2 Fold up the bottom edges of the paper to form 8 tabs.

This makes all tabs the same size.

Step 3 When all the tabs are the same size, crease the paper to hold the tabs in place and staple the sheets together. Turn the paper and label each tab as shown.

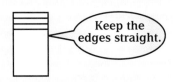

SOUTH ASIA
General Information
Bangladesh
Sri Lanka
Bhutan
Nepal
Pakistan
India

Staple together along the fold.

Reading and Writing As you read, use your foldable to write down the main ideas about each South Asian country. Record the main ideas under each appropriate tab of your foldable.

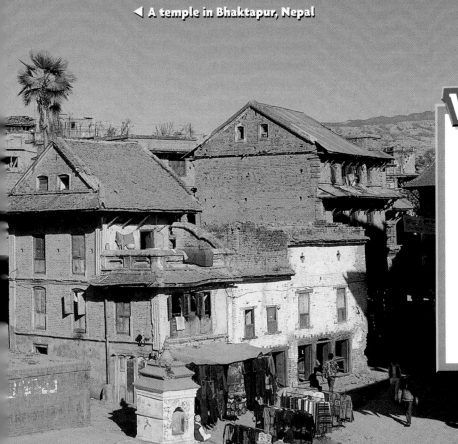

◀ A temple in Bhaktapur, Nepal

Why It Matters

Working Toward Democracy

India is a very large country with a large, diverse population of more than 1 billion people. The country has 18 official languages. For a democracy to be effective, the citizens must be able to educate themselves on important issues. There must be shared values and a respect for differences of opinions. India's democratic government is working toward these goals.

India

Guide to Reading

Main Idea

India is trying to develop its resources and meet the needs of its rapidly growing population.

Terms to Know

- subcontinent
- monsoon
- green revolution
- jute
- cottage industry
- prime minister
- pesticide
- caste
- reincarnation

Reading Strategy

Create a chart like this one. Then fill in at least two key facts about India under each category.

India	
Land	Economy
History	Religion

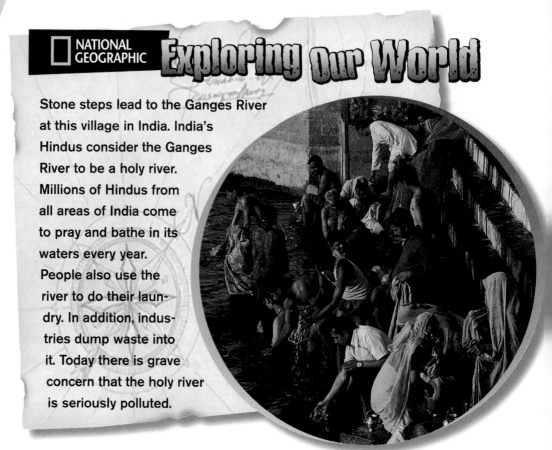

NATIONAL GEOGRAPHIC Exploring Our World

Stone steps lead to the Ganges River at this village in India. India's Hindus consider the Ganges River to be a holy river. Millions of Hindus from all areas of India come to pray and bathe in its waters every year. People also use the river to do their laundry. In addition, industries dump waste into it. Today there is grave concern that the holy river is seriously polluted.

India and several other countries—Pakistan, Bangladesh (BAHNG•gluh•DEHSH), Nepal, Bhutan, Sri Lanka, and the Maldives—make up the South Asian subcontinent. A **subcontinent** is a large landmass that is part of another continent but distinct from it.

India's Land and Economy

Two huge walls of mountains—the **Karakoram** (KAH•rah•KOHR•ahm) **Range** and the **Himalaya** (HIH•muh•LAY•uh)—form India's northern border and separate South Asia from the rest of Asia. (See the map of Asia on page 134.) The tallest mountains in the world, the Himalaya's snowcapped peaks average more than 5 miles (8 km) in height. Mountain ranges also line India's southern coasts.

Most of India is warm or hot all year. The Himalaya block cold northern air from sweeping south into the country. **Monsoons,** or seasonal winds that blow steadily from the same direction for months, also influence the climate. During the rainy season (June through September), southern monsoon winds bring moist air from the Indian Ocean. The map on page 136 shows monsoon patterns for summer and winter.

The Green Revolution   Today, India raises most of the food it needs. In the past it was very different. The world's worst recorded food disaster, known as the **Bengal Famine,** happened in 1943 when the United Kingdom ruled India. An estimated 4 million people died of starvation that year alone. When India won its independence in 1947, government officials turned their attention to improving India's farm output. The green revolution was an effort to use modern techniques and science to increase production of food.

To produce more food, farmers planted more than one crop per year. If the farmers relied only on the monsoon rains to water their crops, they could only plant once per year. The government built dams to collect the water. The dams stored the water and spread it out through irrigation ditches during the dry season. Farmers could then plant twice a year.

The second part of the green revolution was to use improved seeds. New, stronger strains of wheat, rice, and corn were developed that could withstand diseases and droughts and produce more grains. Between 1947 and 1980, farm production improved by more than 30 percent.

Today, India's farmers raise a variety of crops, including rice, wheat, cotton, tea, sugarcane, and jute. Jute is a plant fiber used for making rope, burlap bags, and carpet backing. India is the world's second-largest rice producer, after China.

Industry   Huge factories in India's cities turn out cotton textiles and produce iron and steel. Oil and sugar refineries loom over many urban skylines. Recently, American computer companies have opened offices in India, making it an important source of computer software. Mining is another major industry. India has rich deposits of coal, iron ore, manganese, and bauxite. Its major exports are gems and jewelry.

Many Indian products are manufactured in cottage industries. A cottage industry is a home- or village-based industry in which family members, including children, supply their own equipment to make goods. Items produced in cottage industries include cotton cloth, silk cloth, rugs, leather products, and metalware.

The World's Largest Democracy

India is a representative democracy. The Indian parliament, or congress, has two houses. One house is made up of representatives of the states and is called the Council of States. The other house—called the House of the People—is like our House of Representatives.

NATIONAL GEOGRAPHIC **On Location**

Two Views of India

The growing middle class live comfortable lives in India's suburbs (above), but the poor in India's cities must struggle to survive (left).

Human/Environment Interaction
How would the green revolution benefit India's people?

The most important difference between the two systems is that most of the power to run the government of India is held not by the president, but by the prime minister, who is appointed by the ruling party. The first prime minister of India was Jawaharlal Nehru, who was elected in 1947. While in office, Mr. Nehru pursued peace between India and all other countries. His daughter, Indira Gandhi, was also elected prime minister. Except for a short period, she led India from 1966 until her assassination in 1984. Mrs. Gandhi tried to help India's poor by providing low-cost housing and giving land to those who owned none. She also helped to extend voting rights.

Environmental Challenges   India's economic growth has brought challenges to its environment. Thousands of acres of forests have been cleared for farming. Both water and land have been polluted from burning coal, industrial wastes, and pesticides, or chemicals used to kill insects that destroy crops. The Ganges is considered by many experts to be one of the world's most polluted rivers.

All of these developments have played a part in destroying animal habitats. India's elephants, lions, tigers, leopards, monkeys, and panthers have been greatly reduced in number. The government has set up more than 350 national parks and preserves to save these animals.

✓ Reading Check   **What form of government does India have?**

India's History and People

About 4,000 years ago, the first Indian civilization built well-planned cities along the **Indus River** valley, in present-day Pakistan. In the 1500s B.C., warriors known as **Aryans** (AR•ee•uhns) entered the subcontinent from Central Asia. They set up kingdoms in northern India. Aryan beliefs gradually blended with the practices of the local people to form the religion of **Hinduism.**

NATIONAL GEOGRAPHIC On Location

Leaders of India

Indira Gandhi (left) Mohandas Gandhi (center), and Jawaharlal Nehru (right) were instrumental in bringing democracy to India.

History Which country ruled India before it won its independence?

EXPLORING CULTURE

Music

The tabla is a pair of connected drums from India. The drums are made of wood in the shape of a cylinder. Wooden pegs and leather straps hold the skin tightly onto the right-hand drum. The skin on the left-hand drum is kept slightly loose so that players can push down into it. This creates lower and higher pitches. Although the tabla emerged in India 500 years ago, it is now heard in modern pop and jazz music all over the world.

Looking Closer **Which drum do you think has more variation in sound? Why?**

GO TO

World Music: A Cultural Legacy
Hear music of this region on Disc 2, Track 13.

Over time, Hinduism organized India's society into groups called castes. A caste was a social class based on a person's ancestry. A person was born into a particular caste. People married within their caste, and certain occupations belonged to the specific castes. People could not move from one caste to another. The caste system still influences Indian life, although laws now forbid unfair treatment of one group by another.

Buddhism started in India about 500 B.C., but largely declined there by 300 B.C. You will read more about Buddhism later in this unit. The religion of Islam also influenced India's history. In the A.D. 700s, Muslims from Southwest Asia brought Islam to India. In the 1500s, they founded the **Mogul** (MOH•guhl) **Empire** and ruled India for 200 years.

The British were the last of India's conquerors, ruling from the 1700s to the mid-1900s. They built roads, railroads, and seaports. They also made large profits from the plantations, mines, and factories they set up. An Indian leader named **Mohandas Gandhi** (moh•HAHN•duhs GAHN•dee) led a nonviolent resistance movement to free India from Britain's rule. When India won its independence from the British in 1947, many Muslims were afraid that their voices would not be heard by the Hindu majority. East and West Pakistan were created, one on each side of India, as an independent Muslim homeland. In 1971, East Pakistan became the separate nation of **Bangladesh.**

Religion About 80 percent of India's people are Hindus, or followers of Hinduism. Hindus honor many gods and goddesses, which are often seen as expressions of one eternal spirit. Hinduism teaches that after the body dies, the soul is reborn, often in an animal or human form.

This process, called reincarnation, is repeated until the soul reaches perfection. For this reason, many Hindus believe it is wrong to kill any living creature. Cows are believed to be sacred and are allowed to roam freely.

Islam has over 140 million followers in India. Other religions include Christianity, Sikhism (SEE•KIH•zuhm), Buddhism, and Jainism (JY•NIH•zuhm). Conflict sometimes occurs among members of India's different religious groups. The **Sikhs,** who practice Sikhism, believe in one God as Christians and Muslims do, yet Sikhs also have other beliefs similar to Hindus. Today, many Sikhs would like to form their own independent state.

▲ This statue represents Shiva, one of Hinduism's many deities.

Daily Life More than 1 billion people call India their home. The country has 18 official languages, of which Hindi is the most widely used. English is often spoken in business and government, however. About 70 percent of the people live in farming villages. The government has been working to provide villagers with electricity, drinking water, better schools, and paved roads. Still, many villagers stream to cities to find jobs and a better standard of living.

One of the most popular holidays is **Diwali** (dee•VAH•lee), the Festival of Lights. It is a Hindu celebration marking the coming of winter and the victory of good over evil. Indians also like watching movies. India's movie industry turns out more films than Hollywood.

✓ Reading Check What percentage of India's people live in rural villages?

Section 1 Assessment

Defining Terms
1. Define subcontinent, monsoon, green revolution, jute, cottage industry, prime minister, pesticide, caste, reincarnation.

Recalling Facts
2. Location What two mountain ranges form India's northern border?

3. Culture What is the most widely followed religion in India?

4. History What Indian leader led a movement that brought India its independence in 1947?

Critical Thinking
5. Understanding Cause and Effect How do monsoon winds affect India's climate?

6. Drawing Conclusions What challenges do you think the caste system might have caused in India?

Graphic Organizer
7. Organizing Information India is becoming a more modern country but still has many traditional ways. Create a chart like this one. Then list both modern and traditional aspects of India.

Modern Aspects	Traditional Aspects

Applying Social Studies Skills

8. Analyzing Maps Look at the population density map on page 154. What are the most densely populated areas of India?

Making Connections

CULTURE   GOVERNMENT   PEOPLE   TECHNOLOGY

Shah Jahan of India

Considered one of the world's most beautiful buildings, the Taj Mahal was built by the Muslim emperor Shah Jahan of India. He had it built to house the grave of his beloved wife, Mumtaz Mahal. She died in 1631 shortly after giving birth to their fourteenth child.

The Taj Mahal, Agra, India ▲

Background

While they were married, Mumtaz Mahal and Shah Jahan were constant companions. The empress went everywhere with her husband, even on military expeditions. She encouraged her husband to perform great acts of charity toward the poor. This earned her the love and admiration of the Indian people.

After his wife's death, Shah Jahan ordered the construction of the finest monument ever built. A team of architects, sculptors, calligraphers, and master builders participated in the design. More than 20,000 laborers and skilled craft workers from India, Persia, the Ottoman Empire, and Europe worked together to build the monument. For 22 years they worked to complete the Taj Mahal, which holds a tomb, mosque, rest house, elaborate garden, and arched gateway.

The Mausoleum

The central part of the Taj Mahal is the domed marble mausoleum, or tomb, built on a square marble platform. The central dome is 213 feet (65 m) tall, and four smaller domed chambers surround it. A high minaret, or tower, marks each corner of the platform.

Inside the central chamber, delicately carved marble screens enclose the caskets of Mumtaz Mahal and Shah Jahan. He was buried next to his wife after his death in 1666. Following Islamic tradition, the caskets face east toward Makkah, the religious capital of Islam.

The white marble from which the mausoleum is built seems to change color throughout the day as it reflects light from the sun and moon. Detailed flower patterns are carved into the marble walls and inlaid with colorful gemstones. Verses from Islamic religious writings are etched in calligraphy into the stone archways.

Making the Connection

1. Who is buried in the Taj Mahal?
2. Who built the Taj Mahal and how long did it take?
3. **Understanding Cause and Effect** How did Shah Jahan's feelings for his wife affect the grave site he built for her?

Other Countries of South Asia

Guide to Reading

Main Idea

The other countries of South Asia include once-united Pakistan and Bangladesh, mountainous Nepal and Bhutan, and the island country of Sri Lanka.

Terms to Know

- cyclone
- dzong

Reading Strategy

Create and fill in a chart like this one, listing the main economic activities in these countries of South Asia.

Country	Economic Activity
Pakistan	
Bangladesh	
Nepal	
Bhutan	
Sri Lanka	

NATIONAL GEOGRAPHIC Exploring Our World

Perched on poles planted into the ocean floor, Sri Lankan fishers await their next catch. Although Sri Lanka is trying to build a modern economy, traditional work still goes on. Some people gave up fishing when Sri Lanka seemed ready to become a major tourist destination. However, years of ethnic warfare have kept tourists away and slowed the economy.

Two countries in South Asia—**Pakistan** and **Bangladesh**—are largely Muslim. They share the same religion, but have very different cultures and languages. Two other countries—**Nepal** and **Bhutan**—are landlocked kingdoms. **Sri Lanka** and the **Maldives** are island republics.

Pakistan

Pakistan is about twice the size of California. Towering mountains occupy most of northern and western Pakistan. The world's second-highest peak, K2, rises 28,250 feet (8,611 m) in the Karakoram Range. Another mountain range, the **Hindu Kush,** lies in the far north. Several passes cut through its rugged peaks. The best known is the **Khyber Pass.** For centuries, it has been used by people traveling through South Asia from the north.

Pakistan and India both claim **Kashmir,** a mostly Muslim territory on the border between the two countries. Both countries want to control the entire region, mainly for its vast water resources. This dispute over Kashmir has sparked three wars between Pakistan and India. In fact, it threatens not only the countries of South Asia, but also the rest of the world because both Pakistan and India have nuclear weapons.

Pakistan's People Since independence, Pakistan has had many changes of government. Some of these governments were elected, including a female prime minister, **Benazir Bhutto.** Bhutto was the first woman elected prime minister of an Islamic nation. In other cases, the army seized power from an elected government. The most recent army takeover occurred in 1999, and military leaders still control the country.

About 97 percent of Pakistanis are Muslims. The influence of Islam is seen in large, domed mosques and people bowed in prayer at certain times of the day. Among the major languages are **Punjabi** and **Sindhi.** The official language, **Urdu,** is the first language of only 9 percent of the people. English is widely spoken in government.

Almost 70 percent of Pakistan's people live in rural villages. Most follow traditional customs and live in small homes of clay or sun-dried mud. Pakistanis live in large cities as well. **Karachi,** a seaport on the Arabian Sea, is a sprawling urban area. It has traditional outdoor markets, modern shops, and hotels. In the far north lies **Islamabad,** the capital. The government built this well-planned, modern city to draw people inland from crowded coastal areas.

✓ Reading Check **Why do India and Pakistan both claim Kashmir?**

Bangladesh

Bangladesh, about the size of Iowa, is nearly surrounded by India. Although Bangladesh is a Muslim country like Pakistan, it shares many cultural features with eastern India.

Seeing Bangladesh for the first time, you might describe the country with one word—water. Two major rivers—the **Brahmaputra** (BRAHM•uh•POO•truh) **River** and the **Ganges River**—flow through the lush, low plains that cover most of Bangladesh.

As in India, the monsoons affect Bangladesh. When the monsoons end, cyclones may strike Bangladesh. A cyclone is an intense tropical storm system with high winds and heavy rains. Cyclones, in turn, may be followed by deadly tidal waves that surge up from the **Bay of Bengal.** As deadly as the cyclones and tidal waves may be, it is worse if the rains come too late. When this happens, crops often fail and there is widespread hunger.

A Farming Economy Most people of Bangladesh earn their living by farming. Rice is the most important crop. The fertile soil and plentiful water make it possible for rice to be grown and harvested three times a year. Other crops include sugarcane, jute, and wheat. Cash crops of tea grow in hilly regions in the east. Despite good growing conditions, Bangladesh cannot grow enough food for its people. Its farmers have few modern tools and use outdated farming methods. In addition, the disastrous floods can drown crops and cause food shortages.

School's Out!

Adil Husain is on his way home from middle school. In Pakistan, schooling only goes to grade 10. After grade 10, Adil must decide whether to go to intermediate college (grades 11 and 12) and then the university. Like most Pakistanis, Adil is Muslim. He prays when he hears the call from the mosque. Afterward, he wants to start a game of cricket with his friends. "It's a lot like baseball where teams of 11 players bat in innings and try to score runs. Our rules and equipment are different, though. You should try it!"

Highest Mountain on Each Continent

Analyzing the Graph

Mount Everest is the tallest mountain on the earth.

Place What is the tallest mountain in North America?

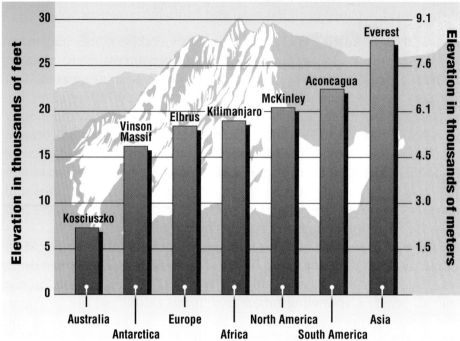

Source: *The World Almanac*, 2000.

The People With about 133.5 million people, Bangladesh is one of the most densely populated countries in the world. It is also one of the poorest countries. More than 80 percent of the people live in rural areas. Because of floods, people in rural Bangladesh have to build their houses on platforms. Many people have moved to crowded urban areas to find work in factories. The largest city is **Dhaka** (DA•kuh), Bangladesh's capital and major port. Most of Bangladesh's people speak **Bengali.**

✓Reading Check What are three ways the climate may cause disasters in Bangladesh?

Nepal

Nepal—about the size of Arkansas—forms a steep stairway to the world's highest mountain range. The Himalaya, dominating about 80 percent of Nepal's land area, are actually three mountain ranges running side by side. Nepal is home to 8 of the 10 highest mountains in the world. **Mount Everest,** the highest, soars 29,035 feet (8,850 m). Nepal's rugged mountains attract thousands of climbers and hikers each year, creating a growing tourist industry.

Nepal's economy depends almost entirely on farming. Farmers grow rice, sugarcane, wheat, corn, and potatoes to feed their families. Most fields are located on the southern plains or on the lower mountain slopes. As the population has increased, Nepalese farmers have moved higher up the slopes. There they clear the forests for new fields and use the cut trees for fuel. Stripped of trees, however, the

Web Activity Visit the *Our World Today: People, Places, and Issues* Web site at owt.glencoe.com and click on **Chapter 5– Student Web Activities** to learn more about Nepal.

slopes erode very easily. During the rainy season, valleys often are flooded, fields destroyed, and rivers filled with mud.

Nepal's People Nepal has 23.5 million people. Most are related to peoples in northern India and Tibet. One group—the **Sherpa**—is known for its skill in guiding mountain climbers. About 85 percent of Nepal's people live in rural villages. A growing number live in **Kathmandu,** Nepal's capital and largest city. Nepal is a parliamentary democracy ruled by a prime minister, who is appointed by Nepal's king.

The founder of Buddhism, **Siddartha Gautama** (sihd•DAHR•tuh GOW•tuh•muh), was born in the Kathmandu region about 563 B.C. Raised as a prince, Gautama gave up his wealth and became a holy man in India. Known as the **Buddha,** or "Enlightened One," he taught that people could find peace from life's troubles by living simply, doing good deeds, and praying. Buddhism later spread to other parts of Asia.

Today Hinduism is the official religion in Nepal, but Buddhism is practiced as well. If you visit Nepal, you will find temples and monuments of both religions scattered throughout the country.

✔️ **Reading Check** What are the two main religions in Nepal?

Bhutan

East of Nepal lies an even smaller kingdom—Bhutan. Bhutan is about half the size of Indiana. The map on page 135 shows you that a small part of India separates Bhutan from Nepal.

As in Nepal, the Himalaya are the major landform of Bhutan. Violent mountain storms are common and are the basis of Bhutan's name, which means "land of the thunder dragon." In the foothills of the Himalaya, the climate is mild. Thick forests cover much of this area. To the south—along Bhutan's border with India—lies an area of plains and river valleys.

More than 90 percent of Bhutan's people are subsistence farmers. They live in the fertile mountain valleys and grow oranges, rice, corn, potatoes, and the spice cardamom. People also herd cattle and yaks, which are a type of oxen. Bhutan is trying to develop its economy, but the mountains slow progress. Building roads is difficult, and there are no railroads. Bhutan has built hydroelectric plants to create electricity from rushing mountain waters. It now exports electricity to India. Tourism is a new industry. However, the government limits the number of tourists in order to protect Bhutan's cultural traditions.

NATIONAL GEOGRAPHIC On Location

Bhutan

This woman makes her living by herding yaks in one of Bhutan's mountain valleys.

Economics What is slowing Bhutan's economic progress?

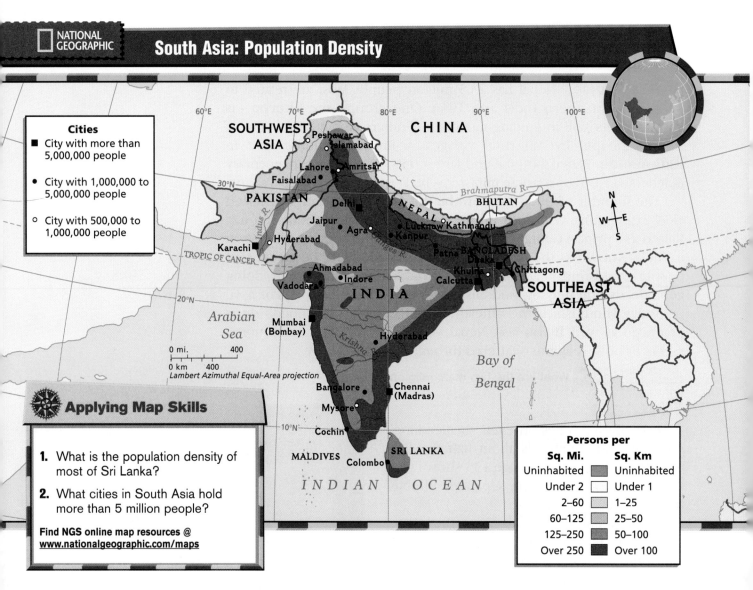

Cities

■ City with more than 5,000,000 people

● City with 1,000,000 to 5,000,000 people

○ City with 500,000 to 1,000,000 people

Applying Map Skills

1. What is the population density of most of Sri Lanka?

2. What cities in South Asia hold more than 5 million people?

Find NGS online map resources @ www.nationalgeographic.com/maps

Persons per		
Sq. Mi.		**Sq. Km**
Uninhabited		Uninhabited
Under 2		Under 1
2–60		1–25
60–125		25–50
125–250		50–100
Over 250		Over 100

Bhutan's People   Bhutan has about 900,000 people. Most speak the Dzonkha dialect and live in rural villages that dot southern valleys and plains. **Thimphu,** the capital, is located in the southern area.

Bhutan was once called the Hidden Holy Land because of its isolation. Most people remain deeply loyal to Buddhism. In Bhutan, Buddhist centers of prayer and study are called dzongs. They have shaped the country's art and culture.

For many years, Bhutan was ruled by strong kings. In 1998 the country began to move toward democracy. At that time, the ruling king agreed to share his power with an elected legislature.

Reading Check **What is the main religion in Bhutan?**

Sri Lanka

Pear-shaped Sri Lanka lies about 20 miles (32 km) off the southeastern coast of India. A little larger than West Virginia, Sri Lanka is an island of white beaches, dense forests, and abundant wildlife. Monsoon winds and heavy rains combine with the island's warm temperatures and fertile soil to make Sri Lanka a good place to farm.

Farmers grow rice and other food crops in lowland areas. In higher elevations, tea, rubber, and coconuts grow on large plantations. The country is one of the world's leading producers of tea and rubber. Sri Lanka is also famous for its sapphires, rubies, and other gemstones. Forests contain many valuable woods, such as ebony and satinwood, and a variety of birds and animals.

Sri Lanka's People For centuries, Sri Lanka prospered because of its location on an important ocean route between Africa and Asia. Beginning in the A.D. 1500s, Sri Lanka—then known as Ceylon—came under the control of European countries. The British ruled the island from 1802 to 1948, when it became independent. In 1972 Ceylon took the name of Sri Lanka, an ancient term meaning "brilliant land." Today, Sri Lanka is a republic with a president who carries out ceremonial duties. Real power is held by a prime minister, who is the head of government.

Sri Lankans belong to two major ethnic groups: the **Sinhalese** (sihng•guh•LEEZ) and the **Tamils** (TA•muhlz). Forming about 74 percent of the population, the Sinhalese live in the southern and western parts of the island. They speak Sinhalese and are mostly Buddhist. The Tamils make up about 18 percent of the population. They live in the north and east, speak Tamil, and are Hindus.

Since 1983 the Tamils and the Sinhalese have fought a violent civil war. The Tamils claim they have not been treated justly by the majority Sinhalese. They want to set up a separate Tamil nation in northern Sri Lanka. Thousands of Sri Lankans have lost their lives in the fighting.

✓Reading Check What are the two main ethnic groups in Sri Lanka?

Section 2 Assessment

Defining Terms
1. Define cyclone, dzong.

Recalling Facts
2. **History** Why has the Khyber Pass been important?
3. **Human/Environment Interaction** What problems are caused by stripping the trees off the slopes in Nepal?
4. **Place** How do Bhutan's people earn a living?

Critical Thinking
5. **Summarizing Information** What were the teachings of the Buddha?
6. **Formulating an Opinion** What could Bangladesh do to increase food production?

Graphic Organizer
7. **Organizing Information** Create a time line like this one. List four events from Sri Lanka's history and their dates.

├──────────┼──────────┼──────────┤

Applying Social Studies Skills

8. **Analyzing Maps** Look at the population density map on page 154 and the physical map on page 134. What is the population density of the southern part of Nepal? The northern part? Explain the difference.

Social Studies Skill

Reading a Circle Graph

Have you ever watched someone dish out pieces of pie? When the pie is cut evenly, everybody gets the same size slice. If one slice is cut a little larger, however, someone else gets a smaller piece.

Learning the Skill

A **circle graph** is like a sliced pie. Often it is even called a pie chart. In a circle graph, the complete circle represents a whole group—or 100 percent. The circle is divided into "slices," or wedge-shaped sections representing parts of the whole.

To read a circle graph, follow these steps:

• Read the title of the circle graph to find out what the subject is.
• Study the labels or the key to see what each "slice" represents.
• Compare the sizes of the circle slices.

Practicing the Skill

Look at the graph below to answer the following questions.

1. What is the subject of the circle graph?
2. Which religion in South Asia has the most followers?
3. What percentage practice Islam?
4. What is the combined percentage of Buddhist and Christian followers?

Applying the Skill

Quiz at least 10 friends about the capitals of India, Pakistan, and Bangladesh. Create a circle graph showing what percentage knew (a) all three capitals, (b) two capitals, (c) one capital, or (d) no capitals.

GO TO

Practice key skills with **Glencoe Skillbuilder Interactive Workbook, Level 1.**

Religions of South Asia

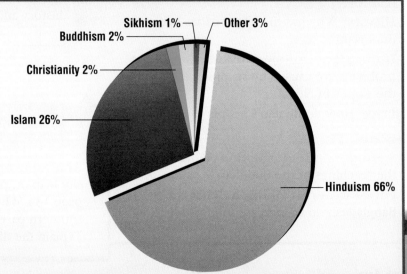

RELIGION	NUMBER OF FOLLOWERS
Hinduism	914,500,000
Islam	366,600,000
Christianity	26,500,000
Buddhism	22,400,000
Sikhism	20,000,000
Other	42,000,000

Sikhism 1% — Other 3%
Buddhism 2%
Christianity 2%
Islam 26%
Hinduism 66%

Source: *Time Almanac,* 2000.

Reading Review

Section 1 | India

Terms to Know
subcontinent
monsoon
green revolution
jute
cottage industry
prime minister
pesticide
caste
reincarnation

Main Idea

India is trying to develop its resources and meet the needs of its rapidly growing population.

✔ Place India is the largest country in South Asia in size and population.

✔ Place The Himalaya and the monsoons affect India's climate.

✔ Economics India's economy is based on both farming and industry.

✔ Culture India has many religions, but the majority of Indians are Hindus.

✔ Human/Environment Interaction The green revolution has greatly improved India's production of food.

Section 2 | Other Countries of South Asia

Terms to Know
cyclone
dzong

Main Idea

The other countries of South Asia include once-united Pakistan and Bangladesh, mountainous Nepal and Bhutan, and the island country of Sri Lanka.

✔ Region Pakistan and India both claim Kashmir. This has led to wars between the two countries.

✔ History Since independence, Pakistan has had many changes of government. Some of the changes were through elections and some through military takeovers.

✔ Place Bangladesh is a densely populated and poor country.

✔ Region The Himalaya are the major landform of Nepal and Bhutan.

✔ Economics Sri Lanka is one of the world's leading producers of tea and rubber.

A teacher and his students have classes outdoors on a pleasant day in Bhutan. ▶

Using Key Terms

Match the terms in Part A with their definitions in Part B.

A.

1. monsoon
2. cyclone
3. green revolution
4. jute
5. subcontinent
6. reincarnation
7. pesticide
8. caste
9. dzong
10. cottage industry

B.

a. social class based on a person's ancestry
b. seasonal wind
c. family members supply their own equipment to make goods
d. large landmass that is part of another continent but distinct from it
e. Buddhist center for prayer and study
f. chemical used to kill insects
g. intense storm system with high winds
h. a government effort to use modern farming methods
i. the belief that after the body dies, the soul is reborn
j. plant fiber used for making rope, burlap bags, and carpet backing

 ## Reviewing the Main Ideas

Section 1 India

11. **Place** What forms a barrier between South Asia and the rest of Asia?
12. **Place** How do the Himalaya affect India's climate?
13. **Economics** What kinds of goods are produced by India's cottage industries?
14. **Culture** What religion do most Indians practice?

Section 2 Other Countries of South Asia

15. **Place** What mountain ranges occupy much of Pakistan?
16. **Human/Environment Interaction** What often happens when the rains come too late in Bangladesh?
17. **Economics** What do most of the people of Bangladesh do for a living?
18. **Human/Environment Interaction** How do mountains hinder economic development in Bhutan?
19. **History** Why was Bhutan once called the Hidden Holy Land?
20. **History** What is the basis of the civil war in Sri Lanka?
21. **Location** How did Sri Lanka's location allow it to prosper for many centuries?

 **South Asia**

Place Location Activity

On a separate sheet of paper, match the letters on the map with the numbered places listed below.

1. Ganges River
2. New Delhi
3. Brahmaputra River
4. Indus River
5. Sri Lanka
6. Himalaya
7. Bangladesh
8. Mumbai
9. Western Ghats
10. Deccan Plateau

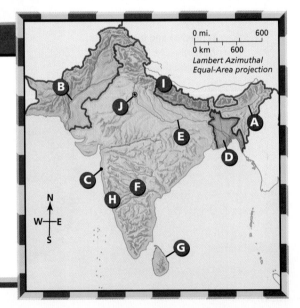

0 mi. 600
0 km 600
Lambert Azimuthal
Equal-Area projection

Our World Today online

Self-Check Quiz Visit the *Our World Today: People, Places, and Issues* Web site at owt.glencoe.com and click on **Chapter 5—Self-Check Quizzes** to prepare for the Chapter Test.

 ## Critical Thinking

22. **Identifying Alternatives** In this chapter you read about South Asia, a region with much poverty. What problems do you think a country faces when it has so many poor people? What are some solutions to this poverty?

23. **Understanding Cause and Effect** Create a chart like this one. List a physical feature of South Asia in the left-hand box. In the right-hand box, explain how that feature affects people's lives.

```
[  ] ────▶ [  ]
```

 ## Current Events Journal

24. **Writing About Religion** Choose Hinduism, Islam, or Buddhism and research its main beliefs and places of worship. Find out the religion's influence in one country today. After your research is complete, create a poster that presents your findings.

 ## Mental Mapping Activity

25. **Focusing on the Region** Create a simple outline map of South Asia, and then label the following:

- Bay of Bengal
- Sri Lanka
- Kashmir
- Pakistan
- Nepal

- Indian Ocean
- Bhutan
- Bangladesh
- Deccan Plateau
- New Delhi

 ## Technology Skills Activity

26. **Using the Internet** Use the Internet to research tourism in one of the following countries: Nepal, India, or Sri Lanka. Create a travel brochure about a trip to the country, featuring information on the equipment and clothing that is needed, the availability of guides, costs, and so on.

The Princeton Review

Standardized Test Practice

Directions: Study the graph below, and then answer the following questions.

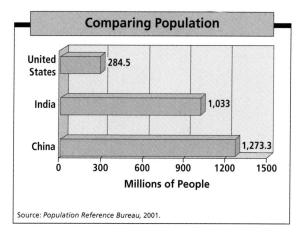

Comparing Population

United States — 284.5
India — 1,033
China — 1,273.3

Millions of People

Source: *Population Reference Bureau, 2001.*

1. **How many people live in India?**
   A 1,033
   B 1,000,033
   C 1,033,000,000
   D 1,033,000,000,000

2. **About how many more people live in India than in the United States?**
   F 2.5 times as many
   G 3.5 times as many
   H 4.5 times as many
   J 5.5 times as many

Test-Taking Tip: You often need to use math skills in order to understand graphs. Look at the information along the sides and bottom of the graph to find out what the bars on the graph mean. Notice that on the graph above, the numbers represent millions of people. Therefore, you need to multiply the number on each bar by 1,000,000 to get the correct answer.

Chapter 6

China and Its Neighbors

The World and Its People NATIONAL GEOGRAPHIC

To learn more about the people and places of China, view **The World and Its People Chapter 24** video.

Our World Today Online

Chapter Overview Visit the **Our World Today: People, Places, and Issues** Web site at owt.glencoe.com and click on **Chapter 6—Chapter Overviews** to preview information about China.

Identifying Main Ideas Study Foldable Make this foldable to help you identify key facts about the people and places of China and its neighbors.

Step 1 Fold the paper from the top right corner down so the edges line up. Cut off the leftover piece.

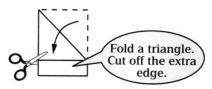

Fold a triangle. Cut off the extra edge.

Step 2 Fold the triangle in half. Unfold.

The folds will form an X dividing four equal sections.

Step 3 Cut up one fold line and stop at the middle. This forms two triangular flaps.

Step 4 Draw an X on one tab and label the other three the following: Flap 1: Mongolia; Flap 2: China; Flap 3: Taiwan.

Step 5 Fold the X flap under the other flap and glue together.

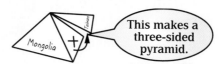

Mongolia

This makes a three-sided pyramid.

Reading and Writing As you read, write main ideas inside the foldable under each appropriate pyramid wall.

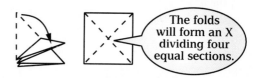

Why It Matters

Opening to Trade

Built to keep out foreigners, the Great Wall of China is the country's best-known monument. Despite this high visibility, much about China remains a mystery to the rest of the world. For centuries, China has worked to protect its culture from outside influences. Recently, however, the need to develop its economy has motivated China to begin opening its doors to trade with other countries.

◀ **Part of the Great Wall of China**

China's Land and Economy

Guide to Reading

Main Idea

China—the third-largest country in the world—has very diverse landforms. China's rapidly growing economy has changed in recent years.

Terms to Know

- dike
- fault
- communist state
- free enterprise system
- invest
- consumer goods
- "one-country, two-systems"

Reading Strategy

Create a diagram like this one. Then list two facts under each heading in the outer ovals.

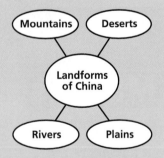

NATIONAL GEOGRAPHIC **Exploring Our World**

Giant pandas look cute and cuddly, but actually they are somewhat hot-tempered. You would be hot-tempered too if your habitat were dwindling in size. Fewer than 1,000 pandas live in the wild, and about 140 live in zoos. The wild pandas make their home on the eastern edge of the Plateau of Tibet. They eat mainly bamboo stems and leaves.

China (officially called the **People's Republic of China**) lies in the central part of eastern Asia. It is the third-largest country in area, after Russia and Canada. China is just slightly larger than the United States.

China's Land and Climate

The physical map on the next page shows the many landforms found within China's vast area. Rugged mountains cover about one-third of the country. Find the Himalaya, Kunlun Shan, Tian Shan, and Altay mountain ranges.

The world's largest plateau is also in China. This high, flat land, commonly called the "Roof of the World," is really the **Plateau of Tibet.** Its height averages about 14,800 feet (4,500 m) above sea level. Scattered shrubs and grasses cover the plateau's harsh landscape. Pandas, golden monkeys, and other rare animals roam the thick forests found at the eastern end of this plateau.

In addition to very high elevations, western China has some extremely low areas. The Turpan Depression, east of the Tian Shan, lies about 505 feet (154 m) *below* sea level. It is partly filled with salt lakes. It also is the hottest area of China. Daytime temperatures can reach as high as 122°F (50°C). Can you find the Turpan Depression on the physical map below? Use the elevation key as a guide.

In the north of China, mountain ranges circle desert areas. One of these areas is the **Taklimakan Desert.** It is an isolated region with very high temperatures. Sandstorms here may last for days and create huge, drifting sand dunes. Farther east lies another desert, the **Gobi.** The Gobi's total area is about 500,000 square miles (1.3 million sq. km), about twice the size of Texas. Instead of sand, the Gobi has rocks and stones. Temperatures here range from -40°F (-40°C) in January to 113°F (45°C) in July.

The map below shows you that plains also run along the coasts of the South China and East China Seas. These fertile plains are rich in mineral resources. The majority of China's people live here. This region is one of the most scenic areas in China.

China: Physical

NATIONAL GEOGRAPHIC

Elevations

Feet	Meters
10,000	3,000
5,000	1,500
2,000	600
1,000	300
0	0

▲ Mountain peak

Applying Map Skills

1. What rivers begin in the high elevations of southwest China?

2. What seas border China?

Find NGS online map resources @ www.nationalgeographic.com/maps

Web Activity Visit the *Our World Today: People, Places, and Issues* Web site at owt.glencoe.com and click on **Chapter 6– Student Web Activities** to learn more about China's rivers.

Rivers Three of China's major waterways—the **Yangtze** (YANG•SEE), **Yellow**, and **Xi** (SHEE) **Rivers**—flow through the plains and southern highlands. They serve as important transportation routes and also as a source of soil. How? For centuries, these rivers have flooded their banks in the spring. The floodwaters have deposited rich soil to form flat river basins that can be farmed. China's most productive farmland is found in valleys formed by these major rivers.

Despite their benefits, the rivers of China also have brought much suffering. The Chinese call the Yellow River "China's sorrow." In the past, its flooding cost hundreds of thousands of lives and caused much damage. To control floods, the Chinese have built dams and dikes, or high banks of soil, along the rivers. Turn to page 167 to learn more about the **Three Gorges Dam**, a project under way on the Yangtze River.

An Unsteady Land In addition to floods, people in eastern China face another danger—earthquakes. Their part of the country stretches along the **Ring of Fire**, a name that describes Pacific coastal areas with volcanoes and frequent earthquakes. Eastern China lies along a fault, or crack in the earth's crust. As a result, earthquakes in this region are common—and can be very violent. Because so many people live in eastern China, these earthquakes can bring great suffering.

✓ Reading Check What problem does China have with its large rivers?

 **Leading Rice-Producing Countries**

 Analyzing the Graph

The most important food crop in Asia is rice.

Economics How many millions of tons of rice does China produce in a year?

Textbook Update

Visit owt.glencoe.com and click on **Chapter 6– Textbook Updates.**

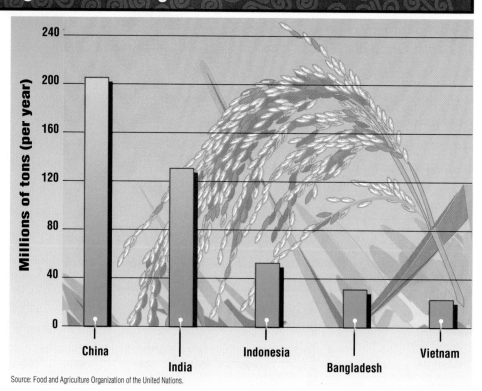

Source: Food and Agriculture Organization of the United Nations.

A New Economy

Since 1949, China has been a communist state, in which the government has strong control over the economy and society as a whole. Government officials—not individuals or businesses—decide what crops are grown, what products are made, and what prices are charged. China discovered that the communist system created many problems. China was still behind other countries in technology, and manufactured goods were of poor quality.

In recent years, China's leaders have begun many changes to make the economy stronger. Without completely giving up communism, the government has allowed many features of the free enterprise system to take hold. In this system, the government allows individuals to choose what jobs they want and where to start their own businesses. Workers can keep the profits they make. Farmers can grow and sell what they wish.

As a result of these and other changes, China's economy has boomed. The total value of goods and services produced in China increased four times from 1978 to 1999. Farm output also rose rapidly. Because of mountains and deserts, only 10 percent of China's land is farmed. Yet China is now a world leader in producing various agricultural products.

Foreign Trade   Eager to learn about new business methods, China has asked other countries to invest, or put money into developing Chinese businesses. Many companies in China are now jointly owned by Chinese and foreign businesspeople. Foreign companies expect two benefits from investing in China. First, they can pay Chinese workers less than they pay workers in their own countries. Second, companies in China have hundreds of millions of possible customers for their goods.

Results of Growth   Because of economic growth, more of China's people are able to get jobs in manufacturing and service industries. Wages have increased, and more goods are available to buy. Some Chinese now enjoy a good standard of living. They can afford consumer goods, or products such as televisions, cars, and motorcycles.

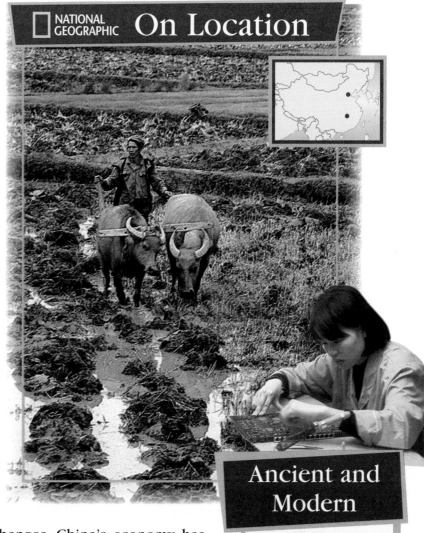

NATIONAL GEOGRAPHIC On Location

Ancient and Modern

In China's rural areas ancient farming methods are still used (left). However, in the industrialized cities, high technology is being developed (right).

Government How has the government affected how things are done in China?

Not everyone has adjusted well to the new economy. Many Chinese find that prices have risen faster than their incomes. Some Chinese have become very rich, while others remain poor.

China's economic growth has also hurt the environment. Many factories dump poisonous chemicals into rivers. Others burn coal, which gives off smoke that pollutes the air. This pollution leads to lung disease, which is the number one cause of death in China.

Hong Kong and Macau   The cities of **Hong Kong** and **Macau** (muh•KOW) are an important part of the economic changes taking place in China. Both cities were once controlled by a European country—Hong Kong by the United Kingdom, and Macau by Portugal. China regained control of Hong Kong in 1997 and of Macau in 1999. Both cities are centers of manufacturing, trade, and finance. Chinese leaders hope that the successful businesses in these cities will help spur economic growth in the rest of the country.

At the same time, foreign companies that are considering investing in these cities must ask themselves whether China will stand by its "one-country, two-systems" pledge. The pledge refers to China's promise to allow Western freedoms and capitalism to exist side by side with Chinese communism. The Time Reports: Focus on World Issues on pages 173–179 looks closely at how free enterprise may affect China's lack of freedom.

✓ Reading Check   **To what does "one-country, two-systems" refer?**

Assessment

Defining Terms
1. Define dike, fault, communist state, free enterprise system, invest, consumer goods, "one-country, two-systems."

Recalling Facts
2. **Place** What are China's two large deserts?
3. **Region** What two very important functions do China's rivers perform?
4. Economics What has caused China's economy to boom?

Critical Thinking
5. Analyzing Information How are China's rivers both a blessing and a disaster?
6. Making Comparisons How is a communist economic system different from a free enterprise system?

7. Analyzing Information What benefits does China receive from foreign investments?

Graphic Organizer
8. Organizing Information Create a diagram like this one. In the proper places on the oval, fill in the physical features you would encounter if you traveled completely around China.

China

Applying Social Studies Skills

9. Analyzing Maps Look at the political map on page 135. What is the capital of China?

The Three Gorges Dam

Since 1919, Chinese officials have dreamed of building a dam across the Yangtze, the third-longest river in the world. Curving through the heart of China, the river provides an important highway for moving people and products from town to town. Yet the Yangtze is unpredictable. For thousands of years, floods have harmed the millions of people who live along its banks. Now construction is under way to build the dam.

The Dam

In 1994 the Chinese government began a 17-year-long project to build a massive dam. It will eventually be 1.5 miles (2.4 km) wide and more than 600 feet (183 m) high. The dam, called the Three Gorges Dam, is being built about halfway between Chongqing and Wuhan (see the map on page 170). The dam will benefit China in several ways. First, it will control water flow and stop floods. Second, its system of locks will allow large ships to travel inland. This will reduce trade and transportation costs for the millions of people who live inland. Third, the dam will create electricity using turbines—water-driven engines.

Controversy

Even with all the proposed benefits, many people within China and elsewhere have questioned the wisdom of building the dam. When completed, the dam will create a deep reservoir nearly 400 miles (644 km) long. This reservoir will flood more than 100 towns and force nearly 1.2 million people to move. Many of these people must leave the farms that their families have worked for centuries. Historians point out that the reservoir will also wash away more than 1,000 important historical sites, including the homeland of the first people to settle the region about 4,000 years ago.

Environmentalists caution that the dam may create pollution and health risks. Industrial sites, once they lie underwater, may leak hazardous chemicals. Sewage from communities surrounding the dam could flow directly into the reservoir and into the Yangtze River. In the past, this problem was less serious because the fast-moving waters of the Yangtze carried waste quickly out to sea.

▶ Making the Connection

1. How have the unpredictable waters of the Yangtze River affected the Chinese?

2. Create a physical map of China showing the major rivers. Mark where the Three Gorges Dam is being built.

3. **Interpreting Points of View** List three reasons in support of constructing the Three Gorges Dam and three reasons against it.

◀ All but the very top of these towering gorges (left) will be deep underwater when the huge dam is completed (far left).

China's People and Culture

Guide to Reading

Main Idea

The arts and ideas of ancient times still influence China today.

Terms to Know

- dynasty
- human rights
- exile
- calligraphy
- pagoda

Reading Strategy

Create a chart like this one. Then list two key facts in the right column for each item in the left column.

China	
History	
Government	
Urban and Rural Life	
Arts	

NATIONAL GEOGRAPHIC Exploring Our World

How do you celebrate the coming of a new year? This costumed figure lives in Tibet. He is a Buddhist monk, or holy man, performing an important ritual celebrating the Tibetan New Year. The mask and colorful robes show that he plays a special role in rituals designed to defeat the forces of evil.

China's population of 1.27 billion is about one-fifth of the world's people. About 92 percent of these people belong to the ethnic group called **Han Chinese.** They have a distinctive culture. The remaining 8 percent belong to 55 other ethnic groups. Most of these groups, such as the **Tibetans,** live in the western part of China. They have struggled to protect their traditions from Han Chinese influences.

China's History

China's civilization is more than 4,000 years old. For many centuries until the early 1900s, rulers known as emperors or empresses governed China. Many lived in the Imperial Palace, located in the heart of Beijing. A **dynasty,** or line of rulers from a single family, would hold power until it was overthrown. Then a new leader would start a new dynasty. Under the dynasties, China built a highly developed culture and conquered neighboring lands.

As their civilization developed, the Chinese tried to keep out foreign invaders. In many ways, this was easy. On most of China's borders, natural barriers such as seas, mountains, and deserts already provided protection. Still, invaders threatened from the north. To defend this area, the Chinese began building the **Great Wall of China** about 2,200 years ago. Over the centuries, the wall was continually rebuilt and lengthened. In time, it snaked more than 4,000 miles (6,437 km) from the Yellow Sea in the east to the deserts of the west. It still stands today.

Culture   Chinese thinkers believed that learning was a key to good behavior. About 500 B.C., a thinker named **Kongfuzi** (KOONG•FOO•DZUH), or Confucius, taught that people should be polite, honest, brave, and wise. Children were to obey their parents, and every person was to respect the elderly and obey the country's rulers. Kongfuzi's teachings shaped China's government and society until the early 1900s.

During Kongfuzi's time, another thinker named **Laozi** (LOW•DZUH) arose. His teachings, called **Daoism** (DAHW•ehzm), stated that people should live simply and in harmony with nature. While Kongfuzi's ideas appealed to government leaders, Laozi's beliefs attracted artists and writers.

Buddhism came to China from Central Asia about A.D. 100. This religion taught that prayer, right thoughts, and good deeds could help people find relief from life's problems. Over time, the Chinese mixed Buddhism, Daoism, and the ideas of Kongfuzi. This mixed spiritual heritage still influences many Chinese people today.

The early Chinese were inventors as well as thinkers. Did you know that they were using paper and ink before people in other parts of the world? Other Chinese inventions included silk, the magnetic compass, printed books, gunpowder, and fireworks. For hundreds of years, China was the most advanced civilization in the world.

Communist China   Foreign influences increasingly entered China during the 1700s and 1800s. Europeans especially wanted to get such fine Chinese goods as silk, tea, and pottery. The United Kingdom and other countries used military power to force China to trade.

In 1911 a Chinese uprising under the Western-educated **Dr. Sun Yat-sen** overthrew the last emperor. China became a republic, or a country governed by elected leaders. Disorder followed until the Nationalist political party took over. The Communist Party gained power as well. After World War II, the Nationalists and the Communists fought for control of China. **General Chiang Kai-shek** (jee•AHNG KY•SHEHK) led the Nationalists. **Mao Zedong** (MOW DZUH•DOONG) led the Communists.

In 1949 the Communists won and set up the People's Republic of China under Mao Zedong and **Zhou Enlai** (JOH ehn•LY). The Nationalists fled to the offshore island of Taiwan. There they set up a rival government.

Reading Check Why was the Great Wall of China built?

Labor Costs

There's a good chance your clothes and shoes were manufactured in China. Some American companies can manufacture their products at much lower costs in China because the wages paid to workers there are low by U.S. standards. These companies pay more and offer better working conditions than Chinese employers. Still, human rights activists are concerned about exploiting workers to make higher profits for U.S. companies.

China's Government and Society

After 1949 the Communists completely changed the mainland of China. All land and factories were taken over by the government. Farmers were organized onto large government farms, and women joined the industrial workforce. Dams and improved agricultural methods brought some economic benefits. Yet many government plans went wrong, and individual freedoms were lost. Many people were killed because they opposed communism.

After Mao Zedong died in 1976, a new Communist leader, **Deng Xiaoping** (DUHNG SHOW•PIHNG), decided to take a new direction. He wanted to make China a more open country. One way to do this was to give people more economic freedom. The government kept tight control over all political activities, however. It continued to deny individual freedoms and acted harshly against any Chinese who criticized its actions. In 1989 thousands of students gathered in Beijing's **Tiananmen** (TEE•EHN•AHN•MEHN) **Square.** The students called for democracy in China. The government answered by sending in tanks and troops. These forces killed hundreds of protesters and arrested thousands more.

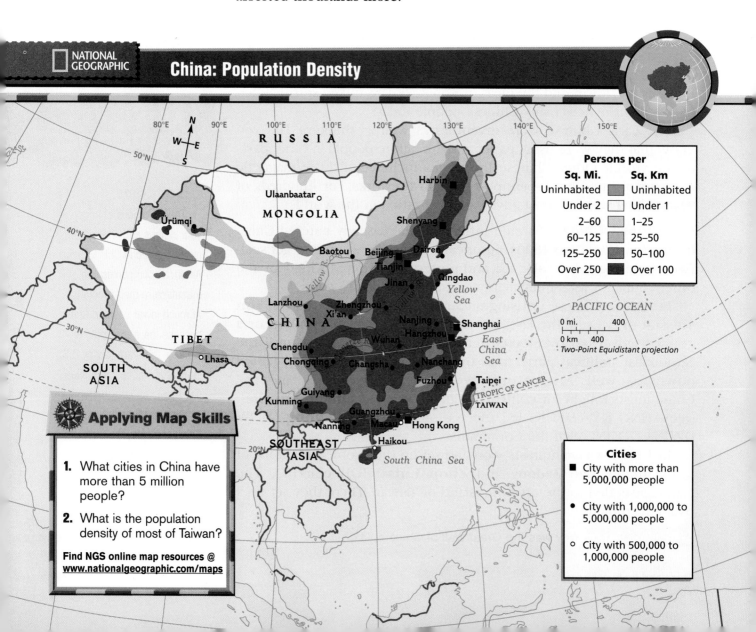

NATIONAL GEOGRAPHIC

China: Population Density

Persons per

Sq. Mi.	Sq. Km
Uninhabited	Uninhabited
Under 2	Under 1
2–60	1–25
60–125	25–50
125–250	50–100
Over 250	Over 100

PACIFIC OCEAN

0 mi.   400
0 km   400
Two-Point Equidistant projection

Cities

■ City with more than 5,000,000 people

● City with 1,000,000 to 5,000,000 people

○ City with 500,000 to 1,000,000 people

Applying Map Skills

1. What cities in China have more than 5 million people?

2. What is the population density of most of Taiwan?

Find NGS online map resources @ www.nationalgeographic.com/maps

Countries around the world have protested the Chinese government's continued harsh treatment of people who criticize it. They say that Chinese leaders have no respect for **human rights**—the basic freedoms and rights, such as freedom of speech, that all people should enjoy. Because of China's actions, some people say that other countries should not trade with China.

China's leaders have also been criticized for their actions in Tibet. Tibet was once a separate Buddhist kingdom. China took control of the area in 1950 and crushed a rebellion there about nine years later. The Tibetan people have demanded independence since then. The **Dalai Lama** (DAH•ly LAH•muh), the Buddhist leader of Tibet, now lives in exile in India. Someone in **exile** is unable to live in his or her own country because of political beliefs. The Dalai Lama travels around the world trying to win support for his people.

Rural Life About 64 percent of China's people live in rural areas. The map on page 170 shows that most Chinese are crowded into the fertile river valleys of eastern China. Families work hard in their fields. They often use hand tools because mechanical equipment is too expensive.

Village life has improved in recent years. Most rural families now live in three- or four-room houses. They have enough food and some modern appliances. Many villages have community centers. People gather there to watch movies and play table tennis and basketball.

Urban Life More than 460 million Chinese people live in cities. China's cities are growing rapidly as people leave farms in hope of finding better-paying jobs. Living conditions in the cities are crowded, but most homes and apartments have heat, electricity, and running water. Many people now earn enough money to buy extra clothes and televisions. They also have more leisure time to attend concerts or Chinese operas, walk in parks, or visit zoos.

✓ Reading Check **Why have people in other countries criticized China's government?**

NATIONAL GEOGRAPHIC On Location

Urban Life

Hundreds of thousands of people use bicycles—not cars—to get around Beijing and other cities.

Place About how many people live in China's cities?

China's Culture

China is famous for its traditional arts. Chinese craft workers make bronze bowls, jade jewelry, decorated silk, glazed pottery, and fine porcelain. The Chinese are also known for their painting, sculpture, and architecture.

The Chinese love of nature has influenced painting and poetry. Chinese artists paint on long panels of paper or silk. Artwork often shows scenes of mountains, rivers, and forests. Artists attempt to portray the harmony between people and nature.

Many Chinese paintings include a poem written in calligraphy, the art of beautiful writing. Chinese writing is different from the print you are reading right now. It uses characters that represent words or ideas instead of letters that represent sounds. There are more than 50,000 Chinese characters, but the average person recognizes only about 4,000 to 5,000.

The Chinese developed the first porcelain centuries ago. Porcelain is made from coal dust and fine, white clay. Painted porcelain vases from early China are considered priceless today.

Most buildings in China's cities are modern. Yet traditional buildings still stand. Some have large, tiled roofs with edges that curve gracefully upward. Others are Buddhist temples with many-storied towers called pagodas. These buildings hold large statues of the Buddha.

Foods   Cooking differs greatly from region to region. In coastal areas, people enjoy fish, crab, and shrimp dishes. Central China is famous for its spicy dishes made with hot peppers. Most Chinese eat very simply. A typical Chinese meal includes vegetables with bits of meat or seafood, soup, and rice or noodles. Often the meat and vegetables are cooked quickly in a small amount of oil over very high heat. This method—called stir-frying—allows the vegetables to stay crunchy.

✓ Reading Check   **Why does it take many years to learn to read and write Chinese?**

Assessment

Defining Terms
1. Define dynasty, human rights, exile, calligraphy, pagoda.

Recalling Facts
2. History Who are two thinkers who influenced life in China?

3. History Who led the Nationalists after World War II? Who led the Communists after World War II? Who won control of China?

4. Culture What scenes are commonly found in Chinese paintings?

Critical Thinking
5. Making Predictions How might the teachings of Kongfuzi prevent rebellions in China?

6. Summarizing Information Why did Europeans want to force China to trade with them?

Graphic Organizer
7. Organizing Information Create a time line like this one. Then list at least five dates and their events in China's history.

Applying Social Studies Skills

8. Analyzing Maps Look at the population density map on page 170. How does the population density in western China differ from that in eastern China?

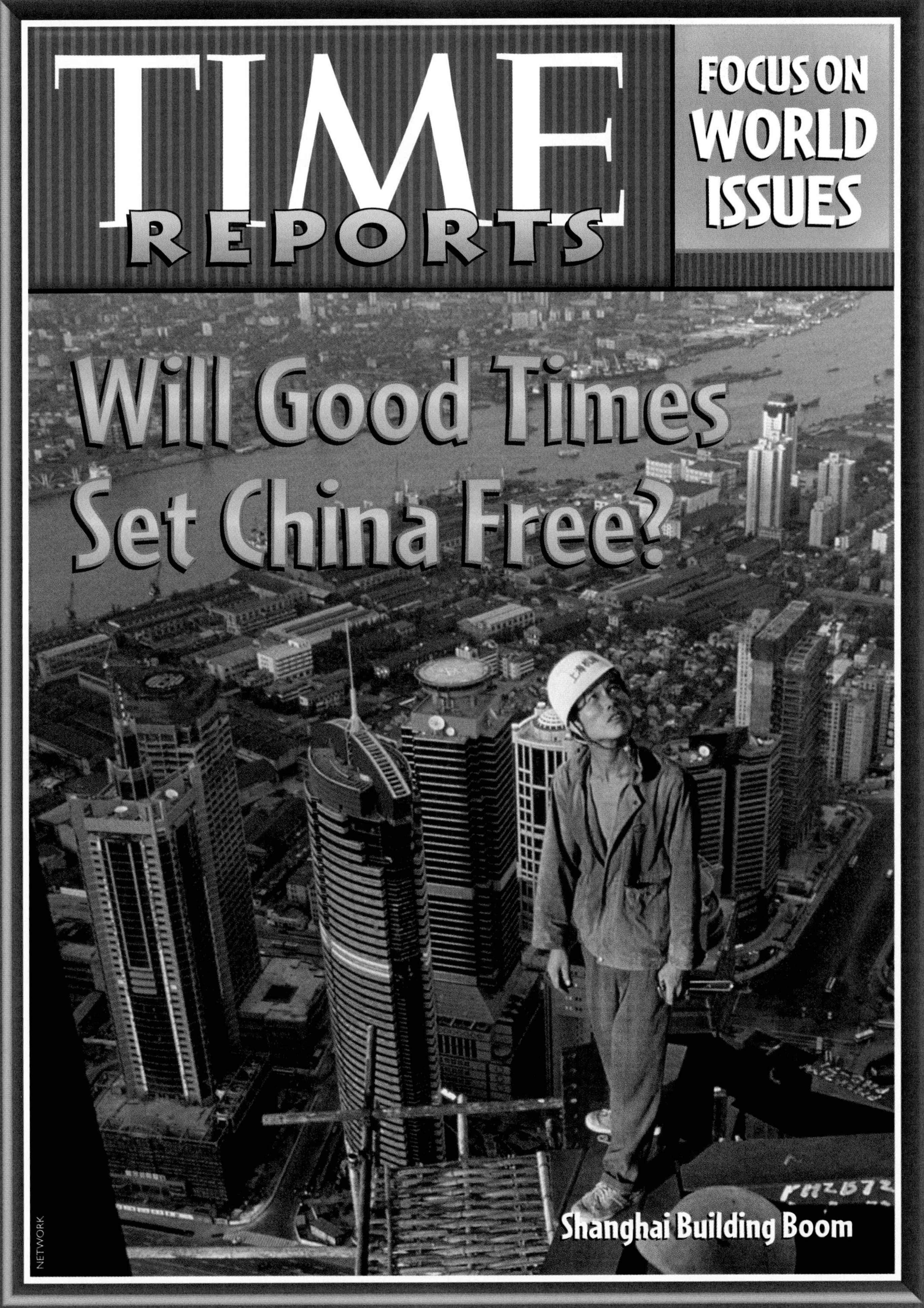

TIME
REPORTS
FOCUS ON WORLD ISSUES
Will Good Times Set China Free?
Shanghai Building Boom
NETWORK

One protester stopped the tanks of the world's biggest army outside Tiananmen Square in 1989.

No Easy Choices

Lloyd Zhao was a wanted man in 2001. He lived in Beijing, China's capital, where he spent hours every day on the Internet. His goal: to keep his religion alive.

The religion, Falun Gong, was barely 10 years old. Some call it a **cult**—a phony religion. Whatever it is, China's Communist leaders fear it. Any large group threatens their hold on power, and Falun Gong was very big. It had several million followers in 2000.

The government outlawed Falun Gong in 1999. It put as many as 6,000 of its followers in jail. But the Internet prevented the government from putting Falun Gong completely out of business.

Its followers used it to stay in touch with each other and with their leaders.

Their effort to outwit the government is a reminder that China is a **police state**, a country whose leaders often crush their opponents.

Because of the Internet, fresh ideas are circulating in China. The government can't control those ideas. It may even stop trying, now that it is preparing to host the 2008 Olympics. Whatever it does, it can't alter the fact that ideas from around the world are changing China forever.

A Closed Society

As recently as the 1970s, China's Communist government banned almost all communication with other countries. It told the Chinese what they could read. It told them where they could live and work. It decided what factories could make and what farms could produce.

It could do this because the government owned all enterprises—factories, farms, stores, and railroads, among other things. And it had the world's largest army to back it up.

During the late 1970s, it became clear that this arrangement wasn't working. China's farms couldn't feed the nation. Its factories produced goods few people wanted.

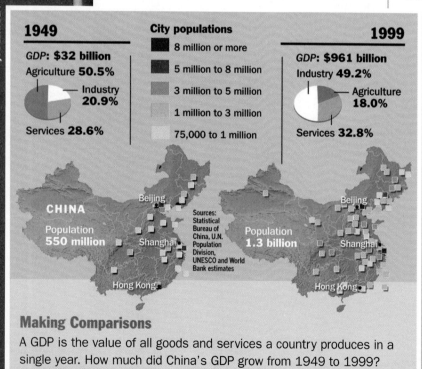

JOE LERTOLA

1949	City populations	1999
GDP: $32 billion	■ 8 million or more	GDP: $961 billion
Agriculture 50.5%	■ 5 million to 8 million	Industry 49.2%
Industry 20.9%	3 million to 5 million	Agriculture 18.0%
Services 28.6%	1 million to 3 million	Services 32.8%
	75,000 to 1 million	

CHINA
Population 550 million
Beijing
Shanghai
Hong Kong

Sources: Statistical Bureau of China, U.N. Population Division, UNESCO and World Bank estimates

Population 1.3 billion
Beijing
Shanghai
Hong Kong

Making Comparisons

A GDP is the value of all goods and services a country produces in a single year. How much did China's GDP grow from 1949 to 1999?

Job-seekers line up in Beijing.

Farmers vote for village leaders.

Britain gave Hong Kong back to China in 1997.

So in 1978 China's government took a big step. It let individuals own their own businesses. It let farmers till their own plots. And it let them keep the money they earned.

Free enterprise worked wonders. China's economy blossomed. Wages went up. Families were able to buy refrigerators, TV sets, computers, and sometimes even automobiles. To bring modern know-how to China, the government let citizens study abroad, hook up to cable TV, and surf the Internet.

New Problems

But with the good news came problems. Better-organized factories and farms needed fewer workers, and millions lost their jobs. Crime increased, and so did shady dealings by government officials. Water and air pollution, already bad, got worse.

The Chinese don't need or want outsiders to help them solve those problems. What they do want is the technology that makes modern

In July 2001 the Chinese celebrated when they learned they would host the 2008 Olympics.

economies run. China must buy that machinery from more industrialized nations.

China pays for those imports with exports—items that carry a "Made in China" label. That trade has connected China's economy to those of nations everywhere.

Globalization has brought China out of its shell. The Internet and the daring of people like Zhao have linked the Chinese to each other and to new ideas. China "watchers" wonder if these changes will bring democracy to China. ◼

EXPLORING THE ISSUE

1. **Making Inferences** Why might China's leaders see the Internet as a threat to their power?

2. **Cause and Effect** China's government let people own farms and businesses and keep the money they earn. How did this decision help make China's economy grow?

175

Limited Freedoms

Young Chinese today enjoy freedoms that their parents could only dream of. They can't criticize government leaders in print, but they can discuss just about anything else. Young people also have the luxury of working, living, and traveling wherever they want.

Such freedoms please people like Lei Xia. Lei, 22, is a guide on a tourist boat that sails up and down China's longest river, the Yangtze. Not long ago she and a friend decided to quit their jobs. They plan to go to Beijing to look for work. "Compared with my parents' generation," Lei said, "maybe I am lucky. I can choose for myself."

Concerts by rocker Cui Jian were once banned.

DAVID J. COULSON—AP

Crushing Dissent

One thing Lei can't choose is China's leaders. The Communist Party officials who run the government will do anything to hold onto power.

They made that clear in 1989. During the spring, students protesting China's lack of democracy camped out on Beijing's main plaza, Tiananmen Square. Ordered to clear them out, soldiers killed at least 300 people. This human rights abuse shocked the world.

► China has almost 2 million people in uniform.

The government continues to crush **dissent**, or opposition. In 1999 it sentenced leaders of a new political party to eight years in jail. Their crime: calling for an end to **one-party rule**. In 2001 the government shut down thousands of cybercafes. **Cybercafes** let people sip tea and surf the Web for as little as 20 cents an hour. Those shut down had let customers tap into anti-government websites.

Baton of Democracy

In June 2001, 20,000 people gathered in a park in Hong Kong, an island city in southeast China. They lighted candles in honor of the people killed in Tiananmen Square. A banner hanging over their heads read, "Pass on the baton of democracy." That's exactly what Chinese leaders don't want their citizens to do. ▪

EXPLORING THE ISSUE

1. **Analyzing** Lei says "Maybe I am lucky" instead of simply "I am lucky." Why might she be unsure?

2. **Explaining** Why do you think China's leaders oppose a two-party system?

The Road to Democracy

People who are free to run their own businesses usually want to choose their own governments. That's what happened in Taiwan, China's island province. Since 1949, Taiwan—a haven for anti-Communists—has gone its own way. It is now a free society with free elections.

But one party ruled it with an iron fist for nearly 40 years. Taiwan's 22 million people became better educated during that time. Taiwan became a showcase for free enterprise. Yet it wasn't until 1986, after those gains were made, that Taiwan's government let new political parties form. In 2000, the Taiwanese elected a member of one of the new parties as their president.

China could follow the same path, experts say. If it does, it could become a full democracy in 40 or 50 years.

Grassroots Democracy

Taiwan prepared its people for democracy slowly. The Taiwanese began electing village, county, and city governments in 1950.

China may be building democracy from the ground up, too. About 900 million farmers live in the nation's 928,000 villages. People in most of those villages have elected their leaders since the 1980s. "Now we have better management," says Li Dongju, whose husband heads a village committee in Hebei Province. "That's because our leaders enjoy the approval of the people."

▶ Li Dongju, who owns an orchard, is a great fan of village elections.

Li and her fellow villagers are learning what it means to be ruled by laws. Laws govern village elections. Laws tell village leaders what they can and can't do. And if the leaders break those laws, another law explains how they can be removed from office.

No laws rule China at the national level, however. Officials in Beijing are free to do just about anything they want. Sometimes visitors to China are arrested as foreign spies or agents. But the Internet and China's new economy are changing everything. "People are thinking rather than blindly following directions," says a Beijing businessman.

Free enterprise has led to democracies in Taiwan, South Korea, and Thailand. It can happen in China too. ◾

EXPLORING THE ISSUE

1. Evaluating Information How can you tell that the United States is a nation ruled by law?

2. Analyzing Information How might mainland China's size make it hard to create a democracy there?

Promoting Democracy: What Can One Person Do?

About 2.1 billion people throughout the world are unable to choose their own governments. Slightly more than half of them live in China.

What can the United States do to urge China's leaders to pass some of their power to ordinary citizens? Confronting China's leaders directly hasn't worked.

What might work, some experts say, is an indirect approach. For example, the U.S. could encourage Asian governments to raise the issue of human rights on their own. Or it could get China to sit down and discuss common problems—like disease control. Human rights might then be slipped into these discussions. That way, China's leaders might be less likely to feel they were being singled out.

Taking Action

You can help promote respect for human rights in China and other countries. For example, you could find out what your representatives in Congress are doing about the issue. Then you could write them, either supporting or disagreeing with their stands.

Since 1961, a group called Amnesty International (AI) has fought to make governments respect their citizens' rights. Its most successful weapon has been the letter-writing campaign. To learn how such campaigns work, go to

Internet portals like sohu.com bring new ideas to China.

GREG GIRARD—CONTACT

AI's Web page **www.amnestyusa.org** and click "Act!" on the top row.

"What you do may seem terribly insignificant," Mohandas Gandhi said, "but it is terribly important that you do it anyway." Gandhi knew what he was talking about. He led the campaign that won India its independence in 1947. ▪

EXPLORING THE ISSUE

1. **Evaluating** China's leaders say that no one outside China has the right to comment on the way they treat their citizens. Why do you agree or disagree?

2. **Cause and Effect** How might letter-writing campaigns make governments think twice about jailing opponents?

REVIEW AND ASSESS

UNDERSTANDING THE ISSUE

1. **Defining Key Terms** Write definitions for the following terms: *cybercafe, cult, police state, free enterprise, globalization, dissent, one-party rule.*

2. **Writing to Inform** In a brief essay, explain how village elections could prepare China's farmers for democracy. Use the terms *free elections, approval by the people,* and *rule of law.*

3. **Writing to Persuade** To get the Chinese to accept free enterprise, one of their leaders once said, "To get rich is glorious." Do you agree with this statement? Write a brief essay to explain your answer.

INTERNET RESEARCH ACTIVITY

4. With your teacher's help, use the Internet to find information about conditions in China today.

Browse the site until you find a topic that interests you. Write a brief report explaining what that topic teaches you about modern-day China.

5. Use the Web to research a specific episode in the history of Chinese immigration to the United States. You might browse through the Internet to find information on Angel Island in San Francisco harbor or search for information on the Chinese immigrant experience in America. Give an oral report of your findings to your class.

BEYOND THE CLASSROOM

6. Visit a library to find books about life in China today. Two places to start: *River Town: Two Years on the Yangtze* by Peter Hessler, and *The Chinese* by Jasper Becker.

▲ **China makes many of the clothes Americans wear.**

Write a short report about a chapter in one of the books.

7. Interview a person in your community who once lived in China. Why did this person move to America? What does he or she think are the most startling differences between the United States and China? What does this person most miss about China? Write a brief portrait of that person. Include his or her answers to the above questions.

GOH CHAI HIN—AFP

U.S. Trade With China
Top Five Imports and Exports Between China and the U.S.

What Americans Buy from China — Billions of dollars

	$0	$5	$10	$15	$20
Toys, games					$19.4 billion
Office machines, automatic data processing machines		$11.0			
Telecommunications, sound & reproduction equipment		$9.9			
Footwear		$9.2			
Electrical machinery and appliances		$9.1			

Source: U.S. Department of Commerce, data for year 2000

What China Buys from the U.S.

	$0	$5
Electrical machinery		$1.7
Transportation equipment, mainly aircraft		$1.7
Office machines, automatic data processing machines		$1.5
Oilseeds		$1.0
Industrial machinery		$0.8

BUILDING SKILLS FOR READING TABLES

1. **Analyzing Data** How do the two "shopping lists" at left differ from one another? How do those differences suggest ways each nation can help the other?

2. **Making Inferences** American factories in China typically pay workers about half what they would pay in the U.S. How does this fact help explain why the U.S. buys more from China than China buys from the U.S.?

FOR UPDATES ON WORLD ISSUES GO TO www.timeclassroom.com/glencoe

China's Neighbors

Guide to Reading

Main Idea

Taiwan and Mongolia have been influenced by Chinese ways and traditions.

Terms to Know

- high-technology industry
- steppe
- nomad
- empire
- yurt

Reading Strategy

Create a diagram like this one. Then write statements that are true of each country under their headings in the outer ovals. Where the ovals overlap, write statements that are true of both countries.

Taiwan   Mongolia

NATIONAL GEOGRAPHIC

Exploring Our World

In the remote, harsh land of western Mongolia, a centuries-old tradition continues. Hunters train eagles to bring their kill back to the human hunter. The people say that female eagles make the best hunters. Because they weigh more than males, they can capture larger prey. Like all eagles, they have superb vision—eight times better than a human's.

Taiwan is an island close to China's mainland, and Mongolia borders China on the north. Throughout history, Taiwan and Mongolia have had close ties to their larger neighbor.

Taiwan

About 100 miles (161 km) off the southeastern coast of China lies the island country of **Taiwan.** It is slightly smaller than the states of Connecticut and Massachusetts put together. Through Taiwan's center runs a ridge of steep, forested mountains. On the east, the mountains descend to a rocky coastline. On the west, they fall away to a narrow, fertile plain. This flat area is home to the majority of the island's people. Like southeastern China, Taiwan has mild winters and hot, rainy summers.

Taiwan's Economy   Taiwan has one of the world's most prosperous economies. Taiwan's wealth comes largely from high-technology industries, manufacturing, and trade with other countries. High-technology industries produce computers and other kinds of electronic equipment. Workers in Taiwan's factories make many different products, including computers, calculators, radios, televisions, and telephones. You have probably seen goods from Taiwan sold in stores in your community.

Taiwan has a growing economic influence on its Asian neighbors. Many powerful companies based in Taiwan have recently built factories in the People's Republic of China and Thailand. Despite their political differences, Taiwan and mainland China have increased their economic ties since the 1990s.

Agriculture also contributes to Taiwan's booming economy. The island's mountainous landscape limits the amount of land that can be farmed. Still, some farmers have built terraces on mountainsides to grow rice. Other major crops include sugarcane, citrus fruits, sweet potatoes, pineapples, tea, and soybeans. In fact, Taiwan's farmers produce enough food not only to feed their own people but also to export.

Taiwan's History and People   For centuries, Taiwan was part of China's empire. Then in 1895, Japan took the island after defeating China in war. The Japanese developed the economy of Taiwan but treated the people very harshly. After Japan's loss in World War II, Taiwan returned to China.

In 1949 the Nationalists under Chiang Kai-shek arrived in Taiwan from the Chinese mainland. Along with them came more than 1.5 million refugees fleeing communist rule. Fearing a communist invasion, the Nationalists kept a large army in the hope of someday retaking the mainland. They also blocked other political groups from sharing in the government.

By the early 1990s, local Taiwanese were allowed more opportunities in government. The virtually one-party system ended, and Taiwan became a democracy. Taiwan still claims to be a Chinese country, but many people would like to declare Taiwan independent. China claims Taiwan as its twenty-third province and believes that it should be under China's control. China has threatened to use force against Taiwan if the island declares its independence.

NATIONAL GEOGRAPHIC **On Location**

Taiwan

Many electronic industries have headquarters in Taiwan.

Place **What kinds of products do high-technology factories in Taiwan produce?**

About 75 percent of Taiwan's 22.5 million people live in urban areas. The most populous city—with 2.6 million people—is the capital, **Taipei.** This bustling center of trade and commerce has tall skyscrapers and modern stores. If you stroll through the city, however, you will see Chinese traditions. Buddhist temples, for example, still reflect traditional Chinese architecture.

✔Reading Check **What is the capital of Taiwan?**

China and Its Neighbors

Ulaanbaatar

Ulaanbaatar in Mongolia began as a Buddhist community in the early 1600s. Today it is a modern cultural and industrial center.

Place Why is Mongolia known as the Land of the Blue Sky?

Mongolia

Landlocked **Mongolia** is a large country about the size of Alaska. Rugged mountains and high plateaus rise in the west and central regions. The bleak landscape of the Gobi spreads over the southeast. The rest of the country is covered by steppes, the dry treeless plains often found on the edges of a desert.

Known as the Land of the Blue Sky, Mongolia boasts more than 260 days of sunshine per year. Yet its climate has extremes. Rainfall is scarce, and fierce dust storms sometimes sweep across the landscape. Temperatures are very hot in the summer. In the winter, they fall below freezing at night.

For centuries, most of Mongolia's people were nomads. Nomads are people who move from place to place with herds of animals. Even today, many Mongolians tend sheep, goats, cattle, or camels on the country's vast steppes. Important industries in Mongolia use products from these animals. Some factories use wool to make textiles and clothing. Others use the hides of cattle to make leather and shoes. Some farmers grow wheat and other grains. Mongolia also has deposits of copper and gold.

Mongolia's History and People Mongolia's people are famous for their skills in raising and riding horses. In the past, they also were known as fierce fighters. In the 1200s, many groups of Mongols joined together under one leader, **Genghis Khan** (JEHNG•guhs KAHN). He led Mongol armies on a series of conquests. The Mongols eventually carved out the largest land empire in history, ruling 80 percent of Eurasia by A.D. 1300. An empire is a collection of different territories under one ruler. The **Mongol Empire** stretched from China all the way to eastern Europe.

During the 1300s, the Mongol Empire weakened and fell apart. China ruled the area that is now Mongolia from the 1700s to the early 1900s. In 1924 Mongolia gained independence and created a strict communist government under the guidance of the Soviet Union. The country finally became a democracy in 1990. Since then, the Mongolian economy has moved slowly from government control to a free enterprise system.

About 85 percent of Mongolia's 2.4 million people are Mongols. They speak the Mongol language. About 60 percent of the people live in urban areas. The largest city is the capital, **Ulaanbaatar** (OO•LAHN•BAH•TAWR). Mongolians in the countryside live on farms. A few still follow the nomadic life of their ancestors. These herder-nomads live in *yurts,* large circle-shaped structures made of animal skins that can be packed up and moved from place to place.

Mongolians still enjoy the sports and foods of their nomadic ancestors. The favorite meal is boiled sheep's meat with rice, washed down with tea. The biggest event of the year is the **Naadam Festival,** held all over the country in mid-summer. It consists of a number of sporting events, including wrestling, archery, and horse racing.

Since before the days of the Mongol Empire, most people in Mongolia have been Buddhists. Buddhism has long influenced Mongolian art, music, and literature. Traditional music has a wide range of instruments and singing styles. In one style of Mongolian singing, male performers produce harmonic sounds from deep in the throat, releasing several notes at once.

For centuries, Buddhist temples and other holy places dotted the country. Under communism, religious worship was discouraged. Many of these historic buildings were either destroyed or left to decay. Today, people are once again able to practice their religion. They have restored or rebuilt many of their holy buildings.

Reading Check What religion do most Mongolians practice?

Assessment

Defining Terms
1. **Define** high-technology industry, steppe, nomad, empire, yurt.

Recalling Facts
2. **Economics** What kinds of products are made in Taiwan?
3. **Government** Why has Taiwan not claimed independence from China?
4. **History** What Mongol warrior conquered much of Eurasia by A.D. 1300?

Critical Thinking
5. **Understanding Cause and Effect** Why did many people flee to Taiwan from China in 1949?
6. **Drawing Conclusions** Why do you think communist leaders discouraged religious worship in Mongolia?

Graphic Organizer
7. **Organizing Information** Create a diagram like this one. Then write either Taiwan or Mongolia in the center oval. Write at least one fact about the country under the headings in each of the outer ovals.

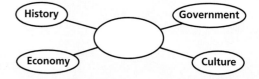

Applying Social Studies Skills

8. **Analyzing Maps** Look at the physical map on page 163. What mountains rise in western Mongolia? What desert is found in southern and southeastern Mongolia?

Critical Thinking Skill

Distinguishing Fact From Opinion

Distinguishing fact from opinion can help you make reasonable judgments about what others say and write. Facts can be proved by evidence such as records, documents, or historical sources. Opinions are based on people's differing values and beliefs.

Learning the Skill

The following steps will help you identify facts and opinions:

- Read or listen to the information carefully. Identify the facts. Ask: Can these statements be proved? Where would I find information to prove them?
- If a statement can be proved, it is factual. Check the sources for the facts. Often statistics sound impressive, but they may come from an unreliable source.
- Identify opinions by looking for statements of feelings or beliefs. The statements may contain words like *should, would, could, best, greatest, all, every,* or *always.*

Practicing the Skill

Read the paragraph below, and then answer the questions that follow.

Anyone who thinks the Internet is not used in China has been asleep at the mouse. China's government-owned factories and political system may seem old-fashioned. When it comes to cyberspace, however, China is moving at Net speed. Internet use is growing explosively. In 1997 only 640,000 Chinese were using the Internet. By 2000, the number had increased to 12.3 million. The Phillips Group estimates that by 2005, the online population should hit 85 million.

1. Identify facts. Can you prove that Chinese Internet use is increasing?
2. Note opinions. What phrases alert you that these are opinions?
3. What is the purpose of this paragraph?

Applying the Skill

Watch a television commercial. List one fact and one opinion that are stated. Does the fact seem reliable? How can you prove the fact?

GO TO

Practice key skills with **Glencoe Skillbuilder Interactive Workbook, Level 1.**

◄ Chinese students attend an Internet exhibit in Beijing.

Section 1 — China's Land and Economy

Terms to Know
dike
fault
communist state
free enterprise system
invest
consumer goods
"one-country, two-systems"

Main Idea
China—the third-largest country in the world—has very diverse landforms. China's rapidly growing economy has changed in recent years.

✓ Place Rugged mountains and harsh deserts cover western China.

✓ Place About 90 percent of China's people live in the lowlands of eastern China.

✓ Human/Environment Interaction China's rivers bring fertile soil along with the danger of flooding to the eastern plains.

✓ Economics China's leaders have changed the economy to give the people more economic freedom. The economy has grown rapidly as a result.

✓ Economics Many companies in China are now jointly owned by Chinese and foreign businesspeople. Foreign companies can pay workers less than they pay workers in their own countries, and they have a great number of possible customers in the Chinese people.

Section 2 — China's People and Culture

Terms to Know
dynasty
human rights
exile
calligraphy
pagoda

Main Idea
The arts and ideas of ancient times still influence China today.

✓ History The ancient teachings of Kongfuzi, Daoism, and Buddhism still influence the people of China.

✓ History For thousands of years, dynasties of emperors ruled China. Today, Communist leaders keep tight control over all areas of political life.

✓ Culture China is famous for the skill of its craft workers and for its distinctive painting and architecture.

Section 3 — China's Neighbors

Terms to Know
high-technology industry
steppe
nomad
empire
yurt

Main Idea
Taiwan and Mongolia have been influenced by Chinese ways and traditions.

✓ Government Taiwan is an island off southeast China. The government of China does not recognize Taiwan as a separate country.

✓ Economics Taiwan's prosperous economy has influenced other Asian economies.

✓ Place Mongolia has rugged terrain and a harsh landscape.

✓ Culture Some people in Mongolia still follow a traditional nomadic lifestyle, and herding remains an important economic activity.

Assessment and Activities

Using Key Terms

Match the terms in Part A with their definitions in Part B.

A.

1. invest
2. dynasty
3. exile
4. high-technology industry
5. dike
6. communist state
7. "one-country, two-systems"
8. calligraphy
9. human rights
10. yurt

B.

a. promise to allow capitalism and communism to exist side-by-side

b. country whose government has strong control over the economy and society

c. high bank of soil along a river to prevent flooding

d. basic freedoms of speech and movement

e. to put money into

f. the art of beautiful writing

g. nomadic tent made of animal skins

h. state of being unable to live in one's own country because of political beliefs

i. line of rulers from the same family

j. business that produces electronic equipment

Reviewing the Main Ideas

Section 1 China's Land and Economy

11. **Place** Where do most of China's people live?
12. **Place** What three major rivers flow through the plains and southern highlands of China?
13. **Human/Environment Interaction** How has the new economy contributed to air pollution in China?
14. **Economics** Give three reasons why China's economy has boomed.

Section 2 China's People and Culture

15. **Culture** What are the ideas of Kongfuzi? Of Laozi?
16. **Government** What kind of government did China have between 1911 and 1949?

Section 3 China's Neighbors

17. **Economics** Why is Taiwan's economy important in Asia?
18. **Place** How does Mongolia's landscape prevent much farming?
19. **Economics** How are Mongolia's main industries related to herding?

 **China**

Place Location Activity

On a separate sheet of paper, match the letters on the map with the numbered places listed below.

1. Plateau of Tibet
2. Yellow River
3. Yangtze River
4. Hong Kong
5. Gobi
6. Beijing
7. Mongolia
8. Shanghai
9. Taklimakan Desert
10. Himalaya

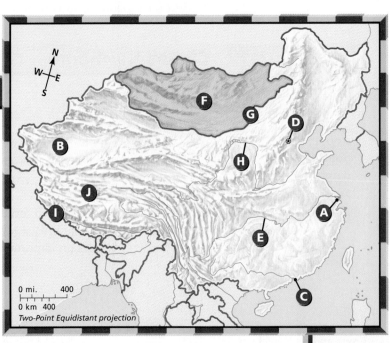

0 mi. 400
0 km 400
Two-Point Equidistant projection

Critical Thinking

20. **Drawing Conclusions** Why do you think China wanted to be isolated from European countries in the 1700s and 1800s?

21. **Organizing Information** Create a chart like the one below. Under each heading, write at least two facts about China.

Land	Economy	History	Government	People

Current Events Journal

22. **Creating a Time Line** Learn more about an event that is happening in China, Mongolia, or Taiwan today. Find out what important events led up to this one and what effect they have had on what is happening now. Create a time line showing your findings. Include illustrations and photos.

Mental Mapping Activity

23. **Focusing on the Region** Create a simple outline map of China and its neighbors, and then label the following:

- Himalaya
- Yellow River
- Taiwan
- Beijing
- Gobi
- Ulaanbaatar
- Yangtze River
- Hong Kong

Technology Skills Activity

24. **Developing a Multimedia Presentation** Using the Internet, research one of the arts of China. You might choose painting, architecture, literature, music, or a craft such as casting bronze or making silk. Create a museum exhibit that presents your findings. Include photographs showing examples of works from different periods in Chinese history.

Standardized Test Practice

Directions: Study the map below, and then answer the questions that follow.

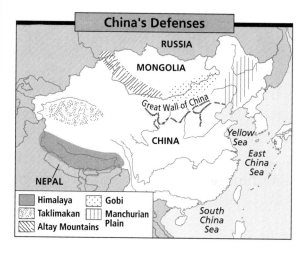

China's Defenses

Legend: Himalaya, Taklimakan, Altay Mountains, Gobi, Manchurian Plain

1. **Where is the Gobi?**

   A Near China's Russian border

   B In the southwestern part of China

   C In the Himalaya

   D Along China's border with Mongolia

2. **Which of the following is a human-made defense?**

   F The Great Wall of China

   G The Gobi

   H The Taklimakan

   J The Himalaya

Test-Taking Tip: Look for key words that will help you find the correct answer. An example is *human-made* in question 2. In this case, all of the answer choices are *natural* defenses of China except for the correct answer. Look at the map closely, using its title, its legend, and the information shown on it to find the correct answer choice.

Soft and sleek, silk is a valuable textile.

The Silk Road

Was there really a road made of silk? Well, not exactly. Silk, however, was one of the main products carried along the Silk Road—a system of trade routes that linked ancient China and the empires of the West. When Chinese silk became fashionable in Rome, the precious cloth traveled the Silk Road.

A Risky Route

The road itself was anything but soft and smooth. Traveling from China, camels laden with silk and other cargo trudged through deserts, including the Taklimakan, a name meaning "go in and you won't come out." Sandstorms and intense heat made passage difficult. Farther along the route, the Pamir mountain range thrust an ice- and snow-covered barrier in the way. The road was dangerous as well. Bandits attacked often, stealing valuable goods.

Few traveled the entire 4,000-mile (6,437-km) series of routes. Instead, merchants bought goods in trading posts and oases along the way and sold them at other markets farther along, much as relay runners pass a baton.

Chinese Secret Agent

Zhang Qian, an agent on a secret mission for Chinese Emperor Wudi, may have started the silk trade. In 139 B.C. invaders swept into China, despite China's Great Wall. Zhang Qian was sent far into Central Asia to find allies to help fight the invaders. He found no allies. Instead, he brought back strong horses for the military, which he had bought with bolts of silk.

Soon the Chinese were trading silk with the Parthian Empire, now present-day Iran. It is said that Rome wanted silk after its soldiers spotted silk banners fluttering above Parthian troops. By the A.D. 100s, China and Rome were trading a variety of goods. From the East came such exotic items as silk, spices, and fruits. Rome paid in glass, wool, and ivory, but mostly in gold.

Ideas also traveled the Silk Road. From India, the religion of Buddhism reached China. Christianity and Islam spread eastward as well. Chinese techniques for making paper and explosives traveled west. Western methods of cloth manufacturing and better gun design went to China. The process for making silk, however, traveled nowhere until much later. The Chinese successfully guarded their secret—that silk was made from the strands of a silkworm's cocoon.

For centuries, goods and ideas traveled between East and West. In the 1300s, however, the Silk Road began to decline as sea routes proved safer than land routes. Nevertheless, even today, parts of the Silk Road are busy with trade—and tourism. In addition to camels, tour buses now travel the caravan routes.

QUESTIONS

1 How is the Silk Road "made of silk"?

2 What were some obstacles along the Silk Road?

A man and his camel travel the Silk Road in China. ▶

NATIONAL GEOGRAPHIC

Silk Road Routes

— Silk Road

RUSSIA

Velikiy Novgorod
Moscow
Istanbul (Constantinople)
Caspian Sea
Aral Sea
Black Sea
Mediterranean Sea
Antioch
IRAN
Baghdad
IRAQ
Samarqand
Pamirs
Taklimakan Desert
MONGOLIA
Anxi
CHINA
Xi'an
INDIA
Arabian Sea
Bay of Bengal
South China Sea

AFRICA

0 mi. 1,000
0 km 1,000
Miller projection

N W E S

Chapter 7

Japan and the Koreas

The World and Its People NATIONAL GEOGRAPHIC

To learn more about the people and places of Japan and the Koreas, view *The World and Its People* **Chapter 25** video.

Our World Today **online**

Chapter Overview Visit the *Our World Today: People, Places, and Issues* Web site at owt.glencoe.com and click on **Chapter 7—Chapter Overviews** to preview information about Japan and the Koreas.

FOLDABLES™
Study Organizer

Compare-Contrast Study Foldable Make this foldable to help you compare and contrast the people and places of Japan and the Koreas.

Step 1 Fold one sheet of paper in half from top to bottom.

Step 2 Fold it in half again, from side to side.

Step 3 Unfold the paper once. Sketch an outline of the Koreas and Japan across both tabs and label them as shown.

Step 4 Cut up the fold of the top flap only.

This cut will make two tabs.

Reading and Writing As you read the chapter, write what you learn about these countries under the appropriate tab. Use your notes to determine how these countries are alike and different.

Why It Matters

Rebuilding

Only about 50 years ago, Japan and Korea were nations largely destroyed by war. The countries recovered to become important centers of technology with prosperous economies. North Korea, under a communist system of government, has not enjoyed economic growth. Today, challenges arise as these nations learn to relate to one another.

◀ **Mount Fuji is the national symbol of Japan.**

Japan

Guide to Reading

Main Idea

Although Japan's people have few mineral resources, they have built a prosperous country.

Terms to Know

- tsunami
- archipelago
- intensive cultivation
- clan
- shogun
- samurai
- constitutional monarchy
- megalopolis

Reading Strategy

Create a chart like this one. In the right column, write a fact about Japan for each topic in the left column.

Japan	Fact
Land	
Economy	
History	
People	

NATIONAL GEOGRAPHIC **Exploring Our World**

Early one morning in 1995, the ground in the Japanese port city of Kobe (KOH•bay) began to shake. The earthquake passed in less than a minute—but the destruction was immense. Buildings and bridges like this one collapsed. Gas lines broke, and the leaking gas caught fire. Thousands of people died, and the damage exceeded $100 billion.

The city of Kobe suffered an earthquake because Japan lies on the **Ring of Fire.** This name refers to an area surrounding the Pacific Ocean where the earth's crust often shifts. Japan experiences thousands of earthquakes a year. People in Japan also have to deal with tsunamis (tsu•NAH•mees). These huge sea waves caused by undersea earthquakes are very destructive along Japan's Pacific coast.

Japan's Land

Japan is an archipelago (AHR•kuh•PEH•luh•GOH), or a group of islands, off the coast of eastern Asia between the Sea of Japan and the Pacific Ocean. Four main islands and thousands of smaller ones make up Japan's land area. The largest islands are **Hokkaido** (hoh•KY•doh), **Honshu, Shikoku** (shee•KOH•koo), and **Kyushu** (KYOO•SHOO).

These islands are actually the peaks of mountains that rise from the floor of the Pacific Ocean. The mountains are volcanic, but many are no longer active. The most famous peak is **Mount Fuji,** Japan's highest mountain and national symbol. Rugged mountains and steep, forested hills dominate most of Japan. Narrowly squeezed between the seacoast and the mountains are plains. The **Kanto Plain** in eastern Honshu is Japan's largest plain. It holds **Tokyo,** the capital, and **Yokohama,** one of Asia's major port cities. You will find most of Japan's cities, farms, and industries on the coastal plains.

No part of Japan is more than 70 miles (113 km) from the sea. In bay areas along the jagged coasts lie many fine harbors and ports. The northern islands catch cold arctic winds and ocean currents. The Pacific Ocean, on the other hand, sends warm ocean currents to the southern part of Japan.

✓Reading Check What are the two major landforms in Japan?

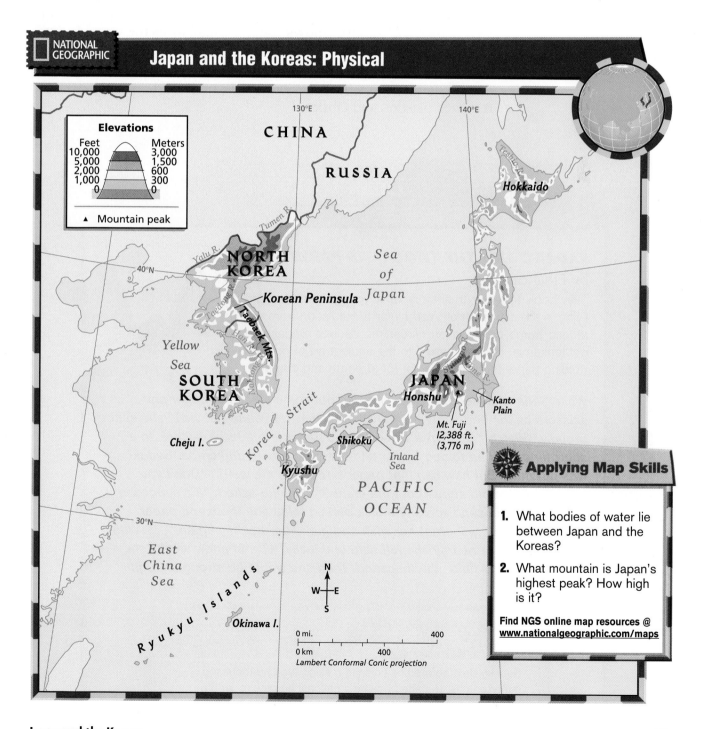

NATIONAL GEOGRAPHIC

Japan and the Koreas: Physical

Elevations
Feet / Meters
10,000 / 3,000
5,000 / 1,500
2,000 / 600
1,000 / 300
0 / 0

▲ Mountain peak

CHINA

RUSSIA

Hokkaido

Tumen R.

Yalu R.

Sea of Japan

NORTH KOREA

Taedong R.

Korean Peninsula

Taebaek Mts.

Han R.

Yellow Sea

Nakdong R.

SOUTH KOREA

Cheju I.

Korea Strait

JAPAN

Honshu

Shinano R.

Tone R.

Kanto Plain

Mt. Fuji
12,388 ft.
(3,776 m)

Shikoku

Inland Sea

Kyushu

PACIFIC OCEAN

East China Sea

Ryukyu Islands

Okinawa I.

0 mi.   400
0 km   400
Lambert Conformal Conic projection

N W E S

Applying Map Skills

1. What bodies of water lie between Japan and the Koreas?

2. What mountain is Japan's highest peak? How high is it?

Find NGS online map resources @ www.nationalgeographic.com/maps

Japan and the Koreas

193

Japan's Economy

Japan's industries have benefited from having highly skilled workers. The people of Japan value hard work, cooperation, and education. After high school graduation, many Japanese students go on to a local university.

Industry Japan has few mineral resources, so it must import raw materials, such as iron ore, coal, and oil. However, Japan is an industrial giant known around the world for the variety and quality of its manufactured goods. Japan's modern factories use new technology and robots to make their products quickly and carefully. These products include automobiles and other vehicles. The graph on page 12 in the Geography Handbook shows you that Japan leads the world in automobile production. Japan's factories also produce consumer goods such as electronic equipment, watches, small appliances, and calculators. Other factories produce industrial goods such as steel, cement, fertilizer, plastics, and fabrics.

Literature

SADAKO AND THE THOUSAND PAPER CRANES
by Eleanor Coerr

This book tells the true story of a young Japanese girl living in the aftermath of World War II. Radiation from the atomic bomb dropped on Hiroshima caused Sadako to get leukemia. Sadako turned to the ancient art of origami (folding paper to make objects) for strength and courage.

While Sadako closed her eyes, Chizuko put some pieces of paper and scissors on the bed . . . "I've figured out a way for you to get well," she said proudly. "Watch!" She cut a piece of gold paper into a large square. In a short time she had folded it over and over into a beautiful crane. Sadako was puzzled. "But how can that paper bird make me well?" "Don't you remember that old story about the crane?" Chizuko asked. "It's supposed to live for a thousand years. If a sick person folds one thousand paper cranes, the gods will grant her wish and make her healthy again." . . . With the golden crane nearby she felt safe and lucky. Why, in a few weeks she would be able to finish the thousand. Then she would be strong enough to go home.

Source: Sadako and the Thousand Paper Cranes by Eleanor Coerr. Copyright 1977. The Putnam Publishing Group.

Analyzing Literature

Sadako died before she finished making the one thousand paper cranes, but she became a national heroine in Japan. What was it about Sadako that made other Japanese people feel connected to her and proud of her?

Agriculture Farmland is very limited. Yet Japan's farmers use fertilizers and modern machinery to produce high crop yields. They also practice intensive cultivation—they grow crops on every available piece of land. You can see crops growing on terraces cut in hillsides and even between buildings and highways. In warmer areas, farmers harvest two or three crops a year. The chief crop is rice, a basic part of the Japanese diet. Other important crops include sugar beets, potatoes, fruits, and tea. Seafood forms an important part of the people's diet. Although Japan's fishing fleet is large, the country imports more fish than any other nation.

Economic Challenges Japan is one of the world's leading exporters. Because of trade restrictions, the country imports few finished goods from other countries, however. This has led to disagreements with trading partners who want to export more goods to Japan.

Another challenge facing Japan is preserving the environment. Air pollution from power plants has produced acid rain. Because of over-fishing, supplies of seafood have dropped. The government has passed laws to limit the amount of fish that can be caught each year.

✓ Reading Check **What are some products made by Japanese manufacturers?**

NATIONAL GEOGRAPHIC **On Location**

Past and Present

Past and present come together in Japan. Here, a priest of the ancient Shinto religion blesses a family's shiny new car.

Place Where was this car probably made? Why?

Japan's History and Government

Japan's history reaches back many centuries. The Japanese trace their ancestry to various clans, or groups of related families, that originally came from the mainland of Asia and lived on the islands as early as the late A.D. 400s.

The Japanese developed close ties with China on the Asian mainland. Ruled by emperors, Japan modeled its society on the Chinese way of life. The Japanese also borrowed the Chinese system of writing and accepted the Buddhist religion brought by Chinese missionaries. Today most Japanese practice Buddhism along with **Shinto,** Japan's own traditional nature religion.

In the 790s, the power of Japanese emperors began to decline. From the late 1100s to the 1860s, Japan was ruled by shoguns, or military leaders, and powerful land-owning warriors known as the

Hard Hats to School?

Okajima Yukiko and Sataka Aya walk along ash-covered sidewalks to Kurokami Junior High School. Why are they wearing hard hats? Their city is near Japan's Mount Oyama Volcano, which has just erupted. Yukiko and Aya have grown up facing the dangers of volcanic eruptions, earthquakes, and tsunamis. At school their first class starts at 8:30 A.M., and their last class ends at 3:40 P.M. Yukiko and Aya must go to school every second Saturday of the month, too.

samurai. Like China, Japan did not want to trade with foreign countries. In 1853 the United States government sent a fleet headed by Commodore Matthew Perry to Japan to demand trading privileges. In response to this action and other outside pressures, the Japanese started trading with other countries.

In the late 1800s, Japanese leaders began to use Western ideas to modernize the country, improve education, and set up industries. By the early 1900s, Japan was the leading military power in Asia.

In the 1930s, Japan needed more resources for its growing population. It took land in China and spread its influence to Southeast Asia. In 1941 Japanese forces attacked the American naval base at **Pearl Harbor** in Hawaii. This attack caused the United States to enter World War II. After four years of fighting, Japan surrendered when the United States dropped atomic bombs on the cities of **Hiroshima** and **Nagasaki.** By that time, many of Japan's cities lay in ruins and the economy had collapsed. With help from the United States, Japan became a democracy and quickly rebuilt its ruined economy.

Government   Japan's democracy is in the form of a constitutional monarchy. The emperor is the official head of state, but elected officials run the government. Voters elect representatives to the national legislature. The political party with the most members chooses a prime minister to lead the government.

Japan has great influence as a world economic power. In addition, it gives large amounts of money to poorer countries. Japan is not a military power, though. Because of the suffering that World War II caused, the Japanese have chosen to keep Japan's military small.

The government of Japan has improved health care and education for its people. Japan has the lowest infant death rate in the world and its literacy rate is almost 100 percent. The crime rate in Japan is very low.

✓Reading Check   What kind of government does Japan have?

Japan's People and Culture

Although about the size of California, Japan has 127.1 million people—nearly one-half the population of the United States. Most of Japan's people belong to the same Japanese ethnic background. Look at the map on page 197 to see where most of Japan's people live. About three-fourths are crowded into urban areas on the coastal plains. The four large cities of Tokyo, Yokohama, Nagoya, and Osaka form a megalopolis, or a huge urban area made up of several large cities and communities near them.

Japan's cities have tall office buildings, busy streets, and speedy highways. Homes and apartments are small and close to one another. Many city workers crowd into subway trains to get to work. Men work long hours and arrive home very late. Women often quit their jobs to raise children and return to work when the children are grown.

You still see signs of traditional life, even in the cities. Parks and gardens give people a chance to take a break from the busy day. It is common to see a person dressed in a traditional garment called a kimono walking with another person wearing a T-shirt and jeans.

Only 22 percent of Japan's people live in rural areas. In both rural and urban Japan, the family traditionally has been the center of one's life. Each family member had to obey certain rules. Grandparents, parents, and children all lived in one house. Family ties still remain strong, but each family member is now allowed more freedom. Many family groups today consist only of parents and children.

Religion   Many Japanese practice two religions—Shinto and Buddhism. Shinto began in Japan many centuries ago. It teaches respect for nature, love of simple things, and concern for cleanliness and good manners. Shinto is different from other religions for several reasons. First, there is no person who founded or started the religion. Shinto did not spread to other areas of the world, but stayed mostly in Japan. Also, there is no collection of writings that make up scripture, such as the Bible or the Quran. Buddhism also teaches respect for nature and the need to achieve inner peace.

Traditional Arts   Japan's religions have influenced the country's arts. Many paintings portray the beauty of nature, often with a few simple brush strokes. Some even include verses of poetry. Haiku (HY•koo) is a well-known type of Japanese poetry that is written according to a very specific formula. Turn to page 199 to learn more about haiku.

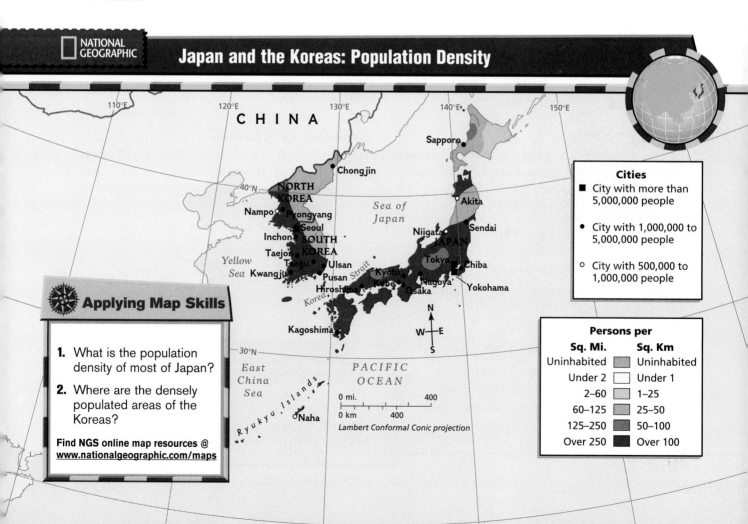

NATIONAL GEOGRAPHIC

Japan and the Koreas: Population Density

Cities

■ City with more than 5,000,000 people

• City with 1,000,000 to 5,000,000 people

○ City with 500,000 to 1,000,000 people

Applying Map Skills

1. What is the population density of most of Japan?

2. Where are the densely populated areas of the Koreas?

Find NGS online map resources @ www.nationalgeographic.com/maps

Persons per	
Sq. Mi.	Sq. Km
Uninhabited	Uninhabited
Under 2	Under 1
2–60	1–25
60–125	25–50
125–250	50–100
Over 250	Over 100

Lambert Conformal Conic projection

0 mi. 400
0 km 400

The Japanese also have a rich heritage of literature and drama. Many scholars believe that the world's first novel came from Japan. The novel is called *The Tale of Genji* and was written by a noblewoman about A.D. 1000. Since the 1600s, Japanese theatergoers have attended the historical plays of the Kabuki theater. In Kabuki plays, actors wearing brilliantly colored costumes perform on colorful stages.

Many of Japan's sports have their origins in the past. A popular sport is sumo, an ancient Japanese form of wrestling. In sumo wrestling, two players try to force their opponent to touch the ground with any part of their body other than the feet. The players may not punch, kick, or pull hair during the competition. Participants in sumo wrestling typically weigh at least 300 pounds (136 kg). Two ancient martial arts—judo and karate—also developed in this area. Today, martial arts are practiced both for self-defense and for exercise.

Modern Pastimes   Along with these traditional arts, the people of Japan enjoy modern pastimes. Many Japanese are enthusiastic about baseball, and the professional leagues in Japan field several teams. Several Japanese players have become stars in the major leagues of the United States. Despite Japan's strong emphasis on education, life is not all work for Japanese young people. They enjoy rock music, modern fashions, television, and movies. Japanese cartoon shows are popular around the world.

✓ Reading Check   **What two religions are found in Japan?**

Assessment

Defining Terms

1. Define tsunami, archipelago, intensive cultivation, clan, shogun, samurai, constitutional monarchy, megalopolis.

Recalling Facts

2. Location Why does Japan experience earthquakes?

3. History Who were the samurai?

4. Culture How have Japan's religions influenced the country's arts?

Critical Thinking

5. Summarizing Information Why do the Japanese not want a large military?

6. Synthesizing Information What three values of the Japanese people have created good workers?

Graphic Organizer

7. Organizing Information Create a diagram like this one. List Japan's economic successes in the large oval and its economic challenges in each of the smaller ovals.

Economic Successes

Challenge   Challenge   Challenge

Applying Social Studies Skills

8. Analyzing Maps Look at the population density map on page 197. Which Japanese cities have more than 5 million people?

Making Connections

CULTURE GOVERNMENT PEOPLE TECHNOLOGY

Haiku

Haiku is a type of poetry that first became popular in Japan during the 1600s. A haiku is a three-line poem, usually about nature and human emotions. The traditional haiku requires 17 syllables—5 in the first line, 7 in the second line, and 5 in the third line. All of the haiku below, written by famous Japanese poets, concern the subject of New Year's Day.*

▲ This Japanese wood-block print shows two girls playing a New Year's game.

For this New Year's Day,
The sight we gaze upon shall be
    Mount Fuji.
        Sôkan

That is good, this too is good,—
New Year's Day
    In my old age.
        Rôyto

New Year's Day;
Whosoever's face we see,
    It is care-free.
        Shigyoku

New Year's Day:
My hovel,
    The same as ever.
        Issa

New Year's Day:
What luck! What luck!
    A pale blue sky!
        Issa

The dawn of New Year's Day;
Yesterday,
    How far off!
        Ichiku

The first dream of the year;
I kept it a secret,
    And smiled to myself.
        Shô-u

*The translations may have affected the number of syllables.
Excerpts from *Haiku, Volume II.* Copyright © 1952 by R.H. Blyth. Reprinted by permission of Hokuseido Press.

► Making the Connection

1. How does the poet Shigyoku think most people react to New Year's Day?

2. From his poem, how can you tell that Ichiku sees the New Year as a new beginning?

3. **Making Comparisons** Compare the two poems by Issa. How does his mood change from one to the other?

Guide to Reading

Main Idea

South Korea and North Korea share the same peninsula and history, but they have very different political and economic systems.

Terms to Know

- dynasty
- monsoon
- famine

Reading Strategy

Create a time line like this one to record four important dates and their events in Korean history.

├───┼───┼───┤

NATIONAL GEOGRAPHIC Exploring Our World

One of Korea's most sacred places is the shrine at Sokkuram. Built in the A.D. 700s, the shrine has 40 statues, including this 11-foot (3.4-m) statue of the Buddha. The original builders created a complex system of stone passages that let air circulate in the shrine. Today air conditioning keeps the statues in good condition.

The **Korean Peninsula** juts out from northern China, between the Sea of Japan and the Yellow Sea. For centuries, this piece of land held a unified country. Today the peninsula is divided into two nations—Communist **North Korea** and non-Communist **South Korea.** For nearly 50 years after World War II, the two governments were bitter enemies. Since the 1990s, they have been drawing closer together.

A Divided Country

The Koreans trace their ancestry to people who settled on the peninsula thousands of years ago. From the 100s B.C. until the early A.D. 300s, neighboring China ruled Korea. When Chinese control ended, separate Korean kingdoms arose throughout the peninsula.

From A.D. 668 to 935, a single kingdom called **Silla** (SHIH•luh) united much of the peninsula. During this time, Korea made many cultural and scientific advances. For example, Silla rulers built one of the world's earliest astronomical observatories in the A.D. 600s.

Other dynasties, or ruling families, followed the Silla. In the 1400s, scholars invented a new way of writing the Korean language. This new system—called *hangul* (HAHN•GOOL)—used less than 30 symbols to write words. This is far fewer than the thousands of characters needed to write Chinese, making the Korean system much easier to learn.

The Korean Peninsula was a stepping stone between Japan and mainland Asia. Trade and ideas went back and forth. In 1910 the Japanese conquered Korea and made it part of their empire. They governed the peninsula until the end of World War II in 1945.

Division and War   After World War II, troops from the Communist Soviet Union took over the northern half of Korea. American troops occupied the southern half. Korea eventually divided along the 38th parallel. A communist state arose in what came to be called North Korea. A noncommunist government controlled South Korea.

In 1950 the armies of North Korea attacked South Korea. They hoped to unite all of Korea under communist rule. United Nations countries, led by the United States, rushed to support South Korea. The **Korean War** finally ended in 1953—without a peace treaty or a victory for either side. By the 1960s, two separate countries had developed on the Korean Peninsula.

After years of bitterness, the two Koreas in the 1990s developed closer relations. In the year 2000, led by a "unification flag," athletes of North and South Korea marched together in the opening ceremony of the Sydney, Australia, Olympics. That same year the leaders of North Korea and South Korea held a meeting for the first time since the division.

✓Reading Check **Why is the Korean Peninsula divided?**

South Korea

South Korea, much of which is covered by mountains, lies at the southern end of the Korean Peninsula. Most South Koreans live in coastal areas where they are affected by monsoons. A monsoon is the

Korean Border

Nearly 50 years after the fighting stopped in Korea, troops still patrol the border between North and South Korea (below left). Seoul, South Korea's modern capital (below right), is less than 25 miles (40 km) from the border.

Location **Where was the line of division drawn between the two countries?**

NATIONAL GEOGRAPHIC On Location

Sonun-sa Temple

Gilded Buddhas such as these at Sonun-sa Temple adorn many Buddhist temples throughout South Korea.

History Which country influenced the religion of South Korea?

seasonal wind that blows over Asia for months at a time. During the summer, a monsoon from the south brings hot, humid weather. In the winter, a monsoon blows in from the north, bringing cold, dry weather.

Manufacturing and trade dominate South Korea's economy. High-technology and service industries have grown tremendously. The country is a leading exporter of ships, cars, textiles, computers, and electronic appliances. In the 1990s, South Korea faced economic difficulties, but it remains one of the economic powers of Asia.

South Korean farmers own their land, although most of their farms are very small. The major crops are rice, barley, onions, potatoes, cabbage, apples, and tangerines. Rice is the country's basic food item. One of the most popular Korean dishes is *kimchi*, a highly spiced blend of vegetables mixed with chili, garlic, and ginger. Many farmers also raise livestock, especially chickens. Some add to their income by fishing.

South Korea's People The people of the two Koreas belong to the same Korean ethnic group. South Korea has nearly 49 million people. More than 80 percent live in cities and towns in the coastal plains. South Korea's capital, **Seoul,** is the largest city.

Most city dwellers live in tall apartment buildings. Many own cars, but they also use buses, subways, and trains to travel to and from work. In rural areas, people live in small, one-story homes made of brick or concrete blocks. A large number of South Koreans have emigrated to the United States since the end of the Korean War.

Buddhism, Confucianism, and Christianity are South Korea's major religions. The Koreans have developed their own culture, but Chinese religion and culture influenced the traditional arts of Korea. In Seoul you will discover ancient palaces modeled after the Imperial Palace in Beijing, China. Historic Buddhist temples—like Sokkuram—dot the hills and valleys of the countryside. Within these temples are beautifully carved figures of the Buddha in stone, iron, and gold. One of the great achievements of early Koreans was pottery. Korean potters still make bowls and dishes that are admired around the world.

Like Japan, Korea has a tradition of martial arts. Have you heard of tae kwon do? This martial art originated in Korea. Those who study it learn mental discipline as well as self-defense.

✓ Reading Check What are the major religions in South Korea?

North Korea

Separated from China by the **Yalu River,** Communist North Korea is slightly larger than South Korea. Like South Korea, monsoons affect the climate, but the central mountains block some of the winter monsoon. The eastern coast generally has warmer winters than the rest of the country.

The North Korean government owns and runs factories and farms. It spends much money on the military. Unlike prosperous South Korea, North Korea is economically poor. Coal and iron ore are plentiful, but industries suffer from old equipment and power shortages.

More than 60 percent of North Korea's rugged landscape is forested. This leaves little land to farm, yet more than 30 percent of North Korea's people are farmers who work on large, government-run farms. These farms do not grow enough food to feed the country. A lack of fertilizer recently produced famines, or severe food shortages.

North Korea's People North Korea has about 22 million people. Nearly 60 percent live in urban areas along the coasts and river valleys. **Pyongyang** is the capital and largest city. Largely rebuilt since the Korean War, Pyongyang has many modern buildings and monuments to Communist leaders. Most of these monuments honor Kim Il Sung, who became North Korea's first ruler in the late 1940s. After Kim's death in 1994, his son Kim Jong Il became the ruler.

The Communist government discourages the practice of religion, although many people still hold to their traditional beliefs. The government also places the needs of the communist system over the needs of individuals and families.

Reading Check **Who controls the economy of North Korea?**

Web Activity Visit the *Our World Today: People, Places, and Issues* Web site at owt.glencoe.com and click on **Chapter 7— Student Web Activities** to learn more about South Korea.

Section 2 Assessment

Defining Terms

1. Define dynasty, monsoon, famine.

Recalling Facts

2. Location Where is the Korean Peninsula?

3. History Why did Korea become divided?

4. Economics What products are made in South Korea?

Critical Thinking

5. Making Comparisons How does the standard of living in South Korea differ from that in North Korea?

6. Summarizing Information What country has had the greatest influence on the culture and arts of South Korea? Explain.

Graphic Organizer

7. Organizing Information Create a diagram like this one. Write facts about each country's economics, government, and natural resources in the outer ovals. Where the ovals overlap, write facts that are common to both countries.

South Korea   North Korea

Applying Social Studies Skills

8. Analyzing Maps Turn to the population density map on page 197. What is the most populous city on the Korean Peninsula? In which country is it located?

Critical Thinking Skill

Making Comparisons

When you make comparisons, you determine similarities and differences among ideas, objects, or events. By comparing maps and graphs, you can learn more about a region.

Learning the Skill

Follow these steps to make comparisons:

- Identify or decide what will be compared.
- Determine a common area or areas in which comparisons can be drawn.
- Look for similarities and differences within these areas.

Practicing the Skill

Use the map and graph below to make comparisons and answer these questions:

1. What is the title of the map? The graph?
2. How are the map and graph related?
3. Which country has the most exports and imports?
4. Does a country's size have any effect on the amount it exports? Explain.
5. What generalizations can you make about this map and graph?

Applying the Skill

Survey your classmates about an issue in the news. Summarize the opinions and write a paragraph comparing the different opinions.

GO TO Practice key skills with **Glencoe Skillbuilder Interactive Workbook, Level 1.**

NATIONAL GEOGRAPHIC

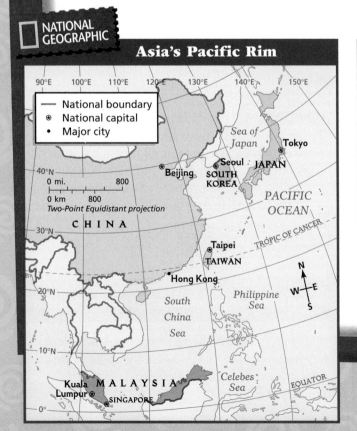

Asia's Pacific Rim

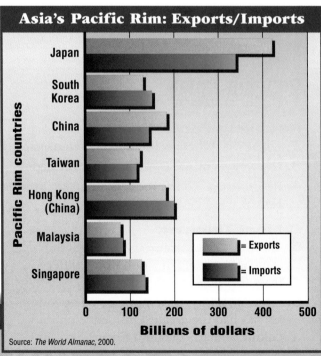

Asia's Pacific Rim: Exports/Imports

Source: *The World Almanac*, 2000.

Reading Review

Section 1   Japan

Terms to Know

tsunami
archipelago
intensive
  cultivation
clan
shogun
samurai
constitutional
  monarchy
megalopolis

Main Idea

Although Japan's people have few mineral resources, they have built a prosperous country.

✓ Location   Japan is an archipelago along the Ring of Fire in the western Pacific Ocean. Volcanoes, earthquakes, and tsunamis may strike these islands.

✓ Economics   Japan is mountainous, but with intensive cultivation its limited farmland is very productive.

✓ Economics   Japan has few resources. Through trade, the use of advanced technology, and highly skilled workers, Japan has built a strong industrial economy.

✓ History   The Japanese people have been strongly influenced by China and by Western countries.

✓ Culture   Most people in Japan live in crowded cities.

✓ Culture   Japanese religion has encouraged a love of nature and simplicity.

Section 2   The Two Koreas

Terms to Know

dynasty
monsoon
famine

Main Idea

South Korea and North Korea share the same peninsula and history, but they have very different political and economic systems.

✓ Culture   The Korean Peninsula lies just south of northern China, and China has had a strong influence on Korean life and culture.

✓ Government   After World War II, the peninsula became divided into two countries, with a communist government in North Korea and a non-communist one in South Korea.

✓ Economics   South Korea has a strong industrial economy.

✓ Culture   Most South Koreans live in cities, enjoying a mix of modern and traditional life.

✓ Government   North Korea has a communist government that does not allow its people many freedoms and spends a great deal of money on the military. It is economically poor.

Because of its beautiful forest-covered mountains, Korea was once known as Land of the Morning Calm. ▶

Using Key Terms

Match the terms in Part A with their definitions in Part B.

A.

1. monsoon
2. tsunami
3. intensive cultivation
4. shogun
5. archipelago
6. dynasty
7. constitutional monarchy
8. clan
9. famine
10. megalopolis

B.

a. group of related families
b. Japanese military leader
c. chain of islands
d. emperor is the official head of state, but elected officials run the government
e. seasonal wind that blows over a continent for months at a time
f. huge wave caused by an undersea earthquake
g. severe food shortage
h. huge supercity
i. ruling family
j. growing crops on every available piece of land

Reviewing the Main Ideas

Section 1 Japan

11. **Human/Environment Interaction** How do Japan's farmers achieve high crop yields?
12. **Economics** What consumer goods and industrial goods are made in Japan?
13. **History** How did Japan change in the late 1800s?
14. **Location** What four cities make up Japan's megalopolis?
15. **Culture** What are three of Japan's traditional arts?

Section 2 The Two Koreas

16. **Location** What large Asian nation lies north of the Korean Peninsula?
17. **History** Why did Korea become divided in 1945?
18. **Location** How do summer and winter monsoons differ in Korea?
19. **Economics** What are the main economic activities in South Korea?
20. **Human/Environment Interaction** Why has North Korea suffered from famine in recent years?

NATIONAL GEOGRAPHIC **Japan and the Koreas**

Place Location Activity

On a separate sheet of paper, match the letters on the map with the numbered places listed below.

1. Mount Fuji
2. Sea of Japan
3. North Korea
4. South Korea
5. Tokyo
6. Honshu
7. Yalu River
8. Seoul
9. Pyongyang
10. Hokkaido

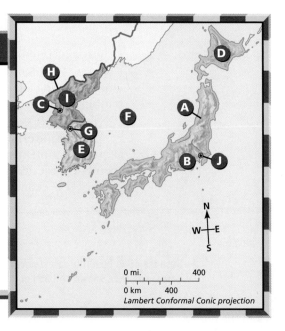

Self-Check Quiz Visit the *Our World Today: People, Places, and Issues* Web site at owt.glencoe.com and click on **Chapter 7—Self-Check Quizzes** to prepare for the Chapter Test.

Critical Thinking

21. **Drawing Conclusions** Why might North Korea find it difficult to change from a communist system to a noncommunist system? Keep in mind the country's location.

22. **Organizing Information** Create a chart like this one. In each column, write two main ideas about Japan, South Korea, and North Korea as they relate to the topics in the first column.

Topic	Japan	South Korea	North Korea
Land			
Economy			
History			
People			

Current Events Journal

23. **Writing a Poem** As you recall, a haiku is a traditional Japanese poem that requires 17 syllables—5 in the first line, 7 in the second line, and 5 in the third line. Write a haiku in which you describe a scene from life in Japan or the Koreas today.

Mental Mapping Activity

24. **Focusing on the Region** Create a map of Japan and the Koreas, and add these labels:

- Honshu
- North Korea
- Korean Peninsula
- Pacific Ocean
- Tokyo
- Sea of Japan

Technology Skills Activity

25. **Using the Internet** Use the Internet to research traditional Japanese culture. You might look at Japanese gardens, Buddhism, literature, or painting. Create a bulletin board display with pictures and write captions that explain what the images show.

Standardized Test Practice

Directions: Read the paragraph below, and then answer the following questions.

In A.D. 1185 Japan's emperor gave political and military power to a shogun, or general. The shogun system proved to be quite strong. Even though the Mongol warrior Kublai Khan tried twice to invade Japan, he did not succeed. On the first invasion in 1274, Japanese warriors and the threat of a storm forced the Mongols to leave. On the second invasion in 1281, 150,000 Mongol warriors came by ship, but a typhoon arose and destroyed the fleet. The Japanese thought of the storm as the kamikaze, or "divine wind." They believed that their islands were indeed sacred.

1. **In what century did shoguns gain political power in Japan?**

   A tenth century

   B eleventh century

   C twelfth century

   D thirteenth century

2. **In what century did the Mongol warrior Kublai Khan try to invade Japan?**

   F tenth century

   G eleventh century

   H twelfth century

   J thirteenth century

Test-Taking Tip: Century names are a common source of error. Remember, in Western societies, a baby's first year begins at birth and ends at age one. Therefore, if you are now 14 years old, you are in your fifteenth year. Using the same type of thinking, what century began in 1201?

Southeast Asia

The World and Its People

NATIONAL GEOGRAPHIC

To learn more about the people and places of Southeast Asia, view *The World and Its People* **Chapter 26** video.

Our World Today online

Chapter Overview Visit the *Our World Today: People, Places, and Issues* Web site at owt.glencoe.com and click on **Chapter 8—Chapter Overviews** to preview information about Southeast Asia.

FOLDABLES™
Study Organizer

Identifying Main Ideas Study Foldable Make this foldable to help you identify key facts about the people and places of Southeast Asia.

Step 1 Fold the paper from the top right corner down so the edges line up. Cut off the leftover piece.

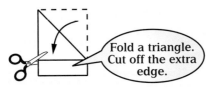

Fold a triangle. Cut off the extra edge.

Step 2 Fold the triangle in half. Unfold.

The folds will form an X dividing four equal sections.

Step 3 Cut up one fold line and stop at the middle. This forms two triangular flaps.

Step 4 Draw an X on one tab and label the other three the following: Flap 1: Mainland Countries; Flap 2: Island Countries; Flap 3: The Philippines.

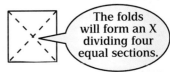

Step 5 Fold the X flap under the other flap and glue together.

This makes a three-sided pyramid.

Reading and Writing As you read, write main ideas inside the foldable under each appropriate pyramid wall.

Why It Matters

A High Price for Prosperity

Some Southeast Asian countries—such as Indonesia, Malaysia, and Singapore—have become very important economic centers, manufacturing goods and exporting natural resources. One possible price paid for this economic prosperity might be in the destruction of the region's most beautiful landscapes.

▲ Outside restaurants are popular in Singapore.

Mainland Southeast Asia

Guide to Reading

Main Idea

The countries of mainland Southeast Asia rely on agriculture as a major source of wealth.

Terms to Know

- monsoon
- precious gems
- deforestation
- socialism
- civil war
- delta

Reading Strategy

Create a chart like this one for each of these countries: Myanmar, Thailand, Laos, Cambodia, and Vietnam. Fill in the right column on each chart with facts about the countries.

Country	
Topic	Key Fact
Land	
Economy	
People	

NATIONAL GEOGRAPHIC *Exploring Our World*

Tattoos and high-heeled shoes in the United States are no match for the fashion statements found in Southeast Asia. This woman belongs to the Padaung ethnic group found in Myanmar and Thailand. A series of brass rings covers her neck. The rings do not stretch the woman's neck but actually push down her collarbone and ribs.

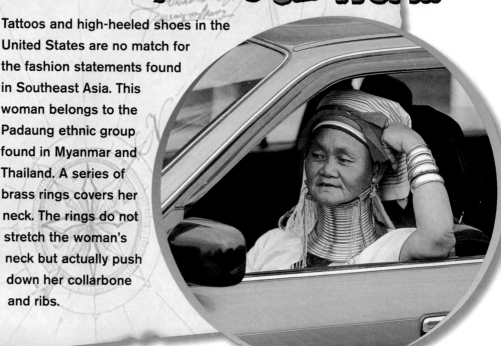

South of China and east of India lies **Southeast Asia.** This region includes thousands of islands and a long arm of land called the **Malay Peninsula.** Several countries lie entirely on the mainland of Southeast Asia. They are Myanmar, Thailand, Laos, Cambodia, and Vietnam.

Myanmar

Myanmar, once called Burma, is about the size of Texas. Rugged, steep mountains sweep through its western and eastern borders. Two wide rivers—the **Irrawaddy** (IHR•ah•WAH•dee) and the **Salween**—flow through vast lowland plains between these mountain ranges.

Monsoons, or seasonal winds that blow over a continent for months at a time, cause the wet summers and dry winters in Myanmar.

About two-thirds of the country's people farm. The main crops are rice, sugarcane, beans, and nuts. Some farmers work their fields with tractors, but most rely on plows drawn by water buffalo.

Factories produce and export such goods as soap, noodles, paper, textiles, and glass bottles. Myanmar also exports precious gems such as rubies, sapphires, and jade. Precious gems are valuable and can be sold for high prices. In addition, the country provides about 75 percent of the world's teakwood. Myanmar's valuable forests are decreasing because of deforestation, or the widespread cutting of trees.

About 75 percent of Myanmar's 47.8 million people live in rural areas. The most densely populated part of the country is the fertile Irrawaddy River valley. Many rural dwellers build their homes on poles above the ground for protection from floods and wild animals.

The capital and largest city, **Yangon** (formerly called Rangoon), is famous both for its modern university and its gold-covered Buddhist temples. Buddhism is the main religion in Myanmar. Most people are of Burman heritage, and Burmese is the main language.

Myanmar was part of British India for many years. It became an independent republic in 1948. Since then, military leaders have turned Myanmar into a socialist country. Socialism is an economic system in which most businesses are owned and run by the government. Some

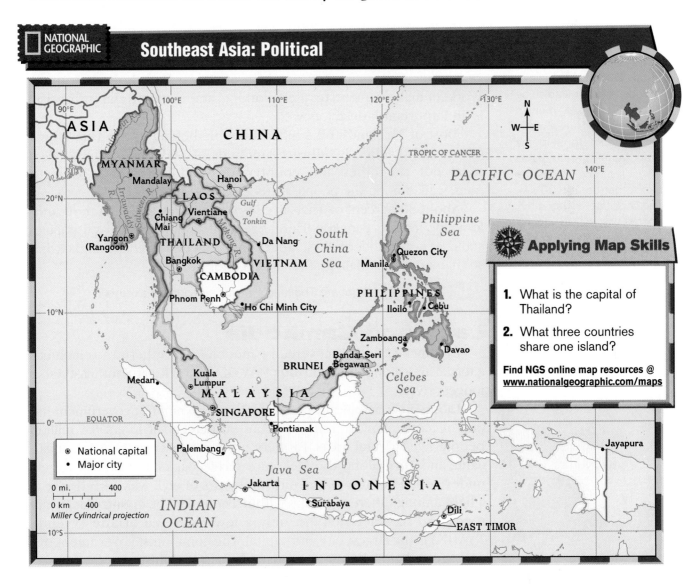

NATIONAL GEOGRAPHIC
Southeast Asia: Political

Applying Map Skills

1. What is the capital of Thailand?

2. What three countries share one island?

Find NGS online map resources @ **www.nationalgeographic.com/maps**

⊛ National capital
• Major city

0 mi.   400
0 km   400
Miller Cylindrical projection

people have struggled to build a democracy in Myanmar. A woman named **Aung San Suu Kyi** (AWNG SAN SOO CHEE) has become a leader in this struggle. In 1991 she was awarded the Nobel Peace Prize for her efforts to bring political changes without violence.

☑ Reading Check Where is Myanmar's most densely populated area?

Thailand

The map on page 211 shows you that **Thailand** looks like a flower on a stem. The "flower" is the northern part, located on the mainland. The "stem" is a narrow strip on the Malay Peninsula. The country's main waterway—the **Chao Phraya** (chow PRY•uh) River—flows through a central plain. Like Myanmar, Thailand has wet summer monsoons and dry winter monsoons.

Once called Siam, *Thailand* means "land of the free." It is the only Southeast Asian country that has never been a European colony. The Thai people trace their independence as a kingdom back to the A.D. 1200s. Thailand still has a king or queen and honors its royal family.

Thailand's main agricultural exports are rubber and teakwood. The government has taken steps to limit deforestation to protect these industries. Thailand is one of the world's leading exporters of tin and gemstones. Most manufacturing is located near **Bangkok,** the capital. Workers make cement, textiles, clothing, and metal products. Tourism is an important industry as well.

Most of Thailand's 62.4 million people belong to the Thai ethnic group and practice Buddhism. Hundreds of Buddhist temples called *wats* dot the cities and countryside. Buddhist monks, or holy men, walk among the people to receive food offerings.

About 80 percent of Thais live in rural villages, although thousands look for jobs in Bangkok. Here, beautiful temples and royal palaces stand next to modern skyscrapers and crowded streets. Bangkok has so many cars that daily traffic jams last for hours.

☑ Reading Check What are Thailand's major agricultural exports?

Laos and Cambodia

Landlocked **Laos** is covered by mountains. Southern Laos includes a fertile area along the **Mekong** (MAY•KAWNG) River, Southeast Asia's longest river.

Laos is an economically poor country. Its communist government has only recently allowed tourism. About 80 percent of Laos's 5.4 million people live in rural areas. Farmers grow rice, sweet potatoes, sugarcane, and corn along the Mekong's fertile banks. Industry is largely undeveloped because of isolation and years of civil war. A civil war is a fight among different groups within a country. The country lacks railroads and has electricity in only a few cities. **Vientiane** (vyehn•TYAHN) is the largest city and capital. The communist government discourages religion, but most Laotians remain Buddhists.

Life as a Monk

After his grandfather died, Nattawud Daoruang became a novice Buddhist monk. "You see," he says, "Thai Buddhists believe they can get to paradise by holding on to a monk's robe. So I became a monk for a month to help my grandfather get to paradise. The novice monks had to get up at 5:00 A.M. and meditate. In the afternoons, we walked around the village with the monks to get food and drink."

Architecture

The temple of Angkor Wat in northwestern Cambodia was built during the 1100s. Dedicated to the Hindu god Vishnu, much of the temple is covered with elaborately carved characters from Hindu legends. The Khmer people designed Angkor Wat to represent the Hindu view of the universe. The moat surrounding the temple stood for the oceans. The tall central tower symbolized Mount Meru, center of the universe and home of the various forms of the Hindu supreme being.

Looking Closer How does the design of Angkor Wat reflect beliefs of the builders?

Cambodia For many years Cambodia was a rich farming country that exported rice and rubber. By the 1980s, its economy was in ruins because of years of civil war and harsh communist rule. Cambodia's few factories produce food items, chemicals, and textiles.

Most of Cambodia's 13.1 million people belong to the **Khmer** (kuh•MEHR) ethnic group. About 80 percent live in rural villages. The rest live in cities such as the capital, **Phnom Penh** (puh•NAWM PEHN). Buddhism is Cambodia's main religion.

In modern times, Cambodia was under French rule, finally becoming independent in 1953. Since the 1960s, it has experienced almost constant warfare among rival political groups. A communist government led by the dictator Pol Pot took control in the mid-1970s, and the people suffered great hardships. Many people from the cities were forced to move to rural areas and work as farmers. More than 1 million Cambodians died. Some fled to other countries. In 1993 Cambodia brought back its king, but rivalry among political groups continues.

✓ Reading Check **Why is Cambodia's economy in ruins?**

Vietnam

Vietnam's long eastern coastline borders the Gulf of Tonkin, the South China Sea, and the Gulf of Thailand. In the north lies the fertile delta of the Red River. A **delta** is an area of land formed by soil deposits at the mouth of a river. In the south you find the wide, swampy delta of the Mekong River. Monsoons bring wet and dry seasons.

Farmers grow large amounts of rice, sugarcane, cassava, sweet potatoes, corn, bananas, and coffee in river deltas. Vietnam's mountain forests provide wood, and the South China Sea yields large catches of fish.

With almost 80 million people, Vietnam has the largest population in mainland Southeast Asia. About 80 percent live in the countryside. The largest urban area is **Ho Chi Minh** (HOH CHEE MIHN) **City,** named for the country's first Communist leader. It used to be called **Saigon** (sy•GAHN). Vietnam's capital, **Hanoi,** is located in the north. Most people are Buddhists and belong to the Vietnamese ethnic group. The rest are Chinese, Cambodians, and other Asian ethnic groups. Vietnamese is the major language, but Chinese, English, and French are also spoken.

The ancestors of Vietnam's people came from China more than 2,000 years ago. From the late 1800s to the mid-1950s, Vietnam was under French rule. Vietnamese Communists drove out the French in 1954. The Communist government controlled northern Vietnam, while an American-supported government ruled the south. In the 1960s, fighting between these two groups led to the **Vietnam War.** During this extended conflict, more than 2.5 million Americans helped fight against the Communists. The United States eventually withdrew its forces in 1973. Within two years, the Communists had captured the south.

In recent years, Vietnam's Communist leaders have opened the country to Western ideas, businesses, and tourists. They also have loosened government controls on the economy. In these two ways, the Communist leaders hope to raise Vietnam's standard of living.

✓ Reading Check What is the largest urban area in Vietnam?

Section 1 Assessment

Defining Terms
1. Define monsoon, precious gems, deforestation, socialism, civil war, delta.

Recalling Facts
2. **Economics** What does Myanmar export?
3. **History** What led to the Vietnam war?
4. **Economics** What has slowed the economies of Laos and Cambodia?

Graphic Organizer
5. **Organizing Information** Create a time line like this one. Then list four events and their dates in Vietnam's history.

Critical Thinking
6. **Summarizing Information** What makes Thailand unique among the countries of Southeast Asia?
7. **Making Predictions** In recent years, the Communist leaders in Vietnam have tried to improve the country's standard of living. How do they hope to do this? Do you think these actions will help? Why or why not?

Applying Social Studies Skills

8. **Analyzing Maps** Look at the political map on page 211. What city is located at 21°N 106°E?

Social Studies Skill

Reading a Contour Map

A trail map would show the paths you could follow if you went hiking in the mountains. How would you know if the trail follows an easy, flat route, though, or if it cuts steeply up a mountain? To find out, you need a **contour map.**

Learning the Skill

Contour maps use lines to outline the shape—or contour—of the landscape. Each contour line connects all points that are at the same elevation. This means that if you walked along one contour line, you would always be at the same height above sea level.

Where the contour lines are far apart, the land rises gradually. Where the lines are close together, the land rises steeply. For example, one contour line may be labeled 1,000 meters (3,281 ft.). Another contour line very close to the first one may be labeled 2,000 meters (6,562 ft.). This means that the land rises 1,000 meters (3,281 ft.) in just a short distance.

To read a contour map, follow these steps:

- Identify the area shown on the map.
- Read the numbers on the contour lines to determine how much the elevation increases or decreases with each line.
- Locate the highest and lowest numbers, which indicate the highest and lowest elevations.
- Notice the amount of space between the lines, which tells you whether the land is steep or flat.

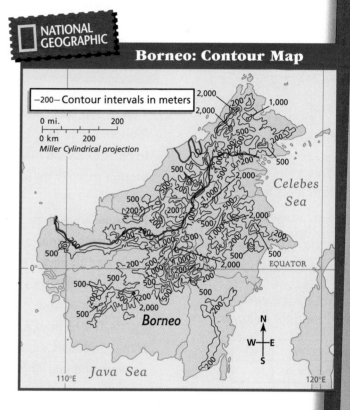

NATIONAL GEOGRAPHIC

Borneo: Contour Map

—200— Contour intervals in meters

0 mi. 200
0 km 200
Miller Cylindrical projection

Celebes Sea

Borneo

EQUATOR

Java Sea

110°E   120°E

N W E S

Practicing the Skill

Study the contour map above, then answer the following questions.

1. What area is shown on the map?
2. What is the lowest elevation on the map?
3. What is the highest elevation on the map?
4. Where is the landscape flattest? How can you tell?
5. How would you describe the physical geography of this island?

Applying the Skill

Turn to page 8 in the Geography Handbook. Use the contour map there to answer the five questions above.

Section 2

Island Southeast Asia

Guide to Reading

Main Idea

The island countries of Southeast Asia have a variety of cultures and economic activities.

Terms to Know

- plate
- strait
- free port
- terraced field

Reading Strategy

Create a chart like this one. As you read, list two facts about each country in the right column.

Country	Facts
Indonesia	
Malaysia	
Singapore	
Brunei	
Philippines	

NATIONAL GEOGRAPHIC Exploring Our World

Villagers in Bali, Indonesia, carry food and gifts to a local Hindu temple. In Bali, it seems as though there is an unending chain of religious festivals. More than 60 festivals a year are dedicated to such events and items as percussion instruments, the birth of a Hindu goddess, woodcarving, and learning.

The island countries of Southeast Asia are Indonesia, Malaysia, Singapore, Brunei (bru•NY), and the Philippines. Indonesia sprawls over an area where two of the earth's tectonic plates meet. Tectonic **plates** are the huge slabs of rock that make up the earth's crust. Indonesia's location on top of these plates causes it to experience earthquakes.

Indonesia

Indonesia is Southeast Asia's largest country. It is an archipelago of more than 13,600 islands. The physical map on page 134 shows you the major islands of Indonesia—**Sumatra, Java,** and **Celebes** (SEH•luh•BEEZ). Indonesia also shares two large islands with other countries. Most of the island of **Borneo** belongs to Indonesia. In addition, Indonesia controls the western half of **New Guinea.** Another country—**Papua New Guinea**—lies on the eastern half.

The volcanoes that formed Indonesia have left a rich covering of ash that makes the soil good for farming. Because Indonesia lies on the Equator, its climate is tropical. Monsoons bring a wet season and a dry season. The tropical climate, combined with fertile soil, has allowed dense rain forests to spread.

Indonesia's Economy Foreign companies build factories on the island of Java because labor is inexpensive. In addition, the island's location makes it easy to ship goods. Agriculture provides work for nearly half of the people of Indonesia. Farmers grow rice, coffee, cassava, tea, coconuts, and rubber trees. Cattle and sheep are also raised.

Indonesia has large reserves of oil and natural gas. Its mines yield tin, silver, nickel, copper, bauxite, and gold. Dense rain forests provide teak and other valuable woods. Some companies that own large tracts of land are cutting down the trees very quickly. The environment suffers from this deforestation. Tree roots help keep the soil in place during heavy rains. When the trees are cut down, the rich soil runs off into the sea.

Indonesia's People Indonesia has about 206 million people—the fourth-largest population in the world. It is also one of the world's most densely populated countries. On Java you will find **Jakarta** (juh•KAHR•tuh), Indonesia's capital and largest city.

Most of Indonesia's people belong to the Malay ethnic group. They are divided into about 300 smaller groups with their own languages. The official language, Bahasa Indonesia, is taught in schools.

Indonesia has more followers of Islam than any other country. Other religions, such as Christianity and Buddhism, are also practiced. On the beautiful island of Bali, Hindu beliefs are held by most of the people.

Thousands of years ago, Malays from mainland Southeast Asia settled the islands that are today Indonesia. Their descendants set up Buddhist and Hindu kingdoms. These kingdoms grew wealthy by controlling the trade that passed through the waterways between the Indian and Pacific Oceans. In the A.D. 1100s, traders from Southwest Asia brought Islam to the region. Four hundred years later, Europeans arrived to acquire the valuable spices grown here. They brought Christianity to the islands. The Dutch eventually controlled most of the islands as a colony. Independence finally came to Indonesia in 1949.

Since the 1960s, unrest and civil war have occurred on several islands. Most recently, the people of East Timor, who are largely Roman Catholic and were once ruled by Portugal, voted for independence. While Indonesia accepted the results of this election, some people in East Timor have not. With so many different ethnic groups, many small political parties arise. As a result, Indonesia's democratic leaders find it difficult to form a government that is strong enough to deal with the challenges facing the country.

✓ **Reading Check** When did Indonesia win its independence?

Exchange of Knowledge

In early times merchants and traders were responsible for the exchange of knowledge as well as goods. Malacca, in Malay, was the richest seaport in the world in the 1500s. Merchants from India, China, and Japan, met Portuguese, British, and Dutch traders. Today, thanks to its geographical location, Singapore has replaced Malacca as the chief center of trade.

East Timor's Challenges

East Timor's road to freedom—finally won on May 20, 2002—was long and difficult. Independence also brings challenges. One of Asia's poorest countries, East Timor suffers from the effects of war and drought. The possibility of wealth from untapped offshore oil and gas fields, however, may brighten East Timor's future.

NATIONAL GEOGRAPHIC On Location

Malaysia

A Malaysian worker taps a rubber tree to get the milky liquid called latex.

Human/Environment Interaction What other Malaysian products are grown for export?

Malaysia

Dense rain forests and rugged mountains make up the landscape of **Malaysia.** The **Strait of Malacca** lies to the west of one part of Malaysia—the Malay Peninsula. A strait is a narrow body of water between two pieces of land. The Strait of Malacca is an important waterway for trade between the Indian Ocean and the Java Sea.

Malaysia is one of the world's leaders in exporting rubber and palm oil. Large amounts of wood are also exported. Malaysia is rich in minerals such as tin, iron ore, copper, and bauxite, as well as in oil and natural gas. Factory workers make high-technology and consumer goods. Malaysia's ports are important centers of trade. **Kuala Lumpur** (KWAH•luh LUM•PUR) is the capital and largest city. The Petronas Towers—the tallest buildings in the world—soar above this city. In contrast, many rural villagers live in thatched-roof homes built on posts a few feet off the ground.

Most of Malaysia's 22.7 million people belong to the Malay ethnic group. Their ancestors came from southern China thousands of years ago. In the 1800s, the British—who then ruled Malaysia—brought in Chinese and South Asian workers to mine tin and to work on rubber plantations. As a result, in marketplaces today you can hear Malay, Chinese, Tamil, and English spoken. Most Malaysians are Muslims, but there are large numbers of Hindus, Buddhists, and Christians.

✓ Reading Check Where are the world's tallest buildings located?

Singapore, Brunei, and the Philippines

Singapore lies off the southern tip of the Malay Peninsula. It is made up of Singapore Island and 58 smaller islands. Singapore is one of the world's smallest countries, yet it has one of the world's most productive economies.

The city of Singapore is the capital and takes up much of Singapore Island. Once covered by rain forests, Singapore Island now has highways, factories, office buildings, and docks.

The city of Singapore has one of the world's busiest harbors. It is a free port, a place where goods can be loaded or unloaded, stored, and shipped again without payment of import taxes. Huge amounts of goods pass through this port. Singapore's many factories make high-tech goods, machinery, chemicals, and paper products. Because of their productive trade economy, the people of Singapore enjoy a high standard of living.

Founded by the British in the early 1800s, Singapore became an independent republic in 1965. Most of the country's 4 million people are Chinese, but Malaysians and Indians make up about 25 percent of the population. These people practice Buddhism, Islam, Christianity, Hinduism, and traditional Chinese religions.

Brunei   On the northern coast of Borneo lies another small nation—**Brunei.** Oil and natural gas exports provide about half of the country's income. Brunei's citizens receive free education and medical care, and low-cost housing, fuel, and food. Today the government is investing in new industries to avoid too much reliance on revenues from fuels. All political and economic decisions are made by Brunei's ruler, or sultan, who governs with a firm hand.

The Philippines   The Philippines is an archipelago of more than 7,000 islands of volcanic mountains and forests. Lava from volcanoes provides fertile soil for agriculture. Farmers have built terraces on the steep slopes of the mountains. Terraced fields are strips of land cut out of a hillside like stair steps. The land can then hold water and be used for farming.

Named after King Philip II of Spain, the Philippines spent more than 300 years as a Spanish colony. As a result of the **Spanish-American War,** the United States controlled the islands from 1898 until World War II. In 1946 the Philippines became an independent democratic republic.

The Philippines is the only Christian country in Southeast Asia. About 90 percent of Filipinos follow the Roman Catholic religion, brought to the islands by Spanish missionaries. The culture today blends Malay, Spanish, and American influences.

Reading Check For whom was the Philippines named and why?

Web Activity Visit the **Our World Today: People, Places, and Issues** Web site at owt.glencoe.com and click on **Chapter 8– Student Web Activities** to learn more about the Philippines.

Assessment

Defining Terms
1. Define plate, strait, free port, terraced field.

Recalling Facts
2. **Location** What five islands are the largest in Indonesia?
3. **Economics** Why do the people of Singapore enjoy a high standard of living?
4. **Culture** What religion do most Filipinos practice?

Critical Thinking
5. **Summarizing Information** How does Brunei's government use its fuel income?
6. **Understanding Culture** Why is it difficult for government officials to rule Indonesia?

Graphic Organizer
7. **Organizing Information** Create a diagram like this one. In the center, list similarities of the countries listed. In the outer ovals, write two ways that the country differs from the others.

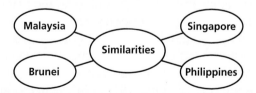

Applying Social Studies Skills

8. **Analyzing Maps** Look at the physical map on page 134. What is the highest point of land in Indonesia?

Shadow Puppets

Late at night, long after dark has fallen on a small stage in Java, a shadow puppet show is about to begin. The glow of a lamp shines behind a wide linen screen. Puppets stand hidden from direct view. The "good" characters are on the right. The "bad" ones are placed on the left. The audience waits anxiously on the other side of the screen. Once the story begins, the performance will continue until dawn.

The Performance

Wayang kulit, the ancient Indonesian shadow puppet theater, dates back at least 1,000 years. Today there are several thousand puppeteers. This makes shadow puppets the strongest theater tradition in Southeast Asia.

Shadow puppets are flat leather puppets, many with movable limbs and mouths, that are operated by sticks. During the show, the puppets cast their shadows onto the screen. The *dalang,* or puppeteer, sits behind the screen and manipulates the figures. He brings each to life in one of the more than 200 traditional puppet stories.

The Stories

Although Islam is now the major religion of Indonesia, much of the traditional shadow puppet theater is based on stories from two ancient Hindu epics from India. At one time the principal purpose of shadow puppetry was to provide moral and religious instruction in Hinduism. Now the stories combine Hindu themes with elements of Buddhism and Islam, as well as Indonesian history and folklore. Often the performance is given in celebration of public or religious holidays or to honor a wedding or birth.

▼ The *dalang* and his orchestra

The Puppeteer

The skill of the *dalang* is critical to the show's success. The *dalang* operates all the puppets, narrates the story, provides sound effects, and directs the gong, drum, and flute orchestra that accompanies the puppet show. The puppeteer changes his voice to create an individual sound for each character. The *dalang* performs without a script or notes, adding jokes and making small changes to suit the crowd and the occasion. Because a shadow puppet show can last as long as nine hours, the *dalang* must have both a tremendous memory and great endurance.

Many *dalang*s carve their own puppets, having learned this art from earlier generations. Each figure must appear in a specific size, body build, and costume. Even the shape of the eyes tells about the figure's character and mood.

→ Making the Connection

1. How do shadow puppets move?
2. What kinds of stories do shadow puppet shows present?
3. **Drawing Conclusions** In what way is the *dalang* a master of many different art forms?

Reading Review

Section 1 | Mainland Southeast Asia

Terms to Know

monsoon
precious gems
deforestation
socialism
civil war
delta

Main Idea

The countries of mainland Southeast Asia rely on agriculture as a major source of income.

✓ **Region** Mainland Southeast Asia includes the countries of Myanmar, Thailand, Laos, Cambodia, and Vietnam.

✓ **Place** These countries have highland areas and lowland river valleys with fertile soil. Monsoons bring heavy rains in the summer.

✓ **History** Thailand is the only country in Southeast Asia free of the influence of colonial rule.

✓ **Economics** Conflict has hurt the economies of Laos, Cambodia, and Vietnam.

Section 2 | Island Southeast Asia

Terms to Know

plate
strait
free port
terraced field

Main Idea

The island countries of Southeast Asia have a variety of cultures and economic activities.

✓ **Region** The island countries of Southeast Asia include Indonesia, Malaysia, Singapore, Brunei, and the Philippines.

✓ **Place** Indonesia—the world's fourth most-populous country—is an archipelago formed by volcanoes.

✓ **Economics** Indonesia has rich supplies of oil, natural gas, and minerals.

✓ **Government** Indonesia's leaders face the challenge of creating a nation out of a land with many different groups and political parties.

✓ **Economics** Malaysia produces palm oil and rubber, among other goods. Its capital, Kuala Lumpur, is a commercial center.

✓ **Economics** The port of Singapore is one of the world's busiest trading centers.

✓ **Culture** The Philippines shows the influence of Malaysian, Spanish, and American culture.

People in Bangkok, Thailand, face traffic snarls and pollution that are among the worst in the world. ▶

Chapter 8

Assessment and Activities

Using Key Terms

Match the terms in Part A with their definitions in Part B.

A.

1. free port
2. delta
3. plates
4. strait
5. deforestation
6. terraced field
7. civil war
8. socialism
9. precious gems
10. monsoon

B.

a. land made from soil deposited at the mouth of a river
b. war fought between groups within a country
c. strip of land cut out of a hillside
d. economic system in which the government owns many businesses
e. stones such as rubies, sapphires, and jade
f. place where shipped goods are not taxed
g. slabs of rock that make up the earth's crust
h. seasonal wind that blows over a continent for months at a time
i. the widespread cutting of trees
j. narrow body of water that runs between two land areas

Reviewing the Main Ideas

Section 1 Mainland Southeast Asia

11. **Economics** Thailand is one of the world's leading exporters of what two items?
12. **History** What countries have poor economies because of recent conflict?
13. **Economics** How is Vietnam trying to improve its economy?

Section 2 Island Southeast Asia

14. **Economics** How do nearly half of the people of Indonesia make a living?
15. **Location** How does location make Indonesia a center of trade?
16. **Government** Why does Indonesia have many political parties?
17. **Place** Why is the strait of Malacca important?
18. **Economics** What economic activity is important in Singapore besides its shipping industry?
19. **Economics** What resources have made Brunei wealthy?
20. **Culture** How does religion show Spanish influence in the Philippines?

NATIONAL GEOGRAPHIC Southeast Asia

Place Location Activity

On a separate sheet of paper, match the letters on the map with the numbered places listed below.

1. Mekong River
2. South China Sea
3. Gulf of Tonkin
4. Hanoi
5. Indonesia
6. Singapore
7. Thailand
8. Vietnam
9. Indian Ocean
10. Philippines

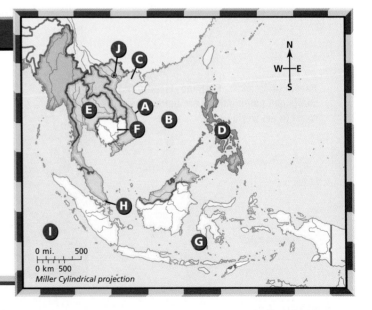

0 mi. 500
0 km 500
Miller Cylindrical projection

Our World Today Online

Self-Check Quiz Visit the *Our World Today: People, Places, and Issues* Web site at owt.glencoe.com and click on **Chapter 8—Self-Check Quizzes** to prepare for the Chapter Test.

Critical Thinking

21. **Predicting Outcomes** Experts believe that Brunei has enough oil reserves to last until 2018. What might happen to the country's economy and standard of living at that time?

22. **Organizing Information** Create a chart like this one. Under each column, write two facts about a country in Southeast Asia. Write about three countries—one from mainland Southeast Asia, Indonesia, and another from island Southeast Asia.

Country	Land	Economy	People

Current Events Journal

23. **Writing a Report** Research the current culture of one of the countries in Southeast Asia. Choose one of the following topics to research: (1) the arts; (2) festivals and holidays; or (3) music and literature. Prepare a written report with illustrations or photos.

Mental Mapping Activity

24. **Focusing on the Region** Draw a map of Southeast Asia, then label the following:

- Borneo
- Irrawaddy River
- Java
- Malay Peninsula
- Philippines
- South China Sea
- Strait of Malacca
- Thailand

Technology Skills Activity

25. **Using the Internet** Use the Internet to learn about the foods in a Southeast Asian country. Find recipes and pictures. Prepare a display that shows a typical meal, or cook the meal yourself and share it with the class.

The Princeton Review

Standardized Test Practice

Directions: Study the graph below, and then answer the questions that follow.

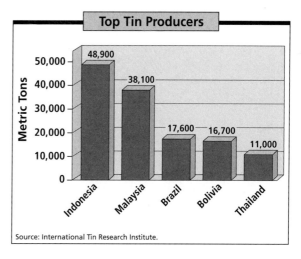

Top Tin Producers

Metric Tons

Indonesia: 48,900
Malaysia: 38,100
Brazil: 17,600
Bolivia: 16,700
Thailand: 11,000

Source: International Tin Research Institute.

1. **About how much tin does Indonesia produce each year?**

   A 48,900 metric tons

   B 48,000,900 metric tons

   C 48.9 million metric tons

   D 48.9 billion metric tons

2. **About how much tin does Brazil produce each year?**

   F 17,600 metric tons

   G 17,600,000 metric tons

   H 17.6 million metric tons

   J 17.6 billion metric tons

Test-Taking Tip: In order to understand any type of graph, look carefully around the graph for keys that show how it is organized. On this bar graph, the numbers along the left side represent the exact number shown. You do not have to multiply by millions or billions to find the number of metric tons.

Unit

4

Woman in Hungary creating folk art

Ancient ruins in Delphi, Greece

Europe

Relatively small as continents go, Europe is rich in history and culture. Like the United States, most nations in Europe are industrialized and have high standards of living. Unlike the United States, however, the people of Europe do not share a common language or government.

▲ The Louvre museum, Paris, France

NGS ONLINE
www.nationalgeographic.com/education

225

REGIONAL ATLAS

Europe

Physical

▲ Mountain peak

0 mi.   500
0 km   500
Lambert Azimuthal Equal-Area projection

ARCTIC CIRCLE

ICELAND

Faroe Is.

Shetland Is.

Orkney Is.

ATLANTIC OCEAN

IRELAND

UNITED KINGDOM

British Isles

MERIDIAN OF GREENWICH (LONDON)

Norwegian Sea

NORWAY

SCANDINAVIA

SWEDEN

FINLAND

Baltic Sea

ESTONIA

LATVIA

LITHUANIA

RUSSIA

RUSSIA

BELARUS

NORTH EUROPEAN PLAIN

POLAND

North Sea

Jutland

DENMARK

NETH.

GERMANY

BELG.

LUX.

CZECH REP.

SLOVAKIA

Carpathian Mountains

UKRAINE

MOLDOVA

Dnieper R.

FRANCE

LIECH.

SWITZ.

ALPS

AUSTRIA

Hungarian Plain

HUNGARY

Bay of Biscay

Mt. Blanc 15,771 ft. (4,807 m)

Matterhorn 14,690 ft. (4,478 m)

SLOV.

CROATIA

ROMANIA

Crimean Peninsula

Black Sea

ANDORRA

Pyrenees

SAN MARINO

BOSN. & HERZG.

SERB. & MONT.

Danube R.

MONACO

Corsica

Adriatic Sea

Apennines

Balkan Peninsula

BULGARIA

PORTUGAL

SPAIN

IBERIAN PENINSULA

Sardinia

ITALY

MACED.

ALBANIA

Aegean Sea

GREECE

Strait of Gibraltar

Sicily

Crete

CYPRUS

MALTA

Mediterranean Sea

RUSSIA

26,247 ft.   8,000 m
19,685 ft.   6,000 m
ALPS
PYRENEES
13,123 ft.   4,000 m
6,562 ft.   2,000 m
LISBON   Sea level   WARSAW

226

UNIT 4

Political

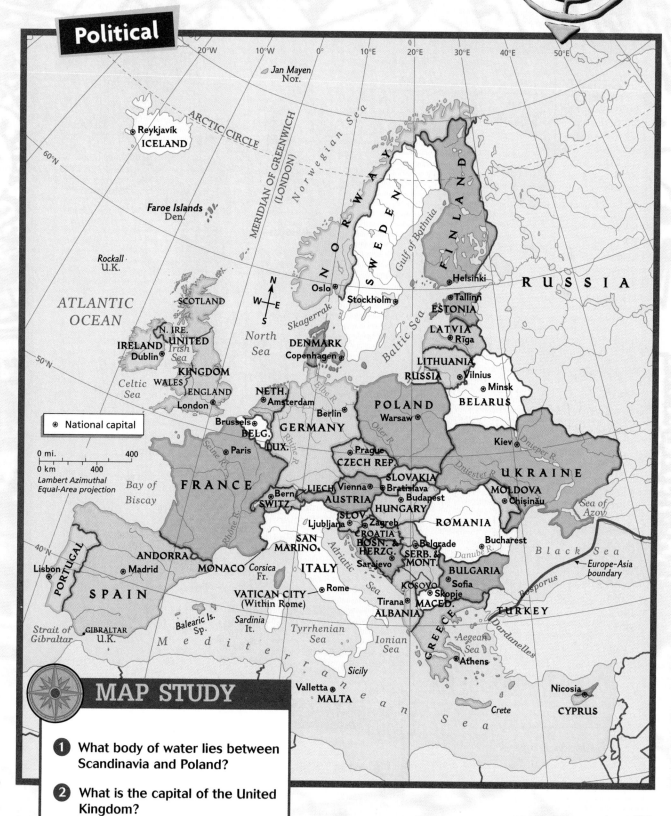

20°W 10°W 0° 10°E 20°E 30°E 40°E 50°E

Jan Mayen
Nor.

ARCTIC CIRCLE

⊛ Reykjavík
ICELAND

Faroe Islands
Den.

Rockall
U.K.

**ATLANTIC
OCEAN**

SCOTLAND

N. IRE.
IRELAND **UNITED**
⊛ Dublin *Irish
Sea*

KINGDOM
WALES

*Celtic
Sea*

ENGLAND
⊛ London

*North
Sea*

NORWAY

SWEDEN

Oslo ⊛
Stockholm ⊛

Gulf of Bothnia

FINLAND

⊛ Helsinki

RUSSIA

⊛ Tallinn
ESTONIA

LATVIA
⊛ Rīga

LITHUANIA
RUSSIA ⊛ Vilnius
⊛ Minsk

BELARUS

Baltic Sea

DENMARK
Copenhagen ⊛

NETH.
⊛ Amsterdam

Berlin ⊛

Brussels ⊛
BELG.
LUX.

GERMANY

Elbe R.
Oder R.
Rhine R.

POLAND
Warsaw ⊛

Dnieper R.
Kiev ⊛

UKRAINE

⊛ National capital

0 mi.   400
0 km   400
Lambert Azimuthal
Equal-Area projection

⊛ Paris
Seine R.

FRANCE

*Bay of
Biscay*

Rhône R.

⊛ Bern
SWITZ.

LIECH.

Prague ⊛
CZECH REP.

Vienna ⊛
AUSTRIA

SLOV.
Ljubljana ⊛

SLOVAKIA
⊛ Bratislava
⊛ Budapest
HUNGARY

Dniester R.

⊛ Chişinău
MOLDOVA

Zagreb ⊛
CROATIA

ROMANIA

*Sea of
Azov*

**SAN
MARINO**

Adriatic

**BOSN. &
HERZG.**
⊛ Sarajevo

**SERB. &
MONT.**

⊛ Belgrade

Bucharest ⊛

Danube R.

Black Sea

PORTUGAL
⊛ Lisbon

Madrid ⊛

ANDORRA

MONACO *Corsica
Fr.*

ITALY

VATICAN CITY
(Within Rome)

⊛ Rome

Sea

KOSOVO
⊛ Skopje
Tirana ⊛ **MACED.**
ALBANIA

BULGARIA
⊛ Sofia

← *Europe-Asia
boundary*

Bosporus

TURKEY

SPAIN

*Strait of
Gibraltar*

GIBRALTAR
U.K.

Balearic Is.
Sp.

Sardinia
It.

*Tyrrhenian
Sea*

Sicily

*Ionian
Sea*

GREECE

*Aegean
Sea*

Dardanelles

⊛ Athens

M e d i t e r r a n e a n

Valletta ⊛
MALTA

Sea

Crete

Nicosia ⊛
CYPRUS

60°N
50°N
40°N

MERIDIAN OF GREENWICH
(LONDON)

Skagerrak

N
W E
S

Norwegian Sea

MAP STUDY

1 What body of water lies between
Scandinavia and Poland?

2 What is the capital of the United
Kingdom?

REGIONAL ATLAS

Europe

Languages

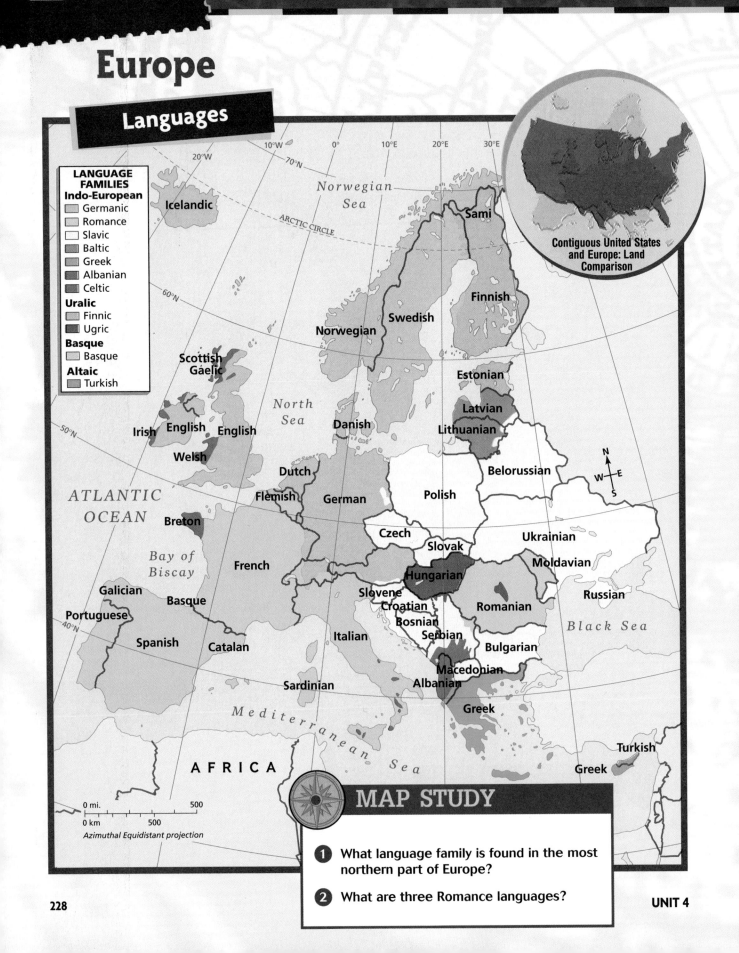

LANGUAGE FAMILIES

Indo-European
- Germanic
- Romance
- Slavic
- Baltic
- Greek
- Albanian
- Celtic

Uralic
- Finnic
- Ugric

Basque
- Basque

Altaic
- Turkish

Contiguous United States and Europe: Land Comparison

Norwegian Sea

ARCTIC CIRCLE

Icelandic

Sami

Finnish

Swedish

Norwegian

Estonian

Latvian

Lithuanian

Scottish Gaelic

North Sea

Danish

Belorussian

Irish

English

English

Polish

Welsh

ATLANTIC OCEAN

Dutch

Flemish

German

Czech

Slovak

Ukrainian

Breton

Hungarian

Moldavian

Bay of Biscay

French

Slovene

Croatian

Romanian

Russian

Galician

Basque

Bosnian

Portuguese

Serbian

Black Sea

Spanish

Catalan

Italian

Bulgarian

Macedonian

Sardinian

Albanian

Greek

Mediterranean Sea

AFRICA

Turkish

Greek

0 mi. 500

0 km 500

Azimuthal Equidistant projection

MAP STUDY

1 What language family is found in the most northern part of Europe?

2 What are three Romance languages?

Fast Facts

COMPARING POPULATION:
United States and Selected Countries of Europe

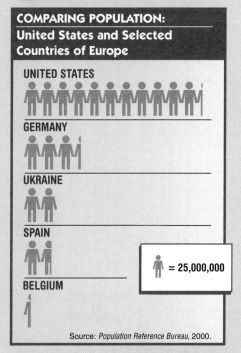

UNITED STATES

GERMANY

UKRAINE

SPAIN

= 25,000,000

BELGIUM

Source: *Population Reference Bureau*, 2000.

RELIGIONS:
Selected Countries of Europe

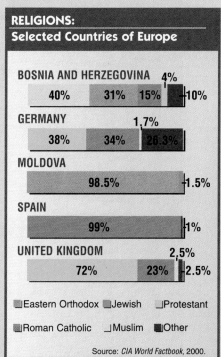

BOSNIA AND HERZEGOVINA   4%

| 40% | 31% | 15% | 10% |

GERMANY   1.7%

| 38% | 34% | 26.3% |

MOLDOVA

| 98.5% | 1.5% |

SPAIN

| 99% | 1% |

UNITED KINGDOM   2.5%

| 72% | 23% | 2.5% |

■Eastern Orthodox ■Jewish □Protestant

■Roman Catholic ⬜Muslim ■Other

Source: *CIA World Factbook*, 2000.

Data Bits

Country	Automobiles per 1,000 people	Telephones per 1,000 people
Austria	469	492
Finland	379	556
France	442	575
Greece	223	516
Ireland	272	411

Population: Urban ▨ vs. Rural ▨

Austria	56%	44%
Finland	63%	37%
France	73%	27%
Greece	65%	35%
Ireland	58%	42%

Source: *World Desk Reference*, 2000.

GRAPHIC STUDY

1. Which two countries have the fewest automobiles per 1,000 people?

2. Roughly, what is the population of Germany? What percentage of the population is Protestant?

REGIONAL ATLAS

Country Profiles

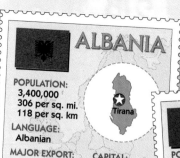

ALBANIA

POPULATION:
3,400,000
306 per sq. mi.
118 per sq. km

LANGUAGE:
Albanian

MAJOR EXPORT:
Asphalt

CAPITAL:
Tirana

MAJOR IMPORT:
Machinery

LANDMASS:
11,100 sq. mi.
28,749 sq. km

ANDORRA

POPULATION:
100,000
578 per sq. mi.
222 per sq. km

LANGUAGES:
Catalan, French,
Spanish

MAJOR EXPORT:
Electricity

CAPITAL:
Andorra la Vella

MAJOR IMPORT:
Manufactured
Goods

LANDMASS:
174 sq. mi.
451 sq. km

AUSTRIA

POPULATION:
8,100,000
250 per sq. mi.
97 per sq. km

LANGUAGE:
German

MAJOR EXPORT:
Machinery

CAPITAL:
Vienna

MAJOR IMPORT:
Petroleum

LANDMASS:
32,378 sq. mi.
83,859 sq. km

BELARUS

POPULATION:
10,100,000
125 per sq. mi.
48 per sq. km

LANGUAGES:
Belarussian, Russian

MAJOR EXPORT:
Machinery

CAPITAL:
Minsk

MAJOR IMPORT:
Fuels

LANDMASS:
80,154 sq. mi.
207,599 sq. km

BELGIUM

POPULATION:
10,300,000
874 per sq. mi.
337 per sq. km

LANGUAGES:
Flemish, French

MAJOR EXPORTS:
Iron and Steel

CAPITAL:
Brussels

MAJOR IMPORT:
Fuels

LANDMASS:
11,787 sq. mi.
30,528 sq. km

BOSNIA and HERZEGOVINA

POPULATION:
3,400,000
172 per sq. mi.
67 per sq. km

LANGUAGE:
Serbo-Croatian

MAJOR EXPORT:
N/A

CAPITAL:
Sarajevo

MAJOR IMPORT:
N/A

LANDMASS:
19,741 sq. mi.
51,129 sq. km

BULGARIA

POPULATION:
8,100,000
189 per sq. mi.
73 per sq. km

LANGUAGE:
Bulgarian

MAJOR EXPORT:
Machinery

CAPITAL:
Sofia

MAJOR IMPORT:
Fuels

LANDMASS:
42,822 sq. mi.
110,909 sq. km

CROATIA

POPULATION:
4,700,000
215 per sq. mi.
83 per sq. km

LANGUAGE:
Serbo-Croatian

MAJOR EXPORT:
Transport
Equipment

CAPITAL:
Zagreb

MAJOR IMPORT:
Machinery

LANDMASS:
21,830 sq. mi.
56,540 sq. km

CYPRUS

POPULATION:
900,000
252 per sq. mi.
97 per sq. km

LANGUAGES:
Greek, Turkish

MAJOR EXPORT:
Citrus Fruits

CAPITAL:
Nicosia

MAJOR IMPORT:
Manufactured
Goods

LANDMASS:
3,571 sq. mi.
9,249 sq. km

CZECH REPUBLIC

POPULATION:
10,300,000
338 per sq. mi.
131 per sq. km

LANGUAGES:
Czech, Slovak

MAJOR EXPORT:
Machinery

CAPITAL:
Prague

MAJOR IMPORT:
Crude Oil

LANDMASS:
30,448 sq. mi.
78,860 sq. km

DENMARK

POPULATION:
5,400,000
325 per sq. mi.
125 per sq. km

LANGUAGE:
Danish

MAJOR EXPORT:
Machinery

CAPITAL:
Copenhagen

MAJOR IMPORT:
Machinery

LANDMASS:
16,637 sq. mi.
43,090 sq. km

ESTONIA

POPULATION:
1,400,000
80 per sq. mi.
31 per sq. km

LANGUAGE:
Estonian

MAJOR EXPORT:
Textiles

CAPITAL:
Tallinn

MAJOR IMPORT:
Machinery

LANDMASS:
17,413 sq. mi.
45,100 sq. km

Countries and flags not drawn to scale

For more information on countries in this region, refer to the Nations of the World Data Bank on pages 690–699.

FINLAND

POPULATION:
5,200,000
40 per sq. mi.
15 per sq. km

LANGUAGES:
Finnish, Swedish

MAJOR EXPORT:
Paper

CAPITAL:
Helsinki

MAJOR IMPORT:
Foods

LANDMASS:
130,560 sq. mi.
338,150 sq. km

FRANCE

POPULATION:
59,200,000
278 per sq. mi.
107 per sq. km

LANGUAGE:
French

MAJOR EXPORT:
Machinery

CAPITAL:
Paris

MAJOR IMPORT:
Crude Oil

LANDMASS:
212,934 sq. mi.
551,499 sq. km

GERMANY

POPULATION:
82,200,000
596 per sq. mi.
230 per sq. km

LANGUAGE:
German

MAJOR EXPORT:
Machinery

CAPITAL:
Berlin

MAJOR IMPORT:
Machinery

LANDMASS:
137,830 sq. mi.
356,980 sq. km

GREECE

POPULATION:
10,900,000
214 per sq. mi.
83 per sq. km

LANGUAGE:
Greek

MAJOR EXPORT:
Foods

CAPITAL:
Athens

MAJOR IMPORT:
Machinery

LANDMASS:
50,950 sq. mi.
131,961 sq. km

HUNGARY

POPULATION:
10,000,000
278 per sq. mi.
108 per sq. km

LANGUAGE:
Hungarian

MAJOR EXPORT:
Machinery

CAPITAL:
Budapest

MAJOR IMPORT:
Crude Oil

LANDMASS:
35,919 sq. mi.
93,030 sq. km

ICELAND

POPULATION:
300,000
8 per sq. mi.
3 per sq. km

LANGUAGE:
Icelandic

MAJOR EXPORT:
Fish

CAPITAL:
Reykjavik

MAJOR IMPORT:
Machinery

LANDMASS:
39,768 sq. mi.
102,999 sq. km

IRELAND

POPULATION:
3,800,000
140 per sq. mi.
54 per sq. km

LANGUAGES:
English, Irish Gaelic

MAJOR EXPORT:
Chemicals

CAPITAL:
Dublin

MAJOR IMPORT:
Foods

LANDMASS:
27,135 sq. mi.
70,280 sq. km

ITALY

POPULATION:
57,800,000
497 per sq. mi.
192 per sq. km

LANGUAGE:
Italian

MAJOR EXPORT:
Metals

CAPITAL:
Rome

MAJOR IMPORT:
Machinery

LANDMASS:
116,320 sq. mi.
301,269 sq. km

LATVIA

POPULATION:
2,400,000
96 per sq. mi.
37 per sq. km

LANGUAGES:
Latvian, Russian

MAJOR EXPORT:
Wood

CAPITAL:
Rīga

MAJOR IMPORT:
Fuels

LANDMASS:
24,942 sq. mi.
64,600 sq. km

LIECHTENSTEIN

POPULATION:
30,000
484 per sq. mi.
186 per sq. km

LANGUAGE:
German

MAJOR EXPORT:
Machinery

CAPITAL:
Vaduz

MAJOR IMPORT:
Machinery

LANDMASS:
62 sq. mi.
161 sq. km

LITHUANIA

POPULATION:
3,700,000
147 per sq. mi.
57 per sq. km

LANGUAGES:
Lithuanian, Polish, Russian

MAJOR EXPORTS:
Foods and Livestock

CAPITAL:
Vilnius

MAJOR IMPORT:
Minerals

LANDMASS:
25,174 sq. mi.
65,201 sq. km

LUXEMBOURG

POPULATION:
400,000
400 per sq. mi.
155 per sq. km

LANGUAGES:
Luxembourgian, German, French

MAJOR EXPORT:
Steel Products

CAPITAL:
Luxembourg

MAJOR IMPORT:
Minerals

LANDMASS:
998 sq. mi.
2,587 sq. km

REGIONAL ATLAS

Country Profiles

MACEDONIA, Former Yugoslav Republic of

POPULATION:
2,000,000
202 per sq. mi.
78 per sq. km

LANGUAGES:
Macedonian, Albanian

MAJOR EXPORT:
Manufactured Goods

MAJOR IMPORT:
Fuels

CAPITAL:
Skopje

LANDMASS:
9,927 sq. mi.
25,711 sq. km

MALTA

POPULATION:
400,000
3,226 per sq. mi.
1,246 per sq. km

LANGUAGES:
Maltese, English

MAJOR EXPORT:
Machinery

MAJOR IMPORT:
Foods

CAPITAL:
Valletta

LANDMASS:
124 sq. mi.
321 sq. km

MOLDOVA

POPULATION:
4,300,000
331 per sq. mi.
128 per sq. km

LANGUAGES:
Moldovan, Russian

MAJOR EXPORT:
Foods

MAJOR IMPORT:
Petroleum

CAPITAL:
Chişinău

LANDMASS:
13,012 sq. mi.
33,701 sq. km

MONACO

POPULATION:
30,000
30,000 per sq. mi.
11,539 per sq. km

LANGUAGE:
French

MAJOR EXPORT:
N/A

MAJOR IMPORT:
N/A

CAPITAL:
Monaco

LANDMASS:
1.0 sq. mi.
2.6 sq. km

NETHERLANDS

POPULATION:
16,000,000
1,015 per sq. mi.
392 per sq. km

LANGUAGE:
Dutch

MAJOR EXPORT:
Manufactured Goods

MAJOR IMPORT:
Raw Materials

CAPITAL:
Amsterdam

LANDMASS:
15,768 sq. mi.
40,839 sq. km

NORWAY

POPULATION:
4,500,000
36 per sq. mi.
14 per sq. km

LANGUAGE:
Norwegian

MAJOR EXPORT:
Petroleum

MAJOR IMPORT:
Machinery

CAPITAL:
Oslo

LANDMASS:
125,050 sq. mi.
323,880 sq. km

POLAND

POPULATION:
38,600,000
309 per sq. mi.
119 per sq. km

LANGUAGE:
Polish

MAJOR EXPORT:
Manufactured Goods

MAJOR IMPORT:
Machinery

CAPITAL:
Warsaw

LANDMASS:
124,807 sq. mi.
323,250 sq. km

PORTUGAL

POPULATION:
10,000,000
282 per sq. mi.
109 per sq. km

LANGUAGE:
Portuguese

MAJOR EXPORT:
Clothing

MAJOR IMPORT:
Machinery

CAPITAL:
Lisbon

LANDMASS:
35,514 sq. mi.
91,981 sq. km

ROMANIA

POPULATION:
22,400,000
243 per sq. mi.
94 per sq. km

LANGUAGES:
Romanian, Hungarian

MAJOR EXPORT:
Textiles

MAJOR IMPORT:
Fuels

CAPITAL:
Bucharest

LANDMASS:
92,042 sq. mi.
238,389 sq. km

SAN MARINO

POPULATION:
30,000
1,304 per sq. mi.
500 per sq. km

LANGUAGE:
Italian

MAJOR EXPORT:
Building Stone

MAJOR IMPORT:
Manufactured Goods

CAPITAL:
San Marino

LANDMASS:
23 sq. mi.
60 sq. km

SERBIA AND MONTENEGRO

POPULATION:
10,700,000
271 per sq. mi.
105 per sq. km

LANGUAGES:
Serbo-Croatian, Albanian

MAJOR EXPORT:
Manufactured Goods

MAJOR IMPORT:
Machinery

CAPITAL:
Belgrade

LANDMASS:
39,448 sq. mi.
102,170 sq. km

SLOVAKIA

POPULATION:
5,400,000
285 per sq. mi.
110 per sq. km

LANGUAGES:
Slovak, Hungarian

MAJOR EXPORT:
Transport Equipment

MAJOR IMPORT:
Machinery

CAPITAL:
Bratislava

LANDMASS:
18,923 sq. mi.
49,011 sq. km

Countries and flags not drawn to scale

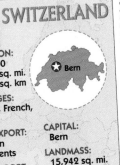
For more information on countries in this region, refer to
the Nations of the World Data Bank on pages 690–699.

SLOVENIA

POPULATION:
2,000,000
256 per sq. mi.
99 per sq. km

LANGUAGES:
Slovene,
Serbo-Croatian

MAJOR EXPORT:
Transport
Equipment

MAJOR IMPORT:
Machinery

CAPITAL:
Ljubljana

LANDMASS:
7,819 sq. mi.
20,251 sq. km

SPAIN

POPULATION:
39,800,000
204 per sq. mi.
79 per sq. km

LANGUAGES:
Spanish, Catalan,
Galician, Basque

MAJOR EXPORTS:
Cars and Trucks

MAJOR IMPORT:
Machinery

CAPITAL:
Madrid

LANDMASS:
195,363 sq. mi.
505,990 sq. km

SWEDEN

POPULATION:
8,900,000
51 per sq. mi.
20 per sq. km

LANGUAGE:
Swedish

MAJOR EXPORT:
Paper Products

MAJOR IMPORT:
Crude Oil

CAPITAL:
Stockholm

LANDMASS:
173,730 sq. mi.
449,961 sq. km

SWITZERLAND

POPULATION:
7,200,000
452 per sq. mi.
174 per sq. km

LANGUAGES:
German, French,
Italian

MAJOR EXPORT:
Precision
Instruments

MAJOR IMPORT:
Machinery

CAPITAL:
Bern

LANDMASS:
15,942 sq. mi.
41,290 sq. km

UKRAINE

POPULATION:
49,100,000
211 per sq. mi.
81 per sq. km

LANGUAGES:
Ukrainian, Russian

MAJOR EXPORT:
Metals

MAJOR IMPORT:
Machinery

CAPITAL:
Kiev

LANDMASS:
233,089 sq. mi.
603,701 sq. km

UNITED KINGDOM

POPULATION:
60,000,000
635 per sq. mi.
245 per sq. km

LANGUAGES:
English, Welsh,
Scottish Gaelic

MAJOR EXPORT:
Manufactured
Goods

MAJOR IMPORT:
Foods

CAPITAL:
London

LANDMASS:
94,548 sq. mi.
244,879 sq. km

VATICAN CITY

POPULATION:
1,000

LANGUAGES:
Italian, Latin

MAJOR EXPORT:
N/A

MAJOR IMPORT:
N/A

CAPITAL:
N/A

LANDMASS:
0.2 sq. mi.
0.4 sq. km

BUILDING CITIZENSHIP

Should students have to wear uniforms? ▼

Participation All citizens are expected to obey the laws of their country. Sometimes, however, the right thing to do is not clear. During World War II, many people in Germany broke the law by helping Jews escape Nazi persecution. During Communist rule, many citizens in Eastern Europe bought and sold goods on the black market.

What do you think would have happened to people helping the Jews if they had been caught?

WRITE ABOUT IT

In the U.S. we work to change laws we believe are unfair. Trying to influence the decisions of our elected leaders is an important part of being an active citizen. Write a letter to your school board explaining why you think students should or should not wear uniforms.

Europe– Early History

The World and Its People NATIONAL GEOGRAPHIC

To learn more about Europe and its people, view **The World and Its People Chapters 10–13** videos.

Our World Today online

Chapter Overview Visit the **Our World Today: People, Places, and Issues** Web site at owt.glencoe.com and click on **Chapter 9–Chapter Overviews** to preview information about the early history of Europe.

Sequencing Events Study Foldable

Make this foldable to help you organize information and sequence events into a flowchart about the early history of Europe.

Step 1 Fold a sheet of paper in half from side to side.

Fold it so the left edge lays about $\frac{1}{2}$ inch from the right edge.

Step 2 Turn the paper and fold it into thirds.

Step 3 Unfold and cut the top layer only along both folds.

This will make three tabs.

Step 4 Turn the paper and label it as shown.

Classical Europe

Medieval Europe

Modern Times

Reading and Writing As you read the chapter, list events that occurred during these three periods in European history under the appropriate tab of your foldable.

▲ Muiderslot Castle in Muiden, Netherlands

Why It Matters

Roots of Western Culture

Our government, economic system, and social systems have grown from the institutions and traditions of Europe. Our laws, family structure, political opinions, and courts have roots in ancient Greek and Roman customs. Medieval Europe saw the growth of cities and the beginning of capitalism. Finally, new Christian churches were founded in Germany, Scotland, and Switzerland.

Classical Europe

Guide to Reading

Main Idea

Ancient Greece and Rome made important contributions to Western culture and civilization.

Terms to Know

- classical
- polis
- democracy
- philosophy
- republic
- consul
- Senate
- emperor

Reading Strategy

Create a chart like the one below. Write one fact that you already know about each category in the "Know" column. After reading the section, write one fact that you have learned about each category in the "Learn" column.

Category	Know	Learn
Greece		
Rome		
Roman law		
Christianity		

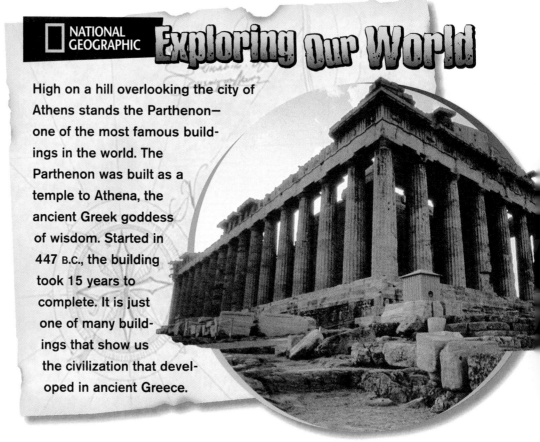

NATIONAL GEOGRAPHIC Exploring Our World

High on a hill overlooking the city of Athens stands the Parthenon—one of the most famous buildings in the world. The Parthenon was built as a temple to Athena, the ancient Greek goddess of wisdom. Started in 447 B.C., the building took 15 years to complete. It is just one of many buildings that show us the civilization that developed in ancient Greece.

When historians talk of Classical Europe they mean the Europe of ancient Greece and Rome. These civilizations flourished from about 800 B.C. to A.D. 400.

The Golden Age of Greece

Much of European, or Western, civilization grew out of the achievements of **Classical Greece.** This Classical period reached its "Golden Age" in the 400s B.C. By that time, the city-state, or polis, had grown from being ruled by a king to the almost direct rule of the people, or democracy. Classical Greece has been called the "cradle of democracy" because we trace the beginnings of our political system to this time.

Athens **Athens** is the best example we have of a democratic city-state. It was the home of the world's first democratic constitution. All free males over the age of 20 had the right to vote and speak freely.

Athenian artists produced famous and influential works of philosophy, literature, and drama. Philosophy is Greek for the "love of wisdom." Two great philosophers, **Socrates** and his student **Plato,** sought to understand and explain human nature. **Aristotle,** a student of Plato's, wrote influential works dealing with social science, politics, literature, ethics, and philosophy. Greek writers and dramatists dealt with these timeless themes in their poems and plays.

Conflict Between the City-States   During this period, city-states such as Athens and **Sparta** wanted to expand their empires. These two rivals often fought against each other. They united temporarily, during the Persian War, when they prevented the Persians from invading Greece. However, from 431–404 B.C. they were again fighting each other. Sparta finally defeated Athens in the Peloponnesian War, which further divided and weakened all of Greece.

Greek Culture Spread   In the 300s B.C., Phillip II of Macedonia and his son, **Alexander the Great** invaded the northern border of Greece. They easily conquered all of Greece. Alexander went on to conquer an empire that included Persia and Egypt and stretched eastward into India. Although his empire barely survived his death, Alexander spread Greek culture everywhere he invaded. Over time Greek customs mixed with Persian and Egyptian culture. The empire's important center was at **Alexandria** in northern Egypt. There a great center of learning, a museum-library, was founded. The empire lasted until it came under attack from the Romans in about 200 B.C.

▲ **Greek theater comedy mask**

✓ Reading Check **Why has Greece been called the "cradle of democracy"?**

The Rise of Rome

According to legend, the city of **Rome** was founded by twin brothers, Romulus and Remus, who had, as infants, been left to die on the banks of the Tiber River. Instead of drowning, they were rescued by a she-wolf,

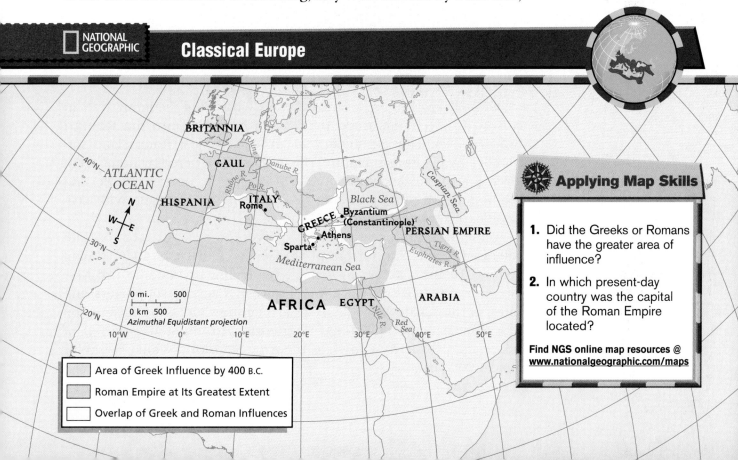

NATIONAL GEOGRAPHIC   **Classical Europe**

BRITANNIA

40°N   ATLANTIC OCEAN

GAUL

Rhine R.

Danube R.

HISPANIA

Rhône R.

Po R.

ITALY

Rome

Caspian Sea

Black Sea

Byzantium (Constantinople)

GREECE

Athens

Sparta

PERSIAN EMPIRE

Tigris R.

Euphrates R.

30°N

Mediterranean Sea

20°N

0 mi.   500
0 km 500
Azimuthal Equidistant projection

AFRICA

EGYPT

Nile R.

Red Sea

ARABIA

10°W   0°   10°E   20°E   30°E   40°E   50°E

Applying Map Skills

1. Did the Greeks or Romans have the greater area of influence?

2. In which present-day country was the capital of the Roman Empire located?

Find NGS online map resources @ www.nationalgeographic.com/maps

⬜ Area of Greek Influence by 400 B.C.

⬜ Roman Empire at Its Greatest Extent

⬜ Overlap of Greek and Roman Influences

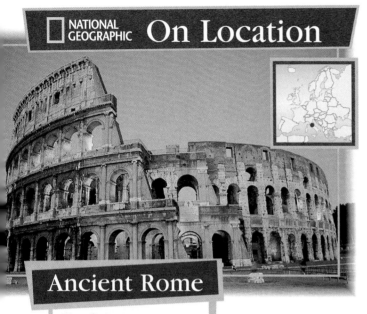

NATIONAL GEOGRAPHIC On Location

Ancient Rome

The Colosseum was built as an arena for gladiator fights.

Place Name some arenas we have for public events today.

who then raised them as her cubs. When grown, the twins built the city on seven hills in central Italy. They fought over who should be king, and Romulus emerged the victor.

Historical Rome What we know for fact is that Rome was settled sometime around 1000 B.C. By about 700 B.C. it had evolved into a major city-state that began to dominate much of the **Italian Peninsula.** Rome lacked natural resources and had only a few good harbors, but it had much more fertile land than Greece. The Romans were therefore a mostly agricultural society and were less likely to live in cities. Italy was more easily invaded than mountainous Greece, so the Romans developed a strong army. Roman art, religion, mythology, and the Latin alphabet were borrowed from the Greeks.

The Roman Republic Rome started as a monarchy but changed to a republic. In a republic, people choose their leaders. Rome was led by two consuls, individuals elected by the people of Rome to represent them. They reported to the Senate. Members of the Senate were landowners who served for life. All of this and more was guaranteed by the system of **Roman law.** The foundation of Roman law was the **Twelve Tables.** The "tables" were actually bronze tablets on which laws regarding wills, courts, and property were recorded. The laws applied to all citizens of Rome, both common and noble. Along with Greek democracy, republican government and Roman law were important contributions to Western civilization and the Modern Age.

✓Reading Check What three important developments from Greece and Rome influenced Western civilization?

From Republic to Empire

From 264 to 146 B.C. a series of wars—including the Punic Wars in North Africa—transformed the Roman Republic into the **Roman Empire.** Eventually the Mediterranean Sea became a "Roman lake" surrounded by the Roman Empire. The peoples conquered by Rome were given Roman citizenship and equality under the Roman law. Beyond the boundaries of its vast empire, Rome opened up trade with civilizations as far away as India and China.

Under the empire, Senators lost power to emperors, or absolute rulers, of Rome. Supporters of the Senate killed the great Roman general **Julius Caesar** in 44 B.C. for trying to become the first emperor. This led to a civil war between Caesar's supporters and those of the Senate. In 31 B.C., Caesar's nephew Octavius became the first Roman emperor, **Caesar Augustus.** He initiated a period of peace and prosperity known as the *Pax Romana,* which lasted for almost 200 years.

Christianity and Rome **Jesus Christ** was born in Palestine under Caesar Augustus and carried out his teaching during the early *Pax Romana.* Two disciples, **Peter** and **Paul,** established the new Christian Church in Rome. Even though the early Christians were cruelly persecuted, Christianity spread over the Roman world. In the A.D. 300s, under the emperors **Constantine I** and **Theodosius I,** Christianity became the official religion of the Roman Empire.

The Decline of the Empire After the period of the *Pax Romana* the Roman Empire began to decline. In A.D. 330, Constantine I moved the capital from Rome in Italy eastward to the newly built city of **Constantinople,** near the Black Sea. Constantine tried to save the empire by reforming the government, but it was too late. Plagues that came in from Asia over trade routes killed numerous people. People fled the cities and escaped to villas, or country estates.

Finally, in the A.D. 400s, the northern frontier defenses crumbled. Rome was left open to invasion by various groups of Germanic peoples. (One group, the Huns, introduced a new technology unknown to the Roman cavalry—the stirrup. The stirrup allowed the Huns to use their weapons while riding horses.) The Germans came to rule over Rome and much of Italy and Europe. The **Eastern Roman Empire,** or Byzantine Empire, did not fall to the Germans but continued on for another 1,000 years until its conquest by the Ottoman Turks in A.D. 1453.

▲ Roman soldier's breastplate

✓ Reading Check How did technology play a role in the decline of the Roman Empire?

Section 1 Assessment

Defining Terms

1. Define classical, polis, democracy, philosophy, republic, consul, Senate, emperor.

Recalling Facts

2. Government In its democratic constitution, what two rights did Athens give all free males over the age of 20?

3. Culture Name four influences that Greece had on Roman culture.

Critical Thinking

4. Analyzing Information Why do you suppose some of Rome's citizens wanted absolute rulers instead of elected senators?

5. Making Connections What is one freedom that American democracy has today that was clearly not recognized in the Roman Empire?

Graphic Organizer

6. Creating Time Lines Create a time line like this. Place the letter of the event next to its date.

A. Greek empire is attacked by Rome

B. Julius Caesar is killed

C. Germans invade Rome

D. Rome is settled

E. Octavius becomes the first Roman emperor

```
                  200 B.C.        31 B.C.
1000 B.C.                  44 B.C.          A.D. 400
```

Applying Social Studies Skills

7. Making Inferences Why do you think the story of Romulus and Remus was created?

Social Studies Skill

Using B.C. and A.D.

Cultures throughout the world have based their dating systems on significant events in their history. For example, Islamic countries use a dating system that begins with Muhammad's flight from Makkah to Madinah. For Western cultures, the most important event was the birth of Jesus.

Learning the Skill

About 515, a Christian monk developed a system that begins dating from *anno Domini,* Latin for "the year of the Lord." Although some historians believe that the monk made a small mistake in his figuring of the exact year of Christ's birth, his system of dating has lasted. Events before the birth of Christ, or "B.C.," are figured by counting backward from A.D. 1. There was no year "0." The year before A.D. 1 is 1 B.C. Notice that "A.D." is written before the date, while "B.C." is written following the date.

Practicing the Skill

Study the time line of Classical Europe to answer the following questions.

1. How old was Plato when he became a student of Socrates?
2. For how long did Alexander the Great rule?
3. How old was Julius Caesar when he was assassinated?
4. Who was emperor nearly 500 years after the rule of Alexander the Great?

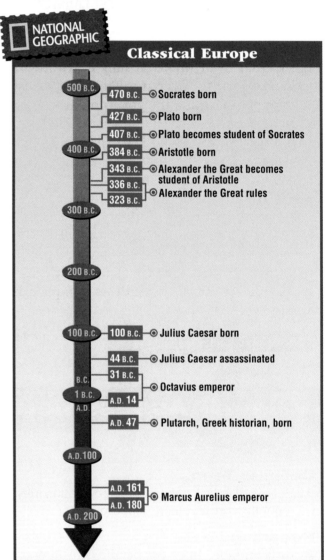

NATIONAL GEOGRAPHIC

Classical Europe

- 500 B.C.
- 470 B.C. — Socrates born
- 427 B.C. — Plato born
- 407 B.C. — Plato becomes student of Socrates
- 400 B.C.
- 384 B.C. — Aristotle born
- 343 B.C. — Alexander the Great becomes student of Aristotle
- 336 B.C. — Alexander the Great rules
- 323 B.C.
- 300 B.C.
- 200 B.C.
- 100 B.C. — Julius Caesar born
- 44 B.C. — Julius Caesar assassinated
- 31 B.C. — Octavius emperor
- B.C. / 1 B.C. / A.D.
- A.D. 14
- A.D. 47 — Plutarch, Greek historian, born
- A.D. 100
- A.D. 161 — Marcus Aurelius emperor
- A.D. 180
- A.D. 200

Applying the Skill

Create a time line using the terms B.M.B. (before my birth) and A.M.B. (after my birth). Fill in the time line with key events that happened before and after you were born. Illustrate the time line with drawings or cutouts from magazines.

GO TO — Practice key skills with **Glencoe Skillbuilder Interactive Workbook, Level 1.**

Guide to Reading

Main Idea

The Middle Ages saw the spread of Christianity, the growth of cities, and the growing powers of kings.

Key Terms

- bishop
- pope
- missionary
- monastery
- convent
- common law
- feudalism
- vassal
- manor
- tenant
- serf
- guild
- apprentice
- charter

Reading Strategy

Create a chart like the one below. Fill in the chief duty or role of each of these members of society.

Lord	
Vassal	
Guild member	
Apprentice	
Serf	

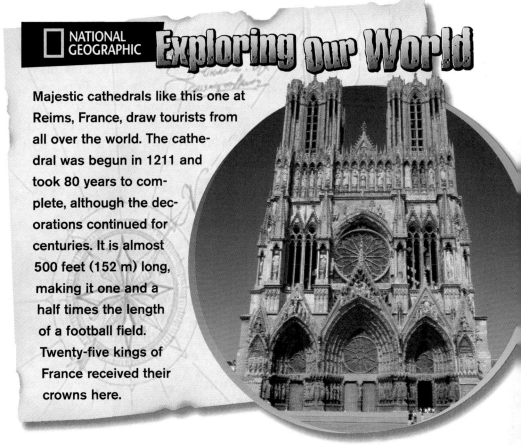

NATIONAL GEOGRAPHIC Exploring Our World

Majestic cathedrals like this one at Reims, France, draw tourists from all over the world. The cathedral was begun in 1211 and took 80 years to complete, although the decorations continued for centuries. It is almost 500 feet (152 m) long, making it one and a half times the length of a football field. Twenty-five kings of France received their crowns here.

With the disappearance of the Roman Empire, a new age, which we call the **Middle Ages,** began. *Medieval* is derived from a Latin word for "Middle Ages." It is a fitting name for the 1,000-year period that lies between Classical and Modern times. Medieval Europe combined characteristics of the Roman Empire with the newer ways of Christianity and the peoples of Europe.

The Rise of Christianity

It was during the Middle Ages that Christianity, in the form of the Roman Catholic Church, became a political power in western Europe. A leader called a **bishop** headed each major Christian community. By the A.D. 500s, the bishops of Rome, now known as **popes,** became the leaders of the Catholic Church. The influence of the Church was so strong at this time that the popes also became important political figures.

In eastern Europe, the Byzantine Empire, started by Constantine I, continued. There, Christianity was known as **Eastern Orthodoxy.** It was not under the leadership of the popes in Rome, but rather the emperors in Constantinople.

241

▲ Jeweled cross and Bible

Spreading the Faith   By A.D. 500, the first Christian Bible was completed. The early popes sent missionaries, teachers of Christianity, to every part of Europe. Many of the missionaries were monks and nuns. Monks were men who devoted their lives to prayer, study, and good works and lived in monasteries. Women who did similarly were nuns and lived in convents. Monks and nuns not only helped the poor and needy, but they were teachers as well. Through its schools, the Christian Church greatly advanced learning in Europe. In the 1100s, the Church also founded the first universities, institutions of higher learning, at Bologna in Italy and Oxford in England.

Beginning in the A.D. 1000s, the Church sponsored a series of holy wars called **Crusades.** It sent armies to capture Jerusalem in Palestine from the Islamic caliphs, or rulers. The Church also crusaded in northern and eastern Europe for about 400 years to spread Christianity.

✓ Reading Check   How did missionaries help spread Christianity?

The Holy Roman Empire

The Germans combined their common law, the unwritten laws that come from local customs, with Roman law and founded kingdoms all over Europe—from Spain to England to Germany and Italy. Many of these kingdoms soon became Christian. The early kings, like the German tribal chiefs before them, were elected by all nobles and knights. Over time, however, the kings became more independent and powerful. The crown was passed down to the next generation, usually the king's first-born son.

Charlemagne   One of the most important German kingdoms was that of the **Franks.** By the A.D. 700s it controlled much of what would become France and Germany. In fact, the name "France" comes from the word *Franks.* In 771 **Charlemagne** was elected king of the Franks. Through war he added more of Germany and parts of Spain and Italy, including Rome, to the kingdom of the Franks.

On Christmas Day in the year 800, Charlemagne knelt before the pope in the Church of St. Peter in Rome and was proclaimed the protector of the Christian Church in the West. He also was crowned the head of the Roman Empire in the West. That empire came to be known as the **Holy Roman Empire.**

After Charlemagne's death in 814, his empire was inherited by his son and grandsons and broke up into several kingdoms. These kingdoms were the foundations for modern Germany, Italy, France, and Spain. At about the same time, several Germanic groups like the Angles, Saxons, Jutes, and Danes helped to found the first English kingdom. "England" gets its name from "Angle land."

✓ Reading Check   What was Charlemagne's role in the spread of Christianity?

Medieval Society

Most people during the Middle Ages were not kings, warriors, traders, or explorers—they were farmers. The medieval political and social system—called feudalism—was based on agriculture. Under feudalism, lords would give land to a noble or knight to work, govern, and defend. In return, those who received the land swore loyalty to the lords and became their **vassals.** Those knights who did not receive land usually served in the armies of the lords, often hoping to be rewarded with land for their service.

The Manor The feudal estate and basic economic unit was called the **manor.** At its heart was usually a manor house or a castle. Most of the population of the manor was made up of common people who farmed and performed other tasks as well. There were two types of farmers. Those who paid rent for their land and then worked the land freely as they pleased were known as **tenants.** The other much larger group was the **serfs.**

Serfs were not as free and were usually poorer than tenant farmers. In return for the use of land, seed, tools, and protection, serfs had to work as ordered by the lords of the manors, whether in the fields or elsewhere. Often the serfs worked on roads, walls, fortifications, and at other hard jobs. Some might become millers making flour out of grain, coopers making barrels and buckets, or blacksmiths making tools, weapons, or horseshoes out of iron and other metals. In times of trouble, male serfs also became foot soldiers who served under the direction of the cavalry of knights.

These were often quite violent times, and the common people rarely strayed too far from the safety of the manor. On occasion the manors might be visited by wanderers with special skills. For example, tinkers made a living by moving from estate to estate, patching pots or fixing other metal objects. Minstrels and other troubadours entertained by playing music, juggling, or acting as comedians or fools.

✓**Reading Check** What did a vassal receive for his service to a lord?

Web Activity Visit the *Our World Today: People, Places, and Issues* Web site at owt.glencoe.com and click on **Chapter 9— Student Web Activities** to learn more about the Crusades.

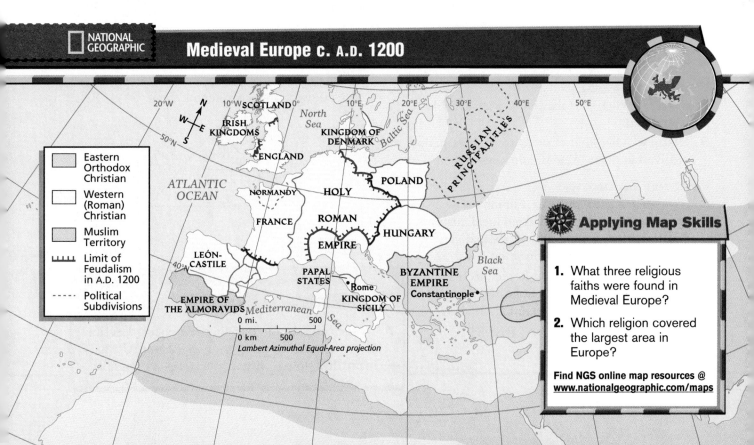

NATIONAL GEOGRAPHIC

Medieval Europe c. A.D. 1200

Legend:
- Eastern Orthodox Christian
- Western (Roman) Christian
- Muslim Territory
- Limit of Feudalism in A.D. 1200
- Political Subdivisions

0 mi. 500
0 km 500
Lambert Azimuthal Equal-Area projection

Applying Map Skills

1. What three religious faiths were found in Medieval Europe?

2. Which religion covered the largest area in Europe?

Find NGS online map resources @ www.nationalgeographic.com/maps

The Growth of Cities

Towns in the Middle Ages were fairly independent and free of the feudal lords' control. Towns served as centers of trade and manufacturing. Their importance increased during the Crusades because the Christian armies needed supplies. By the twelfth century, towns hosted great trade fairs, where merchants from far and wide came together to do business.

Manufacturing came under the control of workers' organizations known as guilds. Different guilds controlled industries such as brewing, cloth making, boat building, and many others. Young workers, called apprentices, spent years learning a trade so that they could join a guild. With experience, the apprentices became journeymen and eventually master craftsmen.

Over time, some towns grew into cities and became political and religious centers as well. The new, more powerful kings and churchmen understood the importance of cities. They built great cathedrals and granted the residents privileges and freedoms in written documents called charters. By doing this, the kings won the support of the townspeople. This support was useful in times of war and for protection against powerful nobles. The kings also raised money by collecting taxes from the towns in return for granting charters.

▲ Stained glass showing a craftsman at work

✓ Reading Check Why did kings want the support of large cities?

Section 2 Assessment

Defining Terms

1. Define bishop, pope, missionary, monastery, convent, common law, feudalism, vassal, manor, tenant, serf, guild, apprentice, charter.

Recalling Facts

2. History When was the first Christian Bible completed?

3. History What kind of work were most people involved in during the Middle Ages?

Critical Thinking

4. Evaluating Information Common laws were unwritten laws that came from local customs. What are the possible difficulties that can arise from having such unwritten laws?

5. Analyzing Information Rome has been known as the Roman Republic, the Roman Empire, and the Holy Roman Empire. How do these labels signify major political changes?

Graphic Organizer

6. Organizing Information Create a triangle like the one below. On the lines, list serfs, vassals, and tenants in the order they would be ranked under a lord in the feudal system. Consider how much a person owned when you rank them.

Lords
_____
_____
_____

Applying Social Studies Skills

7. Describing In three to four sentences, describe life on the manor for a common person. Use as many descriptive words (adjectives) as possible.

The Beginning of Modern Times

Guide to Reading

Main Idea

The study of science, art, and education was renewed in the period following the Middle Ages.

Key Terms

- indulgences
- Protestant
- reform
- Columbian Exchange
- revolution
- divine right of kings
- Parliament
- constitution

Reading Strategy

Create a time line like the one below. As you read, add the following terms to the line in the correct order.

Protestant Reformation
Age of Exploration
American Revolution
Renaissance
French Revolution

NATIONAL GEOGRAPHIC Exploring Our World

From the 1300s to the 1600s, important cultural achievements in the arts and learning spread throughout Europe. Merchant families used their wealth to help artists and scholars explore new ways of thinking. The result was the Renaissance—a French word meaning "rebirth." Michelangelo's statue of David is one of the many masterpieces from this period.

The growth of cities and trade and the gradual breakup of feudalism led to the end of the Middle Ages. Because of the important developments that shaped life as we know it today in the West, historians trace the beginnings of the "Modern Age" from these times. This pre-modern period begins with the **Renaissance** in about A.D. 1350 and ends with the **French Revolution** and Napoleonic Wars in 1815.

The Renaissance

Around 1350 interest in education, art, and science peaked in several parts of Europe, especially in the cities and towns. To many people, this was the beginning of a new golden age like that of ancient Greece and Rome. The Renaissance began in the cities of northern Italy and spread to other cities of Europe.

▲ The Sistine Chapel

Renaissance Artists The artists, scientists, and philosophers of Greece and Rome had taken pride in their ability to think and to appreciate the beauty in human beings and nature. Curiosity and enthusiasm for life were at the heart of the Renaissance. No one better expressed this revived spirit than two Italians, **Leonardo da Vinci** and **Michelangelo Buonarotti.** (See page 250 for some of the achievements of Leonardo da Vinci.)

The painter and sculptor Michelangelo expressed human emotions such as anger, sorrow, and strength in his paintings and sculptures. His most famous work is the mural on the ceiling of the Sistine Chapel in the Vatican Palace in Rome. It covers over 6,000 square feet (1,828 sq. m) and is made up of 145 separate paintings. It took nearly five years to complete.

The Printing Press Writers were also inspired by the ideas of the Renaissance. Until this time, most literature was written in Medieval Latin. To reach a wider audience, writers began to use the language they spoke every day instead of Latin or French, the language of the educated. So, **Geoffrey Chaucer** wrote *The Canterbury Tales* and **William Shakespeare** wrote plays such as *Hamlet* and *Romeo and Juliet* in English. **Miguel de Cervantes** wrote his novel *Don Quixote* in Spanish.

These works were reproduced in many copies for the first time by the printing press, which was invented by **Johannes Gutenberg** around 1450. While the Chinese had developed a printing process, Gutenberg developed the idea of movable type. He has been called the "father of printing in the Western world." The printing press made books more numerous and less expensive, thereby encouraging more people to learn to read and write.

Nationhood During the Renaissance, western European rulers became more powerful. They used their power to unite their countries, creating nations based on a common language and culture. England was strengthened by the first Tudor king, Henry VII and his famous granddaughter Elizabeth I. King Ferdinand and Queen Isabella united Spain by driving out the last of the Muslims and Jews. By the 1450s, the kings of France finally liberated their country from the English.

√ Reading Check How did the printing press make it easier for people to learn to read and write?

The Protestant Reformation

Many of the new ideas given birth in the Renaissance led to questions about religion. Some people believed that Church leaders were more interested in wealth than religion. Others disagreed with corrupt practices of the Church. One of these practices was the selling of documents called indulgences, which freed their owners of punishment for sins they had committed. Because these Christians "protested" Catholic teachings, they came to be called Protestants. The movement to reform, or change, the Catholic Church was called the **Protestant Reformation.**

One of first Protestant leaders to challenge the Catholic Church was **Martin Luther,** a German monk and scholar. In 1520, the pope banished Luther from the Catholic Church for his criticism. Luther organized his own new Christian church, which taught in German, not Latin, from a Bible that Luther himself had translated into German. This split between the Catholics and Protestants led to many long years of religious wars in Europe.

Another early Protestant leader was **John Calvin.** His followers in France were called Huguenots, and in England they were called Puritans. Many came to the Protestant cause seeking not only greater religious freedom, but also political, economic, and intellectual freedom. The Puritans eventually sought freedom in the Americas to practice their own religion.

✓ Reading Check What was the Protestant Reformation?

The Age of Exploration

By the mid-1400s, Europe began to reach out beyond its boundaries in a great age of discovery and exploration. The Portuguese began to sail southward in the Atlantic, down the West African coast, seeking a way to the profitable spice trade in Asia. In 1488, **Bartholomeu Dias** reached the Cape of Good Hope at the southern tip of Africa. Ten years later, **Vasco da Gama** sailed around it to India.

While the Portuguese were searching for a way around Africa, King Ferdinand and Queen Isabella of Spain were trying to find another way to Asia. In 1492, they sent an Italian navigator, **Christopher Columbus,** with three small ships—the *Niña,* the *Pinta,* and the *Santa María*—westward across the Atlantic. Although he never realized it himself, Columbus had landed in a part of the world unknown to Europeans at that time. He called its people "Indians" because he believed he had sailed to the East Indies in Asia.

▲ The Pope acts as money changer while selling indulgences.

The Dutch, English, and French soon joined the Spanish and Portuguese in exploring and settling and trading with the Americas, Asia, and Africa. Eventually—in addition to trade goods—people, diseases, and ideas were distributed around the world in a process called the Columbian Exchange. Europeans unknowingly brought to the Americas diseases such as measles and smallpox, which infected and killed millions of Native Americans. These natives had been used as laborers on plantations and in mines. In their place, traders eventually transported more than 20 million Africans to the Americas as enslaved persons, until the slave trade was outlawed in the early 1800s.

✓ Reading Check Which European nation first began exploration around the coast of Africa?

Revolution

A revolution is a great and often violent change. In the Americas, the colonies won freedom from their European mother countries. In Europe, the people fought for freedom from their kings, queens, and nobles.

The Rule of the People The eighteenth century ended with great changes to Europe and many of its American colonies. The belief in the **divine right of kings**—that European kings and queens ruled by the will of God—was fading. In learning about the examples from ancient Greece and Rome, people came to feel that they should play a greater, more direct role in government. Philosophers such as **John Locke** and **Jean Jacques Rousseau** looked at the nature of man and government. They believed that government should serve them and protect them and their freedom. However, this also meant that they had to take more responsibility for themselves and their own actions.

British Democracy In some cases, revolutionary changes came more peacefully than in others. Over many centuries, Britain had slowly developed a system of shared power and responsibility. The king ruled with the **Parliament,** a popular representative body that gradually took power in the name of the people. Eventually, British kings and queens were forced to accept a **constitution,** a plan for government that shared power but gave most of it to the Parliament.

Literature

THE SCARLET PIMPERNEL
by Baroness Orczy

During and after the French Revolution, many nobles were executed by the lower classes that had rebelled against them. The number of these executions shocked the people of Europe. *The Scarlet Pimpernel* is a novel about an English nobleman who helps aristocrats escape from France. The hero—as true heroes will—minimizes his heroic actions.

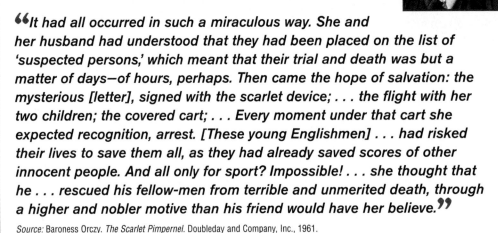

❝*It had all occurred in such a miraculous way. She and her husband had understood that they had been placed on the list of 'suspected persons,' which meant that their trial and death was but a matter of days—of hours, perhaps. Then came the hope of salvation: the mysterious [letter], signed with the scarlet device; . . . the flight with her two children; the covered cart; . . . Every moment under that cart she expected recognition, arrest. [These young Englishmen] . . . had risked their lives to save them all, as they had already saved scores of other innocent people. And all only for sport? Impossible! . . . she thought that he . . . rescued his fellow-men from terrible and unmerited death, through a higher and nobler motive than his friend would have her believe.***❞**

Source: Baroness Orczy. *The Scarlet Pimpernel.* Doubleday and Company, Inc., 1961.

Analyzing Literature

Do you think that the Scarlet Pimpernel's actions were really just for sport? Why or why not? Do you think they were right to try to save the French nobility? Would you?

Democracy in the Americas In the 1770s, the American colonies, beginning with the thirteen British colonies in North America, revolted against British control. The new United States, its **Declaration of Independence, Constitution,** and representative Congress, became a model for many of these revolutions. By the 1830s, most of the Spanish, Portuguese, and British colonies in the Americas south of Canada had also gained their independence.

The French Revolution In the 1780s, revolution erupted in Europe as well, starting with France. The **French Revolution** began in 1789 and went through several stages. When King Louis XVI and Queen Marie Antoinette opposed the revolution and tried to aid the nobility, they were executed. By 1799, **Napoleon Bonaparte**—a military hero of the French Revolution—became the dictator, the absolute leader, of France. He declared himself emperor of a new French Empire in 1804. Eventually, people almost everywhere in Europe reacted against Napoleon and went to war against France. Napoleon was finally defeated in 1815.

The revolution in France stimulated Latin Americans and other European peoples to demand more personal and political control over their lives. Countries such as Greece, Belgium, Italy, and Germany also experienced revolutions.

 **Reading Check** How was the growth of democracy in Britain different from that in France?

Section 3 Assessment

Defining Terms
1. Define indulgences, Protestant, reform, Columbian Exchange, revolution, divine right of kings, Parliament, constitution.

Recalling Facts
2. History What was the movement to reform the Catholic Church called?

3. People Who is known as the "father of printing in the Western world"?

Critical Thinking
4. Examining Results Describe the effects of the process called the Columbian Exchange.

5. Making Connections How might a revolution in one country encourage political changes around the world?

Graphic Organizer
6. Identifying People Create a table like the one below. In the right column, write what each individual is historically recognized for.

Person	Historical Recognition
Michelangelo	
Johannes Gutenberg	
Martin Luther	
Christopher Columbus	
Napoleon Bonaparte	

 Applying Social Studies Skills

7. Identifying Why do you suppose the period known as the Renaissance was considered a "rebirth"?

Leonardo da Vinci

The Italian Leonardo da Vinci is considered one of the greatest artists of the Renaissance. He painted the *Mona Lisa* and the *Last Supper*, two of the world's best-known paintings. He was also a talented architect, engineer, and inventor.

The Artist

Leonardo da Vinci was born in 1452 in a small town near Florence, Italy. As the son of a wealthy man, he received the best education that Florence could offer. Leonardo became known for his ability to create sculptures and paintings that looked almost lifelike. Much of his success in this area came from his keen interest in nature. He also studied human anatomy and used this knowledge to make his figures realistic.

The Inventor

As a child, Leonardo was fascinated with machines and began to draw his own inventions. The first successful parachute jump was made from the top of a French tower in 1783—but Leonardo had sketched a parachute in 1485. He designed flying machines, armored tanks, and aircraft landing gear. He even drew a diver's suit that used tubes and air chambers to allow a swimmer to remain underwater for long periods of time.

Leonardo's Notebooks

Much of what we know about Leonardo comes from the thousands of pages of notes and sketches he kept in his notebooks. He used mirror, or reverse, writing, starting at the right side of the page and moving across to the left. No one is sure why Leonardo wrote this way. Some think he was trying to keep people from reading and stealing his ideas. He may also have been trying

to hide his thoughts from the Roman Catholic Church, whose teachings sometimes conflicted with his ideas. From a practical standpoint, writing in reverse probably helped him avoid smearing wet ink, since he was left-handed.

▲ Leonardo da Vinci

▲ The *Mona Lisa*

▶ Making the Connection

1. What are two of Leonardo's best-known works?

2. Why might Leonardo have written his notebooks in mirror writing?

3. **Understanding Cause and Effect** In what way did Leonardo's interest in the world around him influence his work?

Section 1   Classical Europe

Terms to Know

classical
polis
democracy
philosophy
republic
consuls
Senate
emperor

Main Idea

Ancient Greece and Rome made important contributions to Western culture and civilization.

✓ Government The world's first democratic constitution was written in Athens.

✓ History Alexander the Great conquered all of Greece.

✓ History Rome grew from Republic to Empire.

✓ Religion Christianity spread over the Roman world.

✓ History The Roman Empire was invaded by Germanic peoples and declined.

Section 2   Medieval Europe

Terms to Know

bishop   vassal
pope   manor
missionary   tenant
monastery   serf
convent   guild
common law   apprentice
feudalism   charter

Main Idea

The Middle Ages saw the spread of Christianity, the growth of cities, and the growing powers of kings.

✓ Religion The Roman Catholic Church became a political power in western Europe.

✓ History The first Christian Bible was completed by A.D. 500.

✓ History Charlemagne was crowned head of the Roman Empire in the West and proclaimed Protector of the Christian Church in the West.

✓ Government Feudalism was the medieval political and social system.

Section 3   The Beginning of Modern Times

Terms to Know

indulgences
Protestant
reform
Columbian Exchange
revolution
divine right of kings
Parliament
constitution

Main Idea

The study of science, art, and education was renewed in the period following the Middle Ages.

✓ Culture Important cultural achievements in the arts and learning spread throughout Europe in the period known as the Renaissance.

✓ History Johannes Gutenberg invented the printing press.

✓ Government Countries formed into nations based on a common language and culture.

✓ Religion The Protestant faith emerged in protest to the Roman Catholic Church.

✓ History Christopher Columbus set sail across the Atlantic.

✓ Government Revolution erupted in the Americas and Europe.

Chapter 9

Assessment and Activities

 Using Key Terms

Match the terms in Part A with their definitions in Part B.

A.

1. emperor
2. common law
3. feudalism
4. apprentice
5. indulgences
6. Protestants
7. polis
8. philosophy
9. missionary
10. guild

B.

a. unwritten laws from customs
b. "protested" Catholic teachings
c. freed owners from punishment for sins
d. medieval political and social system
e. absolute ruler
f. young worker learning a trade
g. teacher of Christianity
h. "love of wisdom"
i. workers' organization
j. city-state

 Reviewing the Main Ideas

Section 1 Classical Europe

11. **Government** Where was the first democratic constitution written?
12. **History** Who conquered all of Greece?
13. **Religion** Which religion spread all over the Roman world?
14. **History** Who invaded the Roman Empire?

Section 2 Medieval Europe

15. **Religion** Which religious group became a political power in western Europe?
16. **Economics** Explain the difference between vassals and serfs.
17. **Government** Name the political and social system in medieval Europe.

Section 3 The Beginning of Modern Times

18. **History** What did Johannes Gutenberg invent?
19. **Religion** Which faith emerged out of protest to the Catholic Church?
20. **History** What is Christopher Columbus historically known for?
21. **Government** Where were revolutions taking place?

 **Classical Europe**

Place Location Activity

On a separate sheet of paper, match the letters on the map with the numbered places listed below.

1. Alexandria
2. Macedonia
3. Mediterranean Sea
4. Constantinople
5. Black Sea
6. Greece
7. Athens
8. Rome
9. Tiber River
10. Sparta

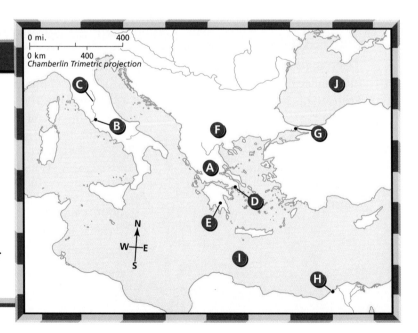

Self-Check Quiz Visit the *Our World Today: People, Places, and Issues* Web site at owt.glencoe.com and click on **Chapter 9—Self-Check Quizzes** to prepare for the Chapter Test.

Critical Thinking

22. **Making Connections** In what ways have our political and social lives today been influenced by ancient Greek and Roman customs?

23. **Drawing Conclusions** Eastern Orthodoxy was ruled by emperors rather than by popes. This made the emperors very powerful. What kinds of problems might have occurred because of this?

Current Events Journal

24. **Writing a Paragraph** Imagine that you are the writer of a tourism book for France. Write a short paragraph describing the cathedral on page 241. Use the information provided in the beginning of Section 2 of this chapter.

Mental Mapping Activity

25. **Identifying People and Places** Create a simple outline map of Europe that includes Germany, Italy, France, Rome, and Greece. Place the letter of the individual's name next to the place from which he originated.

   a. Michelangelo Buonarotti
   b. Alexander the Great
   c. Julius Caesar
   d. Socrates
   e. Charlemagne
   f. Leonardo da Vinci
   g. Christopher Columbus
   h. Napoleon Bonaparte
   i. Martin Luther
   j. Plato

Technology Skills Activity

26. **Using the Internet** Search the Internet for information on the Twelve Tables of Roman law. After reading about the laws, note the ones that you strongly agree or disagree with and tell why. For example, tablet 4 states that "a dreadfully deformed child shall be quickly killed." In our society, this is obviously illegal and inhumane.

Standardized Test Practice

Directions: Read the paragraphs below, and then answer the question that follows.

The ancient Greeks held the Olympic Games in Olympia every four years. The games were a religious festival in honor of Zeus, the Greeks' chief god. Trading and wars stopped while the games took place. The first Greek calendar began with the supposed date of the first Olympic Games in 776 B.C.

Athletes came from all over the Greek-speaking world to compete. Only male athletes, however, were allowed to take part, and women were not permitted even as spectators. Olympic events at first consisted only of a footrace. Later the broad jump, the discus throw, boxing, and wrestling were added. The Greeks crowned Olympic winners with wreaths of olive leaves and held parades in their honor.

1. **From the paragraphs, which of the following statements about Greek culture is correct?**

   F The Greeks stressed group effort over individual achievement.

   G The Greeks believed in one God.

   H The Greeks were not religious.

   J The Greeks encouraged individual glory.

Test-Taking Tip: Read all the choices carefully before choosing the one that correctly describes Greek culture. Eliminate answers that you know are incorrect. For example, all the Olympic events were performed by individuals, not by teams. Therefore, answer F does not describe Greek culture. The question is asking for the statement that DOES describe Greek culture.

253

The World and Its People
NATIONAL GEOGRAPHIC

To learn more about Europe and its people, view **The World and Its People Chapters 10–13** videos.

Our World Today ONLINE

Chapter Overview Visit the **Our World Today: People, Places, and Issues** Web site at owt.glencoe.com and click on **Chapter 10–Chapter Overviews** to preview information about the modern history of Europe.

Summarizing Information Study Foldable

Make the following foldable to help you organize and summarize information about historic events and modern events in Europe, and how they are related.

Step 1 Fold a sheet of paper from side to side, leaving a 2-inch tab uncovered along the side.

Fold it so the left edge lies 2 inches from the right edge.

Step 2 Turn the paper and fold it into thirds.

Step 3 Unfold and cut along the two inside fold lines.

Cut along the two folds on the front flap to make 3 tabs.

Step 4 Label the foldable as shown.

EUROPE: MODERN HISTORY

The Modern Era | Continent Divided | Move Toward Unity

Reading and Writing As you read about the modern history of Europe, write important facts under each appropriate tab of your foldable.

◀ **A modern office building stands next to Billingsgate Fish Market in London, England.**

Why It Matters

The Modern Era

Europe has played a major role in shaping today's world. Industrialization, which started in Europe, is one of the reasons for the high standard of living we experience today. Two world wars, fought largely on European soil, shaped world politics and preserved democracy.

Section 1

The Modern Era

Guide to Reading

Main Idea

Industrialization led not only to a higher standard of living for some, but also to increased tensions in the world. Two world wars changed the balance of power in the world.

Terms to Know

- productivity
- human resources
- textiles
- cottage industry
- union
- strike
- imperialism
- alliance
- communism
- Holocaust
- genocide

Reading Strategy

Create a chart like the one below. Write three statements of fact under the Fact column. In the Opinion column, write three statements that show how you feel about the fact statement.

Fact	Opinion

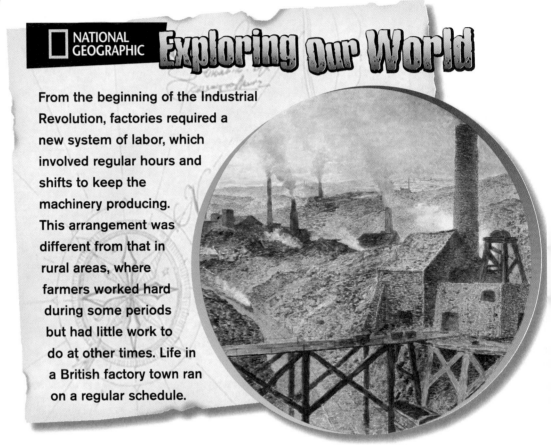

NATIONAL GEOGRAPHIC Exploring Our World

From the beginning of the Industrial Revolution, factories required a new system of labor, which involved regular hours and shifts to keep the machinery producing. This arrangement was different from that in rural areas, where farmers worked hard during some periods but had little work to do at other times. Life in a British factory town ran on a regular schedule.

The **Industrial Revolution** began in Great Britain in the 1700s. It was a time when people used machinery and new methods to increase productivity. Productivity is a measure of how much work can be done in a certain length of time. The changes these machines brought were so great that they led to a revolution in the ways work was done and people lived. Today, machines touch every part of our lives.

A Rapidly Changing World

The Industrial Revolution started in Britain for several reasons. Britain had a ready supply of natural resources such as coal and iron, which were needed to make and run machinery. There was also a plentiful supply of raw materials such as wool and imported cotton, used to make cloth. There was also a source of people—human resources—who could be hired to work the machines. As farmers relied more on machines to plant and harvest crops, fewer people were needed to grow crops. Many people who used to work on the farms went to the cities to find work in factories and shops.

Major Industries   Textiles, or woven cloth, was the first industry to be moved to factories. Before that, spinning and cloth weaving had been a cottage industry carried out by family members working in their own homes. With industrialization, huge quantities of cloth could be produced in factories that employed many workers. Textile mills became even more productive when steam replaced waterpower for running the machinery.

The steam engine was invented by Thomas Newcomen in the early 1700s and was used to pump water out of coal mines. In 1769, James Watt invented a more efficient steam engine, which was used for textile mills, riverboats, and locomotives. Inventions like the railroad stimulated even more industries and growth. By the early 1800s, the Industrial Revolution spread from Britain to much of western Europe and North America.

✓ Reading Check   **How did machinery affect the textile industry?**

Changing Lifestyles

Towns and cities grew, as people's lives changed dramatically. At first, industrial workers, including women and children, had to work hard for long hours often under dangerous conditions. Eventually, the workers formed groups called unions. A union spoke for all the workers in a factory or industry and bargained for better working conditions, higher pay, and a shorter working day. If a factory owner refused these demands, union members often went on strike. That is, they refused to work until their demands were met.

Overall, the Industrial Revolution made life more difficult for people in the short term, but easier in the long run. For example, because manufactured cotton clothing was better and cheaper, people could afford more. They then could afford to change clothes and wash them more often. This new cleanliness helped to reduce sickness and disease, so people generally lived healthier and longer.

The Industrial Revolution also resulted in strong economies in western Europe. It was because of this economic strength that Europe was able to dominate the world in the 1800s and early 1900s.

✓ Reading Check   **How did the Industrial Revolution improve people's lives?**

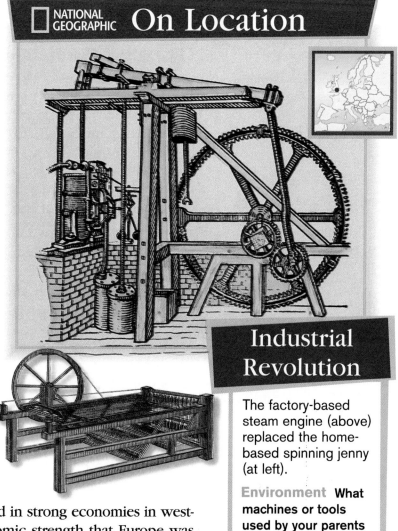

NATIONAL GEOGRAPHIC   On Location

Industrial Revolution

The factory-based steam engine (above) replaced the home-based spinning jenny (at left).

Environment   What machines or tools used by your parents are no longer common in homes today?

Then and Now

The town of Verdun, France, was nearly destroyed during World War II (above). Today it is a thriving commercial center and tourist attraction (right).

Place Name another city that has been rebuilt since WWII.

Rivalry Between Nations

Industrialization created new rivalries among the countries of Europe. Britain, France, Germany, and other European countries competed around the world for markets and resources for their factories. Under a system called imperialism, European countries claimed colonies in Africa and Asia in the late 1800s. European nations built up armies and navies to protect themselves and their empires. Different alliances were formed, whereby various countries agreed to support one another in times of war.

World War I   In 1914, a war broke out in Europe that quickly spread to the European colonies and other areas of the world. It was known as the **Great War,** and later called **World War I.** This war was not like any earlier wars. With the techniques learned in the Industrial Revolution, machines designed for war were mass-produced. Tanks, heavy artillery, machine guns, and airplanes helped to make the war more violent than any before it. In the four years of the war, millions of people were killed or wounded, and many European cities and villages were destroyed.

New Problems Arise   As a result of the war, many once-strong countries became weak. Germany was blamed for starting the war and was asked to pay for much of it. The United States and Japan became great powers. A revolution in Russia in 1917 led to a new political, economic, and social system called communism. Communism was based on the teachings of a German philosopher named Karl Marx. Marx believed that industrialization had created two classes of people—those who owned the means of producing goods and those who worked to produce the goods. He wrote that this system was unfair and needed to be overthrown.

World War II   In the 1930s, worldwide depression severely tested the ability of many governments to provide for their citizens. The problems that were not solved after World War I eventually led to new alliances in Europe. Germany became a dictatorship under Adolf Hitler and his National Socialist German Worker's party. Its members, called **Nazis,** believed in German superiority. By 1939, Germany, Italy, and Japan (the **Axis Powers**) were at war with Britain, France, and China (the **Allies**). In 1941, the United States and Soviet Union joined the Allies in the war that became known as World War II.

▲ Hitler at a Nazi rally, Dortmund, Germany

During the war, Hitler began the Holocaust, which killed over 12 million people. Over 6 million of the victims were Jews. Other persecuted groups included the Romany people (called Gypsies), Serbians, individuals with disabilities, and many other groups that were classified as "undesirable" by the Nazi leaders. The Holocaust is an example of the war crime of genocide, or the mass murder of a people because of race, religion, ethnicity, politics, or culture.

Italy surrendered in 1943. Germany was finally defeated in May 1945, but the Japanese continued to fight. In August, the United States—in an effort to end the war in Asia—dropped two atomic bombs on the Japanese cities of Hiroshima and Nagasaki. From this global conflict, the United States and the Soviet Union emerged as superpowers.

✓ Reading Check What was the Holocaust?

Assessment

Defining Terms

1. Define productivity, human resources, textiles, cottage industry, union, strike, imperialism, alliance, communism, Holocaust, genocide.

Recalling Facts

2. History Where did the Industrial Revolution begin?

3. Government Name the new political, economic, and social system that was envisioned by Karl Marx.

Critical Thinking

4. Comparing and Contrasting How did people's living habits change after the introduction of factories? Do you think people were generally better off?

5. Evaluating Information Why did the new military equipment introduced in World War I change the way wars were fought?

Graphic Organizer

6. Organizing Information Create a diagram like the one below and fill in the names of the countries that made up the two powers fighting each other in World War II.

Axis Powers	Allies

Applying Social Studies Skills

7. Analyzing Maps Refer to the **Reference Atlas** map of the world on pages RA2 and RA3. Which of the Allies was nearest to Japan?

The Holocaust

The Holocaust is one of the most horrifying events in human history. *Holocaust* is a word that means complete and total destruction. Learning about the Holocaust is important so that crimes against humanity such as this will never occur again.

The Final Solution

Adolf Hitler, chancellor of Germany, believed that the Germanic peoples of the world, called Aryans, were a superior race. His goal was to populate Europe with one "master" race of people. During the years before and during World War II, Hitler's government persecuted many racial, religious, and ethnic groups that he considered "undesirable." These included Gypsies, Jehovah's Witnesses, people with disabilities, and political dissidents of all backgrounds.

The chief target of Hitler's plan—which he called his "Final Solution"—was the Jews. Jewish communities throughout Germany and German-controlled territory suffered terribly. Jews, forced to wear identification badges, were blamed for all Germany's economic and social problems.

Between 1939 and 1945, Hitler's Nazi forces attempted to exterminate the Jews in every country Germany invaded, as well as in those countries that were Nazi allies. Jews from Germany, Poland, the Soviet Union, France, Belgium, the Netherlands, Greece, and Hungary were among those killed during the Holocaust.

Mass Murder

In the early years of the war, Jewish people in Eastern Europe were rounded up, gathered together, machine-gunned, and buried in mass graves. Later, millions of Jews were uprooted and forced into concentration camps. Few people survived these. Those too young, sick, or elderly for heavy labor were executed in gas chambers. In all, more than 6 million Jews and about 4 million Gypsies, Poles, Soviet prisoners of war, and others were murdered.

▲ Auschwitz Nazi concentration camp in Oswiecim, Poland

▶ Making the Connection

1. What was the Holocaust?

2. Why did Hitler want to rid Europe of its Jewish people?

3. **Understanding Cause and Effect** How can studying about the Holocaust today help prevent another atrocity from happening in the future?

A Continent Divided

Guide to Reading

Main Idea

Two powers, the democratic United States and the Communist Soviet Union, worked to bring their forms of government to the war-torn nations of Europe.

Terms to Know

- Cold War
- nuclear weapon
- deterrence
- satellite nation
- blockade
- airlift
- glasnost
- perestroika
- capitalism

Reading Strategy

As you read, fill in a time line like the one below with an event that occurred during that year.

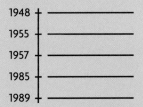

1948
1955
1957
1985
1989

NATIONAL GEOGRAPHIC

Exploring Our World

For nearly 30 years, armed guards patrolled a 103-mile (166-km) wall that divided the German city of Berlin into eastern and western halves. During that time, the citizens of East Berlin were not allowed to travel freely to West Berlin. In late 1989, the wall finally came down. Germans from both parts of the city came together and celebrated.

After World War II, much of Europe was in ruins. The total defeat of Germany, Italy, and Japan left a power gap that would be filled by the rival United States and Soviet Union.

The Cold War

The global competition between the democratic United States and its allies and the Communist Soviet Union and its supporters came to be called the Cold War. It was a dangerous time because by 1948 both sides had nuclear weapons. Nuclear weapons use atomic reactions to release enormous power and cause mass destruction. It was a "cold" war because countries never mobilized armies in an official war.

The Cold War began in Europe. In 1948, the United States started a loan program called the **Marshall Plan** to help rebuild Europe and try to stop the spread of communism. Under the Marshall Plan, factories were rebuilt, mines were reopened, and roads were repaired and replaced. The economies of Western Europe—those countries liberated and occupied by the United States and Great Britain during World War II—began to grow.

Reading Check What was the Cold War?

West European Nations Cooperate

In 1948 under the **Truman Doctrine,** named after U.S. President Harry S Truman, the United States offered military aid to countries such as Greece and Turkey that were engaged in fighting communism inside their borders. In 1949, the North Atlantic Treaty Organization (NATO) was formed to respond to possible attacks by the Soviet Union. In forming NATO, each member country agreed to treat an attack on any other member as an attack on itself. The NATO countries believed that the Soviet Union would not attack Western Europe if Soviet leaders thought such an attack would trigger nuclear war with the United States. This policy is known as deterrence because it is designed to deter, or discourage, an attack.

Eventually, to help themselves, Western European countries began to cooperate with one another. First the small countries Belgium, the Netherlands, and Luxembourg joined together in 1948 to form the **Benelux** trade union, an arrangement for the free movement of money, goods, and people among these nations. Then West Germany,

NATIONAL GEOGRAPHIC

Western and Eastern Europe (c. 1950)

Applying Map Skills

1. Were there more countries in Western Europe or Eastern Europe?

2. Which Eastern European countries were on the border with Western Europe?

Find NGS online map resources@ www.nationalgeographic.com/maps

Western Europe
Eastern Europe

France, and Italy came together with the Benelux countries to form the European Coal and Steel Community. In 1958, this became the **European Economic Community** (or Common Market). The members agreed to free trade among themselves. Workers from one member country could take jobs in any of the other countries. Between 1958 and 1986 Denmark, Great Britain, Ireland, Spain, Portugal, and Greece also joined the Common Market (now known as the **European Union**), which moved toward greater cooperation and economic development.

✓Reading Check **Why did the countries of Western Europe join NATO?**

The Soviet Community

In Eastern Europe, the Soviet Union made satellite nations of those countries surrounding it. Satellite nations are dependent upon a stronger power. They were strictly controlled by the Soviet Union. Bulgaria, Romania, Czechoslovakia, Hungary, and Poland became communist. With these countries, the Soviet Union created the Council for Mutual Economic Assistance, or COMECON, primarily for its own economic benefit.

In 1955, the Soviet Union formed its satellites into an anti-Western military alliance known as the **Warsaw Pact.** It was named after the Polish capital city of Warsaw, where the treaty of alliance was signed.

Yugoslavia and Albania also became communist but refused to be put under Soviet control. During the Cold War, Yugoslavia, India, Egypt, and Indonesia formed the Non-Aligned Community. Its members tried to stay neutral—to not support either side—during the Cold War.

✓Reading Check **In what way was the Warsaw Pact like NATO?**

A Clash Between Superpowers

During the Cold War, there were many "hot spots," or areas of tension and conflict. Some of them were China, Korea, Cuba, and Vietnam. The earliest, however, and one of the most important was Berlin.

Divided Berlin At the end of World War II, the Allies (the United States, the Soviet Union, Great Britain, and France) occupied Germany and its capital city of Berlin. In 1948, to promote peace and German recovery, the United States, Great Britain, and France decided to unite their occupation zones. The Soviet Union was against any plan that would strengthen Germany, its historical enemy. Berlin was located deep in the Soviet zone, but jointly occupied by the four powers. In June 1948, the Soviets blockaded, or closed off, all land and water traffic into Berlin. They hoped this would force the other three powers to leave the city.

In response, the United States and Great Britain began an airlift, or a system of carrying supplies into Berlin by airplane. Day and night the planes flew tons of food, fuel, and raw materials into the city. This heroic effort caused the Soviets to finally end the 11-month blockade of Berlin. That same year, two separate governments were set up—a democratic one for West Germany with Bonn as its new

Exploring Economics

Restructuring

Imagine a family-owned business in which the head of the family makes all the decisions. These include which businesses to begin, who will do which jobs, what products to produce, and how much to pay employees. Then, suddenly the head of the family disappears. Family members now must make all the decisions though they have had no experience. Like this example, for most Soviet citizens, taking control of the local economy was a difficult change.

NATIONAL GEOGRAPHIC On Location

Berlin Airlift

Residents of Berlin (right) wave to an American airlift plane. A line of planes (left) waits to be unloaded at a Berlin airfield.

Technology How did airplane technology affect Soviet military strategy in Germany?

capital and a communist one for East Germany with East Berlin, in the Soviet zone, as its capital. West Berlin remained a democratic stronghold surrounded by communism.

The Berlin Wall Many people in East Germany were unhappy under communist rule. About 3 million people fled to West Berlin in search of political freedom and better living conditions. The East German government wanted to stop these escapes, and in August 1961 the government built a wall between East and West Berlin. The wall, with Soviet soldiers guarding it, became a symbol of the split between Eastern and Western Europe. Many East Germans continued to risk their lives trying to escape over or under the wall.

The Race to Space Part of the Cold War between the United States and the Soviet Union involved the race to explore space. The Soviets first took the lead. In 1957, they launched *Sputnik I,* the first spacecraft to orbit the earth. Four years later Soviet astronaut Yuri Gagarin became the first human being to circle the earth.

Then the United States became the leader. In 1969, Neil Armstrong became the first person to walk on the moon. During the 1970s, the first landings on Venus and Mars were made by U.S. crewless spacecraft. Later, space vehicles explored Jupiter, Saturn, and beyond. The space race brought fame and glory to both powers. Today the United States and Russia are cooperating partners in the international space station project.

✓ Reading Check How did the Soviet Union separate the people of East and West Berlin?

The End of the Cold War

During the cold war, the Soviet Union spent large sums of money on military and space ventures. In spite of plans to improve consumer housing and agriculture, the economies of the Soviet Union and its satellites kept falling further and further behind the United States and its Western European allies.

In 1985, Mikhail Gorbachev became the leader of the Soviet Union. He introduced reforms to try to get the Soviet economy moving again. Under glasnost, or openness, the Soviet people could criticize the system without fear of being punished. Free elections were held in which many noncommunists gained office. Under the policy of perestroika, or restructuring, Gorbachev loosened government control and moved the economy toward capitalism. Under capitalism, most businesses are privately owned, and there is competition to try to create better products at lower prices.

In the late 1980s, as the Soviet Union moved slowly toward democracy and capitalism, tensions within the country and its satellites increased. Some people thought that Gorbachev was moving too quickly with his reforms. Others thought he was not moving fast enough. At the same time, many ethnic groups in the Soviet Union were demanding independence. All of the satellite nations also moved toward freedom from Soviet domination. The Cold War was coming to an end.

✓ **Reading Check** Which Russian leader moved the Soviet Union toward democracy?

Section 2 Assessment

Defining Terms

1. **Define** Cold War, nuclear weapon, deterrence, satellite nation, blockade, airlift, glasnost, perestroika, capitalism.

Recalling Facts

2. **History** What was the purpose of the Marshall Plan?

3. **Place** Which countries were considered satellites of the Soviet Union?

Critical Thinking

4. **Comparing and Contrasting** What are the similarities and differences between a "cold" war and a "hot" war?

5. **Analyzing Information** How did the space race reflect tensions between the United States and the Soviet Union?

Graphic Organizer

6. **Organizing Information** Create a chart like this. Explain how each of the following events intensified the Cold War.

Marshall Plan	
Truman Doctrine	
NATO	
Warsaw Pact	

Applying Social Studies Skills

7. **Analyzing Maps** Look at the map of Western and Eastern Europe on page 262. Name the Western European countries that shared a border with countries in Eastern Europe.

Social Studies Skill •

Reading a Population Map

Population density is the number of people living in a square mile or square kilometer. A **population density map** shows you where people live in a given region. Mapmakers use different colors to represent different population densities. The darker the color, the more dense, or crowded, the population is in that particular area. Cities that are shown by dots or squares also represent different population sizes.

Learning the Skill

To read a population density map, follow these steps:

- Read the title of the map.
- Study the map key to determine what the colors mean.
- On the map, find the areas that have the lowest and highest population density.
- Identify what symbols are used to show how heavily populated the cities are.

Practicing the Skill

Look at the map below to answer the following questions.

1. What color stands for 125–250 people per square mile (50–100 per sq. km)?
2. Which cities have more than 1 million people?
3. Which areas have the lowest population density? Why?

Applying the Skill

Obtain a population density map of your state. What is the population density of your area? What is the nearest city with 1 million people?

GO TO

Practice key skills with **Glencoe Skillbuilder Interactive Workbook, Level 1.**

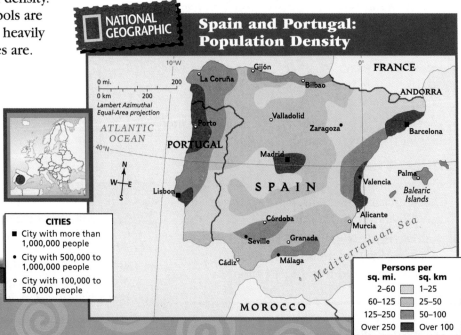

NATIONAL GEOGRAPHIC

Spain and Portugal: Population Density

0 mi. 200
0 km 200
Lambert Azimuthal Equal-Area projection

ATLANTIC OCEAN

FRANCE
ANDORRA

Gijón
La Coruña
Bilbao
Porto
Valladolid
Zaragoza
Barcelona
PORTUGAL
Madrid
Lisbon
Palma
Valencia
Balearic Islands
SPAIN
Alicante
Murcia
Córdoba
Seville
Granada
Cádiz
Málaga
Mediterranean Sea
MOROCCO

CITIES
- ■ City with more than 1,000,000 people
- ● City with 500,000 to 1,000,000 people
- ○ City with 100,000 to 500,000 people

Persons per	
sq. mi.	sq. km
2–60	1–25
60–125	25–50
125–250	50–100
Over 250	Over 100

Moving Toward Unity

Guide to Reading

Main Idea

Although the Cold War is over, many challenges still face the old and new nations of Europe.

Terms to Know

- European Union
- Euro

Reading Strategy

Create a chart like the one below and write one key fact about each topic.

Soviet Union	
Yugoslavia	
European Union	
NATO	
Greenhouse effect	

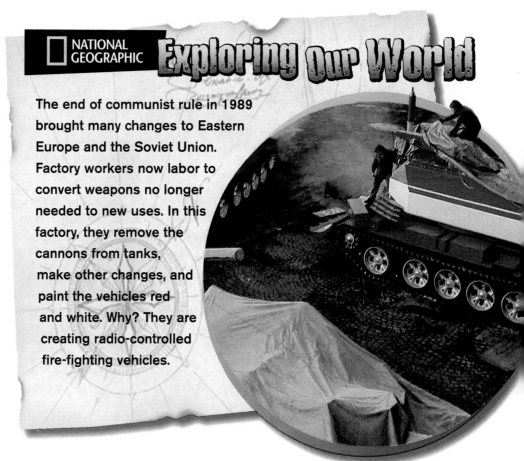

NATIONAL GEOGRAPHIC *Exploring Our World*

The end of communist rule in 1989 brought many changes to Eastern Europe and the Soviet Union. Factory workers now labor to convert weapons no longer needed to new uses. In this factory, they remove the cannons from tanks, make other changes, and paint the vehicles red and white. Why? They are creating radio-controlled fire-fighting vehicles.

In 1989, massive protests occurred in East Germany and the communist government resigned. In Poland, for the first time, a communist government had lost power as a result of a democratic election. The Berlin Wall came down, and West and East Germany reunited in October 1990. By 1991, all of the Soviet satellite nations—even Mongolia in distant Central Asia—had thrown off communist rule in favor of democracy.

The End of the Soviet Union

The movement toward democracy spread to the Soviet Union. In May 1990, **Boris Yeltsin,** a reformer and former Communist, was elected president of Russia, the largest of the 15 Soviet republics. Meanwhile, people in the other 14 republics began to shake off Soviet rule. Freed from daily Russian control, fierce ethnic fighting broke out among various groups in many of these republics.

Standing for Democracy

Russia's president, Boris Yeltsin, stood on a tank in defiance of a communist group that wanted to stop Russia's move to democracy.

Government Why do you think some people did not want Russia to move to democracy?

The Soviet Union officially broke up on Christmas Day 1991. In its place are 15 independent republics stretching across Europe and Asia and moving toward democracy. **Yugoslavia,** a communist country, and democratic **Czechoslovakia** also broke up. After much fighting and civil wars, Yugoslavia became the independent republics Slovenia, Croatia, Bosnia and Herzegovina, Macedonia, and a smaller Yugoslavia made up of Serbia and Montenegro. In 2002, Serbia and Montenegro formed a looser union that ended Yugoslavia. Czechoslovakia peacefully became the Czech Republic and Slovakia. All of these countries today have major problems due to poor economies, ethnic tensions, and a lack of understanding of democracy. You will read more about these eastern European countries in Chapter 12.

✓ Reading Check What happened to the republics of the Soviet Union?

The New Europe

Since the fall of communism and the Soviet Union, there is no longer a political division between western and eastern Europe—although cultural and economic differences remain. As a result of cooperation, Europe is becoming more of an economic power in the world. The Common Market became the European Union (EU) in 1993. At that time, the twelve members included the United Kingdom, Ireland, France, Luxembourg, Spain, Portugal, Denmark, the Netherlands, Belgium, Germany, Italy, and Greece. Austria, Finland, and Sweden joined in 1995. Fourteen other nations, including many from eastern Europe, have applied for membership.

The European Union is moving toward greater unity. Some Europeans would eventually like to see it become a United States of Europe that would include all European countries. In January 2002 most of its members began using a common currency, the Euro, to replace their national currencies. This means that citizens of countries in the EU are using the same type of money to buy goods and services.

The countries of Europe have cooperated in the areas of science and technology, as well as politics and economics. Europe had one of the first treaties on nuclear energy. The European Atomic Energy Community (EURATOM) has wide powers, including the right to conclude contracts, obtain raw materials, and establish standards to protect workers and the general population against the dangers of radiation.

✔ Reading Check **What is the name of the new European Union currency?**

Facing the Region's Problems

There is a growing number of problems in Europe—and the world—that Europeans are actively trying to solve. The income gap between the rich and poor nations of Europe needs to be lessened. In addition, the increasing food and health needs of the people of these countries must be met.

NATO as Peacekeeper Some eastern European countries have joined NATO, and others are on the waiting list. Poland, Hungary, and the Czech Republic are NATO members. Countries that have applied for NATO membership include Estonia, Romania, Bulgaria, and Slovenia. However, the expansion of NATO to Russia's western border is also creating tensions among Russia, the United States, and Europe. NATO is moving beyond its original role as a protector from communist states. It has also taken on European peacekeeping tasks. Since 1990, through NATO, the United States and Europe have supported United Nations efforts in areas such as Iraq and Kuwait and in the former Yugoslav republics of Bosnia and Herzegovina, Serbia and Montenegro, and Macedonia. Sometimes there is not complete agreement among all the members on how NATO forces should be employed.

NATIONAL GEOGRAPHIC **On Location**

The Euro

Ten different national sides of the one Euro coin are shown, along with the front image that does not change.

Place How many nations can you identify by the images chosen to represent the country?

Environmental Issues Yet another important challenge for Europe comes from environmental problems. In France, rivers like the Seine and the Loire are polluted as are the major canals. In the Netherlands, water is not drunk or used on gardens without being filtered first. Nowhere is the problem more acute than in the Rhine River. As the river flows north, it passes through a continuous band of cities and industrial regions. By the time it reaches the Netherlands it is carrying a staggering 25 million tons of industrial water per year. This is all dumped into the North Sea. In its rush to develop and compete, the Soviet Union did damage to the natural environment across Europe and Asia. The resulting air, water, and land pollution from these environmental disasters needs to be cleaned up and reversed.

One pressing issue is the growing greenhouse effect. Heavy use of coal and oil fuels results in a buildup of carbon dioxide gas in the atmosphere. Scientists believe that this may cause the earth's average temperature to rise. Even a slight rise could result in polar ice cap melting and flooding in some coastal regions of the world.

Europeans are realizing that international cooperation and increased contact among peoples hold the key to overcoming the region's—and the world's—most pressing challenges and crises. Because of their history, Europeans have learned the importance of working together to deal with major problems.

✓ Reading Check What are two important issues facing Europeans today?

Assessment

Defining Terms
1. **Define** European Union, Euro.

Recalling Facts
2. **History** What smaller independent republics were formed after the breakup of Yugoslavia?
3. **Economics** What is the European Union trying to achieve?

Critical Thinking
4. **Making Inferences** Why did freedom from Soviet rule lead to ethnic fighting in many former Soviet satellites?
5. **Drawing Conclusions** Do you think Russia will join the European Union? Why or why not?

Graphic Organizer
6. **Organizing Information** Create a list showing some of the problems still facing Europe.

1. _____
2. _____
3. _____
4. _____

Applying Social Studies Skills

7. **Summarizing** Write a paragraph that summarizes the end of the Soviet Union. In your summary be sure to include important events that led to the breakup, any key people who created change, and the final outcome of the breakup.

Chapter 10 Reading Review

Section 1 · The Modern Era

Terms to Know

productivity
human resources
textiles
cottage industry
union
strike
imperialism
alliance
communism
Holocaust
genocide

Main Idea

Industrialization led not only to a higher standard of living for some, but also to increased tensions in the world. Two world wars changed the balance of power in the world.

✓ Economics Machinery made it possible to increase productivity, leading to the Industrial Revolution.

✓ Culture Industry changed the way people worked and lived.

✓ Economics Competition for markets and resources led to imperialism and friction among European countries.

✓ History The two World Wars changed the way wars were fought and created new political power for the United States and the Soviet Union.

Section 2 · A Continent Divided

Terms to Know

Cold War
nuclear weapon
deterrence
satellite nation
blockade
airlift
glasnost
perestroika
capitalism

Main Idea

Two powers, the democratic United States and the Communist Soviet Union, worked to bring their forms of government to the war-torn nations of Europe.

✓ History Competition between the United States and the Soviet Union started a cold war.

✓ Economics Western European countries joined together to form the European Common Market, which moved toward greater cooperation and economic development.

✓ Government The Soviet Union made satellites of its surrounding nations.

✓ History Berlin became a "hot spot" for conflict between the superpowers, symbolized by the Berlin Wall.

✓ Government As its satellites began to rebel, the Soviet Union under Mikhail Gorbachev moved towards a more open system that allowed privately owned businesses.

Section 3 · Moving Toward Unity

Terms to Know

European Union
Euro

Main Idea

Although the Cold War is over, many challenges still face the old and new nations of Europe.

✓ History In 1991, the Soviet Union officially broke up into 15 independent republics.

✓ Economics The fall of the Soviet Union increased Europe's global influence and strengthened the movement towards greater political and economic unity.

✓ Human/Environment Interaction Problems still remain in Europe, including poverty and pollution.

Chapter 10 Assessment and Activities

Using Key Terms

Match the terms in Part A with their definitions in Part B.

A.

1. productivity
2. union
3. imperialism
4. communism
5. genocide
6. Cold War
7. deterrence
8. glasnost
9. capitalism
10. Euro

B.

a. group that bargains for better working conditions
b. openness
c. mass murder of a people because of race, religion, ethnicity, politics, or culture
d. European Union common currency
e. economic system where businesses are privately held
f. countries claim colonies for their resources and markets
g. how much work can be done in a certain length of time
h. conflict between the United States and the Soviet Union
i. political system that called for the overthrow of the industrialized system
j. designed to discourage a first attack

Reviewing the Main Ideas

Section 1 The Modern Era

11. **History** How did the Industrial Revolution change working and living conditions?
12. **Economics** Why did European countries find it necessary to have colonies?
13. **History** What were some of the problems that led to World War II?

Section 2 A Continent Divided

14. **History** What was the Marshall Plan, and why was it important?
15. **History** How did the Truman Doctrine and NATO intensify the Cold War?
16. **Government** What was the Non-Aligned Community, and which nations belonged to it?
17. **History** Why did the Soviet Union build the Berlin Wall?
18. **Government** What were some of the policies that Mikhail Gorbachev introduced?

Section 3 Moving Toward Unity

19. **Government** What are some of the problems facing the new republics formed after the breakup of the Soviet Union?
20. **Economics** What is the Euro?
21. **Human/Environment Interaction** How has Soviet economic development and competition affected the environment?

 **The Allies and Axis Powers**

Place Location Activity

On a separate sheet of paper, match the letters on the map with the numbered places listed below.

1. Germany
2. Italy
3. United Kingdom
4. France
5. China
6. Soviet Union
7. Japan
8. United States

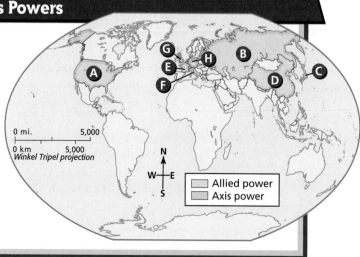

Critical Thinking

22. **Predicting Consequences** What changes will occur in Europe as a result of the European Union and the collapse of the Soviet Union?

23. **Sequencing Events** List five events that made the Cold War "colder."

   1. _____
   2. _____
   3. _____
   4. _____
   5. _____

Current Events Journal

24. **Writing and Evaluating** Write an essay explaining why adopting the Euro as common European currency is a good or bad idea. Find out why the European Union thinks a common currency is important and how Europeans are adapting to it. Use the Technology Skills Activity to provide more information to support your arguments.

Mental Mapping

25. **Focusing on the Region** Draw a simple outline map of Europe and label the following:

   - the United Kingdom
   - Germany
   - Italy
   - France
   - Russia
   - Spain
   - Greece

Technology Skills Activity

26. **Using the Internet** Research the national currencies that are being used in at least five European countries until the Euro is adopted. Note what each country's currency is called and when the country plans to phase it out. Use this information to support your ideas in the Current Events Journal above.

Standardized Test Practice

Directions: Study the map, then answer the question that follows.

Occupation of Germany 1945

North Sea — DENMARK — Baltic Sea

AMERICAN — Bremen — Hamburg

NETHERLANDS — Hannover — Berlin

BRITISH — SOVIET — POLAND

BELGIUM — Bonn — Leipzig

GERMANY — Dresden

Frankfurt

LUX. — AMERICAN — CZECHO-SLOVAKIA

Stuttgart

FRANCE — FRENCH — Munich

AUSTRIA

SWITZERLAND — 0 mi. 200 — 0 km 200

ITALY

1. **In 1945, what country controlled the land surrounding Berlin, Germany's capital?**

   F the United Kingdom
   G the Soviet Union
   H the United States
   J France

Test-Taking Tip: This question asks you to synthesize information on the map with prior knowledge. Notice that the map does not specifically state that the United Kingdom, for example, controlled a portion of Germany. Instead, it refers to this area as "British."

EYE on the Environment

RAIN, RAIN
Go Away

Acid Rain Have you ever sucked on a lemon slice? Yow! Lemons make you pucker up because they are high in acid. Rainwater can be acidic, too. Any form of precipitation that contains high amounts of acid is known as acid rain. In some parts of the world, rain or snow falls that is as acidic as lemon juice.

Why does this happen? When cars and trucks burn gasoline, or when factories and power plants burn coal, sulfur and nitrogen compounds are produced. High in the atmosphere, these gases mix with moisture to form sulfuric acid and nitric acid. These acids make rainwater much more acidic than normal. Acid rain is a problem because it

- harms fish and other animals in lakes and streams;
- damages trees and crops;
- washes nutrients out of soils.

Taking Action Europeans are very concerned about acid rain and its effects. Half of the trees in Germany's Black Forest are sick or dying. Forests in Norway, Austria, Poland, France, and the Czech Republic have also been damaged. In Sweden, 20 percent of the lakes contain few or no fish. The same is true of most lakes in southern Norway.

Many European countries are trying to reduce acid rain by

- installing filters on factory smokestacks;
- putting special exhaust systems on motor vehicles;
- building new factories that do not burn coal.

Acid rain eats away at a statue in Rome.

A German factory spews chemicals that cause acid rain.

Making a Difference

Acid Rain 2000 A project called Acid Rain 2000 is giving students across Europe a chance to study acid rain and its effects. From 2000 to 2005, participating students will be collecting four kinds of environmental data at study sites in Europe.

- **WEATHER** — Each day, students record the wind direction and the acidity of precipitation.
- **PLANTS** — Once a month, students check the condition of trees and other plants at their sites.
- **SOIL** — Once a month, students test the soil at their sites for acid and plant nutrient levels.
- **LICHENS** — Twice a year, students record the condition of plants called lichens. Since lichens die if the air is too polluted, they are good indicators of a site's air quality.

Acid Rain 2000 participants e-mail the data they collect to Northamptonshire Grammar School, near Northampton, England. There, students and staff process the data and publish the project's findings on the Internet. Acid Rain 2000 hopes to show which areas in Europe are most sensitive to acid rain.

Student collects weather data

What Can You Do?

- **Collect Data**
  Although Acid Rain 2000 is a European project, you can collect similar kinds of data at a study site in your community. For more information about how to set up a site and collect data, contact Acid Rain 2000 at www.brixworth.demon.co.uk/acidrain2000

- **Investigate**
  Does acid rain affect your community? If so, what impact has acid rain had on the environment? What are local industries doing to combat the problem? Motor vehicle exhaust contributes to acid rain. What can you do to limit vehicle use on a daily basis?

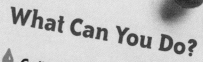

Lichens

Western Europe Today

The World and Its People

NATIONAL GEOGRAPHIC

To learn more about the people and places of western Europe, view *The World and Its People* **Chapters 10–12** videos.

Our World Today online

Chapter Overview Visit the *Our World Today: People, Places, and Issues* Web site at owt.glencoe.com and click on **Chapter 11—Chapter Overviews** to preview information about western Europe.

FOLDABLES™
Study Organizer

Categorizing Information Study Foldable Make this foldable to organize information from the chapter to help you learn more about the people and places of Western Europe.

Step 1 Collect three sheets of paper and place them about 1 inch apart.

Keep the edges straight.

Step 2 Fold up the bottom edges of the paper to form 6 tabs.

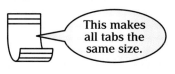
This makes all tabs the same size.

Step 3 When all the tabs are the same size, crease the paper to hold the tabs in place and staple the sheets together. Turn the paper and label each tab as shown.

Staple together along the fold.

WESTERN EUROPE TODAY
United Kingdom and Ireland
France and Benelux Countries
Germany, Switzerland, Austria
The Nordic Countries
Southern Europe

Reading and Writing As you read, use your foldable to write down what you learn about Western Europe. Write facts under each appropriate tab.

Why It Matters

Building a Community

Western Europe has been a center of world trade for hundreds of years. However, in the past the nations of this region often set up trade barriers against one another in order to protect their own industries. Today, as the European Union, these same countries are working together to make their region an even stronger economic power.

◄ **The Eiffel Tower in Paris, France**

The United Kingdom and Ireland

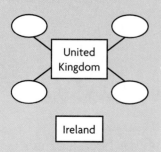
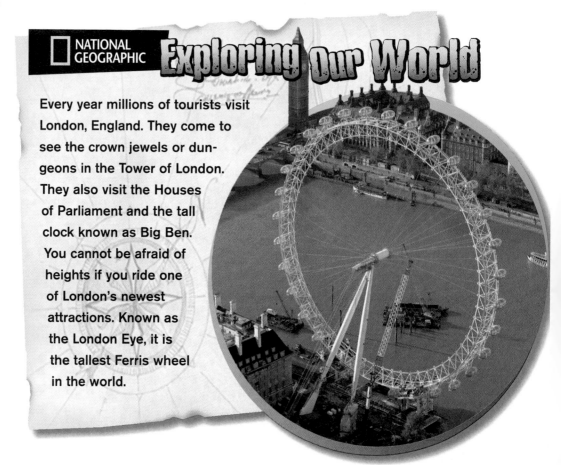

NATIONAL GEOGRAPHIC

Exploring Our World

Every year millions of tourists visit London, England. They come to see the crown jewels or dungeons in the Tower of London. They also visit the Houses of Parliament and the tall clock known as Big Ben. You cannot be afraid of heights if you ride one of London's newest attractions. Known as the London Eye, it is the tallest Ferris wheel in the world.

Also known as the **British Isles,** the countries of the **United Kingdom** and the **Republic of Ireland** lie in the North Atlantic Ocean, west of the European continent. While sharing similar physical characteristics, these two countries differ culturally.

The United Kingdom

About the size of Oregon, the United Kingdom is made up of four regions. **England** dominates the United Kingdom, in population and economic strength. However, **Scotland** and **Wales** are important parts of the United Kingdom. Both were conquered by England centuries ago. Today, movements for independence have grown in both countries, as seen by the popularity of the movie *Braveheart,* a film about the Scottish freedom-fighter William Wallace. In Wales, great effort has been made to keep the Welsh language alive by teaching it in school. (Language is one of the most important ways to keep a people's culture alive.)

The United Kingdom's fourth region—**Northern Ireland**—shares the island of Ireland with the Republic of Ireland.

The Economy   Over 250 years ago, inventors and scientists here sparked the **Industrial Revolution.** Today, the United Kingdom is still a major industrial and trading country. Manufactured goods and machinery are the leading exports, though new computer and electronic industries are gradually replacing older industries. Service industries such as banking, insurance, communications, and health care employ most of the country's people.

Farming is very efficient here. Still, the United Kingdom must import about one-third of its food. Why? A lack of farmland and a limited growing season make it impossible to feed the large population.

The Government   The United Kingdom is a *parliamentary democracy,* a form of government in which voters elect representatives to a lawmaking body called Parliament. It has two houses—the House of Commons and the House of Lords. The political party that has the largest number of members in the House of Commons chooses the government's leader, the prime minister. The House of Lords has little power. Most members of the Lords are nobles who have inherited their titles or who have been given titles by the Queen.

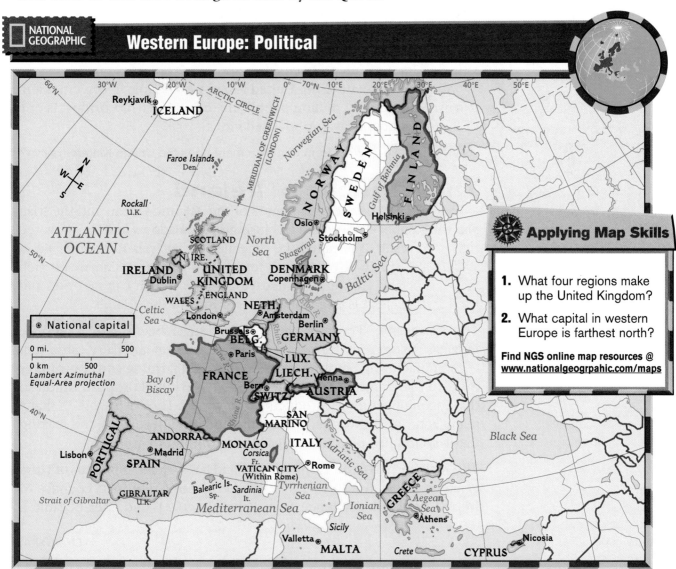

NATIONAL GEOGRAPHIC

Western Europe: Political

⊛ National capital

0 mi.   500
0 km   500
Lambert Azimuthal
Equal-Area projection

Applying Map Skills

1. What four regions make up the United Kingdom?

2. What capital in western Europe is farthest north?

Find NGS online map resources @ www.nationalgeogrpahic.com/maps

Urban and Rural Life

Bright green meadows greet people who venture into Ireland's countryside (above). In London, doubledecker buses show their bright colors (right).

Government Which Ireland is part of the United Kingdom?

The United Kingdom's government is also a **constitutional monarchy,** in which a queen or king is the official head of state. The monarch represents the country at public events but has little power.

The People About 60 million people live in the United Kingdom. The British people speak English, although Welsh and Scottish Gaelic are spoken in some areas. Most people are Protestant Christians, although immigrants practice Islam and other religions.

About 90 percent of the United Kingdom's people live in cities and towns. With nearly 7 million people, the capital city of **London** is one of Europe's most heavily populated cities.

√ Reading Check What form of government does the United Kingdom have?

The Republic of Ireland

Surrounded by the blue waters of the Atlantic Ocean and the Irish Sea, the Republic of Ireland has lush green meadows and tree-covered hills. As a result, it is called the Emerald Isle. At Ireland's center lies a wide, rolling plain covered with forests and farmland. The area is rich in **peat,** or wet ground with decaying plants, which can be dried and used for fuel. Peat is dug from **bogs,** or low swampy lands.

The Economy Potatoes, barley, wheat, sugar beets, and turnips are Ireland's major crops. Farmers raise sheep and beef and dairy cattle, too.

Manufacturing employs more people than farming and contributes more to the country's economy. Ireland joined the European Union so that it could market its products more widely. The Irish work in many manufacturing industries, processing foods and beverages, and making textiles, clothing, pharmaceuticals, and computer equipment.

The Northern Ireland Conflict Ireland has suffered hundreds of years of unrest under British rule. The southern, mostly Catholic, counties of Ireland won independence from Britain in 1921, and later became a republic. The northern counties, where many British Protestants had settled, remained part of the United Kingdom. Peace still did not come to the island. The **Nationalists,** who are typically Catholic, want the six counties of Northern Ireland to be reunited with the Republic. The **Loyalists,** who are typically Protestant, prefer that Northern Ireland remain under British rule. The fighting between these two groups, which the Irish refer to as "the troubles," has led to many deaths.

In 1998 officials of the United Kingdom and the Republic of Ireland met with leaders of the different sides in Northern Ireland. They all signed an agreement to end the political violence. The agreement gave Northern Ireland its own elected assembly, made up of Catholic and Protestant members, to govern Northern Ireland. Although disagreements continue to occur, many hope that the peace will last.

The People The Irish trace their ancestry to the **Celts** who settled Ireland around 500 B.C. Gaelic, a Celtic language, and English are Ireland's two official languages.

Today, Ireland is an urban nation. About 58 percent of the country's people live in cities or towns. Nearly one-third live in or around **Dublin,** the capital. Life often centers on the neighborhood church.

Irish music and folk dancing, are performed around the world. Of all the arts, however, the Irish have had the greatest influence on literature. Playwright **George Bernard Shaw,** poet William Butler Yeats, and novelist James Joyce are some of the country's best-known writers.

✔Reading Check How are Northern Ireland and the Republic of Ireland different?

Section 1 Assessment

Defining Terms
1. Define parliamentary democracy, constitutional monarchy, peat, bog.

Recalling Facts
2. **Region** What are the four regions of the United Kingdom?
3. **Economics** What are the two leading exports of the United Kingdom?
4. **Economics** Why did Ireland join the European Union?

Critical Thinking
5. **Analyzing Information** Why does the House of Lords have little power in the United Kingdom's Parliament?
6. **Understanding Cause and Effect** What disagreement has led to fighting in Northern Ireland?

Graphic Organizer
7. **Organizing Information** Create two diagrams like this one, one for the United Kingdom and one for the Republic of Ireland. Under each heading, list as many facts as you can for both countries.

```
   Land          People
        \        /
         Country
        /        \
  Economy      Government
```

Applying Social Studies Skills
8. **Analyzing Maps** Look at the political map on page 279. What is the capital of the United Kingdom? Of the Republic of Ireland?

Stonehenge

Stonehenge, one of the world's best-known and most puzzling ancient monuments, stands in southern England.

History of Stonehenge

The most noticeable part of Stonehenge is its huge stones set up in four circular patterns. A circular ditch and mound form a border around the site. Shallow dirt holes also circle the stones.

Stonehenge was built over a period of more than 2,000 years. The earliest construction, that of the circular ditch and mound, probably began about 3100 B.C. The outer ring of large pillars, topped with horizontal rocks, was built about 2000 B.C. An inner ring of stone pillars also supports horizontal stones.

There was no local source of stone, so workers carried it from an area that was about 20 miles (32 km) north. The stones are huge—up to 30 feet (9 m) long and 50 tons (45 t) in weight. Before setting the stones in place, workers smoothed and shaped them. They carved joints into the stones so that they would fit together perfectly. Then the builders probably used levers and wooden supports to raise the stones into position.

About 500 years later, builders added the third and fourth rings of stones. This time they used bluestone, which an earlier group of people had transported 240 miles (386 km) from the Preseli Mountains of Wales.

What Does It Mean?

Although much is known about when people built Stonehenge, experts do not agree on who built it. Early theories suggested that an ancient group known as Druids or the Romans built the monument. Now archaeologists believe that the monument was completed long before either of these groups came to the area.

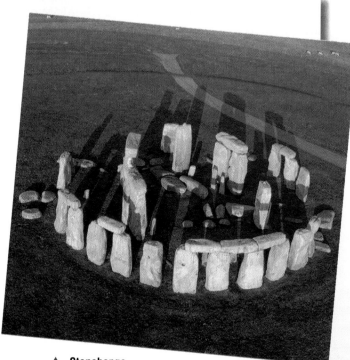

▲ Stonehenge

An even greater mystery is *why* Stonehenge was built. Most experts agree that Stonehenge was probably used as a place of worship. Some believe that the series of holes, stones, and archways were used as a calendar. By lining up particular holes and stones, people could note the summer and winter solstices. They could also keep track of the months. Some scientists think that early people used the site to predict solar and lunar eclipses.

▶ Making the Connection

1. About how old is Stonehenge?

2. From where did the stones used at Stonehenge come?

3. **Sequencing Information** Describe the order in which Stonehenge was built.

France and the Benelux Countries

Guide to Reading

Main Idea

France and the Benelux countries are important cultural, agricultural, and manufacturing centers of Europe.

Terms to Know

- navigable
- republic
- polder
- multinational company
- multilingual

Reading Strategy

Create a diagram like this one. Then list two countries that have major products in these categories. List two products for each country.

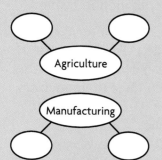

Agriculture

Manufacturing

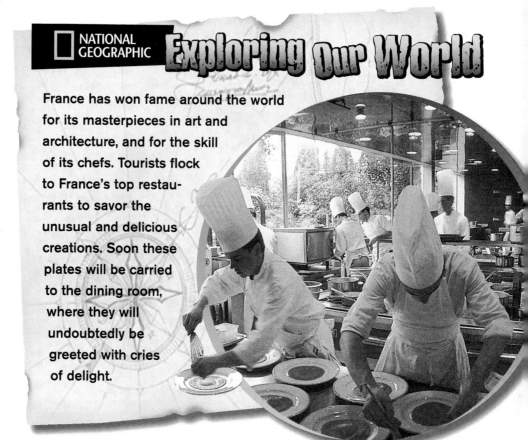

NATIONAL GEOGRAPHIC **Exploring Our World**

France has won fame around the world for its masterpieces in art and architecture, and for the skill of its chefs. Tourists flock to France's top restaurants to savor the unusual and delicious creations. Soon these plates will be carried to the dining room, where they will undoubtedly be greeted with cries of delight.

France and its neighbors in the west of Europe rank as major economic and cultural centers of the world. Today, joined in economic partnership in the European Union, they look forward to a peaceful and prosperous twenty-first century.

France

The largest country in western Europe, **France** is still smaller than the state of Texas. France's landscape is made up of high mountain ranges that separate it from Spain, Italy, and Switzerland, a large flat plain, and several rivers. Most of these rivers are navigable, or wide and deep enough to allow the passage of ships.

Most of France has a climate that is ideal for agriculture. The rich soil in France's flat lowland area makes France an important food producer. In many French towns, you can find open-air markets displaying an abundance of fresh farm produce.

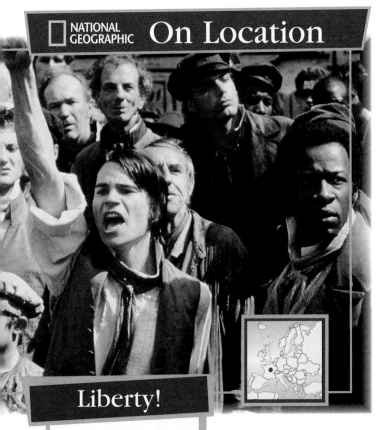

Liberty!

Les Miserables was a popular novel by Victor Hugo before it became a Broadway play. The story portrays the struggle of the French people for "liberty, equality, and fraternity."

History What important **pledge** used by Americans also contains the word *liberty?*

France's Economy France's well-developed economy relies on agriculture and manufacturing. Most people, however, work in service industries such as banking, commerce, communications, and tourism. Tourists from all over the world flock to visit France's historic and cultural sites, such as palaces and museums. They also come to enjoy the blue skies, rocky cliffs, and lovely beaches of France's Mediterranean coast.

France produces more food than any other nation in western Europe. In fact, it ranks as the second-largest food exporter in the world, after the United States. Yet only 5 percent of French workers labor on farms. Their success is a tribute to France's fertile soil, mild climate, and modern farming methods.

French farmers grow grains, sugar beets, fruits, and vegetables. They also raise beef and dairy cattle. In addition, vineyards are a common sight. The grapes are used to make famous French wines. Olives are grown along the warm, dry Mediterranean coast.

France's natural resources include bauxite, iron ore, and coal. France has small petroleum reserves and little hydroelectric power. How does the nation power its industries? About 80 percent of France's electricity comes from nuclear power plants.

Workers produce a variety of manufactured goods, including steel, chemicals, textiles, airplanes, cars, and computers. France is also a leading center of commerce, with an international reputation in fashion.

The French People "Liberté . . . Egalité . . . Fraternité" (Liberty, Equality, Fraternity)—France's national motto—describes the spirit of the French people. Although they have regional differences, the French share a strong national loyalty. Most French trace their ancestry to the Celts, Romans, and Franks of early Europe. They speak French, and about 90 percent of them are Roman Catholic.

France's government is known as the **Fifth Republic.** A republic is a strong national government headed by elected leaders. A powerful president, elected for a five-year term, leads the nation. The French president manages the country's foreign affairs. He or she appoints a prime minister to run the day-to-day affairs of government.

About three-fourths of France's 59.2 million people live in cities and towns. **Paris,** the capital and largest city, with its suburbs has a population of more than 10 million people. The city is home to many universities, museums, and other cultural sites. Outstanding cultural figures who lived in Paris include the writer Victor Hugo and the painters Claude Monet and Pierre-Auguste Renoir. Each year, millions of tourists flock to the City of Light, as Paris is called. They visit such sites as the Eiffel Tower, the cathedral of Notre Dame, and the Louvre (LOOV), one of the world's most famous art museums.

Reading Check What is the main religion in France?

The Benelux Countries

The name *Benelux* comes from combining the first letters of three countries' names: Belgium, the Netherlands, and Luxembourg. These three countries cooperated to form a trade partnership in 1948. These small countries have much in common. Their lands are low, flat, and densely populated. Most people live in cities, work in businesses or factories, and enjoy a high standard of living. All three nations are members of the European Union. They are also parliamentary democracies with constitutional monarchies.

Belgium About the size of Maryland, **Belgium** touches France, Luxembourg, Germany, and the Netherlands. Lying near major industrial regions, Belgium has long been a trade and manufacturing center. Belgian lace, Belgian chocolate, and Belgian diamond-cutting all enjoy a worldwide reputation for excellence. With few natural resources of their own, the Belgian people import metals, fuels, and raw materials. They use these materials to make vehicles, chemicals, and textiles, which are then exported.

Most of the people are Roman Catholic. The country has two main cultural and language groups. The Flemings in the north speak Flemish, a language based on Dutch. The south is home to the French-speaking Walloons. Tensions sometimes arise between the two groups, especially because there is more wealth and industry in the north than in the south. Most Belgians live in crowded urban areas. **Brussels,** the capital and largest city, is an international center for trade.

The Netherlands The **Netherlands**—about half the size of Maine—is one of the most densely populated countries in the world. Sometimes called Holland, its people are known as the Dutch.

Netherlands means "lowlands." True to its name, nearly half of this small, flat country lies below sea level. Without defenses against the sea, high tides would flood much of the country twice a day. To protect and reclaim their land from the sea, the Dutch have developed a simple method of building dikes to keep the sea out, then draining and pumping the wetlands dry. Once run by windmills, pumps are now driven by steam or electricity. These drained lands, called polders, have rich farming soil. The Dutch also build factories, airports, and even towns on them. The Delta Plan Project, completed in 1986, consists of huge barriers that keep the North Sea from overflowing the countryside during storms.

High technology makes small farms so productive that the Dutch can export cheese, vegetables, and flowers. In fact, the Netherlands ranks third in the world—after the United States and France—in the value of its agricultural exports. However, because machines make farming more productive, most people work in service industries, manufacturing, and trade.

Our World Today Online

Web Activity Visit the *Our World Today: People, Places, and Issues* Web site at owt.glencoe.com and click on **Chapter 11— Student Web Activities** to learn more about the Delta Plan Project.

▲ A tulip field in the Netherlands

About 90 percent of the Dutch live in cities and towns. Amsterdam is the capital and largest city. Living in a densely populated country, the Dutch make good use of their space. Houses are narrow but tall, and apartments are often built on canals and over highways. Some of Amsterdam's most famous people are the painters Rembrandt van Rijn and Vincent van Gogh. You may already have read *The Diary of Anne Frank.* This Dutch teenager's autobiography tells how she and her family tried to hide from the German Nazis during World War II.

About two-thirds of the Dutch people are Christian. A small number of immigrants are Muslims. The people of the Netherlands speak Dutch, but most also speak English.

Luxembourg Southeast of Belgium lies **Luxembourg,** one of Europe's smallest countries. The entire country is only about 55 miles (89 km) long and about 35 miles (56 km) wide.

Despite its size, Luxembourg is prosperous. Many multinational companies, firms that do business in several countries, have their headquarters here. It is home to the second-largest steel-producing company in Europe and is a major banking center as well.

Why is Luxembourg so attractive to foreign companies? First, the country is centrally located. Second, most people in this tiny land are multilingual, or able to speak several languages. They speak Luxembourgian, a blend of old German and French; French, the official language of the law; and German, used in most newspapers.

✓ Reading Check **What industries are important in Luxembourg?**

Assessment

Defining Terms
1. Define navigable, republic, polder, multinational company, multilingual.

Recalling Facts
2. Economics Name five of France's agricultural products.
3. Culture What are the two major cultures and languages of Belgium?
4. Human/Environment Interaction How do the Dutch protect their land from the sea?

Critical Thinking
5. Drawing Conclusions France is the second-largest food exporter in the world. Why is that remarkable?
6. Analyzing Information Why do foreign companies come to Luxembourg?

Graphic Organizer
7. Organizing Information Create a diagram like this one. In the center circle list three characteristics that are shared by these countries.

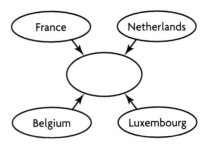

 Applying Social Studies Skills

8. Analyzing Maps Turn to the political map on page 279. What country borders France, Belgium, the Netherlands, and Luxembourg?

Study and Writing Skill

Using Library Resources

Your teacher has assigned a major research report, so you go to the library. As you wander the aisles surrounded by books, you wonder: Where do I start my research? Which reference tools should I use?

Learning the Skill

Libraries contain many resources. Here are brief descriptions of important ones:

- **Encyclopedia:** set of books containing short articles on many subjects arranged alphabetically
- **Biographical Dictionary:** brief biographies listed alphabetically by last names
- **Atlas:** collection of maps and charts
- **Almanac:** reference updated yearly that provides current statistics and historical information on a wide range of subjects
- **Card Catalog:** listing of every book in the library, either on cards or computerized; search for books by author, subject, or title
- **Periodical Guide:** set of books listing topics covered in magazines and newspaper articles
- **Computer Database:** collections of information organized for rapid search and retrieval
- **World Wide Web:** collection of information on the Internet accessed with a Web browser *(Caution: Some information may not be reliable.)*

Practicing the Skill

Suppose you are assigned a research report dealing with Denmark. Read the questions below, and then decide which of the resources on the left you would use to answer each question and why.

1. During which years did Queen Margarethe rule Denmark?
2. What is the current population of Denmark?
3. Besides "The Little Mermaid," what stories did Danish author Hans Christian Andersen write?

Applying the Skill

Using library resources, research the origins and main stories of Icelandic sagas. Find out if the sagas say anything about the land or environment of Iceland. Present the information you find to the class.

◄ The Little Mermaid statue in Copenhagen

Germany, Switzerland, and Austria

Guide to Reading

Main Idea

Germany, Switzerland, and Austria are known for their mountain scenery and prosperous economies.

Terms to Know

- autobahn
- federal republic
- reunification
- neutrality
- continental divide

Reading Strategy

Create a diagram like this one. Under the headings fill each oval with facts about each country. Put statements that are true of all three countries where the ovals overlap.

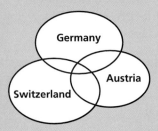

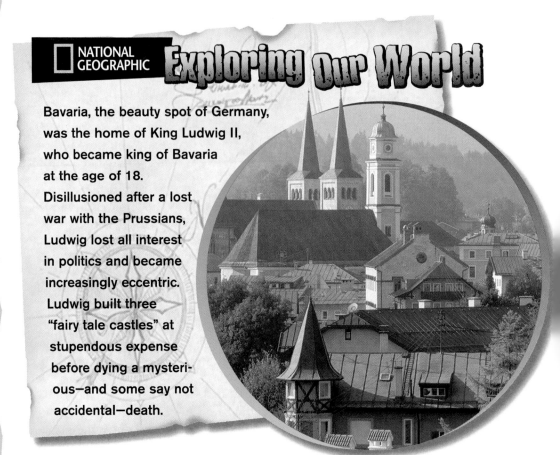

NATIONAL GEOGRAPHIC Exploring Our World

Bavaria, the beauty spot of Germany, was the home of King Ludwig II, who became king of Bavaria at the age of 18. Disillusioned after a lost war with the Prussians, Ludwig lost all interest in politics and became increasingly eccentric. Ludwig built three "fairy tale castles" at stupendous expense before dying a mysterious—and some say not accidental—death.

Today **Germany, Switzerland,** and **Austria** are adjusting to the changes sweeping Europe since the fall of communism. Germany, in particular, is facing challenges of reuniting its eastern and western sectors.

Germany

About the size of Montana, Germany lies in the heart of Europe. Mountains in the south and plains in the north form the physical landscape of Germany. The Alps rise in the southern German state of **Bavaria.** The lower slopes of these mountains—a favorite destination for skiers—are covered with forests.

One of Europe's most important waterways flows through Germany. The **Danube River** winds eastward across southern Germany. Rivers are also important in northern Germany, where they are used to transport raw materials and manufactured goods. The **Rhine River,** in the west, forms part of the border with France.

Because of the rivers and productive land, Germany's northern plain has many cities and towns. **Berlin,** the capital, is the major center of the northeast. To the west lies **Hamburg,** Germany's largest port city, located on the Elbe River.

An Economic and Industrial Power   Germany is a global economic power and a leader in the European Union. In fact, an area in western Germany called the **Ruhr** ranks as one of the world's most important industrial centers. The Ruhr developed around rich deposits of coal and iron ore. Battles have been fought among Europe's leaders for control of this productive area. Factories here produce high-quality steel, ships, cars, machinery, chemicals, and electrical equipment.

The growth of factories, service industries, and high technology in the last decade has used up the supply of workers. Thus, a growing number of immigrants, or guest workers, have come from Turkey, Italy, Greece, and the former Yugoslav republics. Sometimes they are the targets of racist attacks. When the economy takes a downturn and jobs are scarce, native-born people sometimes resent foreign laborers.

Lying just north of the Alps, Germany's **Black Forest** is famous for its beautiful scenery and for its wood products. Is the forest really black? No, but in places the trees grow so close together that the forest appears to be black. The Black Forest has suffered severe damage. Smoke containing sulfur and other chemicals from factories and automobiles makes the rain more acidic. The Germans have not yet found a solution to the acid rain problem because much of the pollution is formed in other countries.

Germany imports about one-third of its food, although it is a leading producer of beer, wine, and cheese. Farmers raise livestock and grow grains, vegetables, and fruits. Superhighways called autobahns, along with railroads, rivers, and canals, link Germany's cities.

Germany's Government   Like the United States, Germany is a federal republic in which a national government and state governments share powers. An elected president serves as Germany's head of state, but he or she carries out only ceremonial duties. The chancellor, chosen by one of the two houses of parliament, is the real head of the government.

One of the challenges of the current government has been reunification—bringing the two parts together under one government. (Remember that Germany was divided into East and West after World War II.) This process has been difficult. Workers in East Germany had less experience and training in modern technology than workers in West Germany. After reunification, many old and inefficient factories in the east could not compete with the more advanced industries in the west and were forced to close. Unlike western Germany, the number of people without jobs has risen in the eastern part.

People   Most of Germany's 82.2 million people trace their ancestry to groups who settled in Europe from about the A.D. 100s to 400s. The people speak German, a language that is related to English. Roman

▲ A cuckoo clock from the Black Forest region of Germany

NATIONAL GEOGRAPHIC On Location

Swiss Chocolate

Switzerland's factories produce some of the best chocolate in the world.

Economics What other products are made in Switzerland?

Catholics and Protestants make up most of the population and are fairly evenly represented. **Munich** (MYOO•nikh), the largest city in southern Germany, is known for its theaters, museums, and concert halls. Berlin has also emerged as a cultural center.

✓ Reading Check How much of its food does Germany import?

Switzerland and Austria

The **Alps** form most of the landscape in Switzerland, Austria, and **Liechtenstein.** That is why they are called the Alpine countries. Liechtenstein is a tiny country—only 60 square miles (155 sq. km)—sandwiched between Switzerland and Austria. The rugged Swiss Alps prevent easy travel between northern and southern Europe. For centuries, landlocked Switzerland guarded the few routes that cut through this barrier.

Switzerland The Swiss have enjoyed a stable democratic government for more than 700 years. Because of its location in the Alps, Switzerland has practiced neutrality—refusing to take sides in disagreements and wars between countries. As a result of this peaceful history, the Swiss city of **Geneva** is today the center of many international organizations.

The Alps make Switzerland the continental divide of central Europe. A continental divide is a high place from which rivers flow in different directions. Several rivers, including the Rhine and the Rhône, begin in the Swiss Alps. Dams built on Switzerland's rivers produce great amounts of hydroelectric power. Most of Switzerland's industries and its richest farmlands are found on a high plateau between two mountain ranges. **Bern,** Switzerland's capital, and **Zurich,** its largest city, are also located on this plateau.

Although it has few natural resources, Switzerland is a thriving industrial nation. Using imported materials, Swiss workers make high-quality goods such as electronic equipment, clocks, and watches. They also produce chemicals and gourmet foods such as chocolate and cheese. Tourism is an important industry, as are banking and insurance.

Zurich and Geneva are important centers of international finance. Recently, Swiss and Austrian banks have been criticized for keeping gold and deposits belonging to Jews sent to concentration camps during the Holocaust. International pressure has resulted in some of these deposits being returned to the heirs.

As you might expect given its location, Switzerland has many different ethnic groups and religions. Did you know that the country has four national languages? They are German, French, Italian, and Romansch. Most Swiss speak German, and many speak more than one language.

Austria   Austria is a landlocked country lying in the heart of Europe. The Alps cover three-fourths of Austria. In fact, Austria is one of the most mountainous countries in the world. Have you ever seen the movie *The Sound of Music*? It took place in Austria's spectacular mountains. Austria's climate is similar to Switzerland's. In winter, low-land areas receive rain, and mountainous regions have snow. Summers are cooler in Austria than they are in Switzerland.

Austria's economy is strong and varied. The mountains provide valuable timber and hydroelectric power. They also attract millions of tourists who enjoy hiking and skiing. Factories produce machinery, chemicals, metals, and vehicles. Farmers raise dairy cattle and other livestock, sugar beets, grains, potatoes, and fruits.

Most of Austria's 8.1 million people live in cities and towns and work in manufacturing or service jobs. The majority of people speak German. About 80 percent of the people are Roman Catholic.

Vienna, on the Danube River, is the capital and largest city. It has a rich history as a center of culture and learning. Some of the world's greatest composers, including Mozart, Schubert, and Haydn, lived or performed in Vienna. The city's concert halls, historic palaces and churches, and grand architecture continue to draw musicians today.

▲ Young people dance in one of Vienna's many ballrooms.

✓ Reading Check   **What economic benefits do Austria's mountains provide?**

Section 3 Assessment

Defining Terms
1. **Define** autobahn, federal republic, reunification, neutrality, continental divide.

Recalling Facts
2. **Human/Environment Interaction** What has damaged the Black Forest?
3. **Culture** What are Switzerland's four languages?
4. **Economics** What types of jobs do most Austrians have?

Critical Thinking
5. **Understanding Cause and Effect** What problems have emerged as a result of German reunification?
6. **Analyzing Information** How have the Alps helped Switzerland maintain its neutrality?

Graphic Organizer
7. **Organizing Information** Create a diagram like the one below. On the lines list two facts about Austria's physical features, two facts about Austria's people, and four facts about Austria's economy.

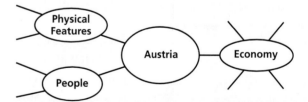

Applying Social Studies Skills

8. **Analyzing Maps** Look at the political map on page 279. The city of Berlin, Germany, is located at what degree of latitude?

The Nordic Countries

Guide to Reading

Main Idea

The Nordic countries have developed diverse economies, and their people enjoy a high standard of living.

Terms to Know

- fjord
- welfare state
- heavy industry
- sauna
- geyser
- geothermal energy

Reading Strategy

Create a chart like this one for each of the following countries: Norway, Sweden, Finland, Denmark, and Iceland. Fill in at least two key facts for (1) the land, (2) the economy, and (3) the people of each country.

Country	(1)
	(2)
	(3)

NATIONAL GEOGRAPHIC Exploring Our World

An ocean of snow covers northern Sweden in winter, when the sun is above the horizon for only a few hours a day and summer seems light-years away. Here, near Sweden's border with Norway, a footbridge used in summer to cross a ravine is almost buried in snow. Instead of using the footbridge, a hardy Swede zips by on his snowmobile.

The northernmost part of Europe is made up of five countries: **Norway, Sweden, Finland, Denmark,** and **Iceland.** People in these countries have standards of living that are among the highest in the world.

Norway

Norway's long, jagged coastline on the Atlantic Ocean includes many **fjords** (fee•AWRDS), or steep-sided, glacial-cut valleys that are inlets of the sea. The fjords provide Norway with sheltered harbors and beautiful scenery that is popular with tourists.

Norway's far northern location results in a mostly cold climate. However, a mild climate is found along Norway's southern and western coasts due to warm winds from the **North Atlantic Current.** As you might expect, most of Norway's 4.5 million people live in the south in close proximity to the coast.

Norway is a wealthy country, partly because of oil and natural gas pumped from beneath the North Sea. Today it is one of the world's largest oil exporters. The seas themselves provide an important export—fish. Warm ocean currents keep most of Norway's harbors ice-free all year.

Norway is a parliamentary democracy. It has a monarchy but is governed by an elected prime minister. In 1994, Norway voted not to join the European Union so that it could keep control of its own economy. EU membership is still hotly debated, however.

The people of Norway greatly value their cultural traditions. Elaborate folk dress is often seen at weddings and village festivals. Norwegians are a very modern people, though. Three-fourths of the population live in urban centers like the capital, **Oslo.** More than one-third own computers. When they are not typing on keyboards, they may be skiing or riding snowmobiles.

✓ Reading Check **What type of government does Norway have?**

Sweden

Like Norway, Sweden is a wealthy, industrial country. Its prosperity comes from abundant natural resources, including iron ore deposits and extensive pine forests. Exports include machinery, motor vehicles, paper products, wood, and electronic products. Only about 8 percent of Sweden's land can be used for farming. Swedish farmers have developed efficient ways to grow crops, and their farms supply most of the nation's food.

Sweden's economic wealth enabled it to become a welfare state —a country that uses high rates of taxation to provide services to people who are sick, needy, jobless, or retired. Sweden is a constitutional monarchy and joined the European Union.

Norway's Economy

A shopper goes from boat to boat looking for bargains in Bergen, Norway's water market (below left). Europe's richest oil and natural gas fields are found in the North Sea (below right).

Human/Environment Interaction **What keeps Norway's harbors ice-free all year?**

NATIONAL GEOGRAPHIC On Location

Most of Sweden's almost 9 million people live in cities in the southern lowlands. Stockholm is the country's capital and largest city. Sweden's high standard of living has attracted more than 1 million immigrants from nearby Norway and Denmark and distant Turkey and Vietnam.

Reading Check What two natural resources have helped make Sweden wealthy?

Finland

Finland holds some of the largest unspoiled wilderness in Europe. Most of Finland's wealth comes from its huge forests of spruce, pine, and birch. Paper and wood products are important exports. In recent years, **heavy industry**—or manufactured goods such as machinery—has driven Finland's economy. The Finns are also leaders in the electronic communications industry. In fact, Finns as young as 10 carry mobile phones to school. In 1995 Finland joined the European Union.

The ancestors of the Finns settled in the region thousands of years ago, probably coming from what is now Siberia in Russia. As a result, Finnish language and culture differ from those of other Nordic countries.

Most of Finland's more than 5 million people live in towns and cities on the southern coast. **Helsinki,** the capital, has over 1 million people, but the city has still kept a small-town atmosphere. For example, there are no high-rise buildings. With snow on the ground for about half of the year, Finns enjoy cross-country skiing. After outdoor activities, many Finns enjoy relaxing in **saunas,** or wooden rooms heated by water sizzling on hot stones.

Reading Check Where did the ancestors of the Finns come from?

Denmark and Iceland

Both Denmark and Iceland are countries whose histories are closely tied to the sea. For centuries, Iceland was ruled by Denmark. For this reason Danish is still widely spoken and understood in Iceland.

Most of Denmark is made up of a peninsula known as Jutland. Denmark also includes nearly 500 islands, only about 100 of which have people living on them. Denmark also rules the large island of Greenland. Throughout history, Denmark's location has made it a link for people and goods between the Nordic countries and the rest of Europe. Ferries and bridges connect Jutland and the islands. A bridge and tunnel now join Denmark's Zealand Island to Sweden.

Denmark has some of the richest farmland in northern Europe. Danish farm products include butter, cheese, bacon, and ham. Royal Copenhagen porcelain, a famous Danish export, is among the finest in the world. The Danes also invented and export the world-famous LEGO® toy building blocks. In 1993 Denmark joined the European Union.

The more than 5 million Danes enjoy a high standard of living. About 85 percent of them live in cities or towns. **Copenhagen,** Denmark's capital, is the largest of the Nordic cities. In Copenhagen's harbor is

Cozy Ballet?

Helle Oelkers (far right) is a member of one of Europe's finest ballet companies—the Royal Danish Ballet. Helle likes to think that her performance encourages audience members to feel *hygge. Hygge* means feeling cozy and snug. She explains, "The greatest compliment a Dane can give is to thank someone for a cozy evening."

a famous attraction: a statue of the Little Mermaid. (See the photo on page 287.) She is a character from a story by the Danish author Hans Christian Andersen. Andersen, who lived and wrote during the 1800s, is one of Denmark's most famous writers.

Iceland   Iceland, an island in the North Atlantic, is a land of hot springs and geysers—springs that shoot hot water and steam into the air. The people of Iceland make the most of this unusual environment. They use geothermal energy, or heat produced by natural underground sources, to heat most of their homes, buildings, and swimming pools.

What makes such natural wonders possible? Sitting on top of a fault line, Iceland is at the mercy of constant volcanic activity. Every few years, one of the country's 200 volcanoes erupts. The volcanoes heat the springs that appear across the length of Iceland.

Iceland's economy depends heavily on fishing. Fish exports provide the money to buy food and consumer goods from other countries. Iceland is concerned that overfishing will reduce the amount of fish available. To reduce its dependence on the fishing industry, Iceland has introduced new manufacturing and service industries.

More than 90 percent of the nearly 290,000 Icelanders live in urban areas. More than half the people live in the capital city of **Reykjavík** (RAY•kyah•veek). The people have a passion for books, magazines, and newspapers. In fact, the literacy rate in Iceland is 100 percent.

 Reading Check   **How do the people of Iceland take advantage of the country's geysers?**

 Section 4 **Assessment**

Defining Terms
1. Define fjord, welfare state, heavy industry, sauna, geyser, geothermal energy.

Recalling Facts
2. Location   Name the five Nordic countries.
3. Economics   What resource produces most of Norway's wealth?
4. History   Why do some Icelanders speak Danish?

Critical Thinking
5. Analyzing Information   How has Denmark's location affected its relationship with the rest of Europe?
6. Understanding Cause and Effect   Why is Finnish culture different from the rest of the Nordic countries?

Graphic Organizer
7. Organizing Information   Create a diagram like the one below that models three effects on Iceland that result because of its location on a fault line.

Fault		Effects on Iceland

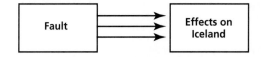 **Applying Social Studies Skills**

8. Analyzing Maps   Study the political map on page 279. Which Nordic capital lies the farthest north? Which Nordic capital lies the farthest south?

Southern Europe

Guide to Reading

Main Idea

The sea has played an important role in each of the countries of southern Europe.

Terms to Know

- dry farming
- parliamentary republic
- sirocco
- coalition government

Reading Strategy

Create a chart like this one for each of the following countries: Spain, Portugal, Italy, and Greece. Fill in at least one key fact about each country for each category listed.

Country	
Land	
Economy	
Government	
People	

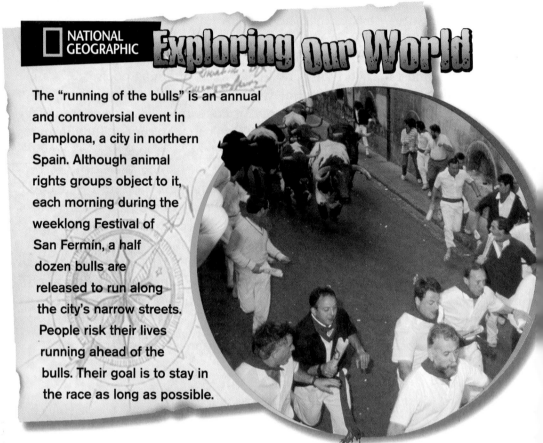

NATIONAL GEOGRAPHIC Exploring Our World

The "running of the bulls" is an annual and controversial event in Pamplona, a city in northern Spain. Although animal rights groups object to it, each morning during the weeklong Festival of San Fermín, a half dozen bulls are released to run along the city's narrow streets. People risk their lives running ahead of the bulls. Their goal is to stay in the race as long as possible.

Spain, **Portugal, Italy,** and **Greece**—and several tiny countries make up southern Europe. A long cultural heritage has produced many of the world's greatest writers, artists, and musicians. As you read in Chapter 9, it was the people of ancient Greece and Rome who played an especially important role in the development of Western civilization.

The Iberian Peninsula

Spain—about twice the size of Oregon—and Portugal—slightly smaller than Indiana—share the **Iberian Peninsula.** Tiny **Andorra,** with only 175 square miles (453 sq. km), perches high in the Pyrenees mountain range near Spain's border with France.

Portugal and most of Spain have mild winters and hot summers. Much of the interior of the peninsula is a dry plateau. In many areas the reddish-yellow soil is poor, and the land is dry-farmed to grow crops such as wheat and vegetables. In dry farming irrigation is not used. Instead the land is left unplanted every few years so that it can store moisture. Some farmers also herd sheep, goats, and cattle.

Growing Economies   Spain and Portugal both belong to the European Union. Once slow in developing manufacturing, the two countries in recent years have worked hard to catch up economically with other European Union nations.

Spain is one of the world's leading producers of olive oil. Portuguese farmers grow potatoes, grains, fruits, olives, and grapes. In addition, Portugal is the world's leading exporter of cork. The cork comes from the bark of certain oak trees, which grow well in central Portugal.

Millions of people travel to the Iberian Peninsula to enjoy the sunny climate, beautiful beaches, and ancient castles and cathedrals. Andorra draws millions of tourists each year who flock to its duty-free shops. Spain and Portugal also depend on the tourist industry. Manufacturing industries benefit both countries' economies as well. Spanish workers mine rich deposits of iron ore and make processed foods, clothing, footwear, steel, and automobiles.

The Government   Both Spain and Portugal are modern democracies. Spain is a constitutional monarchy, in which a king or queen is head of state, but elected officials run the government. Portugal is a **parliamentary republic,** with a president as head of state. A prime minister, chosen by the legislature, is the head of government. Andorra is a semi-independent principality—it is governed by both Spain and France.

The People   Most people in Spain and Portugal are Roman Catholic. Despite similar histories, the people of Spain and Portugal have their cultural differences. Portugal developed a unified culture based on the Portuguese language, while Spain remained a "country of different countries." The Spanish people do not all speak the same language or even have a single culture.

The **Basque** people in the Pyrenees see themselves as completely separate from Spain. They speak Basque, a language unlike any other in the world. Having lived in Spain longer than any other group, many Basques want independence in order to preserve their way of life. Some Basque groups have even used violence against the Spanish government.

Lisbon is Portugal's busy capital, but Portugal is mostly rural. On the other hand, more than three-fourths of Spain's people live in cities and towns. **Madrid,** Spain's capital, has nearly 5 million people and ranks as one of Europe's leading cultural centers. Madrid faces the usual urban challenges of heavy traffic and air pollution. Fast-paced **Barcelona** is Spain's leading seaport and industrial center.

You find some centuries-old traditions even in the modern cities. For example, most Spanish families usually do not eat dinner until 9 or 10 o'clock at night. On special occasions, Spaniards enjoy paella, a traditional dish of shrimp, lobster, chicken, ham, and vegetables mixed with seasoned rice.

Rock and jazz music are popular with young Spaniards and Portuguese, but the people of each region have their own traditional songs and dances as well. Spanish musicians often accompany

Islamic Art
The Muslims brought scientific knowledge to Spain. They introduced methods of irrigation, and new crops. They also brought literature, music, and art.

Islam forbids art that includes human forms. As a result, Muslim artists created complex patterns and elaborate designs. The tile below is an example of the beautiful mosaics seen today the world over.

singers and dancers on guitars, castanets, and tambourines. Spanish dances, such as the bolero and flamenco, and soulful Portuguese music known as fado have spread throughout the world.

Reading Check In what export does Portugal lead the world?

Italy

The Italian peninsula sticks out from Europe into the center of the Mediterranean Sea. The peninsula looks like a boot about to kick a triangle-shaped football. The "football" is **Sicily,** an island that belongs to Italy. Two tiny countries—**San Marino** and **Vatican City**—lie within the Italian "boot."

The Alps tower over northern Italy, while the rumbling of volcanic mountains echoes through the southern part of the peninsula and the island of Sicily. Throughout history, the southern part of Italy has experienced volcanic eruptions and earthquakes.

Most of Italy has a mild climate of sunny summers and rainy winters. In spring and summer, hot dry winds called siroccos blow across Italy from North Africa.

Italy's Economy In the past 50 years, Italy has changed from a mainly agricultural country into one of the world's leading industrial economies. Many of those goods are produced by small, family-owned businesses rather than by large corporations. Italian businesses are known for creating new designs and methods for making products. Italy is a member of the European Union.

Northern Italy is the center of manufacturing. Tourism is also important in northern and central Italy. Resorts in the Alps attract skiers. **Venice,** to the northeast, is built on 117 islands. You find no cars in this city, which is crisscrossed by canals and relies on boats for transportation. In central Italy lies **Rome,** the seat of the ancient Roman Empire. It is from the Vatican City, surrounded by Rome, that the pope guides the Catholic Church. In Rome, you can see ancient ruins and magnificent churches and palaces.

Southern Italy is poorer and less industrialized than northern and central Italy. Unemployment and poverty are common. Many southern Italians have moved to northern Italy or to other parts of Europe.

Italy's Government After World War II, Italy became a democratic republic. Yet democracy did not bring a stable government. Rivalry between the wealthy north and the poorer south has caused political tensions. In addition, many political parties exist, and no single party has been strong enough to gain control. Instead, Italy has seen many coalition governments, where two or more political parties work together to run a country.

A Varied Economy

The canals of Venice draw thousands of tourists (above)—while the city of Milan boasts a fashion industry with stylish models (below).

Economics How has Italy's economy changed in the past 50 years?

Italy's People   About 70 percent of Italy's nearly 58 million people live in towns and cities, and more than 90 percent of Italians work in manufacturing and service industries. Most Italians—more than 95 percent—are Roman Catholic. Celebrating the church's religious festivals is a widely shared part of Italian life. Vatican City in Rome is the headquarters of the Roman Catholic Church. The pope, who is the head of the church, lives and works here. Vatican City has many art treasures as well as the world's largest church, St. Peter's Basilica.

The people of Italy speak Italian, which developed from Latin, the language of ancient Rome. Italian is closely related to French and Spanish. Pasta, made from flour and water, is the basic dish in Italy. Some pasta dishes are spaghetti, lasagna, and ravioli.

✓ Reading Check   Why have coalition governments been necessary in Italy?

Greece

The Greek mainland sits on the southern tip of the Balkan Peninsula, which juts out from Europe into the Mediterranean Sea. Greece also includes 2,000 islands around the mainland. Like other Mediterranean areas, Greece is often shaken by earthquakes. Mountain ranges divide Greece into many separate regions. Historically, this has kept people in one region isolated from people in other regions.

Of the 2,000 Greek islands, only about 170 have people living on them. The largest Greek island, covering more than 3,000 square miles (7,770 sq. km), is **Crete.** Farther east in the Mediterranean is the island country of **Cyprus.** Once under Turkish and then British rule, Cyprus became independent in 1960. For centuries Greeks and Turks have lived on Cyprus, but fighting between the two groups has resulted in a divided country.

EXPLORING CULTURE

Architecture

For 800 years the Leaning Tower of Pisa has stood as a monument to construction mistakes. Begun in 1173, the tower began to tilt even before it was finished. Over time the tower moved even more, until by 1990, it leaned 15 feet (4.5 m) to the south. Fearing the tower would fall over, experts closed it. They added 800 tons (726 t) of lead weights to its base. They also removed 30 tons (27 t) of subsoil from underneath the north side of the tower in hopes that it would sink the opposite way. Visitors have once more returned to the tower.

Looking Closer Why do you think experts fixed the tower's problem but still left it leaning?

Greece's Economy   Greece belongs to the European Union but has one of the least industrialized economies in Europe. Because of the poor, stony soil, most people living in the highlands must graze sheep and goats. Greece must import food, fuels, and many manufactured goods. Farmers cultivate sugar beets, grains, citrus fruits, and tobacco. Greece has important crops of olives, used for olive oil, and grapes, used for wine.

No part of Greece is more than 85 miles (137 km) from the sea. Greece still has one of the largest shipping fleets in the world, including oil tankers, cargo ships, fishing boats, and passenger vessels. Shipping is vital to the economy.

Tourism is another key industry. Each year millions of visitors come to Greece to visit historic sites, such as the Parthenon in the capital city of **Athens** and the temple of Apollo at Delphi. Others come to relax on beaches and to enjoy the beautiful island scenery.

Greece's Government and People   Today, Greece is a parliamentary republic. About 65 percent of Greece's 10.9 million people live in urban areas. The Greeks of today have much in common with their ancestors. They debate political issues with great enthusiasm, and they value the art of storytelling.

More than 95 percent of Greeks are Greek Orthodox Christians. Religion influences much of Greek life, especially in rural areas. Easter is the most important Greek holiday. Traditional holiday foods include lamb, fish, and feta cheese—made from sheep's or goat's milk.

 Greek folk dancers in traditional costumes

✓ Reading Check   **What are two key industries in Greece?**

Section 5 Assessment

Defining Terms
1. **Define** dry farming, parliamentary republic, sirocco, coalition government.

Recalling Facts
2. **Location** What three countries are located on the Iberian Peninsula?
3. **Economics** Which is the more prosperous region in Italy—north or south?
4. **Culture** List four things tourists see in Italy.

Critical Thinking
5. **Analyzing Information** Why is it expected that Greece's economy would be dependent upon the sea?
6. **Understanding Cause and Effect** Why do the Basque people feel separate from the rest of Spain?

Graphic Organizer
7. **Organizing Information** In a chart like this one, list facts about Spain, Portugal, and Greece for each category.

	Spain	Portugal	Greece
Land			
Economy			
Cities			
People			

Applying Social Studies Skills

8. **Analyzing Maps** Turn to the political map on page 279. What body of water touches most of the southern countries of Europe?

Section 1 | **The United Kingdom and Ireland**

Terms to Know

peat   bog

parliamentary democracy

constitutional monarchy

Main Idea

The United Kingdom and Ireland are small in size, but their people have had a great impact on the rest of the world.

✓ Economics The United Kingdom is a major industrial and trading country.

✓ Economics Manufacturing is important to Ireland's economy.

✓ History After years of conflict, a peace plan was adopted in Northern Ireland.

Section 2 | **France and the Benelux Countries**

Terms to Know

navigable   polder

republic

multinational company

multilingual

Main Idea

France and the Benelux countries are important cultural, agricultural, and manufacturing centers of Europe.

✓ Culture Paris is a world center of art, learning, and culture.

✓ Location Belgium's location has made it an international center for trade.

✓ Economics Luxembourg is home to many multinational companies.

Section 3 | **Germany, Switzerland, and Austria**

Terms to Know

autobahn

federal republic

reunification

neutrality

continental divide

Main Idea

Germany, Switzerland, and Austria are known for their mountain scenery and prosperous economies.

✓ Economics The German economy is very strong.

✓ Economics Switzerland produces high-quality manufactured goods.

✓ Economics Austria's economy makes use of its mountainous terrain.

Section 4 | **The Nordic Countries**

Terms to Know

fjord   sauna

geyser

welfare state

heavy industry

geothermal energy

Main Idea

The Nordic countries have developed diverse economies, and their people enjoy a high standard of living.

✓ Region The Nordic countries include Norway, Sweden, Finland, Denmark, and Iceland.

✓ Culture Finnish culture differs from other Nordic countries.

✓ Economics Sweden's prosperity comes from forests and iron ore.

Section 5 | **Southern Europe**

Terms to Know

dry farming

sirocco

parliamentary republic

coalition government

Main Idea

The sea has played an important role in each of the countries of southern Europe.

✓ Location Spain and Portugal occupy the Iberian Peninsula.

✓ Economics Italy is one of the world's leading industrial economies.

✓ Place Greece consists of a mountainous mainland and 2,000 islands.

Assessment and Activities

Using Key Terms

Match the terms in Part A with their definitions in Part B.

A.

1. multilingual
2. heavy industry
3. coalition government
4. neutrality
5. dry farming
6. welfare state
7. polder
8. autobahn
9. constitutional monarchy
10. multinational company

B.

a. land reclaimed from the sea
b. leaving land unplanted to store moisture
c. refusing to take sides
d. country that uses tax money to help people in need
e. company that has offices in several countries
f. government that has a king or queen but is run by elected officials
g. superhighway
h. able to speak several languages
i. two or more political parties working together to run a country
j. production of industrial goods

Reviewing the Main Ideas

Section 1 The United Kingdom and Ireland

11. **Region** What regions make up the United Kingdom?
12. **Culture** Name the major language(s) and religion of the Republic of Ireland.

Section 2 France and the Benelux Countries

13. **Government** What is the Fifth Republic?
14. **Economics** Why do the Dutch reclaim land from the sea?

Section 3 Germany, Switzerland, and Austria

15. **Government** What challenges does the reunification of Germany bring?
16. **Location** Why is Geneva the center of many international organizations?

Section 4 The Nordic Countries

17. **Economics** Why is Norway wealthy?
18. **Culture** What is Iceland's literacy rate?

Section 5 Southern Europe

19. **Culture** Why do the Basque people want independence from Spain?
20. **Economics** How does the rocky landscape influence Greece's economy?

NATIONAL GEOGRAPHIC — Western Europe

Place Location Activity

On a separate sheet of paper, match the letters on the map with the numbered places listed below.

1. Ireland
2. North Sea
3. Belgium
4. Austria
5. Switzerland
6. Spain
7. Norway
8. Portugal
9. Sweden
10. Iceland

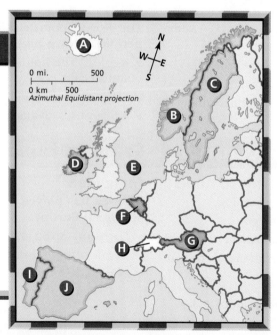

 ## Critical Thinking

21. **Analyzing Information** Why is the name Land of Fire and Ice appropriate for Iceland?

22. **Organizing Information** Create an outline of each country in Section 5. Use the following guide as your base outline.

    I. Name of Country
       A. Land
       B. Economy
       C. People

 ## Current Events Journal

23. **Writing a News Broadcast** Research an important event that recently took place in one of the countries studied in this chapter. Write and present a news broadcast about the event. Make a prediction about how the event might affect life in the future.

 ## Mental Mapping Activity

24. **Focusing on the Region** Create a simple outline map of western Europe, and then label the following:

    • United Kingdom   • Italy
    • France   • Spain
    • Germany   • Switzerland
    • Sweden   • Iceland

 ## Technology Skills Activity

25. **Using a Spreadsheet** List the names of the western European countries in a spreadsheet, beginning with cell A2 and continuing down the column. Find each country's population and record the figures in column B. In column C list each country's area in square miles. Title column D "Population Density" then divide column B by column C to find the population density. Print and share your spreadsheet with the class.

The Princeton Review

Standardized Test Practice

Directions: Study the graph below, and then answer the question that follows.

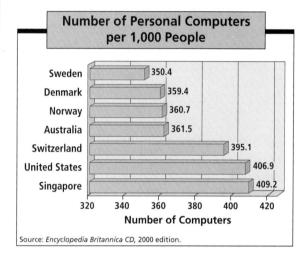

Number of Personal Computers per 1,000 People

Country	Number of Computers
Sweden	350.4
Denmark	359.4
Norway	360.7
Australia	361.5
Switzerland	395.1
United States	406.9
Singapore	409.2

Source: *Encyclopedia Britannica CD,* 2000 edition.

1. **Which Nordic country has the highest number of personal computers per 1,000 people?**

   A Singapore

   B Switzerland

   C Denmark

   D Norway

Test-Taking Tip: Use the information on the graph to help you answer this question. Look carefully at the information on the bottom and the side of a bar graph to understand what the bars represent. The important word in the question is *Nordic.* Other countries may have more personal computers, but which Nordic country listed on the graph has the most personal computers per 1,000 people?

Eastern Europe Today

The World and Its People NATIONAL GEOGRAPHIC

To learn more about the people and places of eastern Europe, view **The World and Its People Chapter 13** video.

Our World Today **online**

Chapter Overview Visit the **Our World Today: People, Places, and Issues** Web site at owt.glencoe.com and click on **Chapter 12—Chapter Overviews** to preview information about eastern Europe.

Compare-Contrast Study Foldable Make the following foldable to help you compare and contrast what you learn about Western Europe and Eastern Europe.

Step 1 Fold a sheet of paper in half from side to side.

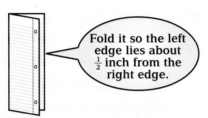

Fold it so the left edge lies about $\frac{1}{2}$ inch from the right edge.

Step 2 Turn the paper and fold it into thirds.

Step 3 Unfold and cut the top layer only along both folds.

This will make three tabs.

Step 4 Label as shown.

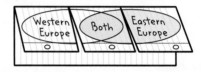

Western Europe | Both | Eastern Europe

Reading and Writing Before you read Chapter 12, record what you learned about Western Europe in Chapter 11 under the Western Europe tab of your foldable. As you read Chapter 12, write what you learn about Eastern Europe under the correct tab. Then list ways these two regions are similar under the middle tab.

Why It Matters

From Communism to Democracy

Since the fall of communism, the countries of eastern Europe have continued to change. The formation of new democratic governments has led to closer ties with other free nations in Europe. The economic influence of eastern Europe grows as the region becomes a new market for western goods. The changes are not occurring smoothly, however, and many challenges have to be met.

◄ Starometske Namesti and Tyn Church in Prague, Czech Republic

Guide to Reading

Main Idea

The countries of East Central Europe are undergoing many changes to their political and economic systems.

Terms to Know

- acid rain
- pope
- landlocked
- bauxite
- spa

Reading Strategy

Create a chart like this one, and write one fact about the people of each country in East Central Europe.

Country	People
Poland	
Estonia	
Latvia	
Lithuania	
Hungary	
Czech Republic	
Slovakia	

NATIONAL GEOGRAPHIC Exploring Our World

The Estonians have a rich tradition of dance, music, and story-telling. Every four years, Estonians stage the Song and Dance Festival to celebrate their people's music. Up to 500,000 people of Estonian descent may take part, many coming from as far away as the United States, Canada, and Australia.

East Central Europe is a region sandwiched between western Europe, the Baltic Sea, the Balkan Peninsula, and the countries of Ukraine and Belarus to the east. **Poland, Estonia, Latvia, Lithuania, Hungary,** the **Czech Republic,** and **Slovakia** make up this region. Though close together in distance, these countries have distinct histories and customs. This region of the world has undergone great changes in recent years.

Poland

Poland is one of the largest countries in eastern Europe. About the size of New Mexico, Poland lies on the huge North European Plain, which stretches from France to Russia. Many Polish people live in this fertile central region. Warm winds blowing across Europe from the Atlantic Ocean bring mild weather year-round.

Poland is dotted with thousands of small farms, on which about 25 percent of Poles work. Farmers here grow more potatoes and rye than most other countries in the world. Other crops include wheat, sugar beets, fruits, and vegetables. Some farmers raise cattle, pigs, and chickens.

Most mining and manufacturing take place in Poland's central and southern regions. The mountains hold copper, zinc, and lead. Poland also has petroleum and natural gas, and produces hydroelectric power. Coal mining is one of the most important industries. Factories process foods and make machines, transportation equipment, and chemicals. Warsaw is Poland's capital and the city of **Gdańsk** (guh•DAHNSK), a Baltic Sea port, is an important shipbuilding center.

NATIONAL GEOGRAPHIC

Eastern Europe: Political

Applying Map Skills

1. What is the capital of Poland?

2. What eastern European countries border the Adriatic Sea?

Find NGS online map resources @ **www.nationalgeographic.com/maps**

National capital
Major city

0 mi.   200
0 km   200
Azimuthal Equidistant projection

Latvia's Economy

Fish caught in the nearby Baltic Sea are sold in Rīga's Central Market.

Economics On what is Latvia's economy based?

Environmental Challenges Under Communist rule, Polish factories caused some of the worst water and air pollution in Europe. Since 1989 the government has moved to clean up the environment. Still, problems continue because Polish factories rely on burning coal. Factory smoke causes acid rain, or rain containing chemical pollutants.

Daily Life About 38.6 million people live in Poland. Almost all of them are ethnic Poles. Poles belong to the large ethnic group called Slavs and speak Polish, a Slavic language.

Poland is more rural than nations in other parts of Europe. About one-third of the people live in the countryside. As Poland's economy changes, more people are moving to cities.

Poles feel a deep loyalty to their country. Religion unites Poles as well. Most are Roman Catholic, and religion has a strong influence on daily life. Poles were very proud in 1978 when Karol Wojtyla (voy•TEE•wah) was named pope, or head of the Roman Catholic Church. Taking the name John Paul II, he was the first Pole ever to become pope.

✔ Reading Check What two beliefs or attitudes unite the Polish people?

The Baltic Republics

The small Baltic republics of Estonia, Latvia, and Lithuania lie on the shores of the Baltic Sea. For most of their history, the Baltic republics have been under Russian control. Though they are independent today, these countries still have large Russian minority populations.

The Baltic republics are located on poor, swampy land. Their well-developed economies are based mainly on dairy farming, beef production, fishing, and shipbuilding. In recent years, increased trade and industry have raised standards of living in this region. All three republics are moving toward free-market economies and closer ties with the countries in western Europe.

There is considerable tension between the native ethnic groups and the large Russian minorities. Many Russians claim that they are now being treated like second-class citizens. Most people in Estonia and Latvia are Protestant Lutherans, while Roman Catholics make up the majority in Lithuania.

✔ Reading Check What two major religions are practiced in the Baltic republics?

Hungary, the Czech Republic, and Slovakia

On the western edge of eastern Europe, you find Hungary, the Czech (CHEHK) Republic, and Slovakia (sloh•VAH•kee•uh). The Czech Republic and Slovakia once were partners in a larger country known as Czechoslovakia.

Hungary Hungary, almost the size of Indiana, is landlocked, meaning that its land does not border a sea or an ocean. Hungary depends on the Danube River for trade and transportation. Its capital, Budapest, is divided in two by this river. Hungary's farmers grow corn, sugar beets, wheat, and potatoes in the country's rich soil. Grapes, grown to make wine, are also an important crop. Hungary has important natural resources such as coal, petroleum, and natural gas. Workers also mine bauxite, a mineral used to make aluminum. Service industries, such as financial services and tourism, now thrive.

About 90 percent of Hungary's 10 million people are descended from the **Magyars,** a nomadic people from Central Asia who came to Hungary about 1,000 years ago. Almost all speak the Hungarian language. About two-thirds are Roman Catholic, while another one-fourth is Protestant. Two-thirds of Hungarians live in towns and cities.

The Czech Republic The Czech Republic is also a landlocked country. Many areas are known for their natural beauty. In the mountains to the north and south, you can visit spas, or resorts with hot mineral springs that people bathe in to improve their health.

The Czechs enjoy a high standard of living compared to other eastern European countries. Large fertile areas make the Czech Republic a major agricultural producer. Farmers grow grains, sugar beets, potatoes,

Music

Early European shepherds were probably the first to play bagpipes. The bag is made from an animal's hide or stomach. To inflate the bag, air is blown through a tube or pumped with a bellows—a small air pump—under one arm. This air escapes through hollow sticks or bones in a controlled way to make different notes. The "drone" pipes produce a steady note, while the "chanter" pipes produce the melody.

Looking Closer Do you think drone or chanter pipes would be more difficult to control? Why?

GO TO

World Music: A Cultural Legacy
Hear music of this region on Disc 1, Track 17.

and other foods. The country has some petroleum and natural gas. Minerals include limestone, coal, and kaolin, a fine clay used for pottery.

Prague (PRAHG), the capital, is a center of service industries, tourism, and high-technology manufacturing. Although manufacturing provides many consumer products, many factories are old, inefficient, and harmful to the environment. The Czechs are trying to modernize them to continue their prosperity.

Today the Czech Republic is a parliamentary democracy, with a powerful president assisted by a prime minister. Two-thirds of the Czech Republic's 10.3 million people live in cities. Prague is often called "the city of a hundred spires" because of its many church steeples. More recently, the Czech Republic has been a leading European center of jazz.

Slovakia Farmers in the lowlands of Slovakia grow barley, corn, potatoes, sugar beets, and wine grapes. The mountains are rich in iron ore, lead, zinc, and copper. Factories produce iron and steel products, cement, plastics, textiles, and processed foods.

Slovaks make up most of the population. They have a language and culture different from the Czechs'. Most Slovaks are Roman Catholic. Nearly 60 percent of Slovakia's 5.4 million people live in modern towns and cities. Tourists visit villages to see people dress in traditional clothes for festivals. You might even see musicians playing folk music on shepherds' flutes and bagpipes.

✓ Reading Check **What crops are common to Hungary, the Czech Republic, and Slovakia?**

Assessment

Defining Terms

1. Define acid rain, pope, landlocked, bauxite, spa.

Recalling Facts

2. Culture What is Poland's language and major religion?

3. Culture What is the difference between the religion of Lithuania and that of the other Baltic republics?

4. Economics What are three of the Czech Republic's natural resources?

Critical Thinking

5. Analyzing Information What has been the cause of considerable tension in the Baltic republics?

6. Analyzing Information Why do the Czechs have a high standard of living?

Graphic Organizer

7. Organizing Information Create a diagram like the one below. Choose one country of East Central Europe and write its name in the center oval. Then add at least two facts under the heading in each outer oval.

```
   People        Land
        Country
       Economy
```

Applying Social Studies Skills

8. Analyzing Maps Refer to the political map on page 307. What do Estonia's and Latvia's capitals have in common that Lithuania's capital does not?

The Balkan Countries

Guide to Reading

Main Idea

The Balkan countries have greatly suffered from ethnic conflicts and economic setbacks.

Terms to Know

- consumer goods
- ethnic cleansing
- refugee
- mosque

Reading Strategy

Create a chart like this one. For each Balkan country, write a fact or cause in the left box. Then write an effect that results from that fact in the right box.

| Cause | → | Effect |

NATIONAL GEOGRAPHIC

Exploring Our World

Traditional dress, folk music, and dancing enliven outdoor festivals in Romania. Many of these traditions come from the Roma people, who have lived here for centuries. If you expect to see folk dress in Romania's capital, however, teenagers there might think you are old-fashioned. These teens in Bucharest listen to rock music and watch TV just as you do.

Europe's **Balkan Peninsula** lies between the Adriatic Sea and the Black Sea. The political map on page 307 shows you that several countries make up this Balkan region. They are **Romania, Bulgaria,** the former **Yugoslav** republics, and **Albania.**

Romania

Romania sits on the northeastern edge of the Balkan Peninsula. The **Carpathian Mountains** take up about one-third of the country's land area. A vast plateau covers central Romania. A coastal region along the Black Sea includes the mouth of the Danube River. Winters can be very cold and foggy, with much snow. Summers are hot and sunny, but rainfall is abundant.

Romania's economic activities include farming, manufacturing, and mining. The forested mountains and central plateau contain deposits of coal, petroleum, and natural gas. Oil derricks rise in the south. Orchards and vineyards stretch along Romania's western, eastern, and southern borders. Farmers also grow grains, vegetables, and herbs here.

Despite abundant resources, Romania's economy has been held back by the communist policies of the past. Under communism, Romania's factories produced steel, chemicals, and machinery. Few **consumer goods**—clothing, shoes, and other products made for people—were manufactured. Romania now has a free-market economy to supply these goods, but aging factories must be updated for Romania's economy to grow. In addition, the country needs to heal an environment widely damaged by air and water pollution.

The Romanians About 56 percent of Romania's people live in towns and cities. Bucharest, the capital and largest city, has more than 2 million people. What does Romania's name tell you about its history? If you guessed that the Romans once ruled this region, you are correct. Romania's history and culture were greatly influenced by the Romans. The Romanian language is closer to French, Italian, and Spanish—which are all based on Latin—than it is to other eastern European languages. In other ways, the Romanians are more like their Slavic neighbors. Many Romanians are Eastern Orthodox Christians.

✓ Reading Check **To what other languages is Romanian related?**

Bulgaria

Mountainous Bulgaria lies south of Romania. Fertile valleys and plains are tucked among Bulgaria's mountains. The coast along the Black Sea has warmer year-round temperatures than the mountainous inland areas.

Bulgaria's economy rests on both agriculture and manufacturing. Wheat, corn, and sugar beets grow in the fertile valleys. Roses are grown in the central Valley of the Roses. Their sweet-smelling oil is used in perfumes. Manufacturing depends on the country's deposits of zinc and coal. Factories produce machinery, metals, textiles, and processed foods. Tourism is growing as visitors flock to Bulgaria's resorts on the Black Sea.

The Bulgarians Most of Bulgaria's 8.1 million people trace their ancestry to the Slavs, Turks, and other groups from Central Asia. Most Slavic people use the **Cyrillic** (suh•RIH•lihk) alphabet, which was first created to write the Slavic language. The Bulgarian language is also written in this Cyrillic alphabet. Most Bulgarians practice the **Eastern Orthodox Christian** religion. About 13 percent of the people are Muslims.

Sofia, with over 1 million people, is the capital and largest city. During the summer, Bulgarians join vacationers from other countries at resorts on the Black Sea coast. Here, modern hotels line wide, sandy beaches.

✓ Reading Check **What alphabet is used in many Slavic languages?**

Transylvania

The region of central Romania known as Transylvania was the setting for English author Bram Stoker's vampire novel *Dracula.* Recently, a doctor noticed that many myths about vampires matched the symptoms of rabies, including pain from bright lights. He found that rabies had spread through the region at the same time that the vampire tales began.

Former Yugoslav Republics

The former Yugoslav republics used to be one large-sized country called **Yugoslavia.** In the early 1990s, long-simmering disputes among ethnic groups boiled to the surface and tore the country apart. Five countries eventually emerged: **Slovenia, Croatia, Bosnia and Herzegovina** (HEHRT•seh•GAW•vee•nah), **Serbia and Montenegro,** and **Macedonia,** also known as the Former Yugoslav Republic of Macedonia (or F.Y.R.O.M.).

After the breakup, Serbia and Montenegro were known as Yugoslavia. Serbia wanted to control the other republics and to protect the Serbs living in them. As a result, wars erupted throughout the 1990s. Some countries forced people of other ethnic groups to leave their homes, a policy called ethnic cleansing. Tens of thousands died or were murdered. Thousands more became refugees, or people who flee to another country to escape danger. These wars left the region badly scarred. By 2002, Serbia's hope of one Yugoslavia had ended. Serbia and Montenegro formed a looser union and dropped the Yugoslav name.

Slovenia Slovenia, in the northwest of the Balkans region, has rugged mountains and fertile, densely populated valleys. Of all the countries of the old Yugoslavia, Slovenia is the most peaceful and prosperous. With many factories and service industries, it also has the region's highest standard of living. About 52 percent of the 2 million Slovenians live in towns and cities. Most are Roman Catholic.

Croatia Croatia spreads along the island-studded coast of the Adriatic Sea. Then it suddenly swings inland, encompassing rugged mountains and a fertile plain. **Zagreb,** the capital and largest city, lies in this inland area. An industrialized republic, Croatia supports agriculture as well. Tourists once flocked to Croatia's beautiful Adriatic beaches, but war has damaged many places.

The Croats, a Slavic group, make up 78 percent of Croatia's 4.7 million people. Another 12 percent are Serbs. Both Croats and Serbs speak the same Serbo-Croatian language, but they use different alphabets. The Croats use the Latin alphabet, the same one that you use for English. The Serbs write with the Cyrillic alphabet. Religion also divides Croats and Serbs. Croats are mainly Roman Catholic, while Serbs are Eastern Orthodox Christians.

NATIONAL GEOGRAPHIC On Location

Vukovar, Croatia

In 1991 Serbs attacked Vukovar in a revolt against Croatian independence. The revolt became a war, which lasted many years and took many lives.

Place What are some ways the war affected Croatia?

Primary Source

ZLATA'S DIARY
by Zlata Filipović

Young Zlata Filipović kept a diary about her experiences in Sarajevo.

"BOREDOM!!! SHOOTING!!! SHELLING!!! PEOPLE BEING KILLED!!! DESPAIR!!! HUNGER!!! MISERY!!! FEAR!!! That's my life! The life of an innocent eleven-year-old schoolgirl!!!! A schoolgirl without a school. A child without games, without friends, without the sun, without birds, without nature, without fruit, without chocolate or sweets, with just a little powdered milk. In short, a child without a childhood. A wartime child. . . . I once heard that childhood is the most wonderful time of your life. And it is. I loved it, and now an ugly war is taking it all away from me. Why? I feel sad. I feel like crying. I am crying."

Taken from *Zlata's Diary: A Child's Life in Sarajevo,* © 1994 by Viking Penguin. Translation copyright Fixotet editions Robert Laffont, 1994.

Analyzing Primary Sources

Making Inferences What little things do you think you would miss the most if war or another tragedy took them from you?

Bosnia and Herzegovina Mountainous and poor, Bosnia and Herzegovina has an economy based mainly on crops and livestock. **Sarajevo** (SAR•uh•YAY•voh), the capital, has the look of an Asian city, with its marketplaces and mosques, or Muslim houses of worship. Many of the Bosnian people are Muslims, followers of the religion of Islam. Others are Eastern Orthodox Serbs or Roman Catholic Croats. Serbs began a bitter war after Bosnia's independence in 1992. The **Dayton Peace Accords** divided Bosnia into two regions under one government in 1995. American and other troops came as peacekeepers.

Serbia and Montenegro Since 2002, Serbia and its reluctant partner Montenegro have formed a loose union. Inland plains and mountains cover the area. The economies of these two republics are based on agriculture and industry. The region's largest city is **Belgrade.** Most of the 10.7 million Serbs and Montenegrins practice the Eastern Orthodox faith.

Serbia has faced growing unrest in some of its local provinces. Muslim Albanians living in the province of **Kosovo** want independence from Serbia. Also living in Kosovo is a smaller group of Eastern Orthodox Serbs. For centuries, Albanians and Serbs here have felt a deep anger toward each other. In 1999 Serb forces tried to push the Albanians out of Kosovo. The United States and other nations bombed Serbia to force it to withdraw its troops. Even with the help of United Nations peacekeeping troops, peace in Kosovo remains shaky.

Macedonia (F.Y.R.O.M.) Macedonia's 2 million people are a mix of different ethnic groups from the Balkans. In **Skopje** (SKAW•pyeh), Macedonia's capital, there is an amazing mix of ancient Christian churches, timeworn Turkish markets, and modern shopping centers. Close to Kosovo, Macedonia handled a huge wave of ethnic Albanian refugees from Kosovo who fled Serb forces in 1999.

✓ Reading Check What nations were formed from the former Yugoslavia?

Albania

Bordering the Adriatic Sea, Albania is slightly smaller than the state of Maryland. Mountains cover most of the country, contributing to Albania's isolation from neighboring countries. A small coastal plain runs along the Adriatic Sea.

Albania is a very poor country. Although the country has valuable mineral resources, it lacks the money to mine them. Most Albanians farm—growing corn, grapes, olives, potatoes, sugar beets, and wheat—in mountain valleys.

The Albanians Almost two-thirds of Albanians live in the country-side. The capital and largest city, **Tirana,** and its suburbs have a population of about 270,000. Although 3.4 million people live in Albania, another 3.2 million Albanians live in nearby countries. These refugees fled Albania to escape the violence that swept the country after communism's fall in the early 1990s.

About 70 percent of Albanians are Muslim. The rest are Christian—either Eastern Orthodox or Roman Catholic. While the Communists opposed religion, Albania's democratic government has allowed people to practice their faith. As a result, many mosques and churches have opened across the country. The most famous Albanian in recent times was the Catholic nun **Mother Teresa,** who served the poor in Calcutta, India.

✓ Reading Check **What is the main religion in Albania?**

Assessment

Defining Terms
1. Define consumer goods, ethnic cleansing, refugee, mosque.

Recalling Facts
2. Place What is the capital of Romania?
3. Economics How are roses used in Bulgaria?
4. History Which of the former Yugoslav republics is most prosperous?

Critical Thinking
5. Drawing Conclusions How do you think people in the Balkans feel about the recent changes in their countries?
6. Understanding Cause and Effect What effect did the policy of ethnic cleansing have on the people of Serbia?

Graphic Organizer
7. Organizing Information Create a chart like the one below and complete it by filling in two facts under each country name.

Romania	Bulgaria	Slovenia	Croatia
Bosnia and Herzegovina	Serbia and Montenegro	Macedonia	Albania

Applying Social Studies Skills

8. Analyzing Maps Study the political map on page 307. What countries are found on the east coast of the Adriatic Sea?

Making Connections

CULTURE GOVERNMENT PEOPLE TECHNOLOGY

Ukrainian Easter Eggs

Ukrainians have a rich folk art tradition that dates back thousands of years. It includes pottery, textiles, and woodworking. The best-known Ukrainian art form, however, is that of *pysanky,* or decorated eggs.

History

Ukrainian Easter eggs are known worldwide for their beauty and skillful designs. Many of the designs date back to a time when people in the region worshiped a sun god. According to legend, the sun god preferred birds over all other creatures. Birds' eggs became a symbol of birth and new life, and people believed the eggs could ward off evil and bring good luck. Eggs were decorated with sun symbols and used in ceremonies that marked the beginning of spring.

When Christianity took hold in Ukraine in A.D. 988, the tradition of decorative eggs continued. The egg came to represent religious rebirth and new life. People decorated eggs in the days before Easter, then gave them as gifts on Easter morning.

Technique

The word *pysanky* comes from Ukrainian words meaning "things that are written upon." This phrase helps explain the wax process used to decorate the eggs. An artist uses a pin or a tool called a *kistka* to "write" a design in hot wax onto the egg. The egg is then dipped into yellow dye, leaving the wax-covered portion of the eggshell white. After removing the egg from the dye, the artist writes with hot wax over another section of the egg. This portion stays yellow as the egg is dipped into a second dye color. The process continues, with the artist adding wax and dipping the egg into a darker and darker color. At the end, the artist removes the wax layers to reveal the multicolored design.

▶ Making the Connection

1. What is *pysanky* and when did it originate?
2. What role does placing wax onto the eggshell play in creating a decorative egg?
3. **Drawing Conclusions** What purposes, other than entertainment, might folk art accomplish?

◀ Ukrainian Easter eggs

Ukraine, Belarus, and Moldova

Guide to Reading

Main Idea

Past ties to Russia have had different effects on the economies and societies of Ukraine, Belarus, and Moldova.

Terms to Know

- steppe
- potash

Reading Strategy

Create a diagram like this one. Under the country name, list at least one fact that shows the Soviet Union's effect on these countries.

Soviet Union's effect on

Ukraine   Moldova

Belarus

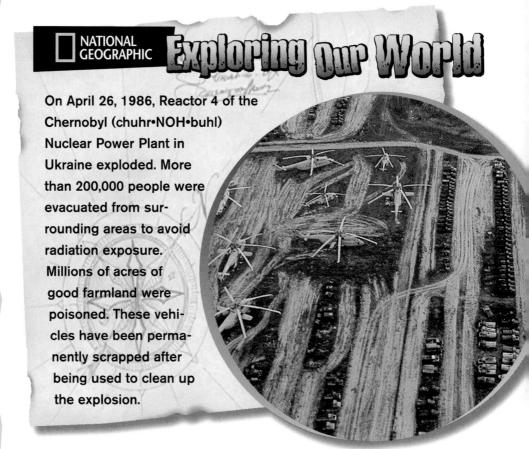

NATIONAL GEOGRAPHIC Exploring Our World

On April 26, 1986, Reactor 4 of the Chernobyl (chuhr•NOH•buhl) Nuclear Power Plant in Ukraine exploded. More than 200,000 people were evacuated from surrounding areas to avoid radiation exposure. Millions of acres of good farmland were poisoned. These vehicles have been permanently scrapped after being used to clean up the explosion.

Ukraine, **Belarus** (BEE•luh•ROOS), and **Moldova** (mawl•DAW•vuh) once belonged to the Soviet Union. When the Soviet Union broke apart in late 1991, Ukraine, Belarus, and Moldova became independent. Since then, they have struggled to build new economies.

Ukraine

Slightly smaller than Texas, Ukraine is by far the largest eastern European country (excluding Russia). The Carpathian Mountains rise along its southwestern border. Farther east, a vast steppe, or gently rolling, partly wooded plain, makes up the country. Numerous rivers twist across the steppe, most of which are too shallow for ships. The most important waterway, the **Dnieper** (NEE•puhr) **River,** has been made navigable so that ships can carry goods to distant markets. The **Crimean Peninsula** juts into the Black Sea. Most of Ukraine has cold winters and warm summers.

Rich dark soil covers nearly two-thirds of Ukraine. Farms are very productive, earning the country the name "breadbasket of Europe." Farmers grow sugar beets, potatoes, and grains and raise cattle and sheep. Factories make machinery, processed foods, and chemicals.

The Ukrainians   Early Slavic groups settled and traded along the rivers of the region. During the A.D. 800s, warriors from Nordic countries united these groups into a large state centered on the city of Kiev (KEE•ihf). A century later, the people of Kiev accepted the Eastern Orthodox faith and built one of Europe's most prosperous civilizations. After 300 years of freedom, the people of Kiev were conquered by Mongols, then Lithuanians and Poles, and finally the Russians.

In the 1930s, Soviet dictator Joseph Stalin brought Ukraine's farms under government control. Millions were murdered or starved in the famine that followed. Millions more died when Germans invaded

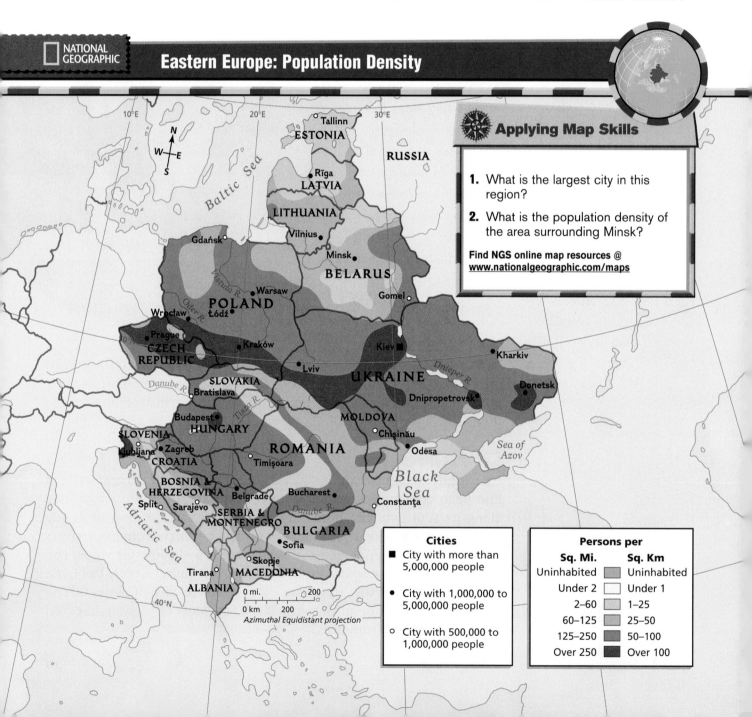

NATIONAL GEOGRAPHIC

Eastern Europe: Population Density

Applying Map Skills

1. What is the largest city in this region?

2. What is the population density of the area surrounding Minsk?

Find NGS online map resources @ www.nationalgeographic.com/maps

Cities

■ City with more than 5,000,000 people

● City with 1,000,000 to 5,000,000 people

○ City with 500,000 to 1,000,000 people

0 mi. 200
0 km 200
Azimuthal Equidistant projection

Persons per	
Sq. Mi.	**Sq. Km**
Uninhabited	Uninhabited
Under 2	Under 1
2–60	1–25
60–125	25–50
125–250	50–100
Over 250	Over 100

Language Families of Europe

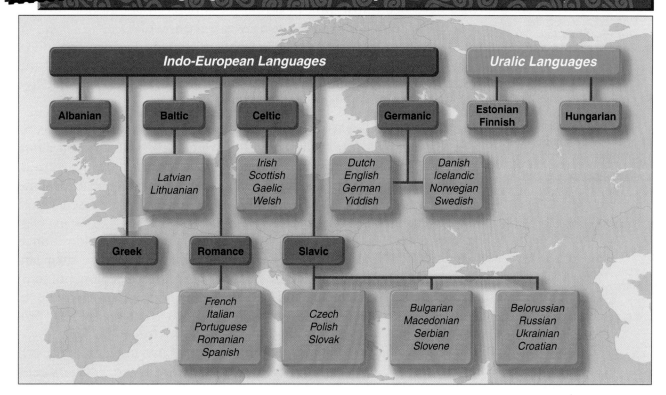

Indo-European Languages

Albanian	Baltic	Celtic	Germanic	

Baltic: Latvian, Lithuanian

Celtic: Irish, Scottish, Gaelic, Welsh

Germanic: Dutch, English, German, Yiddish — Danish, Icelandic, Norwegian, Swedish

Greek | Romance | Slavic

Romance: French, Italian, Portuguese, Romanian, Spanish

Slavic: Czech, Polish, Slovak — Bulgarian, Macedonian, Serbian, Slovene — Belorussian, Russian, Ukrainian, Croatian

Uralic Languages

Estonian Finnish | Hungarian

Ukraine during World War II. Finally in 1991, with the decline of Soviet power, Ukraine once again became a free nation.

Ukraine has over 49 million people. Nearly 75 percent are ethnic Ukrainians. About 22 percent are Russians, who live mainly in eastern areas. Most of the people follow the Eastern Orthodox religion and speak Ukrainian, a Slavic language closely related to Russian.

More than 70 percent of the people live in cities. Kiev, the capital, has more than 2.6 million people. Modern Ukrainians, even teenagers, enjoy listening to folk music played on a stringed instrument called a bandura and watching the acrobatic leaps of the *hopak* dance.

✓Reading Check Why is Ukraine called the "breadbasket of Europe"?

Belarus and Moldova

Belarus, slightly smaller than Kansas, is largely lowlands. Visiting Belarus, you would see wide stretches of birch tree groves, vast forested marshlands, and wooden villages surrounded by fields. Summers are cool and wet, and winters are cold.

Farmers grow potatoes, grains, vegetables, sugar beets, and fruits. Factory workers make equipment, chemicals, and construction materials. Food processing is another important industry. In addition to having petroleum and natural gas, Belarus has potash, a mineral used in fertilizer.

Slavic groups first settled the area that is today Belarus in the A.D. 500s. Surrounded by larger countries, Belarus was under foreign rule

Analyzing the Chart

Seven main language families stem from Indo-European origins.

History From what language family did Ukrainian develop?

Eastern Europe Today

for most of its history. Communist Party leaders control Belarus's government and have maintained close ties with Russia. Foreign companies have been unwilling to do business here. In addition, Belarus is still linked to neighboring Russia's weak economy.

The 10 million people of Belarus are mostly Eastern Orthodox Slavs. Their Belorussian language is closely related to Russian and Ukrainian and is written in Cyrillic. Two-thirds of Belarus's people live in cities. **Minsk,** the largest city, is the capital.

Moldova   Moldova is mostly a rolling hilly plain sliced by rivers. These waterways form valleys that hold rich fertile soil. This soil, along with mild winters and warm summers, provides productive farmland. Farmers grow sugar beets, grains, potatoes, apples, and tobacco. Some grow grapes used to make wine. Factories turn out processed foods, machinery, metals, construction materials, and textiles.

Moldova's flag looks similar to Romania's flag. Why? Moldova once was part of Romania. About two-thirds of the people trace their language and culture to that country. Moldova's eastern region, home to many Russians, Ukrainians, and Turks, recently declared independence. This is an important issue to Moldova because that region also produces about 80 percent of the country's electricity.

Moldova has 4.3 million people. About half live in cities, but much of Moldova's culture is still based on a rural way of life. Villagers celebrate special occasions with lamb, cornmeal pudding, and goat's milk cheese. The main city is the capital, **Chişinău** (KEE•shee•NOW).

 Reading Check   **With what nation does Belarus have close ties?**

Section 3 Assessment

Defining Terms
1. **Define** steppe, potash.

Recalling Facts
2. **Location** Where is the Crimean Peninsula located?
3. **Government** What type of government does Belarus have?
4. **Economics** Name three of Moldova's agricultural products.

Critical Thinking
5. **Categorizing Information** List four agricultural products and three manufactured products of Ukraine.

6. **Understanding Cause and Effect** Why is the culture of Moldova similar to that of Romania?

Graphic Organizer
7. **Organizing Information** Create a time line like this one. Then label five important periods or events in Ukraine's history.

Applying Social Studies Skills

8. **Analyzing Maps** Compare the political and population maps on pages 307 and 318. What is the population density around Ukraine's Dniester River?

TIME REPORTS

FOCUS ON WORLD ISSUES

The European Union: Good for Everyone?

A Common Currency for a Common Market

AVANTIS/ZUMO PRESS/NEWSCOM

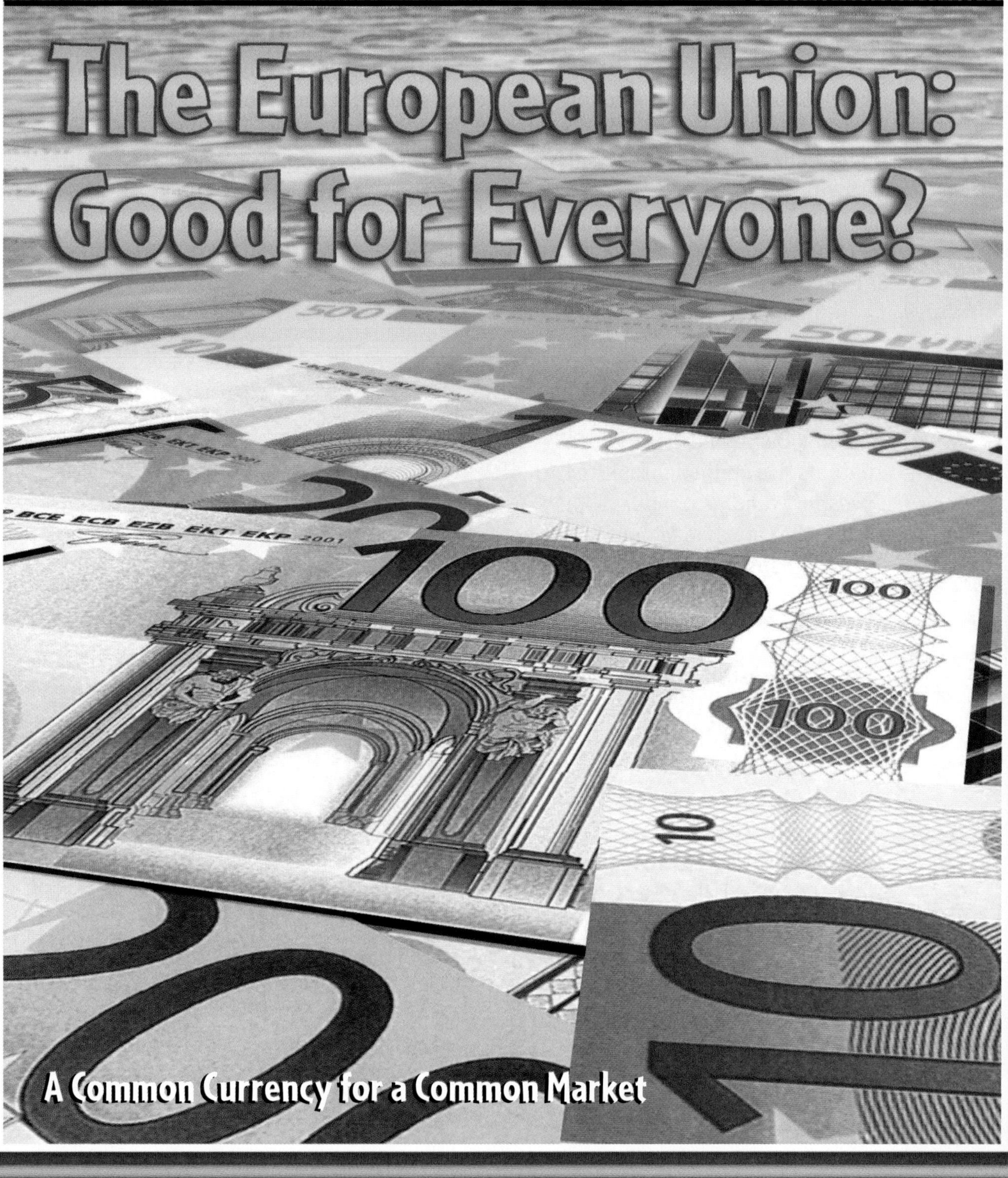

Europeans began using new currency in January 2002. The impact was gigantic—like this mocked-up coin.

Building a United Europe

Damien Barry had a problem. He wanted to work in Paris, France. The trouble was, the French were fussy. French people could work in France. And so could people from 14 other loosely united European nations. All 15 nations belonged to the **European Union**, or the EU. Barry wasn't from an EU nation. He was from Brooklyn, New York.

But that didn't stop him. Ireland is an EU nation. It grants citizenship to anyone with an Irish parent or grandparent. Barry had Irish grandparents. He applied for an Irish passport and got one. Soon after that he had a job in a French bank.

A Big Story for Americans

Barry would never give up his U.S. citizenship. Yet he's not letting go of his Irish passport, either. "It's worth a million dollars to me," he said.

Barry's story suggests how much the EU matters to Americans. The 15 EU nations form the world's largest trading group. That gives them awesome power to control jobs and the price of many things you buy.

That's not all. By 2020, the EU plans to expand to 30 nations. Right now, the U.S. deals one-on-one with countries like Poland and Estonia. In the future, it will have to deal with them through the EU.

What's more, the EU is piecing together a small army. That army will change the U.S. military's role in Europe. "In the next 10 years," TIME magazine said in 2001, "there may be no bigger story than the EU."

Europe in 2002

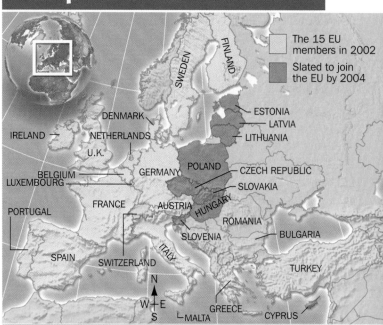

The 15 EU members in 2002

Slated to join the EU by 2004

INTERPRETING MAPS

1. **Categorizing** In what part of Europe—east or west—are most current EU members? In which part are nations that want to join?

2. **Making Inferences** Why might it be hard for all these nations to agree on important issues?

The new euro exchanged for bread.

Farm animals' health is a big EU concern.

This Dutchman is one of hundreds of pro soccer players in the EU.

Common Problems

What is the EU? Simply put, it is a group of nations that have joined forces to solve common problems. Finding a safe way to recycle used batteries is one problem. Convincing Europeans to stop smoking is another. Making sure goods flow freely within Europe is still another. The EU is a **free trade zone**. That means EU nations don't tax goods they import from each other.

The EU hopes to help its members prosper. But it has another goal—bringing peace to a continent with a long history of conflict.

A Heap of Trouble

To get the euro to shoppers by January 2002, the EU sent 56 billion coins to banks in 12 nations. The coins weighed 168,000 tons—24 times more than the Eiffel Tower in Paris, France!

Weak Government

Some people compare the EU with the United States around 1785. The U.S. government had little muscle then. It had no president, no army, no power to raise money. The states had all the money and almost all the power.

In many ways, the EU is like that. Officials at EU offices in Brussels, Belgium, make a lot of decisions, but they have no **authority** to force member nations to give up their armies. They can't even make them stop printing money.

In 1789, America's original 13 states agreed to give up powers like those. They did it by approving the U.S. Constitution.

The EU doesn't have a constitution. Its members are joined by treaties, or written agreements. Without a constitution to guide them, it's hard to get all 15 nations to agree on anything.

New Money

One thing most EU members have agreed on is a common currency, the **euro**. In January 2002, most EU nations replaced their own money with the euro. Three nations—Britain, Denmark and Sweden—chose not to make the switch immediately.

In 2001, Damien Barry got paid in French francs. Now he gets paid in euros. When he goes to Italy and Holland, he no longer carries Italian lira and Dutch guilders. Like his Irish passport, the euro has made his life easier. And it's done the same for the more than 300 million Europeans who use the euro every day.

EXPLORING THE ISSUE

1. **Making Generalizations** Three EU nations refused to replace their currencies with the euro. Why might a nation want to keep its own currency?

2. **Cause and Effect** How might the EU affect your life—today, and in the future?

From Peace to Prosperity

World War II ended in 1945. It was the third time in 75 years that Germany and France had fought each other.

Could another war be prevented? A Frenchman named Jean Monnet thought so. He proposed taking coal and steel production out of the hands of individual countries. Without fuel and steel, he said, nations couldn't wage war.

U.S. President John F. Kennedy shares a smile with Jean Monnet.

UPI/CORBIS-BETTMANN

In 1951, six nations accepted Monnet's proposal. They were Belgium, West Germany, Italy, Luxembourg, the Netherlands, and France. They set up an organization that told each nation how much coal and steel it could produce.

A Free Trade Zone

That was a big step. But there were bigger ones to come. In 1957, all six agreed to stop taxing goods they imported from each other. Those taxes

How the EU Grew

1951: France, Germany, Italy, the Netherlands, Belgium and Luxembourg agree to pool their coal, iron ore, and steel industries.

1973: Denmark, Ireland, and the United Kingdom join.

1981: Greece joins.

1986: Spain and Portugal join.

1995: Austria, Finland, and Sweden join.

acted like walls, stopping goods from moving between nations. By removing those walls, the six nations created a **common market**.

Common markets were nothing new. The United States had had one for more than 150 years. California never taxed beef "imported" from Texas, for instance. Free trade was new for Europe, however. And it helped businesses there grow.

Growing Pains

Nine nations eventually joined the original six. Looking ahead, the EU expects to let about 15 other nations join by 2020.

Getting 30 nations to work together won't be easy. But no one doubts that the EU's impact on the world is going to grow. Somewhere Jean Monnet, who died in 1979, must be smiling. ▪

EXPLORING THE ISSUE

1. **Explaining** In what ways might the simple fact of the EU's existence promote peace?

2. **Cause and Effect** How might a common market help businesses grow?

A Model for Change

▲ **U.S. President George W. Bush confers with a high EU official.**

Hungary had a bumpy 50 years after World War II. This East European nation suffered under Communist rule from 1948 to 1990. Now Hungarians have a democratic government. Individuals there can own their own businesses again. Those changes put Hungary on track to joining the EU.

Qualifying for entry wasn't easy. Hungary's government had to budget its spending. It had to sell factories and land it owned to private citizens. Thousands of workers lost their jobs.

Creating Jobs

Hungarians were willing to make the sacrifices, because they wanted to join the EU. Once in, they would be able to sell what they made to other

The Top Two Output of goods and services in 1999, in trillions of dollars

$8.5 — European Union

$9.3 — United States

Analyzing information One of every 10 people in the world lives in the U.S. or in EU nations. But every year U.S. and EU workers together create more than half the world's goods and services. Why do you think this is so?

EU members. Those sales would create jobs at home and make lives easier for Hungarians.

Would Hungary have changed if the EU didn't exist? Certainly. But chances are it wouldn't have changed so fast—and so completely. The EU has served as a model for Hungary and other former Communist nations to follow.

The EU has been especially good for the United States. Every day the U.S. and the EU nations sell each other goods worth $2 billion. In 1999, EU citizens bought $32 billion worth of goods from Texas and California alone. That money paid the salaries of at least 327,000 Texans and Californians.

Ads for Democracy

The U.S. and the EU compete with each other. They often disagree on major issues. But they are firm friends, and both are good advertisements for democracy and free trade. The prospect of joining their "club" has spurred Hungary and other nations to change—and to change quickly. ▪

EXPLORING THE ISSUE

1. Comparing How are the U.S. and the EU alike?

2. Making Inferences Why might a Hungarian worker be both for and against change?

325

Resolving Differences: What Can One Person Do?

The EU and the United States are good friends. But friends have their differences. Here are four:

1. The U.S. doesn't trade with Libya, Iran, and Cuba. It tried to get EU nations to do the same, but the EU refused. Companies in EU nations want to be free to sell goods to anyone.

2. The EU puts limits on some U.S. companies that do business in Europe. Some Americans don't think the EU should be able to tell U.S. companies how to run their businesses.

3. Another dispute involved food. U.S. food companies wanted to grow **genetically altered crops** that resisted disease. So their scientists invented new types of crops. Many Europeans are afraid that those crops might pose health risks. Some EU countries won't even allow those products inside their borders.

4. Global warming is another sticking point. Scientists fear that gases from factories and automobiles keep the earth's heat from escaping into space.

SCIENCE SOURCE/PHOTO RESEARCHERS

▲ Genetically altered foods are creating an EU controversy.

EU nations and the U.S. can't agree on the best way to solve the problem.

Choose one of the four problems. Research each side's argument. Then create a solution to the problem—one you think both sides might accept.

Make your views public. Put them in a letter. Send the letter to your representatives in Congress. You might even want to send a copy to the European Union's ambassador to the United States. Address: Ambassador, Delegation of the European Commission to the United States, 2300 M Street, NW, Washington, D.C., 20037. ■

EXPLORING THE ISSUE

1. Analyzing Information What might all four disputes have to do with each side's view of its "rights?"

2. Making Predictions How might these disputes affect parts of the world outside the U.S. and the EU?

REVIEW AND ASSESS

UNDERSTANDING THE ISSUE

1. Defining Key Terms Write definitions for the following terms: *European Union, import tax, free trade zone, common market, authority, genetically-altered crops,* and *euro.*

2. Writing to Inform Write a short article about the European Union, explaining how it could affect the lives of your fellow students. Use as many words as you can from the above list.

3. Writing to Persuade Write a letter to an imaginary friend in Denmark. Convince your friend that all European countries should use the euro.

INTERNET RESEARCH ACTIVITY

4. Use Internet resources to find information about the European Union. Read about the EU's three main governing bodies. Choose one and write a brief description of it in your own words. Then decide, with your classmates, how those bodies work together and which ones have the most power.

5. With your teacher's help, use Internet resources to research the symbols of the European Union: the flag, the anthem, and Europe Day. How is the EU's flag like—and different from—the first U.S. flag? Download the EU anthem. Why do you think the EU chose it? How is Europe Day like Independence Day in the U.S.? Put your answers in a 250-word essay.

BEYOND THE CLASSROOM

6. Research the history of the U.S. dollar. How hard was it to get Americans to accept U.S. currency in 1792? Ask your parents about the U.S.

The EU flag: golden stars in a blue sky represent the nations

$2.00 bill. How did they react to its introduction? How is the dollar like the euro? Explain your answers in an article appropriate for a school newspaper.

7. Divide the class into three teams. Debate this resolution: "It is unfair for the EU to let only European citizens work in EU countries." A panel of student judges will decide which team has the most convincing arguments.

It's All About Jobs!

Where U.S. Exports Create Jobs
(Top 10 States with Jobs Created by Exports to Europe)

State	Thousands of Jobs
1 California	250,270
2 Washington	157,500
3 New York	119,500
4 Illinois	92,200
5 Texas	77,100
6 Massachusetts	71,700
7 New Jersey	70,400
8 Ohio	66,200
9 Michigan	56,400
10 Pennsylvania	52,000

Where Europe's Money Creates Jobs
(Top 10 States with Jobs Created by European Companies)

State	Thousands of Jobs
1 California	292,000
2 New York	246,900
3 Texas	235,300
4 N. Carolina	182,000
5 Pennsylvania	173,300
6 Florida	171,100
7 Illinois	169,300
8 New Jersey	164,400
9 Michigan	163,300
10 Ohio	154,200

Source: European-American Business Council. Note: "Europe" refers to the 15 EU members plus four members of a related group, the European Free Trade Association (Iceland, Liechtenstein, Norway, and Switzerland).

BUILDING SKILLS FOR READING TABLES

1. Analyzing Data Using an almanac, find the 10 states with the largest populations. How many of those states are listed among the top 10 on each graph? What relationships do you see between state populations and jobs supported by exports? How might state populations influence the number of jobs European companies create in the U.S.? Put your answers in a short report.

2. Making Inferences European companies create more jobs in Florida and North Carolina than exports do. How might you explain this?

FOR UPDATES ON WORLD ISSUES GO TO www.timeclassroom.com/glencoe

Study and Writing Skill

Taking Notes

Effective note taking involves more than just writing facts in short phrases. It involves breaking up information into meaningful parts so that it can be remembered.

Learning the Skill

To take good notes, follow these steps:

- Write key points and important facts and figures quickly and neatly. Use abbreviations and phrases.
- Copy words, statements, or diagrams from the board or your research.
- Ask the teacher to repeat important points you do not understand.
- When studying textbook material, organize your notes into an outline (see page 654) or concept map that links important information.
- For a research report, take notes on cards. Note cards should include the title, author, and page number of sources.

Practicing the Skill

Suppose you are writing a research report on eastern Europe. First, identify main idea questions about this topic, such as "Who has ruled Poland?" or "What economic activities are found in the Czech Republic?" Then research each question.

Using this textbook as a source, read the material on pages 307 and 309–310 and prepare notes like this:

Main Idea: What economic activities are found in Poland?
1. Agriculture:
2. Mining:
3. Manufacturing:
Main Idea: What economic activities are found in the Czech Republic?
1.
2.
3.

Applying the Skill

In an encyclopedia or on the Internet, find information about Poland's coal industry and the environmental consequences of burning coal. Take notes by writing the main idea and supporting facts. Then rewrite the article using only your notes.

◀ A Czech teenager displays Soviet souvenirs for tourists who flock to Prague.

Section 1 | East Central Europe

Terms to Know
acid rain
pope
landlocked
bauxite
spa

Main Idea
The countries of East Central Europe are undergoing many changes to their political and economic systems.

✓ Culture The Poles feel deep loyalty to their country and the Catholic Church.

✓ Economics The Baltic countries of Estonia, Latvia, and Lithuania are moving toward free-market economies and closer ties with western Europe.

✓ Economics The Czech Republic is prosperous but must modernize its factories.

Section 2 | The Balkan Countries

Terms to Know
consumer goods
ethnic cleansing
refugee
mosque

Main Idea
The Balkan countries have greatly suffered from ethnic conflicts and economic setbacks.

✓ Culture Romanian history and culture were greatly influenced by the Romans. The Romanian language is not like other Eastern European languages, however in other ways, Romanians are like their Slavic neighbors.

✓ History Ethnic conflicts have torn apart the former Yugoslav republics.

✓ Economics Albania is rich in minerals but is too poor to mine them.

Section 3 | Ukraine, Belarus, and Moldova

Terms to Know
steppe
potash

Main Idea
Past ties to Russia have had different effects on the economies and societies of Ukraine, Belarus, and Moldova.

✓ Geography Ukraine's rich soil allows it to grow large amounts of food.

✓ Culture The government of Belarus has maintained ties with Russia.

✓ Culture Moldova's eastern region, home to many Russians, Ukranians, and Turks, has declared its independence.

Sculptures and chandeliers ▶
made of salt decorate this
room in a Polish salt mine.

Assessment and Activities

Using Key Terms

Match the terms in Part A with their definitions in Part B.

A.

1. spa
2. ethnic cleansing
3. acid rain
4. pope
5. steppe
6. landlocked
7. mosque
8. consumer goods
9. refugee
10. bauxite

B.

a. head of the Roman Catholic Church
b. resort with hot mineral springs
c. a mineral that is mined and used to make aluminum
d. having no access to the sea
e. products made for people
f. Muslim house of worship
g. rain containing chemical pollutants
h. forcing people from other ethnic groups to leave their homes
i. person who must flee to another country to escape danger or disaster
j. gently rolling, partly wooded plain

Reviewing the Main Ideas

Section 1 East Central Europe

11. **Economics** What is one of Poland's most important industries?
12. **History** What has raised standards of living in the Baltic republics?
13. **Place** What is the capital of the Czech Republic?

Section 2 The Balkan Countries

14. **Economics** What factors are holding back Romania's economy?
15. **Culture** What is the main religion of Bulgaria?
16. **History** What caused Yugoslavia to fall apart?
17. **History** Over 3 million Albanians live outside of Albania in nearby countries. Explain why.

Section 3 Ukraine, Belarus, and Moldova

18. **Place** What is the capital of Ukraine?
19. **Culture** To what languages is Belorussian similar?
20. **History** Why does Moldova's flag look similar to Romania's flag?

Eastern Europe

Place Location Activity

On a separate sheet of paper, match the letters on the map with the numbered places listed below.

1. Danube River
2. Black Sea
3. Croatia
4. Albania
5. Latvia
6. Hungary
7. Warsaw
8. Carpathian Mountains
9. Baltic Sea
10. Ukraine

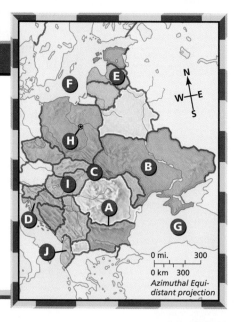

Critical Thinking

21. **Understanding Cause and Effect** What has led to the unrest in parts of Serbia and Montenegro?

22. **Categorizing Information** In a chart like the one below, identify eastern European countries that are succeeding and ones that continue to struggle economically. Include one fact that explains each situation.

Countries That Are Succeeding	Countries That Are Struggling

Current Events Journal

23. **Writing a Poem** A cinquain is a poem with five lines. The first line is a one-word title. Line two has two words that describe the title. Line three has three action words that describe the title. Line four is a four-word phrase that expresses a feeling about the subject. Line five is one word that is a synonym or restatement of the title. Write a cinquain about a current issue in one eastern European country.

Mental Mapping Activity

24. **Focusing on the Region** Create a map of eastern Europe, then label the following:

- Poland
- Albania
- Czech Republic
- Serbia & Montenegro
- Hungary
- Lithuania
- Black Sea
- Danube River
- Ukraine
- Adriatic Sea

Technology Skills Activity

25. **Building a Database** Create a database of eastern European countries. Include fields for capital, size, population, government, and products. After analyzing your database, predict which countries have a good chance of improving their standard of living.

Standardized Test Practice

Directions: Study the map below, and then answer the question that follows.

European Union 2000

- Existing members
- Applying for membership

1. **Which of the following nations in eastern Europe has applied for membership in the European Union?**

   A Poland

   B Spain

   C Ireland

   D Germany

Test-Taking Tip: Notice that the question asks you to base your answer on location. Three of the choices deal with countries in western Europe. You should use the process of elimination to find the correct answer.

Unit 5

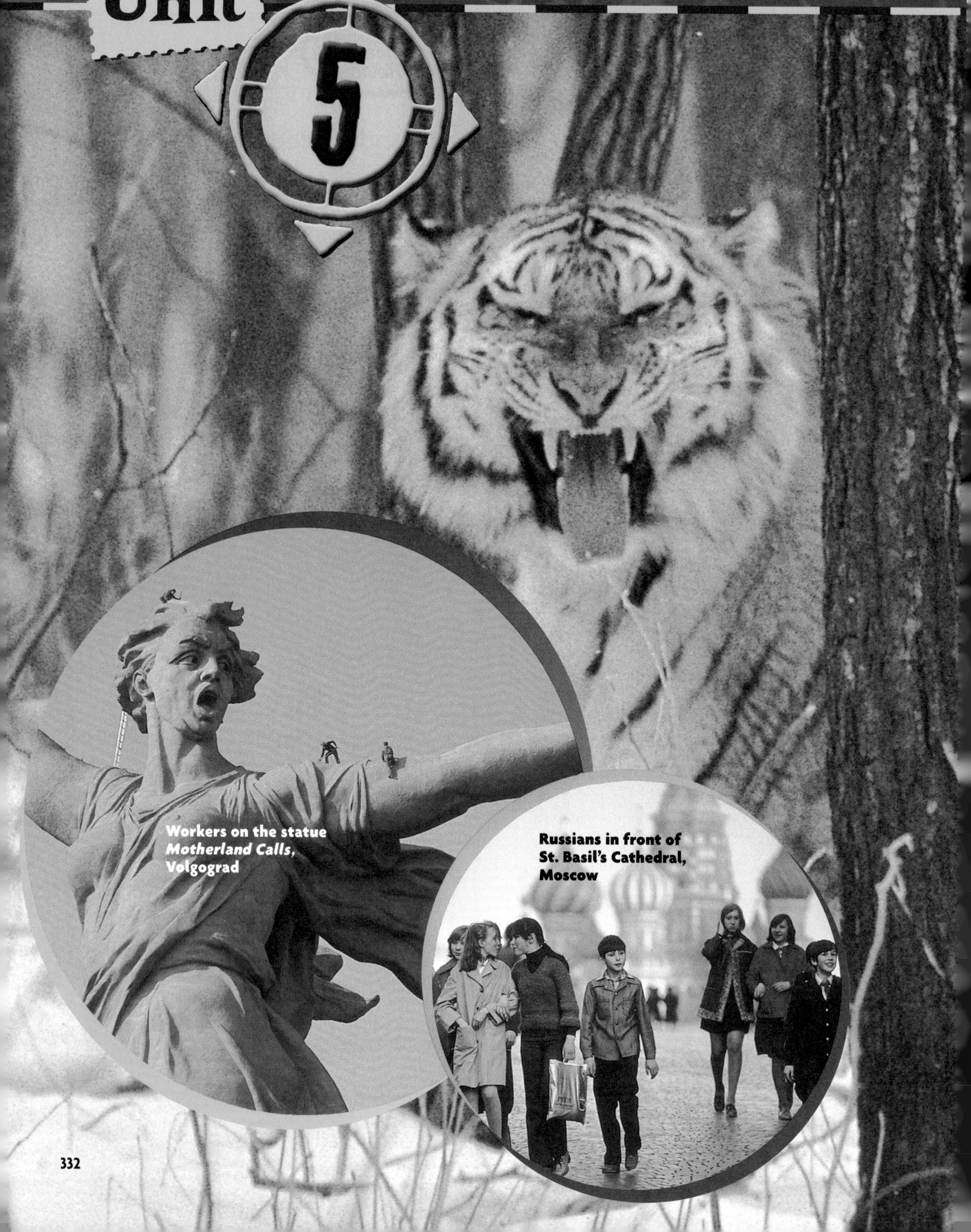

Workers on the statue *Motherland Calls*, Volgograd

Russians in front of St. Basil's Cathedral, Moscow

NATIONAL
GEOGRAPHIC

Russia and the Eurasian Republics

If you had to describe Russia in one word, that word would be BIG! Russia is the largest country in the world in area. Its almost 6.6 million square miles (17 million sq. km) are spread across two continents—Europe and Asia. As you can imagine, such a large country faces equally large challenges. In 1991, Russia emerged from the Soviet empire as an independent country. Since then it has been struggling to unite its many ethnic groups, set up a demo-cratic government, and build a stable economy.

▲ Siberian tiger in a forest in eastern Russia

NGS ONLINE
www.nationalgeographic.com/education

333

Russia and the Eurasian Republics

Physical

GREENLAND

ICELAND

ATLANTIC OCEAN

ARCTIC OCEAN
North Pole

Wrangel I.

Chukchi Peninsula

East Siberian Sea

North Land

New Siberian Islands

Laptev Sea

Kara Sea

Novaya Zemlya

Barents Sea

Klyuchevskaya Sopka 15,584 ft. (4,750 m)

KAMCHATKA PENINSULA

EUROPE

Baltic Sea

Kola Peninsula

RUSSIA

NORTH EUROPEAN PLAIN

Moscow

URAL MOUNTAINS

Ob R.

WEST SIBERIAN PLAIN

CENTRAL SIBERIAN PLATEAU

SIBERIA

Verkhoyansk Range

Lena R.

Kolyma R.

Kolyma Range

Sea of Okhotsk

Sakhalin Island

Don R.

Volga R.

Kama R.

Ural R.

Irtysh R.

Yenisey R.

Ob R.

RUSSIA

Lake Baikal

Sayan Mts.

Yablonovyy Range

Stanovoy Range

Amur R.

Mt. Elbrus 18,510 ft. (5,642 m)

Caucasus Mts.

GEORGIA
T'bilisi

ARMENIA
Yerevan

Baku

AZERBAIJAN

Caspian Sea

KAZAKHSTAN

Astana

Aral Sea

THE STEPPES

Lake Balkhash

Sea of Japan

TURKMENISTAN

UZBEKISTAN

Ashgabat

Tashkent

Bishkek

KYRGYZSTAN

Dushanbe

TAJIKISTAN

ASIA

Sea level

0 mi. 1,000

0 km 1,000

Two-Point Equidistant projection

⊛ National capital
▲ Mountain peak

N
W E
S

TROPIC OF CANCER

PACIFIC OCEAN

26,247 ft.	NORTH EUROPEAN PLAIN			0 mi. 500				8,000 m
19,685 ft.		URAL MOUNTAINS		0 km 500		KAMCHATKA PENINSULA		6,000 m
13,123 ft.				SAYAN MOUNTAINS		STANOVOY RANGE		4,000 m
6,562 ft.	MOSCOW		IRTYSH RIVER		LAKE BAIKAL		SEA OF OKHOTSK	2,000 m

Political

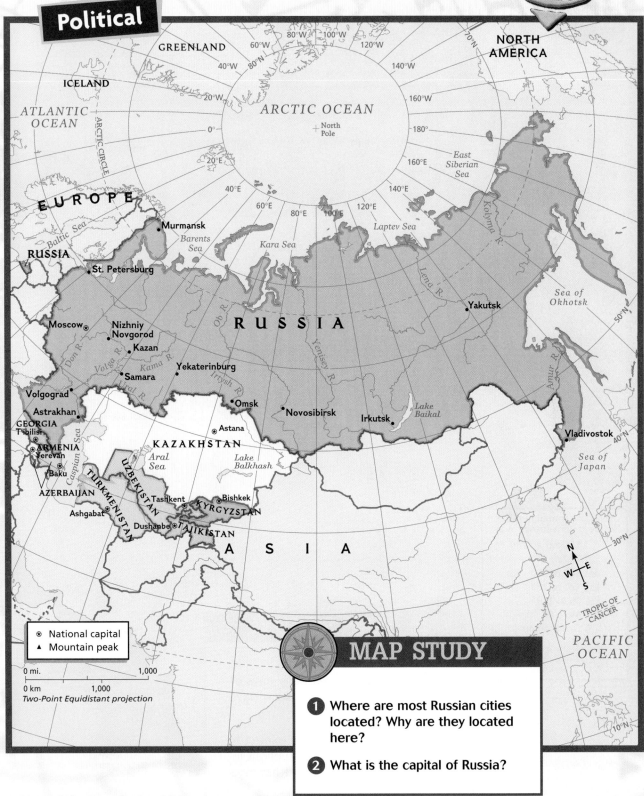

GREENLAND

ICELAND

ATLANTIC
OCEAN

ARCTIC CIRCLE

80°W 100°W

60°W

40°W 80°N

20°W

0°

20°E

40°E

60°E

80°E 100°E

120°W

140°W

160°W

180°

160°E

140°E

120°E

70°N

NORTH
AMERICA

ARCTIC OCEAN

+ North
Pole

East
Siberian
Sea

E U R O P E

RUSSIA

Baltic Sea

• Murmansk

Barents
Sea

Kara Sea

Laptev Sea

Kolyma R.

Sea of
Okhotsk

• St. Petersburg

Ob R.

Lena R.

Moscow ⊛

Nizhniy
Novgorod

Don R.

Volga R.

• Kazan

Kama R.

Yekaterinburg

Irtysh R.

Yenisey R.

R U S S I A

• Yakutsk

50°N

Volgograd •

• Samara

Ural R.

Astrakhan •

GEORGIA
T'bilisi •

Caspian Sea

⊛ Astana

KAZAKHSTAN

• Omsk

• Novosibirsk

Irkutsk •

Lake
Baikal

Amur R.

40°N

Vladivostok •

Sea of
Japan

ARMENIA
Yerevan •

• Baku

AZERBAIJAN

Aral
Sea

Lake
Balkhash

UZBEKISTAN

TURKMENISTAN

Tashkent ⊛

• Bishkek

KYRGYZSTAN

Ashgabat •

Dushanbe ⊛

TAJIKISTAN

A S I A

30°N

TROPIC OF
CANCER

N
W   E
S

PACIFIC
OCEAN

10°N

Legend:
- ⊛ National capital
- ▲ Mountain peak

0 mi.   1,000

0 km   1,000

Two-Point Equidistant projection

MAP STUDY

1 Where are most Russian cities located? Why are they located here?

2 What is the capital of Russia?

Russia

The Russian Winter

Average annual number of days with snow cover

- More than 240
- 200 to 240
- 160 to 200
- 120 to 160
- 80 to 120
- 40 to 80
- Less than 40

7 — Daily average hours of sunshine in January

0 mi. — 1,000
0 km — 1,000
Two-Point Equidistant projection

Contiguous United States and Russia: Land Comparison

MAP STUDY

1 On average, how many days of snow cover does Moscow have per year?

2 Which city would you expect to have more hours of sunlight in June—Vladivostok or Khatanga?

Fast Facts

COMPARING POPULATION:
United States and Russia

UNITED STATES

RUSSIA

= 25,000,000

Source: *Population Reference Bureau, 2000.*

COMPARING AREA AND POPULATION:
Russia East and West of the Ural Mountains

AREA

West of Urals 24.4%

East of Urals 75.6%

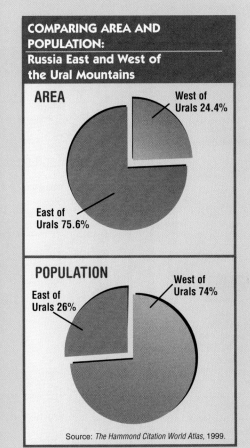

POPULATION

West of Urals 74%

East of Urals 26%

Source: *The Hammond Citation World Atlas, 1999.*

Russia
Data Bits

	Automobiles per 1,000 people	120
	Telephones per 1,000 people	183
VOTE	Democratic elections	Yes

Ethnic Makeup

Tatar 4%
Chuvash 1%
Ukrainian 3%
Other 10%
Russian 82%

World Ranking

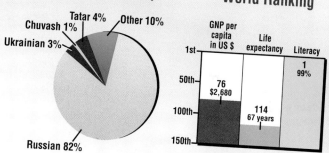

GNP per capita in US $ — 76 $2,680
Life expectancy — 114 67 years
Literacy — 1 99%
1st, 50th, 100th, 150th

Population: Urban vs. Rural

76%   24%

Source: *World Desk Reference, 2000.*

GRAPHIC STUDY

1 Which country, Russia or the U.S., has a **higher** population?

2 What percentage of Russia's people live west of the Ural Mountains?

Country Profiles

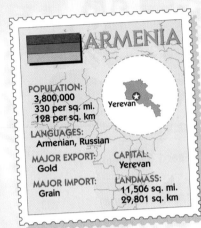

ARMENIA

POPULATION:
3,800,000
330 per sq. mi.
128 per sq. km

LANGUAGES:
Armenian, Russian

MAJOR EXPORT:
Gold

MAJOR IMPORT:
Grain

CAPITAL:
Yerevan

LANDMASS:
11,506 sq. mi.
29,801 sq. km

AZERBAIJAN

POPULATION:
8,100,000
242 per sq. mi.
94 per sq. km

LANGUAGES:
Azeri, Russian,
Armenian

MAJOR EXPORT:
Petroleum

MAJOR IMPORT:
Machinery

CAPITAL:
Baku

LANDMASS:
33,436 sq. mi.
86,599 sq. km

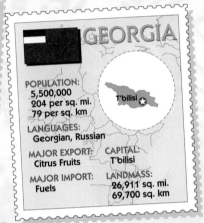

GEORGIA

POPULATION:
5,500,000
204 per sq. mi.
79 per sq. km

LANGUAGES:
Georgian, Russian

MAJOR EXPORT:
Citrus Fruits

MAJOR IMPORT:
Fuels

CAPITAL:
T'bilisi

LANDMASS:
26,911 sq. mi.
69,700 sq. km

KAZAKHSTAN

POPULATION:
14,800,000
14 per sq. mi.
5 per sq. km

LANGUAGES:
Kazakh, Russian

MAJOR EXPORT:
Petroleum

MAJOR IMPORT:
Machinery

CAPITAL:
Astana

LANDMASS:
1,049,039 sq. mi.
2,717,011 sq. km

KYRGYZSTAN

POPULATION:
5,000,000
65 per sq. mi.
25 per sq. km

LANGUAGES:
Kirghiz, Russian

MAJOR EXPORT:
Cotton

MAJOR IMPORT:
Grain

CAPITAL:
Bishkek

LANDMASS:
76,641 sq. mi.
198,500 sq. km

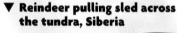

▼ **Reindeer pulling sled across the tundra, Siberia**

RUSSIA

POPULATION:
144,400,000
22 per sq. mi.
9 per sq. km

LANGUAGES:
Russian, Local
Languages

MAJOR EXPORT:
Petroleum

MAJOR IMPORT:
Machinery

CAPITAL:
Moscow

LANDMASS:
6,592,819 sq. mi.
17,075,401 sq. km

Countries and flags not drawn to scale

For more information on countries in this region, refer to the Nations of the World Data Bank on pages 690–699.

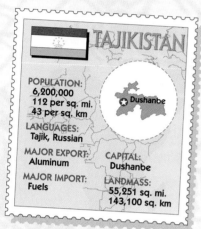

TAJIKISTAN

POPULATION:
6,200,000
112 per sq. mi.
43 per sq. km

LANGUAGES:
Tajik, Russian

MAJOR EXPORT:
Aluminum

MAJOR IMPORT:
Fuels

CAPITAL:
Dushanbe

LANDMASS:
55,251 sq. mi.
143,100 sq. km

Dushanbe

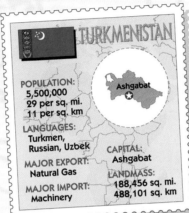

TURKMENISTAN

POPULATION:
5,500,000
29 per sq. mi.
11 per sq. km

LANGUAGES:
Turkmen,
Russian, Uzbek

MAJOR EXPORT:
Natural Gas

MAJOR IMPORT:
Machinery

CAPITAL:
Ashgabat

LANDMASS:
188,456 sq. mi.
488,101 sq. km

Ashgabat

UZBEKISTAN

POPULATION:
25,100,000
145 per sq. mi.
56 per sq. km

LANGUAGES:
Uzbek,
Russian, Tajik

MAJOR EXPORT:
Cotton

MAJOR IMPORT:
Machinery

CAPITAL:
Tashkent

LANDMASS:
172,741 sq. mi.
447,399 sq. km

Tashkent

BUILDING CITIZENSHIP

Initiative Under communism, the government is the main employer. Many people no longer had a steady income when the Soviet Union broke apart. The government could no longer take care of them. People had to figure out on their own how to solve the problem of making enough money to feed their families. In other words, they had to show initiative.

Describe a time when you showed initiative to solve a problem you faced.

WRITE ABOUT IT

You can develop initiative with practice. Use the problem-solving process to identify a business you could start alone or with some friends. Gather information, list and consider your options, consider the advantages and disadvantages, and make a decision.

Teens washing cars to earn money ▼

Chapter 13

Russia and Its Neighbors

The World and Its People

NATIONAL GEOGRAPHIC

To learn more about Russia and its neighboring countries, view *The World and Its People* **Chapters 14** and **18** videos.

Our World Today
ONLINE

Chapter Overview Visit the *Our World Today: People, Places, and Issues* Web site at owt.glencoe.com and click on **Chapter 13—Chapter Overviews** to preview information about Russia.

Compare-Contrast Study Foldable Make this foldable and use it to help you organize what you learn about Russia and its neighbors.

Step 1 Stack four sheets of paper, one on top of the other. On the top sheet of paper, trace a large circle.

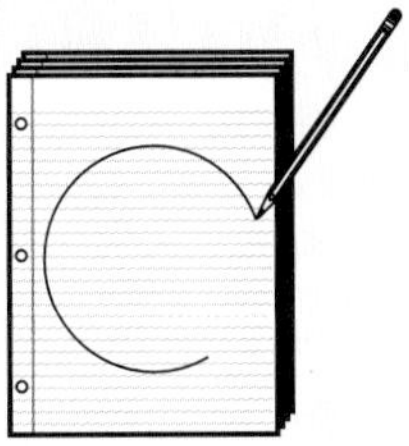

Step 2 With the papers still stacked, cut along the circle line you traced.

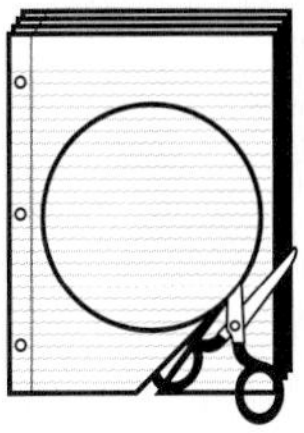

Step 3 Staple the paper circles together at one point around the edge.

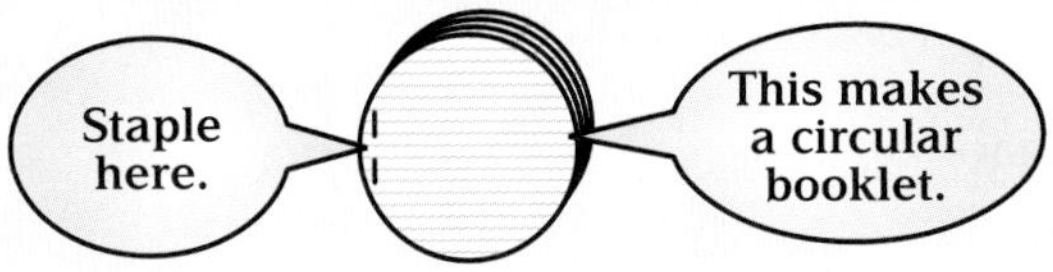

Staple here.

This makes a circular booklet.

Step 4 Label the front circle **Russia**. Take notes on the pages that open to the right. Turn the book upside down and label the back **Southern Neighbors**. Take notes on the pages that open to the right.

Russia

Reading and Writing As you read the chapter, write facts about the people and places of Russia and its neighbors in the appropriate places of your circular foldable booklet.

Why It Matters

Rebuilding a Country

For nearly half a century two superpowers—the United States and the Soviet Union—dominated world politics. With the failure of communism, the Soviet Union broke up into 15 independent nations. Russia, which was the heart of the Soviet Union, is now trying to rebuild its government and economy. Russia's efforts to create a democratic government may show whether democracy is a workable system, not just for some but for all people.

◀ Cathedral of St. Basil the Blessed, Red Square, Moscow, Russia

The Russian Land

Guide to Reading

Main Idea

Most people in Russia live west of the Urals, where the climate is mild. The people of Russia have met the challenges created by the country's gigantic size and harsh climates.

Terms to Know

- Eurasia
- urban
- suburb
- consumer goods
- rural
- tundra
- taiga
- steppe
- communism
- heavy industry
- light industry

Reading Strategy

Create a chart like this one. Give a specific name of each type of physical feature listed.

Russia	
Lakes	
Mountains	
Plateaus	
Rivers	

NATIONAL GEOGRAPHIC Exploring Our World

Siberian tigers hunt in the eastern forests of Russia—sometimes even climbing trees to find food. Only a few hundred now live in the wild, though. The animals they hunt—elk, deer, and wild boar—are dwindling, and the tigers themselves are hunted by people. Poachers who kill the tigers illegally can sell a skin for $15,000. Russia is trying to enforce laws to save these animals.

Russia is the world's largest country. Nearly twice as big as the United States, Russia is called a Eurasian country because its lands lie on two continents—Europe and Asia. The western part of Russia borders eastern European countries such as **Belarus** and **Ukraine.** As you move east, you run into the **Ural Mountains**—the dividing line between Europe and Asia. The rest of Russia stretches across Asia to the Pacific Ocean. Along the way, it shares borders with 14 other countries. Russia is so wide that it includes 11 time zones from east to west. When it is 12:00 P.M. (noon) in eastern Russia and people are eating lunch, people in western Russia are still sound asleep at 1:00 A.M.

Russia's Climate

As you can see from the map on page 334, Russia's southern border is in the middle latitudes while the north reaches past the **Arctic Circle.** Russia has a long coastline along the **Arctic Ocean.** Ice makes shipping difficult or impossible most of the year. Even many of Russia's ports on the Baltic Sea and Pacific Ocean are closed by ice part of the year.

Russia's gigantic size and harsh climates make transportation difficult within the country as well. If you visited Russia, you would discover that, unlike in the United States, railroads, rivers, and canals still are important means of getting around. With about 54,000 miles (about 87,000 km) of track, railroads are the leading movers of people and goods in Russia.

European Russia   Two mountain ranges rise in western Russia—the Urals and the Caucasus (KAW•kuh•suhs). The Ural Mountains, very old and worn down, do not reach very high. Their length is extensive, though, running from the Arctic Ocean to Russia's southern boundary. The Ural Mountains form the geographical boundary between the continents of Europe and Asia.

In general, the part of Russia that lies in Europe is warmer than Asian Russia. Large plains cover European Russia, which lies west of the Ural Mountains. The plains have a mild climate, and about 75 percent of the population lives in European Russia. This region holds the national capital, **Moscow,** and other important cities such as **St. Petersburg** and **Volgograd.**

Russia's urban, or city, areas are large and modern with stone or concrete buildings and wide streets. Tall buildings hold apartments for

NATIONAL GEOGRAPHIC On Location

Urban vs. Rural

The GUM state department store (left) is very similar to modern malls in the United States. In some rural areas, however, people still live without heat, electricity, or plumbing (below).

Culture How are cities in Russia like urban areas in the U.S.?

Vegetation in Russia

The tundra (left) is found in northern Russia. South of the tundra is the huge expanse of taiga woods (center). The steppes dominate southwest Russia (right).

Place What type of vegetation grows on the taiga?

hundreds of small families. Many of these apartments are small and cramped, however. When people in cities relax, they spend time with their families and friends, take walks through parks, or attend concerts, movies, and the circus.

Russian cities have changed in recent years. New cars speed down Moscow's streets. More people are building large houses outside the city limits, where few people lived before. As a result, Russia is developing its first **suburbs,** or smaller communities that surround a city.

Large numbers of Russia's city dwellers remain poor. These people lack the money to buy the **consumer goods** that are now more and more available. Consumer goods are items sold directly to the public, such as clothes, radios, and automobiles. The poor must stand in long lines to get the food they need to survive. Some of them resent the success of the newly wealthy people.

In Russia's **rural** areas, or countryside, most people live in houses built of wood. As in the United States, the quality of health care and education is often lower in rural areas than in the cities. Over the years, many people have left rural areas to find work in Russia's cities.

Tundra In the northern part of Siberia is a large, treeless region where only the top few inches of the ground thaw during the summer. This is called **tundra.** The few people who live there make their living by fishing, hunting seals and walruses, or herding reindeer. Because there are few trees, many of the huts are made of walrus skins. For two months out of the year, while the sun stays visible for most of the night, temperatures may rise to 70°F (21°C). Because the distances are great and the land is usually covered in ice and snow, people may use helicopters for travel.

South of the tundra is the world's largest forest, the taiga (TY•guh). The forest of evergreen trees stretches about 4,000 (6,436 km) miles across the country in a belt 1,000 to 2,000 (1,609 to 3,218 km) miles wide. Like the tundra, few people live in this area. Those who do support themselves by lumbering or hunting. This area is so sparsely populated that forest fires sometimes burn for weeks before anyone notices.

Inland Water Areas   Russia touches many inland bodies of water. In the southwest, it borders on the **Black Sea.** Through this sea, Russia can reach the Mediterranean Sea. If you look at the physical map on page 334, you will find another large sea in southwestern Russia—the **Caspian Sea.** About the size of California, the Caspian Sea is actually the largest inland body of water in the world. Like the Great Salt Lake in Utah, the Caspian Sea has salt water, not freshwater. Russia shares this sea with four other countries—Azerbaijan, Iran, Turkmenistan, and Kazakhstan.

High in the Central Siberian Plateau is **Lake Baikal.** This is the world's deepest freshwater lake. In fact, Lake Baikal holds almost 20 percent of the world's supply of unfrozen freshwater. It is also the world's oldest lake, dating back nearly 30 million years. Some of the plant and fish species in the lake can be traced to prehistoric times. No other lake in the world has so many unusual and rare species. As a result, Lake Baikal is a huge natural laboratory that attracts scientists from all over the world. Tourists travel by train to see the lake's shimmering blue waters.

Unfortunately, a large paper mill nearby has polluted the Lake Baikal region. The paper mill is a major source of jobs and wealth. An important problem for this region is how to both save the lake and keep the badly needed industry.

Russia has several important rivers. The **Volga**—the longest river in Europe—is an important transportation route. It and other rivers of European Russia are connected by canals. These transport people and goods from one city to another.

The Caucasus and Central Asia   To the south of European Russia lay the high, rugged **Caucasus Mountains.** This is a fertile region of valleys where many non-Russian people live. Thirteen different ethnic groups live in the eight countries, or republics, that border Russia in this region. Many, like the Armenians, have cultures far older than that of the Russians.

Central Asia is a land of desert and grassland steppes occupied by Turkish- and Persian-speaking peoples. These people were forced to settle down on the land as farmers by the conquering Russians. Fewer than 10 percent of the people here even speak Russian. You will learn more about the countries in these regions in Section 3.

✓ Reading Check   **What is an important problem for the Lake Baikal region?**

Fighting Pollution

In the United States, our government passes strong laws to prevent or limit pollution. However, some companies that contribute to the pollution problem fight the laws because making their companies pollution-free is very expensive. Fortunately, our government is strong and able to enforce the anti-pollution laws. In Russia, however, the new government is not strong enough to enforce the anti-pollution laws it has passed and many companies are still polluting areas such as Lake Baikal.

Russia's Economic Regions

Russia has large deposits of coal, oil, and natural gas. It also has many minerals, including nickel, iron ore, tin, and gold. The southwestern area can produce rich yields of grains. Russia's fishing industry is among the largest in the world. The vast forests of Siberia provide plenty of timber. Many people who live in Russia have jobs relating to the rich resources found there.

Even with these resources, Russia's economy is not strong. Under communism, it was very difficult for individual people to start their own companies or run their own farms. The state wanted people to work for government factories and on large, government-owned farms because the government had strong control over the economy and society as a whole. The Russian people have not had very much experience in creating jobs, starting businesses, and making money. You will learn more about the changing Russian economy in Chapter 14. For now, we will look at what Russia's economy produces.

NATIONAL GEOGRAPHIC

Russia: Economic Activity

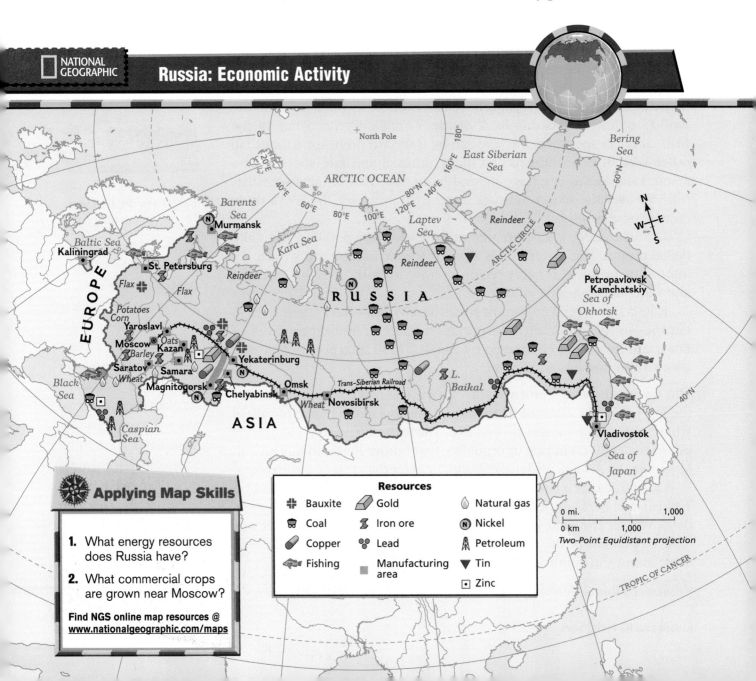

Resources

- ✚ Bauxite
- 🜚 Coal
- ⬮ Copper
- 🐟 Fishing
- ▬ Gold
- ⚒ Iron ore
- ⚬⚬ Lead
- ■ Manufacturing area
- 💧 Natural gas
- Ⓝ Nickel
- ⚑ Petroleum
- ▼ Tin
- ⊡ Zinc

0 mi. 1,000
0 km 1,000
Two-Point Equidistant projection

🧭 Applying Map Skills

1. What energy resources does Russia have?

2. What commercial crops are grown near Moscow?

Find NGS online map resources @ www.nationalgeographic.com/maps

Russia is divided into four different economic regions: the Moscow Region, Port Cities, the Volga and Urals Region, and Siberia. Today people living in these regions face many challenges.

The Moscow Region About 800 years old, Moscow is the political and cultural center of Russia. The largest city in Russia, Moscow is the country's economic center and largest transportation hub as well. Look at the economic activity map on page 346. Many of Russia's manufacturing centers are located in the western part of Russia. In the past, many of the country's factories focused on heavy industry, or making goods such as machinery, mining equipment, and steel. In recent years, more factories have shifted to light industry, or the making of such goods as clothing, shoes, furniture, and household products. Today factories make more consumer goods.

Some farming also takes place in the Moscow region. Farmers raise dairy cattle, barley, oats, potatoes, corn, and sugar beets. Other crops include flax, which is used to make textiles. Railroads and canals criss-cross Moscow and the area to carry raw materials.

Port Cities Russia has two important northwestern ports—**Kaliningrad** and **St. Petersburg.** Look at the map on page 346. Do you see that Russia owns a small piece of land on the Baltic Sea separated from the rest of the country? The port of Kaliningrad is located on this land. This city is Russia's only Baltic port that remains free of ice year-round. Russian officials, hoping to increase trade here, have eliminated all taxes on foreign goods brought to this city. However, companies that deliver goods to this port must still transport their goods another 200 miles (322 km) through other countries to reach the nearest inland part of Russia. In summer when St. Petersburg's port is not frozen, ships can travel another 500 miles (805 km) north to reach this city.

St. Petersburg, once the capital of Russia, is another important port and a cultural center. The city was built by Czar Peter the Great in the early 1700s on a group of more than 100 islands connected by bridges. Large palaces stand gracefully on public squares. (You will read more about this beautiful city when we look at Russian culture.) Factories in St. Petersburg make light machinery, textiles, and scientific and medical equipment. Located on the **Neva River** near the Gulf of Finland, the city is also a shipbuilding center.

Murmansk, in Russia's far north, and **Vladivostok,** in the east, are other important port cities. Vladivostok is Russia's largest port on the Pacific Ocean. Trade in these port cities brings needed goods to the Russian people.

Comparing Latitudes

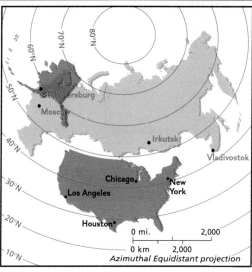

Azimuthal Equidistant projection

Analyzing the Map

As you can see in this map, ports in the contiguous United States are at much lower latitudes than Russian ports and are not affected by freezing weather.

Location Look at the political map of the United States on page 475. What body of water is close to the same latitude as St. Petersburg?

Siberia   This region has the largest supply of minerals in Russia, including iron ore, uranium, gold, diamonds, and coal. Huge deposits of oil and natural gas lie beneath the frozen ground of northern Siberia. About two-thirds of Siberia is covered with trees that could support a timber industry. The people here also fish, hunt, and herd reindeer.

Tapping all of these resources is very difficult, however. Siberia is mostly undeveloped because of its harsh, cold climate. Another problem is size—it can take eight or more days to travel across all of Russia by train. Finding a way to develop the remote resources of Siberia is very important for Russia's economic future. Many of the minerals and fuels of western Russia have been used up. The industrial centers there need the resources from Siberia.

The Volga and Urals Region   Tucked between the Moscow area and Siberia lies the industrial region of the Volga River and Ural Mountains. The Volga River carries almost one-half of Russia's river traffic. It provides water for irrigation and for hydroelectric power— the power generated by fast-flowing water. The region is also home to Russia's most productive farmlands.

The Ural Mountains are rich in minerals. Workers here mine copper, gold, lead, nickel, and bauxite, a mineral used to make aluminum. The mountains have energy resources of coal, oil, and natural gas.

✓ Reading Check   **Why are Siberia's mineral resources important?**

Assessment

Defining Terms
1. **Define** Eurasia, urban, suburb, consumer goods, rural, tundra, taiga, steppe, communism, heavy industry, light industry.

Recalling Facts
2. **Culture** What is the political and cultural center of Russia?
3. **Location** What mountain range separates Europe and Asia?

Critical Thinking
4. **Analyzing Information** Why is Kaliningrad such an important port in Russia?
5. **Drawing Conclusions** How have economic changes affected the lives of people since the collapse of the Soviet Union?

Graphic Organizer
6. **Organizing Information** Create a chart like the one below and list the most important method of transportation for the locations listed.

Place	Transportation Method
Volga	
Kaliningrad	
Siberia	
Moscow	

Applying Social Studies Skills

7. **Analyzing Maps** Turn to the economic activity map on page 346. In what part of Russia does most of the manufacturing take place?

Technology Skill

Evaluating a Web Site

Why Learn This Skill?

The Internet has become a valuable research tool. It is convenient to use, and the information contained on the Internet is plentiful. However, some Web site information is not necessarily accurate or reliable. When using the Internet as a research tool, the user must distinguish between quality information and inaccurate or incomplete information.

▲ **The Peace Corps Web site is government sponsored.**

Learning the Skill

There are a number of things to consider when evaluating a Web site. Most important is to check the accuracy of the source and content. The author and publisher or sponsor of the site should be clearly indicated, and the user must also determine the usefulness of the site. The information on the site should be current, and the design and organization of the site should be appealing and easy to navigate.

To evaluate a Web site, ask yourself the following questions:

- Are the facts on the site documented?
- Does the site contain a bibliography?
- Is the author clearly identified?
- Does the site explore the topic in-depth or only scratch the surface?
- Does the site contain links to other useful and up-to-date resources?
- Is the information easy to access? Is it properly labeled?

Practicing the Skill

Visit the Peace Corps Web site listed below and answer the following questions.

1. Who is the author or sponsor of the Web site?
2. What links does the site contain? Are they appropriate to the topic?
3. What sources were used for the information contained on the site?
4. Does the site explore the topic in-depth? Why or why not?
5. Is the design of the site appealing? Why or why not?

Applying the Skill

Locate two other Web sites about Russia. Evaluate them for accuracy and usefulness, and then compare them to the Peace Corps site listed below.

(**www.peacecorps.gov/kids/world/europemed/russia.html**)

The People of Russia

Guide to Reading

Main Idea

Russia's people enjoy a culture rich in art, music, literature, and religion.

Terms to Know

- Slav
- majority culture
- minority culture

Reading Strategy

Create a chart like this one. List one or two examples for Russia beside each heading.

Art	
Music	
Literature	
Religion	

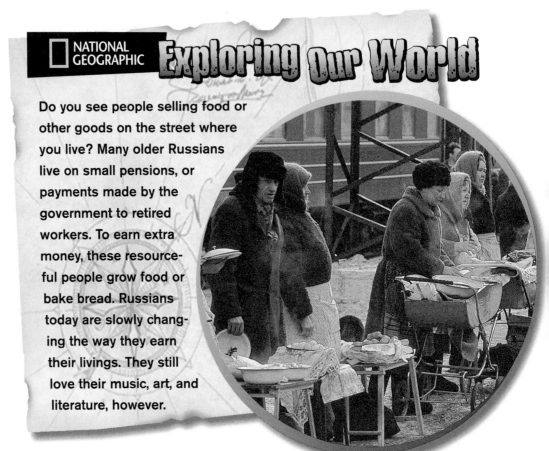

NATIONAL GEOGRAPHIC Exploring Our World

Do you see people selling food or other goods on the street where you live? Many older Russians live on small pensions, or payments made by the government to retired workers. To earn extra money, these resourceful people grow food or bake bread. Russians today are slowly changing the way they earn their livings. They still love their music, art, and literature, however.

Russia is one of the most populous countries in the world, with more than 145 million people. It is a huge land, with about 100 different ethnic groups. Most of the people, however, are Russian.

Ethnic Groups

Russians, along with Ukrainians and Belorussians, are part of a larger group of people called **Slavs.** Hundreds of years ago, the Slavs migrated from northeastern Europe to settle in western Russia. In Russia today, more than 80 percent of the people are Slavs. Slavs are the **majority culture.**

Besides the Slavs, many other ethnic groups live in Russia. Each group has its own distinctive language and culture. Some groups have a Christian heritage, while others are Islamic, Buddhist, or Jewish. These cultures are called **minority cultures** because they are not the group that controls most of the wealth and power in the society.

When the Soviet Union existed, the central government kept tight control over its people. When the Soviet Union fell apart, many old feuds and remembered wrongs came to the surface. Fighting broke out among many of the people who had been enemies in the past and whose differences had never been resolved. Like many other areas of the world, one of the biggest challenges facing people in this region is to learn how to cooperate and how to protect people who belong to minority cultures.

Reading Check What is the largest, most powerful ethnic group in Russia?

Culture in Russia

Russia has a rich tradition of art, music, and literature. Russians view these cultural achievements with pride. Some of their most beloved works of art are based on religious, historical, or folk themes.

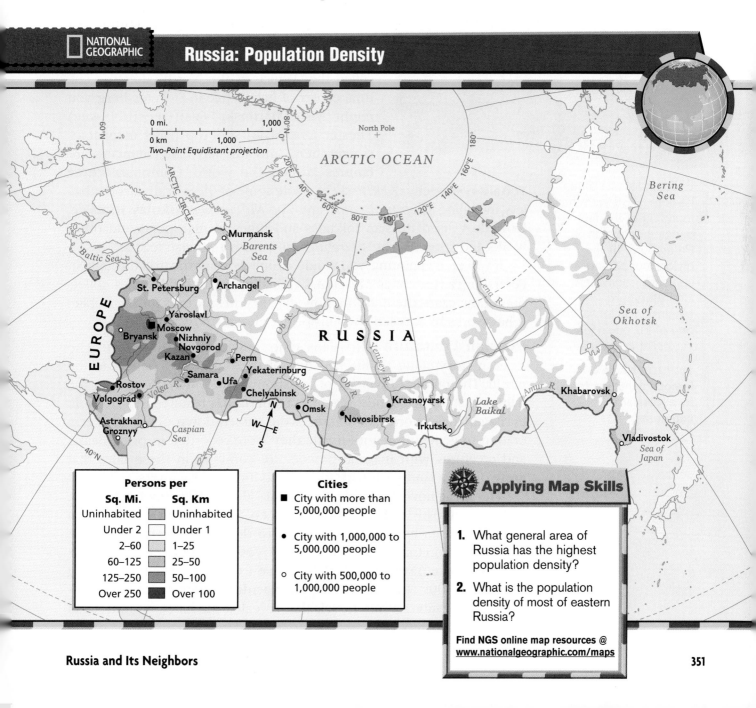

NATIONAL GEOGRAPHIC

Russia: Population Density

Persons per

Sq. Mi.		Sq. Km
Uninhabited		Uninhabited
Under 2		Under 1
2–60		1–25
60–125		25–50
125–250		50–100
Over 250		Over 100

Cities

■ City with more than 5,000,000 people

● City with 1,000,000 to 5,000,000 people

○ City with 500,000 to 1,000,000 people

Applying Map Skills

1. What general area of Russia has the highest population density?

2. What is the population density of most of eastern Russia?

Find NGS online map resources @ www.nationalgeographic.com/maps

Russia and Its Neighbors

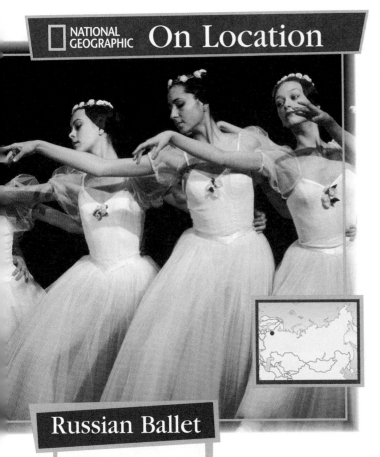

NATIONAL GEOGRAPHIC On Location

Russian Ballet

Ballet in Russia dates back to 1738 with the founding of the first dancing school in St. Petersburg for the children of palace servants. Today, the St. Petersburg Mariinsky Ballet, is one of Europe's leading ballet companies.

Culture Name a famous ballet written by a Russian composer.

Art and Music If you wanted to see Russian culture at its best, you would probably go to St. Petersburg. It was founded in 1703 by **Peter the Great.** Peter came to the Russian throne in 1682. Nearly 7 feet tall, he had boundless energy. During his reign, he took an 18-month tour of England and visited the Netherlands. Unlike most tourists, Peter stopped at shipyards, factories, and laboratories. He learned carpentry and enough skill in dentistry and surgery to want to practice on others. When he returned home, Peter forced the Russian nobility to adopt the ways of western Europe. In fact, those who refused to study math and geometry were not allowed to get married!

Peter built St. Petersburg to face Europe as a "window on the West." The city has so many beautiful museums and statues that it is also called the "Venice of the North" after the cultural center of Italy. While in St. Petersburg, you might visit Mariinsky (MAH•ree•IHN•skee) Theater. One of Russia's top ballet companies dances here, and Russian ballet dancers are famous around the world. Composer **Peter Tchaikovsky** (chy•KAWF•skee) wrote some of the world's favorite ballets, including *Sleeping Beauty* and *The Nutcracker.* **Nikolay Rimsky-Korsakov** used Russian folktales and tunes in his operas and other works. **Igor Stravinsky's** *Firebird* came from a Russian legend.

If you enjoy painting, you would definitely want to stroll through St. Petersburg's Hermitage Museum. Originally built to hold the art collection of the czars, the museum now displays these works for the public. It has works by Russian and European painters and sculptors.

Russian Literature Russians enjoy all kinds of literature. The Russian storytelling tradition is one of the oldest and richest in the world. The cultural diversity of Russia's peoples, stretching from the Baltic coast to the farthest reaches of Siberia, has resulted in a rich folk heritage. These stories, or *skazki,* were passed down orally from generation to generation, until finally they were recorded in print. Beasts and creatures with magical powers are common in these tales that grew out of a land with dark forests and long, cold winters.

The great novels and plays of Russia also reflect Russian themes. These are mostly political, however. **Leo Tolstoy's** novel *War and Peace* recounts how Russians rallied to defeat the French emperor Napoleon Bonaparte. **Fyodor Dostoyevsky** (FEE•uh•dor DAHS•tuh•YEHF•skee) wrote many novels that explored Russian life during the late 1800s. In the 1970s, **Alexander Solzhenitsyn** (SOHL•zhuh•NEET•suhn) wrote novels that revealed the harsh conditions of Communist society.

Religion in Russia   In spite of Communist laws in the past forbidding the practice of religion, the Russian Orthodox Church is very popular. Russian Orthodox is a Christian faith. It is headed by a figure called the *patriarch.* The word *patriarch* is a Greek term for "father."

Russian Orthodoxy was responsible for a special alphabet called Cyrillic. According to legend, this alphabet was developed by St. Cyril, an Orthodox priest, to help the Slavs read and write their own language. He invented new letters for sounds in the Slavic language that were not present in the Greek or Latin languages.

While more than 70 percent of the Russian population is Russian Orthodox, this is by no means the only religion in Russia. Many Muslims (followers of Islam), Roman Catholics, Protestants, and Buddhists today live within Russia's boundaries. Many of the Jews that at one time lived in Russia have now emigrated to other areas. Today, fewer than 1 million Jews live in Russia.

Family Life in Russia   About three-fourths of Russians live in cities, mostly in large apartment blocks. The typical Russian apartment is small, often with just a bedroom, living room, kitchen, and bathroom for a family of four. The living room may also be used as a bedroom.

It is very hard to find housing in the cities in Russia. For this reason, many generations may share the same home. This can be very helpful, however, because many Russian mothers work outside of the

Primary Source

ALEXANDER SOLZHENITSYN

(1918–)

For many years Russian author Alexander Solzhenitsyn was the voice of protest for his people, speaking out through his novels about injustices in the Soviet Union's Communist system. Since the people could not "see" freedom for themselves, he used his great literary talent to bring truth to as many people as possible.

"The sole substitute for an experience which we have not ourselves lived through is art and literature," he wrote. **"Wherever else it fails, art always has won its fight against lies, and it always will."**

Source: Nobel Lecture, 1972 by Alexander Isayevich Solzhenitsyn

Analyzing Primary Sources

1. What does Solzhenitsyn mean when he says that literature can substitute for an experience we have not had? Do you agree?

2. Describe an event you "experienced" through art. This might include a scary story or a powerful scene from a film.

▲ A babushka selling Russian folk art

home. The grandmother, or *babushka*, may take care of the cooking, cleaning, and caring for the young children. Shopping for food can take a long time because it sometimes means waiting in long lines.

There are many similarities between American and Russian women. As in the United States, women and men have equal rights in Russia. But also like the United States, women are still expected to do most of the household work. Women in Russia earn less than men for the same work. There is an active women's movement in Russia, and many women have begun to demand more help in the home from Russian men.

Celebrations and Food Russians enjoy small family get-togethers as well as national holidays. New Year's Eve is the most festive nonreligious holiday for Russians. Children decorate a fir tree and exchange presents with others in their families. Russians also celebrate the first day of May with parades and speeches. May Day honors Russian workers.

If you have dinner with a Russian family, you might begin with a big bowl of *borscht,* a soup made from beets, or *shchi,* a soup made from cabbage. Next, you might have meat turnovers, called *piroshki.* For the main course, you are likely to eat meat, poultry, or fish with boiled potatoes. On special occasions, Russians like to eat caviar. This delicacy is eggs of the sturgeon, a fish from the Caspian Sea.

 Reading Check **What are some ways that Russian and American cultures are similar?**

Section 2 Assessment

Defining Terms
1. Define Slav, majority culture, minority culture.

Recalling Facts
2. Culture What Russian composer wrote the world famous ballets *Sleeping Beauty* and *The Nutcracker*?

3. Government What did the Communist government try to do to religion in Russia?

4. Culture What is the major religion of Russia?

Critical Thinking
5. Synthesizing Information How would your family's living conditions change if you lived in a typical Russian apartment?

6. Making Predictions Art ideas are frequently drawn from life. What themes do you think you will see in future Russian arts?

Graphic Organizer
7. Organizing Information Create a diagram like this one, and list two facts for each topic in the four outer ovals.

Music — Russian Life — Family
Religion — Russian Life — Food

Applying Social Studies Skills

8. Analyzing Maps Look at the population density map on page 351. What city is located at about 55°N latitude and 38°E longitude? What is the estimated population of this city?

Making Connections

Count Leo Tolstoy

Count Leo Tolstoy (1828–1910) was a famous Russian novelist. Two of his epic works are *War and Peace* and *Anna Karenina.* What is not generally known is that Tolstoy also wrote for children. He wrote: "[These writings] will be used to teach generations of all Russian children, from the czar's to the peasant's, and from these readers they will receive their first poetic impressions, and having written these books, I can now die in peace."

Russian literature, even stories for children, contains more suffering and tragedy than American children would appreciate. The stories also celebrate qualities such as helpfulness, compassion, mercy, and justice. These are the values needed to survive difficult times. This story is an example of just such literature.

▲ Russian grandfather

The Grandfather and His Little Grandson
by Count Leo Tolstoy (1828–1910)

The grandfather had become very old. His legs would not carry him, his eyes could not see, his ears could not hear, and he was toothless. And when he ate, he was untidy. His son and the son's wife no longer allowed him to eat with them at the table and had him take his meals near the stove. They gave him his food in a cup. Once he tried to move the cup closer to him and it fell to the floor and broke. The daughter-in-law scolded the old man, saying that he damaged everything around the house and broke their cups, and she warned him that from that day on she would give him his food in a wooden dish. The old man sighed and said nothing.

One day the old man's son and his wife were sitting in their hut, resting. Their little son was playing on the floor. He was putting together something out of small bits of wood. His father asked him: "What are you making, Misha?" And Misha said: "I'm making a wooden bucket. When you and Mommie get old, I'll feed you out of this wooden bucket."

The young peasant and his wife looked at each other and tears appeared in their eyes. They were shamed to have treated the old man so unkindly, and from that day they again ate with him at the table and took better care of him.

Source: "The Grandfather and His Little Grandson" from *A Harvest of Russian Children's Literature,* edited by Miriam Morton. Copyright © 1967. University of California Press (Berkeley and Los Angeles, CA)

Making the Connection

1. What reasons did Leo Tolstoy give for writing stories for children?

2. What do you think the young peasant and his wife learned from their son?

3. **Making Comparisons** Compare this story with one you have learned. How are they different? How are they the same?

Russia's Southern Neighbors

Guide to Reading

Main Idea

Many different ethnic groups live in the Eurasian Republics.

Terms to Know

- ethnic conflict
- nomad
- oasis
- elevation

Reading Strategy

Create a chart like this one. In the left column, list the Eurasian Republics. In the right column, list the dominant religion in the republic.

Republic	Religion

NATIONAL GEOGRAPHIC

Exploring Our World

For centuries, the city of Bukhara was a stop along an ancient trading route called the Silk Road, which stretched from China to Europe. In the past, precious Chinese silk was carried on the backs of camels. Silk is still sold in Bukhara's markets. Today, however, international trade communications occur through a system of fiber-optic cables laid along the ancient tracks of the Silk Road camels.

The Eurasian Republics are made up of the three republics of the **Caucasus** and the five republics of **Central Asia.** The peoples of this region have all been ruled at one time or another by Arabs, Turks, Persians, and Russians. Disagreements among these countries and among the ethnic groups within the countries have sparked violent conflicts. Most of the conflicts are disagreements over who owns the land.

The Caucasus

The republics of the Caucasus include **Armenia, Georgia,** and **Azerbaijan.** These republics became independent in 1991 for the first time in centuries. Since then, the region has struggled to develop its own industries and businesses.

Making this effort more difficult is the conflict between ethnic Armenia and neighboring Azerbaijan. Many Christian Armenians live in Azerbaijan, but want to become part of Armenia. This dispute led

Azerbaijan to cut off needed supplies of fuel and other resources to Armenia. The economy of both countries has been seriously hurt by this conflict.

Azerbaijan is split in two by Armenian territory. The Azeris are Muslims who speak their own language called Azeri. The Azeris are excellent weavers and craftspeople and are productive farmers. Their hand-knotted rugs are highly valued for their rich color and intricate designs.

Ethnic conflicts have also erupted in Georgia and hurt its efforts to move toward democracy. **Ethnic conflicts** are disagreements or fights between two or more groups who differ from each other. The majority culture of dominant Georgians is Christian and the people speak Georgian, a distinctive language. They are skilled farmers whose products make up one-third of the country's goods. Georgia also has natural resources such as copper and coal.

✓ Reading Check What republics are included in the republics of the Caucasus?

The Central Asian Republics

The Central Asian Republics include **Kazakhstan, Kyrgyzstan** (KIHR•gih•STAN), **Tajikistan, Turkmenistan,** and **Uzbekistan.** All five Central Asian Republics are Islamic countries. (See the map on page 359.)

Kazakhstan The Kazakh ancestors were horse-riding warriors called the **Mongols.** They were **nomads** who moved from place to place and did not have permanent homes. They were forced to give up their nomadic ways when they became part of the Soviet Union. Many Russians moved to this area in the mid-1900s to enforce Communist laws. The environment has been badly damaged, because many factories were built quickly. Nuclear, chemical, and industrial waste has polluted the land and water of this region.

Uzbekistan Uzbekistan is slightly larger than California in area. Uzbekistan is one of the world's largest cotton producers. This boom in cotton, unfortunately, has had disastrous effects on the environment. Large farms needing irrigation have nearly drained away the rivers flowing into the **Aral Sea.** Receiving less freshwater, the sea has steadily shrunk, and its salt level has increased. Fish and wildlife have died, and salt particles have polluted the air and soil. To create prosperity, Uzbek leaders want to

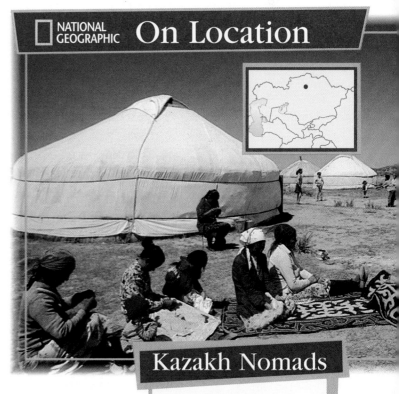

NATIONAL GEOGRAPHIC On Location

Kazakh Nomads

The traditional home of Kazakhs—called a yurt—can be easily taken apart and moved.

Culture Why would the Mongols and early Kazakh people need a house that could be moved?

Architecture

In 1983, two towns from opposite sides of the world—Boulder, Colorado, and Dushanbe, Tajikistan—became "sister cities." From Dushanbe came the largest gift ever from any former Soviet nation to the United States. The gift was a beautiful *chaikhona,* or teahouse. In Tajikistan, a *chaikhona* serves as a community center where men gather to drink beverages, tell stories, hear news, and socialize. It often becomes the very heart of a city. In Boulder, men and women of all ages come here to appreciate this part of Tajik culture.

Looking Closer What are some gathering places in your community?

use newly discovered deposits of oil, gas, and gold. This would create more pollution in the region.

Most of Uzbekistan's roughly 25 million people are Uzbeks who generally live in fertile valleys and oases. An oasis is a green area in a desert watered by an underground spring. **Tashkent,** the capital, is the largest city and industrial center in Central Asia. About 2,000 years ago, the oases of Tashkent, Bukhara, and Samarqand were part of the busy trade route—the **Silk Road**—that linked China and Europe.

Kyrgyzstan The lofty **Tian Shan** (tee•AHN SHAHN) mountain range makes up most of Kyrgyzstan. The climate depends on an area's height above sea level, or elevation. Lower valleys and plains have warm, dry summers and chilly winters. Higher areas have cool summers and bitterly cold winters. The harsh climate and lack of fertile soil hinder farmers, but they manage to grow cotton, vegetables, and fruits. Many also raise sheep or cattle. Kyrgyzstan has few industries, but it does have valuable deposits of mercury and gold.

More than half of the people belong to the Kyrgyz ethnic group. Differences among clans, or family groups, often separate one part of the country from another. Kyrgyzstan is a bilingual country—one that has two official languages. These are Kirghiz, related to Turkish, and Russian. About 40 percent of the people live in cities, such as the capital, **Bishkek.**

Tajikistan **Tajikistan** is also very mountainous. Ismail Samani Peak, once called Communist Peak, is the highest mountain in Central Asia and is located here.

Agriculture is the most important activity in Tajikistan. Farmers grow cotton, rice, and fruits in fertile river valleys. Mountain streams provide water for irrigation as well as for hydroelectric power.

The largest city is **Dushanbe** (doo•SHAM•buh), the capital. Most of Tajikistan's people are Tajiks, who are related to the Persians. Another 25 percent are Uzbeks, a group related to the Turks. In 1992 a bitter civil war broke out between rival clans. Many people were killed, and the economy was severely damaged. Despite a peace agreement in 1997, tensions still remain high.

Turkmenistan   **Turkmenistan** is larger than neighboring Uzbekistan, but it has far fewer people. Why? Most of this vast land—about 85 percent of it—is part of a huge desert called the **Garagum** (GAHR•uh•GOOM). Making up the northern and central region, *Garagum* means "black sand."

Despite the harshness of the land and climate, agriculture is the leading economic activity. Raising livestock is another important activity. However, not enough produce is available to feed everyone in the country, and much food has to be imported.

NATIONAL GEOGRAPHIC

The Eurasian Republics: Political

Applying Map Skills

1. What country is split into two parts by Armenia?

2. What is the capital of Georgia?

Find NGS online map resources @ www.nationalgeographic.com/maps

⊕ National capital
• Major city

0 mi.   400
0 km   400
Two-Point Equidistant projection

Turkmenistan is important to world energy markets because it contains one of the world's largest reserves of natural gas. Estimates of the total gas resource range as high as 535 trillion cubic feet. Turkmenistan is hoping that its oil and natural gas resources will give it a brighter future.

Turkmenistan is a one-party state dominated by its president and his closest advisers. In spite of having a democratic constitution, the current president retains absolute control over the country. In 1991, the Communist Party was renamed the Democratic Party.

More than 75 percent of Turkmenistan's roughly 4.5 million people belong to the Turkmen ethnic group. **Ashgabat,** the capital, is the country's largest city and leading economic and cultural center. More than one-half of Turkmenistan's people live in rural areas, though. Turkmen villages usually are located near oases formed by mountain streams.

The Turkmen people were nomads who raised camels and other livestock in the desert. Under Soviet rule, the nomads were forced to settle on farms. In the 1950s, Soviet engineers built a large irrigation and shipping canal. This technological feat greatly increased the land area used for growing cotton. However, as in Uzbekistan, cotton growing has helped to dry up the Aral Sea.

✓ Reading Check **Why do most Turkmen live along the southern border?**

Section 3 Assessment

Defining Terms
1. Define ethnic conflict, nomad, oasis, elevation.

Recalling Facts
2. Culture What are the different ethnic groups that make up the Central Asian Republics?
3. Economics What conflict is making economic development more difficult for the republics of the Caucasus?

Critical Thinking
4. Summarizing Information How has religion played a part in the ethnic conflicts in the Caucasus?
5. Understanding Cause and Effect What are the results of diverting the rivers that empty into the Aral Sea?

Graphic Organizer
6. Organizing Information Create a chart like the one shown below. Fill in the chart with information about Georgia and Uzbekistan that you learned in this section.

	Georgia	Uzbekistan
Ethnic group		
Natural resources		
Economic activity		

Applying Social Studies Skills

7. Analyzing Maps Look at the political map chart on page 335. Which of the Eurasian Republics do not share a border with Russia?

Section 1 | The Russian Land

Terms to Know

Eurasia
urban
suburb
consumer goods
rural
tundra
taiga
steppe
communism
heavy industry
light industry

Main Idea

Most people in Russia live west of the Urals, where the climate is mild. The people of Russia have met the challenges created by the country's gigantic size and harsh climates.

✓ Location Spanning two continents—Europe and Asia—Russia is the world's largest country.

✓ Region 75 percent of the people live in the western half of Russia, which is mostly a lowland plain.

✓ Economics The people of Siberia support themselves by fishing, hunting, herding reindeer, and lumbering.

✓ Movement Inland waterways, such as rivers and canals, and railroads are important for moving goods through Russia.

Section 2 | The People of Russia

Terms to Know

Slav
majority culture
minority culture

Main Idea

Russia's people enjoy a culture rich in art, music, literature, and religion.

✓ Culture Russia is a huge, populous country with about 100 different ethnic groups.

✓ Culture The city of St. Petersburg in Russia has many beautiful museums and statues.

✓ Religion The Russian Orthodox Church has provided the people of Russia with religious beliefs, beautiful churches, and the Cyrillic alphabet.

Section 3 | Russia's Southern Neighbors

Terms to Know

ethnic conflict
nomad
oasis
elevation

Main idea

Many different ethnic groups live in the Eurasian Republics.

✓ Economics The Caucasus Republics have struggled to develop their own industries and businesses, but are facing many ethnic conflicts.

✓ Environment The Central Asian Republics have a great deal of pollution in their land, air, and water.

✓ Environment The Aral Sea is drying up because farms are draining away the rivers that feed it to irrigate their crops.

Camels walk where fish once swam in the Aral Sea. ▶

Chapter 13 Assessment and Activities

Using Key Terms

Match the terms in Part A with their definitions in Part B.

A.

1. majority culture
2. light industry
3. rural
4. steppe
5. consumer goods
6. taiga
7. hydroelectric power
8. ethnic conflict
9. heavy industry
10. urban

B.

a. huge, subarctic evergreen forest
b. relating to cities
c. dry, treeless grassland
d. the most powerful culture group
e. a disagreement or fight between two or more groups who differ from one another
f. production of consumer goods
g. electricity generated by water
h. production of industrial goods
i. relating to the countryside
j. products for personal use, such as clothing

Reviewing the Main Ideas

Section 1 The Russian Land

11. **Culture** What is the political and cultural center of Russia?
12. **Economics** Why is Kaliningrad so important to Russia?
13. **Movement** What river carries almost half of Russia's river traffic?
14. **Location** What part of Russia has the most minerals?

Section 2 The People of Russia

15. **Culture** What is the majority culture in Russia?
16. **Culture** In which city would you find the Hermitage Museum and many of Russia's museums, theaters, and ballets?
17. **History** What happened to religion during the Communist rule of Russia?

Section 3 Russia's Southern Neighbors

18. **Economics** What has seriously hurt the the economies of Armenia and Azerbaijan?
19. **Culture** Why have the republics been in conflict?

 **Russia—A Eurasian Country**

Place Location Activity

On a separate sheet of paper, match the letters on the map with the numbered places listed below.

1. Ural Mountains
2. Kamchatka Peninsula
3. Lake Baikal
4. Volga River
5. Moscow
6. Lena River
7. West Siberian Plain
8. Caspian Sea
9. Caucasus Mountains
10. Sakhalin Island

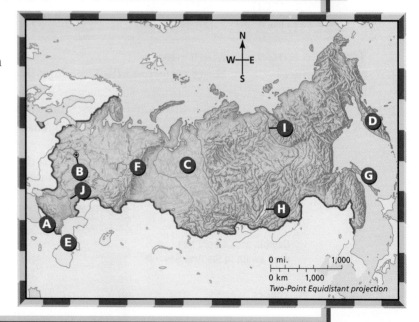

Self-Check Quiz Visit the *Our World Today: People, Places, and Issues* Web site at owt.glencoe.com and click on **Chapter 13—Self-Check Quizzes** to prepare for the Chapter Test.

Critical Thinking

20. **Making Generalizations** How have recent changes in Russia affected its economy?

21. **Organizing Information** Create a chart like this one. Then list the similarities and differences between the economies of two Central Asian Republics.

Country	Similarities	Differences

Current Events Journal

22. **Writing a Paragraph** Read a newspaper or magazine article about Russia or one of its neighbors and imagine yourself in the scene. Write a description using vivid words to portray the sights, sounds, and people you would encounter. What is the location? What can you hear? What can you see? Describe what is happening.

Mental Mapping Activity

23. **Focus on the Region** Create a simple outline map of Russia and label the following:

- Black Sea
- Caspian Sea
- Aral Sea
- Volga River
- Ural Mountains
- St. Petersburg
- Moscow
- Lake Baikal

Technology Skills Activity

24. **Using the Internet** Search the Internet for information on the problems facing the Aral Sea. Find out what distinctive creatures live in the Aral Sea that cannot be found anywhere else in the world. Find a map that shows what countries border the Aral Sea.

The Princeton Review

Standardized Test Practice

Directions: Read the paragraph below, and then answer the following question.

For 70 years, the Communist government of the Soviet Union stopped at nothing to industrialize the country. Soviet leaders gave little thought to the health of the people or to the land they were ruining. From St. Petersburg to Vladivostok, across more than 8 million square miles, the Russian environment shows decay. Today Russia's great rivers are sewers of chemicals and human waste. Air in more than 200 cities is at least five times more polluted than standards allow, putting millions at risk of lung diseases. Tons of nuclear waste lie under Arctic waters, and the use of toxic fertilizers has poisoned the soil.

Adapted from "The U.S.S.R.'s Lethal Legacy" by Mike Edwards, *National Geographic*, August 1994.

1. **Which of the following best explains why the Russian environment is so polluted today?**

   F Soviet leaders did not care about the health of the people.

   G Soviet leaders wanted to industrialize the country at all costs.

   H Farmers were careless about the chemicals they spread on their fields.

   J Eastern Russia is not as polluted as western Russia.

Test-Taking Tip: Notice that the question asks for the *best* explanation. Read all the choices carefully before choosing the best one. Although all of the answer statements might be true, you need to find the best answer to the question.

The World and Its People

NATIONAL GEOGRAPHIC

To learn more about Russia's land and economy, view *The World and Its People* Chapter 15 video.

Our World Today online

Chapter Overview Visit the *Our World Today: People, Places, and Issues* Web site at owt.glencoe.com and click on **Chapter 14—Chapter Overviews** to preview information about Russia.

Categorizing Information Study Foldable When you group information into categories, it is easier to make sense of what you are learning. Make this foldable to help you learn about Russia's past and present.

Step 1 Fold one sheet of paper in half from top to bottom.

Step 2 Fold it in half again, from side to side.

Step 3 Unfold the paper once. Cut up the inside fold of the top flap only.

This cut will make two tabs.

Step 4 Turn the paper and sketch a map of the USSR and Russia on the front tabs. Label your foldable as shown.

Past
USSR

Present
Russia

Reading and Writing As you read the chapter, write under the appropriate flaps of your foldable what you learn about the former USSR and present-day Russia.

Why It Matters

A New Government

Russia is a land rich in natural resources but with a troubled political history. The various peoples in Russia have had little experience with "hands-on" government. This experience is needed for a stable democracy to work. On the other hand, a strong, central government is needed to create policies to prevent continued air and water pollution and to build up the economy. How will Russia meet both of these aims? The answer is important to us all.

◄ **Statue of Vladimir Lenin at the Exhibition of Economic Achievement Moscow, Russia**

A Troubled History

Guide to Reading

Main Idea

The harsh rule of powerful leaders has often sparked violent uprisings in Russia.

Terms to Know

- czar
- serf
- industrialize
- communist state
- Cold War
- glasnost

Reading Strategy

Create a chart like this one. List three main czars and important things to remember about them.

Czar	Importance

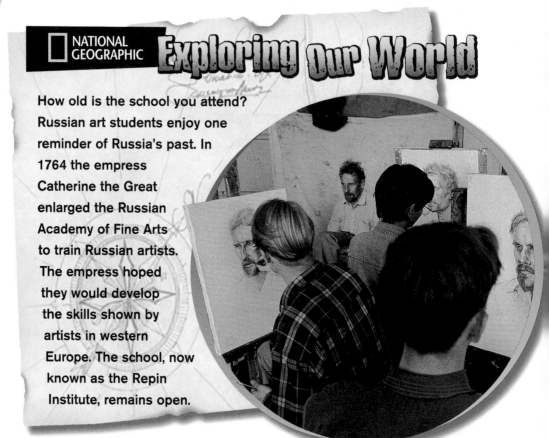

NATIONAL GEOGRAPHIC

Exploring Our World

How old is the school you attend? Russian art students enjoy one reminder of Russia's past. In 1764 the empress Catherine the Great enlarged the Russian Academy of Fine Arts to train Russian artists. The empress hoped they would develop the skills shown by artists in western Europe. The school, now known as the Repin Institute, remains open.

As you read in Chapter 13, Russia is the largest country in the world. Early in its history, however, it was a small territory on the edge of Europe. Strong rulers gradually expanded the Russian territory. Their harsh rule led to unrest, eventually resulting in two major upheavals—one in 1917, the other in 1991.

The History of Russia

To understand the challenges facing Russia today, let us go back through Russia's history. Modern Russians descend from early groups of Slavs who settled along the rivers of what is today Ukraine and Russia. During the A.D. 800s, these early Slavs built a civilization around the city of **Kiev,** today the capital of **Ukraine.** This civilization was called **Kievan Rus** (kee•AY•vuhn ROOS). By the 1000s, the ruler and people of Kievan Rus had accepted Eastern Orthodox Christianity. They prospered from trade with the Mediterranean world and western Europe.

In the 1200s, the **Mongols** swept in from Central Asia and conquered Kiev. Under their rule, Kiev lost much of its wealth and power. Meanwhile, Moscow became the center of a new Slavic territory, called **Muscovy** (muh•SKOH•vee). In 1480 **Ivan III,** a prince of Muscovy, drove out the Mongols and made the territory independent.

Rise of the Czars   Muscovy slowly developed into the country we know today as Russia. Russian rulers expanded their power, built up their armies, and seized land and other resources. They called themselves czars, or emperors. (Sometimes you will see this word written *tsar.*) They had complete and total control over the government. As a citizen of Muscovy, you would have feared **Czar Ivan IV,** who ruled during the 1500s. Known as "Ivan the Terrible," Ivan IV used a secret police force to tighten his iron grip on the people and control their lives.

The czars gradually conquered nearby territories. As a result, many non-Russian peoples became part of the growing Russian Empire. (Russia today still suffers from ethnic tensions caused by these conquests.) Czars, such as **Peter the Great** and **Catherine the Great,** pushed the empire's borders southward and westward. They also tried to make Russia modern and more like Europe. As explained in Chapter 13, Peter built a new capital—St. Petersburg—in the early 1700s. Built close to Europe near the Baltic coast, St. Petersburg was designed like a European city with its elegant palaces, public squares, and canals. If you had been a Russian noble at this time, you would have spoken French as well as Russian. You also would have put aside traditional Russian dress, worn European clothes, and attended fancy balls and parties.

Early Russia

Ivan III, or "Ivan the Great," (left) ruled Muscovy until 1505. His grandson, Ivan IV, also known as "Ivan the Terrible," (right) used a secret police force to control the people of Muscovy.

History Who drove the Mongols out of Kiev?

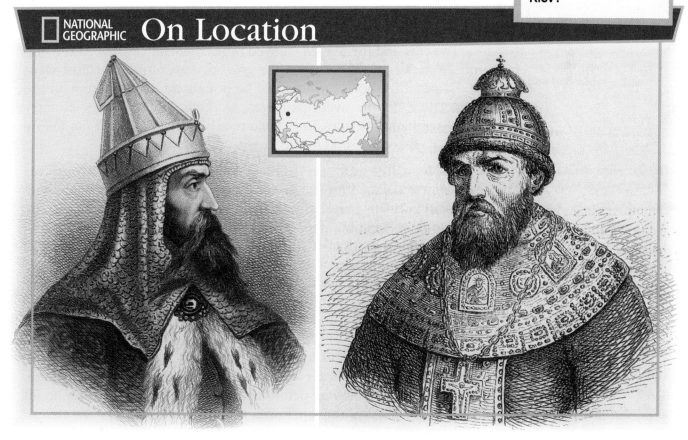

NATIONAL GEOGRAPHIC On Location

Kievan Territory
1360–1524
1524–1689
1689–1917
1917–1945
— Boundary of the Soviet Union in 1945
— Present-day Russian boundary

ALASKA (Sold to U.S. in 1867)

Chukchi Sea

East Siberian Sea

Bering Sea

North Pole

ARCTIC OCEAN

Barents Sea

Kara Sea

Laptev Sea

Sea of Okhotsk

ARCTIC CIRCLE

R U S S I A

L. Baikal

Black Sea

Caspian Sea

Aral Sea

0 mi. 1,000
0 km 1,000
Two-Point Equidistant projection

Applying Map Skills

1. During which time period was the most land added to Russia?

2. Was Russia's land area larger in 1945, or is it larger today?

Find NGS online map resources @ www.nationalgeographic.com/maps

The czar and the nobles enjoyed rich, comfortable lives. At the bottom of society, however, were the great masses of people. Most were serfs, or farm laborers, who could be bought and sold along with the land. These people lived hard lives, working on the nobles' country estates or in city palaces. Few could read or write. They did not follow Western customs, but kept the Russian traditions.

Dramatic Changes In 1812, a French army led by **Napoleon Bonaparte** invaded Russia. Brave Russian soldiers and the fierce winter weather finally forced the French to retreat. The year 1812 became a symbol of Russian patriotism. Have you ever heard the *1812 Overture,* with its dramatic ending that includes the ringing of bells and the bursts of cannon fire? Written by the Russian composer **Peter Tchaikovsky,** this musical masterpiece celebrates the Russian victory over Napoleon.

In the late 1800s, Russia entered a period of economic and social change. The Russian Empire expanded southward into the Caucasus Mountains and eastward into Central Asia and toward China and the Pacific Ocean. In 1861 **Czar Alexander II,** known as the Czar-Liberator, freed the serfs from being tied to the land. His new law did little to lift them out of poverty, though. Russia began to industrialize, or change its economy to rely more on manufacturing and less on farming.

Railroads, including the famous **Trans-Siberian Railroad,** spread across the country. It linked Moscow in the west with Vladivostok on Russia's Pacific coast.

✓ Reading Check What was the name of the civilization that early Slavs built in the area that is today Ukraine?

The Soviet Era

In 1914, World War I broke out in Europe. Russian and German armies met and fought bloody battles in eastern Europe. Unprepared for war, Russia suffered many defeats and had few victories. As the fighting dragged on, shortages of food in Russian cities caused much starvation. The Russian people blamed the czars for their troubles.

The Russian Revolution In 1917, political leaders, soldiers, and factory workers forced **Czar Nicholas II** to give up the throne. Later that year, a political rebel named **Vladimir Lenin** (VLAH•deh•meehr LEN•in) led a second revolution and seized control. He and his followers set up a communist state. A communist state is a country whose government has almost total control over the economy and society as a whole. Fearing invasion, the Communists moved Russia's capital from coastal St. Petersburg to Moscow.

Growth of Soviet Power By 1922, after a brutal civil war, Russia's Communist leaders were securely in power. In that year, they formed the **Union of Soviet Socialist Republics (USSR),** or the Soviet Union. This vast territory included Russia and most of the conquered territories of the old Russian Empire. After Lenin died in 1924, Communist Party officials disagreed over who was to lead the country.

Within a few years, **Joseph Stalin** won out over the others and became the Soviet Union's leader. Under Stalin's orders, the government took complete control over the economy. Those who opposed Stalin's actions were killed or sent to remote prison camps deep in the vast forests of icy Siberia. Millions of people were brutally murdered or forced into slave labor under Stalin's rule.

Web Activity Visit the **Our World Today: People, Places, and Issues** Web site at owt.glencoe.com and click on **Chapter 14— Student Web Activities** to learn more about the Russian Revolution.

NATIONAL GEOGRAPHIC On Location

Collective Farm Workers

In an effort to increase food production, Stalin combined small peasant farms into large collectives, shown in this painting by Alexander Volkov. Peasant resistance turned the experiment into a disaster.

Government How do you think Stalin dealt with the farmers who resisted?

EXPLORING CULTURE

Art

Peter Carl Fabergé was no ordinary Russian jeweler. His successful workshop designed extravagant jeweled flowers, figures, and animals. He is most famous for crafting priceless gold Easter eggs for the czar of Russia and other royalty in Europe and Asia. Each egg was unique and took nearly a year to create. Lifting the lid of the egg revealed a tiny surprise. One egg Fabergé created (shown here) held an intricate ship inside.

Looking Closer Why do you think Fabergé's workshop closed after the Russian Revolution of 1917?

Fabergé egg ▲

In 1941, Nazi Germany invaded the Soviet Union, bringing the country into World War II. During the conflict, the Soviets joined with Great Britain and the United States to defeat the Germans. Millions of Russians—soldiers and civilians—died in what Russians call the Great Patriotic Fatherland War.

A Superpower   When World War II ended, Stalin wanted to protect the Soviet Union from any more invasions. He sent troops to set up communist governments in neighboring Eastern European countries. These countries formed what was called an "iron curtain" behind which corruption and brutality were the norm. Stalin and the leaders who followed him spent large amounts of money on the military and weapons. The Soviet Union became one of the two most powerful nations in the world. The other superpower—the United States—opposed Soviet actions. These two nations engaged in the Cold War, competing for world influence without breaking out in actual fighting. They even competed in areas outside the world. Both the Soviet Union and the United States launched rockets in a bid to be first in outer space.

The Cold War Years   From 1940 to the late 1980s—the Cold War years—the Soviet economy faced many problems. The government factories and businessess had no competition and no one was allowed to make a profit. Factories became inefficient and produced poor-quality goods. The government cared more about making tanks and airplanes for military purposes than cars and refrigerators. As a result, people had few consumer goods to buy. Food often became scarce, and people often waited in long lines to buy bread, milk, and other necessary items. As during World War I, the Russian people and those living in Soviet-controlled areas became very unhappy.

The Soviet Union had another challenge. This vast empire included not only Russians but also people from many other ethnic groups. Instead of being scattered throughout the country, people in each of these other groups generally lived together in the same area. They resented the control of the government in Moscow, which they believed favored Slavic Russians. They wanted to leave the Soviet Union and form their own countries.

Soviet Collapse In 1985, **Mikhail Gorbachev** (GAWR•buh•CHAWF) became the leader of the Soviet Union. Gorbachev hoped to lessen the government's control of the economy and society. He allowed farmers and factory managers to make many of their own decisions. He allowed people to speak freely about the government and important issues, a policy called glasnost, or "openness." Instead of strengthening the country, however, his policies only made people doubt the communist system even more. People's demands for more and more changes eventually led to the collapse of both communism and the Soviet Union.

▲ Mikhail Gorbachev tried to lessen the Russian government's control of the economy and society.

By late 1991, each of the 15 republics that made up the Soviet Union had declared its independence. The Soviet Union no longer existed. Russia emerged as the largest and most powerful of those republics. Although a rough road lay ahead, many Russians were thrilled by the end of communism and the chance to enjoy freedom.

✓ Reading Check **What were three reasons for the breakup of the Soviet Union?**

Section 1 Assessment

Defining Terms
1. Define czar, serf, industrialize, communist state, Cold War, glasnost.

Recalling Facts
2. History Why did Peter the Great build a new capital of Russia?
3. History Who led the 1917 revolution in Russia?
4. History What happened to the Soviet Union in 1991?

Critical Thinking
5. Understanding Cause and Effect What problems were created when Mikhail Gorbachev allowed the policy of glasnost in Russia?
6. Analyzing Information How did glasnost weaken the communist system?

Graphic Organizer
7. Organizing Information In a chart like this one, write facts that show the contrast between the nobles and the serfs of Russia.

Nobles	Serfs

Applying Social Studies Skills

8. Creating Mental Maps Create your own map of early Russian territory. Label where Kievan Rus was located. Then label where Peter the Great moved the capital.

Making Connections

CULTURE   GOVERNMENT   PEOPLE   TECHNOLOGY

Cooperative Space Ventures

The space age officially began in 1957, when Russia launched *Sputnik I*. It was the first artificial satellite to orbit the earth.

The Space Race

The Russians sent the first man into space in 1961, when cosmonaut Yuri Gagarin orbited the earth. A few weeks later, Alan Shepard made the United States's first spaceflight. John Glenn was the first astronaut to orbit the earth in 1962. After this, the "space race" between the United States and Russia was of global importance. It was feared that one country could dominate the world if it had the right equipment in space.

Over the years, both Russia and the United States launched many spacecraft. In 1986, the Russian space station *Mir*, which means "peace," began to orbit the earth. This was the first permanently staffed laboratory in space. Astronauts from more than a dozen countries were invited to participate on the space station *Mir*. The astronauts and Russian cosmonauts performed many experiments about the effects of weightlessness.

In 1993 the United States and Russia decided to work jointly to build an International Space Station. On November 2, 2000, the International Space Station had its first permanent human inhabitants. The crew was made up of both Russian cosmonauts and American astronauts. Finally, on March 22, 2001, after 15 years of use, *Mir* was allowed to plummet back to Earth.

▶ Making the Connection

1. What country launched the space age?
2. How have the United States and Russia cooperated on space ventures?
3. **Making Predictions** What space technology do you think we will see in the future? What social consequences might result from this?

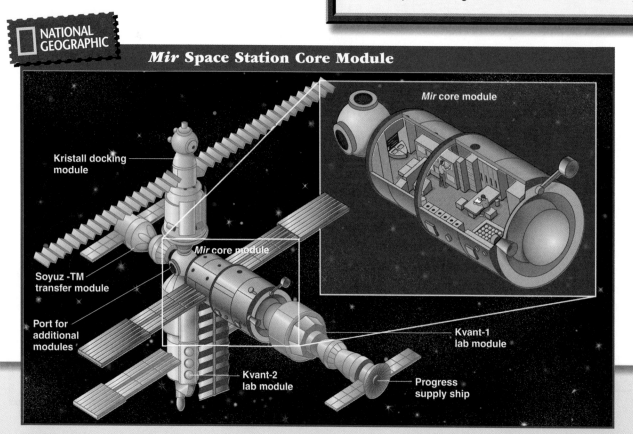

NATIONAL GEOGRAPHIC

Mir **Space Station Core Module**

Mir core module

Kristall docking module

Mir core module

Soyuz -TM transfer module

Port for additional modules

Kvant-1 lab module

Kvant-2 lab module

Progress supply ship

Guide to Reading

Main Idea

Russia has a rich cultural heritage but faces challenges in adopting a new economic system and government.

Terms to Know

- free market economy
- nuclear energy
- federal republic

Reading Strategy

Create a diagram like this one. Then write four challenges facing Russia.

Challenges Facing Russia

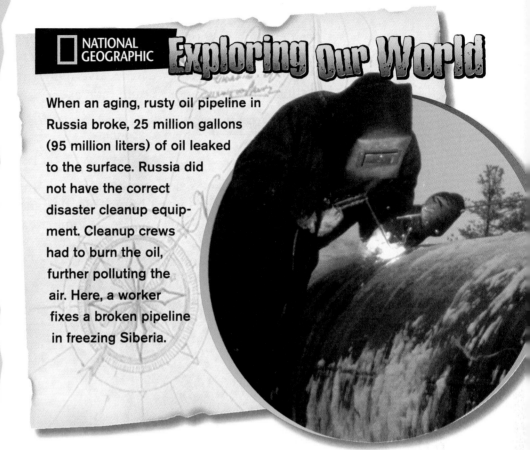

NATIONAL GEOGRAPHIC *Exploring Our World*

When an aging, rusty oil pipeline in Russia broke, 25 million gallons (95 million liters) of oil leaked to the surface. Russia did not have the correct disaster cleanup equipment. Cleanup crews had to burn the oil, further polluting the air. Here, a worker fixes a broken pipeline in freezing Siberia.

The Russian people have been working to move away from the strict, tightly controlled rules of the czars and Communists. They are moving toward a democratic government and an economy based on competition and private ownership. They have found that these changes do not come easily.

From Communism to Free Enterprise

The fall of communism turned Russia's economy upside down. The new Russian government turned to a free market economy, the system followed in the United States. Under a free market economy, the people, not the government, decide what businesses to start and run. Today Russian factory managers can decide what products to make. People can open businesses—such as restaurants, stores, or computer companies—and choose their own careers.

The Russian people gained freedom, but staying free is hard work. People now can make their own decisions, but those decisions do not always lead to success. Businesses can fail. People may become unemployed. Under communism, everybody had jobs. Workers today can lose their jobs when business is poor.

In addition, the government no longer sets prices for food and other goods. When prices were set low, the Russian people could afford the goods, but they often faced shortages. Without government controls, prices have risen. Higher prices make it harder to buy necessities such as food and clothing. It is hoped that, in the future, factories will start producing more goods. Now that manufacturers and producers know they can receive higher prices and more profits for their goods, supplies should increase.

✓ Reading Check **Who decides what businesses to start and run in a free market economy?**

Environmental Issues

Pollution is a serious problem in Russia today. Forest lands in Russia have been cut without replanting seedlings to hold the soil. This is causing serious soil erosion in some areas. Chemical fertilizers have been heavily used to increase crop production. These chemicals can build up in the soil over time and destroy its ability to grow food.

NATIONAL GEOGRAPHIC On Location

Food Shortages

Under communism, prices were low, but people faced shortages and had to wait in line to buy goods.

Economics Why can many Russians today not afford goods and services?

Air Pollution The government built power plants to make nuclear energy, or energy from controlled atomic reactions. Nuclear power plants can leave dangerous by-products in the air. Air pollution from heavy industry is also particularly bad. Gases are given off by coal-fired electric plants, vehicles, and other forms of transportation in major cities. It has caused many people to suffer from lung diseases. Rising numbers of people have cancer, and people in Russia are dying sooner.

Water Pollution Chemicals that are used in agriculture and industry often end up in rivers and lakes. Poor sewer systems also pollute water systems in Russia. Yet another source of water pollution is the chemical weapons that were developed by the Soviet Union during the Cold War. Many of these are buried in dumps throughout Russia. Age is causing the weapons to deteriorate and some of the chemicals are finding their way into groundwater or nearby waterways.

✓ Reading Check What pollution problems do nuclear power plants create?

NATIONAL GEOGRAPHIC On Location

Chechnya

These Chechens are standing in line to receive humanitarian aid after Russian rebels fought in this area.

Culture Why might Chechens want to be independent from Russia?

Political Challenges

Today Russia is a democracy, a government in which people freely elect their leaders. Russia, like the United States, is also a **federal republic.** This means that power is divided between national and state governments. A president is elected to lead the nation.

A Russian president has stronger powers than an American president. For example, the Russian president can issue orders that become laws even if they are not passed by the legislature. Russia's first two presidents—**Boris Yeltsin** (BOH•rehs YEHL•tzehn) and **Vladimir Putin** (VLAH•deh•meehr POO•tihn)—used their powers to help develop and strengthen Russia's economy.

In adjusting to a new form of government, Russians face important political challenges. Democracy is built on the idea of the rule of law. In the past, Russian leaders did what they wanted to do. In the new system, they must learn to follow the law. Also, past governments punished people who criticized their decisions. In a democracy, officials have to respect other opinions.

Another challenge results from the fact that Russia is home to many different ethnic groups. Some of these groups want to form their own countries. Among these groups are the **Chechens** (CHEH•chehnz), who live in **Chechnya** (CHEHCH•nee•uh) near the Caspian Sea and Caucasus Mountains in southern Russia. Russian troops have fought Chechen forces to keep Chechnya a part of Russia.

✓ Reading Check   Why is being home to different ethnic groups a challenge for Russia?

Russia—Past and Present

The Challenge of Change

Russia and the Eurasian Republics are presently facing many challenges. The most important of these have been discussed in the last two chapters. One of these is the change from a communist to a democratic government. Another challenge involves creating a free market economy. People must be given an opportunity to learn, work, and raise families. Third, if the region is to have peace, trust must grow among the different ethnic groups. Finally, the land, air, and waterways must be cleaned up and preserved for future generations.

The free countries of the world have many reasons to want Russia to succeed. If the Russian people cannot live under a democratic government, will they try to turn back to communism? Is an unstable Russia dangerous for Europe? A strong, free Russia may guide neighboring countries toward democracy. Russia has seen two revolutions in the last century, one violent and one peaceful. The problem faced by the world is how to best help this region reach its goals so that there will be no more violent Russian revolutions.

✔️ Reading Check **What are some of the challenges facing Russia and the Eurasian Republics?**

Assessment

Defining Terms

1. Define free market economy, nuclear energy, federal republic.

Recalling Facts

2. Environment How has the government added to the problem of pollution?

3. Environment What are three sources of water pollution in Russia?

4. Government Why has there been fighting with Chechnya?

Critical Thinking

5. Synthesizing Information After years of living under Communist rule, why is it hard for Russians to live under democracy?

6. Making Predictions Russia is now a federal republic, and government officials need to learn how to make decisions in a democracy. How do you think that will affect life in Russia in the future?

Graphic Organizer

7. Organizing Information Create a diagram like this one, and list two facts about Russia for each topic in the four outer ovals. Facts should reflect the conditions in Russia today.

```
Government            Culture
        Russian
         Life
Economy             Environment
```

Applying Social Studies Skills

8. Comparing Governments Compare Russia's government during communism and since it has changed to a democratic government. List similarities and differences.

TIME REPORTS

FOCUS ON WORLD ISSUES

The New Russia

SOVFOTO/EASTFOTO

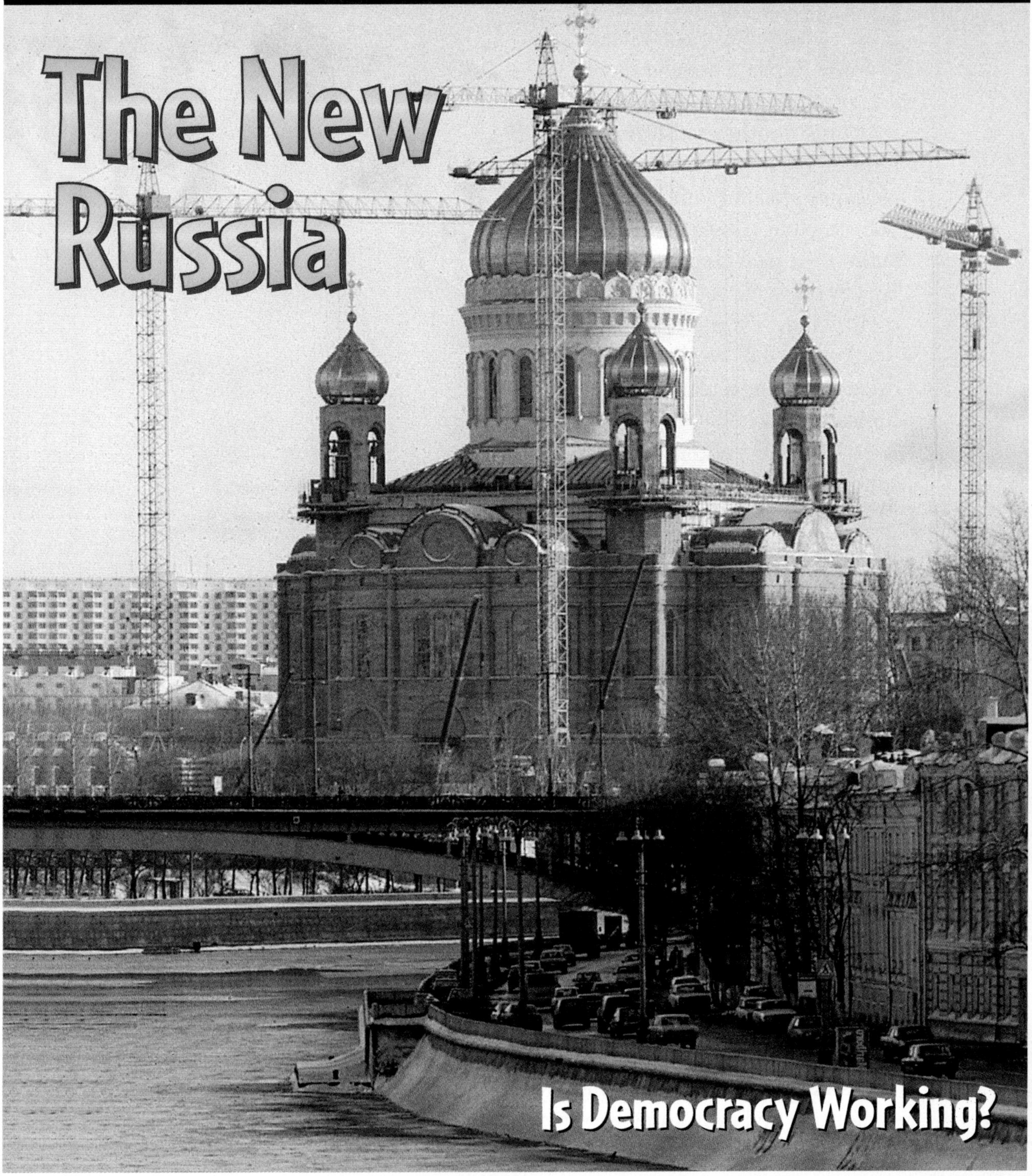

Is Democracy Working?

SOVFOTO/EASTFOTO

CHRISTOPHER MORRIS/ BLACK STAR

In Moscow a cathedral destroyed by the Communists has been rebuilt. Older Russians are especially pleased.

Inventing a Nation

Daniel Strigin lives in Moscow, the capital of Russia. There he shares a tiny, three-room apartment with his mother, his grandmother, his wife—and the parts of a one-seat airplane. Strigin, 30, is what Russians call a **kulibini**—a part-time inventor. By day he works as a computer technician. By night and on weekends, he works on his dream of flying a plane he built himself.

Strigin is one of tens of thousands of *kulibini* in Russia. "There is something in the Russian man's soul," he says, "that pushes him to invent."

The impulse to invent is something Russia needs badly today. Its Communist government collapsed in 1991. Since then the country of 146 million people has been struggling to remake itself as a democracy.

Remarkable Gains

So far Russia has made impressive strides:

- Russians now elect their leaders, something they had never been allowed to do before.

- Russians now own factories, shops, restaurants, and other **enterprises**, or businesses. Before 1991, the government owned everything.

Prefab housing goes up in Provideniya, a port city west of Alaska in Russia's Far East.

WOLFGANG KAEHLER/CORBIS

- Russia has shrunk its borders. Once it had been the Soviet Union's leading power. But that union fell to pieces. The republics that were part of it went their own way. Now all 15 of those former republics, including Russia, are independent nations.

Russia still has a long way to go. Its elected leaders sometimes act illegally to silence their critics. The government still owns all the nation's land. Steel companies and other huge businesses ended up in the hands of a few powerful people. Criminal gangs and dishonest public officials thrive. And in Chechnya, part of the Russian Federation, rebels have been at

A man votes near a statue of Vladimir Lenin, the first Communist dictator.

A man dressed as a bear advertises a new restaurant in St. Petersburg.

Billboards near St. Basil Cathedral are evidence of the new Russian economy.

war with the government since 1994. The **Russian Federation** is Russia's official name.

Misery For Many

The reforms caused great hardship. In the shift to privately owned enterprises, thousands of farms and factories failed. Millions lost their jobs, and the government had no money with which to help them. In 1999, 55 million people—one out of three Russians—scraped by on less than $6 a month.

Millions landed on their feet, however. "Everyone willing to work hard has a chance nowadays," said a restaurant owner. "Not everyone is prepared to do that. I haven't taken a day off since I opened this place."

EXPLORING THE ISSUE

1. **Making Inferences** Why might younger Russians find it easier than older Russians to learn to rely on themselves?

2. **Compare and Contrast** Suppose all the states in the United States became independent nations. How might that situation be like—and unlike—what happened to the Soviet Union?

Misplaced Trust

Russia's Communist government didn't ask Russians to plan their lives. "You did what was expected of you," said one woman. "We didn't think to ask questions or doubt the [Communist] system. Now," she added, "I can't imagine being so trusting."

Today's Russia faces an uncertain future. Despite its stockpile of nuclear weapons, it is no longer a military superpower. But it remains the world's largest country, and its people are well educated. Its natural resources—oil, lumber, and minerals—are plentiful. And many of its privately owned factories have at last figured out how to make first-rate products.

"A Russian is **inventive**," says one of Daniel Strigin's *kulibini* friends, "because he has to find solutions in bad conditions."

Bad conditions haunt today's Russia. Time will tell whether its people have the will—and the inventiveness—to overcome them.

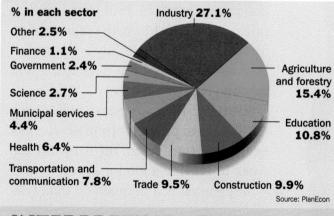

Where Russians Work

% in each sector

Other **2.5%**
Finance **1.1%**
Government **2.4%**
Science **2.7%**
Municipal services **4.4%**
Health **6.4%**
Transportation and communication **7.8%**
Trade **9.5%**
Construction **9.9%**
Education **10.8%**
Agriculture and forestry **15.4%**
Industry **27.1%**

Source: PlanEcon

INTERPRETING GRAPHS

Explaining How does this graph tell you that about one in four Russians makes or sells industrial products?

Are Russians Better Off?

Valentina Fedotova cries when she tells her story. In 1946, she was a student nurse in the Ukrainian city of Kiev. One day the secret police arrested her. They never told her why. After a four-minute "trial," she was shipped off to Russia's brutally cold Siberia. There she spent 10 years in a labor camp, working year-round mining gold. After 10 years, she was freed. But her sentence required her to stay in Siberia for 10 more years. By

▼ Many older Russians long for earlier times. This war veteran's hero is Joseph Stalin, a brutal dictator.

CHRISTOPHER MORRIS/BLACK STAR

EXPLORING THE ISSUE

1. **Finding the Main Idea** How does Fedotova's story illustrate the main point of this article?

2. **Contrasting** Elected leaders are less likely than dictators to arrest and imprison people without cause. Why do you think this is so?

the time those years were up, Fedotova was a broken woman. She never left the far east, where she now lives alone.

Millions of people who lived through the Soviet era, from 1917 to 1991, have similar stories. The Communist government headed by Joseph Stalin between 1924 and 1953 imprisoned, executed, or starved to death millions of people. Prisoners in labor camps built canals, railroads, hydroelectric stations, mines, and other industries.

The Price of Freedom

The freedom that followed the Soviet Union's collapse in 1991 came with a price. People had to take responsibility for their lives. "In today's Russia," a businessman says, "you have to learn to rely on yourself."

Even self-reliant Russians often suffer. Thanks to private enterprise, Russia's economy is growing. Yet it is still too weak to provide a job for everyone.

But Russia is a democracy today. Its government is no longer free to destroy lives like Valentina Fedotova's. That fact alone, most Russians believe, makes them better off today than ever. ■

The Road to Somewhere

As the Soviet Union was ending in 1991, protesters gathered in Moscow's Red Square. One man held a sign that said, "70 Years to Nowhere." The sign spoke of the past—the years of Communist rule that had led to a dead end.

What about the next 70 years? They should bring fairer courts, for one thing. Russian judges are used to taking the government's side. Soon juries will be deciding many cases, making courts more even-handed.

Jobs and Health

Tomorrow's Russians will be wealthier and healthier than today's. Today hospital patients must supply their own food, sheets, and medicine. Life expectancy for men today is only 59 years—down from 64 in 1989.

But Russia's healthcare system is getting stronger, along with the nation's economy. A stronger economy will mean more jobs and less poverty. Steady jobs should persuade Russian men to stop abusing alcohol. That drug is shortening their lives.

How quickly will those changes come? It all depends on how quickly Russians change the way they think. Russians don't yet have democracy in their hearts. They are not used to voting or taking part in community affairs, either as volunteers or as elected officials. They tend to think it is more important to help themselves than their neighbors.

EAST NEWS/GETTY IMAGES/NEWS.COM

▲ **Russian students hope to enjoy freedoms their parents never knew.**

Self-Serve Government

Government workers think the same way. Few see themselves as **public servants**. Many of them serve themselves first. People must pay money "under the table" to get driver's licenses, fair treatment by police, and permits to build houses.

Today "70 Years to Somewhere" could be Russia's slogan. It's just far too early to say what that somewhere will be. ▪

EXPLORING THE ISSUE

1. Explaining Why might a stronger Russian economy lead to better health?

2. Problem Solving What could the United States do to help Russians learn to put "democracy in their hearts"?

Helping Russia Rebuild: What Can One Person Do?

I n July 2000, former hockey star Mike Gartner made boys in two Russian hockey clubs very happy. One club was in Penza, a town outside Moscow. The other club was far to the east in Novokuznetsk, a city in Siberia. Gartner gave each club something it couldn't afford—hockey equipment worth thousands of dollars.

Goals & Dreams

Gartner heads the Goals & Dreams program of the National Hockey League Players' Association (NHLPA). "We're not doing this to try to make future NHL hockey players," Gartner said. "The goal is to try to make kids better people."

That's also the goal of the head of Novokuznetsk's hockey. "We are working toward a healthier lifestyle for our youth," he said.

That's not easy in a nation as hardpressed as Russia. The Novokuznetsk club gives its members free food and medical care. But it has no money left over to buy equipment.

Encouraging Words

You can help Russians simply by supporting efforts like the NHLPA's. You don't have to send sports equipment. You don't have to send money. Just send those groups a letter, letting them know how much you appreciate their efforts. Groups that provide assistance to others

▲ Former pro hockey star Mike Gartner meets members of a Russian hockey club.

gain strength just from knowing that others care.

Many groups are helping Russia today. One is the Eurasia Foundation, based in Washington, D.C. The World Wildlife Federation is another.

And don't forget Goals & Dreams. "We have stacks of letters from kids and families thanking us," Mike Gartner said. "This is a great job—kind of like being Santa Claus."

REVIEW AND ASSESS

The old and new reflect Russia's future. ▶

SERGEI GUNEYEV/TIMEPIX

UNDERSTANDING THE ISSUE

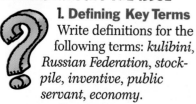

1. Defining Key Terms Write definitions for the following terms: *kulibini, Russian Federation, stockpile, inventive, public servant, economy.*

2. Writing to Inform Pretend you are in a Russian middle school. Write a letter to an American friend explaining Russia's problems. Use at least five of the key terms listed above.

3. Writing to Persuade "In today's Russia, you have to learn to rely on yourself." Write a letter to an imaginary Russian friend. Explain why self-reliance is a good thing.

INTERNET RESEARCH ACTIVITY

4. Russian army units have "adopted" a few thousand of the 1 million to 2 million Russian kids who have no home. Children as young as 11 live on army bases, wear uniforms, and attend school. They are not sent to war. Elsewhere children do fight wars. To learn about them, with your teacher's help, browse the Internet for information. List ways that real child soldiers seem like, and are different from, Russian kids in uniform. Compare your list with those of your classmates.

5. Since 1999, the Library of Congress has brought Russian officials to the United States to see democracy at work. Browse the Internet to find out more about this Library of Congress program. In a short essay, describe the program and explain how it might benefit both Russians and Americans. Put your answers in a 250-word essay.

BEYOND THE CLASSROOM

6. Visit your school or local library to learn about the Soviet Union. Working in groups, find out what it was like to live under a Communist government. What basic freedoms did Russians not have? Discuss your findings with your classmates.

7. Research another nation that has exchanged one-party rule for democracy. What might Russians learn from the other nation's experience? Put your findings in a short report.

RUSSIA'S 11 TIME ZONES

The earth is divided into 24 time zones, one for each hour of the day. Russia spans 11 time zones, stretching nearly halfway around the globe. We've labeled Russia's time zones from A to K. There's an hour's difference between each zone. It's always later in the East, where the sun rises, than in the West.

BUILDING MAP READING SKILLS

1. Interpreting Maps If it's 9:00 A.M. in Kaliningrad, what time is it in Moscow? What time is it in Tura, Chita, Vladivostok, and Magadan? Suppose it is 2:00 A.M., January 20, in Tomsk. What time and day is it in Samara?

2. Transferring Data Across the top of a sheet of paper, write the name of one city in each time zone, from Kaliningrad to Anadyr. Draw a clock beneath each name. Set the sixth clock at midnight. Draw the correct time on the 10 other clocks.

FOR UPDATES ON WORLD ISSUES GO TO www.timeclassroom.com/glencoe

Critical Thinking Skill

Understanding Cause and Effect

Understanding cause and effect involves considering *why* an event occurred. A *cause* is the action or situation that produces an event. What happens as a result of a cause is an *effect*.

▲ Revolutionary leaders and philosophers Lenin, Engels, and Marx

Learning the Skill

To identify cause-and-effect relationships, follow these steps:

- Identify two or more events or developments.
- Decide whether one event caused the other. Look for "clue words" such as *because, led to, brought about, produced, as a result of, so that, since,* and *therefore.*
- Look for logical relationships between events, such as "She overslept, and then she missed her bus."
- Identify the outcomes of events. Remember that some effects have more than one cause, and some causes lead to more than one effect. Also, an effect can become the cause of yet another effect.

Practicing the Skill

For each number below, identify which statement is the cause and which is the effect.

1. (A) Russia's capital was moved from coastal St. Petersburg to Moscow in the heart of the country.
   (B) The capital of Russia was threatened by an outside invasion.

2. (A) Revolutionary leaders seized control of the Russian government.
   (B) During World War I, shortages of food in Russian cities caused much starvation.
   (C) Discontent grew among the Russian people.

3. (A) The Soviet government kept prices for goods and services very low.
   (B) Many goods and services were in short supply in the Soviet Union.

Applying the Skill

In your local newspaper, read an article describing a current event. Determine at least one cause and one effect of that event. Show the cause-and-effect relationship in a diagram like the one here:

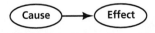

Cause ⟶ Effect

GO TO

Practice key skills with **Glencoe Skillbuilder Interactive Workbook, Level 1.**

Section 1   A Troubled History

Terms to Know
czar
serf
industrialize
communist state
Cold War
glasnost

Main Idea
The harsh rule of powerful leaders has often sparked violent uprisings in Russia.

✓ History   Emperors ruled the Russian Empire from 1480 to 1917.

✓ History   The czars expanded Russian territory to reach from Europe to the Pacific.

✓ Government   Under the Communists, Russia became part of the Soviet Union.

✓ History   In 1991 the Soviet Union broke apart, and Russia became an independent republic.

Section 2   A New Russia

Terms to Know
free market
  economy
nuclear energy
federal republic

Main Idea
Russia has a rich cultural heritage but faces challenges in adopting a new economic system and government.

✓ Economics   The change to a free market economy has been a challenge for Russians as they face rising unemployment and rising prices.

✓ Government   Russians have had to learn how to live in a democracy.

✓ Government   Some non-Russian ethnic groups want to create independent nations outside of Russia.

Young people strolling and singing in St. Petersburg ▶

Russia–Past and Present

Using Key Terms

Match the terms in Part A with their definitions in Part B.

A.

1. Cold War
2. nuclear energy
3. czar
4. industrialize
5. serf
6. federal republic
7. free market economy

B.

a. to change an economy to manufacturing
b. a farm laborer bound to the land he worked
c. period of tension without actual fighting
d. power is divided between national and state governments with a president who leads the nation
e. former emperor of Russia
f. power from a controlled atomic reaction
g. people decide what businesses to start and run

 ## Reviewing the Main Ideas

Section 1 A Troubled History

8. **History** Which czar used a secret police force to maintain strict control over the people?
9. **History** When was the Union of Soviet Socialist Republics formed?
10. **Government** Why did Stalin send people to Siberia?
11. **History** What took place in the Soviet Union in 1991?

Section 2 A New Russia

12. **Economics** What type of economic system has the new Russian government adopted?
13. **History** Who were Russia's first two presidents?
14. **Government** What are some changes Russians face as a result of their changing government?
15. **Human/Environment Interaction** What are some environmental problems facing Russia today?
16. **Government** Why do free countries of the world want Russia's new government to succeed?

 **Russia–Past and Present**

Place Location Activity

On a separate sheet of paper, match the letters on the map with the numbered places listed below.

1. St. Petersburg   5. Vladivostok
2. Caspian Sea   6. Sea of Japan
3. Moscow   7. Irkutsk
4. Baltic Sea   8. Omsk

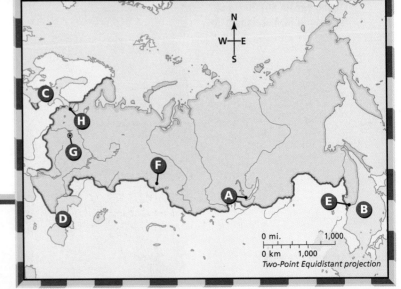

Self-Check Quiz Visit the *Our World Today: People, Places, and Issues* Web site at owt.glencoe.com and click on **Chapter 14—Self-Check Quizzes** to prepare for the Chapter Test.

Critical Thinking

17. **Understanding Cause and Effect** How did World War I help lead to the Russian Revolution?

18. **Organizing Information** Create a diagram like this one. Complete it with four characteristics of the free enterprise system in Russia.

Free Enterprise System in Russia

Current Events Journal

19. **Writing a News Article** Learn about the nuclear disaster at Chernobyl. Find out what happened and its effects on the local people, animals, and land. Write a news article that might have appeared a few days after the explosion. Then write a follow-up article on how the explosion is still affecting Russia and neighboring countries today.

Mental Mapping Activity

20. **Focusing on the Region** Create a simple outline map of Russia and label the following:

- Arctic Ocean
- Bering Sea
- Pacific Ocean
- St. Petersburg
- Vladivostok
- Siberia
- Moscow
- Baltic Sea

Technology Skills Activity

21. **Developing Multimedia Presentations** Choose an environmental or political problem that the Russian people have faced in the last 10 years. Research your choice and create a multimedia presentation on this problem. Include information on when, what, where, and other interesting facts. Use pictures, maps, and time lines to make your presentation more visual.

The Princeton Review

Standardized Test Practice

Directions: Read the paragraph below, and then answer the question that follows.

You may be surprised to know that Kazakhstan was—and still is—important to the exploration of outer space. The Russian space center Baikonur (by•kuh•NOOR) lies in south-central Kazakhstan. During the Soviet period, Baikonur was used for many space launches. Several historic "firsts in space" occurred here. For example, the first satellite was launched in 1957. The first crewed flight took place when cosmonaut Yuri Gagarin orbited the earth in 1961. In addition, the flight of the first woman in space, Valentina Tereshkova, was launched in 1963. After the Soviet collapse, the Russian-owned center remained in independent Kazakh territory.

1. **The Soviet space program at Baikonur holds great importance, mostly because**

   F it is located in south-central Kazakhstan.

   G it provides jobs for the people who live near the launch site.

   H many "first in space" flights were launched from it.

   J Valentina Tereshkova was the first woman in space.

Test-Taking Tip: When a question uses the word *most* or *mostly*, it means that more than one answer may be correct. Your job is to pick the *best* answer. For example, Baikonur's location in Kazakhstan may be important to the people who live near it, which is answer G. Another answer, however, provides a more general reason for Baikonur's importance.

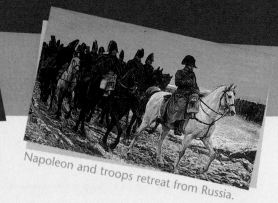

Napoleon and troops retreat from Russia.

RUSSIA'S STRATEGY:
Freeze Your Foes

Winter weather can cancel school, bring flu outbreaks, and stop traffic. It can even change history. Such was the case when French ruler Napoleon Bonaparte thought he had conquered the Russian Empire.

In fact, Napoleon did not want to conquer Russia. His real enemy was Great Britain. Napoleon wanted Russia and other countries to stop trading with Great Britain. Yet Russia's czar, Alexander I, refused. By 1812, Napoleon was determined to change Alexander's mind. In June, leading an army of more than half a million soldiers, Napoleon invaded Russia. To reach Moscow and the czar, Napoleon had to fight his way across the Russian countryside.

By the time Napoleon's battle-weary forces reached Moscow, supplies were scarce. All along the route, Russians had burned villages as they retreated, leaving no food or shelter. Reaching Moscow, Napoleon found the city in flames and nearly empty of people. The czar had moved to St. Petersburg. Napoleon took Moscow without a fight, but most of the city was in ashes.

Winter Wins a War

With winter approaching, Napoleon waited in Moscow for Alexander I to offer peace. The czar remained silent, however. With dwindling supplies and many of his troops lacking winter clothes, Napoleon was forced to retreat. He tried to take a new way back, but the Russians made Napoleon use the same ruined route he had used before. Armed bands of Russians attacked at every turn. Starving and desperate to escape the bitter cold, several of Napoleon's soldiers threw themselves into burning buildings. Most of Napoleon's troops never made it out of Russia.

History Repeats

More than a century later, during World War II, Russia's winter was again a mighty foe. On June 22, 1941, Adolf Hitler's German army invaded Russia, then part of the Soviet Union. As the German army fought its way to Moscow, Soviet leader Joseph Stalin issued his own "scorched-earth policy." Soviet citizens burned anything of use to the invaders. By December, German troops were within sight of the Kremlin, Moscow's government center, when winter struck.

Snow buried the invaders. Temperatures fell below freezing. Grease in guns and oil in vehicles froze solid. German soldiers suffered frostbite and died. The Soviets were better clothed and had winterized their tanks and trucks. Stalin's troops pushed back the German army. Once again the Russians triumphed with help from "General Winter."

QUESTIONS

1 After Napoleon conquered Moscow in 1812, why did he retreat?

2 How did Russia's winter affect fighting in World War II?

German prisoners of Russia's winter ▶

NATIONAL GEOGRAPHIC

Average Winter Temperatures

EUROPE

⊕ Moscow

RUSSIA

N
W   E
S

ASIA

Napoleon's Advance,
June–October 1812

German Forces Front
Line, December 1941

< -40°F
-40° to -31°F
-30° to -21°F
-20° to -11°F
-10° to 0°F

0° to 10°F
11° to 20°F
21° to 30°F
> 30°F

Unit 6

Waterfront of Cape Town, South Africa

Woman making butter in Chad

Africa South of the Sahara

The region of Africa south of the Sahara is home to more than 2,000 ethnic groups. Its hot, humid forests and dry grasslands support a variety of wild animals. Both people and animals face tough challenges in this region. The people are struggling to build stable governments and economies. The animals are threatened with extinction as human activities destroy natural habitats.

NGS ONLINE
www.nationalgeographic.com/education

Africa South of the Sahara

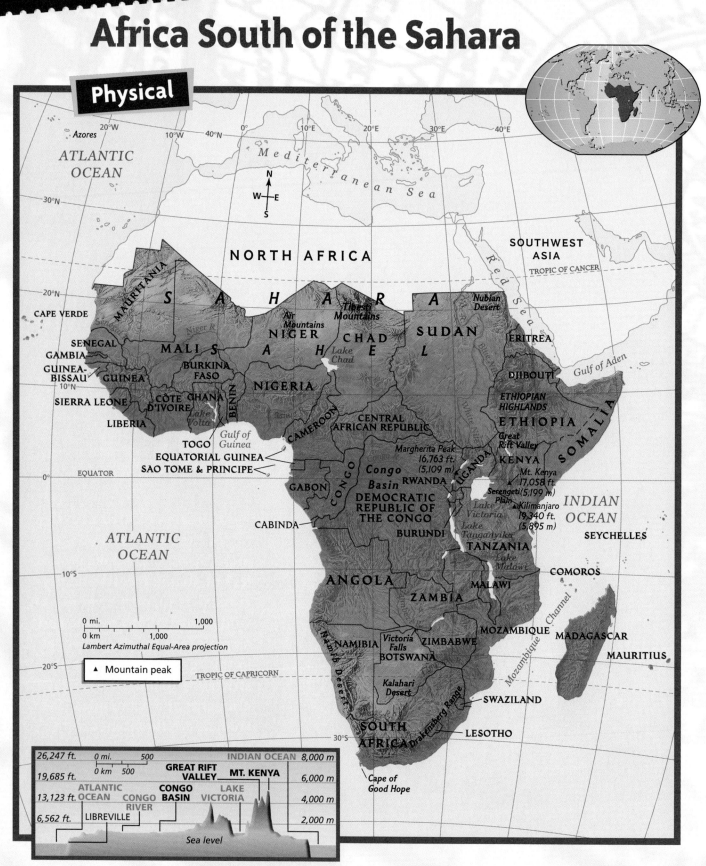

Physical

ATLANTIC OCEAN

Mediterranean Sea

Azores

20°W

10°W

40°N

10°E

20°E

30°E

40°E

30°N

N
W E
S

NORTH AFRICA

SOUTHWEST ASIA

TROPIC OF CANCER

Red Sea

CAPE VERDE

20°N

MAURITANIA

S A H A R A

Air Mountains

Tibesti Mountains

Nubian Desert

SENEGAL

MALI

NIGER

CHAD

SUDAN

Niger R.

ERITREA

Lake Chad

GAMBIA

GUINEA-BISSAU

BURKINA FASO

S A H E L

DJIBOUTI

10°N

GUINEA

NIGERIA

Gulf of Aden

SIERRA LEONE

CÔTE D'IVOIRE

GHANA

BENIN

Lake Volta

ETHIOPIAN HIGHLANDS

LIBERIA

CAMEROON

CENTRAL AFRICAN REPUBLIC

ETHIOPIA

Blue Nile

Great Rift Valley

TOGO

Gulf of Guinea

EQUATORIAL GUINEA

SAO TOME & PRINCIPE

CONGO

Congo Basin

Margherita Peak
16,763 ft.
(5,109 m)

UGANDA

KENYA

SOMALIA

White Nile

RWANDA

Mt. Kenya
17,058 ft.

0°

EQUATOR

GABON

DEMOCRATIC REPUBLIC OF THE CONGO

Serengeti Plain (5,199 m)

INDIAN OCEAN

Lake Victoria

Kilimanjaro
19,340 ft.
(5,895 m)

CABINDA

BURUNDI

Lake Tanganyika

ATLANTIC OCEAN

TANZANIA

SEYCHELLES

BURUNDI

Lake Malawi

COMOROS

10°S

ANGOLA

ZAMBIA

MALAWI

Mozambique Channel

MOZAMBIQUE

MADAGASCAR

0 mi. 1,000

0 km 1,000

Lambert Azimuthal Equal-Area projection

Namib Desert

NAMIBIA

Victoria Falls

ZIMBABWE

MAURITIUS

▲ Mountain peak

BOTSWANA

TROPIC OF CAPRICORN

20°S

Kalahari Desert

Drakensberg Range

SWAZILAND

SOUTH AFRICA

LESOTHO

30°S

Cape of Good Hope

26,247 ft.	0 mi. 500	INDIAN OCEAN	8,000 m
19,685 ft.	0 km 500	GREAT RIFT VALLEY MT. KENYA	6,000 m
13,123 ft.	ATLANTIC OCEAN CONGO RIVER	CONGO BASIN LAKE VICTORIA	4,000 m
6,562 ft.	LIBREVILLE		2,000 m
	Sea level		

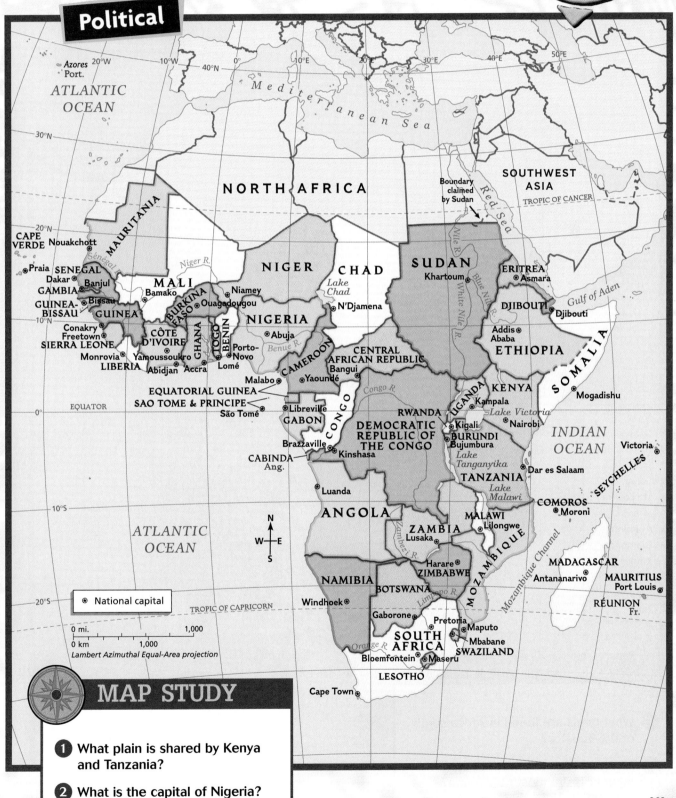
Political

ATLANTIC OCEAN

Azores Port.

Mediterranean Sea

NORTH AFRICA

Boundary claimed by Sudan

SOUTHWEST ASIA

TROPIC OF CANCER

Red Sea

CAPE VERDE

Nouakchott

MAURITANIA

Praia
SENEGAL
Dakar
GAMBIA
Banjul
GUINEA-BISSAU
Bissau

Senegal R.

Niger R.

MALI
Bamako

NIGER
Niamey

CHAD

Lake Chad

SUDAN
Khartoum

Blue Nile R.
White Nile R.
Nile R.

ERITREA
Asmara

DJIBOUTI
Djibouti

Gulf of Aden

GUINEA
Conakry
Freetown
SIERRA LEONE
Monrovia
LIBERIA

BURKINA FASO
Ouagadougou

CÔTE D'IVOIRE
Yamoussoukro
Abidjan

GHANA
TOGO
BENIN
Accra
Lomé
Porto-Novo

NIGERIA
Abuja

Benue R.

N'Djamena

CENTRAL AFRICAN REPUBLIC
Bangui

Addis Ababa

ETHIOPIA

SOMALIA

Mogadishu

EQUATORIAL GUINEA
SAO TOME & PRINCIPE
São Tomé

CAMEROON
Malabo
Yaoundé

GABON
Libreville

CONGO
Brazzaville

CABINDA Ang.

Congo R.

DEMOCRATIC REPUBLIC OF THE CONGO
Kinshasa

RWANDA
Kigali
BURUNDI
Bujumbura

UGANDA
Kampala

Lake Victoria

KENYA
Nairobi

INDIAN OCEAN

Victoria
SEYCHELLES

EQUATOR

Lake Tanganyika

TANZANIA

Dar es Salaam

Lake Malawi

COMOROS
Moroni

Luanda

ANGOLA

ZAMBIA
Lusaka

MALAWI
Lilongwe

MOZAMBIQUE

Mozambique Channel

MADAGASCAR
Antananarivo

MAURITIUS
Port Louis

Zambezi R.

Harare
ZIMBABWE

RÉUNION Fr.

ATLANTIC OCEAN

NAMIBIA

BOTSWANA

Windhoek

Limpopo R.

Gaborone

Pretoria
Maputo
Mbabane
SWAZILAND

Orange R.

SOUTH AFRICA

Bloemfontein
Maseru
LESOTHO

Cape Town

⊛ National capital

TROPIC OF CAPRICORN

0 mi.   1,000
0 km   1,000

Lambert Azimuthal Equal-Area projection

N W E S

MAP STUDY

1 What plain is shared by Kenya and Tanzania?

2 What is the capital of Nigeria?

393

NATIONAL GEOGRAPHIC REGIONAL ATLAS

Africa South of the Sahara

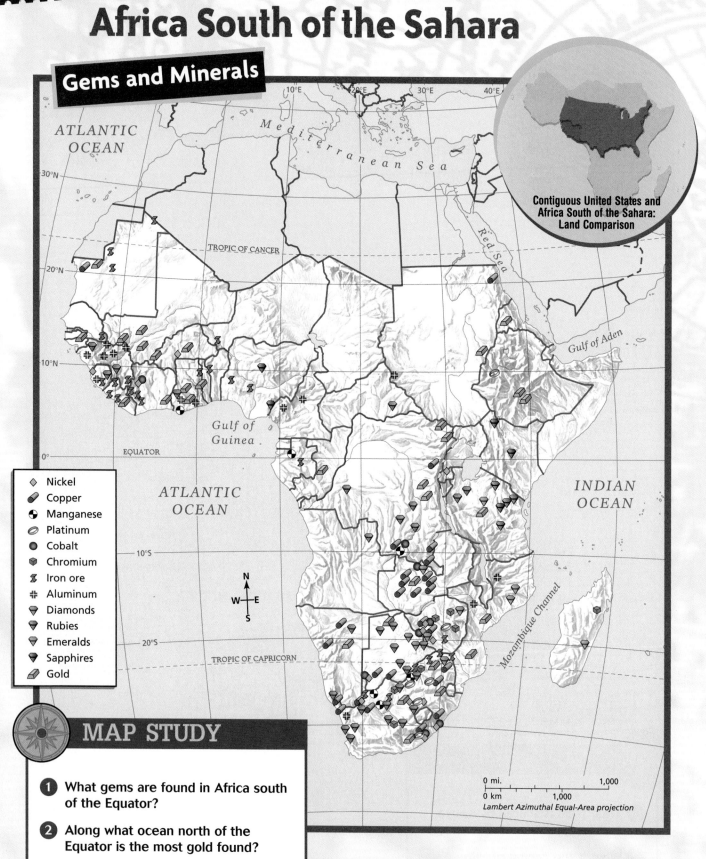

Gems and Minerals

Contiguous United States and Africa South of the Sahara: Land Comparison

ATLANTIC OCEAN

Mediterranean Sea

TROPIC OF CANCER

Red Sea

Gulf of Aden

Gulf of Guinea

EQUATOR

ATLANTIC OCEAN

INDIAN OCEAN

Mozambique Channel

TROPIC OF CAPRICORN

Legend:
- ◇ Nickel
- ⬮ Copper
- ◕ Manganese
- ⬭ Platinum
- ● Cobalt
- ⬡ Chromium
- ⟆ Iron ore
- ✤ Aluminum
- ▽ Diamonds
- ▽ Rubies
- ▽ Emeralds
- ▽ Sapphires
- ⬦ Gold

0 mi. 1,000
0 km 1,000
Lambert Azimuthal Equal-Area projection

MAP STUDY

1. What gems are found in Africa south of the Equator?

2. Along what ocean north of the Equator is the most gold found?

UNIT 6

Fast Facts

COMPARING POPULATION:
United States and Selected Countries of Africa South of the Sahara

UNITED STATES

NIGERIA

DEMOCRATIC REPUBLIC OF THE CONGO

SOUTH AFRICA

KENYA

 = 30,000,000

SENEGAL

Source: *Population Reference Bureau*, 2000.

SELECTED RURAL AND URBAN POPULATIONS:
Africa South of the Sahara

	Rural	Urban
WEST AFRICA		
Niger	83%	17%
Cape Verde	56%	44%
CENTRAL AFRICA		
Angola	68%	32%
Central African Republic	61%	39%
EAST AFRICA		
Rwanda	95%	5%
Djibouti	17%	83%
SOUTHERN AFRICA		
Lesotho	84%	16%
South Africa	55%	45%

Source: *Population Reference Bureau*, 2000.

Data Bits

Country	Automobiles per 1,000 people	Telephones per 1,000 people
Ghana	5	6
Mauritania	8	5
Sudan	10	4
Tanzania	1	3
Zambia	17	9

Religions

Country	Islam	Christian	Traditional Beliefs	Other
Ghana	11%	43%	38%	8%
Mauritania	100%	—	—	—
Sudan	70%	9%	20%	1%
Tanzania	33%	33%	30%	4%
Zambia	0.5%	63%	36%	.5%

Source: *World Desk Reference*, 2000.

GRAPHIC STUDY

1. Which country has the highest percentage of Christians? Which has the lowest percentage?

2. Of the African countries shown in the chart at lower left, which is least urbanized? Which is most urbanized?

REGIONAL ATLAS

Country Profiles

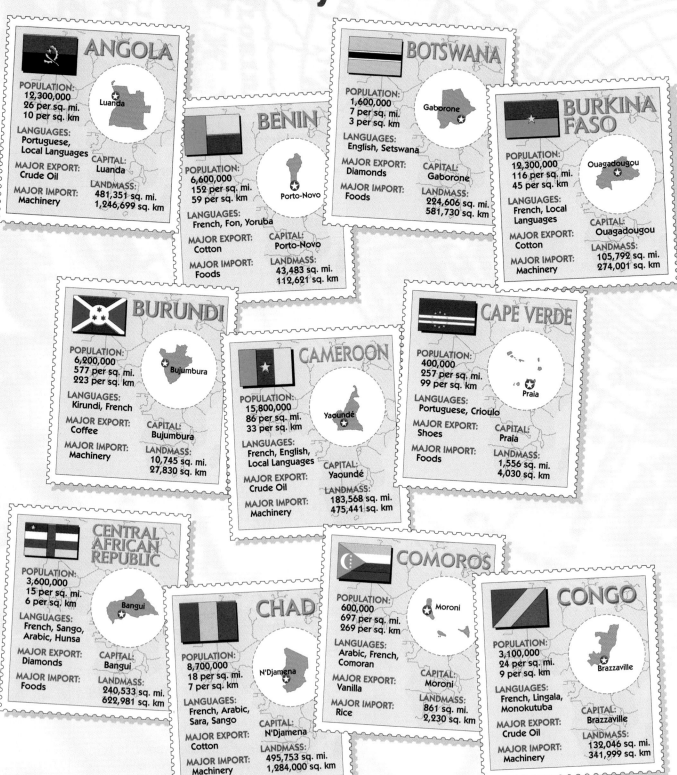

ANGOLA

POPULATION:
12,300,000
26 per sq. mi.
10 per sq. km

LANGUAGES:
Portuguese,
Local Languages

MAJOR EXPORT:
Crude Oil

MAJOR IMPORT:
Machinery

CAPITAL:
Luanda

LANDMASS:
481,351 sq. mi.
1,246,699 sq. km

Luanda

BENIN

POPULATION:
6,600,000
152 per sq. mi.
59 per sq. km

LANGUAGES:
French, Fon, Yoruba

MAJOR EXPORT:
Cotton

MAJOR IMPORT:
Foods

CAPITAL:
Porto-Novo

LANDMASS:
43,483 sq. mi.
112,621 sq. km

Porto-Novo

BOTSWANA

POPULATION:
1,600,000
7 per sq. mi.
3 per sq. km

LANGUAGES:
English, Setswana

MAJOR EXPORT:
Diamonds

MAJOR IMPORT:
Foods

CAPITAL:
Gaborone

LANDMASS:
224,606 sq. mi.
581,730 sq. km

Gaborone

BURKINA FASO

POPULATION:
12,300,000
116 per sq. mi.
45 per sq. km

LANGUAGES:
French, Local
Languages

MAJOR EXPORT:
Cotton

MAJOR IMPORT:
Machinery

CAPITAL:
Ouagadougou

LANDMASS:
105,792 sq. mi.
274,001 sq. km

Ouagadougou

BURUNDI

POPULATION:
6,200,000
577 per sq. mi.
223 per sq. km

LANGUAGES:
Kirundi, French

MAJOR EXPORT:
Coffee

MAJOR IMPORT:
Machinery

CAPITAL:
Bujumbura

LANDMASS:
10,745 sq. mi.
27,830 sq. km

Bujumbura

CAMEROON

POPULATION:
15,800,000
86 per sq. mi.
33 per sq. km

LANGUAGES:
French, English,
Local Languages

MAJOR EXPORT:
Crude Oil

MAJOR IMPORT:
Machinery

CAPITAL:
Yaoundé

LANDMASS:
183,568 sq. mi.
475,441 sq. km

Yaoundé

CAPE VERDE

POPULATION:
400,000
257 per sq. mi.
99 per sq. km

LANGUAGES:
Portuguese, Crioulo

MAJOR EXPORT:
Shoes

MAJOR IMPORT:
Foods

CAPITAL:
Praia

LANDMASS:
1,556 sq. mi.
4,030 sq. km

Praia

CENTRAL AFRICAN REPUBLIC

POPULATION:
3,600,000
15 per sq. mi.
6 per sq. km

LANGUAGES:
French, Sango,
Arabic, Hunsa

MAJOR EXPORT:
Diamonds

MAJOR IMPORT:
Foods

CAPITAL:
Bangui

LANDMASS:
240,533 sq. mi.
622,981 sq. km

Bangui

CHAD

POPULATION:
8,700,000
18 per sq. mi.
7 per sq. km

LANGUAGES:
French, Arabic,
Sara, Sango

MAJOR EXPORT:
Cotton

MAJOR IMPORT:
Machinery

CAPITAL:
N'Djamena

LANDMASS:
495,753 sq. mi.
1,284,000 sq. km

N'Djamena

COMOROS

POPULATION:
600,000
697 per sq. mi.
269 per sq. km

LANGUAGES:
Arabic, French,
Comoran

MAJOR EXPORT:
Vanilla

MAJOR IMPORT:
Rice

CAPITAL:
Moroni

LANDMASS:
861 sq. mi.
2,230 sq. km

Moroni

CONGO

POPULATION:
3,100,000
24 per sq. mi.
9 per sq. km

LANGUAGES:
French, Lingala,
Monokutuba

MAJOR EXPORT:
Crude Oil

MAJOR IMPORT:
Machinery

CAPITAL:
Brazzaville

LANDMASS:
132,046 sq. mi.
341,999 sq. km

Brazzaville

Countries and flags not drawn to scale

For more information on countries in this region, refer to the Nations of the World Data Bank on pages 690–699.

CONGO,
Democratic Republic of the

POPULATION:
53,600,000
59 per sq. mi.
23 per sq. km

LANGUAGES:
French, Lingala, Kingwana

MAJOR EXPORT:
Diamonds

MAJOR IMPORT:
Manufactured Goods

CAPITAL:
Kinshasa

LANDMASS:
905,351 sq. mi.
2,344,859 sq. km

Kinshasa

CÔTE D'IVOIRE

POPULATION:
16,400,000
132 per sq. mi.
51 per sq. km

LANGUAGES:
French, Dioula

MAJOR EXPORT:
Cocoa

MAJOR IMPORT:
Foods

CAPITALS:
Yamoussoukro, Abidjan

LANDMASS:
124,502 sq. mi.
322,460 sq. km

Yamoussoukro
Abidjan

DJIBOUTI

POPULATION:
600,000
67 per sq. mi.
26 per sq. km

LANGUAGES:
French, Arabic

MAJOR EXPORTS:
Hides and Skins

MAJOR IMPORT:
Foods

CAPITAL:
Djibouti

LANDMASS:
8,958 sq. mi.
23,201 sq. km

Djibouti

EQUATORIAL GUINEA

POPULATION:
500,000
46 per sq. mi.
18 per sq. km

LANGUAGES:
Spanish, French, Fang, Bubi, Ibo

MAJOR EXPORT:
Petroleum

MAJOR IMPORT:
Machinery

CAPITAL:
Malabo

LANDMASS:
10,830 sq. mi.
28,050 sq. km

Malabo

ERITREA

POPULATION:
4,300,000
95 per sq. mi.
37 per sq. km

LANGUAGES:
Afar, Amharic, Arabic, Tigre

MAJOR EXPORT:
Livestock

MAJOR IMPORT:
Processed Foods

CAPITAL:
Asmara

LANDMASS:
45,405 sq. mi.
117,599 sq. km

Asmara

ETHIOPIA

POPULATION:
65,400,000
153 per sq. mi.
59 per sq. km

LANGUAGES:
Amharic, Tigrinya, Orominga

MAJOR EXPORT:
Coffee

MAJOR IMPORTS:
Foods and Livestock

CAPITAL:
Addis Ababa

LANDMASS:
426,371 sq. mi.
1,104,301 sq. km

Addis Ababa

GABON

POPULATION:
1,200,000
12 per sq. mi.
5 per sq. km

LANGUAGES:
French, Local Languages

MAJOR EXPORT:
Crude Oil

MAJOR IMPORT:
Machinery

CAPITAL:
Libreville

LANDMASS:
103,347 sq. mi.
267,669 sq. km

Libreville

GAMBIA

POPULATION:
1,400,000
321 per sq. mi.
124 per sq. km

LANGUAGES:
English, Mandinka, Fula

MAJOR EXPORT:
Peanuts

MAJOR IMPORT:
Foods

CAPITAL:
Banjul

LANDMASS:
4,363 sq. mi.
11,300 sq. km

Banjul

GHANA

POPULATION:
19,900,000
216 per sq. mi.
83 per sq. km

LANGUAGES:
English, Local Languages

MAJOR EXPORT:
Gold

MAJOR IMPORT:
Machinery

CAPITAL:
Accra

LANDMASS:
92,100 sq. mi.
238,539 sq. km

Accra

GUINEA

POPULATION:
7,600,000
80 per sq. mi.
31 per sq. km

LANGUAGES:
French, Local Languages

MAJOR EXPORT:
Bauxite

MAJOR IMPORT:
Petroleum Products

CAPITAL:
Conakry

LANDMASS:
94,927 sq. mi.
245,861 sq. km

Conakry

GUINEA-BISSAU

POPULATION:
1,200,000
86 per sq. mi.
33 per sq. km

LANGUAGES:
Portuguese, Crioulo, Fula

MAJOR EXPORT:
Cashews

MAJOR IMPORT:
Foods

CAPITAL:
Bissau

LANDMASS:
13,946 sq. mi.
36,120 sq. km

Bissau

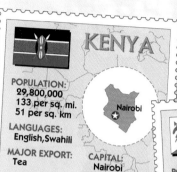

National Geographic

Country Profiles

KENYA
POPULATION:
29,800,000
133 per sq. mi.
51 per sq. km

LANGUAGES:
English, Swahili

MAJOR EXPORT:
Tea

MAJOR IMPORT:
Machinery

CAPITAL:
Nairobi

LANDMASS:
224,081 sq. mi.
580,370 sq. km

LESOTHO
POPULATION:
2,200,000
188 per sq. mi.
73 per sq. km

LANGUAGES:
English, Sesotho,
Zulu, Xhosa

MAJOR EXPORT:
Clothing

MAJOR IMPORT:
Corn

CAPITAL:
Maseru

LANDMASS:
11,718 sq. mi.
30,350 sq. km

LIBERIA
POPULATION:
3,200,000
74 per sq. mi.
29 per sq. km

LANGUAGES:
English, Local
Languages

MAJOR EXPORT:
Diamonds

MAJOR IMPORT:
Natural Gas

CAPITAL:
Monrovia

LANDMASS:
43,000 sq. mi.
111,370 sq. km

MADAGASCAR
POPULATION:
16,400,000
72 per sq. mi.
28 per sq. km

LANGUAGES:
French, Malagasy

MAJOR EXPORT:
Coffee

MAJOR IMPORT:
Machinery

CAPITAL:
Antananarivo

LANDMASS:
226,656 sq. mi.
587,039 sq. km

MALAWI
POPULATION:
10,500,000
230 per sq. mi.
89 per sq. km

LANGUAGES:
Chewa, English

MAJOR EXPORT:
Tobacco

MAJOR IMPORT:
Foods

CAPITAL:
Lilongwe

LANDMASS:
45,745 sq. mi.
118,480 sq. km

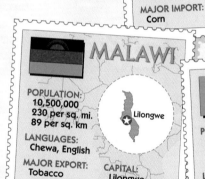

MALI
POPULATION:
11,000,000
23 per sq. mi.
9 per sq. km

LANGUAGES:
French, Bambara

MAJOR EXPORT:
Cotton

MAJOR IMPORT:
Machinery

CAPITAL:
Bamako

LANDMASS:
478,838 sq. mi.
1,240,190 sq. km

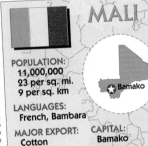

MAURITANIA
POPULATION:
2,700,000
7 per sq. mi.
3 per sq. km

LANGUAGES:
Hasaniya Arabic,
Wolof

MAJOR EXPORT:
Fish

MAJOR IMPORT:
Foods

CAPITAL:
Nouakchott

LANDMASS:
395,954 sq. mi.
1,025,521 sq. km

MAURITIUS
POPULATION:
1,200,000
1,523 per sq. mi.
588 per sq. km

LANGUAGES:
English, Creole,
Bhojpuri, French

MAJOR EXPORT:
Sugar

MAJOR IMPORT:
Foods

CAPITAL:
Port Louis

LANDMASS:
788 sq. mi.
2,041 sq. km

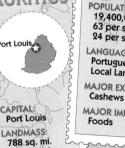

MOZAMBIQUE
POPULATION:
19,400,000
63 per sq. mi.
24 per sq. km

LANGUAGES:
Portuguese,
Local Languages

MAJOR EXPORT:
Cashews

MAJOR IMPORT:
Foods

CAPITAL:
Maputo

LANDMASS:
309,494 sq. mi.
801,590 sq. km

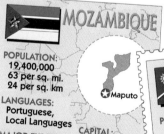

NAMIBIA
POPULATION:
1,800,000
6 per sq. mi.
2 per sq. km

LANGUAGES:
English, Afrikaans, Local Languages

MAJOR EXPORT:
Diamonds

MAJOR IMPORT:
Construction
Materials

CAPITAL:
Windhoek

LANDMASS:
318,259 sq. mi.
824,291 sq. km

NIGER
POPULATION:
10,400,000
21 per sq. mi.
8 per sq. km

LANGUAGES:
French, Hausa,
Djerma

MAJOR EXPORT:
Uranium Ore

MAJOR IMPORT:
Manufactured
Goods

CAPITAL:
Niamey

LANDMASS:
489,189 sq. mi.
1,267,000 sq. km

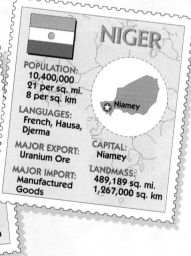

Countries and flags not drawn to scale

For more information on countries in this region, refer to the Nations of the World Data Bank on pages 690–699.

NIGERIA

POPULATION:
126,600,000
355 per sq. mi.
137 per sq. km

LANGUAGES:
English, Hausa,
Yoruba, Igbo

MAJOR EXPORT:
Petroleum

MAJOR IMPORT:
Machinery

CAPITAL:
Abuja

LANDMASS:
356,668 sq. mi.
923,770 sq. km

★ Abuja

RWANDA

POPULATION:
7,300,000
718 per sq. mi.
277 per sq. km

LANGUAGES:
Kinyarwanda,
French, English

MAJOR EXPORT:
Coffee

MAJOR IMPORT:
Foods

CAPITAL:
Kigali

LANDMASS:
10,170 sq. mi.
26,340 sq. km

★ Kigali

SAO TOME and PRINCIPE

POPULATION:
200,000
539 per sq. mi.
208 per sq. km

LANGUAGES:
Portuguese, Crioulo

MAJOR EXPORT:
Cocoa

MAJOR IMPORT:
Textiles

CAPITAL:
São Tomé

LANDMASS:
371 sq. mi.
961 sq. km

✪ São Tomé

SENEGAL

POPULATION:
9,700,000
128 per sq. mi.
49 per sq. km

LANGUAGES:
French, Wolof,
Pulaar, Diola

MAJOR EXPORT:
Fish

MAJOR IMPORT:
Foods

CAPITAL:
Dakar

LANDMASS:
75,954 sq. mi.
196,721 sq. km

★ Dakar

SEYCHELLES

POPULATION:
100,000
575 per sq. mi.
222 per sq. km

LANGUAGES:
English, French,
Creole

MAJOR EXPORT:
Fish

MAJOR IMPORT:
Foods

CAPITAL:
Victoria

LANDMASS:
174 sq. mi.
451 sq. km

Victoria ✪

SIERRA LEONE

POPULATION:
5,400,000
195 per sq. mi.
75 per sq. km

LANGUAGES:
English, Mende,
Temne, Krio

MAJOR EXPORT:
Diamonds

MAJOR IMPORT:
Foods

CAPITAL:
Freetown

LANDMASS:
27,699 sq. mi.
71,740 sq. km

Freetown ✪

SOMALIA

POPULATION:
7,500,000
31 per sq. mi.
12 per sq. km

LANGUAGES:
Somali, Arabic

MAJOR EXPORT:
Livestock

MAJOR IMPORT:
Textiles

CAPITAL:
Mogadishu

LANDMASS:
246,201 sq. mi.
637,661 sq. km

✪ Mogadishu

SOUTH AFRICA

POPULATION:
43,600,000
92 per sq. mi.
36 per sq. km

LANGUAGES:
Afrikaans,
English, Zulu

MAJOR EXPORT:
Gold

MAJOR IMPORT:
Transport Equip.

CAPITALS:
Pretoria, Cape Town,
Bloemfontein

LANDMASS:
471,444 sq. mi.
1,221,038 sq. km

Pretoria
★ Bloemfontein
Cape Town

SUDAN

POPULATION:
31,800,000
33 per sq. mi.
13 per sq. km

LANGUAGES:
Arabic, Nubian,
Ta Bedawie

MAJOR EXPORT:
Cotton

MAJOR IMPORT:
Petroleum
Products

CAPITAL:
Khartoum

LANDMASS:
967,494 sq. mi.
2,505,809 sq. km

★ Khartoum

SWAZILAND

POPULATION:
1,100,000
165 per sq. mi.
64 per sq. km

LANGUAGES:
English, Swazi

MAJOR EXPORT:
Soft Drink
Concentrates

MAJOR IMPORT:
Machinery

CAPITAL:
Mbabane

LANDMASS:
6,703 sq. mi.
17,361 sq. km

Mbabane ✪

TANZANIA

POPULATION:
36,200,000
99 per sq. mi.
38 per sq. km

LANGUAGES:
Swahili, English

MAJOR EXPORT:
Coffee

MAJOR IMPORT:
Machinery

CAPITAL:
Dar es Salaam

LANDMASS:
364,900 sq. mi.
945,087 sq. km

Dar es Salaam

Country Profiles

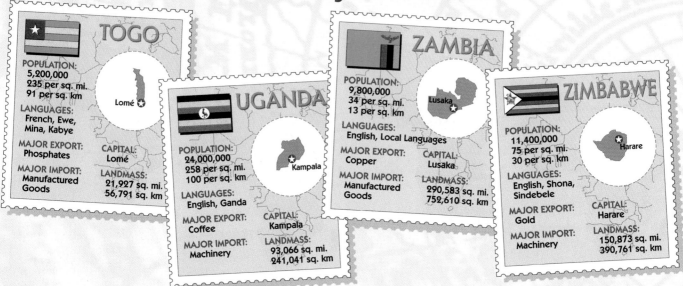

TOGO

POPULATION:
5,200,000
235 per sq. mi.
91 per sq. km

LANGUAGES:
French, Ewe,
Mina, Kabye

MAJOR EXPORT:
Phosphates

CAPITAL:
Lomé

MAJOR IMPORT:
Manufactured
Goods

LANDMASS:
21,927 sq. mi.
56,791 sq. km

Lomé

UGANDA

POPULATION:
24,000,000
258 per sq. mi.
100 per sq. km

LANGUAGES:
English, Ganda

MAJOR EXPORT:
Coffee

CAPITAL:
Kampala

MAJOR IMPORT:
Machinery

LANDMASS:
93,066 sq. mi.
241,041 sq. km

Kampala

ZAMBIA

POPULATION:
9,800,000
34 per sq. mi.
13 per sq. km

LANGUAGES:
English, Local Languages

MAJOR EXPORT:
Copper

CAPITAL:
Lusaka

MAJOR IMPORT:
Manufactured
Goods

LANDMASS:
290,583 sq. mi.
752,610 sq. km

Lusaka

ZIMBABWE

POPULATION:
11,400,000
75 per sq. mi.
30 per sq. km

LANGUAGES:
English, Shona,
Sindebele

MAJOR EXPORT:
Gold

CAPITAL:
Harare

MAJOR IMPORT:
Machinery

LANDMASS:
150,873 sq. mi.
390,761 sq. km

Harare

Countries and flags not drawn to scale

BUILDING CITIZENSHIP

Closing the Door on Racism By 1994, South Africa's racist policy of apartheid was officially over. Nelson Mandela became the first black person to be elected president of South Africa. Just three years earlier he had been released from jail after spending 27 years there for antiapartheid activities. When he became president, he created a panel to grant pardons to both blacks and whites who had admitted to committing political crimes in the past. Mandela believed that only by "closing the door" on the past could the country move on to its future.

Why do you think Nelson Mandela was willing to pardon people?

The flags of African countries often represent the history or culture of the country. For example, the "y" shape in the South African flag symbolizes a divided people going forward in unity. Research the flag of an African country and write a paragraph about the meaning of the flag.

▼ **Women at an antiapartheid rally in South Africa**

View over terraced fields
and small settlements,
Kabale, Uganda

East and Central Africa

The World and Its People NATIONAL GEOGRAPHIC

To learn more about the people and places of East and Central Africa, view **The World and Its People Chapters 20** and **21** videos.

Our World Today **online**

Chapter Overview Visit the *Our World Today: People, Places, and Issues* Web site at <u>owt.glencoe.com</u> and click on **Chapter 15—Chapter Overviews** to preview information about East and Central Africa.

Compare-Contrast Study Foldable Make this foldable to compare and contrast traditional and modern cultures in East and Central Africa.

Step 1 Fold one sheet of paper in half from side to side.

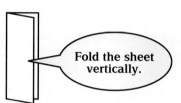

Fold the sheet vertically.

Step 2 Fold again, one inch from the top. (Tip: The middle knuckle of your index finger is about one inch long.)

Step 3 Open and label as shown.

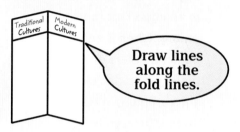

Traditional Cultures   Modern Cultures

Draw lines along the fold lines.

Reading and Writing As you read this chapter, record information in the two columns of your foldable chart. Be sure to write the information you find in the appropriate column of your foldable.

▲ Elephants on the Serengeti Plain in Tanzania

Why It Matters

Rich in Heritage

Some of Africa's most important early civilizations flourished in the location that is now part of East and Central Africa. These societies grew to become large and complex as they developed the skills to master the region's difficult environment. They were successful farmers, herders, metalworkers, artisans, and merchants. Today, the people of this region are facing difficult challenges just to survive.

Section 1
East Africa: Kenya and Tanzania

Guide to Reading

Main Idea

Both Kenya and Tanzania are countries of diverse landscapes and peoples.

Terms to Know

- coral reef
- poaching
- free enterprise system
- cassava
- sisal
- habitat
- eco-tourist

Reading Strategy

Create a chart like this one. Then list facts about the land, economy, and people of Kenya and Tanzania.

Fact	Kenya	Tanzania
Land		
Economy		
People		

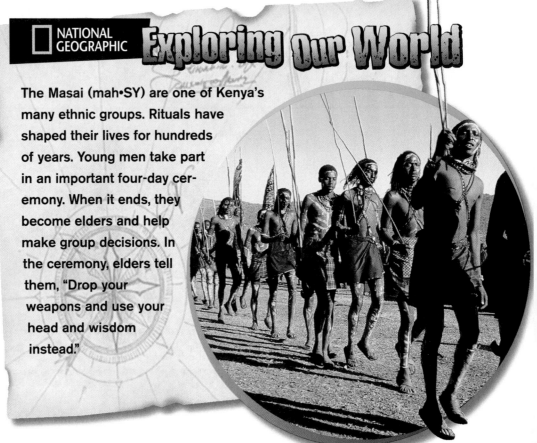

NATIONAL GEOGRAPHIC Exploring Our World

The Masai (mah•SY) are one of Kenya's many ethnic groups. Rituals have shaped their lives for hundreds of years. Young men take part in an important four-day ceremony. When it ends, they become elders and help make group decisions. In the ceremony, elders tell them, "Drop your weapons and use your head and wisdom instead."

Both traditional and modern cultures meet in **Kenya.** The **Masai** follow ways of life similar to their ancestors, while city dwellers live in apartments and work in offices. **Tanzania** has one of the largest wild animal populations in the world.

Kenya

Kenya is about two times the size of Nevada. The country's Indian Ocean coastline has stretches of white beaches lined with palm trees. Offshore lies a coral reef, a natural formation at or near the water's surface that is made of the skeletons of small sea animals. Lions, elephants, rhinoceroses, and other wildlife roam an upland plain. Millions of acres are set aside by the government to protect plants and wildlife. Still, in recent years there has been heavy poaching, the illegal hunting of protected animals.

In the western part of the country are highlands and the **Great Rift Valley.** (See photo on page 59.) This valley is really a fault—a crack in the earth's crust. The Great Rift Valley begins in southeastern Africa and stretches about 3,000 miles (4,825 km) north to the Red Sea. Lakes have formed in many places and volcanoes also dot the area. One of them—**Mt. Kenya**—rises 17,058 feet (5,199 m) high. It is in the Great Rift Valley that fossils of the earliest human ancestors, dating back about 4 million years, have been found.

Kenya's Economy   Kenya has a developing economy based on a free enterprise system. In this economic system, people can start and run businesses with limited government involvement. Kenya's capital, **Nairobi** (ny•ROH•bee), has become a center of business and commerce for all of East Africa. Foreign companies have set up regional headquarters in this city.

Many Kenyans remain poor, however. Farmers raise corn, bananas, cassava, and sweet potatoes. Cassava is a plant whose roots are ground to make porridge. Some larger farms raise coffee and tea for export. In recent years, the weather has not been good for crops. Also,

Literature

A CHANGING KENYA

As developing countries modernize, traditional ways of life often change. In the following poem, Kenyan poet and playwright Micere Githae Mugo expresses the challenge of living in a changing world.

WHERE ARE THOSE SONGS?
by Micere Githae Mugo

Where are those songs
my mother and yours
always sang
fitting rhythms
to the whole
vast span of life?
. . . .
I have forgotten
my mother's song
my children
will never know.

This I remember:
Mother always said
sing child sing
make a song
and sing
beat out your own rhythms
the rhythms of your life
but make the song soulful
and make life
sing
. . . .

From "Where are those Songs" by Micere G. Mugo. Reprinted by permission of the author.

Analyzing Literature

What do you think the poet's mother meant when she said to "beat out your own rhythms"?

Nairobi, Kenya

Like most cities, Kenya's capital has crowded markets, high-rise office buildings, and elegant mansions. Many city workers maintain close ties to relatives in the countryside.

Place About how many people live in Nairobi?

corrupt practices of government officials have hurt the economy.

One of the fastest-growing industries in Kenya is tourism. Thousands of tourists visit each year. Visitors often take tours called safaris in jeeps and buses to see the country's wildlife in its natural surroundings.

Kenya's People The people of Kenya believe in *harambee,* which means "pulling together." The spirit of *harambee* has led the different ethnic groups to build schools and clinics in their communities. They have raised money to send good students to universities.

History and Government During the A.D. 700s, Arab traders from Southwest Asia settled along the coast where a blending of cultures took place. The **Swahili** language came about from this blending. The name *Swahili* comes from an Arabic word meaning "of the coast." The language includes features of several African languages as well as Arabic. Today Swahili is one of Kenya's two official languages. English is the other.

The British made Kenya a colony in 1920 after World War I. They took land from the Africans and set up farms to grow coffee and tea for export. By the 1940s, Kenya's African groups like the **Mau Mau** fought in violent civil wars to end British rule. Kenya finally won its independence in 1963 and became a republic. The country's first president, **Jomo Kenyatta** (JOH•moh kehn•YAHT•uh), won respect as an early leader in Africa's movement for freedom. Under Kenyatta, Kenya enjoyed economic prosperity and had a stable government. In recent years, the economy has weakened. In response, many Kenyans have demanded democratic changes.

Kenya Today Kenya's roughly 30 million people are divided among 40 different ethnic groups. The **Kikuyu** (kee•KOO•yoo) people are Kenya's main group, making up less than one-fourth of the population. Most Kenyans live in rural areas where they struggle to grow crops. In recent years, large numbers of people have moved to cities in search of a better life.

About one-third of Kenya's people live in cities. Nairobi is the largest city, with about 2 million people. **Mombasa** (mohm•BAH•sah) is Kenya's chief port on the Indian Ocean. This city has the best harbor in East Africa, making it an ideal site for oceangoing trade.

✔ Reading Check What city is Kenya's chief port?

Tanzania

Tourists flock to Tanzania's **Serengeti** (SEHR•uhn•GEH•tee) **Plain,** famous for its wildlife preserve, huge grasslands, and patches of trees and shrubs. To the north, near the Kenyan border, a snowcapped mountain called **Kilimanjaro** towers over this region. It is the highest point in Africa. The Great Rift Valley cuts two gashes through Tanzania, one in the center of the country and the other along the western border. Unusual fish swim in the deep, dark waters of Lake Tanganyika (TAN•guhn•YEE•kuh). Lake Victoria, also in Tanzania, is Africa's largest lake and one of the sources of the Nile River.

Tanzania's Economy   More than 80 percent of all Tanzanians work in farming or herding. Important export crops are coffee and sisal, a plant fiber used to make rope and twine. Do you enjoy eating baked ham? If so, you might have tasted the spice called cloves, often used to flavor ham. The islands of Zanzibar and Pemba, off the coast of Tanzania, produce more cloves than any other place in the world.

NATIONAL GEOGRAPHIC

East and Central Africa: Political

Applying Map Skills

1. What is the capital of Tanzania?

2. What countries in East and Central Africa include islands?

Find NGS online map resources @ www.nationalgeographic.com/maps

⊛ National capital
• Major city

0 mi.   1,000
0 km   1,000
Azimuthal Equidistant projection

Tourism is a fast-growing industry in Tanzania. The government has set aside several national parks to protect the habitats of the country's wildlife. A habitat is the type of environment in which a particular animal species lives. Serengeti National Park covers about 5,600 square miles (14,504 sq. km). Lions and wild dogs hunt among thousands of zebras, wildebeests, and antelopes. The park attracts many eco-tourists, or people who travel to another country to view its natural wonders.

Tanzania's leaders are also taking steps to preserve farmland. In recent years, many trees have been cut down. Without trees, the land cannot hold soil or rainwater in place. As a result, the land dries out, and soil blows away. To prevent the land from becoming desert, the government of Tanzania has announced a new policy. For every tree that is cut down, five new trees should be planted.

History and Government Tanzania's roughly 32 million people include more than 120 different ethnic groups. Each group has its own language, but most people also speak Swahili. The two main religions are Christianity and Islam. In 1964 the island country of Zanzibar united with the former German colony of Tanganyika to form Tanzania. Since then, Tanzania has been one of Africa's more politically stable republics. During the 1960s, Tanzania's socialist government controlled the economy. By the 1990s, however, it had moved toward a free market system. In taking this step, Tanzania's leaders hoped to improve the economy and reduce poverty. Meanwhile, the government also moved toward more democratic elections with more than one political party.

✓ Reading Check Why is sisal important to Tanzania?

Section 1 Assessment

Defining Terms
1. Define coral reef, poaching, free enterprise system, cassava, sisal, habitat, eco-tourist.

Recalling Facts
2. Place Describe the Great Rift Valley.
3. Culture What are Kenya's official languages?
4. Culture What are the two major religions of Tanzania?

Critical Thinking
5. Evaluating Information Why might two countries such as Tanganyika and Zanzibar unite?
6. Analyzing Information Why would the government of Tanzania put so much effort into preserving its national parks?

Graphic Organizer
7. Organizing Information Review the information about the history and government of Kenya. Then, on a time line like the one below, label four important events and their dates in Kenya's history.

|——————|——————|——————|

Applying Social Studies Skills
8. Analyzing Maps Study the political map on page 407. Name the four bodies of water that border Tanzania. On what body of water is Dar es Salaam located?

Critical Thinking Skill

Making Predictions

Predicting consequences is obviously difficult and sometimes risky. The more information you have, however, the more accurate your predictions will be.

Learning the Skill

Follow these steps to learn how to better predict consequences:

- Gather information about the decision or action you are considering.
- Use your knowledge of history and human behavior to identify what consequences could result.
- Analyze each of the consequences by asking yourself: How likely is it that this will occur?

Practicing the Skill

Study the graph below, and then answer these questions:

1. What is measured on this graph? Over what time period?
2. In what year did the fewest tourists visit Kenya?
3. What trend does the graph show?
4. Do you think this trend is likely to continue?
5. On what do you base this prediction?
6. What are three possible consequences of this trend?

Applying the Skill

Analyze three articles in your local newspaper. Predict three consequences of the actions in each of the articles. On what do you base your predictions?

GO TO

Practice key skills with **Glencoe Skillbuilder Interactive Workbook, Level 1.**

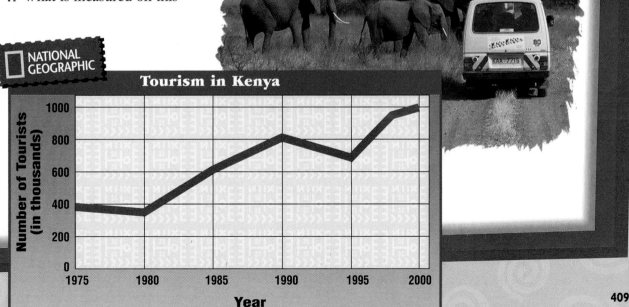

NATIONAL GEOGRAPHIC

Tourism in Kenya

Number of Tourists (in thousands): 0, 200, 400, 600, 800, 1000

Year: 1975, 1980, 1985, 1990, 1995, 2000

Source: *Europa Year Book; Yearbook of Tourism Statistics, 1978–1999.*

Other Countries of East Africa

Guide to Reading

Main Idea

The other countries of East Africa have all been scarred by conflict in recent years.

Terms to Know

- drought
- plate
- clan
- endangered species
- genocide
- refugee

Reading Strategy

Create a chart like this one. Choose three countries of East Africa. Write the cause of conflict in each country next to that country's name. Then write the effects of that conflict.

Country	Cause of conflict	Effects of conflict

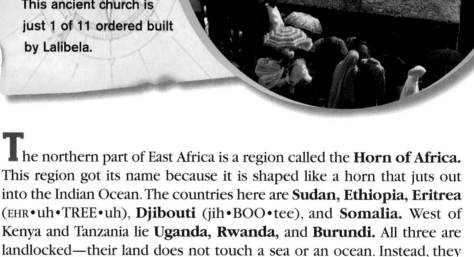

NATIONAL GEOGRAPHIC

Exploring Our World

In the late 1100s and early 1200s, a king named Lalibela ruled Ethiopia. He had his subjects build Christian churches by carving them out of solid rock. First they cut a huge rectangular trench into the ground. Then they carved the rock inside that trench to form the church. This ancient church is just 1 of 11 ordered built by Lalibela.

The northern part of East Africa is a region called the **Horn of Africa.** This region got its name because it is shaped like a horn that juts out into the Indian Ocean. The countries here are **Sudan, Ethiopia, Eritrea** (EHR•uh•TREE•uh), **Djibouti** (jih•BOO•tee), and **Somalia.** West of Kenya and Tanzania lie **Uganda, Rwanda,** and **Burundi.** All three are landlocked—their land does not touch a sea or an ocean. Instead, they use three large lakes for transportation and trade.

Sudan

Sudan is the largest country in Africa—about one-third the size of the continental United States. Nomads raise camels and goats in the north. Most of Sudan's people live along the Nile River or one of its tributaries. Like the Egyptians, they use water from the Nile to irrigate their fields. Farmers grow sugarcane, grains, nuts, dates, and cotton—the country's leading export. Oil fields in the south offer another possibility of income.

Sudan's People and History   In ancient times, Sudan was the center of a powerful civilization called Kush. The people of Kush had close cultural and trade ties with the Egyptians to the north. Kushites traded metal tools for cotton and other goods from India, Arabia, and China. They built a great capital at Meroë (MAR•oh•EE). In Meroë were huge temples, stone palaces, and small pyramids. Kush began to lose power around A.D. 350.

During the A.D. 500s, missionaries from Egypt brought Christianity to the area. About 900 years later, Muslim Arabs entered northern Sudan and converted its people to Islam. From the late 1800s to the 1950s, the British and the Egyptians together ruled the entire country. Sudan became an independent nation in 1956. Since then, military leaders generally have ruled Sudan. In the 1980s, the government began a "reign of terror" against the southern Christian peoples. The fighting has disrupted the economy and caused widespread hunger, especially in the south. A recent drought—a long period of extreme dryness and water shortages—made the situation worse. Millions of people have starved to death, and major outbreaks of diseases have swept through the country. The war continues despite occasional peace talks aimed at granting the south greater independence. Time Reports: Focus on World Issues on pages 415-421 looks at Sudanese refugees.

Reading Check   **What is the main export of Sudan?**

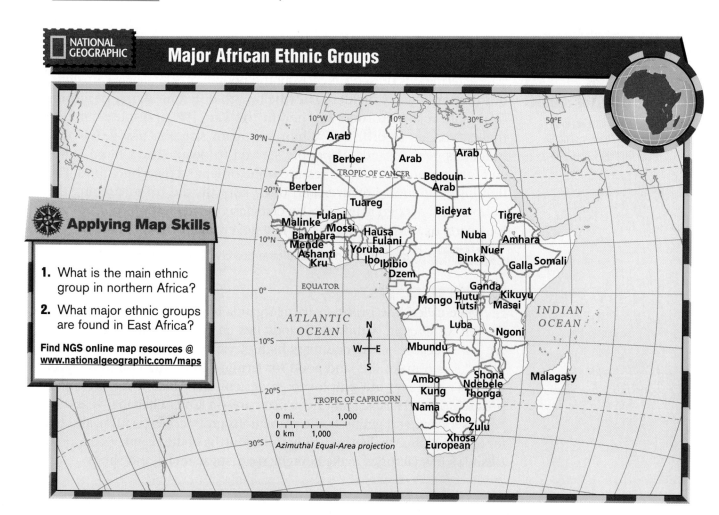

NATIONAL GEOGRAPHIC

Major African Ethnic Groups

Applying Map Skills

1. What is the main ethnic group in northern Africa?

2. What major ethnic groups are found in East Africa?

Find NGS online map resources @ www.nationalgeographic.com/maps

Ethiopia and Eritrea

Landlocked Ethiopia is almost twice the size of Texas. Famine brought Ethiopia to the world's attention in the 1980s. At that time, a drought turned fields once rich in crops into seas of dust. Despite food aid, more than 1 million Ethiopians died.

Since 1974 Ethiopia has been trying to build a democratic government. This goal is hindered by warfare with neighboring **Eritrea,** a small country that broke away from Ethiopia in 1993. Eritrean women formed about one-third of the army that won the war. After the war ended, the new government passed laws that gave Eritrean women more rights than they had ever had before.

Muslims now form about 45 percent of Ethiopia's roughly 65 million people. About 40 percent of Ethiopians are Christians. Others practice traditional African religions. Almost 80 languages are spoken in Ethiopia. Amharic, similar to Hebrew and Arabic, is the official language.

✔ Reading Check What crisis brought Ethiopia to the world's attention?

Djibouti and Somalia

The tiny country of Djibouti is one of the hottest, driest places on the earth. This country lies at the northern tip of the Great Rift Valley, where three of the earth's plates join. Plates are huge slabs of rock that make up the earth's crust. In Djibouti, two of these plates are pulling away from each other. As they separate, fiery hot rock rises to the earth's surface, causing volcanic activity.

Somalia Shaped like the number seven, Somalia is almost as large as Texas. Like Eritrea and Djibouti, much of Somalia is hot, dry country where farming is difficult. Nearly all the people of Somalia are Muslims, but they are deeply divided. They belong to different clans, or groups of people related to one another. In the late 1980s, disputes among these clans led to civil war. When a drought struck a few years later, hundreds of thousands of people starved to death. The United States and other countries tried to restore some order and distribute food. The fighting continued, however, and often kept the aid from reaching the people who needed it. Even today, clan-based armed groups control various parts of Somalia. There is no real government that is in charge.

✔ Reading Check What kind of conflict led to civil war in Somalia?

Uganda

Once called "the pearl of Africa," Uganda is a fertile, green land about the size of Oregon. Although Uganda lies on the Equator, temperatures are mild because of the country's high elevation. Uganda's rich soil and plentiful rain make the land good for farming. More than 80 percent of Uganda's workers are employed in agriculture. Most work on small, privately owned farms. They grow plantains—a kind of banana—cassava, potatoes, corn, and grains. Some plantations grow coffee, cotton, and tea for export. Coffee makes up nearly three-fourths of the country's exports. Uganda's few factories make cement, soap, sugar, metal, and shoes.

▲ In May 2000, this woman voted in Ethiopia's second-ever democratic election.

Kampala, Uganda

Public transportation in Kampala includes passenger boats on Lake Victoria.

Culture How does this compare to public transportation where you live?

Uganda's People About two-thirds of Ugandans are Christians. The remaining one-third practice Islam or traditional African religions. At one time there were large numbers of Hindus and Sikhs from South Asia living in the country. A dictator, Idi Amin, drove them out in 1972. Recently, the government invited them back, and many have returned.

The people of Uganda belong to more than 40 different ethnic groups. They have a rich cultural heritage of songs, folktales, and poems. In the past, these were passed by word of mouth from one generation to the next. Today this heritage has been preserved in books.

History and Government The dictator Idi Amin's cruel rule hurt Uganda throughout much of the 1970s. Civil wars forced Amin out of power. The role of local chiefs or kings has been reduced since independence to try to lessen tribal fighting.

Ugandans now have one of the fastest-growing economies in Africa. Uganda also enjoys a stable government. It is a republic with an elected president and legislature. Still, the future is clouded. Uganda, along with other African countries, faces a new threat: the disease called AIDS. Hundreds of thousands of Ugandans have died from it, and many more are infected with HIV, the virus that causes AIDS.

Reading Check What kind of government does Uganda have today?

Rwanda and Burundi

Rwanda and Burundi are located in inland East Africa. High elevation gives both Rwanda and Burundi a moderate climate even though they lie near the Equator. Gorillas live within rain forests here. Because of illegal hunting, or poaching, gorillas are now an endangered species. An endangered species is a plant or an animal threatened with extinction.

Web Activity Visit the *Our World Today: People, Places, and Issues* Web site at owt.glencoe.com and click on **Chapter 15— Student Web Activities** to learn more about Rwanda's mountain gorillas.

Coffee is the main export crop in Burundi and Rwanda. The people who live along Lake Kivu and Lake Tanganyika also fish. Because both countries are landlocked, they have trouble getting their goods to foreign buyers. Few paved roads and no railroads exist. Most goods must be transported by road to Lake Tanganyika, where boats take them to Tanzania or the Democratic Republic of the Congo.

Ethnic Conflict Rwanda and Burundi are among the most densely populated countries in Africa. Two ethnic groups form most of the population of both countries—the **Hutu** and the **Tutsi.** The Hutu make up 80 percent or more of the population in both Rwanda and Burundi. The two groups have vied for political and economic power, with the Tutsi traditionally controlling the governments and economies.

The constant power struggle erupted into a full-scale civil war and genocide in the 1990s. Genocide is the deliberate destruction of a group of people because of their race or culture. A Hutu-led government in Rwanda killed hundreds of thousands of Tutsi people. Two million more became refugees, or people who flee to another country to escape persecution or disaster. The fighting between the Hutu and Tutsi has lessened, but both countries face many challenges as they try to rebuild with the help and cooperation of the international community.

✔ Reading Check   **Which ethnic group makes up the majority of the population in Rwanda and Burundi?**

Section 2 Assessment

Defining Terms
1. Define drought, plate, clan, endangered species, genocide, refugee.

Recalling Facts
2. History To what nation did Eritrea once belong?
3. Government Describe the current political situation in Somalia.
4. Region What endangered species lives in the forests of Rwanda and Burundi?

Critical Thinking
5. Evaluating Information How could a deadly epidemic, such as AIDS, affect a country's economy?
6. Analyzing Information How do ethnic differences create problems for the countries of East Africa?

Graphic Organizer
7. Organizing Information Create a diagram like the one below. Then write two facts about Sudan under each of the category headings in the outer ovals.

People — Sudan — Economy
History

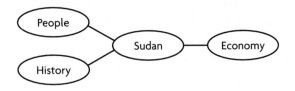

Applying Social Studies Skills

8. Analyzing Maps Study the political map on page 407. What are the four landlocked countries of East Africa?

TIME REPORTS

Refugees On the Move

DEREK HUDSON/CORBIS SYGMA

The Lost Boys of Sudan

At a refugee camp, boys collected sticks and reeds to make huts, then cooked a rare meal of beans.

The Lost Boys of Sudan

In November 1987, William Deng was tending cattle several miles from his village in southern Sudan. Two brothers and some cousins were with him. One afternoon they heard distant gunfire but ignored it. "The next morning," William said, "we saw the smoke. I climbed a tree and saw that my whole village was burned."

They raced to the village. There they learned that government troops had swept through. "Nobody was left standing," William said. "Some were wounded; some were killed. My father was dead. So we just ran away. I was 5."

The boys headed toward Ethiopia. Crossing marshlands and desert, they joined thousands of other Sudanese, mostly boys. They walked for two months. They ate berries, dried leaves, birds, and mice—anything they could find. Thousands died. "You think that maybe later that will be you," said one boy.

Into Ethiopia

The survivors finally reached a refugee camp in Ethiopia. **Refugees** are people forced by fear to find refuge, or shelter, outside their countries.

In 1991 Ethiopia closed its camps. Soldiers forced all of the "Lost Boys of Sudan," as they came to be called, back to their homeland.

After a year in Sudan, 10,000 of the boys fled to a refugee camp in Kenya. And there they stayed— some for as long as 10 years.

All Too Common

Sadly, William's experience is not unique. Throughout Africa south of the Sahara, millions of people have had to flee their homes. Most live in crowded camps set up by groups such as the United Nations. There they wait—until it is safe to go home, or until another country lets them stay.

Africans on the Move

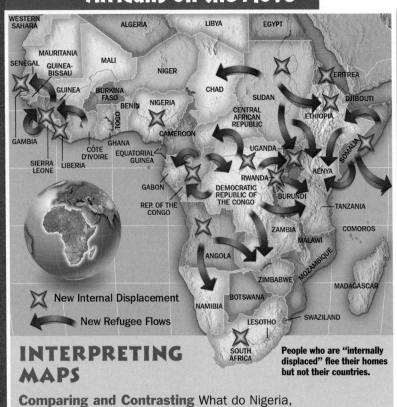

New Internal Displacement

New Refugee Flows

INTERPRETING MAPS

Comparing and Contrasting What do Nigeria, Eritrea, and South Africa have in common?

People who are "internally displaced" flee their homes but not their countries.

Older boys "adopted" younger ones.

Some drew on clay.

The youngest wore the faces of suffering.

TOP PHOTOS: DEREK HUDSON/CORBIS SYGMA

The Lost Boys are victims of a war that began in 1983. The government has been heartless. U.S. President George W. Bush explained in 2001, "Some 2 million Sudanese have lost their lives; 4 million more have lost their homes." The Sudan, he concluded, is a "disaster area for human rights."

Human rights include the right to safety, to food, and to shelter, among other things. In democracies, they also include the rights of citizens to choose their own leaders.

Defining Refugees' Rights

In 1951, members of the United Nations agreed to guarantee basic human rights to refugees. They signed a **convention**, or special document, that gives refugees a unique legal status, or position. That status gives them the right to **asylum**, or safety, in foreign countries. It also gives them the right to be treated like any other foreign resident of their host country.

The convention defines refugees as people who leave their countries to flee **persecution**. Persecution is unfair treatment based on such characteristics as race, religion, or ethnic background. Recently the United Nations expanded this definition. Today people whose governments can't protect them from the dangers of war are also entitled to refugee status, the UN says.

The boys made huts to protect themselves from rain and wild animals.

DEREK HUDSON/CORBIS SYGMA

EXPLORING THE ISSUE

1. **Analyzing Information** Why do you think thousands of the Lost Boys traveled together instead of alone?

2. **Making Predictions** How might your life change if you no longer had the basic human rights listed here?

Environmental Refugees

People who flee natural disasters, such as floods and famines, aren't refugees. They are "displaced persons" or "environmental refugees." Immigrants aren't refugees, either. Immigrants may leave their countries to get an education or find a better job. Refugees like the Lost Boys of Sudan have little choice. They flee their countries to find safety.

Africa's Troubled Past

Refugees have existed in many places, not only countries in Africa. Yet rarely has the flow of refugees been as widespread as it is today. Worldwide, about 35 million people were on the move in the year 2000. Some 21 million of them were **internally displaced persons (IDPs)**— people who flee to safety inside their own countries. About 14 million more were refugees seeking freedom from war and persecution outside their countries. During the year 2000, Africa alone held more than 3 million refugees and at least 11 million IDPs.

The Impact of Violence

The presence of these uprooted people is reshaping Africa. Away from their villages, refugees no longer grow crops, worsening food shortages. The crush of refugees drains the resources of the already poor countries that host them. And refugees spread AIDS, a disease that by 2001 had killed the parents of 12 million African children.

Many experts trace Africa's refugee problem back to the late 1800s. That's when European nations began to carve the continent into colonies. Africa's

▲ War chased thousands of terrified Burundians into the Congo in 1995.

2,000 ethnic groups speak around one thousand languages. But European colonizers failed to respect those differences. They set up boundaries that split individual tribes into many pieces. Other borders forced traditional enemies such as Rwanda's Hutu and Tutsi tribes to share the same space.

How America Is Different

Americans living in Britain's 13 colonies didn't face such problems in 1776. When the U.S. was born, most Americans spoke English and shared similar values. They were ready to rule themselves as a democracy.

Africa's colonies became independent nearly two centuries later. But they contained groups that had little interest in working together. That made it hard for democracy to take root. In many nations, armed groups muscled their way to power. Such struggles for control turned millions of Africans into refugees. ▪

EXPLORING THE ISSUE

1. **Making Inferences** In what ways might Africa's refugee problem hurt all Africans?

2. **Analyzing Information** How might the refugee problem keep governments from building roads and providing services such as education and health care?

Struggling to Survive

When refugees enter another country, they may face dangers. In 1997, for example, soldiers rounded up refugees who had lived many years in Tanzania. They forced the refugees into camps. "I never thought the Tanzanian government would do this to us," said a woman who fled Burundi in 1971. "I am now held in a refugee camp, but my children are still outside. They have no money to come here."

Even refugees allowed to stay in private homes face risks. "We are frequently arrested by the police," said an Ethiopian who fled to Kenya. "They require bribes before they will release us."

Refugee Children

For children, refugee life brings special problems. Those separated from their parents must fend for themselves. Some girls and boys are forced to become soldiers. Others must work for little or no pay. Abby, 14, fled the war in Sierra Leone. Now she lives in a refugee camp in Guinea. "In the morning," she said, "I fetch water, sweep, and pray. Then I go find a job for the day. I usually pound rice. I get no food,

only [a tiny amount of money]. I will be in the sun until evening. I feel pain all over my body. I don't go to school. I live with my grandmother, and she is very old. I need to take care of her."

Many of the overcrowded camps are dirty and unhealthy. But until their countries become safe again, the refugees have few choices. Either they stay in the camps, or they return home to the horrors they fled. ▨

SEBASTIAO SALGADO/AMAZONAS/CONTACT

▲
A refugee camp at Benako, Tanzania, overflowed with Rwandans in 1994.

EXPLORING THE ISSUE

1. **Drawing Conclusions** What do you think makes many refugee camps so difficult to live in?

2. **Problem Solving** Gather information from the text to identify problems faced by refugees in camps. Consider changes that might be made to the camps to improve life there.

Helping Refugees: What Can One Person Do?

▲ Joseph Maker found a warm welcome in Houston, Texas.

Countries that offer asylum to refugees are known as **host countries.** Starting life over in a host country can be hard for refugees and their families. William Deng and Joseph Maker are among more than 3,000 Lost Boys of Sudan who have found refuge in the United States. William lives in Grand Rapids, Michigan. Joseph lives in Houston, Texas, as do 170 other Lost Boys. In both places, local people taught them how to take buses, shop, and even use faucets and refrigerators. Many of those helpers were volunteers.

Eventually, many refugees who get such help grow to love their adopted country. They come to love it as much as or even more than people born there. ▪

◄ These Michigan second graders are learning English at school.

Strangers in a New Land

Joseph Maker and fellow Lost Boys felt they had landed on another planet when they reached Houston, their new home. They had to be taught to use electricity, running water, air conditioners, flush toilets, stoves, and telephones.

Packaged foods baffled them. At the refugee camp in Kenya, they had eaten the same meal—beans and lentils—every day for nine years. In Houston they discovered junk food—and the fear of getting fat. "I've heard in America, people can become big," said Joseph's friend James Thon Aleer.

The newcomers also had to learn new ways to act. In Sudan, it's disrespectful to look into the eyes of the person you're speaking to. In America, it's impolite to look away. Joseph and his friends adopted the American way, something they think helped all of them get jobs.

One thing they picked up quickly was American humor. The tag on James Thon Aleer's key chain says "Don't Mess with Texas."

EXPLORING THE ISSUE

1. **Categorizing** If you were an African refugee in your community, what things might confuse you the most?

2. **Problem Solving** How might volunteers help refugees adapt to life in your community?

REVIEW AND ASSESS

UNDERSTANDING THE ISSUE

1. Defining Key Terms Write definitions for the following terms: *refugee, refugee camp, internally displaced person, human rights, asylum, convention,* and *persecution.*

2. Writing to Inform Write a 250-word article about the refugee issue that a school newspaper could publish. Use the key terms listed above.

3. Writing to Persuade "There is no greater sorrow on Earth than the loss of one's native land." A Greek thinker wrote those words about 2,500 years ago. Is his statement as true today? Write a short essay to explain.

INTERNET RESEARCH ACTIVITIES

4. With your teacher's help, use Internet resources to find information on issues involving refugees in the world, particularly in Africa today. Write a brief report on current "refugee hot spots" in Africa south of the Sahara. Be prepared to report on these problems in class.

5. With your teacher's help, choose two nations in Africa south of the Sahara. Then use Internet resources to learn more about these countries. Concentrate your research on refugee problems in your chosen nations and report your findings to the class.

BEYOND THE CLASSROOM

6. Visit your school or local library to find books in which young refugees share their experiences. (Enter the key words "refugee children" on the Amazon.com website, and you can find some titles to start with.) Bring those books to class to share with your classmates.

▲ This Hmong boy from Laos now lives in Wisconsin.

STEVE LISS

7. Research a refugee problem outside Africa. List ways that it is like—and different from—refugee problems in Africa. Report your findings to your classmates.

8. Work in groups to come up with ways young people could make it easier for newcomers to your community. Put your suggestions on a poster. Include phone numbers of groups that provide services for refugees. Display the poster for all students to see.

Where the World's Refugees Come From

(Top 15 Sources of Refugees as of January 1, 2001)

Palestinians	4,000,000
Afghanistan	3,600,000
Sudan	460,000
Iraq	450,000
Burundi	420,000
Angola	400,000
Sierra Leone	400,000
Burma	380,000
Somalia	370,000
Dem. Rep. of Congo	350,000
Eritrea	350,000
Croatia	315,000
Vietnam	300,000
Bosnia and Herzegovina	250,000
El Salvador	230,000
Total:	**12,275,000**

Around the world in 2001, more than 14 million people lived as refugees. This table lists the 15 nations that most of them fled. Besides refugees, at least 21 million others are internally displaced persons (IDPs). IDPs seek safety inside their nations' borders but far from their homes.

Source: U.S. Committee for Refugees

BUILDING SKILLS FOR READING TABLES

1. Categorizing Use your text to categorize the 15 nations listed here under one of five regions: Africa, Central America, Eastern Europe, Southeast Asia, and Southwest Asia.

2. Analyzing Data Which of the above regions have the greatest and smallest number of refugees?

3. Transferring Data Create a bar graph based on the country data in this table.

FOR UPDATES ON WORLD ISSUES GO TO www.timeclassroom.com/glencoe

Guide to Reading

Main Idea

Central Africa has rich natural resources that are largely undeveloped because of civil war and poor government decisions.

Terms to Know

- savanna
- canopy
- hydroelectric power
- tsetse fly
- deforestation

Reading Strategy

Create a chart like this one. Choose two countries of Central Africa. Then list two facts about the people of each country.

Country	Fact #1	Fact #2

NATIONAL GEOGRAPHIC

Exploring Our World

In the 2000 Olympic Summer Games, the gold medal in men's soccer went to Cameroon's team, the Indomitable Lions. The streets of Cameroon's capital, Yaoundé, and other cities were jammed with wildly excited fans screaming with joy. Cameroon's president even declared the following Monday a national holiday to celebrate the victory.

Central Africa includes seven countries. They are the **Democratic Republic of the Congo, Cameroon,** the **Central African Republic, Congo, Gabon** (ga•BOHN), **Equatorial Guinea,** and **São Tomé** (sow too•MAY) **and Príncipe** (PREEN•see•pee). Africa's second-longest river—the **Congo River**—flows through the center of the Democratic Republic of the Congo in the very heart of Africa.

Democratic Republic of the Congo

One-fourth the size of the United States, the Democratic Republic of the Congo has only about 23 miles (37 km) of coastline. Most of its land borders other African countries—nine in all.

High, rugged mountains rise in the eastern part of the country. Here you will find four large lakes—Lake Albert, Lake Edward, Lake Kivu, and Lake Tanganyika. **Lake Tanganyika** is the longest freshwater lake in the world. It is also the second deepest, after Russia's Lake Baikal. Savannas, or tropical grasslands with few trees, cover the highlands in the far north and south of the country. In these areas, lions and leopards stalk antelopes and zebras for food.

One of the world's largest rain forests covers the center of the Democratic Republic of the Congo. The treetops form a canopy, or an umbrella-like forest covering so thick that sunlight rarely reaches the forest floor. More than 750 different kinds of trees grow here. The rain forests are being destroyed at a rapid rate, however, as they are cleared for timber and farmland.

The mighty Congo River—about 2,800 miles (4,506 km) long—weaves its way through the country on its journey to the Atlantic Ocean. The river current is so strong that it carries water about 100 miles (161 km) into the ocean. The Congo River and its tributaries, such as the Kasai River, provide hydroelectric power, or electricity generated by flowing water. In fact, these rivers produce more than 10 percent of all the world's hydroelectric power. The Congo is also the country's highway for trade and travel.

The Economy The Democratic Republic of the Congo has the opportunity to be a wealthy nation. The country exports gold, petroleum, diamonds, and copper. It is Central Africa's main source of industrial diamonds, as shown on the graph below. These diamonds are used in making strong industrial tools that cut metal. The country's factories make steel, cement, tires, shoes, textiles, processed foods, and beverages.

The Democratic Republic of the Congo has not been able to take full advantage of its rich resources, however. Why? One reason is the difficulty of transportation. Many of the minerals are found deep in the country's interior. Lack of roads and the thick rain forests make it hard to reach these areas. Another reason is political unrest. For many years, power-hungry leaders kept the nation's wealth for themselves. Then a

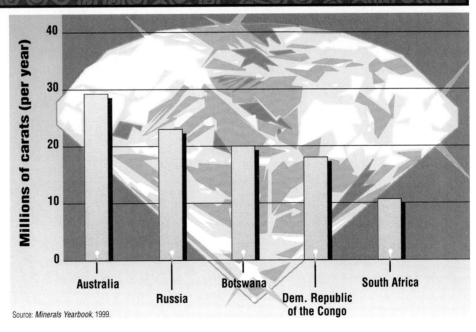

Leading Diamond-Producing Countries

Analyzing the Graph

Three of the world's top diamond-producing countries are in Africa south of the Sahara.

Economics What two countries produce the most diamonds in Africa?

Visit owt.glencoe.com and click on **Chapter 15— Textbook Updates.**

Source: *Minerals Yearbook,* 1999.

civil war broke out in the late 1990s. This war has hurt efforts to develop the country's economy.

The People of the D.R.C.   The Democratic Republic of the Congo's roughly 52 million people consist of more than 200 different ethnic groups. One of these groups is the Kongo people, after whom the country is named. The country's official language is French, but many people speak local languages, such as Lingala or Kingwana. More than 75 percent of Congolese are Christians. Most of these are Roman Catholic.

Most people in the Democratic Republic of the Congo live in rural areas. Less than one-third are city dwellers. Still, **Kinshasa,** the capital, has about 5 million people. After years of civil war, life in the country is still unsettled. The economy has nearly collapsed, and many people in the cities are without work.

In rural areas people follow traditional ways of life. They plant seeds, tend fields, and harvest crops. Most of the harvest goes to feeding the

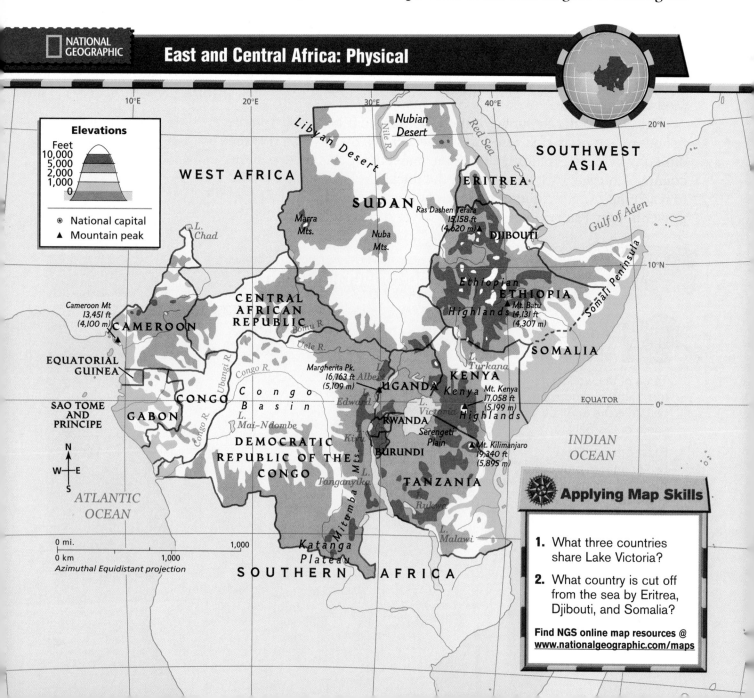

NATIONAL GEOGRAPHIC

East and Central Africa: Physical

Elevations

Feet
10,000
5,000
2,000
1,000
0

⊛ National capital
▲ Mountain peak

10°E   20°E   30°E   40°E   20°N

Nubian Desert

Libyan Desert

Nile R.

Red Sea

SOUTHWEST ASIA

WEST AFRICA

SUDAN

ERITREA

Marra Mts.

Nuba Mts.

Ras Dashen Terara 15,158 ft (4,620 m) ▲

DJIBOUTI

Gulf of Aden

L. Chad

10°N

CENTRAL AFRICAN REPUBLIC

Bomu R.

Uele R.

Cameroon Mt 13,451 ft (4,100 m) ▲

CAMEROON

Ethiopian Highlands ▲ Mt. Batu 14,131 ft (4,307 m)

ETHIOPIA

Somali Peninsula

EQUATORIAL GUINEA

Ubangi R.

Congo R.

Margherita Pk. 16,763 ft (5,109 m)

L. Albert

UGANDA

KENYA

L. Turkana

SOMALIA

SAO TOME AND PRINCIPE

CONGO

Congo Basin

L. Edward

Mt. Kenya 17,058 ft (5,199 m) ▲

Kenya Highlands

EQUATOR   0°

GABON

Congo R.

L. Mai-Ndombe

RWANDA

L. Victoria

DEMOCRATIC REPUBLIC OF THE CONGO

Kirt

Serengeti Plain

▲ Mt. Kilimanjaro 19,340 ft (5,895 m)

INDIAN OCEAN

BURUNDI

N
W   E
S

Mitemba Mts.

L. Tanganyika

TANZANIA

L. Rukwa

ATLANTIC OCEAN

0 mi.   1,000
0 km   1,000
Azimuthal Equidistant projection

Katanga Plateau

L. Malawi

SOUTHERN   **AFRICA**

Applying Map Skills

1. What three countries share Lake Victoria?

2. What country is cut off from the sea by Eritrea, Djibouti, and Somalia?

Find NGS online map resources @ www.nationalgeographic.com/maps

family. Any extra goes to the local market—or to the boats moving along the rivers—to sell or trade for goods the people need.

History and Government The Congo region was first settled about 10,000 years ago. The Bantu people—ancestors of most of the Congolese people today—moved here from Nigeria around the A.D. 600s and 700s. Several powerful kingdoms arose in the savannas south of the rain forests. The largest of these kingdoms was the **Kongo.**

In the late 1400s, European traders arrived in Central Africa. During the next 300 years, European and African agents enslaved many people from the Congo region. Most of these Africans were transported to the Americas.

Like many other African nations, the Democratic Republic of the Congo was once a European colony. It became independent in 1960, and was renamed **Zaire.** A harsh dictator named Mobutu Sese Seko ruled Zaire until civil wars in neighboring Rwanda and Burundi sparked a civil war in Zaire. In 1997, Mobutu's government was finally overthrown, and again the country was given a new name. Zaire became the Democratic Republic of the Congo and another dictator took power. The country has not yet been able to organize a stable government.

✓Reading Check What kind of government does the Democratic Republic of the Congo now have?

NATIONAL GEOGRAPHIC **On Location**

Market Day

This marketplace in Kinshasa, Democratic Republic of the Congo, is bustling with activity.

Culture How is this market different from where your family shops? How is it similar?

Cameroon and the Central African Republic

Find Cameroon and the Central African Republic on the map on page 424. These countries lie just north of the Equator. Most people in the Central African Republic and Cameroon farm for a living. A few large plantations raise cacao, cotton, tobacco, and rubber for export. Some people herd livestock in areas that are safe from tsetse flies. A parasite that is often transmitted by the bite of the tsetse (SEET•see) fly causes a deadly disease called sleeping sickness. Turn to page 428 to find out more about sleeping sickness.

These two countries are only beginning to industrialize, or base their economies more on manufacturing and less on farming. Cameroon has had greater success in this effort. It has coastal ports and forest products, petroleum, and bauxite. The Central African Republic can claim only diamond mining as an important industry.

Rural Living

A row of thatch houses stands in a village in rural Democratic Republic of the Congo (right). This village in Gabon (above) boasts a very different type of house.

Culture Why might house styles differ from country to country?

A colony of France from 1910 until 1960, the Central African Republic recognizes French as its official language. Yet most of its people speak Sango, the national language of the Central African Republic, to ease communication among the many ethnic groups. Cameroon, divided between the British and the French until 1960, uses both English and French as its official languages.

✓ Reading Check **Why has Cameroon had greater success than the Central African Republic in industrializing?**

Congo and Gabon

Congo and Gabon both won their independence from France in 1960. In Congo, a plain stretches along the Atlantic coast and rises to low mountain ranges and plateaus. Here the Congo River supports most of the country's farmlands and industries. To the north, a large swampy area along the Ubangi River supports dense vine thickets and tropical trees. Both the Ubangi and Congo Rivers provide Congo with hydroelectric power. They also make Congo the door to the Atlantic Ocean for trade and transport.

More than half of Congo's and Gabon's people farm small plots of land. Both countries' economies rely on exports of lumber. They are beginning to depend more on rich offshore oil fields, however, for their main export. Congo also exports diamonds. Gabon suffers from deforestation, or the cutting of too many trees too quickly. Gabon also has valuable deposits of manganese and uranium.

Only about 1.2 million people live in Gabon—mainly along rivers or in the coastal capital, **Libreville.** Congo's 3.1 million people generally live along the Atlantic coast or near the capital, **Brazzaville.**

✓ Reading Check **What two exports are most important to Congo and Gabon?**

Island Countries

Once a Spanish colony, Equatorial Guinea won its independence in 1968. Equatorial Guinea includes land on the mainland of Africa and five islands. Today the country is home to about 500,000 people. Most live on the mainland, although the capital and largest city—**Malabo** (mah•LAH•boh)—is on the country's largest island.

Farming, fishing, and harvesting wood are the country's main economic activities. For many years, timber and cacao grown in the islands' rich volcanic soil were the main exports. Oil was recently discovered and now leads all other exports.

The island country of São Tomé and Príncipe gained its independence from Portugal in 1975. The Portuguese had first settled here about 300 years earlier. At that time, no people lived on the islands. Today about 150,000 people live here, with almost all living on the main island of São Tomé.

São Tomé and Príncipe are volcanic islands. As a result, the soil is rich and productive. Farmworkers on the islands grow various crops, including coconuts and bananas for export. The biggest export crop, however, is cacao, which makes cocoa and chocolate.

 **Reading Check** Which of these island countries is also located on the African mainland?

Section 3 Assessment

Defining Terms
1. **Define** savanna, canopy, hydroelectric power, tsetse fly, deforestation.

Recalling Facts
2. **Economics** Why has the Democratic Republic of the Congo not taken full advantage of its resources?
3. **Place** How has Cameroon's location helped it prosper?
4. **Economics** What natural resource was recently discovered in Equatorial Guinea?

Critical Thinking
5. **Evaluating Information** Why do you think Europeans wanted to colonize parts of Africa, such as the Congo?
6. **Understanding Cause and Effect** How could furniture buyers in the United States affect lumber exports in Central Africa?

Graphic Organizer
7. **Organizing Information** Complete a chart like this with one fact about each country.

Country	Fact
Democratic Republic of the Congo	
Cameroon	
Central African Republic	
Congo	
Gabon	
Equatorial Guinea	
São Tomé & Príncipe	

 Applying Social Studies Skills

8. **Analyzing Maps** Study the political map on page 407. The Ubangi River forms part of the boundaries of which countries?

Making Connections

CULTURE GOVERNMENT PEOPLE TECHNOLOGY

Battling Sleeping Sickness

Since the 1300s, people in Africa south of the Sahara have battled a disease now commonly called sleeping sickness. Yet it was not until the early 1900s that scientists began to understand the disease and that it was transmitted through the bite of an infected tsetse fly.

The Tsetse Fly

Found only in parts of Africa, the tsetse fly is the common name for any of about 21 species of flies that can transmit sleeping sickness. The flies are larger than the houseflies common to the United States. Tsetse flies thrive in forests and in areas of thick shrubbery and trees near lakes, ponds, and rivers.

Although the bite of a tsetse fly is painful, the bite itself is not necessarily harmful. What gives the tsetse fly its dreadful reputation is the disease-causing parasite it may carry.

Sleeping Sickness

The World Health Organization (WHO) estimates that more than 60 million people in Africa are at risk of being infected with sleeping sickness. As many as 500,000 people carry the disease. If left untreated, the disease leads to a slow breakdown of bodily functions and, eventually, death. Sleeping sickness need not be fatal. When the disease is treated in its early stages, most people recover. Treatment is expensive, however, and many of those infected lack medical care. Even if they are cured, they may become infected again.

Governments Work Together

To prevent the spread of sleeping sickness requires a united action on the part of the governments of the many African nations affected by the disease. Thirty-seven African countries lie within the African tsetse belt. This belt covers a total of 10 million sq. km in an area stretching from Senegal to South Africa. African leaders have met in conferences and passed a resolution calling for the eradication of the sleeping-sickness-causing tsetse flies from the continent. Perhaps working together to fight a common enemy will encourage the governments to consult on other regional issues, as well.

▶ Making the Connection

1. Where do tsetse flies live?
2. What causes sleeping sickness?
3. **Drawing Conclusions** Why is treatment of infected humans only part of the solution to eliminating sleeping sickness?

◀ Children in Central Africa have learned to report bites of the tsetse fly.

Reading Review

Section 1 — East Africa: Kenya and Tanzania

Terms to Know

coral reef

poaching

free enterprise system

cassava

sisal

habitat

eco-tourist

Main Idea

Both Kenya and Tanzania are countries of diverse landscapes and peoples.

✓ Place The western part of Kenya is marked by highlands and the wide Great Rift Valley.

✓ Economics Many people in Kenya are farmers. Coffee and tea are grown for export. Tourism is also a major industry in Kenya.

✓ Culture Kenya's people come from many different cultures. They speak Swahili and English.

✓ Economics Farming and tourism are the main economic activities in Tanzania.

✓ Government Tanzania's government has been stable and democratic.

Section 2 — Other Countries of East Africa

Terms to Know

drought

plate

clan

endangered species

genocide

refugee

Main Idea

The other countries of East Africa have all been scarred by conflict in recent years.

✓ History Sudan has been torn by a civil war between the northern Muslim Arabs and the southern Christian peoples.

✓ Human/Environment Interaction In the past, drought has caused severe famine in Ethiopia.

✓ History Fighting between rival clans and drought have caused suffering in Somalia.

✓ History Rwanda and Burundi have suffered a brutal civil war between the Hutu and the Tutsi ethnic groups.

Section 3 — Central Africa

Terms to Know

savanna

canopy

hydroelectric power

tsetse fly

deforestation

Main Idea

Central Africa has rich natural resources that are largely undeveloped because of civil war and poor government decisions.

✓ Movement The Congo River—the second-largest river in Africa—provides transportation and hydroelectric power.

✓ Place The Democratic Republic of the Congo has many resources but has not been able to take full advantage of them.

✓ Government A recent civil war in the Democratic Republic of the Congo overthrew a harsh ruler, but an elected government is not yet in place.

✓ Culture Although English and French are the official languages of the Central African Republic, most of its people speak Sango. Sango is the national language and eases communication among the many ethnic groups.

✓ Economics The economies of Congo and Gabon rely on exports of lumber.

Using Key Terms

Match the terms in Part A with their definitions in Part B.

A.

1. canopy
2. tsetse fly
3. poaching
4. endangered species
5. habitat
6. savanna
7. hydroelectric power
8. drought
9. refugee
10. eco-tourist

B.

a. electricity created by flowing water

b. extended period of extreme dryness

c. tropical grassland with few trees

d. person who flees to another country for safety

e. hunting and killing animals illegally

f. topmost layer of a rain forest

g. people who travel to view natural wonders

h. environment in which an animal species normally lives

i. insect whose bite can cause sleeping sickness

j. plant or animal in danger of dying out

Reviewing the Main Ideas

Section 1 East Africa: Kenya and Tanzania

11. **Culture** What does the term *harambee* mean to Kenyans?
12. **Economics** In which part of Tanzania are cloves produced?
13. **Region** Name some of the animals that live on the Serengeti Plain.

Section 2 Other Countries of East Africa

14. **Economics** How do Burundi and Rwanda get their goods to foreign buyers?
15. **Culture** What two ethnic groups are fighting in Rwanda and Burundi?
16. **Place** What is the largest country in Africa?

Section 3 Central Africa

17. **Economy** How has transportation affected the economy of the Democratic Republic of the Congo?
18. **Human/Environment Interaction** How has being a coastal country helped Cameroon become more industrialized?
19. **Culture** What is the official language of the Central African Republic? Why?
20. **History** Who originally settled São Tomé and Príncipe?

East and Central Africa

Place Location Activity

On a separate sheet of paper, match the letters on the map with the numbered places listed below.

1. Congo River
2. Cameroon
3. Kilimanjaro
4. Central African Republic
5. Sudan
6. Tanzania
7. Democratic Republic of the Congo
8. Somalia
9. Rwanda
10. Gabon

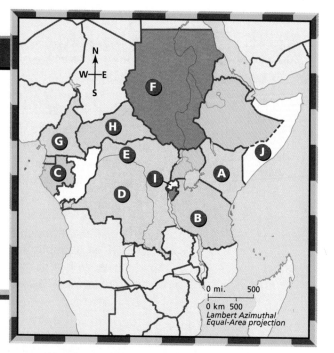

Critical Thinking

21. **Analyzing Information** Much of East Africa depends on agriculture as a main economic activity. Why is a good transportation system important to an agricultural society?

22. **Sequencing Information** In a diagram like the one below, describe and put in order the steps that can lead to the creation of a desert.

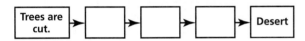

| Trees are cut. | → | | → | | → | | → | Desert |

Current Events Journal

23. **Summarizing Information** Look in the daily newspaper or in a weekly newsmagazine for an article about one of the countries in East or Central Africa. Then create a poster labeled "News Update—Africa Desk." On the poster provide the following information in written or visual form: (1) the topic of the article, (2) the conflict or problem, (3) the major players or factors, (4) possible solutions or consequences, and (5) your comments or thoughts.

Mental Mapping Activity

24. **Focusing on the Region** Create a simple outline map of Africa, then label the following:

- Sudan
- Indian Ocean
- Kenya
- Tanzania
- Democratic Republic of the Congo
- Lake Victoria
- Uganda
- Congo
- Nile River

Technology Skills Activity

25. **Developing a Multimedia Presentation** Choose one of Africa's endangered animals and create a multimedia presentation about it. Include pictures or video clips of the animal, maps of its habitat area, and the steps being taken to protect this animal.

Standardized Test Practice

Directions: Study the map below, and then answer the following question.

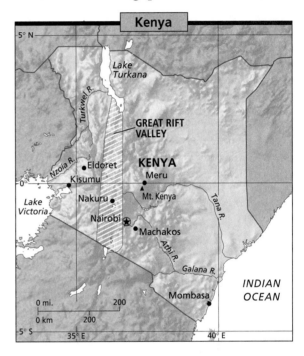

Kenya

1. **About how many miles is it from Nairobi to Mombasa?**

   F 100 miles

   G 200 miles

   H 300 miles

   J 400 miles

Test-Taking Tip: Look carefully at the map key to understand its *scale*, or distance from one point to another. If you find it hard to judge distances by eye, use a small piece of scrap paper to measure the units described in the key.

The World and Its People NATIONAL GEOGRAPHIC

To learn more about the people and places of West Africa, view *The World and Its People* Chapter 19 video.

Our World Today ONLINE

Chapter Overview Visit the *Our World Today: People, Places, and Issues* Web site at owt.glencoe.com and click on **Chapter 16—Chapter Overviews** to preview information about West Africa.

FOLDABLES™
Study Organizer

Summarize Information Study Foldable Make this foldable to determine what you already know, to identify what you want to know, and to summarize what you learn about West Africa.

Step 1 Fold a sheet of paper into thirds from top to bottom.

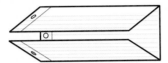

Step 2 Turn the paper horizontally, unfold, and label the three columns as shown.

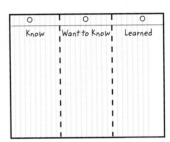

Reading and Writing Before you read the chapter, write what you already know about West Africa under the "Know" tab. Write what you want to know about West Africa under the "Want to Know" tab. Then, as you read the chapter, write what you learn under the "Learned" tab. Be sure to include information you wanted to know (from the second column).

Why It Matters

Cultural Roots

Many African Americans today can trace their roots to West Africa. Enslaved peoples were carried from the "slave coast" to the Americas in the 1600s and 1700s. Liberia was founded as a haven for returning Africans. West Africa also includes Nigeria, the continent's most populous country.

◄ **The Central Mosque of Kano, Nigeria**

Guide to Reading

Main Idea

A large, oil-rich country, Nigeria has more people than any other African nation.

Terms to Know

- mangrove
- savanna
- harmattan
- subsistence farm
- cacao
- compound
- civil war

Reading Strategy

Create a chart like the one below. Then list two facts about Nigeria in each category.

Nigeria	Fact #1	Fact #2
Land		
Economy		
People		

NATIONAL GEOGRAPHIC Exploring Our World

In 1991 Abuja replaced Lagos as Nigeria's capital. The new city was built in an undeveloped region in central Nigeria. Today Abuja has new buildings and a network of roads linking it with other parts of the country. It also has schools. These children prepare to pray at the Islamic Academy in Abuja.

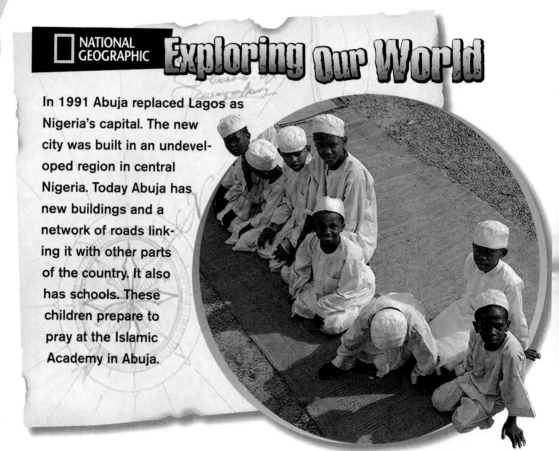

The West African country of **Nigeria** takes its name from the **Niger River,** which flows through western and central Nigeria. One of the largest nations in Africa, Nigeria is more than twice the size of California.

From Tropics to Savannas

Nigeria has a long coastline on the Gulf of Guinea, an arm of the Atlantic Ocean. Along Nigeria's coast, the land is covered with mangrove swamps. A mangrove is a tropical tree with roots that extend both above and beneath the water. As you travel inland, the land becomes vast tropical rain forests. Small villages appear in only a few clearings. The forests gradually thin into savannas in central Nigeria. Savannas are tropical grasslands with only a few trees. Highlands and plateaus also make up this area. Most of the country has high average temperatures and seasonal rains. The grasslands of the far north have a dry climate. In the winter months, a dusty wind called the harmattan blows south from the Sahara.

✓ Reading Check What kinds of vegetation are found in Nigeria?

Economic Challenges

Nigeria is one of the world's major oil-producing countries. More than 90 percent of the country's income comes from oil exports. The government has used money from oil to build highways, schools, skyscrapers, and factories. These factories make food products, textiles, chemicals, machinery, and vehicles. Still, more than one-third of Nigeria's people lack jobs and live in poverty.

Nigeria began to experience economic troubles during the 1980s. As a result of falling world oil prices, Nigeria's income dropped. At the same time, many people left their farms in search of better-paying jobs in the cities. In addition, a few years of low rainfall meant smaller harvests. As a result, food production fell. Nigeria—which had once exported food—had to import food to feed its people.

Despite oil resources, Nigeria's people mainly work as farmers. Most have **subsistence farms,** or small plots that grow just enough

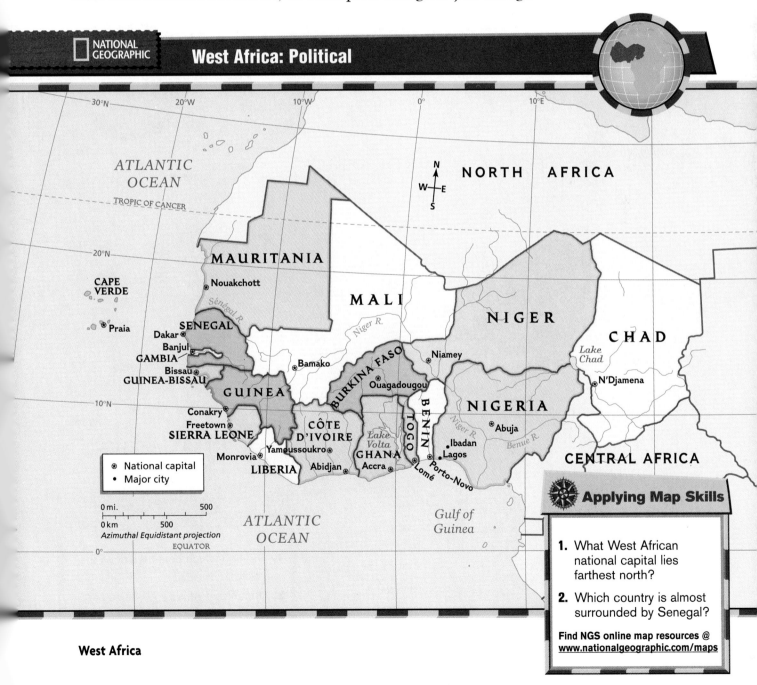

West Africa: Political

NATIONAL GEOGRAPHIC

Applying Map Skills

1. What West African national capital lies farthest north?

2. Which country is almost surrounded by Senegal?

Find NGS online map resources @ www.nationalgeographic.com/maps

West Africa

to feed their families. Some work on larger farms that produce such cash crops as rubber, peanuts, palm oil, and cacao. The **cacao** is a tropical tree whose seeds are used to make chocolate and cocoa. Nigeria is a leading producer of cacao beans.

√ Reading Check How has Nigeria's government used money from oil exports?

Nigeria's People

About 126.6 million people live in Nigeria—more people than in any other country in Africa. The map on page 444 shows that most of the people live along the coast and around the city of **Kano** in the north.

One of the strongest bonds that Africans have is a sense of belonging to a group or family. Nigeria has about 250 ethnic groups. The four largest are the Hausa (HOW•suh), Fulani (foo•LAH•nee), Yoruba (YAWR•uh•buh), and Ibo (EE•boh). Nigerians speak many different

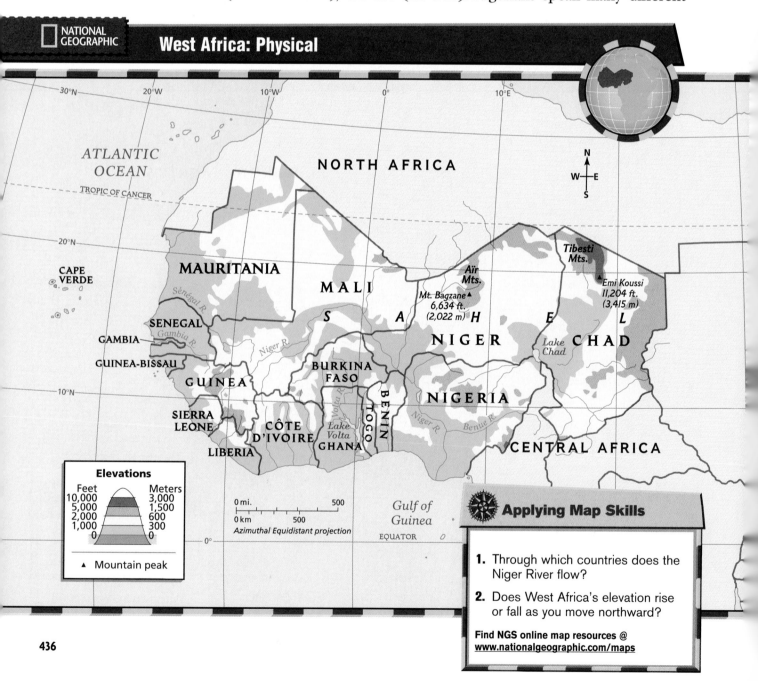

NATIONAL GEOGRAPHIC

West Africa: Physical

ATLANTIC OCEAN

NORTH AFRICA

TROPIC OF CANCER

CAPE VERDE

MAURITANIA

MALI

Tibesti Mts.

Aïr Mts.

Emi Koussi 11,204 ft. (3,415 m)

Mt. Bagzane 6,634 ft. (2,022 m)

SENEGAL

Sénégal R.

S   A   H   E   L

GAMBIA

Gambia R.

Niger R.

NIGER

Lake Chad

CHAD

GUINEA-BISSAU

GUINEA

BURKINA FASO

BENIN

NIGERIA

10°N

SIERRA LEONE

CÔTE D'IVOIRE

Lake Volta

TOGO

GHANA

Niger R.

Benue R.

CENTRAL AFRICA

LIBERIA

Elevations

Feet	Meters
10,000	3,000
5,000	1,500
2,000	600
1,000	300
0	0

▲ Mountain peak

0 mi. 500
0 km 500
Azimuthal Equidistant projection

Gulf of Guinea

EQUATOR

Applying Map Skills

1. Through which countries does the Niger River flow?

2. Does West Africa's elevation rise or fall as you move northward?

Find NGS online map resources @ www.nationalgeographic.com/maps

Nigeria's Economy

Nigerian women canoe past an oil refinery in the Niger River delta (left). Cacao pods are harvested in Nigeria (above).

Human/Environment Interaction What are Nigeria's important cash crops?

African languages. They use English in business and government affairs, though. About one-half of Nigeria's people are Muslim, and another 40 percent are Christian. The remaining 10 percent practice traditional African religions.

About 60 percent of Nigerians live in rural villages. The typical family lives in a **compound,** or a group of houses surrounded by walls. Usually the village has a weekly market run by women. The women sell locally-grown products such as meat, cloth, yams, nuts, and palm oil. The market also provides a chance for friends to meet.

Long-standing rural ways are changing, however. Many young men now move to the cities to find work. The women stay in the villages to raise children and to farm the land. The men, when they are able, return home to see their families and to share the money they have made.

Nigeria's largest city is the port of **Lagos,** the former capital. Major banks, department stores, and restaurants serve the 11 million people who live in Lagos and its surrounding areas. **Ibadan** (EE•bah•DAHN), Kano, and Abuja (ah•BOO•jah) lie inland. **Abuja,** the present capital, is a planned city that was begun during the 1980s.

Nigerians take pride in both old and new features of their culture. Artists make elaborate wooden masks, metal sculptures, and colorful cloth. In the past, Nigerians passed on stories, sayings, and riddles by word of mouth from one generation to the next. In 1986 Nigerian writer **Wole Soyinka** (WAW•lay shaw•YIHNG•ka) became the first African to win the Nobel Prize in literature.

History and Government The earliest known inhabitants of the area were the **Nok** people. They lived between the Niger and Benue Rivers between 300 B.C. and A.D. 200. The Nok were known as skilled metalworkers and traders.

Over the centuries, powerful city-states and kingdoms became centers of trade and the arts. People in the north came in contact with Muslim cultures and adopted the religion of Islam. People in the south developed cultures based on traditional African religions.

During the 1400s, Europeans arrived in Africa looking for gold and Africans to take overseas as enslaved laborers. In 1884 European leaders divided most of Africa into colonies. The borders of these colonies, however, often sliced through ethnic lands. As a result, many ethnic groups found their members living in two or more separate territories. By the early 1900s, the British had taken control of Nigeria.

In 1960 Nigeria finally became an independent country. Ethnic, religious, and political disputes soon tore it apart, however. One ethnic group, the Ibo, tried to set up its own country. A civil war—a fight between different groups within a country—resulted. In this bloody war, starvation and conflict led to 2 million deaths. The Ibo were defeated, and their region remained part of Nigeria.

Nigeria has faced the challenge of building a stable government. Military leaders have often ruled the country. In 1999 Nigerians were able to vote for a president in free elections. Nigerians are continung to work towards greater national unity, but face enormous problems.

Reading Check **What are the four largest ethnic groups in Nigeria?**

Assessment

Defining Terms

1. Define mangrove, savanna, harmattan, subsistence farm, cacao, compound, civil war.

Recalling Facts

2. Place Describe the changes in Nigeria's physical geography as you move from the coast inland.

3. Place What is the capital of Nigeria?

4. Culture How many ethnic groups are represented by the people of Nigeria?

Critical Thinking

5. Understanding Cause and Effect Why did a drop in oil prices cause economic troubles in Nigeria in the 1980s?

6. Drawing Conclusions Why do you think an ethnic group, such as the Ibo, would want to set up their own country?

Graphic Organizer

7. Organizing Information On a time line like the one below, place the following events and their dates in order: Nigeria becomes independent. Nok people work in metal and trade for goods. Free elections are held. British take control of Nigeria.

Applying Social Studies Skills

8. Analyzing Maps Study the physical map on page 436. Into what larger body of water does the Niger River empty?

Critical Thinking Skill

Drawing Inferences and Conclusions

Suppose your teacher brought to class a colorful wooden mask, and a classmate said, "Wow. That's from Nigeria." You might infer that your classmate has an interest in African art and, therefore, recognizes the mask as coming from Nigeria.

Yoruba wood masks ▲

Learning the Skill

To *infer* means to evaluate information and arrive at a conclusion. When you make inferences, you "read between the lines" or draw conclusions that are not stated directly in the text. You must use the available facts *and* your own knowledge and experience to form a judgment or opinion about the material.

Use the following steps to help draw inferences and make conclusions:

- Read carefully for stated facts and ideas.
- Summarize the information and list the important facts.
- Apply related information that you may already know to make inferences.
- Use your knowledge and insight to develop some conclusions about these facts.

Practicing the Skill

Read the passage below, and then answer the questions that follow.

Nigerian art forms reflect the people's beliefs in spirits and nature. Yoruba masks are carved out of wood, reflecting the forces of nature and gods. The masks are used in ceremonies to help connect with the spirit of their ancestors. The masks also appear at funerals in order to please the spirits of the dead. Of all the Yoruba masks, the helmet masks of the Epa cult are the most spectacular.

1. What topic is the writer describing?
2. What facts are presented?
3. What can you infer about the role of masks in Nigerian life?
4. What do you already know about religious ceremonies?
5. What conclusion can you make about traditional religions in Nigeria?

Applying the Skill

Study the photos of Nigerians on page 437. What can you infer about life in Nigeria from the photographs? What evidence supports this inference, or conclusion?

GO TO

Practice key skills with **Glencoe Skillbuilder Interactive Workbook, Level 1.**

Other Countries of West Africa

Guide to Reading

Main Idea

The Sahel countries face a continuing struggle to keep grasslands from turning into desert, but the coastal countries receive plenty of rainfall.

Terms to Know

- overgraze
- drought
- desertification
- bauxite
- phosphate

Reading Strategy

Create five charts like this one, filling in at least one key fact about five West African countries for each category.

Country	
Land	
Economy	
Culture	

NATIONAL GEOGRAPHIC Exploring Our World

Slowly but surely, the desert is creeping into grassy inland areas of West Africa north of Nigeria. Over the past 100 years, a stretch of the Sahara about 100 miles (161 km) wide has swallowed parts of countries in West Africa. This growing desert is like an invading army slowly taking over the countries of the vast Sahel.

Five countries—**Mauritania** (MAWR•uh•TAY•nee•uh), **Mali** (MAH•lee), **Burkina Faso** (bur•KEE•nuh FAH•soh), **Niger** (NY•juhr), and **Chad**—are located in an area known as the **Sahel.** The word *Sahel* comes from an Arabic word that means "border." In addition to the Sahel countries, West Africa includes 11 coastal countries. One country—**Cape Verde**—is a group of islands in the Atlantic Ocean. The other countries, including **Togo** and **Benin,** stretch along the Gulf of Guinea and the Atlantic coast.

Land and History of the Sahel

The Sahel receives little rainfall, so only short grasses and small trees can support grazing animals. Most people have traditionally herded livestock. Their flocks, unfortunately, have overgrazed the land in some places. When animals overgraze land, they strip areas so bare that plants cannot grow back. Then bare soil is blown away by winds.

In the Sahel, dry and wet periods usually follow each other. When the seasonal rains do not fall, drought takes hold. A drought is a long period of extreme dryness and water shortage. The latest drought occurred in the 1980s. Rivers dried up, crops failed, and millions of animals died. Thousands of people died of starvation. Millions of others fled to more productive southern areas. Overgrazing and drought have led to desertification where grasslands have become deserts.

History   From the A.D. 500s to 1500s, three great African empires—**Ghana** (GAH•nuh), **Mali,** and **Songhai** (SAWNG•hy)—arose in the Sahel. These empires controlled the trade in gold, salt, and other goods between West Africa and the Arab lands of North Africa and Southwest Asia. To learn more about the salt trade, turn to page 450.

In the early 1300s, Mali's most famous ruler, **Mansa Musa,** made a journey in grand style to Makkah. This is the holy city of Islam located in the Arabian Peninsula. A faithful Muslim, Mansa Musa made his capital, Tombouctou (TOH•book•TOO), a leading center of Islamic learning. People came from all over the Muslim world to study there.

Invaders from North Africa defeated Songhai—the last of the great empires—in the late 1500s. During the 1800s, the Sahel region came under French rule. The French created five colonies in the area. In 1960 these five colonies became the independent nations of Mauritania, Mali, Upper Volta (now Burkina Faso), Niger, and Chad.

✓ Reading Check   **What has caused the desertification of the Sahel?**

The People of the Sahel

The Sahel countries are large in size but have small populations. If you look at the population map on page 444, you will see that most people live in the southern areas of the Sahel. Rivers flow here, and the land can be farmed or grazed. Yet even these areas do not have enough water and fertile land to support large numbers of people.

Clothing

To protect themselves from the hot Saharan sun, the Tuareg people wear layers of clothing under their long flowing robes. These loose cotton clothes help slow the evaporation of sweat and conserve body moisture. As a sign of respect for their superiors, Tuareg men cover their mouths and faces with veils. Women usually wear veils only for weddings. The veils are made of blue cloth dyed from crushed indigo. The blue dye easily rubs off onto the skin, earning the men the nickname "the Blue Men of the Desert."

Looking Closer How is the clothing of the Tuareg appropriate for the land in which they live?

Today most people in the Sahel live in small towns. They are subsistence farmers who grow grains, such as millet and sorghum (SAWR•guhm). For years, many people were nomads. Groups such as the Tuareg (TWAH•rehg), for example, would cross the desert with herds of camels. The Fulani herded cattle, goats, and sheep. The recent droughts forced many of them to give up their traditional way of life and move to the towns. Here they often live in crowded camps of tents.

Two countries in the Sahel, Mali and Burkina Faso, are among the poorest countries in the world. The landlocked countries of the Sahel have problems getting their products to overseas markets.

The people of the Sahel practice a mix of African, Arab, and European traditions. Most are Muslims and follow the Islamic religion. They speak Arabic as well as a variety of African languages. In many of the larger cities, French is also spoken.

✓ Reading Check **Why have many people in the Sahel given up the nomadic lifestyle?**

The Coastal Countries

If you look at the map on page 444, you will see the Cape Verde Islands off the Atlantic Coast. Skipping to **Senegal,** follow the countries in order around the coast: **Gambia, Guinea-Bissau, Guinea, Sierra Leone, Liberia, Côte d'Ivoire, Ghana,** Togo, and Benin.

Sandy beaches, thick mangrove swamps, and rain forests cover the shores of West Africa's coastal countries. Highland areas with grasses and trees lie inland. Several major rivers flow from these highlands to the coast. They include the Senegal, Gambia, Volta, and Niger Rivers. Rapids and shallow waters prevent large ships from traveling far inland.

Because they border the ocean, the coastal countries receive plenty of rainfall. Warm currents in the Gulf of Guinea create a moist, tropical rain forest climate in most coastal lowlands year-round. For many years, tropical disease, thick rain forests, and river rapids kept European explorers from entering the interior.

Damage to rain forests is also a problem along the densely settled West African coast, where forests have been cleared to make space for palm, coffee, cacao, and rubber plantations, as well as for many small farms. Population pressure on cultivated land is increasing rapidly. Dense clusters of settlements ring port cities such as **Abidjan** (Côte d'Ivoire), **Accra** (Ghana), and Lagos and **Port Harcourt** (Nigeria). Oil discoveries in eastern Nigeria and Gabon are now attracting even more people to the West African coast.

Despite rich agricultural resources, coastal West African countries import more in industrial goods than they export in natural products. Why? Agricultural products often rise and fall in price suddenly, and their value is not equal to finished goods. To meet their countries' needs, governments have to borrow money from other countries or international organizations.

✓ Reading Check **What is attracting people to the West African coast?**

Exploring Economics

Monoculture

The economies of some West African countries, such as Côte d'Ivoire, depend upon the production of one or two major crops. This practice is called monoculture. While this has the advantage of being able to produce enough product to export, it also has disadvantages. If worldwide demand for the product drops, the price also drops. A major drought or epidemic could destroy harvests and wipe out the nation's only major source of income.

History of the Coastal Countries

Ancient Ghana, an empire located at the headwaters of the Senegal and Niger Rivers, flourished between A.D. 700 and 1200. The people of Ghana knew how to make iron weapons, which they used to conquer neighboring groups of farmers and herders. Located on trade routes that connected the gold mines in the rain forests of West Africa with the copper and salt mines in the Sahara, Ghana prospered by taxing the goods that moved north and south along these trade routes. Ghana also had major deposits of gold.

The capital city of ancient Ghana covered a square mile and housed 30,000 people. The kingdom had a well-developed bureaucracy, controlled a large population, and could field an army of 200,000 warriors—at a time when major European battles involved only small numbers of soldiers. The wealth of the king's court was legendary. Europeans called Ghana "the land of gold." This empire fell into decay in the 1200s.

In later times, the powerful and wealthy kingdoms of **Ashanti** and **Abomey** ruled West Africa's coastal region. These kingdoms were centers of trade, learning, and the arts. From the late 1400s to the early 1800s, Europeans set up trading posts along the West African coast. From these posts they traded with Africans for gold, ivory, and other goods that people in Europe wanted.

Effects of Slave Trade Many African states had sold people as slaves to Europeans and Asians long before the Portuguese reached Africa. Most of these slaves were prisoners of war captured in local battles. Only after the development of European sailing ships did the trade in human beings become a major source of income for the kings of West African states, however.

The Europeans also enslaved and forced millions of Africans to migrate to the Americas to work on plantations and in mines. This trade in human beings, which also took place among African countries, was a disaster for West Africa. The removal of so many young and skilled people devastated West African families, villages, and economies.

The French, British, and Portuguese eventually divided up the coastal region and set up colonies to obtain the region's rich resources. In 1957 Ghana became the first country in Africa to become independent. By the late 1970s, no West African country was under European rule.

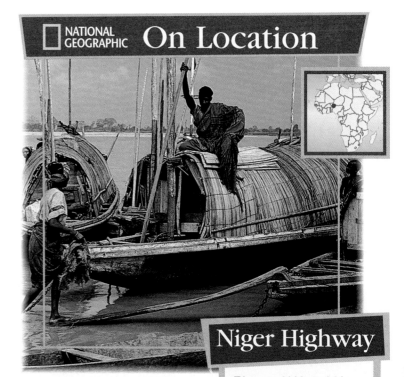

NATIONAL GEOGRAPHIC On Location

Niger Highway

Rivers of West Africa provide not only water but transportation. Here, freight boats on the Niger River deliver goods to Benin's people.

Place What prevents large ships from traveling far inland on West Africa's rivers?

✓ Reading Check What two things allowed ancient Ghana to prosper economically?

West Africa: Population Density

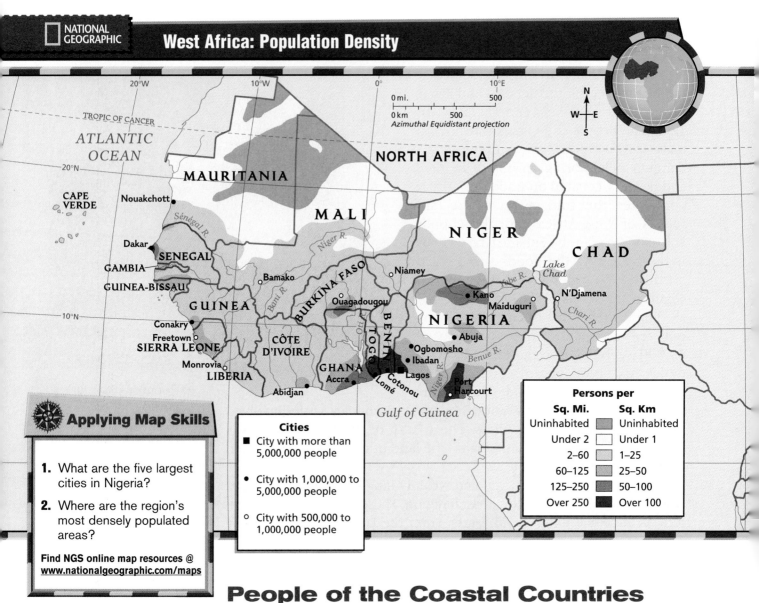

TROPIC OF CANCER

ATLANTIC OCEAN

NORTH AFRICA

0 mi. 500
0 km 500
Azimuthal Equidistant projection

MAURITANIA

MALI

NIGER

CHAD

CAPE VERDE

Nouakchott

Sénégal R.

Dakar

SENEGAL

GAMBIA

GUINEA-BISSAU

GUINEA

Conakry

Freetown
SIERRA LEONE

Monrovia

LIBERIA

Bamako

Niger R.

Bani R.

BURKINA FASO

Ouagadougou

CÔTE D'IVOIRE

GHANA

Accra

Abidjan

Niamey

Oti R.

BENIN

TOGO

Lomé

Cotonou

Lagos

NIGERIA

Kano

Maiduguri

Abuja

Ogbomosho

Ibadan

Benue R.

Port Harcourt

Niger R.

Lake Chad

Yobe R.

N'Djamena

Chari R.

Gulf of Guinea

Cities
■ City with more than 5,000,000 people

● City with 1,000,000 to 5,000,000 people

○ City with 500,000 to 1,000,000 people

Persons per	
Sq. Mi.	**Sq. Km**
Uninhabited	Uninhabited
Under 2	Under 1
2–60	1–25
60–125	25–50
125–250	50–100
Over 250	Over 100

Applying Map Skills

1. What are the five largest cities in Nigeria?

2. Where are the region's most densely populated areas?

Find NGS online map resources @ www.nationalgeographic.com/maps

People of the Coastal Countries

People in coastal West Africa cherish family ties. Some practice traditional African religions, whereas others are Christian or Muslim. Local African languages are spoken in everyday conversation. Reflecting the region's colonial histories, languages such as French, English, and Portuguese are used in business and government. Most of the people in Gambia, Senegal, and Guinea work in agriculture. Guinea is also rich in bauxite and diamonds. **Bauxite** is a mineral used to make aluminum. Guinea has about 25 percent of the world's reserves of bauxite. Senegal is an important source of phosphate. **Phosphate** is mineral salt that has phosphorus, which is used in fertilizers.

Liberia is the only West African nation that was never a colony. African Americans freed from slavery founded Liberia in 1822. **Monrovia,** Liberia's capital, was named for James Monroe—the president of the United States when Liberia was founded. From 1989 to 1996, a civil war cost many lives and destroyed much of the country's economy.

Like Liberia, Sierra Leone was founded as a home for people freed from slavery. The British ruled Sierra Leone from 1787 to 1961. Most of the land is used for farming, but the country also has mineral resources, especially diamonds.

Côte d'Ivoire has a French name that means "ivory coast." From the late 1400s to the early 1900s, a trade in elephant ivory tusks in Côte d'Ivoire brought profits to European traders. Today the ivory trade is illegal, and the country protects its few remaining elephants. The port of Abidjan is the largest urban area and economic center. It has towering office buildings and wide avenues. Abidjan is the official seat of government think, but **Yamoussoukro** (YAH•moo•SOO•kroh), some 137 miles (220 km) inland, has been named the new capital.

Ghana's people belong to about 100 ethnic groups. The Ashanti and the Fante are the largest. Many groups still keep their local kings, but these rulers have no political power. The people respect these ceremonial rulers and look to them to keep traditions alive. About 35 percent of Ghana's people live in cities. Accra, on the coast, is the capital and largest city. A giant dam on the Volta River provides hydroelectric power to urban areas. The dam also has created **Lake Volta,** one of the world's largest artificial lakes.

Web Activity Visit the *Our World Today: People, Places, and Issues* Web site at owt.glencoe.com and click on **Chapter 16— Student Web Activities** to learn more about Liberia.

✓ Reading Check **What are the capitals of Ghana and Côte d'Ivoire?**

Section 2 Assessment

Defining Terms

1. Define overgraze, drought, desertification, bauxite, phosphate.

Recalling Facts

2. History What three great empires ruled in the Sahel from A.D. 500 to 1500?

3. Economics What types of goods must West African countries import?

4. History Which West African country was never a colony?

5. Government How much political power do the local kings in Ghana have?

Critical Thinking

6. Making Predictions What problems do you think will arise as people move from the Sahel to more productive areas?

7. Drawing Conclusions Why do governments of coastal West African countries have to borrow money?

8. Analyzing Information What West African products do you use?

Graphic Organizer

9. Organizing Information On a diagram like this one, record at least three different facts about the three ancient African empires of Ghana, Mali, and Songhai.

Ghana	Mali	Songhai

Applying Social Studies Skills

10. Analyzing Maps Study the population density map on page 444. Why would you expect the heavy population centers to be located along the coast?

Making Connections

CULTURE GOVERNMENT PEOPLE TECHNOLOGY

Great Mosque of Djenné

In the West African city of Djenné (jeh•NAY), Mali, stands a huge structure built entirely of mud. It is the Great Mosque of Djenné, and it covers an area the size of a city block. Considered one of Africa's greatest architectural wonders, the existing Great Mosque is actually the third mosque to occupy the location.

Djenné

Located between the Sahara and the African savanna, the city of Djenné was an important crossroads on a trade route connecting northern and southern Africa. Caravans and boats carried gold, salt, and other goods through the city.

During the A.D. 1200s, the ruler of Djenné ordered the construction of the first Great Mosque. Having recently converted to Islam, he had his palace torn down to make room for the huge house of worship. The city became an important Islamic religious center. Over the years, political and religious conflicts led to a decline in the city. People abandoned the Great Mosque, and a second, much smaller one replaced it. Then in 1906, builders began to raise a new Great Mosque. Today the Great Mosque is once more an important part of the religious life of the people of Djenné.

The Great Mosque

Built facing east toward Makkah, the holy city of Islam, the Great Mosque of Djenné is constructed from the same sun-dried mud bricks as most of the rest of the city. The mud walls of the mosque vary in thickness between 16 and 24 inches (41 and 61 cm), providing insulation to keep the interior cool. Roof vents can be removed at night to allow cooler air inside.

With its five stories and three towers, or minarets, the mosque rises above the surrounding buildings. Inside the mosque, the main prayer hall is open to the sky. Although the mosque contains loudspeakers used to issue the call to prayer, there are few other modern improvements.

Maintaining the Mosque

Rain, wind, and heat can damage mud structures, and without care the Great Mosque would soon deteriorate. Each spring the people of Djenné plaster the mosque from top to bottom with fresh mud. It is a great festival day, and nearly everyone volunteers. Workers climb up the sides of the mosque on wooden rods permanently mounted to the walls. They dump mud and water onto the walls, then smooth it with their bare hands. The townspeople know that, with such care, the Great Mosque will remain a place of worship for generations to come.

→ Making the Connection

1. When was the first Great Mosque built?
2. What elements of the Great Mosque help keep the inside cool?
3. **Making Comparisons** In what way is the Great Mosque like the other buildings in Djenné? In what way is it different?

Section 1 | Nigeria

Terms to Know

mangrove
savanna
harmattan
subsistence farm
cacao
compound
civil war

Main Idea

A large, oil-rich country, Nigeria has more people than any other African nation.

✓ Place Nigeria's major landforms are coastal lowlands, savannas, highlands, plateaus, and partly dry grasslands.

✓ Economics More than 90 percent of Nigeria's income comes from oil exports.

✓ Culture Nigeria has about 250 ethnic groups. The four largest ethnic groups are the Hausa, Fulani, Yoruba, and Ibo.

Section 2 | Other Countries of West Africa

Terms to Know

overgraze
drought
desertification
bauxite
phosphate

Main Idea

The Sahel countries face a continuing struggle to keep grasslands from turning into desert, but the coastal countries receive plenty of rainfall.

✓ Region The Sahel countries are Mauritania, Mali, Niger, Chad, and Burkina Faso.

✓ Region The Sahel receives little rainfall, so only short grasses and small trees can support grazing animals.

✓ Human/Environment Interaction Overgrazing and drought have caused many grassland areas in this region to become desert.

✓ Region The 11 countries that make up coastal West Africa are Senegal, Gambia, Guinea, Guinea-Bissau, Cape Verde, Liberia, Sierra Leone, Côte d'Ivoire, Ghana, Togo, and Benin.

✓ Economics West Africa's coastal countries import more in industrial goods than they export in natural products.

The port of Abidjan, Côte d'Ivoire ▶

Assessment and Activities

 ## Using Key Terms

Match the terms in Part A with their definitions in Part B.

A.

1. overgraze
2. harmattan
3. drought
4. mangrove
5. compound
6. phosphate
7. desertification
8. cacao
9. subsistence farm
10. savanna

B.

a. process in which deserts expand
b. a group of houses surrounded by a wall
c. a dusty wind that blows south from the Sahara
d. mineral salt used in fertilizers
e. tropical tree whose seeds are used to make cocoa and chocolate
f. tropical grassland with scattered trees
g. produces enough to support a family's needs
h. extended period of extreme dryness
i. when animals strip the land so bare that plants cannot grow
j. tropical tree with roots above and beneath the water

 ## Reviewing the Main Idea

Section 1 Nigeria

11. **Economics** What is Nigeria's major export?
12. **History** Why have there been so many conflicts in Nigeria since 1960?
13. **Culture** Who was the first African to win the Nobel Prize in literature?
14. **Culture** What are the four largest ethnic groups in Nigeria?

Section 2 Other Countries of West Africa

15. **Region** What is the meaning of the word *Sahel?*
16. **History** Who was Mansa Musa?
17. **Culture** What religion do most people of the Sahel follow?
18. **History** What has led to desertification in the Sahel?
19. **Movement** Why are ships unable to sail very far inland in coastal West Africa?
20. **History** What was the slave trade?
21. **History** What early kingdom was called "the land of gold"?
22. **Culture** What are the largest ethnic groups in Ghana?

 **West Africa**

Place Location Activity

On a separate sheet of paper, match the letters on the map with the numbered places listed below.

1. Gulf of Guinea
2. Nigeria
3. Niger River
4. Liberia
5. Cape Verde
6. Lagos
7. Mali
8. Ghana
9. Chad
10. Monrovia

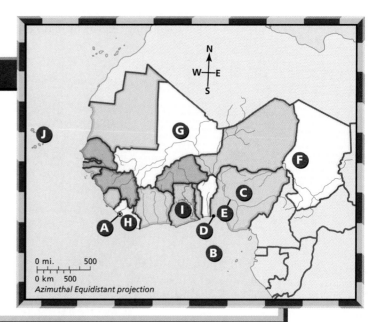

Our World Today Online

Self-Check Quiz Visit the *Our World Today: People, Places, and Issues* Web site at <u>owt.glencoe.com</u> and click on **Chapter 16—Self-Check Quizzes** to prepare for the Chapter Test.

Critical Thinking

23. **Evaluating Information** What do you feel is the major challenge facing the countries of West Africa today? Explain your answer.

24. **Sequencing Information** After reviewing the entire chapter, choose what you feel are five of the most important events in the history of West Africa. Place those events and their dates on a time line like this one.

Current Events Journal

25. **Writing a Poem** Read a newspaper or magazine article about one of the countries of West Africa. Imagine that you are there and write an "I am ..." poem. Begin each line with the words "I am ..." and then complete it with a description, action, or emotion that you feel relates to the subject. Share your poem with the rest of the class.

Mental Mapping Activity

26. **Focusing on the Region** Create a simple outline map of West Africa, then label the following:

- Niger River
- Senegal
- Atlantic Ocean
- Côte d'Ivoire
- Gulf of Guinea
- Chad
- Tropic of Cancer
- Mali
- Nigeria
- Mauritania
- Niger
- Liberia

Technology Skills Activity

27. **Using the Internet** Conduct a search for information about one of the ancient empires or kingdoms of West Africa. Look for maps, pictures, and descriptions of the important rulers. Then write a report using the information you found. Share your report with the rest of the class.

Standardized Test Practice

Directions: Study the graph, and then answer the question that follows.

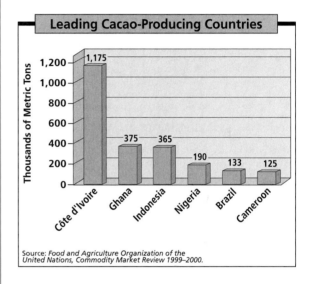

Leading Cacao-Producing Countries

Source: *Food and Agriculture Organization of the United Nations, Commodity Market Review 1999–2000.*

1. **What countries on the graph are leading cacao-producing countries from West Africa?**

   A Ghana, Indonesia, and Nigeria

   B Côte d'Ivoire, Ghana, and Indonesia

   C Côte d'Ivoire, Nigeria, and Cameroon

   D Côte d'Ivoire, Ghana, and Nigeria

Test-Taking Tip: The important words in this question are "from West Africa." You need to use information on the graph as well as information you learned in Chapter 16 to answer this question. As with any graph, read the title bar and information along the side and bottom of the graph first. Then analyze and compare the sizes of the bars to one another.

People trade salt and other goods at a market on the Niger River in ancient Africa.

PLEASE PASS THE SALT:
Africa's Salt Trade

Passing the salt at dinner may not be a big deal, but in parts of Africa, salt built empires. How did such a basic substance come to play such an important role in Africa?

Good as Gold

Salt is essential for life. Every person contains about 8 ounces (227 g) of salt—enough to fill several saltshakers. Salt helps muscles work, and it aids in digesting food. In hot climates, people need extra salt to replace the salt lost when they sweat. In tropical Africa, salt has always been precious.

Salt is plentiful in the Sahara and scarce in the forests south of the Sahara (in present-day countries such as Ghana and Côte d'Ivoire). These conditions gave rise to Africa's salt trade. Beginning in the A.D. 300s, Berbers drove camels carrying European glassware and weapons from Mediterranean ports into the Sahara. At the desert's great salt deposits, such as those near the ancient sites of Terhazza and Taoudenni, they traded European wares for salt.

The salt did not look like the tiny crystals in a saltshaker. It was in the form of large slabs, as hard as stone. The slabs were pried from hardened salt deposits that were left on the land long ago when landlocked seas evaporated. The salt slabs were loaded onto camels, and the animals were herded south. To people in the south, salt was literally worth its weight in gold. The slabs were cut into equal-sized blocks and exchanged for gold and other products such as ivory and kola nuts. Salt was also traded for enslaved people.

Rise and Decline

Before camels arrived in Africa from Asia in A.D. 300, only a trickle of trade, mostly carried by human porters, made it across the blistering desert. In time, caravans of thousands of camels loaded with tons of salt arrived at southern markets.

Local kings along the trade routes put taxes—payable in gold—on all goods crossing their realms. The ancient empires of Mali, Ghana, and Songhai rose to great power from wealth brought by the salt trade.

Trade routes also provided avenues for spreading ideas and inventions. By the A.D. 800s, Arab traders brought to Africa a system of weights and measures, a written language, and the concept of money. They also brought a new religion: Islam.

Today trucks have replaced many of the camels. Salt no longer dominates trade in the region. However, salt is still important, and the salt trade continues in Mali and in the markets of other West African nations.

QUESTIONS

1 What goods were exchanged in the salt trade?

2 How did the salt trade affect regions south of the Sahara?

A present-day salt caravan in Niger ▶

NATIONAL
GEOGRAPHIC

Salt Trade Routes

ATLANTIC OCEAN

Mediterranean Sea

0 mi. 1,000

0 km 1,000

Taghaza

SAHARA

Taoudenni

ASIA

Niger R.

Red Sea

N
W E
S

NIGER

Lake Chad

CÔTE D'IVOIRE   GHANA

Songhai · Salt deposit
Mali · ---▸ Trade routes
Ghana · — Present boundaries

The World
and Its People

NATIONAL
GEOGRAPHIC

To learn more about the people and
places of South Africa and its
neighbors, view **The World and Its
People Chapter 22** video.

Our World Today
online

Chapter Overview Visit the **Our World
Today: People, Places, and Issues**
Web site at owt.glencoe.com and click on
Chapter 17—Chapter Overviews to
preview information about South Africa and
its neighbors.

Why It Matters

New Challenges

In spite of its beautiful scenery and hospitable people, for years South Africa was virtually isolated from the world community because of its racist policies. Today, after decades of struggling for justice and equality, South Africa's new leaders have challenges of a different kind. Poverty and the spread of AIDS plague the lives of many of this nation's people.

FOLDABLES™
Study Organizer

Categorizing Information Study Foldable Make the following foldable to help you organize data about historic and modern events that have occurred in the countries of southern Africa.

Step 1 Fold a sheet of paper from side to side, leaving a 2-inch tab uncovered along the side.

Fold it so the left edge lays 2 inches from the right edge.

Step 2 Turn the paper and fold it into thirds.

Step 3 Unfold the paper and cut along the two inside fold lines.

Cut along the two folds on the front flap to make 3 tabs.

Step 4 Label the foldable as shown.

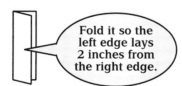

Southern Africa

| Republic of South Africa | Inland Southern Africa | Atlantic/ Indian Ocean Countries |

Reading and Writing As you read the chapter, write information under each appropriate tab of your foldable to record past and present events that have affected the countries and cultures of southern Africa.

◄ Table Mountain overlooks Cape Town, South Africa.

The New South Africa

Guide to Reading

Main Idea
Rich in resources, South Africa has recently seen major social and political changes.

Terms to Know
- industrialized country
- Boer
- apartheid
- township
- enclave

Reading Strategy
Create a time line like this one. Then list five key events and their dates in South Africa's history.

NATIONAL GEOGRAPHIC Exploring Our World

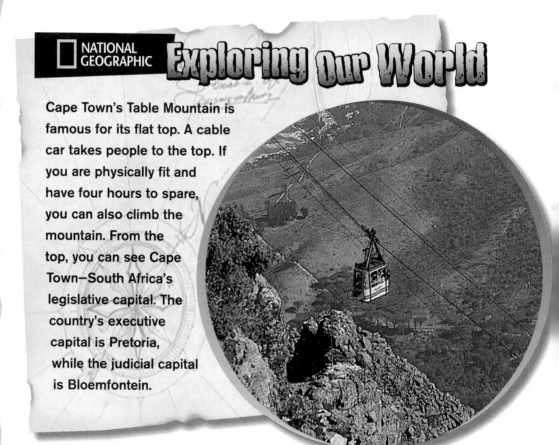

Cape Town's Table Mountain is famous for its flat top. A cable car takes people to the top. If you are physically fit and have four hours to spare, you can also climb the mountain. From the top, you can see Cape Town—South Africa's legislative capital. The country's executive capital is Pretoria, while the judicial capital is Bloemfontein.

South Africa (officially called the **Republic of South Africa**) is a land of beautiful scenery and great mineral wealth. It is also a land of great change. Here you will find the continent's biggest mammal, the African elephant, and smallest mammal, the miniature shrew. To protect these creatures, the government has set aside land as national parks.

A Land Rich in Resources

South Africa borders the Atlantic Ocean on the west and the Indian Ocean on the south and east. The **Namib Desert** reaches into the northwest. The **Cape of Good Hope** is the southernmost point of Africa.

South Africa is the most industrialized country in Africa. An **industrialized country** is one in which a great deal of manufacturing occurs. Not all South Africans benefit from this prosperous economy, however. In rural areas, many people continue to depend on subsistence farming and live in poverty.

South Africa's economy is supported in several ways. In terms of mineral resources, South Africa is one of the richest countries in the world. It is the world's largest producer and exporter of gold. South Africa also has large deposits of diamonds, chromite, platinum, and coal. The country also exports machinery, chemicals, clothing, and processed foods. Among the crops cultivated are corn, wheat, fruits, cotton, sugarcane, and potatoes. Ranchers on the high plains raise sheep, cattle for beef, and dairy cows.

√ Reading Check How have South Africa's resources helped its economy?

South Africa's History and People

About 43.6 million people live in South Africa. Black African ethnic groups make up about 78 percent of the population. Most trace their ancestry to Bantu-speaking peoples who settled throughout Africa between A.D. 100 and 1000. The largest groups in South Africa today are the **Sotho** (SOO•too), **Zulu,** and **Xhosa** (KOH•suh).

In the 1600s, the Dutch settled in South Africa. They were known as the Boers, a Dutch word for farmers. German, Belgian, and French settlers joined them. Together these groups were known as Afrikaners

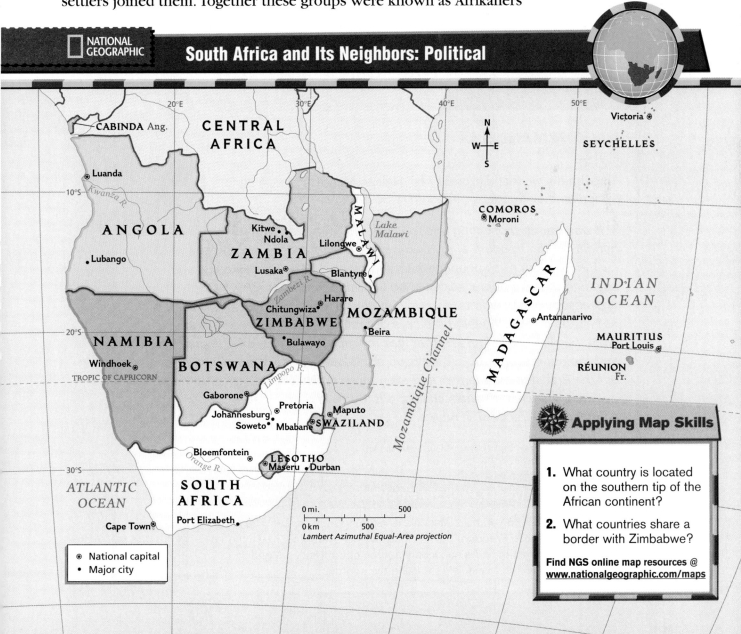

NATIONAL GEOGRAPHIC

South Africa and Its Neighbors: Political

⊛ National capital
• Major city

0 mi. 500
0 km 500
Lambert Azimuthal Equal-Area projection

Applying Map Skills

1. What country is located on the southern tip of the African continent?

2. What countries share a border with Zimbabwe?

Find NGS online map resources @ www.nationalgeographic.com/maps

and spoke their own language—Afrikaans (A•frih•KAHNS). They pushed Africans off the best land and set up farms and plantations. They brought many laborers from India to work on sugar plantations.

The British first came to South Africa in the early 1800s. Later, the discovery of diamonds and gold attracted many more British settlers. Tensions between the British and the Afrikaners resulted in the 1902 defeat of the Afrikaners in the Boer War. In 1910 Afrikaner and British territories became the Union of South Africa. It was part of the British Empire and was ruled by whites. Black South Africans founded the **African National Congress** (ANC) in 1912 in hopes of gaining power.

In 1948 the whites set up a system of apartheid, or "apartness." Apartheid (uh•PAHR•TAYT) made it illegal for different races and ethnic groups to mix, thus limiting the rights of blacks. For example, laws forced black South Africans to live in separate areas, called "homelands." People of non-European background were not even allowed to vote.

For more than 40 years, people inside and outside South Africa protested against the practice of apartheid. Many black Africans were jailed for their actions in the long struggle for justice and equality.

Primary Source

NELSON MANDELA
(1918–)
Through his work and experiences, Nelson Mandela has come to symbolize the struggle for freedom in South Africa.

"It was during those long and lonely years [in prison] that my hunger for the freedom of my own people became a hunger for the freedom of all people, white and black. I knew as well as I knew anything that the oppressor must be liberated just as surely as the oppressed. A man who takes away another man's freedom is a prisoner of hatred, he is locked behind the bars of prejudice and narrow-mindedness. When I walked out of prison, that was my mission, to liberate the oppressed and the oppressor both. . . . We have not taken the final step of our journey, but the first step on a longer and even more difficult road. For to be free is not merely to cast off one's chains, but to live in a way that respects and enhances the freedom of others."

From *The Long Walk to Freedom*, the autobiography of Nelson Mandela.

Analyzing Primary Sources

What do you think Mandela is referring to in his title *The Long Walk to Freedom*? What is the walk a symbol for? Use the Internet to find a secondary source about Nelson Mandela. Research Mandela's life using the source.

The United Nations declared that apartheid was "a crime against humanity." Many countries cut off trade with South Africa. Finally, in 1991 apartheid ended. South Africa held its first democratic election in April 1994, electing **Nelson Mandela** as the first black president.

The People   South Africa has 11 official languages, including Afrikaans, English, Zulu, and Xhosa. About two-thirds of South Africans are Christians while the rest practice traditional African religions.

One of the challenges facing South Africa today is to develop a better standard of living for its poorer people. Most European South Africans live in modern homes and enjoy a high standard of living. Most black African, Asian, and mixed-group South Africans live in rural areas and crowded townships, or neighborhoods outside cities. The government has introduced measures to improve education and basic services.

Another challenge facing South Africa is the AIDS epidemic. Millions of people throughout Africa have been infected with the virus that causes AIDS. South Africa is one of the countries hit hardest.

Lesotho and Swaziland   Within South Africa lie two other African nations—**Lesotho** (luh•SOO•too) and **Swaziland.** These tiny kingdoms are enclaves—small countries located inside a larger country. Both are poor countries that depend heavily on South Africa. Lesotho's only natural resource is water, some of which it sells to South Africa. Many of Lesotho's and Swaziland's people are engaged in subsistence farming. Others work in mines in South Africa.

 **Reading Check**   How do Lesotho and Swaziland earn money from South Africa?

Section 1 Assessment

Defining Terms

1. Define industrialized country, Boer, apartheid, township, enclave.

Recalling Facts

2. Place Name the largest mammal and the smallest mammal in Africa.

3. Government Who is Nelson Mandela?

4. Culture What challenges face South Africa?

Critical Thinking

5. Drawing Conclusions How did the rest of the world view apartheid?

6. Analyzing Information Why do you think workers in Lesotho and Swaziland travel to South Africa to work in mines?

Graphic Organizer

7. Organizing Information In a chart like the one below, write the resources and products of South Africa in the two boxes.

South Africa	
Resources	Products

 Applying Social Studies Skills

8. Analyzing Maps Study the political map on page 455. What are the three national capitals of the Republic of South Africa?

Making Connections

Mining and Cutting Diamonds

A diamond is a mineral made entirely of carbon. It is the hardest known substance on the earth and the most popular gemstone. Most diamonds formed billions of years ago deep inside the earth's mantle. There, intense pressure and heat transformed carbon into diamond crystal.

Mining

There are two major techniques used for mining diamonds: open pit and underground mining. In open pit mining, the earth is dug out in layers, creating a series of steps or roads that circle down into a pit. After drills and explosives loosen the rock containing diamonds, shovels and trucks remove it. When the pit becomes too deep to reach easily, underground mining may begin.

Underground mining requires sinking a shaft into the ground and tunneling to the rock. Explosives blast the rock loose, and the resulting rubble is crushed and carried to the surface for further processing.

To remove the diamonds, the crushed rock is mixed with water and placed in a washing pan. Heavier minerals, such as diamonds, settle to the bottom, while lighter wastes rise to the top and overflow. Next, the heavier mixture travels to a grease table. Diamonds cling to the grease while other wetted minerals flow past. Workers continue the sorting and separating by hand.

Cutting

The newly mined diamond resembles a piece of glass, not a sparkling jewel. To enhance their brilliance and sparkle, gem-quality diamonds are precisely cut and polished. The cutter uses high-speed diamond-tipped tools to cut facets, or small flat surfaces, into the stone. One of the most popular diamond cuts is the brilliant cut, which has 58 facets. The job of the cutter requires extreme skill, because the angles of the facets must be exactly right to bring out the diamond's beauty.

Making the Connection

1. Of what are diamonds made?
2. Why are gemstone diamonds cut and polished?
3. **Making Comparisons** How are open pit and underground diamond mining techniques alike? How are they different?

NATIONAL GEOGRAPHIC

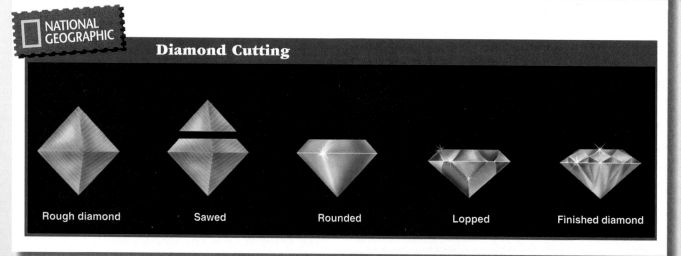

Diamond Cutting

Rough diamond   Sawed   Rounded   Lopped   Finished diamond

Inland Southern Africa

Guide to Reading

Main Idea

Most of inland southern Africa is rich in resources and home to a wide variety of ethnic groups.

Terms to Know

- copper belt
- sorghum

Reading Strategy

Create a chart like this one. Then list the main economic activities of each country.

Country	Economic Activities
Zambia	
Malawi	
Zimbabwe	
Botswana	

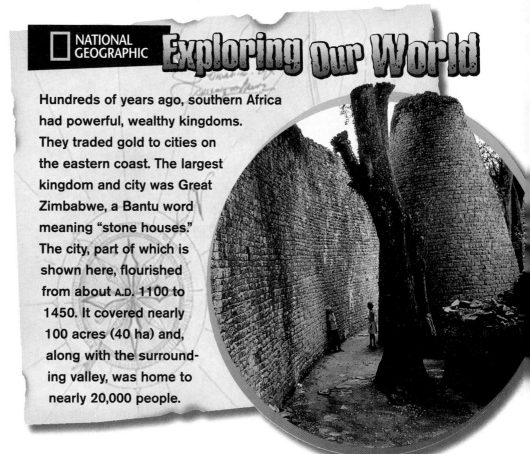

NATIONAL GEOGRAPHIC Exploring Our World

Hundreds of years ago, southern Africa had powerful, wealthy kingdoms. They traded gold to cities on the eastern coast. The largest kingdom and city was Great Zimbabwe, a Bantu word meaning "stone houses." The city, part of which is shown here, flourished from about A.D. 1100 to 1450. It covered nearly 100 acres (40 ha) and, along with the surrounding valley, was home to nearly 20,000 people.

The four countries of inland southern Africa include **Zambia, Malawi** (mah•LAH•wee), **Zimbabwe,** and **Botswana** (baht•SWAH•nah). They have several things in common. First, they all are landlocked. A high plateau dominates much of their landscape and gives them a mild climate. In addition, about 70 percent of the people practice subsistence farming in rural villages. Thousands move to cities each year to look for work.

Zambia

Zambia is slightly larger than Texas. The **Zambezi** (zam•BEE•zee) **River**—one of southern Africa's longest rivers—crosses the country. The Kariba Dam—one of Africa's largest hydroelectric projects—spans the Zambezi River. Also along the Zambezi River are the spectacular Victoria Falls, named in honor of British Queen Victoria, who ruled in the 1800s. The falls are known locally as *Mosi oa Tunya*—or "smoke that thunders."

A large area of copper mines, known as a copper belt, stretches across northern Zambia. One of the world's major producers of copper, Zambia relies on it for more than 80 percent of its income. As a result, when world copper prices go down, Zambia's income goes down too. As copper reserves dwindle, the government has encouraged city dwellers to return to farming. Zambia must import much of its food.

Once a British colony, Zambia gained its independence in 1964. The country's 9.8 million people belong to more than 70 ethnic groups and speak many languages. English is the official language. Those who live in urban areas such as **Lusaka,** the capital, work in mining and service industries. Villagers grow corn, rice, and other crops to support their families. Their main food is porridge made from corn.

✓ Reading Check What happens to Zambia when copper prices go down?

Malawi

If you travel through narrow Malawi, you see green plains and grasslands in western areas. Vast herds of elephants, zebras, and antelope roam national parks and animal reserves here.

The Great Rift Valley runs through eastern Malawi. In the middle of it lies beautiful **Lake Malawi.** This lake holds about 500 fish species, more than any other inland body of water in the world. Malawi is also famous for its more than 400 orchid species.

Malawi has few mineral resources and little industry. Tobacco, tea, sugar, coffee, and peanuts are exported. Farmers also grow sorghum, a tall grass whose seeds are used as grain and to make syrup. Donations, loans, and foreign aid help support Malawi's people.

Bantu-speaking people arrived in the area about 2,000 years ago, bringing with them knowledge of iron working. The most famous European explorer to reach Malawi was the Scottish missionary **David Livingstone** during the mid-1800s. Today most people in Malawi are Protestant Christians as a result of the teachings of missionaries.

In 1964 the British colony became independent. Malawi has recently returned to democratic government after a long period of rule by a dictator. After years of harsh government, the works of many modern writers emphasize themes such as human rights and abuse of power.

Malawi is one of the most densely populated countries in Africa. It has about 229 people per square mile (88 people per sq. km). Jobs are scarce, so thousands of men seek work in South Africa and Zambia.

✓ Reading Check What types of landforms cover western Malawi?

Zimbabwe

Crossing Zimbabwe, you might think you were in the western United States. The vast plateau is studded with large outcrops of rock. The **Limpopo River** winds through southern lowlands. The Zambezi River crosses the north.

Mining gold, copper, iron ore, and asbestos provides most of the country's income. Some large plantations grow coffee, cotton, and

Exploring GOVERNMENT

Democracy in Action

AIDS is a serious problem for many African countries. The governments of some countries, such as Zimbabwe, do not have the resources to deal with the disease. Other countries, such as Botswana, are working with the international community to combat the spread of HIV, the virus that causes AIDS. Because of Botswana's stable, democratic government, clinics have been established, roads are well maintained, and medical supplies can be quickly distributed. Botswana has a good chance for a successful treatment program.

Music

The talking drum is a popular instrument in Africa south of the Sahara. Animal skins cover both ends and are held together by strings. While holding the drum under the arm, the musician strikes one skin with a curved wooden mallet. By squeezing down on the strings with the arm, the skins are stretched tighter and the pitch of the drum becomes higher. The loosening and tightening of the strings give the drum its characteristic "talking" sound.

Looking Closer How is the talking drum different from the traditional drums in the United States?

GO TO

World Music: A Cultural Legacy
Hear music of this region on Disk 2, Track 1.

tobacco. Europeans own many of the large plantations, while many Africans work only small plots. Since the 1980s, the government has tried to redistribute land to Africans. Progress has been slow, and with President Robert Mugabe's support, protesters recently took over large and small European-owned farms to force changes.

Another serious challenge to Zimbabwe's economy comes from the spread of AIDS. People who have the disease often cannot work to support their families. Many children have been orphaned by AIDS. The government lacks the means to effectively deal with the AIDS crisis.

Zimbabwe takes its name from an ancient African city and trading center—**Great Zimbabwe.** This remarkable stone fortress was built by an ethnic group called the Shona in the A.D. 1100s to 1400s. The **Shona** and the **Ndebele** (ehn•duh•BEH•leh) ruled large stretches of south-central Africa until the late 1800s. In the 1890s, the British controlled the area and called it Rhodesia. They named it after **Cecil Rhodes,** a British businessman who expanded British rule in Africa.

Europeans ran Rhodesia and owned all the best farmland. In response, the Africans organized into political groups and fought European rule. In 1980 free elections brought an independent African government to power. The country was renamed Zimbabwe.

Today Zimbabwe has about 12 million people. Most of them belong to the Shona and Ndebele ethnic groups. About a fourth of the population is Christian. Many residents practice traditional African religions. The largest city is **Harare** (hah•RAH•ray), the capital.

✓ Reading Check How has AIDS affected Zimbabwe's economy?

Web Activity Visit the ***Our World Today: People, Places, and Issues*** Web site at owt.glencoe.com and click on **Chapter 17— Student Web Activities** to learn more about Zimbabwe.

Botswana

Botswana lies in the center of southern Africa. The vast **Kalahari Desert** spreads over southwestern Botswana. This hot, dry area has rolling, red sands and low, thorny shrubs. The Okavango River in the northwest forms one of the largest swamp areas in the world. This area of shifting streams has much wildlife.

Botswana's national emblem (as well as its basic monetary unit) is a one-word motto—*Pula*—meaning "rain." In Botswana, there is never much of it. From May to October, the sun bakes the land. Droughts strike often, and many years can pass before the rains fall again.

Botswana is rich in mineral resources. Diamonds account for more than 75 percent of Botswana's export income. Thousands of tourists visit Botswana's game preserves every year. Farming is difficult, and the country grows only about 50 percent of its food needs. It must import the rest. To earn a living, many people work in South Africa for several months a year.

After nearly 80 years of British colonial rule, Botswana became independent in 1966. Today it has one of Africa's strongest democracies. Many of Botswana's people are Christians, although a large number practice traditional African religions. The official language is English, but 90 percent of the population speak an African language called Setswana. **Gaborone** is the capital and largest city.

✓ Reading Check **What is Botswana's biggest source of export income?**

Section 2 Assessment

Defining Terms
1. **Define** copper belt, sorghum.

Recalling Facts
2. **Economics** What is Zambia's most important export?
3. **Place** What makes Lake Malawi unique?
4. **Culture** Where did Zimbabwe get its name?

Graphic Organizer
5. **Organizing Information** Choose two of the countries in this section. Put the name and three facts about each country in the outer ovals. Where the ovals overlap, put facts that are true of both countries.

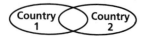

Critical Thinking
6. **Synthesizing Information** Imagine that someone from Great Zimbabwe traveled to that country today. What do you think he or she would describe as the greatest difference between then and now?
7. **Analyzing Information** Why do you think the people of Botswana chose *Pula*, or "rain," as their motto?

Applying Social Studies Skills

8. **Analyzing Maps** Study the political map on page 455. What five African nations does the Tropic of Capricorn cross?

Technology Skill

Developing Multimedia Presentations

Your geography homework is to make a presentation about Botswana. You want to make your presentation informative but also interesting and fun. How can you do this? One way is to combine several types of media into a **multimedia presentation.**

Learning the Skill

A multimedia presentation involves using several types of media, including photographs, videos, or sound recordings. The equipment can range from simple cassette players to overhead projectors to VCRs to computers and beyond. In your presentation on Botswana, for example, you might show photographs of cheetahs in the Okavango River delta or women fishing. You could also play a recording of local music or find a video of people working in diamond mines. If you have the proper equipment, you can then combine all these items on a computer.

Computer multimedia programs allow you to combine text, video, audio, art or graphics, and animation. The tools you need include computer graphic and drawing programs, animation programs that make certain images move, and systems that tie everything together. Your computer manual will tell you which tools your computer can support.

Practicing the Skill

Use the following questions as a guide when planning your presentation:

1. Which forms of media do I want to include? Video? Sound? Animation? Photographs? Graphics?
2. Which of the media forms does my computer support?
3. Which kinds of media equipment are available at my school or local library?
4. What types of media can I create to enhance my presentation?

Applying the Skill

Plan and create a multimedia presentation on a country discussed in this unit. List three ideas you would like to cover. Use as many multimedia materials as possible and share your presentation with the class.

▼ Various equipment is needed to make multimedia presentations. For example, photographs and videos of cheetahs will make your report on Botswana more interesting.

Atlantic and Indian Ocean Countries

Guide to Reading

Main Idea

Africa's Atlantic and Indian Ocean countries are struggling to develop their economies.

Terms to Know

- exclave
- slash-and-burn farming
- deforestation
- cyclone

Reading Strategy

Create a diagram like this one. Choose one Atlantic and one Indian Ocean country. Write facts about each in the outer ovals under its heading. Where the ovals overlap, write statements that are true of both countries.

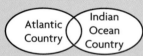

Atlantic Country | Indian Ocean Country

NATIONAL GEOGRAPHIC *Exploring Our World*

Ostriches, lions, and elephants have found a way of surviving in the Namib Desert located along Namibia's Atlantic Ocean coast. Most nights a damp fog forms over the ocean. This fog floats inland, carrying moisture as far as 60 miles (97 km). Some of the hardy animals here survive by eating moistened tree leaves or finding small water holes.

Angola and **Namibia** have long coastlines on the Atlantic Ocean and are known as southern Africa's Atlantic countries. **Mozambique** and four island countries—**Madagascar** (MA•duh•GAS•kuhr), **Comoros** (KAH•muh•ROHZ), **Seychelles** (say•SHEHL), and **Mauritius** (maw•RIH•shuhs)—form southern Africa's Indian Ocean region.

Angola

Angola is almost twice the size of Texas. Angola also includes a tiny exclave called **Cabinda.** An exclave is a small part of a country that is separated from the main part. Hilly grasslands cover northern Angola. The southern part of the country is a rocky desert. In Cabinda, rain forests thrive.

Angola's main economic activity is agriculture. About 85 percent of the people make their living from subsistence farming. Some farmers grow coffee and cotton for export. Angola's main source of income, however, is oil. Oil deposits off the coast of Cabinda account for 90 percent of Angola's export earnings. Other important industries include diamond mining, fish processing, and textiles. Still, Angola is not a wealthy country. Different groups have struggled for control of the country, which has hurt the economy.

Most of Angola's people trace their ancestry to the Bantu-speaking peoples who spread across much of Africa many centuries ago. In the 1400s, the Kongo kingdom ruled a large part of northern Angola.

From the 1500s until its independence in 1975, Angola was a colony of Portugal. Portugal is still an important trading partner, and Portuguese is the official language. Bantu and other African languages are also widely spoken. Almost 50 percent of Angolans practice the Roman Catholic faith brought to Angola by the Portuguese.

After Angola gained its independence, civil war broke out among different political and ethnic groups. The fighting has lasted more than 25 years and continues to bring great suffering to the people.

✓ Reading Check Why is Angola's economy weak, even though it has rich resources?

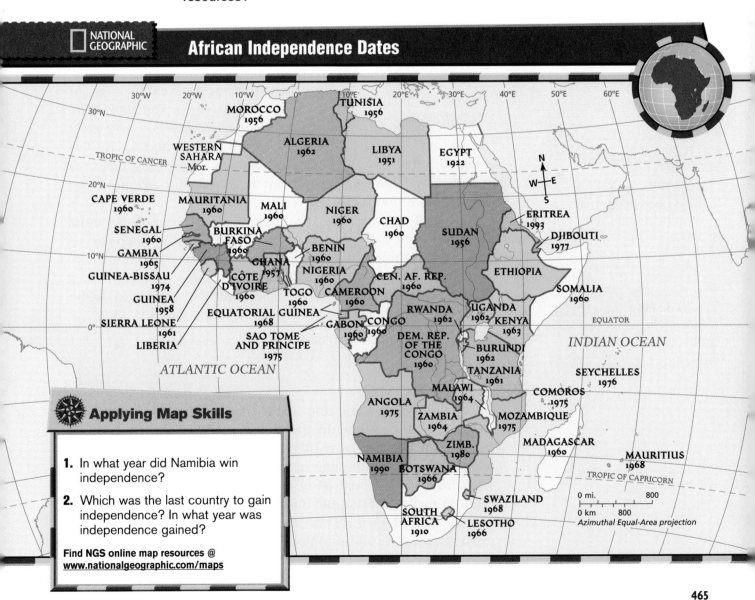

NATIONAL GEOGRAPHIC
African Independence Dates

Applying Map Skills

1. In what year did Namibia win independence?

2. Which was the last country to gain independence? In what year was independence gained?

Find NGS online map resources @ www.nationalgeographic.com/maps

Maputo, Mozambique

A high-rise building is being constructed in Maputo. Hotels and industrial projects are helping the city's economy to grow.

Economics What slowed industrial growth in Maputo in the 1980s and 1990s?

Namibia

Namibia is one of Africa's newest countries. Namibia became independent in 1990 after 75 years of rule by the Republic of South Africa. Before that it was a colony of Germany.

A large plateau runs through the center of the country. This area of patchy grassland is the most populous section of Namibia. The rest is made up of deserts. The **Namib Desert,** located along Namibia's Atlantic coast, is a narrow ribbon of towering dunes and rocks. Tourists come from all over the world to "sand-board" down these dunes. The Kalahari Desert stretches across the southeastern part of the country. As you might guess, most of Namibia has a hot, dry climate.

Namibia has rich deposits of diamonds, copper, gold, zinc, silver, and lead. It is a leading producer of uranium, a substance used for making nuclear fuels. The economy depends on the mining, processing, and exporting of these minerals.

Despite this mineral wealth, most of Namibia's people live in poverty. The income from mineral exports goes to a small group of Namibia's people and to the foreign companies that have invested in Namibia's mineral resources. As a result, half of Namibia's people depend on subsistence farming, herding, and working in food industries.

Only 1.8 million people live in Namibia. It is one of the most sparsely populated countries in Africa. In fact, in the language of Namibia's Nama ethnic group, *namib* means "the land without people." Most Namibians belong to African ethnic groups. A small number are of European ancestry. Namibians speak African languages, while most of the white population speaks Afrikaans and English.

√Reading Check When did Namibia become an independent country?

Mozambique

Sand dunes, swamps, and fine natural harbors line Mozambique's long Indian Ocean coastline. In the center of this Y-shaped country stretches a flat plain covered with grasses and tropical forests.

Most people in Mozambique are farmers. Some practice slash-and-burn farming—a method of clearing land for planting by cutting and burning forest. Slash-and-burn farming along with commercial logging has caused deforestation, or cutting down of forests. Deforestation can, in turn, lead to flooding during the rainy season. Such floods drove over one million people from their homes in early 2000.

Mozambique also experiences deadly cyclones. A cyclone is an intense storm system with heavy rain and high circular winds.

Mozambique's major crops are cashews, cotton, sugarcane, tea, coconuts, and tropical fruits. The main source of income, however, comes from its seaports. South Africa, Zimbabwe, Swaziland, and Malawi all pay to use the docks at Maputo, the capital, and other ports.

During the 1980s and early 1990s, a fierce civil war slowed industrial growth. In recent years, however, foreign companies have begun to invest in metal production, natural gas, fishing, and transportation services.

Most of Mozambique's 19.4 million people belong to one of 16 major African ethnic groups. A former colony of Portugal, Mozambique's official language is Portuguese, but most people speak African languages. About half of the people practice traditional African religions. Most of the rest are Muslim or Christian.

Reading Check What is a negative result of slash-and-burn farming?

Madagascar

The island of Madagascar broke away from the African mainland about 160 million years ago. As a result, it has many plants and animals that are not found elsewhere. It produces most of the world's vanilla beans. The main cash crop is coffee, and rice is also grown. About 80 percent of the island has been slashed and burned by people who must farm and herd to survive. The government has taken steps to save what forests are left.

Only about 22 percent of Madagascar's people are city dwellers. **Antananarivo** (AHN•tah•NAH•nah•REE•voh), the capital, lies in the central plateau. Called "Tana" for short, this city is known for its colorful street markets, where craftspeople sell a variety of products.

Music revolves around dance rhythms that reflect Madagascar's Southeast Asian and African heritage. The people are known for their rhythmic style of singing accompanied only by hand clapping.

Reading Check Why does Madagascar have wildlife that appears nowhere else on the earth?

Small Island Countries

Far from Africa in the Indian Ocean are three other island republics—Comoros, Seychelles, and Mauritius. The people of these countries have many different backgrounds.

Comoros Volcanoes formed Comoros thousands of years ago. Dense tropical forests cover Comoros today. Most of the approximately 600,000 people are farmers. The main crops are rice, vanilla, cloves, coconuts, and bananas. Even though agriculture employs 80 percent of the workforce, Comoros cannot grow enough food for its growing population. The government is trying to encourage industry, including tourism.

The people of Comoros are a mixture of Arabs, Africans, and people from Madagascar. They speak Arabic, French, and Comoran. Most

▲ This ring-tailed lemur lives on the island of Madagascar.

practice Islam. Once ruled by France, the people of Comoros declared their independence in 1974. Since then, they have suffered from fighting among political groups for control of the government.

Seychelles   About 90 islands form the country of Seychelles. About 40 of the islands are granite with high green peaks. The rest are small, flat coral islands with few people. Nearly 90 percent of the country's roughly 80,000 people live on Mahé, the largest island.

Seychelles was not inhabited until the 1700s. Under French and then British rule, it finally became independent in 1976. Most of the country's people are of mixed African, European, and Asian descent. Coconuts and cinnamon are the chief cash crops. Fishing and tourism are important industries as well.

Mauritius   Like Comoros, the islands of Mauritius were formed by volcanoes. Sugar is its main agricultural export. Clothing and textiles account for about half of the country's export earnings. Tourism is an important industry, too.

Mauritians come from many different backgrounds. About 70 percent are descendants of settlers from India. The rest are of African, European, or Chinese ancestry. Because of this varied ethnic heritage, the foods of Mauritius have quite a mix of ingredients. You can sample Indian chicken curry, Chinese pork, African-made roast beef, and French-style vegetables.

√ Reading Check   What natural force created the islands of Comoros and Mauritius?

Section 3 Assessment

Defining Terms

1. Define exclave, slash-and-burn farming, deforestation, cyclone.

Recalling Facts

2. Economics   What is Angola's main source of income?

3. Place   What two deserts can be found in Namibia?

4. Location   Where are most of the world's vanilla beans grown?

Critical Thinking

5. Understanding Cause and Effect   Why is Namibia one of the most sparsely populated countries in Africa?

6. Evaluating Information   How do the foods of Mauritius show its heritage?

Graphic Organizer

7. Organizing Information   Create a diagram like the one below. Then write facts about Madagascar that fit the category heading in each of the outer ovals.

Land — Madagascar — People
Economy — Culture

Applying Social Studies Skills

8. Analyzing Maps   Study the map of African independence dates on page 465. Which southern African country first achieved independence?

Section 1	The New South Africa

Terms to Know
industrialized
   country
Boer
apartheid
township
enclave

Main Idea
Rich in resources, South Africa has recently seen major social and political changes.

✓ Economics Because of its abundant mineral resources, South Africa has the most industrialized economy in Africa.

✓ Government In 1994 South Africa held its first democratic election in which people from all ethnic groups could vote.

✓ Government South Africa is working to improve the lives of its poorer citizens.

Section 2	Inland Southern Africa

Terms to Know
copper belt
sorghum

Main Idea
Most of inland southern Africa is rich in resources and home to a wide variety of ethnic groups.

✓ Economics Zambia is one of the world's largest producers of copper.

✓ Economics Zimbabwe has many mineral resources and good farmland.

✓ Economics Mining and tourism earn money for Botswana, but many of its people work in South Africa for several months a year.

Section 3	Atlantic and Indian Ocean Countries

Terms to Know
exclave
slash-and-burn
   farming
deforestation
cyclone

Main Idea
Africa's Atlantic and Indian Ocean countries are struggling to develop their economies.

✓ Economics Angola's main source of income is oil.

✓ Culture Few Namibians benefit from the country's rich mineral wealth. Most live in poverty.

✓ Human/Environment Interaction Slash-and-burn farming in Mozambique has led to deforestation and flooding. Neighboring countries pay fees for the use of Mozambique's ports.

✓ Location Madagascar's island location has resulted in many plants and animals found nowhere else in the world.

✓ Economics Comoros continues to be a mainly agricultural economy, but Mauritius has succeeded in developing a variety of industries.

✓ Economics Seychelles's beaches and tropical climate draw many tourists.

Using Key Terms

Match the terms in Part A with their definitions in Part B.

A.

1. copper belt
2. cyclone
3. exclave
4. slash-and-burn farming
5. township
6. apartheid
7. Boer
8. deforestation
9. sorghum
10. enclave

B.

a. separating racial and ethnic groups
b. storm with high circular winds
c. widespread cutting of trees
d. small nation located inside a larger country
e. large area of copper mines
f. small part of a nation separated from the main part of the country
g. areas of forest are cleared by burning
h. tall grass used as grain and to make syrup
i. settlement outside cities in South Africa
j. Dutch farmer in South Africa

Reviewing Main Ideas

Section 1 The New South Africa

11. **Location** What is the southernmost point of Africa?
12. **History** When was South Africa's first election allowing all people to vote?
13. **Economics** What is Lesotho's only important natural resource?

Section 2 Inland Southern Africa

14. **Place** What river crosses Zambia?
15. **Economics** Where is the copper belt?
16. **Economics** How are the people of Malawi supported?
17. **History** What was Great Zimbabwe?
18. **History** Who ruled Botswana for nearly 80 years?

Section 3 Atlantic and Indian Ocean Countries

19. **History** What European country colonized Angola?
20. **Culture** What does *namib* mean?
21. **Culture** What is the official language of Mozambique?
22. **Economics** What is Madagascar's main cash crop?

 **South Africa and Its Neighbors**

Place Location Activity

On a separate sheet of paper, match the letters on the map with the numbered places listed below.

1. Madagascar
2. Lake Malawi
3. Zambezi River
4. Kalahari Desert
5. Angola
6. Zimbabwe
7. Pretoria
8. Mozambique
9. South Africa
10. Namibia

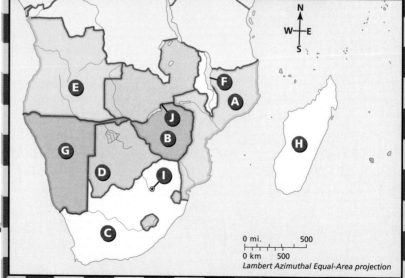

0 mi. 500
0 km 500
Lambert Azimuthal Equal-Area projection

Self-Check Quiz Visit the *Our World Today: People, Places, and Issues* Web site at owt.glencoe.com and click on **Chapter 17–Self-Check Quizzes** to prepare for the Chapter Test.

Critical Thinking

23. **Supporting Generalizations** What facts support the statement "South Africa has the most industrialized economy in Africa"?

24. **Evaluating Information** Many countries of southern Africa are hoping to build and improve their industries. On a chart like the one below, list the positive and negative aspects of industrialization under the correct headings.

Industrialization	
Positives	Negatives

Current Events Journal

25. **Writing a Myth** Throughout history, different cultures have created myths or stories to explain events in nature, such as thunder and lightning or an eclipse. Research one country of southern Africa and choose an aspect of nature that influences life in that country today. Write a story that might explain its occurrence.

Mental Mapping Activity

26. **Focusing on the Region** Create a simple outline map of Africa, then label the following:

- Atlantic Ocean
- Pretoria
- Madagascar
- Angola
- South Africa
- Cape Town
- Namib Desert
- Mozambique
- Indian Ocean
- Kalahari Desert

Technology Skills Activity

27. **Using the Internet** The Zulu are a well-known ethnic group in Africa. Research this group on the Internet. Write a speech answering these questions: Who are the Zulu? How have they affected the history of southern Africa? Where do they live today?

The Princeton Review

Standardized Test Practice

Directions: Study the map below, and then answer the question that follows.

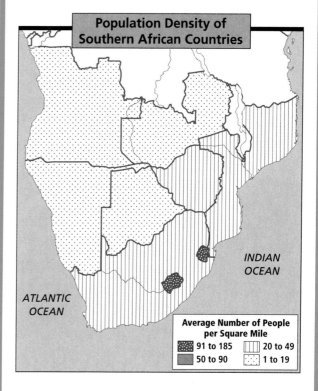

Population Density of Southern African Countries

INDIAN OCEAN

ATLANTIC OCEAN

Average Number of People per Square Mile
- 91 to 185
- 50 to 90
- 20 to 49
- 1 to 19

1. **Which country has the fewest people per square mile?**

   A Lesotho

   B Egypt

   C South Africa

   D Namibia

Test-Taking Tip: This question involves both recalling where countries are located as well as using the legend. Start with the answer choices. Think about what you learned about each country. You may be able to get rid of wrong answer choices simply by recalling these facts.

Unit 7

◀ **Skier in Idaho's stretch of the Rocky Mountains**

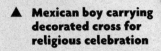

▲ **Mexican boy carrying decorated cross for religious celebration**

Farm on the Manitoba plains

North America and Middle America

Northland America is a large continent that stretches from above the Arctic Circle to a few degrees north of the Equator. Home to many cultures and peoples, it is a land of stark contrasts. The United States and Canada, in the northern part of the continent, share a similar history and a common language. Mexico and Central America also have a common language and history, and—with the islands of the Caribbean— make up a region known as Middle America. Middle America is not technically a separate continent, but it can be defined as a culture region.

NGS ONLINE
www.nationalgeographic.com/education

473

REGIONAL ATLAS

The United States and Canada

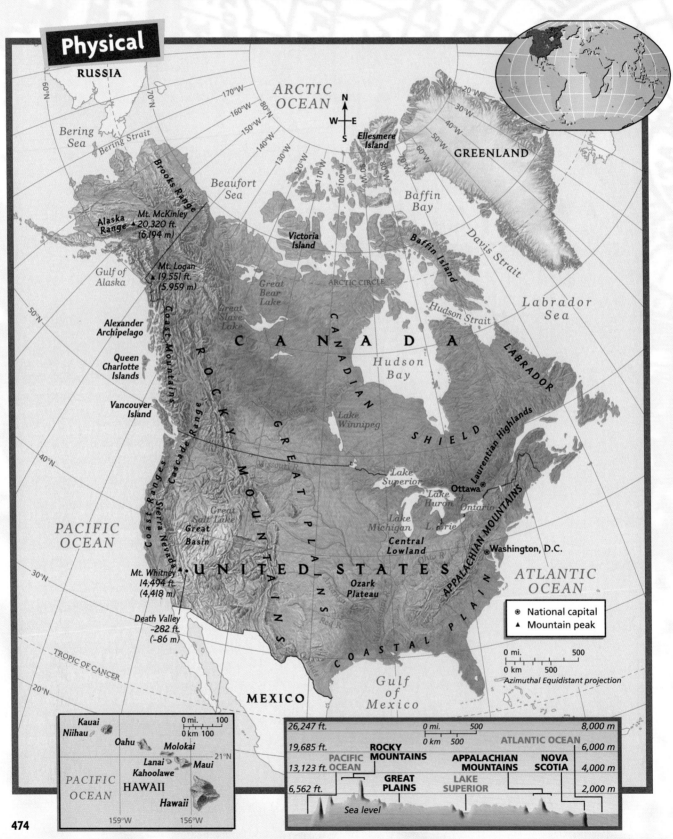

Physical

RUSSIA

ARCTIC OCEAN

Bering Sea

Bering Strait

Ellesmere Island

GREENLAND

Beaufort Sea

Brooks Range

Alaska Range
Mt. McKinley ▲ 20,320 ft. (6,194 m)

Gulf of Alaska

Mt. Logan ▲ 19,551 ft. (5,959 m)

Baffin Bay

Baffin Island

Davis Strait

Victoria Island

ARCTIC CIRCLE

Great Bear Lake

Hudson Strait

Labrador Sea

Alexander Archipelago

Great Slave Lake

C A N A D A

C A N A D I A N

S H I E L D

LABRADOR

Queen Charlotte Islands

Coast Mountains

Vancouver Island

R O C K Y

Lake Winnipeg

Hudson Bay

Laurentian Highlands

Missouri R.

Lake Superior

Ottawa ⊛

Cascade Range

M O U N T A I N S

G R E A T

Lake Huron

L. Ontario

APPALACHIAN MOUNTAINS

Coast Ranges

Sierra Nevada

Great Salt Lake

Great Basin

P L A I N S

Lake Michigan

L. Erie

Central Lowland

⊛ Washington, D.C.

PACIFIC OCEAN

Mt. Whitney ▲ 14,494 ft. (4,418 m)

U N I T E D   S T A T E S

Ozark Plateau

Ohio R.

ATLANTIC OCEAN

Death Valley -282 ft. (-86 m)

Red R.

C O A S T A L   P L A I N

⊛ National capital
▲ Mountain peak

TROPIC OF CANCER

MEXICO

Gulf of Mexico

0 mi.   500
0 km   500

Azimuthal Equidistant projection

Kauai
Niihau

0 mi.   100
0 km   100

Oahu

Molokai

21°N

Lanai
Kahoolawe

Maui

PACIFIC OCEAN

HAWAII

Hawaii

159°W   156°W

26,247 ft.

19,685 ft.

PACIFIC OCEAN

ROCKY MOUNTAINS

0 mi.   500
0 km   500

ATLANTIC OCEAN

8,000 m

6,000 m

13,123 ft.

APPALACHIAN MOUNTAINS

NOVA SCOTIA

4,000 m

6,562 ft.

GREAT PLAINS

LAKE SUPERIOR

2,000 m

Sea level

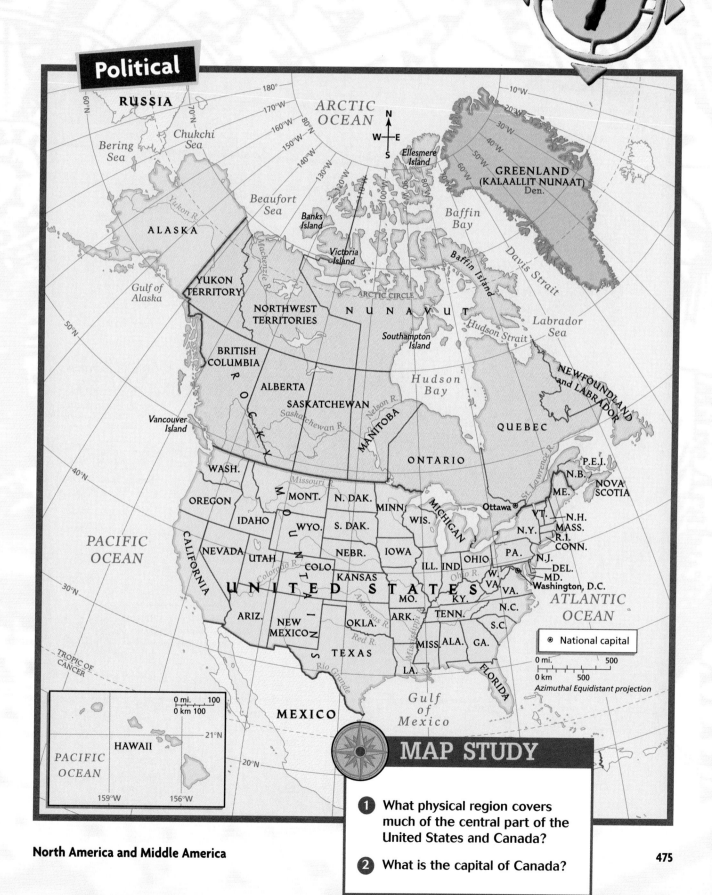

Political

RUSSIA

ARCTIC OCEAN

Bering Sea

Chukchi Sea

ALASKA

Beaufort Sea

Banks Island

Ellesmere Island

GREENLAND (KALAALLIT NUNAAT) Den.

Baffin Bay

Gulf of Alaska

YUKON TERRITORY

NORTHWEST TERRITORIES

Victoria Island

ARCTIC CIRCLE

NUNAVUT

Baffin Island

Davis Strait

Labrador Sea

BRITISH COLUMBIA

Mackenzie R.

ALBERTA

SASKATCHEWAN

Southampton Island

Hudson Strait

Hudson Bay

NEWFOUNDLAND and LABRADOR

Vancouver Island

Saskatchewan R.

Nelson R.

MANITOBA

ONTARIO

QUEBEC

P.E.I.

N.B.

NOVA SCOTIA

St. Lawrence R.

ME.

WASH.

Missouri R.

MONT.

N. DAK.

MINN.

MICHIGAN

Ottawa ⊛

VT.

N.H.

OREGON

IDAHO

WYO.

S. DAK.

WIS.

N.Y.

MASS.

R.I.

CONN.

PACIFIC OCEAN

CALIFORNIA

NEVADA

UTAH

COLO.

Colorado R.

NEBR.

KANSAS

IOWA

ILL. IND

MO.

OHIO

KY.

PA.

N.J.

DEL.

MD.

W. VA.

VA.

Washington, D.C.

N.C.

ATLANTIC OCEAN

UNITED STATES

ARIZ.

NEW MEXICO

OKLA.

Arkansas R.

Red R.

ARK.

TENN.

MISS. ALA.

GA.

S.C.

Mississippi R.

TEXAS

LA.

FLORIDA

Rio Grande

⊛ National capital

MEXICO

Gulf of Mexico

0 mi. 500
0 km 500

Azimuthal Equidistant projection

HAWAII

PACIFIC OCEAN

0 mi. 100
0 km 100

21°N

159°W 156°W

North America and Middle America

MAP STUDY

1 What physical region covers much of the central part of the United States and Canada?

2 What is the capital of Canada?

Middle America

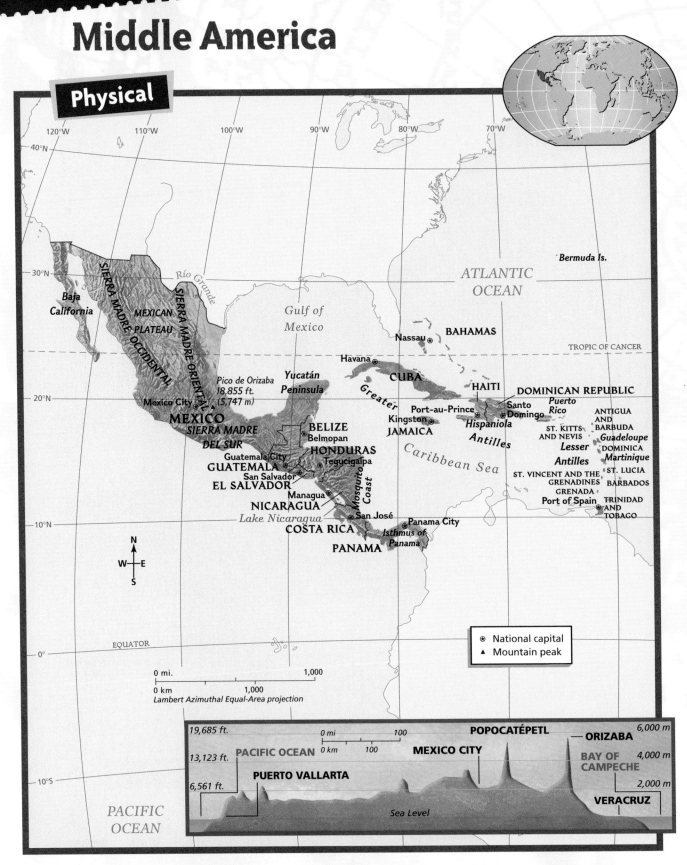

Physical

MEXICO

Baja California

SIERRA MADRE OCCIDENTAL

MEXICAN PLATEAU

SIERRA MADRE ORIENTAL

Río Grande

Gulf of Mexico

ATLANTIC OCEAN

Bermuda Is.

Pico de Orizaba 18,855 ft. (5,747 m)

Yucatán Peninsula

Mexico City

SIERRA MADRE DEL SUR

BELIZE
Belmopan

Guatemala City
GUATEMALA

San Salvador
EL SALVADOR

HONDURAS
Tegucigalpa

Managua

NICARAGUA

Lake Nicaragua

COSTA RICA

San José

PANAMA

Panama City

Isthmus of Panama

Mosquito Coast

Nassau

BAHAMAS

TROPIC OF CANCER

Havana

CUBA

HAITI

DOMINICAN REPUBLIC

Greater

Port-au-Prince

Santo Domingo

Puerto Rico

Kingston

JAMAICA

Hispaniola

Antilles

Caribbean Sea

Lesser Antilles

ANTIGUA AND BARBUDA

ST. KITTS AND NEVIS

Guadeloupe

DOMINICA

Martinique

ST. LUCIA

ST. VINCENT AND THE GRENADINES

BARBADOS

GRENADA

Port of Spain

TRINIDAD AND TOBAGO

EQUATOR

⊛ National capital
▲ Mountain peak

0 mi. 1,000
0 km 1,000
Lambert Azimuthal Equal-Area projection

PACIFIC OCEAN

19,685 ft.

13,123 ft.

6,561 ft.

PACIFIC OCEAN

PUERTO VALLARTA

0 mi 100
0 km 100

MEXICO CITY

POPOCATÉPETL

ORIZABA

BAY OF CAMPECHE

VERACRUZ

Sea Level

6,000 m

4,000 m

2,000 m

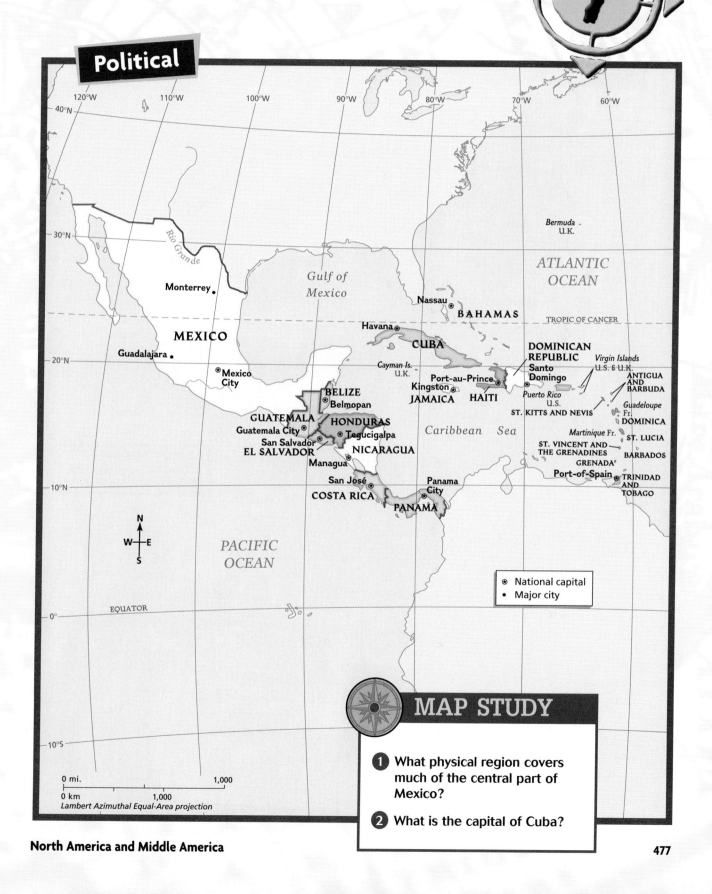
Political

40°N
120°W   110°W   100°W   90°W   80°W   70°W   60°W

30°N

Rio Grande

Monterrey •

Gulf of Mexico

MEXICO

Guadalajara •

20°N

⊛Mexico City

Bermuda
U.K.

ATLANTIC OCEAN

Nassau ⊛

BAHAMAS   TROPIC OF CANCER

Havana ⊛

CUBA

Cayman Is.
U.K.

Port-au-Prince ⊛

Kingston ⊛

JAMAICA   HAITI

DOMINICAN REPUBLIC
Santo Domingo

Virgin Islands
U.S. & U.K.

Puerto Rico
U.S.

ST. KITTS AND NEVIS

ANTIGUA AND BARBUDA

Guadeloupe
Fr. **DOMINICA**

BELIZE
• Belmopan

GUATEMALA
Guatemala City ⊛

San Salvador ⊛
EL SALVADOR

HONDURAS
⊛ Tegucigalpa

NICARAGUA

Managua ⊛

Caribbean   Sea

Martinique Fr. **ST. LUCIA**

ST. VINCENT AND THE GRENADINES

BARBADOS

GRENADA

Port-of-Spain ⊛   **TRINIDAD AND TOBAGO**

10°N

San José ⊛

COSTA RICA

Panama City ⊛

PANAMA

N
W—E
S

PACIFIC OCEAN

0°   EQUATOR

⊛ National capital
• Major city

10°S

0 mi.   1,000
0 km   1,000
Lambert Azimuthal Equal-Area projection

MAP STUDY

1 What physical region covers much of the central part of Mexico?

2 What is the capital of Cuba?

North America and Middle America

Major Languages

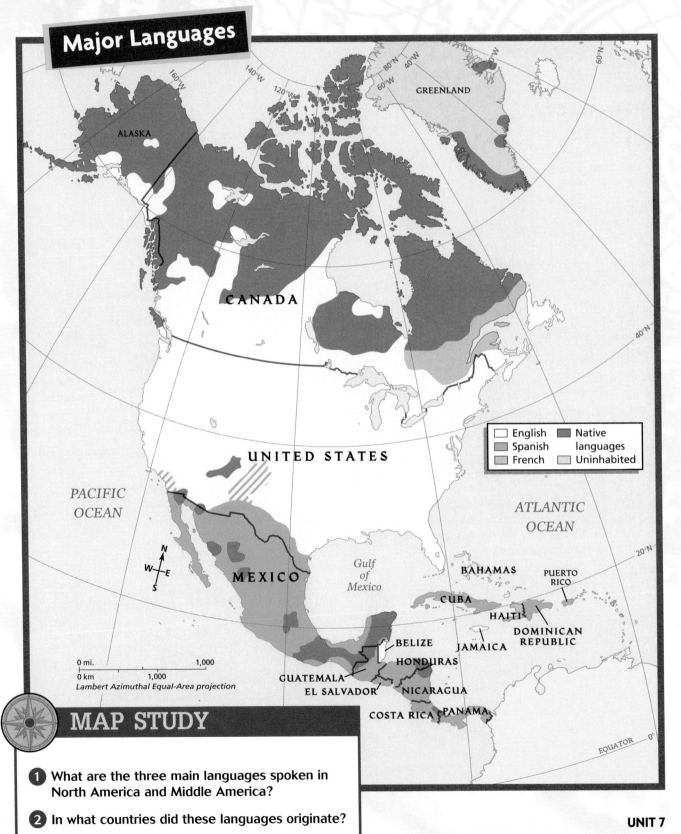

Legend:
- English
- Spanish
- French
- Native languages
- Uninhabited

ALASKA

GREENLAND

CANADA

UNITED STATES

PACIFIC OCEAN

ATLANTIC OCEAN

MEXICO

Gulf of Mexico

BAHAMAS

PUERTO RICO

CUBA

HAITI

DOMINICAN REPUBLIC

JAMAICA

BELIZE

HONDURAS

GUATEMALA

EL SALVADOR

NICARAGUA

COSTA RICA

PANAMA

EQUATOR

0 mi. 1,000
0 km 1,000
Lambert Azimuthal Equal-Area projection

MAP STUDY

1 What are the three main languages spoken in North America and Middle America?

2 In what countries did these languages originate?

Fast Facts

COMPARING POPULATION:
United States, Canada, Mexico, and Guatemala

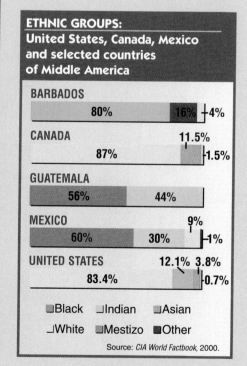

UNITED STATES

CANADA

MEXICO

GUATEMALA

= 25,000,000

Source: *Population Reference Bureau*, 2000.

ETHNIC GROUPS:
United States, Canada, Mexico and selected countries of Middle America

BARBADOS
80%   16%   4%

CANADA
11.5%
87%   1.5%

GUATEMALA
56%   44%

MEXICO
9%
60%   30%   1%

UNITED STATES
12.1% 3.8%
83.4%   0.7%

■ Black   ⌐ Indian   ■ Asian
⌐ White   ■ Mestizo   ■ Other

Source: *CIA World Factbook*, 2000.

Mexico

Data Bits

🚗	Automobiles per 1,000 people	93
📱	Telephones per 1,000 people	96
VOTE	Democratic elections	Yes

Religions

Other 4%
Protestant 6%
Roman Catholic 90%

World Ranking

	GNP per capita in US $	Life expectancy	Literacy
1st			
50th	61 $3,700	62 72 years	94 90%
100th			
150th			

Population: Urban ▓ vs. Rural ░

75%   25%

Source: *World Desk Reference*, 2000.

GRAPHIC STUDY

1. What is life expectancy in Mexico?

2. What countries have a majority of mestizos (people of mixed European and Native American ancestry)?

Country Profiles

ANTIGUA and BARBUDA

POPULATION:
100,000
588 per sq. mi.
227 per sq. km

LANGUAGE:
English

MAJOR EXPORT:
Petroleum Products

MAJOR IMPORTS:
Foods and
Livestock

CAPITAL:
St. John's

LANDMASS:
170 sq. mi.
440 sq. km

BAHAMAS

POPULATION:
300,000
56 per sq. mi.
22 per sq. km

LANGUAGES:
English, Creole

MAJOR EXPORT:
Pharmaceuticals

MAJOR IMPORT:
Foods

CAPITAL:
Nassau

LANDMASS:
5,359 sq. mi.
13,880 sq. km

BARBADOS

POPULATION:
300,000
1807 per sq. mi.
698 per sq. km

LANGUAGE:
English

MAJOR EXPORT:
Sugar

MAJOR IMPORTS:
Manufactured
Goods

CAPITAL:
Bridgetown

LANDMASS:
166 sq. mi.
430 sq. km

BELIZE

POPULATION:
300,000
34 per sq. mi.
13 per sq. km

LANGUAGE:
English

MAJOR EXPORT:
Sugar

MAJOR IMPORT:
Machinery

CAPITAL:
Belmopan

LANDMASS:
8,865 sq. mi.
22,960 sq. km

CANADA

POPULATION:
31,000,000
8 per sq. mi.
3 per sq. km

LANGUAGES:
English, French

MAJOR EXPORT:
Newsprint

MAJOR IMPORT:
Crude Oil

CAPITAL:
Ottawa

LANDMASS:
3,849,670 sq. mi.
9,970,645 sq. km

COSTA RICA

POPULATION:
3,700,000
188 per sq. mi.
72 per sq. km

LANGUAGE:
Spanish

MAJOR EXPORT:
Coffee

MAJOR IMPORT:
Raw Materials

CAPITAL:
San José

LANDMASS:
19,730 sq. mi.
51,100 sq. km

CUBA

POPULATION:
11,300,000
264 per sq. mi.
102 per sq. km

LANGUAGE:
Spanish

MAJOR EXPORT:
Sugar

MAJOR IMPORT:
Petroleum

CAPITAL:
Havana

LANDMASS:
42,803 sq. mi.
110,860 sq. km

DOMINICA

POPULATION:
100,000
345 per sq. mi.
133 per sq. km

LANGUAGES:
English, French

MAJOR EXPORT:
Bananas

MAJOR IMPORT:
Manufactured
Goods

CAPITAL:
Roseau

LANDMASS:
290 sq. mi.
751 sq. km

DOMINICAN REPUBLIC

POPULATION:
8,600,000
457 per sq. mi.
176 per sq. km

LANGUAGE:
Spanish

MAJOR EXPORT:
Ferronickel

MAJOR IMPORT:
Foods

CAPITAL:
Santo Domingo

LANDMASS:
18,815 sq. mi.
48,731 sq. km

EL SALVADOR

POPULATION:
6,400,000
788 per sq. mi.
304 per sq. km

LANGUAGE:
Spanish

MAJOR EXPORT:
Coffee

MAJOR IMPORT:
Raw Materials

CAPITAL:
San Salvador

LANDMASS:
8,124 sq. mi.
21,041 sq. km

GRENADA

POPULATION:
100,000
763 per sq. mi.
295 per sq. km

LANGUAGES:
English, French

MAJOR EXPORT:
Bananas

MAJOR IMPORT:
Foods

CAPITAL:
St. George's

LANDMASS:
133 sq. mi.
339 sq. km

Countries and flags not drawn to scale

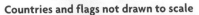

For more information on countries in this region, refer to the Nations of the World Data Bank on pages 690–699.

GUATEMALA

POPULATION:
13,000,000
309 per sq. mi.
119 per sq. km

Guatemala City

LANGUAGES:
Spanish, Mayan Languages

MAJOR EXPORT:
Coffee

CAPITAL:
Guatemala City

MAJOR IMPORT:
Petroleum

LANDMASS:
42,042 sq. mi.
108,889 sq. km

HAITI

POPULATION:
7,000,000
723 per sq. mi.
279 per sq. km

Port-au-Prince

LANGUAGES:
French, Creole

MAJOR EXPORT:
Manufactured Goods

CAPITAL:
Port-au-Prince

MAJOR IMPORT:
Machinery

LANDMASS:
10,714 sq. mi.
27,750 sq. km

HONDURAS

POPULATION:
6,700,000
155 per sq. mi.
60 per sq. km

Tegucigalpa

LANGUAGE:
Spanish

MAJOR EXPORT:
Bananas

CAPITAL:
Tegucigalpa

MAJOR IMPORT:
Machinery

LANDMASS:
43,278 sq. mi.
112,090 sq. km

JAMAICA

POPULATION:
2,600,000
613 per sq. mi.
237 per sq. km

Kingston

LANGUAGES:
English, Creole

MAJOR EXPORT:
Alumina

CAPITAL:
Kingston

MAJOR IMPORT:
Machinery

LANDMASS:
4,243 sq. mi.
10,989 sq. km

MEXICO

POPULATION:
99,600,000
132 per sq. mi.
51 per sq. km

Mexico City

LANGUAGES:
Spanish, Native American Languages

MAJOR EXPORT:
Crude Oil

CAPITAL:
Mexico City

MAJOR IMPORT:
Machinery

LANDMASS:
756,062 sq. mi.
1,958,201 sq. km

NICARAGUA

POPULATION:
5,200,000
104 per sq. mi.
40 per sq. km

Managua

LANGUAGE:
Spanish

MAJOR EXPORT:
Coffee

CAPITAL:
Managua

MAJOR IMPORT:
Manufactured Goods

LANDMASS:
50,193 sq. mi.
130,000 sq. km

PANAMA

POPULATION:
2,900,000
100 per sq. mi.
38 per sq. km

Panama City

LANGUAGE:
Spanish

MAJOR EXPORT:
Bananas

CAPITAL:
Panama City

MAJOR IMPORT:
Machinery

LANDMASS:
29,158 sq. mi.
75,519 sq. km

PUERTO RICO *

POPULATION:
3,900,000
1,129 per sq. mi.
436 per sq. km

San Juan

LANGUAGES:
Spanish, English

MAJOR EXPORT:
Pharmaceuticals

CAPITAL:
San Juan

MAJOR IMPORT:
Chemical Products

LANDMASS:
3,456 sq. mi.
8,951 sq. km

* U.S. Commonwealth

ST. KITTS and NEVIS

POPULATION:
40,000
288 per sq. mi.
111 per sq. km

Basseterre

LANGUAGE:
English

MAJOR EXPORT:
Machinery

CAPITAL:
Basseterre

MAJOR IMPORT:
Electronic Goods

LANDMASS:
139 sq. mi.
360 sq. km

ST. LUCIA

POPULATION:
200,000
837 per sq. mi.
323 per sq. km

Castries

LANGUAGES:
English, French

MAJOR EXPORT:
Bananas

CAPITAL:
Castries

MAJOR IMPORT:
Foods

LANDMASS:
239 sq. mi.
619 sq. km

Country Profiles

ST. VINCENT and the GRENADINES

POPULATION:
100,000
662 per sq. mi.
256 per sq. km

LANGUAGES:
English, French

MAJOR EXPORT:
Bananas

MAJOR IMPORT:
Foods

CAPITAL:
Kingstown

LANDMASS:
151 sq. mi.
391 sq. km

★ Kingstown

TRINIDAD and TOBAGO

POPULATION:
1,300,000
656 per sq. mi.
253 per sq. km

LANGUAGE:
English

MAJOR EXPORT:
Petroleum

MAJOR IMPORT:
Machinery

CAPITAL:
Port-of-Spain

LANDMASS:
1,981 sq. mi.
5,131 sq. km

Port-of-Spain

UNITED STATES

POPULATION:
284,500,000
77 per sq. mi.
30 per sq. km

LANGUAGE:
English

MAJOR EXPORT:
Machinery

MAJOR IMPORT:
Crude Oil

CAPITAL:
Washington, D.C.

LANDMASS:
3,717,796 sq. mi.
9,629,091 sq. km

Washington, D.C.

VIRGIN ISLANDS (U.S.)

POPULATION:
120,917
902 per sq. mi.
349 per sq. km

LANGUAGE:
English

MAJOR EXPORT:
Chemical Products

MAJOR IMPORT:
Crude Oil

CAPITAL:
Charlotte Amalie

LANDMASS:
134 sq. mi.
347 sq. km

Charlotte Amalie

Countries and flags not drawn to scale

BUILDING CITIZENSHIP

Building Better Communities Many people in the United States try to improve their communities or help other people by volunteering in nonprofit and civic organizations. Nonprofit organizations may be churches, boys and girls clubs, sporting and leisure clubs, homeless shelters, or other social agencies. They supply services and assistance that the government or business is unable to provide. More than half of all Americans volunteer at least three hours per week in nonprofit organizations.

Tell about a nonprofit organization supported by you or someone you know.

WRITE ABOUT IT

Imagine that you are the head of a nonprofit organization and you have been invited to your school to tell the students about it. Knowing that your efforts may inspire students to participate, write a short talk describing the organization's activities.

▲ **Students participating in a service activity**

Spider monkey and Mayan ruins, Mexico ▲

18 Canada

The World and Its People

NATIONAL GEOGRAPHIC

To learn more about Canada's people and places, view **The World and Its People Chapter 5** video.

Our World Today online

Chapter Overview Visit the **Our World Today: People, Places, and Issues** Web site at owt.glencoe.com and click on **Chapter 18—Chapter Overviews** to preview information about Canada.

FOLDABLES™
Study Organizer

Compare-Contrast Study Foldable Make this foldable to help you analyze the similarities and differences between the landforms, climate, and cultures of northern and southern Canada.

Step 1 Mark the midpoint of the side edge of a sheet of paper.

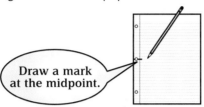

Draw a mark at the midpoint.

Step 2 Turn the paper and fold the outside edges in to touch at the midpoint.

Step 3 Turn and label your foldable as shown.

Northern Canada

Southern Canada

Reading and Writing As you read the chapter, collect and write information under the appropriate tab that will help you compare and contrast northern and southern Canada.

Why It Matters

Sharing a Border

The boundary line between Canada and the United States has formed the longest unprotected border in the world. Citizens of these countries have been allowed to travel freely between them, which is symbolic of the free trade between these nations.

◄ **Vancouver, British Columbia**

Canada's Major Regions

Guide to Reading

Main Idea

Canada is a vast country with many landforms, climates, resources, and peoples.

Terms to Know

- province
- Nunavut
- indigenous
- glacier
- tundra
- service industry
- overfishing

Reading Strategy

Create a chart like this one and list Canada's provinces in the left column. In the right column, list the main economic activities in each province.

Province	Economic Activities

NATIONAL GEOGRAPHIC

Exploring Our World

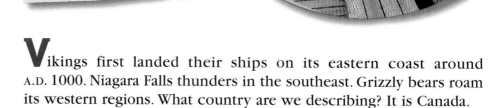

Have you ever seen a lumbering, snarling grizzly bear up close and personal? Many tourists come to Banff National Park in western Canada hoping to spot such a creature. Located in the Rocky Mountains, Banff is Canada's oldest, best-loved, and busiest national park. More than 4 million visitors a year are drawn to its spectacular mountain scenery.

Vikings first landed their ships on its eastern coast around A.D. 1000. Niagara Falls thunders in the southeast. Grizzly bears roam its western regions. What country are we describing? It is Canada.

Landforms and Climate

The map on the next page shows you that Canada is located north of the contiguous United States. Between Canada and the United States lies the world's longest undefended border. The friendship developed between the two countries has allowed thousands of people to cross this border every day. Canada is the world's second-largest country in land area. Only Russia is larger. Instead of being made up of states, Canada has 10 **provinces,** or regional political divisions. It also includes 3 territories. The newest territory, **Nunavut** (NOO•nuh•vuht), in the north of Canada was created in 1999 as a homeland for Canada's **indigenous,** or native, people. You will learn more about Nunavut in the next section.

Thousands of years ago, huge glaciers, or giant sheets of ice, covered most of Canada. The weight of these glaciers pushed much of the land down and created a large, low basin. Highlands rose on the western, eastern, and northern edges of this basin. Water filled the land that was pushed very low. As a result, Canada today has many lakes and inland waterways—more than any other country in the world.

The map below shows you that Canada generally has a cool or cold climate. In Canada's far north, people shiver in the cold tundra climate. A tundra is a treeless plain where the soil beneath the first few inches is permanently frozen. Farther south, between 70°N and 50°N latitude, Canadians experience cool summers and long, cold winters. Southeastern Canada has a milder climate and, as you would expect, most Canadians live in this area. The warmest temperatures are found along Canada's southwest coastline.

√ Reading Check Why do you think most of Canada's people live in the southern part of the country?

The Economy of Canada

Canada's economy is similar to that of the United States. Canada is known for fertile farmland, rich natural resources, and skilled workers. Manufacturing, farming, and service industries are the country's major economic activities. Service industries are businesses, such as banking and education, which provide services to people instead of producing goods. Fishing, mining, and lumbering are also important to the Canadian economy.

Canada, like the United States, has a free market economy in which people start and run businesses with limited government involvement. Canada's government, however, plays a more direct role in the

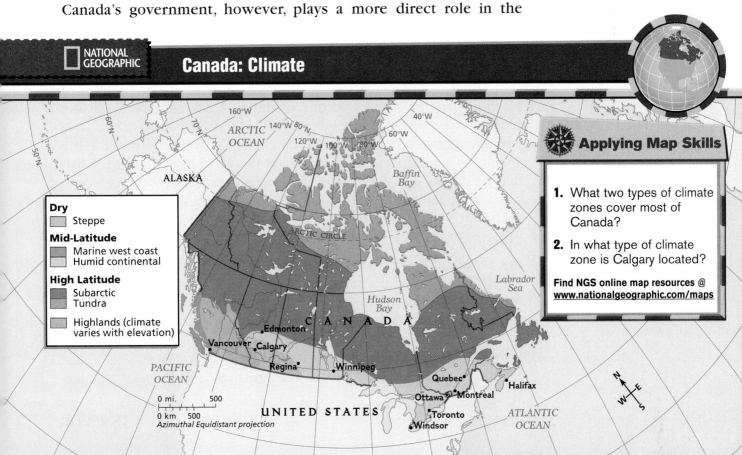

NATIONAL GEOGRAPHIC

Canada: Climate

Applying Map Skills

1. What two types of climate zones cover most of Canada?

2. In what type of climate zone is Calgary located?

Find NGS online map resources @ www.nationalgeographic.com/maps

Dry
Steppe
Mid-Latitude
Marine west coast
Humid continental
High Latitude
Subarctic
Tundra
Highlands (climate varies with elevation)

0 mi.   500
0 km   500
Azimuthal Equidistant projection

1

1

National Geographic On Location

Nighttime Harvest

At harvest time in southern Saskatchewan, the work goes on around the clock. The farms in the Prairie Provinces are large and depend on machinery.

Economics What other economic activities take place in the Prairie Provinces?

Canadian economy. For example, Canada's national and provincial governments provide health care for citizens. Broadcasting, transportation, and electric power companies are heavily regulated. These public services might not have been available in Canada's remote areas without government assistance.

About $1 billion worth of trade passes between Canada and the United States each day. In 1992 Canada, the United States, and Mexico entered into the North American Free Trade Agreement (NAFTA) to remove trade barriers among the three countries. This agreement took effect in 1994. Some Canadians do not support NAFTA because they fear that their economy is too dependent on the United States. They worry that the American economy is so large that it will dominate the partnership.

Reading Check What is a free market economy?

Canada's Industries

Fishing, manufacturing, farming, ranching, and energy production are Canada's important industries. As you would expect, geography plays a major role in where industries are located. Factors such as nearness to the ocean, location along the border with the United States, and oil and coal deposits determine where industries, jobs, and people can be found.

Fishing Fishing has been the major industry in **Newfoundland** and the **Maritime Provinces.** The Grand Banks, off the coast of Newfoundland, is known as one of the best fishing grounds in the world. These waters have been overfished, however. **Overfishing** results when too many fish are taken without giving time for the species to reproduce. The government now regulates how many fish may be caught in these waters. **Halifax** is a major shipping center in this region. Its harbor remains open in winter when ice closes most other eastern Canadian ports.

Manufacturing Manufacturing and service industries are dominant in Canada's largest province, **Quebec.** Almost one-fourth of Canadians live in Quebec, where agriculture and fishing are also important. **Montreal,** an important port on the St. Lawrence River, is Canada's second-largest city and a major financial and industrial center. The city of **Quebec,** founded by the French in 1608, is the capital of the province. Many historic sites and a European charm make it popular with tourists. Canada's second-largest province is **Ontario,** but it has the most people and greatest wealth. Ontario produces more than half of Canada's manufactured goods. Southern Ontario also has fertile land and a growing season long enough for commercial farming. Farmers here grow grains, fruits, and vegetables and raise beef and dairy cattle.

Toronto, the capital of Ontario, is Canada's largest city. It is also the country's chief manufacturing, financial, and communications center. **Ottawa,** the capital city of Canada, lies in Ontario near the border with Quebec. Many Canadians work in government offices in Ottawa.

Farming, Ranching and Energy Production   Farming and ranching are major economic activities in the Prairie Provinces of **Manitoba, Saskatchewan,** and **Alberta.** Canada is one of the world's biggest producers of wheat, most of which is exported to Europe and Asia. Some of the world's largest reserves of oil and natural gas are found in Alberta and Saskatchewan. These resources contribute to Canada's wealth. Huge pipelines carry the oil and gas to other parts of Canada and to the United States. Canada is the fifth-largest producer of energy in the world.

Thick forests blanket much of **British Columbia.** The province helps make Canada the world's leading producer of newsprint, the type of paper used for printing newspapers. Fishing and tourism are also strong economic activities in British Columbia. Fishing fleets sail out into the Pacific Ocean to catch salmon and other kinds of fish. **Vancouver** is a bustling trade center and the nation's main Pacific port.

✔ Reading Check Which city is Canada's communications center?

Assessment
Section 1

Defining Terms
1. **Define** province, Nunavut, indigenous, glacier, tundra, service industry, overfishing.

Recalling Facts
2. **History** What is unusual about the border between Canada and the United States?
3. **Place** Who lives in the north of Canada?
4. **Economics** Which province is the world's leading producer of newsprint?

Critical Thinking
5. **Analyzing Information** What province in Canada has the most people and the most economic activity?
6. **Drawing Conclusions** Explain why some Canadians fear NAFTA.

Graphic Organizer
7. **Organizing Information** Create a chart like this one. Then list each province, the resources found in it, and major cities located there, if any.

Province	Resources	Cities

Applying Social Studies Skills

8. **Analyzing Maps** Look at the political map on page 475. Name the Canadian provinces that border the United States.

Social Studies Skill

Reading a Time Zones Map

The earth revolves 360° in 24 hours. The earth's surface has been divided into 24 time zones. Each time zone represents 15° longitude, or the distance that the earth rotates in 1 hour.

Learning the Skill

The Prime Meridian, or 0° longitude, is the starting point for figuring out time around the world. Traveling west from 0° longitude, it becomes 1 hour earlier. Traveling east, it becomes 1 hour later. The international date line is set at the 180° line of longitude. Traveling west across this imaginary line, you add a day. Traveling east, you subtract a day. To read a time zones map:

- Choose a place for which you already know the time and locate it on the map.
- Locate another place and determine if it is east or west of the first place.
- Calculate the time by either adding (going east) or subtracting (going west) an hour for each time zone.
- Determine whether you have crossed the International Date Line, and identify the day of the week.

Practicing the Skill

1. On the map below, if it is 4 P.M. in Rio de Janeiro, what time is it in Honolulu?
2. If it is 10:00 A.M. in Tokyo on Tuesday, what day and time is it in Moscow?

Applying the Skill

Imagine you have a friend living in Rome, Italy. What time (your time) would you call if you wanted to talk to your friend after 7:00 P.M.?

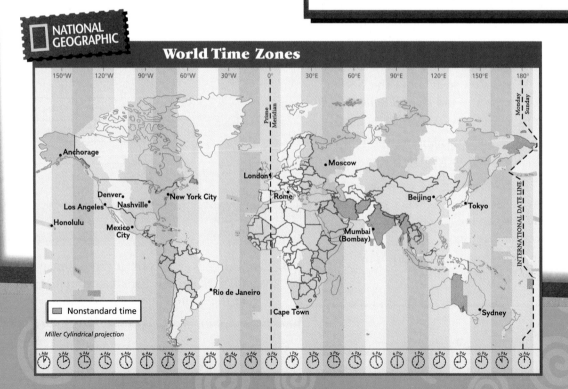

NATIONAL GEOGRAPHIC

World Time Zones

Nonstandard time

Miller Cylindrical projection

The Canadians

Guide to Reading

Main Idea

Canadians of many different backgrounds live in towns and cities close to the United States border.

Terms to Know

- colony
- dominion
- parliamentary democracy
- prime minister
- bilingual
- autonomy

Reading Strategy

Create a chart like this one and give at least two facts about Canada for each topic.

History		
Population		
Culture		

NATIONAL GEOGRAPHIC

Exploring Our World

Arrêt or Stop? People living in Quebec need to know both words when they cross the street. Canada has two official languages— English and French. All government documents are printed in both languages. In Quebec, even the school system is divided into French and English.

Like the United States, Canada's population is made up of many different cultures. The largest group of Canadians has a European heritage, but the country is home to people from all continents. Yet unlike the United States, Canada has had difficulty achieving a strong sense of being one nation. The country's vast distances and largely separate cultures have made Canadians feel more closely attached to their own region or culture than to Canada as a whole.

Canada's History

Inuit and other Native Americans lived for thousands of years in Canada before European settlers arrived. Some lived in coastal fishing villages. Others were hunters and gatherers constantly on the move. Still others founded permanent settlements. The first Europeans in Canada were Viking explorers who landed in about A.D. 1000. They lived for a while on the Newfoundland coast but eventually left.

In the 1500s and 1600s, both Britain and France claimed areas of Canada. French explorers, settlers, and missionaries founded several cities. The most important were Quebec and Montreal. For almost 230

Clothing

The Inuit of the Canadian Arctic designed their clothes for protection from the harsh climate. Traditional clothing was made up of a caribou or sealskin parka, pants, mittens, and boots. In winter the Inuit wore their furs facing toward the skin. This created air pockets that trapped warm air close to the body. On top they wore another layer with the fur facing outward. The clothing flapped as the wearer moved, creating a breeze that kept the person from overheating while running or working.

Looking Closer How does traditional clothing protect the Inuit from the harsh climate?

years, France ruled the area around the St. Lawrence River and the Great Lakes. This region was called New France.

During the 1600s and 1700s, England and France fought each other for territory around the globe. Eventually, by 1763, the British gained control of all of Canada. Tragically, European warfare and diseases were destroying the Native American cultures during this time.

From Colony to Nation For about 100 years, Great Britain held Canada as a colony. A **colony** is an overseas territory with political and economic ties to the parent country. While Canada was ruled by Great Britain, English and French areas were kept separate. Each region had its own colonial government. In 1867 the different colonies of Canada became one nation known as the Dominion of Canada. As a **dominion,** Canada had its own government to run local affairs. Great Britain, however, still controlled Canada's relations with other countries.

The new Canadian government promised continued protection for the French language and culture in Quebec. Yet many English-speaking Canadians did not always keep this promise. French speakers often claimed that they were treated unfairly because of their heritage. Canada often was torn apart by disputes between the two ethnic groups.

During the 1900s, Canadians fought side by side with the British and Americans in two World Wars. Canada's loyal support in these conflicts gradually led to the nation's full independence. In 1982 Canadians peacefully won the right to change their constitution without British approval. Today, only one major link between Canada and Britain remains. The British king or queen still reigns as king or queen of Canada, but this is a ceremonial position with no real power.

Canada's Government The Canadians have a British-style parliamentary democracy. In a **parliamentary democracy,** voters elect representatives to a lawmaking body called Parliament. These representatives then choose an official called the prime minister to head the government. Since the British king or queen visits Canada only once in a while, a Canadian official called the governor-general carries out most of the government's ceremonial duties.

✓ Reading Check **What was the result of Great Britain keeping the French and British areas separate?**

A Bilingual Country

Canada's history of being colonized by both France and Britain means that today, two European languages and cultures exist side by side. About one-fourth of the Canadians are descended from French-speaking settlers. (By comparison, in the United States, only 1 person out of 20 claims French ancestry.) Most of these people live in Quebec. In Quebec, the French, not the British, are the majority ethnic group.

The people of Quebec have long refused to give up their French language and customs. They did not want to "become English." As a result, Canada today is a bilingual country, with two official languages.

Primary Source

A DECLARATION OF FIRST NATIONS

"We the Original Peoples of this land know the Creator put us here. The Creator gave us laws that govern all our relationships to live in harmony with nature and mankind.
The Laws of the Creator defined our rights and responsibilities.
The Creator gave us our spiritual beliefs, our languages, our culture, and a place on Mother Earth, which provided us with all our needs. We have maintained our Freedom, our Languages, and our Traditions from time immemorial.
We continue to exercise the rights and fulfill the responsibilities and obligations given to us by the Creator for the land upon which we were placed.
The Creator has given us the right to govern ourselves and the right to self-determination.
The rights and responsibilities given to us by the Creator cannot be altered or taken away by any other Nation."

Copyright © Assembly of First Nations National Indian Brotherhood 2001

Analyzing Primary Sources

The U.S. Declaration of Independence states that ". . . all men are created equal, that they are endowed by their Creator with certain **unalienable** Rights. . ." Look up the meaning of *unalienable*. Then, identify the line in the Declaration of First Nations that supports this statement.

Time to Play

Fifteen-year-old Natalie Menard has been playing ice hockey since she was five years old. Natalie also enjoys visiting her cousin Angela, who lives in Toronto, Ontario. More than 6 miles (10 km) of covered walkways and underground tunnels in downtown Toronto connect subways with shops, offices, hotels, and restaurants. Natalie and Angela walk from place to place without even thinking of the weather.

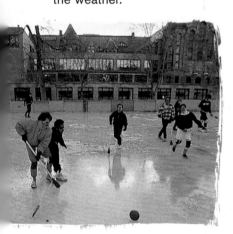

An official language is one that is recognized by the government as being a legal language for conducting government business. Government documents and publications in Canada are printed in English and French. Traffic signs are also printed in both languages. School students learn to speak both languages. Of course, some areas of the country favor one language over the other. What language do you think is more popular in Quebec?

For many years, many French-speaking people have wanted Quebec to secede, or withdraw, from Canada. They would like Quebec to become an independent country, apart from the rest of the Canadian provinces. They do not believe that French culture can be protected in a largely English-speaking country. So far, they have been defeated in two very important votes on this issue. However, Canada's future as a united country is still uncertain.

✓ Reading Check **What are Canada's two official languages?**

Nunavut, A New Territory

As you have already learned, the first peoples of Canada were Inuit and other Native Americans. In recent years, the Canadian government has given the first peoples more control over their land. In 1999, the new territory of Nunavut was created for the Inuit. *Nunavut* is an Inuit word that means "our land." The Inuit now control the government and mineral rights in this new territory. In this way, most of the Inuit living in Canada have autonomy, the right to govern themselves. When issues involve other nations, however, the national government of Canada still makes the decisions.

Nunavut is almost three times the size of the state of Texas. Part of it lies on the North American continent, but more than half of Nunavut is made up of hundreds of islands in the Arctic Ocean. As large as it is, Nunavut does not include all of Canada's Inuit people. Many live in Quebec, Labrador, and the Northwest Territories.

The population of Nunavut is also different from the rest of Canada because of its age. More than 60 percent of the population is under the age of 25. Finding jobs to take care of the young population is difficult because there is not much industry in this region. The government is the largest employer, but there are not enough jobs. People often must hunt and fish to make sure they have enough food and warm clothes to stay alive. Nunavut must develop an economy that will grow along with its population so that its citizens will not have to depend on government welfare.

✓ Reading Check **For whom was Nunavut created?**

A Growing Ethnic Diversity

Like the United States, Canada has opened its doors to a great many immigrants. Canada opened its doors to Ukrainians, who settled in the Prairie Provinces about 100 years ago. Many other settlers came from Italy, Hungary, and other European countries.

In the 1960s, Canada welcomed refugees and other people who lost their homes due to war or natural disasters. Many of these people came from Asia, especially China, Southeast Asia, and India. Cities such as Vancouver on the west coast have sizeable Asian populations. Many Africans have also migrated to Canada.

Canada also has a long history of religious diversity. Most Canadians are Roman Catholic or Protestant. Still, many also follow Judaism, Buddhism, Hinduism, or Islam.

Food, Sports, and Recreation   Because Canada has such ethnic diversity, people here can enjoy a variety of tasty foods. People from many different groups have settled in cities such as Toronto. You can walk down the street and sample the foods of Ukraine, Greece, Italy, the Caribbean, and Asia all in the same day.

Canadians enjoy a variety of activities, especially outdoor sports. You will find local parks and national parks crowded with people exercising and having fun. Many young Canadians enjoy playing ice hockey. They also take part in other winter sports, including skiing, skating, curling, and snowboarding. During the summer, they might go sailing on Lake Ontario. Professional football and hockey are popular spectator sports. Many Canadian sports fans also flock to see the major league baseball games played in Toronto's and Montreal's large indoor stadiums.

Reading Check   What groups make up Canada's diverse population?

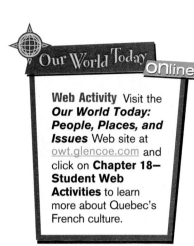

Web Activity Visit the *Our World Today: People, Places, and Issues* Web site at owt.glencoe.com and click on **Chapter 18— Student Web Activities** to learn more about Quebec's French culture.

Assessment

Defining Terms

1. Define colony, dominion, parliamentary democracy, prime minister, bilingual, autonomy.

Recalling Facts

2. History Who were the first peoples to live in Canada?

3. Government What is the new territory that was created in 1999, and what does its name mean?

4. Culture Name four activities enjoyed by Canadians.

Critical Thinking

5. Analyzing Information What is the link between Canada and Great Britain?

6. Summarizing Information What are two reasons for Canada's ethnic diversity?

Graphic Organizer

7. Organizing Information Create a diagram like this one. List two examples under each heading in the outer ovals.

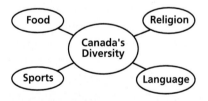

Applying Social Studies Skills

8. Analyzing Maps Study the major languages map on page 478. How does the language pattern in Canada differ from that of the United States?

Matthew Coon Come: Man With a Mission

The National Chief of the Mistissini First Nation of the Cree, Ne-Ha-Ba-Nus, "the one who wakes up with the sun," is also known as Matthew Coon Come. Matthew's goal is to preserve the right of the native peoples to choose whether they will live in the cities or the wilderness.

A Proud Chief

In 1990 Matthew Coon Come led a fight against a proposed hydroelectric project, which would have flooded Cree lands in Quebec. He helped organize a canoe trip to get publicity for Cree leaders. The trip was from James Bay, through Lake Erie, down the Hudson River, and finally to New York City. The strategy was brilliantly effective. Coon Come gained much-needed worldwide attention and made his plea directly to New Yorkers, who cancelled their plans to buy power from the proposed project.

As Grand Chief of the Crees of northern Quebec, Matthew Coon Come has become a foe of industry and politicians who want to separate Quebec from Canada. He has stated that even if Quebec secedes, the Native Americans living in Quebec choose to stay part of Canada. Coon Come speaks for only 12,000 Cree, Inuit, Nadkapi, and Innu people, but they live on two-thirds of the land area of Quebec.

What if the Native American peoples who control two-thirds of the territory should elect to leave the new Quebec and rejoin Canada? Would this mean a problem for the newly formed Quebec nation? No one is really sure what would happen in that case.

For recognition of his leadership in environmental, human rights, and tribal communities, Matthew Coon Come has received numerous awards.

▲ Matthew Coon Come

▶ Making the Connection

1. Why was Matthew Coon Come so opposed to the proposed hydroelectric project?

2. How will the Native Americans be affected if Quebec is successful in seceding from Canada?

3. **Synthesizing Information** Matthew Coon Come has his Christian name and his Cree name. They represent the two worlds he lives in. Develop a new name for yourself and explain what it represents.

Section 1 — Canada's Major Regions

Terms to Know
province
Nunavut
indigenous
glacier
tundra
service industry
overfishing

Main Idea
Canada is a vast country with many landforms, climates, resources, and peoples.

✓Region Canada, the second-largest country in the world, is rich in natural resources.

✓Economics Canada's economy is rich in fertile farmland, natural resources, and skilled workers.

✓Economics One of the best fishing grounds in the world is found in the Grand Banks off the coast of Newfoundland.

✓Place Quebec is the largest province.

✓Culture Quebec and Ontario have Canada's largest cities and most of its people.

✓Region Most of Canada has a cool or cold climate. Milder temperatures are found in the southern part of the country.

Section 2 — The Canadians

Terms to Know
colony
dominion
parliamentary democracy
prime minister
bilingual
autonomy

Main Idea
Canadians of many different backgrounds live in towns and cities close to the United States border.

✓History Inuit and other Native Americans were the first Canadians. French and British settlers later built homes in Canada. Large numbers of immigrants have recently come from Asia and eastern Europe.

✓Government Canada's government is a parliamentary democracy headed by a prime minister.

✓Culture Some people in French-speaking Quebec want to separate from the rest of Canada.

✓Culture Canada's native peoples have recently been given more autonomy to govern themselves.

Horseshoe Falls, Canada—
one of the two waterfalls
that makes up Niagara Falls ▶

chapter 18

Assessment and Activities

Using Key Terms

Match the terms in Part A with their definitions in Part B.

A.

1. province
2. glacier
3. Nunavut
4. indigenous
5. overfishing
6. dominion
7. service industry
8. bilingual
9. prime minister
10. parliamentary democracy

B.

a. having or speaking two languages
b. giant sheet of ice
c. businesses that provide services to people
d. voters elect representatives to a lawmaking body called Parliament
e. government leader chosen by Parliament
f. Canada's newest territory created in 1999
g. too many fish are taken without giving time for the species to reproduce
h. regional political division
i. native
j. nation that has its own government to run local affairs

Reviewing the Main Ideas

Section 1 Canada's Major Regions

11. **Economics** What kind of economy does Canada have?
12. **Region** Which three provinces are good agricultural areas?
13. **Location** Which province is the most heavily populated?
14. **Government** Describe two ways in which Canada's government plays a role in the nation's economy.
15. **History** Who are the main inhabitants of many of Canada's northern regions?
16. **Economics** What are three economic activities of British Columbia?
17. **Human/Environment Interaction** Explain why Canada's government must regulate how many fish can be caught in the Grand Banks.

Section 2 The Canadians

18. **Culture** Why do some of Quebec's people want independence from Canada?
19. **Culture** What are Canada's two official languages?
20. **History** Who were the first people of Canada?

 **Canada**

Place Location Activity

On a separate sheet of paper, match the letters on the map with the numbered places listed below.

1. Hudson Bay
2. Nunavut
3. British Columbia
4. Ottawa
5. Quebec (province)
6. St. Lawrence River
7. Rocky Mountains
8. Winnipeg
9. Ontario
10. Nova Scotia

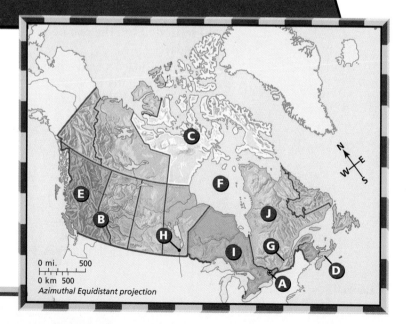

Azimuthal Equidistant projection

Self-Check Quiz Visit the *Our World Today: People, Places, and Issues* Web site at <u>owt.glencoe.com</u> and click on **Chapter 18—Self-Check Quizzes** to prepare for the Chapter Test.

Critical Thinking

21. **Making Comparisons** Compare the climates of western and eastern Canada.

22. **Analyzing Information** Why do most Canadians live in the south of Canada?

23. **Categorizing Information** Choose one of Canada's provinces or territories. Complete a chart like the one below with at least two facts or examples under the headings.

Province or Territory	Landforms	Resources
	Major Cities	Products

Current Events Journal

24. **Writing a Speech** List what you know about Matthew Coon Come. Take the role of either an environmentalist, a Native American, a vice president of the hydroelectric project, a school teacher, or a parent of five children living in Montreal. Write a short speech explaining your views of the hydroelectric project or the secession of Quebec.

Mental Mapping Activity

25. **Focusing on the Region** Create a simple outline map of Canada. Refer to the map on page 475, and then label the following on your map:

- Arctic Ocean
- Quebec (province)
- Pacific Ocean
- Ontario
- Atlantic Ocean
- British Columbia
- Rocky Mountains
- Nunavut
- Hudson Bay
- Ottawa

Technology Skills Activity

26. **Using the Internet** Access the Internet and search for information on the Inuit and the new territory of Nunavut. Create an illustrated time line that shows the steps leading to the creation of the new territory.

The Princeton Review

Standardized Test Practice

Directions: Study the graph below, and then answer the questions that follow.

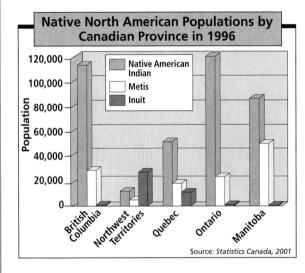

Native North American Populations by Canadian Province in 1996

Source: *Statistics Canada, 2001*

1. **Which of the following provinces has the largest Native American population?**

   A Northwest Territories

   B Manitoba

   C Ontario

   D British Columbia

2. **In what part of Canada does most of the Inuit population live?**

   F The northern part of the country

   G Along the U.S. border

   H Near the Atlantic coast

   J In cities such as Toronto and Ontario

Test-Taking Tip: Sometimes you cannot answer a question directly from the information in a map or a graph. In these cases, you have to make an *inference*, or draw a conclusion, that is supported by information in the map or graph. The clues may also help you get rid of wrong choices.

Chapter 19

The United States

The World and Its People
NATIONAL GEOGRAPHIC

To learn more about the people and places of the United States, view *The World and Its People* **Chapter 4** video.

Our World Today **online**

Chapter Overview Visit the *Our World Today: People, Places, and Issues* Web site at owt.glencoe.com and click on **Chapter 19—Chapter Overviews** to preview information about the United States.

Identifying Main Ideas Study Foldable
Asking yourself questions as you read helps you to focus on main ideas of the material and better understand it. Make this foldable and use it as a journal to record and answer your own questions about the United States.

Step 1 Fold a sheet of paper in half from top to bottom.

Step 2 Then fold the paper in half from side to side.

Step 3 Label the foldable as shown.

Reading and Writing Before you read the chapter, list questions you have about the land, people, and economy of the United States. Then, as you read the chapter, write down more questions that occur to you on the pages of your journal. Be sure to review your questions and fill in all the correct answers.

◀ **Statue of Liberty in New York Harbor, New York**

Why It Matters

Leading the Free World
The United States is the most powerful nation in the world. It has the world's largest economy and is a leading representative democracy. Immigrants from nearly every other nation of the world have moved here in order to enjoy the freedom the United States Constitution provides.

A Vast, Scenic Land

Guide to Reading

Main Idea

The United States has a great variety of landforms and climates.

Terms to Know

- contiguous
- megalopolis
- coral reef

Reading Strategy

Create a chart like the one below. Fill in details about each of the seven physical regions of the United States.

Region	Details

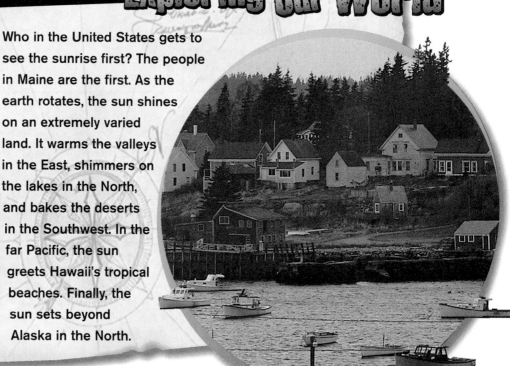

Who in the United States gets to see the sunrise first? The people in Maine are the first. As the earth rotates, the sun shines on an extremely varied land. It warms the valleys in the East, shimmers on the lakes in the North, and bakes the deserts in the Southwest. In the far Pacific, the sun greets Hawaii's tropical beaches. Finally, the sun sets beyond Alaska in the North.

The United States stretches 2,807 miles (4,517 km) across the middle part of North America. The 48 states in this part of the country are **contiguous,** or joined together inside a common boundary. These states touch the Atlantic Ocean, the Gulf of Mexico, and the Pacific Ocean. Our neighbors are Canada to the north and Mexico to the south.

Two states lie apart from the other 48. Alaska—the largest state—lies in the northwestern portion of North America. Hawaii is in the Pacific Ocean about 2,400 miles (3,862 km) southwest of California.

From Sea to Shining Sea

The United States is ranked as the fourth-largest country in the world. Only Russia, Canada, and China are larger. Like a patchwork quilt, the United States has regional patterns of different landscapes. You can see swamps and deserts, tall mountains and flat plains.

The contiguous states have five main physical regions: the Coastal Plains, the Appalachian Mountains, the Interior Plains, the Mountains and Basins, and the Pacific Coast. Alaska and Hawaii each has its own set of physical landforms.

The Coastal Plains   A broad lowland runs along the eastern and southeastern coasts of the United States. The eastern lowlands are called the Atlantic Coastal Plain. The lowlands in the southeast border the Gulf of Mexico and are called the Gulf Coastal Plain. Find these coastal plains on the map below. Then look at the population map on page 522. What large cities lie in the Atlantic Coastal Plain?

Boston, New York City, Philadelphia, Baltimore, and Washington, D.C., all lie in the Atlantic Coastal Plain. These cities and their suburbs form an almost continuous line of settlement. Geographers call this kind of huge urban area a megalopolis.

The Gulf Coastal Plain is wider than the Atlantic plain. Soils in this region are better than those along the Atlantic coast. Texas and Louisiana both have rich deposits of oil and natural gas. The large cities

NATIONAL GEOGRAPHIC

The United States: Physical

Elevations

Feet		Meters
10,000		3,000
5,000		1,500
2,000		600
1,000		300
0		0

⊕ National capital
▲ Mountain peak

Applying Map Skills

1. What river forms part of the boundary between the United States and Mexico?

2. What is the tallest mountain in the 50 states?

Find NGS online map resources @
www.nationalgeographic.com/maps

NATIONAL GEOGRAPHIC On Location

City and Country

The Interior Plains of the United States include industrial cities of the North, such as Chicago (above), and the agricultural lands of the Great Plains, like this area in Texas (above right).

Region What river divides much of the Interior Plains?

of the Gulf Coastal Plain include Houston and New Orleans, which are shown on the map on page 522.

The Appalachian Mountains Along the western edge of the Atlantic Coastal Plain rise the Appalachian (A•puh•LAY•chuhn) Mountains. The Appalachians are the oldest mountains on the continent. How can you tell? Their rounded peaks show their age. Erosion has worn them down over time. The highest peak, Mount Mitchell in North Carolina, reaches 6,684 feet (2,037 m).

The Interior Plains When you cross the Appalachians heading west, you enter the vast Interior Plains. This region has two parts. East of the Mississippi River are the Central Lowlands. Here you will find grassy hills, rolling flatlands, and thick forests. The land is fertile, and farms are productive. This area also contains important waterways.

The Great Lakes—the largest group of freshwater lakes in the world—lie in the Central Lowlands. Glaciers formed Lake Superior, Lake Michigan, Lake Huron, Lake Erie, and Lake Ontario in the distant past. The waters of these connected lakes flow into the St. Lawrence River, which empties into the Atlantic Ocean.

West of the Mississippi River stretch the Great Plains. The landscape in many places is blanketed with neat fields of grain and grassy pastures and takes on a checkerboard pattern. The Great Plains are about 500 miles (805 km) wide and stretch west to the Rocky Mountains, north into Canada, and south to the Mexican border. The rich grasslands of the Great Plains once provided food for millions of buffalo and the Native Americans who lived there. Today, farmers grow grains and ranchers raise cattle on the Great Plains.

Mountains and Plateaus The Rocky Mountains begin in Alaska and run all the way south to Mexico. Running along these mountains is a ridge called the Continental Divide. This ridge separates rivers that flow west—toward the Pacific Ocean—from those that flow east—toward the Mississippi River. Many important rivers begin in the high, snowy peaks of the Rockies. The Rio Grande, as well as the Missouri, Platte, Arkansas, and Red Rivers all flow east. The Colorado, Snake, and Columbia Rivers flow west.

Between the Rockies and the Pacific Coast are plateaus, canyons, and deserts. Plateaus are areas of flat land that rise above the land around them. A canyon is a deep valley with steep sides. The most famous of these is the Grand Canyon in Arizona.

The Pacific Coast Near the Pacific Coast rise two other mountain ranges, the Cascade Range and the Sierra Nevada. Find these ranges on the map on page 503. Even in a place as far south as California, the tops of these high mountains remain covered with snow year-round.

To the west of these Pacific ranges lie fertile valleys. The Willamette Valley in Oregon and the Central Valley in California both produce abundant crops. Many of the fruits and vegetables you eat may come from these valleys.

Alaska Mountain ranges form a semicircle over the northern, eastern, and southern parts of Alaska. Mount McKinley—the tallest mountain in North America—stands 20,320 feet (6,194 m) high in the Alaska Range. The northern part of the state borders on the frigid Arctic Ocean, and you can almost see Russia from Alaska's western shores. Most people in Alaska live along the southern coastal plain or in the central Yukon River valley.

Hawaii Eight large islands and more than 120 smaller islands make up Hawaii, our island state in the Pacific Ocean. Volcanoes on the ocean floor erupted and formed these islands. Some of the islands have coral reefs, formed by the skeletons of small sea animals. These structures lie just above or submerged just below the surface of the water.

Reading Check **What is the Continental Divide?**

NATIONAL GEOGRAPHIC **On Location**

Mt. McKinley

Fierce winds and high altitude make McKinley one of the coldest mountains in the world. The summit is below 0°F (−17.8°C) almost all of the time.

Location Why do you think it gets colder as the altitude increases?

A Variety of Climates

Because the United States is such a large country, you probably expect it to have a variety of climates. You are right! Most of the country lies squarely in the middle latitude region—from 23°30′N to 66°30′N latitude. This means the region has the greatest variety of climates. With Alaska and Hawaii, our country also has high-latitude and tropical climates.

In spite of the great variety of climates in the United States, generally the climate is mild, or temperate. That means most places are not usually too hot or too cold. In general, the farther north you travel, the summers grow shorter and cooler and the winters become longer and colder. As you travel south, the summers become longer and hotter and the winters shorter and milder. Being located in the middle of North America gives the United States an average warmer climate than Canada, and an average cooler climate than Mexico.

It is important to remember that climate also changes with elevation. The higher the land, the cooler the temperatures. The highest mountain peaks are snow-covered year-round, while the low valleys and deserts stay much warmer year-round.

As you would expect, the climate in much of Alaska is similar to that of northern Canada. Hawaii and Florida have warm tropical climates with heavy rainfall much of the year.

✓ Reading Check In general, what is the climate of the U.S.? What does this mean?

Section 1 Assessment

Defining Terms

1. **Define** contiguous, megalopolis, coral reef.

Recalling Facts

2. **Place** How does the United States rank in size among all the countries of the world?

3. **History** Which region once supported millions of buffalo and the Native Americans who depended on them?

4. **Place** What is the largest group of freshwater lakes in the world?

Critical Thinking

5. **Understanding Cause and Effect** How were the Hawaiian Islands formed?

6. **Drawing Conclusions** What challenges do you think result from the distance between Alaska, Hawaii, and the other states?

Graphic Organizer

7. **Organizing Information** Create a diagram like this one to compare the Atlantic and Gulf Coastal Plains. In the separate outer parts of the ovals, write the qualities that make each region different. In the overlapping area, write the characteristics that the two areas share.

Atlantic Coastal Plain ⬭⬭ Gulf Coastal Plain

Applying Social Studies Skills

8. **Analyzing Maps** Look at the physical map on page 503 and the population map on page 522. At what elevations do the cities with more than 5 million people lie?

An Economic Leader

Guide to Reading

Main Idea

The powerful United States economy runs on abundant resources and the hard work of Americans.

Terms to Know

- free enterprise system
- fossil fuel
- landfill
- recycling
- free trade

Reading Strategy

Create a chart like this one. List at least five challenges the United States faces in the twenty-first century.

Challenges

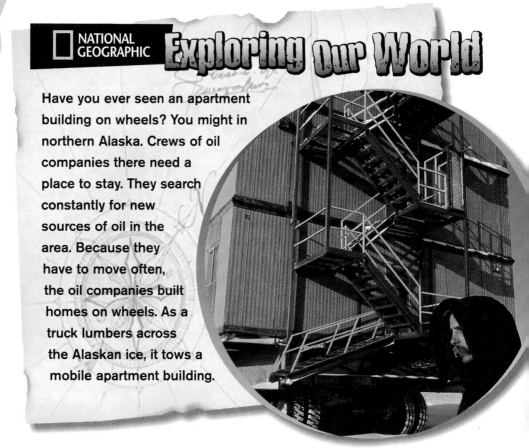

NATIONAL GEOGRAPHIC

Exploring Our World

Have you ever seen an apartment building on wheels? You might in northern Alaska. Crews of oil companies there need a place to stay. They search constantly for new sources of oil in the area. Because they have to move often, the oil companies built homes on wheels. As a truck lumbers across the Alaskan ice, it tows a mobile apartment building.

The United States has a large, energetic, and growing economy. Fueling all of this economic activity is freedom. The **free enterprise system** is built on the idea that individual people have the right to run businesses to make a profit with limited government interference and regulation. Americans are free to start their own businesses and to keep the profits they earn. They are free to work in whatever jobs they want—and for whatever employers they want. This has helped create great economic success.

The World's Economic Leader

The United States is rich in resources and has a hardworking labor force. As a result, the country has built the world's largest economy—in terms of how much money is made from the sale of its goods and services. In fact, the American economy is larger than the next two largest economies—China's and Japan's—combined.

Farms in the United States produce about one-half of the world's corn and about one-tenth of its wheat. American farmers raise about 20 percent of the world's beef, pork, and lamb. The country exports

Analyzing the Graph

The economy of the United States is divided into the four main areas shown here.

Economics What percentage of the U.S. economy is not made up of service/information industry?

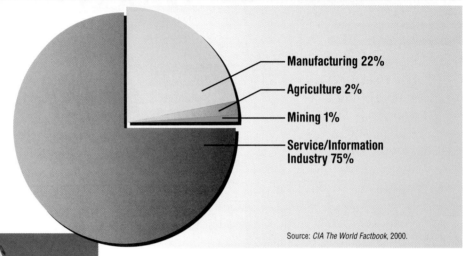

Manufacturing 22%

Agriculture 2%

Mining 1%

Service/Information Industry 75%

Source: *CIA The World Factbook*, 2000.

▲ Filmmaking is a service industry.

more food than any other nation. Yet agriculture is only a small part of the American economy. It makes up about 2 percent of the value of all goods produced in the country.

The United States has rich mineral resources. About one-fifth of the world's coal and copper and one-tenth of the world's petroleum come from the United States. The country also has large amounts of iron ore, zinc, lead, silver, gold, and many other minerals. Mining, though, makes up little more than 1 percent of the nation's economy.

American factory workers build cars and airplanes. They make computers and appliances. They process foods and make medicines. Manufacturing accounts for about one-fifth of the American economy.

By far, the largest part of the economy is services. A service industry is a business that provides services to people instead of producing goods. Banking and finance are services. So is entertainment—and people all over the world buy American movies and CDs. The United States is a leader in tourism, another service industry. Computer-based, online services have also emerged as an important American service industry.

✓ Reading Check What is the largest part of the United States economy?

In the Twenty-First Century

The American economy, although strong, faces challenges in the twenty-first century. One challenge is how to clean up pollution and trash. Americans burn fossil fuels—coal, oil, and natural gas—to power their factories and run their cars. When sulfur oxides from coal-burning power plants and nitrogen oxides from cars combine with moisture in the air, they form acids. When acidic moisture falls to Earth as rain or snow, it is called acid rain. Many scientists believe that acid rain harms trees, rivers, and lakes.

The fast-paced American way of life creates another problem. People generate huge amounts of trash. **Landfills,** the areas where trash companies dump the waste they collect, grow higher and higher each year. Many communities now promote **recycling,** or reusing materials instead of throwing them away. Recycling cuts down on the amount of trash.

Quality Schools   The ability to develop new technology has been a major source of strength for the American economy. Researchers work constantly to find new products to make people's lives easier, healthier—and more fun. Quality schools that produce educated and creative people have helped the country become a world leader in satellites, computers, health care, and many other fields. Keeping our number one position will require just as much creative thinking and hard work. You will need to learn and use these new technologies to stay productive in your future jobs.

Health Care   Some of the best health care in the world is available in the United States. New medical technologies are constantly being developed to improve the level of health care. These new technologies help to save lives, but they also raise medical costs overall.

The rising cost of medical care is a problem that affects the delivery of health care in the United States. Employers of temporary, contract,

Factories

A worker inspects computer components. Along with agriculture, America's economy is strong in technology, science, education, and medicine.

Economics **What percentage of the American economy is manufacturing?**

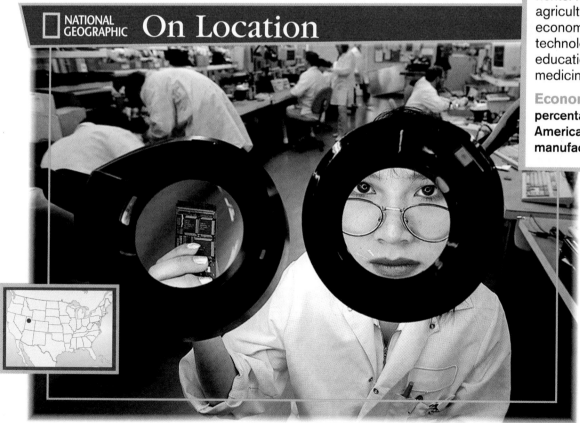

NATIONAL GEOGRAPHIC On Location

and part-time workers do not usually offer health insurance as a benefit. Approximately one out of five people in the United States does not have medical insurance. The majority of these uninsured are children of lower income families. Because they do not have insurance, they might not be able to get some forms of medical care. For example, some preventative services are not available to people without medical insurance.

Some people think that the United States should have a national program of health insurance. In this type of program, the government provides some level of medical care at an affordable rate. Critics of nationalized health care point to countries such as Canada, where people may have lengthy waiting periods for operations and where there is little choice, if any, of doctors or hospitals.

World Trade The United States leads the world in the value of all its imports and exports. Millions of Americans depend on trade for their jobs. American leaders have worked hard to promote free trade. Free trade means taking down trade barriers such as tariffs or quotas so that goods flow freely between countries. In 1992 the United States, Mexico, and Canada entered into the North American Free Trade Agreement (NAFTA). This agreement, which took effect in 1994, promised to remove all barriers to trade among those three countries.

✓ Reading Check How might burning fossil fuels harm the environment in the United States?

2 Assessment

Defining Terms
1. Define free enterprise system, fossil fuel, landfill, recycling, free trade.

Recalling Facts
2. **Economics** What is a major challenge of the American economy in the twenty-first century?
3. **Culture** What problem has been created by the fast-paced American way of life?
4. **Economics** What was the goal of the North American Free Trade Agreement (NAFTA)?

Critical Thinking
5. **Analyzing Information** Describe two characteristics of the United States that have helped it become a world leader.
6. **Understanding Cause and Effect** What reasons can you give for the fast-paced American way of life causing more trash?

Graphic Organizer
7. **Organizing Information** Create a diagram like this one. In the outer ovals, write one specific example under each heading.

Applying Social Studies Skills

8. **Interpreting Graphs** Study the graph of the American economy on page 508. Then determine what part of the American economy is two times larger than mining.

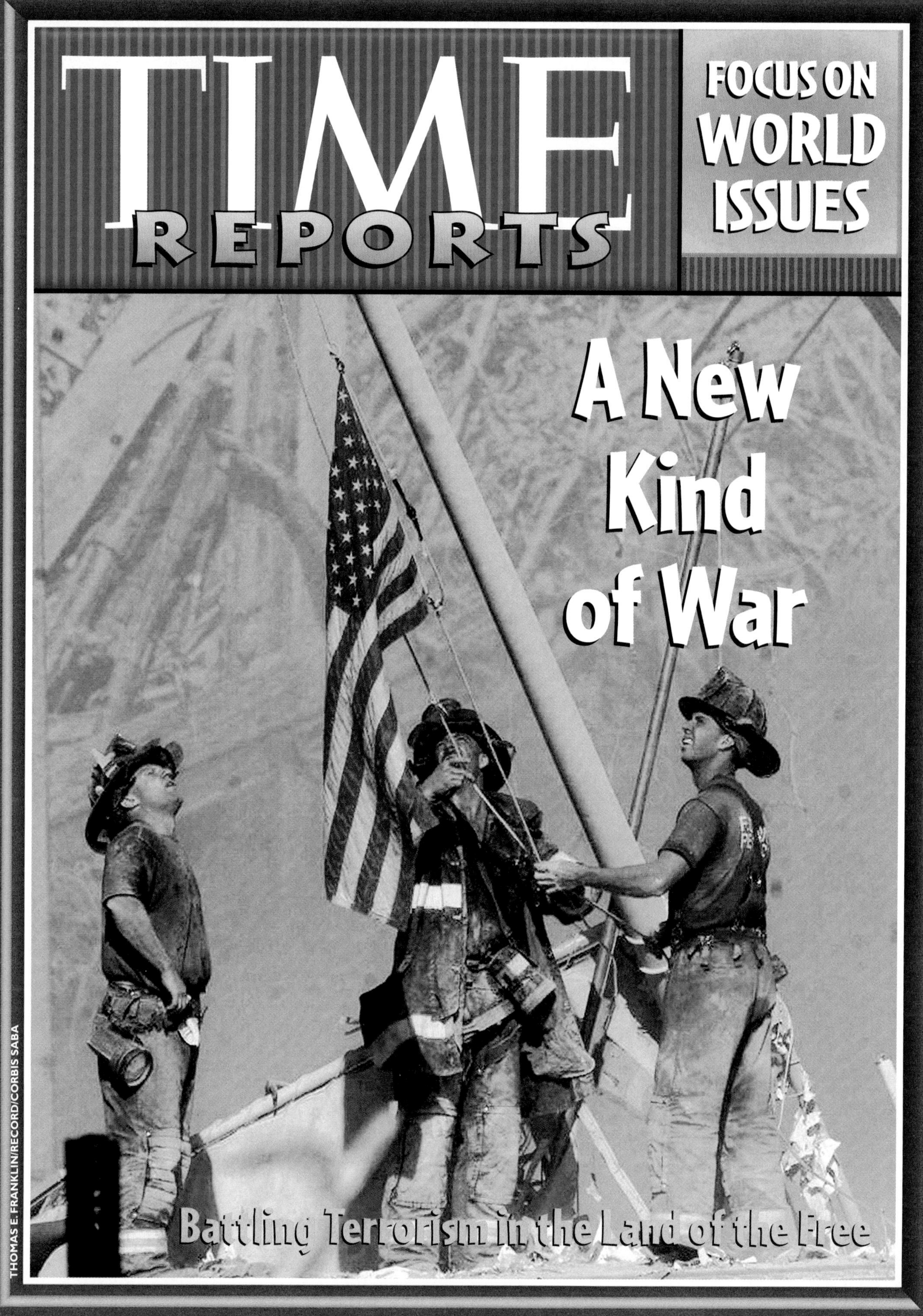

TIME
REPORTS
FOCUS ON WORLD ISSUES
A New Kind of War
Battling Terrorism in the Land of the Free
THOMAS E. FRANKLIN/RECORD/CORBIS SABA

A Day for Heroes

September 11, 2001, was a day John Jonas will never forget. A New York City firefighter, Jonas was the captain of Ladder Company 6. That morning two **hijacked** commercial airplanes were deliberately flown into the World Trade Center's twin towers.

Jonas and five other men in his company rushed to the scene. An hour later they were walking down the fire stairs of the south tower. An older woman they were rescuing told them she couldn't go on. They told her they wouldn't leave her. And just then the building collapsed around them in clouds of dust. "I'm thinking," Jonas said later, "'I can't believe this is how it ends for me.'"

But life didn't end either for Jonas or for the five firefighters with him. Nor did it end for Josephine Harris, the woman they saved. Above and below them and in the nearby north tower, more than 2,600 people died. But somehow the part of the stairway they were on didn't collapse. "It was a freak of timing," said Jonas. Another minute,

▲ New York City firefighters battle the World Trade Center disaster on September 11, 2001.

either way, and the crumpling building would have crushed them like the others.

More than 300 rescue workers lost their lives trying to save others that day. All of them were heroes.

No Ordinary Crime

On September 11, hijackers also flew an airplane into the Pentagon—the headquarters of the U.S. military. Another 125 people died. Other hijackers seized a fourth airplane, but the passengers heroically fought back. Instead of hitting a building, the plane crashed in a field in Pennsylvania leaving no survivors.

Clearly, the horrible crashes were not ordinary crimes. They were acts of **terrorism**. Terrorism is the illegal use of violence against people or property to make a point. The point may involve a particular belief, such as religion or politics.

EXPLORING THE ISSUES

1. **Interpreting Points of View** Do you agree with the article's view of heroes? Why or why not?

2. **Making Inferences** How is the violence of terrorism different from the violence of war?

1998
Terrorists destroy the U.S. Embassy in Nairobi, Kenya.

2000
Terrorists in Yemen batter a U.S. warship, the USS Cole.

2001
Suicide pilots level the World Trade Center's twin towers.

Al-Qaeda

The terrorists who hijacked the airplanes belonged to a group called al-Qaeda (al•KAY•dah). The group was founded by Osama bin Laden, a wealthy Saudi Arabian.

Al-Qaeda was created to fight the Russian invasion of Afghanistan. After the Russians left Afghanistan, al-Qaeda members changed their goals. They wanted to force all non-Muslims out of the Middle East. They hated the U.S. troops based in Saudi Arabia and the Jewish people living in Israel.

Al-Qaeda's members also believed Muslims were being changed too much by modern ideas. They hated freedom of religion and wanted strict religious leaders to control Muslim countries.

Al-Qaeda's beliefs were not shared by all Muslims. The attacks on the United States horrified people around the world, including millions of Muslims who live in the Middle East, the United States, and elsewhere.

Closing in on America

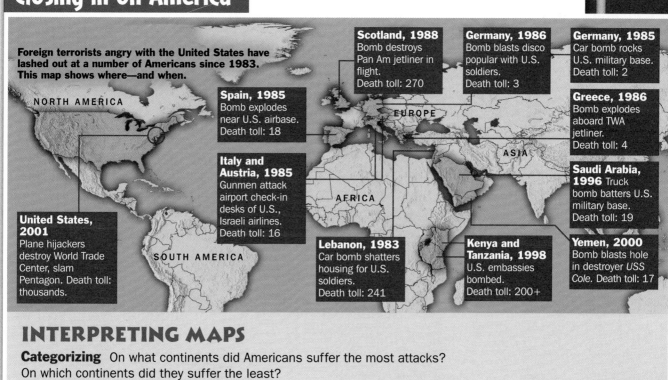

Foreign terrorists angry with the United States have lashed out at a number of Americans since 1983. This map shows where—and when.

Scotland, 1988 Bomb destroys Pan Am jetliner in flight. Death toll: 270

Germany, 1986 Bomb blasts disco popular with U.S. soldiers. Death toll: 3

Germany, 1985 Car bomb rocks U.S. military base. Death toll: 2

Spain, 1985 Bomb explodes near U.S. airbase. Death toll: 18

Greece, 1986 Bomb explodes aboard TWA jetliner. Death toll: 4

Italy and Austria, 1985 Gunmen attack airport check-in desks of U.S., Israeli airlines. Death toll: 16

Saudi Arabia, 1996 Truck bomb batters U.S. military base. Death toll: 19

United States, 2001 Plane hijackers destroy World Trade Center, slam Pentagon. Death toll: thousands.

Lebanon, 1983 Car bomb shatters housing for U.S. soldiers. Death toll: 241

Kenya and Tanzania, 1998 U.S. embassies bombed. Death toll: 200+

Yemen, 2000 Bomb blasts hole in destroyer USS Cole. Death toll: 17

INTERPRETING MAPS

Categorizing On what continents did Americans suffer the most attacks? On which continents did they suffer the least?

513

Behind the Hatred

What makes the United States the target of so much deadly anger? One answer is its support for Israel. Israel was founded in 1948. Soon afterwards, an Arab-Israeli war forced about 750,000 Palestinian Arabs from their homes.

Today many of those Palestinians live in refugee camps. So do their children and their grandchildren. Those 4 million Palestinians want a nation of their own. Israel has offered to exchange land for a promise of peace. But so far the Palestinians have rejected that offer.

U.S. Troops in Saudi Arabia

Another source of anger is the presence of U.S. troops in Saudi Arabia. The Saudi government asked the United States to station troops there. But the holy cities of Makkah (Mecca) and Madinah (Medina) are in Saudi Arabia. To many Muslims, U.S. troops on Saudi soil are an insult to Islam.

To terrorist Osama bin Laden and his followers, the solution to these problems was violence. In 1996, he urged Muslims to kill U.S. troops in Saudi Arabia. In 1998, he called for attacks on American **civilians**. Civilians are people

▲ **Fanatics hail terrorist Osama bin Laden in 2001.**

not in the armed forces or diplomatic services. By the end of 2001, several thousand people had been killed.

The United States responded to September 11 with a determination and resolve bin Laden surely didn't expect. "Our war on terror begins with al-Qaeda," President George W. Bush said. "It will not end until every terrorist group of global reach has been found, stopped, and defeated."

EXPLORING THE ISSUES

1. Drawing Conclusions What are some reasons many Americans support Israel's presence in the Middle East?

2. Making Inferences Why are acts of terror against civilians often effective?

War on All Fronts

On September 11, 2001, President George W. Bush spoke to the nation and announced a war on terrorism. He warned the world, "We will make no distinction between the terrorists who committed these acts and those who harbor them."

Al-Qaeda had followers all over the world, but it was based in Afghanistan. The Taliban, a strict religious party that controlled Afghanistan's government, protected Osama bin Laden and al-Qaeda. They refused to help the U.S. So in October, the President ordered the U.S. military to attack Afghanistan.

Aid for Children

The U.S. was not at war with the Afghan people, but with the Taliban and al-Qaeda. During the attack, U.S. planes dropped food and medicine to the men, women, and children in the civilian population.

Nations around the world backed the United States and began arresting al-Qaeda members hiding in their countries. Some sent troops to help the Americans fight in Afghanistan.

A few weeks after the attack began, the Taliban government collapsed. With the aid of the U.S. and its allies, the Afghan people created a new government. Meanwhile, American and allied troops began hunting for al-Qaeda forces in Afghanistan. The U.S. also sent troops to the Philippines, Yemen, and the nation of Georgia to train local troops to fight terrorists.

MARK RICHARDS

▲ Safety checks help prevent terrorism. But the cost—less freedom—worries many Americans.

Liberty and Security

At home, federal agencies stepped up their efforts to find terrorists. President Bush ordered banks to hold money belonging to groups linked to terrorists. Congress passed a new antiterrorist law making it easier to tap phones, intercept e-mail, and search homes.

Some people worried that the new antiterrorist law would chip away at our **liberties**—such as freedom of speech and the right to privacy. For this reason, Congress set a five-year time limit on parts of the new law. ▨

EXPLORING THE ISSUE

1. **Analyzing Information** Shortly after September 11, 2001, President Bush said, "No one should be singled out for unfair treatment or unkind words because of their ethnic background or religious faith." What do you think he meant by that statement?

2. **Problem Solving** What liberties, if any, might you be willing to give up in order to ensure national security?

Stopping Terrorism: What Can One Person Do?

The rescue workers who responded to the attacks on the World Trade Center and the Pentagon were true heroes. In the months that followed, Americans honored them for their courage and sacrifice.

The response of Americans to tragedy showed the world the nation's hidden strengths—its people. Wherever they lived, Americans reacted. They gave blood. They held candlelight **vigils** to honor the victims. They flew flags to show their unity. They cut deeply into their budgets, contributing more than $200 million in the first week to help victims' families.

They all made it clear, as a girl from Ohio told TIME For Kids, that no terrorist can weaken the nation's spirit. "They bent steel," said Danielle, 12, of the World Trade Center murderers, "but they can't break the U.S."

Be a Local Hero

Wherever you live, you can help keep that spirit alive. And you can do it even years after the disasters of September 2001 took place.

Learn all you can about terrorism. Learn what it is, why it exists, and how people at all levels of government are fighting it.

Then join that fight any way you can. With posters and letters, report successful efforts to combat this evil. Raise money for groups that help out the victims of terrorism everywhere.

▲ Terrorist attacks in September 2001 trigger a burst of patriotism everywhere in America.

Finally, refuse to give in to fear. Terrorists use fear as a weapon. If you can keep fear from changing your life, you will have taken a big bite out of terrorism.

The novelist Stephen King—who often writes about human fears—agreed. "If everybody continues working," he said, "they [the terrorists] don't win."

EXPLORING THE ISSUE

1. **Problem Solving** What might people do to stop the fear of terrorism from keeping them from doing what they want to do?

2. **Summarizing the Main Idea** Write a new title for this piece. Share it with your classmates. Explain why you think your title fits the story.

REVIEW AND ASSESS

UNDERSTANDING THE ISSUE

1. Defining Key Terms In your own words, define the following terms: *terrorism, hijacker, the Pentagon, al-Qaeda, Taliban, liberty, security, principles, ideals,* and *vigil.*

2. Writing to Inform In a 300-word article, describe a terrorist act you heard or read about. Describe how you reacted when you heard about it.

3. Writing to Persuade What do you think Americans should know about terrorism? Put your answer in a 250-word letter to the editor of your local newspaper. Support your answer with facts. Use at least five of the terms listed above.

INTERNET RESEARCH ACTIVITY

4. Use Internet resources to find information on what individuals and organizations are doing today to help victims of terrorism. Use what you learn to write a report on current efforts, and share it with the class.

5. With your teacher's help, use Internet resources to learn more about how the tragedies of September 11 resulted in an increase of visible patriotism in the United States. Focus your research on finding personal stories of how the attacks increased individual Americans' beliefs in and loyalty to the United States. Prepare a brief report on your findings.

BEYOND THE CLASSROOM

6. Study the map on page 513. Research one of the terrorist attacks noted there. What does the attack tell you about the goals, thoughts, and methods terrorists have? Describe the attack and answer those questions in a brief oral report.

▲ **Muslims mourn victims of the World Trade Center attack.**

7. Visit your school or local library. Research a country, such as Israel, Northern Ireland, or Bosnia where the people have suffered from terrorist attacks. Find out what programs have been started by groups or individuals to bring an end to the violence. Present your findings to the whole class.

TIME/CNN Poll

Fighting Terrorism: How Far Would You Go?

What are Americans willing to do to fight terrorism? These pie graphs show what a TIME/CNN Poll found out.

To prevent terrorist attacks, would you favor or oppose the government doing each of the following?

	Favor		Oppose
1. Allow police to wire-tap phone conversations of suspected terrorists without a court's okay:	68%		29%
2. Let courts jail, for as long as they want, people suspected of links to terrorist groups:	59%		38%
3. Let police intercept e-mail messages sent by anyone in the United States and scan them for suspicious words or phrases:	55%		42%
4. Require everyone in the United States to carry an identification card issued by the Federal Government:	57%		41%
5. Let police stop people on the street and search them:	29%		69%

Source: TIME magazine, October 8, 2001; Gray slices indicate respondents who were not sure.

BUILDING GRAPH READING SKILLS

1. Analyzing the Data The U.S. Constitution bars the government from making "unreasonable searches and seizures" of citizens. Which of the graphs show how Americans think about this right? What's your thinking on this issue?

2. Making Inferences The U.S. Constitution bars the government from taking away a person's "life, liberty, or property" without a fair trial. What do most people who took part in this poll seem to think about this right? Why do you think they hold that view?

FOR UPDATES ON WORLD ISSUES GO TO www.timeclassroom.com/glencoe

517

Critical Thinking Skill

Mental Mapping

Think about how you get from place to place each day. In your mind you have a picture—or **mental map**—of your route. If necessary, you could probably create sketch maps like the one below of many familiar places.

Learning the Skill

To develop your mental mapping skills, follow these steps.

- When a country or city name is mentioned, find it on a map to get an idea of where it is and what is near it.
- Create a sketch map of it and include a compass rose to determine the cardinal directions.
- As you read or hear information about the place, try to picture where on your sketch you would fill in this information.
- Compare your sketch to an actual map of the place. Change your sketch if you need to, thus changing your mental map.

Practicing the Skill

Study the sketch map at the right. Picture yourself standing *in* the map, then answer the following questions.

1. If you were facing north, looking at the Chicago Cultural Center, what route would you take to reach the Chicago Harbor?

2. You are at the Sears Tower, one of the tallest buildings in the world. About how many miles would you have to walk to get to Medinah Temple?
3. If you met your friend at the cultural center, would it be too far to walk to the Art Institute? Should you take a taxi? Explain.

Applying the Skill

Think about your own neighborhood. Create a sketch map of it from your mental map. Which neighborhood streets or roads did you include? What are the three most important features on your map?

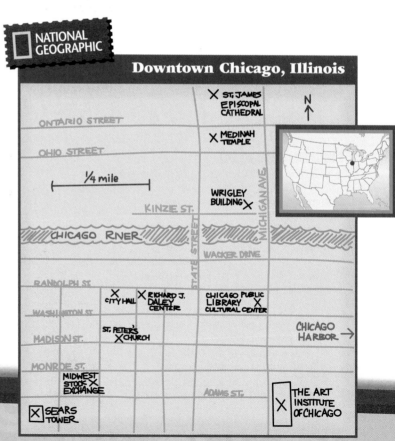

NATIONAL GEOGRAPHIC

Downtown Chicago, Illinois

The Americans

Guide to Reading

Main Idea

The United States has attracted people from all over the world who have created a land of many cultures.

Terms to Know

- representative democracy
- federal republic
- amend
- ethnic group
- rural
- urban
- suburb

Reading Strategy

Create a diagram like this one. In each outer oval, write under the heading one fact about American society as it relates to the topic given.

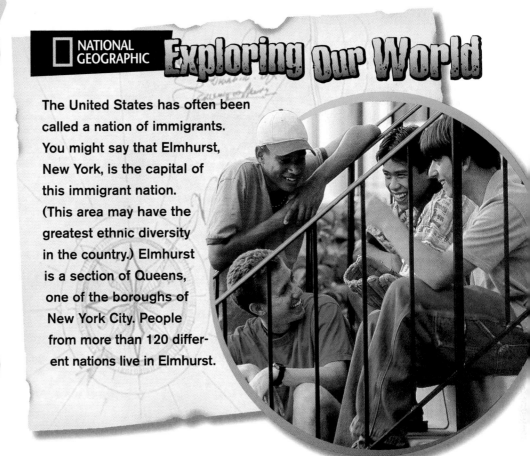

NATIONAL GEOGRAPHIC Exploring Our World

The United States has often been called a nation of immigrants. You might say that Elmhurst, New York, is the capital of this immigrant nation. (This area may have the greatest ethnic diversity in the country.) Elmhurst is a section of Queens, one of the boroughs of New York City. People from more than 120 different nations live in Elmhurst.

The United States is full of people from many different lands. What attracts people to the United States? One attraction is the freedom that Americans enjoy. Economic opportunity is another. The United States gives people in many other lands hope that they and their children can enjoy better lives.

A Rich History

The first Americans lived all over the continent. Over time, they developed different ways of life using local resources. Groups fished, planted corn and beans, or followed buffalo herds. Around A.D. 1500, Europeans began to explore the Americas. The raw materials they saw—forests, animal furs, and rich soils—soon led them to set up colonies, or overseas settlements that are tied to a parent country. The French built trading posts around the Great Lakes and interior river valleys. The Spanish built towns and missions in Florida and Georgia and from Texas to California. British and northern European colonists settled along the Atlantic coast from Georgia to Massachusetts.

Analyzing the Diagram

The United States government has three main branches.

Government Which branch makes the laws?

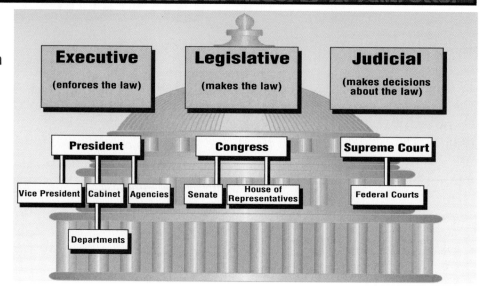

A Federal Republic By the mid-1700s, many people living in the British colonies were frustrated with British policies that infringed on their rights. In 1775, 13 colonies rebelled. On July 4, 1776, they declared independence and created the United States of America. In 1788, they adopted a new constitution that is still used today.

The United States is a republic. A republic is a form of government in which there is no king. The head of a republic is usually a president. In a republic, power belongs to the citizens who vote. A republic is a type of representative democracy. Voters elect representatives to make laws for the benefit of the people they represent. The United States is also a federal republic. A federal government is divided between a central or national government, and individual state governments. As you can see from the chart above, the national government of the United States is also divided into three branches.

The Constitution There are many different types of democracies, and you will be studying some of these in other units. One thing that most democracies have, however, is some kind of constitution, or document that identifies the rights and responsibilities of the people. The Constitution of the United States, written in 1787, is a world-famous document. It has been used as a model by many countries.

There are many reasons for the success of this document, but experts believe that the most important reason is that the Constitution can be changed, when necessary, to meet the changing needs of the country's people. Through a process called amending, the people of the United States have a peaceful way to change the basic laws of their government. U.S. citizens can adapt their laws to situations that did not exist when the original Constitution was written. Americans can also correct injustices that arise from unequal treatment of people. Examples include the Thirteenth Amendment, which abolished slavery, and the Fifteenth Amendment, which gave the right to vote to all men regardless of color or race. (Sometimes, it took more than one amendment to correct the problem. The Nineteenth Amendment gave the right to vote to women also.) The Constitution is a remarkable document that grows with the needs of the country.

A Period of Growth From 1800 to 1900, the United States grew from the 13 states along the Atlantic coast to include 45 states that reached to the Pacific Ocean. Settlers cleared forests, farmed, and often fought with Native Americans who were being pushed out of the way. Farmers grew corn in the Midwest and cotton in the South. Some people mined gold and silver in the Rocky Mountains and California.

In 1861, the nation experienced a crisis. Several Southern states seceded, or withdrew, from the national government over issues of slavery and states' rights. For four years, the North and the South fought a bitter civil war. In the end, the Southern states were brought back into the Union, and slavery was abolished. Afterward, the country began a period of great industrial and economic growth. Factories sprang up, especially in the Northeast and Midwest. Railroads reached out to bring faraway places into a national marketing system. This economic growth attracted another great wave of immigrants.

A World Leader During the early 1900s, the United States became one of the leading economies in the world. Automobiles rolled off assembly lines, electricity became common, and other technologies—the telephone and radio, for example—entered daily life.

The world plunged into two World Wars in the first half of the twentieth century. The United States took part in these wars. Our country's leaders urged the world's people to fight for freedom. American factories built tanks and airplanes, while American soldiers helped win the wars.

After World War II, the United States enjoyed great influence around the world. American companies shipped their products to all continents.

The Founders

With very few exceptions, the world knew only monarchies and absolute rulers when courageous leaders such as Thomas Jefferson (left), George Washington (center), and James Madison (right) risked their lives and fortunes to spearhead the drive for an independent United States. Jefferson was the chief author of the Declaration of Independence that the Continental Congress formally issued on July 4, 1776. Washington led the new nation's army in the Revolution, chaired the Constitutional Convention, and became the first president under the Constitution. Madison is considered the master builder of the Constitution and later served as president.

Beliefs Why do you think that the Founders were willing to risk their lives and fortunes to establish the United States?

NATIONAL GEOGRAPHIC On Location

American leaders worked to establish democracy and free enterprise in other countries. American culture spread around the globe.

At home, however, tensions existed among groups within American society. Many of the Americans who had fought in the two World Wars or had taken care of the home front were women, African Americans, Hispanic Americans, and Native Americans. After World War II, these groups became more active in seeking equal rights. Many people, including such leaders as Martin Luther King, Jr., developed methods that led to civil change. The poems on page 526 describe two views of Americans struggling to be accepted.

✓ Reading Check How did a strong economy in the U.S. help spread American culture?

One Out of Many

About 285 million people live in the United States, making it the third most populous country in the world after China and India. Compared with people in most other countries, Americans enjoy a

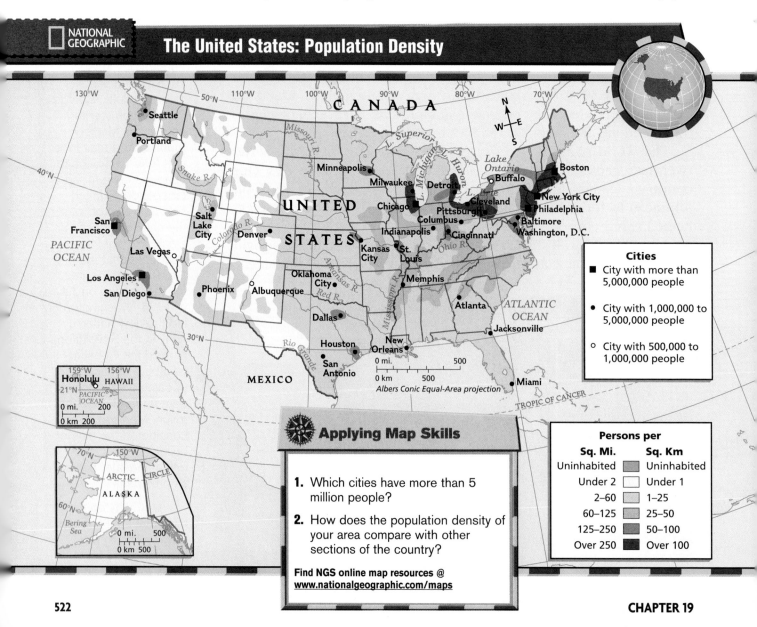

NATIONAL GEOGRAPHIC

The United States: Population Density

Cities
- ■ City with more than 5,000,000 people
- ● City with 1,000,000 to 5,000,000 people
- ○ City with 500,000 to 1,000,000 people

Albers Conic Equal-Area projection

Applying Map Skills

1. Which cities have more than 5 million people?

2. How does the population density of your area compare with other sections of the country?

Find NGS online map resources @ www.nationalgeographic.com/maps

Persons per	
Sq. Mi.	**Sq. Km**
Uninhabited	Uninhabited
Under 2	Under 1
2–60	1–25
60–125	25–50
125–250	50–100
Over 250	Over 100

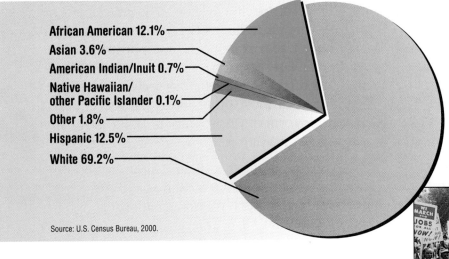

African American 12.1%
Asian 3.6%
American Indian/Inuit 0.7%
Native Hawaiian/
other Pacific Islander 0.1%
Other 1.8%
Hispanic 12.5%
White 69.2%

Source: U.S. Census Bureau, 2000.

Analyzing the Graph

There are six main ethnic groups in the United States.

Culture What percentage of the population is made up of Asians?

▲ The African American ethnic group has struggled to achieve equality with other Americans.

very high standard of living. Americans, on the average, can expect to live about 76 years. Medical advances help people live longer than their grandparents could expect to live.

Almost three-fourths of the people in our country are descended from European ethnic groups. An **ethnic group** is a group of people who share a culture, language, or history. African American ethnic groups form about 12 percent of the population. Hispanics, who trace their heritages to the countries of Latin America and Spain, are the fastest-growing ethnic group. Today, many immigrants to the United States come from China, India, other Asian countries, and the Pacific Islands. The smallest ethnic groups have lived in the country the longest—Native Americans who live in Alaska.

Language The main language of the United States is English, but you can hear many different languages spoken on American streets. One of the concerns about language has centered around how to teach children who come to school not knowing English. Schools play an important role in developing good citizens. Learning a common language, sharing in national holidays, and being taught together with children from different countries help to create a feeling of patriotism and belonging. People who are against teaching immigrant children in their native languages believe that bilingual education discourages that sense of being an American. Other people think that it is just as important to preserve the culture and language of the immigrants, and that it is better for the students to learn in their own language, at least for some period of time. This debate has not been settled to most people's satisfaction one way or the other. It is likely that the debate will continue for years to come.

Religion Religion has always been an important influence on American life. One of the first laws passed by the new country stated that "Congress shall make no laws respecting an establishment of

Rights and Responsibilities

The Constitution of the United States protects certain rights of citizens. Many of those rights, such as freedom of religion and speech, are listed in the Bill of Rights. Along with our rights as citizens, however, we also have responsibilities. Attending school so that we will be informed and effective citizens is one of our responsibilities. Obeying school rules and local state laws is one of our most important responsibilities.

religion, or prohibiting the free exercise thereof. . . ." In other words, this law said that the government could not say which religion people should follow. It also said that public or taxpayer money should not be used to support the goals of specific religions. The judicial branch of government has drawn a very sharp line between government and religion.

This does not mean that religion is not popular in the United States. About 80 percent of Americans consider themselves religious, and almost 50 percent attend some religious service on a regular basis. This rate is higher than in most other industrialized countries. Most Americans follow some form of Christianity. Judaism, Islam, Buddhism, and Hinduism are also important religions in our country.

Mobility   Americans have always been a mobile people, moving from place to place. At one time, our nation was made up entirely of rural, or countryside, areas. Now we are a nation of urban, or city, dwellers. To find more room to live, Americans move from cities to the suburbs, or smaller communities surrounding a larger city. They also move from one region to another to seek a better climate or better jobs. Since the 1970s, the fastest-growing areas in the country have been in the South and Southwest—often called the Sunbelt.

Security   Americans normally feel safe in their own country. After the terrorists attacks of September 11, 2001, this feeling of security was endangered. President Bush responded by creating the Office of Homeland Security on October 8, 2001. This office set up a national strategy to fight terrorism. In late October, Congress also passed a new antiterrorism law to help police track down terrorists.

Then, on November 25, 2002, the president signed the Homeland Security Act into law. The act established a new cabinet department— the **Department of Homeland Security**—to coordinate government agencies charged with protecting the nation from terrorist attacks.

Many Americans also feared that terrorists or hostile governments would use *bioterrorism*. Bioterrorism is the use of living organisms, such as bacteria or viruses, as weapons. To deal with bioterrorism, the government created the **Office of Public Health Preparedness** and collected medicine in case of an attack.

√ Reading Check   **What ethnic group is the fastest growing?**

American Culture

American artists and writers have developed distinctly American styles. The earliest American artists used materials from their environments to create works of art. Native Americans carved wooden masks or made beautiful designs on pottery from clay found in their areas. Later artists were attracted to the beauty of the American land. Winslow Homer painted the stormy waters of the North Atlantic. Georgia O'Keeffe painted the colorful cliffs and deserts of the Southwest. Thomas Eakins painted scenes of city life.

Two themes are common to American literature. One theme focuses on the rich diversity of the people in the United States. The poetry of Langston Hughes and the novels of Toni Morrison portray the triumphs and sorrows of African Americans. The novels of Amy Tan examine the lives of Chinese Americans. Oscar Hijuelos and Sandra Cisneros write about the country's Hispanics.

A second theme focuses on the landscape and history of particular regions. Mark Twain's books tell about life along the Mississippi River in the mid-1800s. Nathaniel Hawthorne wrote about the people of New England. Willa Cather and Laura Ingalls Wilder portrayed the struggles people faced in settling the Great Plains. William Faulkner examined life in the South.

Sports and Recreation Many Americans spend their leisure time at home, watching television, playing video games, or using a computer. Many also pursue active lives outdoors. They bike and hike, ski and skate, shoot baskets and kick soccer balls. Many enjoy spectator sports such as baseball and football. Stock-car races and rodeos also attract large crowds. Millions each year travel to national parks, or areas set aside to protect wilderness and wildlife and to offer recreation.

✓Reading Check **What are two common themes in American literature?**

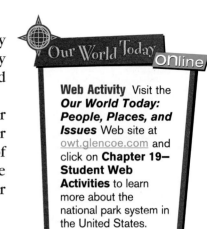

Web Activity Visit the *Our World Today: People, Places, and Issues* Web site at owt.glencoe.com and click on **Chapter 19— Student Web Activities** to learn more about the national park system in the United States.

Assessment

Defining Terms

1. Define representative democracy, federal republic, amend, ethnic group, rural, urban, suburb.

Recalling Facts

2. History When was the war to free the colonies fought?

3. History Give two reasons people from other countries are attracted to the United States.

4. Culture What theme do the works of Langston Hughes and Toni Morrison share?

Critical Thinking

5. Drawing Conclusions Because the United States is a nation of immigrants, bilingual education is an issue. How would you resolve the issue of educating non-English-speaking students? Would you help these students preserve their native language and culture, or would you immerse these students in English-language education? Explain your reasoning.

6. Drawing Conclusions Why did the United States take part in two World Wars during the first half of the twentieth century?

Graphic Organizer

7. Organizing Information Create a diagram like the one below. At the tops of the three arrows, complete the diagram by listing three reasons that Americans today are moving more frequently than ever.

Americans are on the move.

Applying Social Studies Skills

8. Analyzing Maps According to the population density map on page 522, what are the two largest cities in the northwest United States?

Making Connections

Americans All

Native Americans and African Americans endured many years of injustice. Even so, the pride and determination of these Americans remained strong. Read the poems by Native American poet Simon J. Ortiz and African American poet Langston Hughes to see how they express these feelings.

▲ Picking cotton near Dallas, Texas, 1907

Survival This Way
by Simon J. Ortiz (1941–)

Survival, I know how this way.
This way, I know.
It rains.
Mountains and canyons and plants
grow.
We travelled this way,
gauged our distance by stories
and loved our children.
We taught them
to love their births.
We told ourselves over and over
again, "We shall survive
this way."

"Survival This Way" by Simon J. Ortiz. Reprinted by permission of the author.

I, Too
by Langston Hughes (1902–1967)

I, too, sing America.

I am the darker brother.
They send me to eat in the kitchen
When company comes,
But I laugh,
And eat well,
And grow strong.

Tomorrow,
I'll be at the table
When company comes.
Nobody'll dare
Say to me,
"Eat in the kitchen,"
Then.

Besides,
They'll see how beautiful I am
And be ashamed—

I, too, am America.

"I, Too" from *Collected Poems* by Langston Hughes. Copyright © 1994 by the Estate of Langston Hughes. Reprinted by permission of Alfred A. Knopf, a Division of Random House, Inc.

▲ Native Americans on the Great Plains, 1891

► Making the Connection

1. How does the poem "Survival This Way" tell how Native Americans feel about their children?

2. What does Langston Hughes mean by the phrase "I, too, sing America"?

3. **Making Comparisons** In what way do both poems convey a message of hope?

Reading Review

| Section 1 | A Vast, Scenic Land |

Terms to Know
contiguous
megalopolis
coral reef

Main Idea
The United States has a great variety of landforms and climates.

✓ Region The United States has five main physical regions: the Coastal Plains, the Appalachian Mountains, the Interior Plains, the Mountains and Basins region, and the Pacific Coast. Alaska and Hawaii make up two additional regions.

✓ History Forty-eight of the United States are contiguous, joined together inside a common boundary between the Atlantic and Pacific Oceans.

✓ Economics The Central Lowlands area is well suited to agriculture, as are western coastal valleys.

✓ Place The high Rocky Mountains have a ridge called the Continental Divide, which separates rivers that flow east from rivers that flow west.

| Section 2 | An Economic Leader |

Terms to Know
free enterprise system
fossil fuel
landfill
recycling
free trade

Main Idea
The powerful United States economy runs on abundant resources and the hard work of Americans.

✓ Economics Because of many natural resources and a hardworking labor force, the United States has the world's most productive economy.

✓ Economics Service industries contribute the most to the American economy, followed by manufacturing, agriculture, and mining.

✓ Economics The economy of the United States faces many challenges in the twenty-first century. Some of those challenges include cleaning up pollution and trash.

✓ Economics Creativity and hard work are needed to continue to develop new technologies and help the American economy grow.

| Section 3 | The Americans |

Terms to Know
representative democracy
federal republic
amend
ethnic group
rural
urban
suburb

Main Idea
The United States has attracted people from all over the world who have created a land of many cultures.

✓ Culture The American people are immigrants or the descendants of immigrants who came from all over the world.

✓ Government The United States is a republic. A republic is a type of representative democracy.

✓ Culture Ethnic groups in America are descendants of five main peoples: Europeans, Africans, Hispanics, Asians and Pacific Islanders, and Native Americans.

✓ Culture American arts celebrate the country's ethnic and regional diversity.

Assessment and Activities

Using Key Terms

Match the terms in Part A with their definitions in Part B.

A.

1. contiguous
2. megalopolis
3. free enterprise system
4. fossil fuel
5. suburb
6. amend
7. recycling
8. free trade
9. ethnic group
10. representative democracy

B.

a. oil, natural gas, and coal
b. smaller community surrounding a city
c. areas joined inside a common boundary
d. reusing materials
e. limited government control over the economy
f. group of people who share a common culture, language, and history
g. huge urban area
h. peaceful way to change laws
i. goods flow freely between countries
j. voters choose government leaders

Reviewing the Main Ideas

Section 1 A Vast, Scenic Land

11. **Region** How does the climate of the United States compare to those of Mexico and Canada?
12. **Region** What are the five main physical regions of the U.S.?
13. **Place** What cities make up the huge urban area along the East Coast of the United States?

Section 2 An Economic Leader

14. **History** What countries are part of the free trade agreement that took effect in 1994? What is the agreement called?
15. **Economics** What type of economic system does the United States have?
16. **Human/Environment Interaction** What is happening to America's landfills?

Section 3 The Americans

17. **Government** What is the most important reason for the success of the Constitution?
18. **Culture** Which ethnic group is growing the fastest in the United States?
19. **Place** Which parts of the United States have the fastest-growing populations?

The United States

Place Location Activity

On a separate sheet of paper, match the letters on the map with the numbered places listed below.

1. Rocky Mountains
2. Mississippi River
3. Appalachian Mountains
4. Washington, D.C.
5. Chicago
6. Lake Superior
7. Ohio River
8. Gulf of Mexico
9. Texas
10. Los Angeles

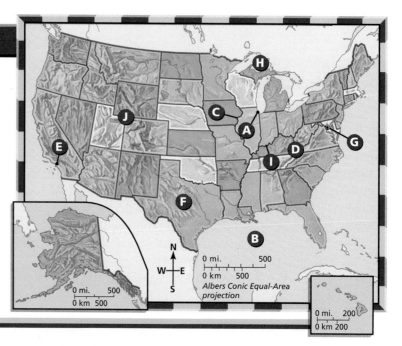

0 mi. 500
0 km 500
Albers Conic Equal-Area projection

0 mi. 500
0 km 500

0 mi. 200
0 km 200

 ## Critical Thinking

20. **Understanding Cause and Effect** What physical features of the Interior Plains have affected the economy of that region?

21. **Categorizing Information** Create a diagram like the one below. In the outer ovals, write two facts about the United States under each heading.

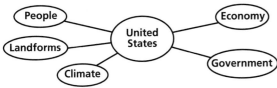

  ## Current Events Journal

22. **Writing a Paragraph** Write a paragraph describing the recycling efforts of your school and community. Explain how you can help with these efforts.

Mental Mapping Activity

23. **Focusing on the Region** Create a simple outline map of the United States. Refer to the map on page 503, and then label the following:

- Appalachian Mountains
- Great Lakes
- Alaska
- Rocky Mountains
- Hawaii
- Mississippi River
- Pacific Ocean
- Atlantic Ocean
- Gulf of Mexico
- Great Plains

 ## Technology Skills Activity

24. **Using the Internet** Search the Internet to find out where different ethnic groups have settled in your state. Create a state map and label where the groups are located.

Standardized Test Practice

Directions: Study the graph, and then answer the following questions.

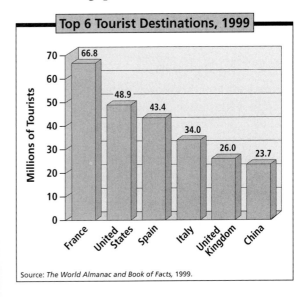

Top 6 Tourist Destinations, 1999

Source: *The World Almanac and Book of Facts,* 1999.

1. **According to the graph, about how many tourists visited the United States in 1999?**

   A 48.9

   B 66.8

   C 48,900

   D 48,900,000

2. **Which country on the graph had the least number of tourists?**

   F France

   G China

   H Spain

   J Italy

Test-Taking Tip: A common error when reading graphs is to overlook the information on the bottom and the side of the graph. Check these areas of the graph to see what the numbers mean.

529

TOO MUCH
Trash

Tons of Trash If you are an average American, you throw away about 4 pounds (2 kg) of trash each day. Not much, right? Think again. That is 1,460 pounds (663 kg) a year. By age 13, you have produced almost 10 tons (9 t) of trash!

Americans create more than one-third of the world's trash—200 million tons (181 million t) each year. That is enough to fill a line of garbage trucks that would circle the earth eight times!

What happens to trash?

State Recycling Rates
- 30% or greater
- 20–29%
- 10–19%
- Less than 10%
- Unavailable

Source: U.S. EPA Municipal Solid Waste Handbook—Internet Version.

- ⊛ Most ends up in landfills.

- ⊛ Some is burned in incinerators.

- ⊛ Some is dumped into lakes, rivers, and oceans.

All of these disposal methods create pollution and harm living things. When landfills fill up, new ones must be created. However, sites for new landfills are getting hard to find. Would you want to live near one?

The Three R's Surprisingly, the solution to too much trash is simple. We need to produce less waste. How? By following the three R's—reduce, reuse, and recycle.

- ⊛ REDUCE the amount of trash you throw away each day.

- ⊛ REUSE products and containers.

- ⊛ RECYCLE some of your trash. About 80 percent of household trash can be recycled.

If we reduce, reuse, and recycle, we can win the war against trash.

Trash piles up at a landfill in New Jersey.

Making a Difference

Dig It! You are at the ball game. You toss part of your hot dog into a trash can. Eventually, the hot dog goes to a landfill. How long will it take the hot dog to decay in a landfill?

In 1973 archaeologist William Rathje began the Garbage Project. He wanted to "dig up" facts about the trash Americans throw away. Rathje and his University of Arizona students spent years studying garbage from landfills across the United States. His research results are full of surprises. In some landfills, team members found foods such as steaks and hot dogs that were 15 to 20 years old! Lack of air, light, and moisture prevents wastes from breaking down. Another surprise: About a third of the trash in landfills is paper. Yard waste, food scraps, plastics, construction materials, and furniture are some of the other items we throw away.

William Rathje

Recycling to the Max Linda Munn and her husband, Frank Schiavo, are teachers in California. They have not set out a curbside garbage can in more than 20 years. That is because they recycle or compost almost everything they use. They produce only about two handfuls of trash a week—and that goes to a recycling center, too.

What Can You Do?

✷ **Make Toys From Trash**
Create toys from discarded clean paper, cardboard, or plastic. Have a class contest and award prizes to students who reuse trash in the most creative ways.

✷ **Campaign Against Waste**
Investigate products used every day. Which ones have too much packaging—layers of plastic or paper thrown away once the product is used? Can you think of ways to eliminate the excess? Identify companies that make these products and send them a letter or an e-mail outlining your packaging changes.

REDUCE | REUSE | RECYCLE

SANTA MONICA RECYCLE!

CANS, GLASS & PLASTICS
458-8527

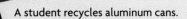
A student recycles aluminum cans.

The World and Its People
NATIONAL GEOGRAPHIC

To learn more about the people and places of Mexico, view **The World and Its People Chapter 6** video.

Our World Today online

Chapter Overview Visit the **Our World Today: People, Places, and Issues** Web site at owt.glencoe.com and click on **Chapter 20—Chapter Overviews** to preview information about Mexico.

Categorizing Information Study Foldable

When you group information into categories on a table, it is easier to study characteristics of items. Make this foldable to help you describe Mexico's land, economy, and government—past and present.

Step 1 Fold a sheet of paper into thirds from top to bottom.

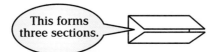

This forms three sections.

Step 2 Open the paper and refold it into fourths from side to side.

Fold it in half, then in half again.

This forms four sections.

Step 3 Unfold, turn the paper, and draw lines along the folds.

Step 4 Label your table as shown.

	Past	Present
Mexico's Land		
Mexico's Economy		
Mexico's Government		

Reading and Writing As you read the chapter, record key facts about Mexico's land, economy, and government in the appropriate places on your table foldable.

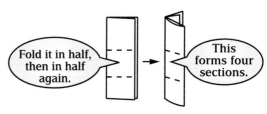

Why It Matters

Moving Forward

Mexico is a country working hard to catch up with the more industrialized countries of the world. Today, Mexico is an important trading partner of the United States. However, a rapidly growing population and a rapidly developing economy have made it difficult for Mexico to support all of its people.

◄ **The Lighthouse of Commerce and the Cathedral of Monterrey, Monterrey, Mexico.**

Mexico's Land and Economy

Guide to Reading

Main Idea

Mexico's mountainous landscape and varied climate create different economic regions.

Terms to Know

- land bridge
- peninsula
- latitude
- altitude
- hurricane
- vaquero
- maquiladora
- subsistence farm
- plantation
- industrialize
- service industry
- NAFTA

Reading Strategy

Create a chart of Mexico's economic regions like this one. List the main economic activity of each region.

Region of Mexico	Economic Activity
Northern	
Central	
Southern	

NATIONAL GEOGRAPHIC Exploring Our World

Mexican farmer Dionisio Pulido was plowing his cornfield one day. Suddenly his son heard a rumble in the ground. Then white smoke began to spew into the air. When they awoke the next day, they saw a volcano 30 feet (9 m) high. Today, more than 50 years later, the volcano named Paricutín soars nearly 9,200 feet (2,800 m) above sea level.

Paricutín and other volcanoes are scattered throughout Mexico because the country sits where three plates in the earth's crust collide. Sometimes the movement of these plates brings disastrous results. Hot magma, or melted rock, shoots through a volcano. The ground shifts violently in an earthquake. Do you see why Native Americans once called Mexico "the land of the shaking earth"?

Bridging Two Continents

Mexico forms part of a land bridge, or narrow strip of land that joins two larger landmasses. This land bridge connects North America and South America. Look at the map on page 535. You can see that Mexico borders the southern United States.

Physical geographers, people who study continents and landforms, think of Mexico as part of North America. It is also true, however, that cultural geographers think of Mexico as being part of Latin America.

For cultural geographers, language, customs, religion, and history are important areas of study. Both groups are correct. Mexico is a Latin (Spanish-speaking) country on the continent of North America. Mexico's culture is closely tied to Central and South America. Its location in North America makes it an important trading partner to the United States and Canada. It is a country that bridges two continents.

The Pacific Ocean borders Mexico on the west. Extending south along this western coast is **Baja** (BAH•hah) **California.** It is a long, narrow peninsula, or piece of land with water on three sides. On Mexico's eastern side, the **Gulf of Mexico** and the **Caribbean Sea** border the coasts. Between the Gulf and the Caribbean Sea is another peninsula—the **Yucatán** (YOO•kah•TAHN) **Peninsula.**

Mexico is a rugged land. If you were to see it from space, you might think that the country looked like a crumpled piece of paper with deep folds. Towering mountain ranges and a huge, high plateau occupy the center of the country.

The Sierra Madre   Three different mountain ranges in Mexico make up the **Sierra Madre** (SYEHR•rah MAH•thray), or "mother range." Because of

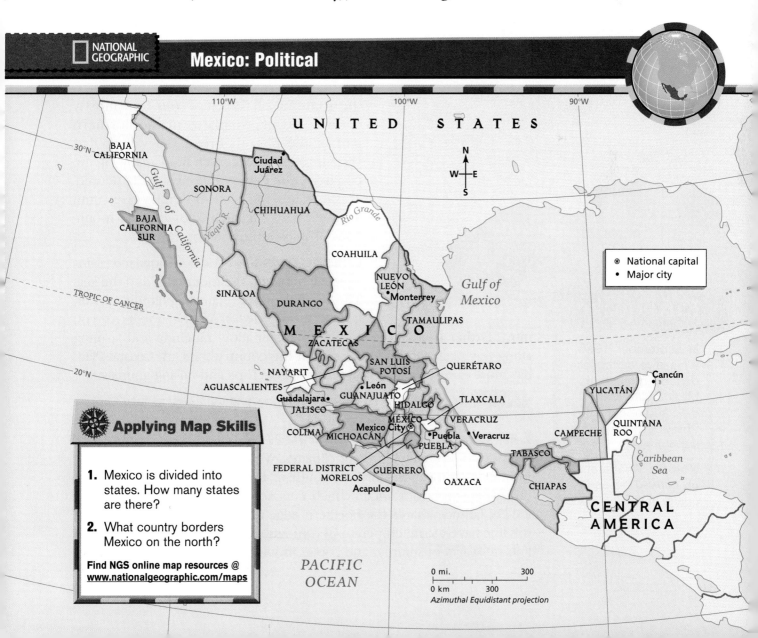

⊛ National capital
• Major city

Applying Map Skills

1. Mexico is divided into states. How many states are there?

2. What country borders Mexico on the north?

Find NGS online map resources @ www.nationalgeographic.com/maps

0 mi. 300
0 km 300
Azimuthal Equidistant projection

the rugged terrain, few people live in the Sierra Madre. The mountains are rich in resources, though. They hold copper, zinc, silver, and timber.

Many of Mexico's mountains are volcanoes. **Popocatepetl** (POH•puh•KAT•uh•PEHT•uhl), or "El Popo," as Mexicans call it, erupted violently centuries ago. In December 2000, El Popo erupted again, hurling molten rock into the sky. About 30,000 people from surrounding areas were forced to temporarily leave their homes. Tens of millions of people live 50 miles (80 km) or less from the mountain and could face even worse eruptions in the future.

Mexicans face another danger from the land. Earthquakes can destroy their cities and homes. A 1985 earthquake killed nearly 10,000 people in Mexico's capital, **Mexico City,** even though the earthquake's center was about 185 miles (298 km) away. You can understand why there are so many earthquakes here if you remember that Mexico is one of the countries that border the "Ring of Fire." This ring includes three-fourths of the world's active volcanoes. Earthquakes in the zone are common due to movement of the huge Pacific plate deep under the earth's crust.

NATIONAL GEOGRAPHIC **On Location**

Earthquake

In September, 1985, a massive earthquake destroyed much of Mexico City. Many died and about 250,000 people were left homeless.

Region What probably caused the earthquake in Mexico City?

The Plateau of Mexico The Sierra Madre surround the large, flat center of the country, the **Plateau of Mexico.** You find mostly deserts and grassy plains in the northern part of the plateau. Broad, flat valleys that slice through the center hold many of the country's chief cities and most of its people. To the south, the plateau steadily rises until it meets with the high, snowcapped mountains of the southern Sierra Madre.

Coastal Lowlands Mexico's lowland plains squeeze between the mountains and the sea. The Pacific Coastal Plain begins with a hot, largely empty desert in the north. As you move farther south, better soil and rainfall allow ranching and farming along this plain. On the other side of the country, the Gulf Coastal Plain has more rain and fertile soil for growing crops and raising animals.

✓ Reading Check What two dangers from the land do Mexicans face?

Land of Many Climates

Mexico has many different climates. Why? As you read in Chapter 2, latitude—or location north or south of the Equator—affects temperature. The Tropic of Cancer, which cuts across the center of Mexico at 23½°N latitude, marks the northern edge of the Tropics. Areas south of this line have warm temperatures throughout the year. Areas north of this line are warm in summer and cooler in winter.

Mexico's Altitude Zones

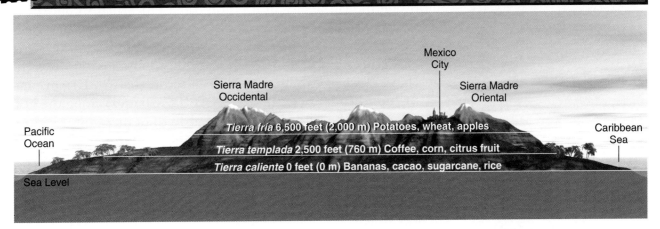

Mexico City

Sierra Madre Occidental

Sierra Madre Oriental

Pacific Ocean

Caribbean Sea

Tierra fría 6,500 feet (2,000 m) Potatoes, wheat, apples

Tierra templada 2,500 feet (760 m) Coffee, corn, citrus fruit

Tierra caliente 0 feet (0 m) Bananas, cacao, sugarcane, rice

Sea Level

Analyzing the Diagram

Mexico has zones of different climates that result from different altitudes.

Location In which altitude zone is Mexico City located?

Altitude, or height above sea level, affects temperature in Mexico as well. The higher the altitude, the cooler the temperatures—even within the Tropics. The diagram above shows that Mexico's mountains and plateau create three altitude zones. You could travel through all of these zones in a day's trip across the Sierra Madre.

Because the coastal lowlands are near sea level, they have high temperatures. Mexicans call this altitude zone the *tierra caliente* (tee•AY•rah kah•lee•AYN•tay), or "hot land." Moving higher in altitude, you find the *tierra templada* (taym•PLAH•dah), or "temperate land." Here the climate becomes more moderate. In the highest zone, the climate becomes even cooler. Mexicans call this the *tierra fría* (FREE•ah), or "cold land."

Rainfall varies throughout Mexico. Baja California and northern Mexico receive very little precipitation. Other regions receive more, mostly in the summer and early fall. From June to October, Mexico can be hit by hurricanes. These fierce tropical storms with high winds and heavy rains form over the warm waters of the Atlantic or Pacific Oceans. They can strike Mexico with fury.

✓Reading Check What is Mexico's warmest altitude zone?

Mexico's Economic Regions

Mexico's physical geography and climate together give Mexico three distinct economic regions: the North, Central Mexico, and the South. Large stretches of northern Mexico are too dry and rocky to farm. By building canals to carry water to their fields, people without irrigation can grow cotton, fruits, grains, and vegetables.

Northern Mexico Did you know that the skills used by American cowhands originated in Mexico? Mexican cowhands, called vaqueros (vah•KEHR•ohs), developed the tools and techniques for herding, roping, and branding cattle. Vaqueros in northern Mexico carry on this work today.

▲ A skilled seamstress makes clothing in a maquiladora in northern Mexico.

Northern Mexico has seen an economic boom. **Monterrey,** Mexico's main producer of steel and cement, has long been an important industrial city. In this and other cities, many companies from the United States and elsewhere have built maquiladoras (mah•KEEL•ah•DOHR•as), or factories that assemble parts made in other countries. As a result, thousands of Mexicans have flocked to cities such as **Tijuana** (tee•WAH•nah) and **Ciudad Juárez** (see•ooh•DAHD HWAH•rayz) along the border with the United States. The growth in these border cities has raised the standard of living in northern cities through factory work and increased trade. However, this quick growth has also brought concerns about damaging the environment, pollution, and dangers to the health and safety of workers.

Central Mexico More than half of Mexico's people live in the central region, the country's heartland. Why do they call this area home? The climate is one reason. Although central Mexico lies in the Tropics, its high elevation keeps it from being hot and humid. Temperatures are mild, and the climate is pleasant year-round. A second reason is the fertile soil created by volcanic eruptions over the centuries. This allows good production for farming and ranching.

Large industrial cities such as Mexico City and **Guadalajara** also prosper in central Mexico. More than 18 million people live in Mexico City and its suburbs, making it one of the largest cities in the world. Mexico City has been the largest city in the Americas since before the Spanish arrived in the early 1500s.

Southern Mexico The South is the poorest economic region of the country. The mountains towering in the center of this region have poor soil. Subsistence farms, or small plots where farmers grow only enough food to feed their families, are common here. In contrast, the coastal lowlands of this area have good soil and plentiful rain. Wealthy farmers grow sugarcane or bananas on plantations, large farms that raise a single crop for sale.

Both coasts also have beautiful beaches and a warm climate. Tourists from all over the world flock to such resort cities as **Acapulco** and **Puerto Vallarta** on the Pacific coast and **Cancún** on the Yucatán Peninsula.

√ Reading Check How does the economic region of northern Mexico differ from that of southern Mexico?

Mexico's Economy Today

With many resources and workers, Mexico has a growing economy. Did you know that Mexico's economy ranks among the top 15 in the world? As in the past, agriculture is important. Farmers raise food to feed people at home—and also to ship around the world. Corn, beans, wheat, and rice are the main crops grown for food. Exports include coffee, cotton, vegetables, fruit, livestock, and tobacco.

In recent years, Mexico has industrialized, or changed its economy to rely less on farming and more on manufacturing. Factories in

Mexico now make cars, consumer goods, and steel. The labels on your clothing might even say "Made in Mexico."

Mexico has large deposits of petroleum and natural gas in the Gulf of Mexico and along the southern coast. As a result, Mexico is among the world's major oil-producing nations.

Mexico is also home to important service industries such as banking and tourism. Service industries, you recall, are businesses that provide services to people rather than producing goods.

NAFTA In 1994, Mexico, the United States, and Canada became partners in NAFTA, the North American Free Trade Agreement. Under this agreement, most goods traded between these countries would be free of tariffs, or special taxes. This means a homemaker in Canada would probably choose to buy a tablecloth made in Mexico rather than to pay more for a taxed tablecloth produced in Europe.

Some Americans have been afraid that belonging to NAFTA would mean American jobs would "go south." They feared that the lower rate of pay for labor in Mexico would encourage many manufacturers to move their businesses to Mexico rather than keep them in the United States. The debate about the overall effect of NAFTA is still going on. However, as one supporter of NAFTA stated, "As Americans, it is far better to have prosperous than struggling neighbors, and NAFTA contributes to this objective."

✓Reading Check **Why are some Americans afraid jobs will "go south"?**

Assessment

Defining Terms
1. Define land bridge, peninsula, latitude, altitude, hurricane, vaquero, maquiladora, subsistence farm, plantation, industrialize, service industry, NAFTA.

Recalling Facts
2. History How did the vaqueros of Mexico influence American ranching?
3. Location Why is Mexico a land bridge?
4. Economics Why have many Mexicans moved to the cities of the north?

Critical Thinking
5. Understanding Cause and Effect How has NAFTA affected the people in Canada and the people in Mexico? Do you think NAFTA has been good or bad for the people in border cities of the United States? Explain.

6. Analyzing Information Explain why Mexico is part of both North America and Latin America.

Graphic Organizer
7. Organizing Information Create a diagram like this one, and then list two facts that explain the large population of central Mexico.

```
────────────►┌──────────────┐
             │     High     │
             │  Population of│
────────────►│Central Mexico│
             └──────────────┘
```

Applying Social Studies Skills

8. Analyzing Diagrams Study Mexico's altitude zones on page 537. At which elevation do you think most people live? Why do they live here?

Making Connections

CULTURE GOVERNMENT PEOPLE TECHNOLOGY

Good Neighbors

Every 60 seconds, goods worth more than $500,000 cross the U.S.-Mexican border. Goods worth nearly $750,000 cross the U.S.-Canadian border. Those numbers are double what they were in 1994. What made them grow, experts say, was the North American Free Trade Agreement, or NAFTA.

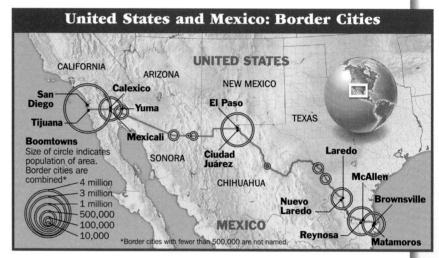

United States and Mexico: Border Cities

Boomtowns
Size of circle indicates population of area. Border cities are combined*
— 4 million
— 3 million
— 1 million
— 500,000
— 100,000
— 10,000

*Border cities with fewer than 500,000 are not named.

Map by Joe Lertola for TIME

Pros and Cons

NAFTA is an agreement that allows for the free movement of goods across the international borders of Canada, the United States, and Mexico. The free movement of goods means people are able to import goods from other countries without paying import taxes, called *tariffs*.

Since NAFTA took effect, millions of new jobs have been created in all three countries, especially Mexico. On the negative side, some U.S. and Canadian companies moved their factories to Mexico where wages are low. As a result, many U.S. and Canadian workers lost their jobs.

The United States and Mexico are interdependent. Mexicans depend on U.S. companies to create jobs in Mexico. The United States depends on the ability of Mexican workers to produce goods cheaply.

U.S. border communities depend on Mexico, too. From El Paso alone, Mexican factories buy $8 billion to $10 billion worth of goods and services every year. U.S. companies often try to make workers' lives easier by building safe and affordable housing.

Mexico's Future

Mexico's economy must get stronger. In 2001, one of every two people in Mexico was poor.

One in five was desperately poor, scraping by on less than $2 per day. Farmworkers earned only about $5 per day. That isn't nearly enough to support a family.

Jobs in border factories, or maquiladoras, that opened after NAFTA was signed pay on the average of $10 per day. This is still a small amount compared to U.S. jobs, but it is much better than before.

Mexico's president Vicente Fox is working closely with U.S. officials to ease such problems as immigration and trade restrictions. He would like to see the interdependence created by NAFTA continue to grow and benefit all its members.

→ Making the Connection

1. List the combined border cities on the map above in order of size, from largest to smallest.

2. **Drawing Conclusions** Which border cities would you expect to be connected by bridges? Why?

Mexico's History

Guide to Reading

Main Idea

Mexico's culture reflects a blend of its Native American and Spanish past.

Terms to Know

- jade
- obsidian
- maize
- hieroglyphics
- mural
- colony
- hacienda

Reading Strategy

Create a chart like this one, then provide one example of how Native Americans and Europeans influenced Mexican culture.

Ethnic Groups	Influence on Mexican Culture
Native Americans	
Europeans	

NATIONAL GEOGRAPHIC — Exploring Our World

Thousands of people visit an ancient temple in Mexico's Yucatán Peninsula on the spring and fall equinoxes. On those two days, the setting sun casts a shadow on the stairs of the temple's north face. The area that is not shadowed looks like the ancient Native American god called Kukulcan, or the Feathered Serpent, going down the temple stairs.

The first people to arrive in Mexico were the ancestors of today's Native Americans. Mexico's Native American heritage has shaped the country's culture. So has Mexico's European heritage, brought by the Spaniards who conquered the area in the 1500s.

Native American Civilizations

Native Americans came to Mexico thousands of years ago. From about 1200 B.C. to the A.D. 1500s, these people built a series of brilliant, highly advanced civilizations on Mexican soil. Of these, the Olmec, Mayan, and Aztec civilizations are the best-known. Look at the map on page 542 to see where the Olmec, Mayan, and Aztec civilizations thrived.

The Olmecs The **Olmecs** built the first civilization in the Americas around 1200 B.C. They decorated their cities with large carved stone statues, some standing about 10 feet (3 m) high and weighing over 20 tons (18 t). They also carved smaller and more personal objects like jewelry out of jade, a local shiny stone that comes in many shades of green as well as other colors. All these items were carved with obsidian, a hard, black glass created by the volcanoes in the area, because the Olmecs had no metals.

The Olmecs were the first to grow maize, or corn, to feed their many people. In addition to cities and ceremonial centers, they built large drainage systems to direct rainwater away from their fields and settlements. The Olmecs lasted longer than any other Native American civilization, finally disappearing in about 400 B.C.

The Maya   The people called the **Maya** lived in the rain forests of the Yucatán Peninsula and surrounding areas from about A.D. 250 to 900. Religion held Mayan society together. Mayan priests needed to measure time accurately to hold religious ceremonies at the correct moment. They studied the heavens and developed a calendar of 365 days.

The Maya built huge stone temples in the shape of pyramids with steps. One of these structures, the temple of **Kukulcan,** showed careful planning. Each side of Kukulcan had 91 steps, totaling 364. The platform at the temple's top made one more step for a grand total of 365—just like the days in the year.

The Maya also developed hieroglyphics, a form of writing that uses signs and symbols. They had a complex number system. Artists decorated temples and tombs with elaborate murals, or wall paintings.

Around A.D. 900, Mayan civilization declined. Why? Historians do not know. Some suggest that the Maya overused the land and could not grow enough food. Others suggest that warfare or the spread of disease caused their decline. The Maya did not disappear, however. Their descendants still live in the same area and speak the Mayan language.

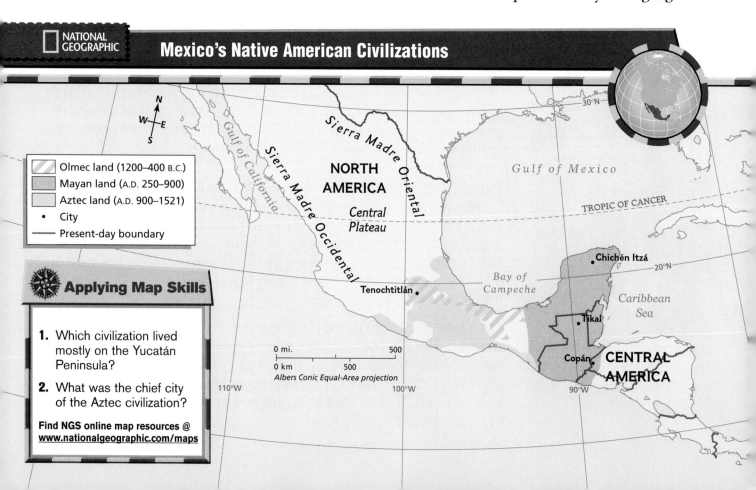

NATIONAL GEOGRAPHIC

Mexico's Native American Civilizations

Olmec land (1200–400 B.C.)
Mayan land (A.D. 250–900)
Aztec land (A.D. 900–1521)
• City
— Present-day boundary

NORTH AMERICA
Central Plateau
Gulf of California
Sierra Madre Occidental
Sierra Madre Oriental
Gulf of Mexico
TROPIC OF CANCER
Bay of Campeche
Chichén Itzá
Tenochtitlán
Tikal
Caribbean Sea
Copán
CENTRAL AMERICA

0 mi. 500
0 km 500
Albers Conic Equal-Area projection

30°N
20°N
110°W   100°W   90°W

Applying Map Skills

1. Which civilization lived mostly on the Yucatán Peninsula?

2. What was the chief city of the Aztec civilization?

Find NGS online map resources @ www.nationalgeographic.com/maps

The Aztec Around A.D. 1200, a people called the Mexica moved into central Mexico from the north. The Spanish later called these people the Aztec. The **Aztec** conquered a large empire in central Mexico. Their capital, **Tenochtitlán** (tay•NAWCH• teet•LAHN) was magnificent. Today Mexico City—Mexico's capital—stands on this ancient site.

Tenochtitlán was originally built on two islands in the middle of Lake Texcoco. Long dikes connected it to land. The city had huge stepped pyramids. Merchants traded gold, silver, and pottery in busy marketplaces. Farmers grew their crops in structures called "floating gardens," or rafts filled with mud. The rafts eventually sank to the lake bottom and piled up, forming fertile islands.

The Aztec people and many of their traditions survive today in Mexico. The food, crafts, and language of Mexico have roots in Aztec culture. Even the name of the country comes from the word the Aztec called themselves—the *Mexica*. The flag of modern Mexico honors this ancient civilization. In the center of the flag is the Aztec symbol of an eagle with a snake in its beak.

✓ **Reading Check** **What Native American cultures flourished in Mexico?**

Spanish Mexico

In 1519 Mexico's history changed dramatically. A Spanish army led by **Hernán Cortés** landed on Mexico's Gulf coast. He and about 600 soldiers marched to Tenochtitlán, which they had heard was filled with gold.

But how did just 600 Spaniards conquer the heart of the Aztec Empire, which contained about 6 million people? There are several reasons for their success. First, the Spanish made treaties with the Indians who had been conquered by the Aztec. These treaties brought the Spanish thousands of warriors and many needed supplies. Second, Cortés's men had steel swords, muskets, and cannons, while the Aztec had only wooden weapons with jade and obsidian blades and points. Third, the Spanish had horses, which were unknown to the Indians until then. In fact, when Cortés was on his horse, the Aztec thought that he and his horse were one creature. They thought he was one of their most important gods, *Quetzlcoatl,* and that he had returned to punish them. So, of course, many were afraid of him.

Spain made Mexico a colony, or an overseas territory, because Mexico's rocky land held rich deposits of gold and silver. Many Spanish settlers came to live in Mexico. Some raised cattle on large ranches called haciendas (ah•see•AYN•dahs). Others started gold and silver mines. The Spaniards made Native Americans work on the ranches and

**NATIONAL GEOGRAPHIC On Location**

Teotihuacán

Hundreds of years before the appearance of the Aztec or the Spanish, Native Americans built monuments, such as the Temple of the Sun in the city of Teotihuacán. Teotihuacán was located near what is now Mexico City.

History How was the small Spanish army able to defeat the Aztec?

Web Activity Visit the *Our World Today: People, Places, and Issues* Web site at owt.glencoe.com and click on **Chapter 20— Student Web Activities** to learn more about Mexico's history.

in the mines. Thousands of Native Americans died from mistreatment. Many thousands more died of diseases such as the common cold and smallpox, which they caught from the Europeans. Spanish priests came to Mexico and in their own way tried to improve the lives of the Native Americans. Because of their work, many Native Americans accepted the priests' teachings. Today about 90 percent of Mexico's people follow the Roman Catholic religion.

Reading Check Why was Mexico a valuable colony for Spain?

Independence and Revolution

The people of Mexico eventually rebelled against Spanish rule and gained independence in 1821. In 1824 they set up a republic with an elected president. Soon after independence, Mexico lost some valuable territory to the United States. The territory included Texas and California.

For many decades, rich families, army officers, and Catholic Church leaders held most of the power and wealth in Mexico. In 1910 Mexican peasants revolted. **Emiliano Zapata,** who commanded a rebel army, stated the goals of this revolution. He wanted to give to the poor "the lands, woods, and water that the landlords or bosses have taken from us." Zapata's forces swooped down and seized many large haciendas. They divided the land among the poor. In the northwestern provinces along the U.S. border, **Francisco "Pancho" Villa** also tried to help the poor, mostly Indian peasants.

Reading Check Who led the 1910 revolution?

Section 2 Assessment

Defining Terms
1. Define jade, obsidian, maize, hieroglyphics, mural, colony, hacienda.

Recalling Facts
2. History Describe three achievements of the ancient Maya.

3. History Which European country conquered and colonized Mexico?

4. History What were Emiliano Zapata's goals?

Critical Thinking
5. Sequencing Information Put the following events in the correct chronological order: Cortés conquers the Aztec, Mexico wins independence from Spain, the Mexica move into central Mexico, Zapata leads a revolution.

6. Understanding Cause and Effect How did the arrival of Europeans affect the Native Americans in Mexico?

Graphic Organizer
7. Organizing Information Create a chart like this one. In each column list the major advancements of each civilization.

Olmec	Maya	Aztec

Applying Social Studies Skills

8. Analyzing Maps Refer to the map of Mexico's Native American civilizations on page 542. Which Native American group settled the farthest south?

Social Studies Skill

Reading a Physical Map

A map that shows the different heights of the land is called a physical map. Physical maps use colors and shading to also show relief—or how flat or rugged the land surface is. Colors are also used to show the land's elevation—or height above sea level. Green often shows the lowest elevations (closest to sea level). Yellows, oranges, browns, and reds usually mean higher elevations. Sometimes the highest areas, such as mountain peaks, are white.

Learning the Skill

To read a physical map, apply these steps:

- Read the map title to identify the region shown on the map.
- Use the map key to find the meaning of colors and symbols.
- Identify the areas of highest and lowest elevation on the map.
- Find important physical features, including mountains, rivers, and coastlines.
- Mentally map the actual shape of the land.

Practicing the Skill

Look at the map to answer the following:

1. What country is shown on the map?
2. What mountain ranges are labeled?
3. What is the elevation of the green areas on the map (in feet and meters)?

4. What color on the map means 2,000–5,000 feet (600–1,500 m)?
5. Briefly describe the physical landscape of the area shown on the map, moving from west to east.

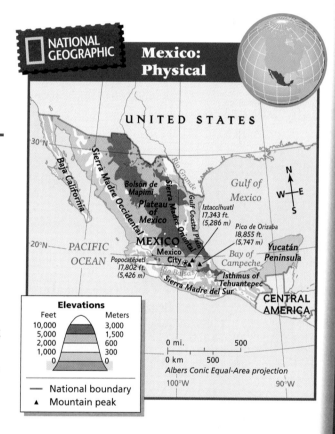

NATIONAL GEOGRAPHIC

Mexico: Physical

Applying the Skill

Look at the physical map of Canada on page 474. Describe the physical landscape of the country, moving from east to west.

GO TO

Practice key skills with **Glencoe Skillbuilder Interactive Workbook, Level 1.**

Mexico Today

Guide to Reading

Main Idea

Mexicans enjoy a rich and lively culture but face many serious challenges.

Terms to Know

- plaza
- adobe
- federal republic
- migrant worker
- national debt
- smog

Reading Strategy

Create a diagram like this one. In each of the smaller ovals, write a feature of Mexican culture. Add as many smaller ovals as you need.

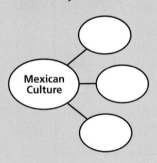

Mexican Culture

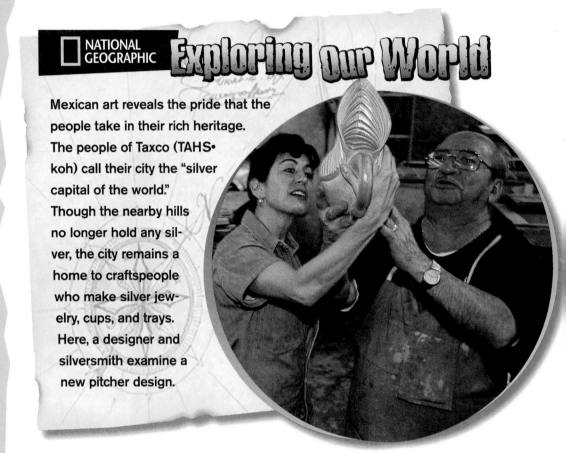

NATIONAL GEOGRAPHIC Exploring Our World

Mexican art reveals the pride that the people take in their rich heritage. The people of Taxco (TAHS•koh) call their city the "silver capital of the world." Though the nearby hills no longer hold any silver, the city remains a home to craftspeople who make silver jewelry, cups, and trays. Here, a designer and silversmith examine a new pitcher design.

Mexico—the third-largest country in area in Latin America, after Brazil and Argentina—has a large and dynamic population. About 75 percent of all Mexicans live in the country's bustling cities.

Mexico's Cities and Villages

In the center of Mexico's cities, you often find large plazas, or public squares. Around each city's plaza stand important buildings such as a church and a government center. When you look at the buildings, you can see the architectural style of Spanish colonial times. Newer sections of the cities have a mix of towering glass office buildings and modern houses. In the poorer sections of town, people build small homes out of whatever materials they can find. This may include boards, sheet metal, or even cardboard.

Rural villages also have central plazas. Streets lead from the plazas to residential areas. Many homes are made of adobe (uh•DOH•bee), or sun-dried clay bricks. The roofs might be made of straw or of colored tile, in the Spanish style.

✓**Reading Check** What do you find in the center of Mexico's cities and villages?

Mexican Culture

Mexican artists and writers have created many national treasures. In the early 1900s, Mexican painters produced beautiful murals—just as Native American painters had done centuries before. Among the most famous of these mural painters were José Clemente Orozco, David Alfero Sequieros, and Diego Rivera. Rivera's wife, Frida Kahlo, became well-known for her paintings, which revealed her inner feelings. Modern writers such as Carlos Fuentes and Octavio Paz have written poems and stories that reflect the values of Mexico's people.

Food If you have tasted Mexican food, you know that it is a rich blend of flavors. Corn—first grown in Mexico—continues to be an important part of the Mexican diet. Chocolate, tomatoes, beans, squash, and chilies were all Native American foods as well. When the Spanish came, they brought beef, chicken, cheese, and olive oil, which Mexicans added to their cooking.

Today, Mexicans combine these different cooking traditions in popular foods such as tacos and enchiladas. Both dishes combine a flat bread called a tortilla with meat or beans, vegetables, cheese, and spicy chilies.

Celebrations Throughout the year, Mexicans enjoy several special celebrations called fiestas (fee•EHS•tuhs). These special days include parades, fireworks, music, and dancing. Mariachi (MAHR•ee•AH•chee) bands may play such traditional instruments as the violin, guitar, horn, and bass at fiestas. More likely, however, you will hear the fast-paced rhythms and singing of Latino bands, which have influenced the United States.

National holidays include **Independence Day** (September 16) and **Cinco de Mayo** (May 5). This holiday celebrates the day in 1862 that Mexicans defeated an invading French army in battle. November 2

Art

Mexican artist Diego Rivera is one of the most famous mural painters of the twentieth century. He believed that art belonged to the people. In Mexico City, Rivera's murals line the courtyard of the Ministry of Education building and cover the walls of the National Palace. With their characteristically vivid colors and distinctive style, Rivera's murals tell the story of the work, culture, and history of the Mexican people.

Looking Closer How did Rivera's work support his belief that art belongs to the people?

Mexico Through the Centuries ▶

NATIONAL GEOGRAPHIC On Location

Fiestas

On September 16, parades celebrate the women and men who helped win Mexican independence.

Culture What is the purpose of fiestas?

is a special religious celebration called the **Day of the Dead.** On this day, families gather in cemeteries where they honor their departed loved ones by laying down food and flowers.

✓ Reading Check What are some important celebrations in Mexico?

Mexico's Government

Mexico, like the United States, is a federal republic, where power is divided between national and state governments, and a strong president leads the national government. Mexico's national government differs in that it has much more power than the state governments. The president of Mexico is head of the executive branch of government. He or she can serve only one six-year term but has more power than the legislative and judicial branches.

For many decades, one political party, called the **Party of Institutional Revolution (PRI),** led Mexico. All the presidents and most other elected officials came from this party. In recent years, economic troubles and the people's lack of political power led to growing frustration. In the year 2000, the newly elected president of Mexico, **Vicente Fox,** came from a different political party—for the first time in more than 70 years.

Mexico's government faces many difficult challenges. People in Mexico are demanding more political freedom to make decisions that affect their everyday lives. Traffic in illegal drugs is protected by corrupt government officials. Nearly 40 percent of Mexico's 100 million people live below the poverty line. To fight the country's pressing problems, from poverty to drugs, a strong central government is needed. To increase democracy in Mexico, however, Fox must give power back to local and state agencies. Fox will have to help his country find the balance between these two levels of government.

✓ Reading Check What form of government does Mexico have?

Mexico's Challenges

Mexico has tried to use its resources to improve the lives of its people. These actions have had strong effects on Mexican life—and created some challenges for the future.

Population Mexico's population has increased rapidly in recent decades. Because many people have moved to the cities to find jobs, the cities have grown quickly. Many people have been forced to take jobs that pay low wages. As a result, thousands of people crowd together in poor sections of the cities.

Those Mexicans who cannot find any work in their country may become migrant workers. These are people who travel from place to place when extra workers are needed to plant or harvest crops. They legally and sometimes illegally cross Mexico's long border to work in the United States. Though the pay is low, the migrant workers can earn more in the United States than in Mexico.

Another issue with people involves the descendents of the ancient Maya Indians. The present-day Maya live in the southernmost state of Mexico called **Chiapas.** This state is one of the poorest states in Mexico. Over 75 percent of the people there live below the poverty level. Most of the wealth in Chiapas is concentrated in a very small number of ranching families who are of Spanish descent. Diseases and illness that result from the poverty and a lack of health care cause many thousands of deaths every year. Believing that the Mexican government would never help improve their situation, many Maya rebelled. The Zapatista Army they formed is still fighting for independence from the central government.

Foreign Debt For decades, the Mexican government refused to let foreign companies build factories in Mexico. Leaders feared that the companies would take their profits to their own country, thus draining

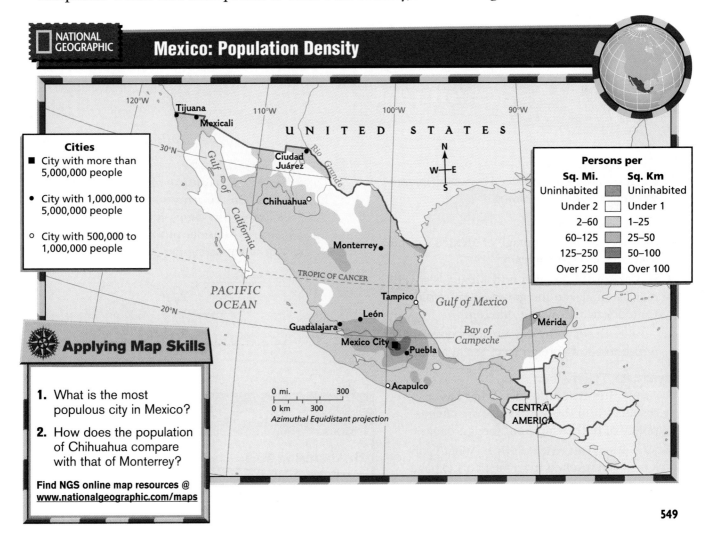

NATIONAL GEOGRAPHIC

Mexico: Population Density

Cities
- ■ City with more than 5,000,000 people
- ● City with 1,000,000 to 5,000,000 people
- ○ City with 500,000 to 1,000,000 people

Persons per	
Sq. Mi.	**Sq. Km**
Uninhabited	Uninhabited
Under 2	Under 1
2–60	1–25
60–125	25–50
125–250	50–100
Over 250	Over 100

UNITED STATES

PACIFIC OCEAN

Gulf of Mexico

Bay of Campeche

CENTRAL AMERICA

TROPIC OF CANCER

0 mi. 300
0 km 300
Azimuthal Equidistant projection

Applying Map Skills

1. What is the most populous city in Mexico?

2. How does the population of Chihuahua compare with that of Monterrey?

Find NGS online map resources @ www.nationalgeographic.com/maps

money out of Mexico. In the 1990s, the government changed this policy. Mexican officials were still concerned that money would be lost, but hoped that the new factories would create more jobs for Mexicans.

To help its economy grow, Mexico borrowed money from foreign banks. The government then had to use any money it earned in taxes to pay back the loans. As a result, Mexico's leaders did not have enough funds to spend on the Mexican people when the economy began to struggle. Many Mexicans grew angry. Yet, if the government did not make the loan payments, banks would refuse to lend more money for future plans. Because there are still loans to be repaid, Mexicans will face this situation for many years. The problem of repaying a national debt, or money owed by the government, is one that is being faced by many countries in the world today.

Pollution   As Mexico's population boomed, its cities grew very large. At the same time, the economy industrialized. Both of these changes contributed to rising pollution in Mexico.

The mountains that surround Mexico City trap the exhaust fumes from hundreds of thousands of cars. People wake each day to a thick haze of fog and chemicals called smog. Many people wear masks when they leave their homes to go to work or school. In northern Mexico, many factories release dangerous chemicals into the air or water. One environmental group says that the Rio Grande is now one of the most polluted rivers in North America.

√Reading Check   **What challenges does Mexico face?**

Section 3 Assessment

Defining Terms
1. Define plaza, adobe, federal republic, migrant worker, national debt, smog.

Recalling Facts
2. Culture What percentage of Mexico's population lives in urban areas?

3. Government Explain how Mexico's government is similar to the government of the United States. How is it different?

4. Government Why did Mexico's government refuse to allow foreign factories in Mexico?

Critical Thinking
5. Analyzing Information What has resulted from the Mexican government's policy of borrowing from foreign banks?

6. Summarizing Information What problems have resulted from Mexico's expanding population?

Graphic Organizer
7. Organizing Information Create a diagram like this one. On the arrows list three factors that have led to the smog problem of Mexico City. Be sure to consider physical characteristics of the area when listing the factors.

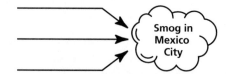

Applying Social Studies Skills

8. Analyzing Maps Look at the population density map on page 549. What is the population of Guadalajara? What is the population density of the area surrounding Mérida?

| Section 1 | Mexico's Land and Economy |

Terms to Know
land bridge
peninsula
latitude
altitude
hurricane
vaquero
maquiladora
subsistence farm
plantation
industrialize
service industry
NAFTA

Main Idea

Mexico's mountainous landscape and varied climate create different economic regions.

✓Location Mexico is part of a land bridge that connects North and South America.

✓Movement Mexico's economy is growing; and many people are moving to the northern cities.

✓Location Much of Mexico lies in the Tropics, but the climate in some areas is cool because of high elevation.

✓Economics Landforms and climate combine to create three economic zones in Mexico.

| Section 2 | Mexico's History |

Terms to Know
jade
obsidian
maize
hieroglyphics
mural
colony
hacienda

Main Idea

Mexico's culture reflects a blend of its Native American and Spanish past.

✓History Mexico's Native American civilizations—the Olmec, Maya, and Aztec—made many contributions to Mexico's culture.

✓Culture Mexico's people reflect the country's Native American and Spanish roots.

✓History The Spanish ruled Mexico from the 1500s to 1821, when Mexico won its independence.

✓History The poor people in Mexico revolted against the rich and powerful church and military leaders in 1910.

| Section 3 | Mexico Today |

Terms to Know
plaza
adobe
federal republic
migrant worker
national debt
smog

Main Idea

Mexicans enjoy a rich and lively culture but face many serious challenges.

✓Culture About 75 percent of Mexicans live in cities today.

✓Environment Mexico is suffering from air pollution.

✓Culture Challenges facing Mexico include problems caused by population growth, foreign investment and debt, and pollution.

✓Government Mexico's government is a federal republic.

Chapter 20

Assessment and Activities

Using Key Terms

Match the terms in Part A with their definitions in Part B.

A.

1. altitude
2. hurricane
3. vaquero
4. maquiladora
5. jade
6. adobe
7. plaza
8. smog
9. mural
10. subsistence farm

B.

a. factory that assembles parts from other countries

b. cowhand

c. sun-dried clay bricks

d. wall painting

e. height above sea level

f. fog mixed with smoke

g. produces only enough to support a family's needs

h. fierce tropical storm

i. public square

j. shiny stone that comes in many shades of green

Reviewing the Main Ideas

Section 1 Mexico's Land and Economy

11. **Location** How does Mexico's latitude affect its climate?
12. **Economics** What are Mexico's major exports?
13. **Movement** How have maquiladoras affected northern Mexico's cities?

Section 2 Mexico's History

14. **History** What was the capital city of the Aztec civilization?
15. **History** What effects did Spanish conquest have on Native Americans?
16. **History** When did Mexico win its independence from Spain?

Section 3 Mexico Today

17. **Government** What are people in Mexico demanding from the Mexican government?
18. **Culture** What does Cinco de Mayo celebrate?

Mexico

Place Location Activity

On a separate sheet of paper, match the letters on the map with the numbered places listed below.

1. Sierra Madre Occidental
2. Mexico City
3. Plateau of Mexico
4. Yucatán Peninsula
5. Baja California
6. Rio Grande
7. Gulf of Mexico
8. Guadalajara
9. Monterrey
10. Sierra Madre del Sur

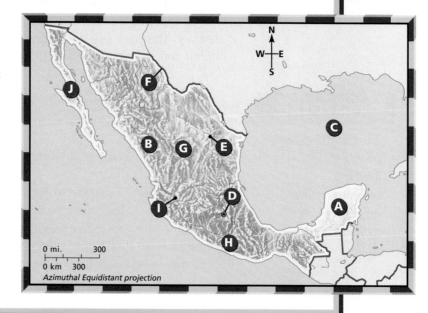

Our World Today Online

Self-Check Quiz Visit the *Our World Today: People, Places, and Issues* Web site at owt.glencoe.com and click on **Chapter 20—Self-Check Quizzes** to prepare for the Chapter Test.

Critical Thinking

19. **Understanding Cause and Effect** Why have Mexico's leaders encouraged free trade agreements with other countries?

20. **Analyzing Information** If you were Mexico's president, what would you do to rid Mexico of the problems of illegal drugs and poverty?

Current Events Journal

21. **Writing a Brochure** In this chapter, you learned about Mexico's economic and environmental challenges. Research the economy of your city or town. Include the types of industries, trading partners, and environmental effects of the industries. Share the information you gather by making a brochure. Include a map of your area.

Mental Mapping Activity

22. **Focusing on the Region** Create a simple outline map of Mexico. Refer to the physical map on page 476 and then label the following:

- Pacific Ocean
- Gulf of Mexico
- Yucatán Peninsula
- Baja California
- Mexico City
- Rio Grande
- Caribbean Sea

Technology Skills Activity

23. **Developing a Multimedia Presentation** Imagine that you work for Mexico's Economic Development Office. Create a multimedia presentation to present to a group of foreign investors. Use a software application such as PowerPoint® to showcase positive features like climate, resources, and labor supply. Your goal is to show investors that Mexico is a good place for them to invest their money.

The Princeton Review

Standardized Test Practice

Directions: Read the paragraph below, and then answer the question that follows.

The Aztec civilization was organized into classes. At the top was the emperor. His power came from his control of the army and the religious beliefs of the people. Next came the nobles, followed by commoners. Commoners included priests, merchants, and artists. Below commoners were the serfs, or workers who farmed the nobles' fields. Slaves, the lowest class, included criminals and people in debt, as well as female and child prisoners of war. Male prisoners of war were sacrificed to the Aztec gods. The Aztec believed that live human sacrifices were needed to keep the gods pleased and to prevent floods and other disasters.

1. **Which of the following statements is an opinion about the information given above?**

   F The Aztec civilization was organized into classes.

   G Male prisoners of war were sacrificed to the Aztec gods.

   H Slaves included children.

   J The Aztec should not have sacrificed people to the gods.

Test-Taking Tip: This question asks you to identify an opinion. An opinion is a person's belief. It is not a proven fact (such as answer F). Opinions often contain subjective words, such as *easier, best,* or *should.*

553

Central America and the West Indies

The World and Its People

NATIONAL GEOGRAPHIC

To learn more about the people and places of Central America and the West Indies, view *The World and Its People* **Chapter 7** video.

Our World Today ONLine

Chapter Overview Visit the *Our World Today: People, Places, and Issues* Web site at owt.glencoe.com and click on **Chapter 21—Chapter Overviews** to preview information about Central America and the West Indies.

Why It Matters

Building Trust

Ever since the Monroe Doctrine warned European nations against interfering with the affairs of countries in the Americas, U.S. presidents have worked to develop a special relationship with our near neighbors.

FOLDABLES™
Study Organizer

Compare-Contrast Study Foldable Make this foldable to help you determine how Central America and the West Indies are similar and different.

Step 1 Fold one sheet of paper in half from top to bottom.

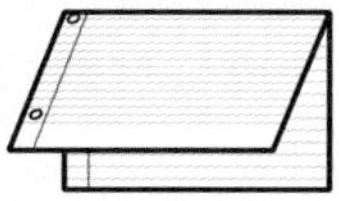

Step 2 Fold it in half again, from side to side.

Step 3 Unfold the paper once. Sketch an outline of Central America and the West Indies across both tabs and label them as shown.

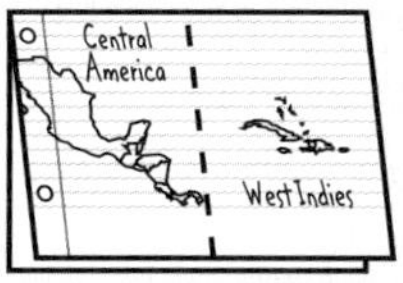

Central America
West Indies

Step 4 Cut up the fold of the top flap only.

Central America
West Indies

This cut will make two tabs.

Reading and Writing As you read the chapter, write facts under the appropriate tabs of your foldable. Use what you write to compare and contrast the people and places of Central America and the West Indies.

Central America

Guide to Reading

Main Idea

Central America is made up of seven nations that are home to a variety of peoples, exotic animals, and diverse landforms.

Terms to Know

- isthmus
- hurricane
- plantation
- subsistence farm
- canopy
- eco-tourist
- literacy rate
- republic
- parliamentary democracy

Reading Strategy

Create a chart like this one, listing each country in Central America and two key facts about each country.

Country	Key Facts

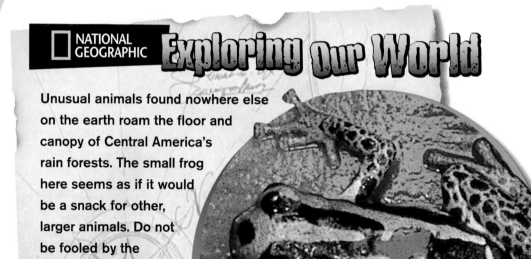

NATIONAL GEOGRAPHIC Exploring Our World

Unusual animals found nowhere else on the earth roam the floor and canopy of Central America's rain forests. The small frog here seems as if it would be a snack for other, larger animals. Do not be fooled by the enlargement of the photo, however. Many frogs like this one hold a deadly poison in their skin, which would quickly kill anything that tried to eat them.

Central America, part of Middle America, is an isthmus, or a narrow piece of land that links two larger areas of land—North America and South America. Most of the countries on the isthmus have two coastlines—one on the Pacific Ocean and one on the Caribbean Sea. This narrow region is actually part of North America. Seven countries make up Central America: Belize, Guatemala, El Salvador, Honduras, Nicaragua, Costa Rica, and Panama.

A Rugged Land

Like Mexico, Central America sits where plates in the earth's crust meet. The collision of these plates produces volcanoes and earthquakes in the region. The Central Highlands, which curve like a backbone through inland Central America, are actually a chain of volcanic mountains. Because of their ruggedness, the Central Highlands are difficult to cross. This causes serious problems for transportation and communication and has also kept many of the region's people isolated from one another. The volcanoes of the Central Highlands do bring some benefits to farmers, though. Volcanic material has made the soil very fertile.

Central America is mostly tropical, although the mountainous areas are cool year-round. The Caribbean Lowlands have a hot, tropical rain forest climate throughout the year. Here you can expect about 100 inches (254 cm) of rain each year. Breezes from the Caribbean Sea provide some cooling relief. These breezes can be replaced by deadly hurricanes during the summer and fall, though. Hurricanes are fierce storms with winds of more than 74 miles (119 km) per hour.

✓ Reading Check How have the volcanoes in Central America been helpful?

The Economy

The economies of the Central American countries depend on farming and harvesting wood from their rain forests. Central America has two kinds of farms. Wealthy people and companies own plantations— commercial farms that grow crops for sale. Major crops include coffee, bananas, cotton, and sugarcane. Plantations export their harvest to the United States and other parts of the world. Farmers in Guatemala and Costa Rica also grow flowers and ornamental plants for export.

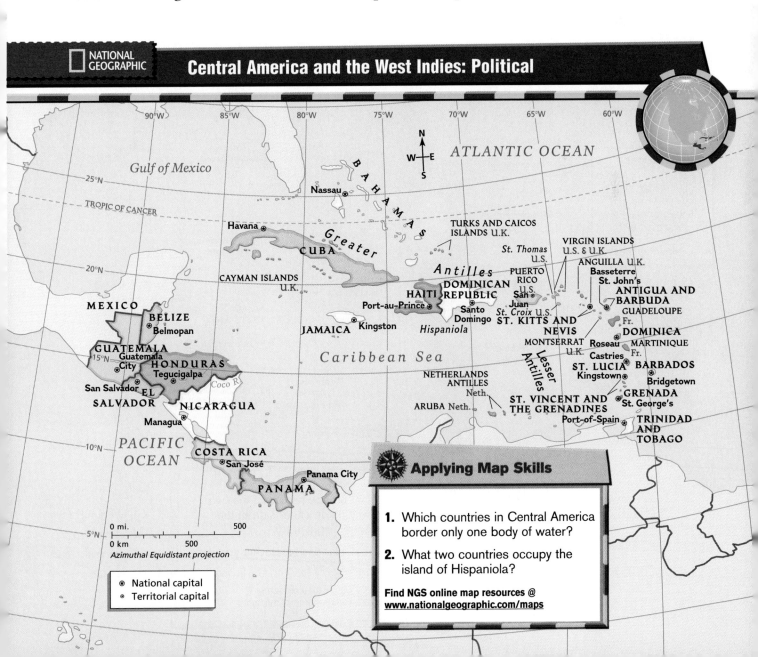

NATIONAL GEOGRAPHIC

Central America and the West Indies: Political

Applying Map Skills

1. Which countries in Central America border only one body of water?

2. What two countries occupy the island of Hispaniola?

Find NGS online map resources @ www.nationalgeographic.com/maps

⊛ National capital
◦ Territorial capital

Azimuthal Equidistant projection

Many farms in Central America are not plantations but subsistence farms, or small plots of land where poor farmers grow only enough food to feed their families. Subsistence farmers typically raise livestock and grow corn, beans, and rice.

Rain Forests Under Central America's green canopy, or topmost layer of the rain forest, which shades the forest floor, valuable resources and ancient ruins of past empires can be found. The dense forests offer expensive woods—mahogany and rosewood, for example. Unusual animal and plant species also thrive here. Scientists research the plants to develop new medicines.

Both local and foreign-owned companies have set up large-scale operations in the rain forests. Lumber companies cut down and export the valuable trees. Other companies and local farmers also cut or burn the trees to clear land for farming. Without trees to provide nutrients and hold the soil in place, rains wash it and its nutrients away. As a result, the land soon becomes poor. The businesses and farmers then move on, clearing trees from another piece of land.

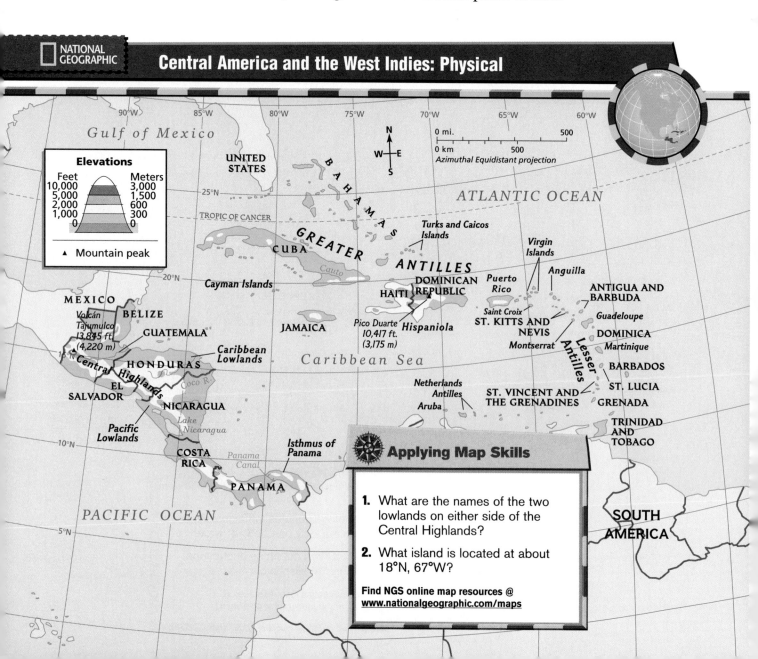

NATIONAL GEOGRAPHIC

Central America and the West Indies: Physical

Applying Map Skills

1. What are the names of the two lowlands on either side of the Central Highlands?

2. What island is located at about 18°N, 67°W?

Find NGS online map resources @ www.nationalgeographic.com/maps

Many Central Americans worry about the rapid destruction of the rain forests. Some countries are responding to this crisis by helping workers replant cleared areas. Costa Rica has set aside one-fourth of its forests as national parks. It uses the rain forests to attract eco-tourists, or people who travel to other countries to enjoy natural wonders.

Industry Missing from the skylines of most major Central American cities are the smokestacks of industry. The few industries that exist generally focus on preparing foods. In Guatemala, Honduras, and Nicaragua, some factories produce clothing for export.

Guatemala, which has some oil reserves, exports crude oil. Costa Rica produces computer chips, other electronic goods, and medicines. With its varied economy, Costa Rica enjoys one of the highest standards of living in Latin America. It also has one of the highest literacy rates, or percentage of people who can read and write.

The Panama Canal The economy in Panama is based on farming—as the economy is throughout Central America—but Panama also earns money from its canal. The **Panama Canal** stretches across the narrow **Isthmus of Panama.** Ships pay a fee to use the canal to shorten travel time between the Atlantic and Pacific Oceans. Turn to page 562 to see how the canal works.

The United States built the canal and owned it for more than 80 years. Panama was given final control of the canal on December 31, 1999. Panama hopes to use this waterway to build its economy. Nearly half of Panama's 2.9 million people live and work in the canal area.

✓Reading Check What are the major crops grown on Central America's plantations?

The History and People of Central America

Native Americans settled Central America thousands of years ago. The Olmecs were the first civilization in the area, from about 1200 B.C. to 400 B.C. The Maya flourished in the rain forests of the north from about A.D. 250 to 900. Look at the Native American civilizations map on

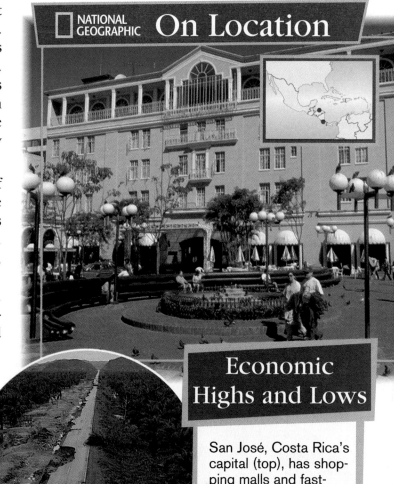

NATIONAL GEOGRAPHIC **On Location**

Economic Highs and Lows

San José, Costa Rica's capital (top), has shopping malls and fast-food chains like many North American cities. In 1998 Hurricane Mitch caused massive mudslides that buried whole villages and destroyed crops in Honduras (bottom).

Issues During what seasons do hurricanes strike Central America?

page 542. In **Tikal** (tee•KAHL), Guatemala, and **Copán** (koh•PAHN), Honduras, the Maya created impressive temples and sculptures. Before Columbus arrived, Tikal was the site of the highest structure in the Americas, a 212-foot (64.6-m) temple rising from the floor of the rain forest. The Maya were a very highly developed civilization. Their religion focused on the careful study of time and the stars, astronomy, and mathematics. The Maya developed a calendar and kept records on stone slabs. Then the Maya mysteriously left their cities. Many of their descendants still live in the area today.

In the 1500s, Spaniards established settlements in Central America. For the next 300 years, Spanish landowners forced Native Americans to work on plantations. The two cultures gradually blended. Native Americans started to speak the Spanish language and follow the Roman Catholic faith. Native Americans taught the Spanish about local plants for medicines and how to trap animals for food and hides.

Most Central American countries gained independence from Spain by 1821. The two exceptions are Panama and Belize. Panama was part of the South American country of Colombia for decades. In 1903 the United States helped Panama win its independence in exchange for the right to build the Panama Canal. Belize, a British colony until 1981, was the last Central American country to gain independence.

After Independence Most Central American countries faced constant strife after they became independent. A small number of people in each country held most of the wealth and power. Rebel movements arose as poor farmers fought for changes that would give them land and better lives. Civil wars raged in Nicaragua, El Salvador, and Guatemala as recently as the 1980s and 1990s.

In Guatemala from 1960 to 1996, government military forces fought rebel groups living in the highlands. About 150,000 people died, and the civil war severely weakened Guatemala's economy. Tens of thousands of Guatemalans left the country to look for work in the United States.

In contrast, Costa Ricans have enjoyed peace. A stable democratic government rules, and the country has avoided conflict for most of its history. As a result of these peaceful relations, the country has no army—only a police force to maintain law and order.

Today each country in Central America has a democratic government, with voters choosing government officials. Six countries are also **republics,** with elected presidents as head of the government. Belize is a British-style **parliamentary democracy,** in which an elected legislature chooses a prime minister to head the government.

Daily Life Nearly 38 million people live in Central America. About one-third of this number live in Guatemala, the most heavily populated country in the region. Only about 300,000 people live in Belize, the region's least populous country. Spanish is the official language

Teen Scene

What a Catch!

The deep blue waters of Lake Nicaragua are home to the world's only freshwater sharks and swordfish. Now the lake holds one less swordfish. Amadeo Robelo, who lives in Granada, Nicaragua, just spent three hours battling the powerful fish. Amadeo enjoys fishing with his father on weekends. His father wants Amadeo to become part of Nicaragua's middle class—something new in a region where you are either one of the few with wealth or one of the many who live in poverty.

throughout the region, except for English-speaking Belize. Many Central Americans also speak Native American languages, such as Mayan. Guatemala's population, for instance, is largely Native American and has more than 20 different Native American languages. Most Central Americans follow the Roman Catholic religion.

About 50 percent of all Central Americans live on farms or in small villages. At least one major city, usually the capital, is densely populated in each country. Guatemala's capital, **Guatemala City,** ranks with **San José,** Costa Rica, as one of the most populous cities in Central America. People living in urban areas hold manufacturing or service industry jobs, or they work on farms outside the cities. Those living in coastal areas may harvest shrimp, lobster, and other seafood to sell in city markets or for export.

Whether rural or urban, most people enjoy a major celebration called **Carnival.** This festival comes before Lent, a solemn period of prayer and soul-searching before the Christian celebration of Easter. During Carnival—and at other times—bands play salsa, a mixture of Latin American popular music, jazz, and rock. Do you like baseball? It is a national sport in Nicaragua and very popular in Panama, too. Most people throughout the region also enjoy *fútbol,* or soccer.

✓ Reading Check **Why is the government of Belize different from that of other countries in Central America?**

Assessment

Defining Terms
1. Define isthmus, hurricane, plantation, subsistence farm, canopy, eco-tourist, literacy rate, republic, parliamentary democracy.

Recalling Facts
2. Economics What is the difference between plantation and subsistence farming?
3. Culture What are the major religion and language of Central America?
4. Place Which country in Central America is the most heavily populated? The most sparsely populated?

Critical Thinking
5. Making Comparisons How have the differences in government stability affected the citizens of Guatemala and Costa Rica?

6. Analyzing Cause and Effect Explain why rain forest soil does not keep its nutrients long.

Graphic Organizer
7. Organizing Information Create a diagram like this one. On the lines list the major products and industries of Central America.

Major products and industries

Applying Social Studies Skills

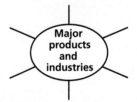

8. Analyzing Maps Refer to the political map on page 557. Which countries of Central America border Mexico? Which border the Pacific Ocean?

The Panama Canal Locks

Before the Panama Canal was built, ships had to sail around the southern tip of South America to go from the Atlantic Ocean to the Pacific Ocean and vice versa. The canal provides a shortcut that reduces that trip by about 7,000 miles (11,270 km).

Digging the Canal

The first attempts to build a canal across Panama were begun in 1881 by a private French company. Huge expenses, poor planning, and the effects of diseases such as malaria and yellow fever stopped construction. In 1904 the United States government took over. Doctors had recently learned that bites from infected mosquitoes caused malaria and yellow fever. Workers drained swamps and cleared brush to remove the mosquitoes' breeding grounds. Then the digging began. The canal's course ran through hills of soft volcanic soil. Massive landslides regularly occurred before the 50-mile (80-km) canal was completed in 1914.

An Engineering Masterpiece

To move ships through the canal, engineers designed three sets of locks—the largest concrete structures on the earth. They allow ships to move from one water level to another by changing the amount of water in the locks. Together, the locks can raise or lower ships about 85 feet (26 m)—the height of a seven-story building. The diagram below shows you how these locks work.

→ Making the Connection

1. Why was a canal through Panama desirable?
2. What function do locks perform?
3. **Understanding Cause and Effect** How did medical advances affect the building of the Panama Canal?

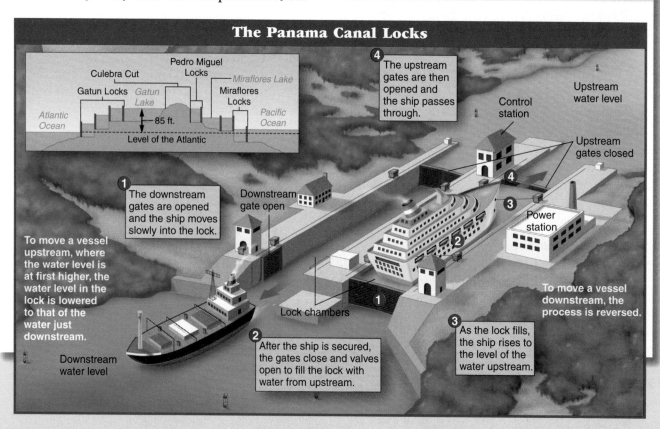

The Panama Canal Locks

Culebra Cut
Pedro Miguel Locks
Gatun Locks — Gatun Lake
Miraflores Lake
Miraflores Locks
Atlantic Ocean
Pacific Ocean
85 ft.
Level of the Atlantic

4 The upstream gates are then opened and the ship passes through.

Upstream water level

Control station

Upstream gates closed

1 The downstream gates are opened and the ship moves slowly into the lock.

Downstream gate open

Power station

To move a vessel upstream, where the water level is at first higher, the water level in the lock is lowered to that of the water just downstream.

Lock chambers

To move a vessel downstream, the process is reversed.

Downstream water level

2 After the ship is secured, the gates close and valves open to fill the lock with water from upstream.

3 As the lock fills, the ship rises to the level of the water upstream.

The West Indies

Guide to Reading

Main Idea

The islands of the West Indies rely on tourism to support their economies.

Terms to Know

- commercial crop
- bauxite
- cooperative
- communist state
- embargo
- free trade zone
- commonwealth

Reading Strategy

Create a diagram like this one. In each oval, list a country in the West Indies and features that are specific to it. Where the ovals overlap, list features that are similar to both countries.

Country 1 / Country 2

NATIONAL GEOGRAPHIC

Exploring Our World

The warm waters of the Caribbean Sea lure millions of tourists to the West Indies every year. Some tourists go scuba diving so they can see the colorful fish, which swim in the islands' clear waters. Others shop at the local stores, buying hand-crafted goods. This diver uses a metal detector to look for objects from a Spanish ship that sank in the 1600s.

A number of islands dot the Caribbean Sea and form groups known as archipelagos that make up the West Indies. Study the map on page 557 to become familiar with the names of these island groups.

Mountaintop Islands

When you look at the islands of the West Indies, you are really looking at the tops of mountains. Many West Indian islands are part of an underwater chain of mountains formed by volcanoes. A typical volcanic island has central highlands ringed by coastal plains. The volcanic soil in the highlands is rich.

Other islands are limestone mountains pushed up from the ocean floor by pressures under the earth's crust. Limestone islands generally are flatter than volcanic islands. The sandy soil found on many limestone islands is not good for farming.

Climate The West Indies lie in the Tropics. Most islands have a fairly constant tropical climate. Sea and wind, more than elevation, affect the climate here. Northeast breezes sweep across the Caribbean Sea,

563

taking on the temperature of the cooler water beneath them. When the winds blow onshore, they keep temperatures pleasant.

For half the year, hurricanes threaten the West Indies. The word *hurricane* comes from the Taíno, an early Native American people who lived on the islands. They worshiped a god of storms named Hurakan. (Two other terms that we have borrowed from the Taíno include *canoe* and *hammock,* a bed made by stringing ropes or plant fibers between two trees.)

✓ Reading Check **What formed the islands of the West Indies?**

The Economy of the West Indies

Farming and tourism are the main economic activities in the West Indies. Wealthy landowners grow crops such as sugarcane, bananas, coffee, and tobacco for export. Many laborers work on the plantations that grow these commercial crops. Commercial crops are grown to sell, not to be eaten by the grower. The Caribbean islands face an economic danger by depending on one commercial crop. If the crop fails, no income is earned. If too much of the crop is produced worldwide, overall prices fall and the economy is in serious trouble. Most of the islands do not have large amounts of minerals, although several islands have some resources. Jamaica, for example, mines bauxite, a mineral used to make aluminum. The country of Trinidad and Tobago exports oil products. In Puerto Rico, companies make chemicals and machinery. Haiti and the Dominican Republic have textile factories that make cloth.

✓ Reading Check **What are the two major industries in the West Indies?**

History and Culture

When Christopher Columbus reached San Salvador in the Bahamas in 1492, who met him? As you probably guessed, it was a Native American group—the Taíno. The Taíno and other Native Americans lived on the islands long before the coming of Europeans.

The Spaniards established the first permanent European settlement in the Western Hemisphere in 1496. That settlement is now the city of Santo Domingo, capital of the Dominican Republic. During the next 200 years, the Spaniards, the English, the French, and the Dutch also founded colonies, or overseas settlements, on many of the islands. They found the soil and climate perfect for growing sugarcane.

By the mid-1600s, most Native Americans had died from European diseases and harsh treatment. The Europeans then brought enslaved Africans to work on sugar plantations. When the slave trade ended in the early 1800s, plantation owners still in need of workers brought them from Asia, particularly India. The Asians agreed to work a set number of years in return for free travel to the West Indies and low wages.

Independence During the 1800s and 1900s, many Caribbean islands won their freedom from European rule. The first to become independent were the larger island countries, such as Haiti, the Dominican Republic, and Cuba. Later, smaller islands such as Barbados

and Grenada became nations. Many countries—like Haiti and the Dominican Republic—are republics. Others—like Jamaica and the Bahamas—are British-style parliamentary democracies.

Daily Life About 60 percent of West Indians live in cities and villages. The other 40 percent live and work in the countryside. Many islanders have jobs in the hotels or restaurants that serve the tourist industry.

If you visit the Caribbean, you are likely to hear lively music. The bell-like tones of the steel drum, developed in Trinidad, are part of the rich musical heritage of the region. Enslaved Africans created a kind of music called calypso. Jamaica's reggae and calypso music combine African rhythms and American popular music. Cuban salsa blends African rhythms, Spanish styles, and jazz.

On several islands, you will hear a different sound—the crack of a baseball bat. People in Puerto Rico, the Dominican Republic, and Cuba have a passion for baseball. Soccer and cricket are other popular sports.

Schoolgirls on Barbados walk past vast sugar plantations that European countries started in the colonial period (left). A steel-drum band entertains tourists in Trinidad (above).

Region What attracts so many tourists to the islands of the West Indies?

✓ Reading Check Where was the first permanent European settlement in the West Indies?

Island Profiles

The islands of the West Indies share many similarities, but they also have differences. Some of these differences can be seen in Cuba, Haiti, the Dominican Republic, and Puerto Rico.

Bee Hummingbird

How small is this bird? The bee hummingbird of Cuba measures only 2 inches (5.1 cm) from head to tail. That is small enough to make it the tiniest bird in the world. The bird's wings move so fast—80 beats per second—that the human eye cannot see them. At two grams, the bee hummingbird weighs less than a penny.

Cuba   One of the world's top sugar producers, Cuba lies about 90 miles (145 km) south of Florida. Most farmers work on cooperatives, or farms owned and operated by the government. In addition to growing sugarcane, they grow coffee, tobacco, rice, and fruits. In **Havana,** Cuba's capital and the largest city in the West Indies, workers make food products, cigars, and household goods.

Cuba is a communist state, where the government has strong control over the economy and society. Cuba is the only communist state in the Western Hemisphere. A dictator, President Fidel Castro, leads the country. Cuba won its independence from Spain in 1898. The country had a democratic government, although military leaders sometimes seized power. In 1959 Castro led a revolution that took control of the government. He set up a communist state and turned to the Soviet Union for support. When he seized property belonging to American companies, the United States government responded. It put in place an embargo, or a ban on trade, against Cuba.

Cuba lost its major source of aid when the Soviet Union collapsed in 1991. Since then, Cuba's leaders have been unable to meet even the basic needs of their people. Some Americans believe that the United States should help the Cuban people by ending the trade embargo with Cuba. Other Americans, especially Cuban Americans, many of whom fled Castro's regime, strongly oppose reopening trade. They believe that such trade would only help an oppressive Communist dictator stay in power.

Haiti   On the western half of the island of Hispaniola, you will find the country of Haiti. Led by a former slave, Francois-Dominique Toussaint-Louverture, Haiti fought for and won its independence from France in 1804. It was the second independent republic in the Western Hemisphere (after the United States). It became the first nation in the history of the world to be founded by former slaves. About 95 percent of Haiti's 7 million people are of African ancestry.

Civil war has left Haiti's economy in ruins, and most Haitians are poor. In the 1980s, the staple of the people's diet was the Creole pig. Creole pigs were also important as a source of income. When it came time to send a child to school, a pig could be sold to pay for books, fees, and a uniform. When planting season came, selling a pig paid for seeds and tools. Having a pig around often meant the difference between life and death when health emergencies arose. Creole pigs were known as the "peasants' savings bank." In 1983 a swine flu outbreak forced the government to destroy the population of Creole pigs to prevent an epidemic in the Americas. A plan to substitute pigs from the United States failed. Unlike the Creole pigs, which had foraged for their own food, the U.S. breeds required expensive feed, which Haitian farmers could not afford. The imported pigs required housing with cement floors, while most Haitian peasants lived in homes with dirt floors. The U.S. pigs were not used to Haiti's tropical climate. Within a short time, most of the U.S. pigs had died. Today, the Haitian government, with U.S. help, has begun a program to help communities replace the destroyed Creole pig population and try to lessen the poverty of the families.

Dominican Republic   The Dominican Republic shares the island of Hispaniola with Haiti, filling the eastern part. Though they share the same island, the two countries have different histories and little contact. Haiti was a French colony. The Dominican Republic was settled by Spaniards, who brought enslaved Africans to work on sugar plantations. Sugar is still an important crop to the Dominicans. Tourism is growing, too, and many Dominicans sell goods in the country's free trade zone. Free trade zones are areas where people can buy goods from other countries without paying taxes.

Puerto Rico   To be or not to be a state in the United States, that is the question that Puerto Ricans ask themselves every few years. The last time they voted on the question, they said no. How did Puerto Rico become part of the United States? The island was a Spanish colony from 1508 to 1898. After the Spanish-American War in 1898, the United States won control of Puerto Rico. Since 1952 the island has been a commonwealth, or a partly self-governing territory, under American protection. By law, Puerto Ricans are citizens of the United States. They can come and go from the island to the United States as they wish. Today around 3 million Puerto Ricans live in the United States. The island itself holds about 3.9 million and boasts a high standard of living compared to most other Caribbean islands.

Web Activity Visit the *Our World Today: People, Places, and Issues* Web site at owt.glencoe.com and click on **Chapter 21— Student Web Activities** to learn more about Puerto Rico.

Reading Check   **What is a commonwealth?**

Assessment

Defining Terms

1. Define commercial crop, bauxite, cooperative, communist state, embargo, free trade zone, commonwealth.

Recalling Facts

2. Economy What are the two main economic activities in the West Indies?

3. History Name four groups who have influenced the culture of the Caribbean region.

4. Government How is Cuba different from every other country in the Western Hemisphere?

Critical Thinking

5. Drawing Conclusions Explain why you think Puerto Ricans might be satisfied remaining a commonwealth, rather than becoming a state.

6. Making Predictions What is the danger of a country's depending on only one crop?

Graphic Organizer

7. Organizing Information Complete a chart like the one below with facts about Haiti and the Dominican Republic.

Country	Haiti	Dominican Republic
Location		
Colonized by		
Economy		

Applying Social Studies Skills

8. Cause and Effect Create a cause and effect chart that explains why the U.S. pigs did not adapt to Haiti.

Social Studies Skill

Interpreting an Elevation Profile

You have learned that differences in land elevation are often shown on physical or relief maps. Another way to show elevation is on **elevation profiles.** When you view a person's profile, you see a side view. An elevation profile is a diagram that shows a side view of the landforms in an area.

Learning the Skill

Suppose you could slice right through a country from top to bottom and could look at the inside, or *cross section.* The cross section, or elevation profile, below pictures the island of Jamaica. It shows how far Jamaica's landforms extend below or above sea level.

Follow these steps to understand an elevation profile:

- Read the title of the profile to find out what country you are viewing.
- Look at the line of latitude written along the bottom of the profile. On a separate map, find the country and where this line of latitude runs through it.
- Look at the measurements along the sides of the profile. Note where sea level is located and the height in feet or meters.
- Now read the labels on the profile to identify the heights of the different landforms shown along with their elevation.
- Compare the highest and lowest points.

Practicing the Skill

Use the elevation profile below to answer the following questions.

1. At what elevation is Kingston?
2. What are the highest mountains, and where are they located?
3. Where are the lowest regions?
4. Along what line of latitude was this cross section taken?

Applying the Skill

Turn to page 13 in the Geography Handbook. Use the elevation profile of Africa to answer questions 2–4 above about *that* continent.

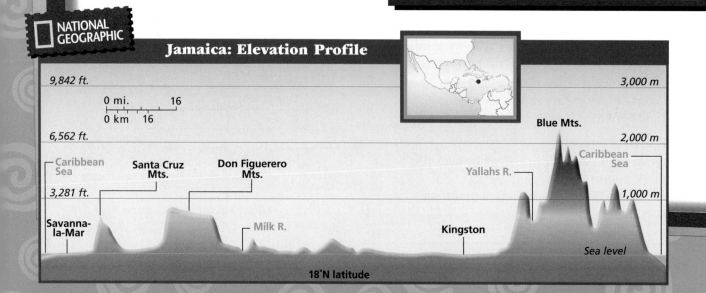

Jamaica: Elevation Profile

Section 1 | Central America

Terms to Know

isthmus
hurricane
plantation
subsistence farm
canopy
eco-tourist
literacy rate
republic
parliamentary
  democracy

Main Idea

Central America is made up of seven nations that are home to a variety of peoples, exotic animals, and diverse landforms.

✓ Region Central America includes seven countries: Belize, Guatemala, Honduras, El Salvador, Nicaragua, Costa Rica, and Panama.

✓ Region Volcanic mountains run down the center of Central America with coastal lowlands on either side.

✓ Economics Most people in the region farm—either on plantations or subsistence farms.

✓ Culture Most countries in Central America have a blend of Native American and Spanish cultures.

The Panama Canal ▶

Section 2 | The West Indies

Terms to Know

commercial crop
bauxite
cooperative
communist state
embargo
free trade zone
commonwealth

Main Idea

The islands of the West Indies rely on tourism to support their economies.

✓ History Christopher Columbus landed in this region in 1492.

✓ History Most of the islands were at one time colonies of European countries.

✓ Economics Farming and tourism are the major economic activities in the West Indies.

✓ Culture The cultures of the West Indies mix Native American, European, African, and Asian influences.

✓ Government Most governments in the West Indies are democratic, but a dictator rules Communist Cuba.

Using Key Terms

Match the terms in Part A with their definitions in Part B.

A.

1. isthmus
2. literacy rate
3. plantation
4. eco-tourist
5. commercial crop
6. bauxite
7. commonwealth
8. embargo
9. free trade zone
10. republic

B.

a. large commercial farm
b. mineral ore from which aluminum is made
c. ban on trade
d. narrow piece of land connecting two larger pieces of land
e. area where people can buy goods from other countries without paying taxes
f. person who travels to another country to enjoy its natural wonders
g. country with an elected president
h. percentage of adults who can read and write
i. partly self-governing territory
j. product grown to sell rather than to eat

Reviewing the Main Ideas

Section 1 Central America

11. **Region** What seven countries make up Central America?
12. **Economics** Why are the Central American rain forests being destroyed?
13. **History** In what Central American countries did the Maya live?
14. **Culture** What percentage of Central Americans live on farms or in small villages?

Section 2 The West Indies

15. **Economics** What two activities form the basis of the West Indian economies?
16. **Region** Many of the Caribbean islands were formed by what type of tectonic activity?
17. **Culture** What types of music can you find in the West Indies?
18. **History** What was the first nation in the world to be founded by formerly enslaved people?
19. **Economics** Why are commercial crops sometimes a risky business?

 **NATIONAL GEOGRAPHIC** **Central America and the West Indies**

Place Location Activity

On a separate sheet of paper, match the letters on the map with the numbered places listed below.

1. Guatemala
2. Caribbean Sea
3. Cuba
4. Puerto Rico
5. Costa Rica
6. Panama
7. Bahamas
8. Haiti
9. Jamaica
10. Honduras

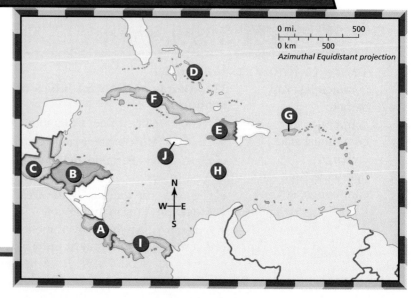

Self-Check Quiz Visit the *Our World Today: People, Places, and Issues* Web site at owt.glencoe.com and click on **Chapter 21—Self-Check Quizzes** to prepare for the Chapter Test.

Critical Thinking

20. **Analyzing Information** Explain why Cuba's location is an important factor in the United States's relationship with that nation.

21. **Categorizing Information** Create a diagram like this with details about the people, history, and economy of a country in Chapter 21.

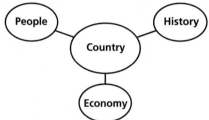

Current Events Journal

22. **Writing an Itinerary** Write an itinerary, or travel plan, for a cruise ship that makes five stops in the Caribbean. Include a map showing the route and descriptions of the sites and activities at each stop.

Mental Mapping Activity

23. **Focusing on the Region** Create an outline map of Central America and the West Indies, and then label the following:

- Pacific Ocean
- Cuba
- Caribbean Sea
- Puerto Rico
- Guatemala
- Jamaica
- Panama
- Bahamas

Technology Skills Activity

24. **Building a Database** Create a database about Central America, using the Country Profiles in the Unit 7 Regional Atlas for your database. Make a record for each country. Each record should have a field for the following: population, landmass, and capital city. Sort the records from largest to smallest for population. What generalizations can you make based on these data?

Standardized Test Practice

Directions: Study the map below, and then answer the question that follows.

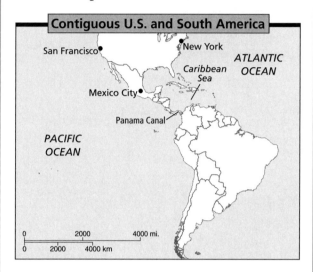

1. **Which of the following was true before the Panama Canal was completed?**

    A A ship sailing from New York to California had to travel nearly 10,000 additional miles.

    B A ship sailing from New York to California had to travel nearly 5,000 additional kilometers.

    C The completion of the canal increased trade between Mexico City and San Francisco.

    D Mexico City was extremely far away from New York City.

Test-Taking Tip: The scale shows you the actual distance between places on a map. Use your finger or a piece of paper to mark off the distance of the scale. Then, use your finger or piece of paper to gauge the distance between two places on the map.

Unit

8

Peruvian Indian woman and child

Peaks of the Andes, Chile

South America

South America is a continent in change. It contains 350 million people. Today, the larger countries of this region are diversifying their economies and attempting to meet the demands of rapidly growing populations. Many are becoming urban nations as people leave the countryside for the cities. Other countries are burdened by social stress and political turmoil.

◄ Rio de Janeiro, Brazil

South America

Physical

80°W 70°W 60°W 50°W

10°N

Caribbean Sea

Caracas ⊛
VENEZUELA

LLANOS
SURINAME
Georgetown ⊙
GUYANA
Paramaribo
⊙ Cayenne
FRENCH GUIANA

Bogotá ⊛
COLOMBIA

GUIANA HIGHLANDS

EQUATOR
0°
Galápagos Is.
Quito ⊛
ECUADOR

Marajó I.

A M A Z O N

La Montaña
B A S I N

S E L V A S

CATINGAS

PERU

10°S

A N D E S

Lima ⊛

B R A Z I L

MATO GROSSO PLATEAU

BRAZILIAN

PACIFIC OCEAN

Lake Titicaca
La Paz ⊛
BOLIVIA
Sucre ⊛

⊛ Brasília

Pantanal

HIGHLANDS

20°S

Altiplano

PARAGUAY

TROPIC OF CAPRICORN

Atacama Desert

GRAN CHACO

Salado R.

⊛ Asunción

Cape São Tomé

Paraguay R.

30°S

CHILE

Aconcagua
22,834 ft.
(6,960 m) ▲

N
W E
S

Santiago ⊛

P A M P A S

Buenos ⊛
Aires

Uruguay R.

⊛ Montevideo
URUGUAY

Río de la Plata

ATLANTIC OCEAN

ARGENTINA

40°S

Isla Grande
de Chiloé

Valdés
Peninsula

A N D E S

P A T A G O N I A

⊛ National capital
⊙ Territorial capital
▲ Mountain peak

0 mi. 1,000

0 km 1,000
Lambert Azimuthal Equal-Area projection

50°S

Strait of Magellan
Tierra
del Fuego
Cape Horn

Falkland Is.
(Islas Malvinas)

South Georgia I.

26,247 ft. | 0 mi. 500 | 8,000 m

19,685 ft. | 0 km 500 | 6,000 m

ANDES | BRAZILIAN HIGHLANDS | 4,000 m
13,123 ft. | MATO GROSSO PLATEAU

AMAZON BASIN

6,562 ft. | 2,000 m

LIMA | Sea level | SALVADOR

UNIT 8

Political

90°W 80°W 70°W 60°W 50°W 40°W 30°W

10°N

Caracas

VENEZUELA
Georgetown SURINAME
Medellín GUYANA Paramaribo
Bogotá Cayenne
Cali FRENCH GUIANA
COLOMBIA Fr.

EQUATOR Quito
0° Galápagos Is. ECUADOR Negro R. Amazon R.
Ecua. Manaus

Madeira R.

Recife

PERU BRAZIL

10°S Lima

Salvador
PACIFIC Lake
OCEAN Titicaca Brasília
La Paz
BOLIVIA ATLANTIC
Sucre OCEAN

20°S PARAGUAY Belo Horizonte
Paraná R.

TROPIC OF CAPRICORN Rio de Janeiro
Asunción São Paulo
Paraguay R. Curitiba
CHILE

Porto Alegre
30°S ARGENTINA
Valparaíso URUGUAY
Santiago Rosario Montevideo
Buenos Río de la Plata
Aires

N
40°S W E
S
0 mi. 1,000
0 km 1,000
Lambert Azimuthal Equal-Area projection

⊗ National capital
⊙ Territorial capital Falkland Is.
• Major city U.K.

50°S
South Georgia I.
U.K.

MAP STUDY

1 What huge lowland area lies in northern Brazil?

2 What is the capital of Chile?

South America

Urban Population Growth

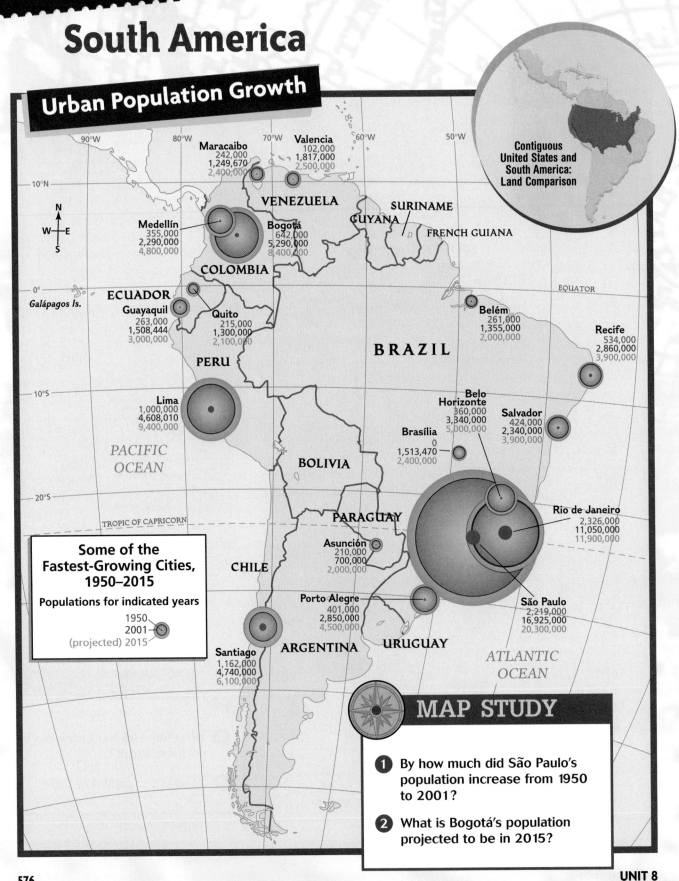

Contiguous
United States and
South America:
Land Comparison

Maracaibo
242,000
1,249,670
2,400,000

Valencia
102,000
1,817,000
2,500,000

VENEZUELA

SURINAME

GUYANA

FRENCH GUIANA

Medellín
355,000
2,290,000
4,800,000

Bogotá
642,000
5,290,000
8,400,000

COLOMBIA

ECUADOR

Galápagos Is.

Guayaquil
263,000
1,508,444
3,000,000

Quito
215,000
1,300,000
2,100,000

PERU

BRAZIL

EQUATOR

Belém
261,000
1,355,000
2,000,000

Recife
534,000
2,860,000
3,900,000

Lima
1,000,000
4,608,010
9,400,000

Belo Horizonte
360,000
3,340,000
5,000,000

Salvador
424,000
2,340,000
3,900,000

Brasília
0
1,513,470
2,400,000

PACIFIC OCEAN

BOLIVIA

TROPIC OF CAPRICORN

PARAGUAY

Rio de Janeiro
2,326,000
11,050,000
11,900,000

Some of the Fastest-Growing Cities, 1950–2015

Populations for indicated years

1950
2001
(projected) 2015

Asunción
210,000
700,000
2,000,000

CHILE

Porto Alegre
401,000
2,850,000
4,500,000

São Paulo
2,219,000
16,925,000
20,300,000

Santiago
1,162,000
4,740,000
6,100,000

ARGENTINA

URUGUAY

ATLANTIC OCEAN

MAP STUDY

1 By how much did São Paulo's population increase from 1950 to 2001?

2 What is Bogotá's population projected to be in 2015?

Fast Facts

COMPARING POPULATION:
United States and Selected Countries of South America

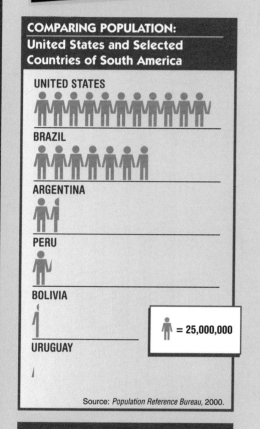

UNITED STATES

BRAZIL

ARGENTINA

PERU

BOLIVIA

= 25,000,000

URUGUAY

Source: *Population Reference Bureau*, 2000.

ETHNIC GROUPS:
Selected Countries of South America

ARGENTINA
97% ‖3%

BOLIVIA
55% | 30% | 15%

BRAZIL
6%
55% | 38% ‖1%

PERU
45% | 37% | 15% ‖3%

URUGUAY
8%
88% ‖4%

■Black ▯Indian ▨African/European
▯White ▨Mestizo ■Other

Source: *Population Reference Bureau*, 2000.

Data Bits

Country	Automobiles per 1,000 people	Telephones per 1,000 people
Chile	71	180
Colombia	19	148
Ecuador	40	75
Suriname	59	130
Venezuela	68	116

Population: Urban ▨ vs. Rural ▯

Chile	84%	16%
Colombia	73%	27%
Ecuador	58%	42%
Suriname	50%	50%
Venezuela	93%	7%

Source: *World Desk Reference*, 2000.

GRAPHIC STUDY

❶ Which country has the fewest automobiles per 1,000 people, and which has the fewest telephones?

❷ What country has a majority of Indians?

REGIONAL ATLAS

Country Profiles

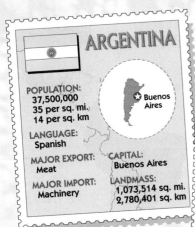

ARGENTINA

POPULATION:
37,500,000
35 per sq. mi.
14 per sq. km

LANGUAGE:
Spanish

MAJOR EXPORT:
Meat

MAJOR IMPORT:
Machinery

CAPITAL:
Buenos Aires

LANDMASS:
1,073,514 sq. mi.
2,780,401 sq. km

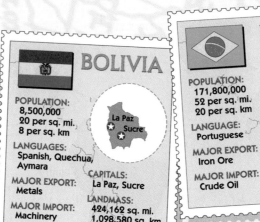

BOLIVIA

POPULATION:
8,500,000
20 per sq. mi.
8 per sq. km

LANGUAGES:
Spanish, Quechua,
Aymara

MAJOR EXPORT:
Metals

MAJOR IMPORT:
Machinery

CAPITALS:
La Paz, Sucre

LANDMASS:
424,162 sq. mi.
1,098,580 sq. km

BRAZIL

POPULATION:
171,800,000
52 per sq. mi.
20 per sq. km

LANGUAGE:
Portuguese

MAJOR EXPORT:
Iron Ore

MAJOR IMPORT:
Crude Oil

CAPITAL:
Brasília

LANDMASS:
3,300,154 sq. mi.
8,547,399 sq. km

CHILE

POPULATION:
15,400,000
53 per sq. mi.
20 per sq. km

LANGUAGE:
Spanish

MAJOR EXPORT:
Copper

MAJOR IMPORT:
Machinery

CAPITAL:
Santiago

LANDMASS:
292,135 sq. mi.
756,630 sq. km

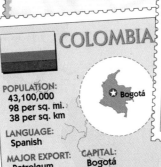

COLOMBIA

POPULATION:
43,100,000
98 per sq. mi.
38 per sq. km

LANGUAGE:
Spanish

MAJOR EXPORT:
Petroleum

MAJOR IMPORT:
Machinery

CAPITAL:
Bogotá

LANDMASS:
439,734 sq. mi.
1,138,911 sq. km

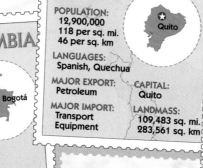

ECUADOR

POPULATION:
12,900,000
118 per sq. mi.
46 per sq. km

LANGUAGES:
Spanish, Quechua

MAJOR EXPORT:
Petroleum

MAJOR IMPORT:
Transport
Equipment

CAPITAL:
Quito

LANDMASS:
109,483 sq. mi.
283,561 sq. km

FRENCH GUIANA*

POPULATION:
200,000
6 per sq. mi.
2 per sq. km

LANGUAGE:
French

MAJOR EXPORT:
Shrimp

MAJOR IMPORT:
Foods

CAPITAL:
Cayenne

LANDMASS:
34,749 sq. mi.
90,000 sq. km

* Territory of France

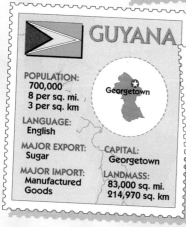

GUYANA

POPULATION:
700,000
8 per sq. mi.
3 per sq. km

LANGUAGE:
English

MAJOR EXPORT:
Sugar

MAJOR IMPORT:
Manufactured
Goods

CAPITAL:
Georgetown

LANDMASS:
83,000 sq. mi.
214,970 sq. km

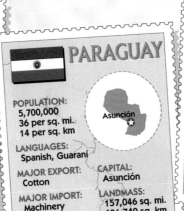

PARAGUAY

POPULATION:
5,700,000
36 per sq. mi.
14 per sq. km

LANGUAGES:
Spanish, Guaraní

MAJOR EXPORT:
Cotton

MAJOR IMPORT:
Machinery

CAPITAL:
Asunción

LANDMASS:
157,046 sq. mi.
406,749 sq. km

PERU

POPULATION:
26,100,000
53 per sq. mi.
20 per sq. km

LANGUAGES:
Spanish, Quechua,
Aymara

MAJOR EXPORT:
Copper

MAJOR IMPORT:
Machinery

CAPITAL:
Lima

LANDMASS:
496,224 sq. mi.
1,285,217 sq. km

Countries and flags not drawn to scale

For more information on countries in this region, refer to the Nations of the World Data Bank on pages 690–699.

SURINAME

POPULATION:
400,000
6 per sq. mi.
3 per sq. km

LANGUAGE:
Dutch

MAJOR EXPORT:
Bauxite

CAPITAL:
Paramaribo

MAJOR IMPORT:
Machinery

LANDMASS:
63,039 sq. mi.
163,271 sq. km

URUGUAY

POPULATION:
3,400,000
50 per sq. mi.
19 per sq. km

LANGUAGE:
Spanish

MAJOR EXPORT:
Wool

CAPITAL:
Montevideo

MAJOR IMPORT:
Machinery

LANDMASS:
68,498 sq. mi.
177,410 sq. km

VENEZUELA

POPULATION:
24,600,000
70 per sq. mi.
27 per sq. km

LANGUAGE:
Spanish

MAJOR EXPORT:
Petroleum

CAPITAL:
Caracas

MAJOR IMPORT:
Raw Materials

LANDMASS:
352,143 sq. mi.
912,050 sq. km

BUILDING CITIZENSHIP

Public and Private Needs
More than one-third of the area of Brazil is covered by a rain forest. This fragile ecosystem is home to millions of plant, animal, and insect species. Some of the plants are important sources of medicines. According to scientists, more than 50 percent of the world's species live in the rain forest.

The rain forest is also a major source of timber, minerals, fruits, and vegetables. Building roads and clearing land to reach these resources has led to major destruction of the rain forest habitat. The government of Brazil has tried to set aside large portions of the rain forest as preserves while allowing development of its natural resources, as income for its citizens.

Because of its effect on climate, the rain forest is important not just to Brazil but to the whole world. Who should have more say about how much of the rain forest is preserved—Brazil or the United Nations?

WRITE ABOUT IT

Imagine that a new golf course is being built in your city and the area where it is being built includes natural wetlands where birds and animals live. Write a letter to the city council outlining what steps you think the golf course developers should take to protect the wetlands.

Brazilian rain forest

Brazil and Its Neighbors

The World and Its People

NATIONAL GEOGRAPHIC

To learn more about the people and places of Brazil and its neighbors, view **The World and Its People Chapter 8** video.

Our World Today online

Chapter Overview Visit the **Our World Today: People, Places, and Issues** Web site at owt.glencoe.com and click on **Chapter 22—Chapter Overviews** to preview information about Brazil and its neighbors.

FOLDABLES™
Study Organizer

Summarizing Information Study Foldable Make this foldable and use it to organize note cards with information about the people and places of Brazil and its neighbors.

Step 1 Fold a 2-inch tab along the long edge of a sheet of paper.

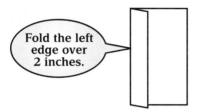

Fold the left edge over 2 inches.

Step 2 Fold the paper in half so the tab is on the inside.

The tab can't be seen when the paper is folded.

Step 3 Open the paper pocket foldable, turn it, and glue the edges of the pockets together.

Glue here.

Glue here.

Step 4 Label the pockets as shown.

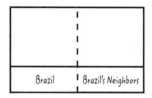

Brazil | Brazil's Neighbors

Reading and Writing As you read the chapter, summarize key facts about Brazil and its neighbors on note cards or on quarter sheets of notebook paper. Organize your notes by placing them in your pocket foldable inside the appropriate pockets. (Keep your pocket foldable to use with Chapter 23.)

Why It Matters

Preserving the Environment

The Amazon rain forest—sometimes called the "lungs of the planet" because of the huge amounts of oxygen given off by its trees—is home to up to 30 percent of the animal and plant life on Earth. Destroying these trees may cause the extinction of many wildlife species and damage to the earth's environment—on which we all depend for our own lives. This is just one of many issues facing the people and government of Brazil.

◄ **The Amazon River, Brazil**

Brazil

Guide to Reading

Main Idea

Brazil is a large country with many resources, a lively culture, and serious economic challenges.

Terms to Know

- basin
- *selva*
- escarpment
- favela
- deforestation
- republic

Reading Strategy

Create a chart like the one below and fill in at least one key fact about Brazil in each category.

Brazil	
Land	
Climate	
History	
Economy	
Government	
People	

NATIONAL GEOGRAPHIC

Exploring Our World

Some of the world's largest fresh-water fish swim in the mighty Amazon River in Brazil. Called pirarucu (pih•RAHR•uh•KEW), these fish can grow up to 15 feet (4.6 m) long. What a catch! The people who catch these huge fish often make the fish scales into souvenir key chains for tourists.

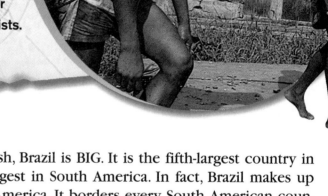

Like the pirarucu fish, Brazil is BIG. It is the fifth-largest country in the world and the largest in South America. In fact, Brazil makes up almost half of South America. It borders every South American country except Chile and Ecuador.

Brazil's Land

Because Brazil covers such a large area, it has many different types of landforms and climates. The map on page 583 shows you that Brazil has narrow coastal plains, highland areas, and lowland river valleys.

The **Amazon River** is the world's second-longest river, winding almost 4,000 miles (6,437 km) from the Andes mountain ranges to the Atlantic Ocean. On its journey to the Atlantic, the Amazon drains water from a wide, flat basin. A basin is a low area surrounded by higher land. In the **Amazon Basin,** rainfall can reach as much as 120 inches (305 cm) a year. These rains support the growth of thick tropical rain forests, which Brazilians call *selvas.* Turn to page 600 to learn more about this rain forest, which covers one-third of Brazil.

Brazil has lowlands along the Paraná River and the São Francisco River. The Brazilian Highlands cover about half of the country, then drop sharply to the Atlantic Ocean. This drop is called the **Great Escarpment.** An escarpment is a steep cliff between higher and lower land.

✓ Reading Check What is significant about the Amazon River?

Brazil's Economy

How do Brazilians earn a living? Agriculture, mining, and forestry have been important for centuries. The Amazon Basin has been a mysterious region whose secrets were guarded by the Native Americans

NATIONAL GEOGRAPHIC

Brazil and Its Neighbors: Physical

Elevations

Feet	Meters
10,000	3,000
5,000	1,500
2,000	600
1,000	300
0	0

▲ Mountain peak

Applying Map Skills

1. Which area of Brazil—the north or the south—has the highest elevation?

2. Name two rivers that flow into the Amazon River.

Find NGS online map resources @ www.nationalgeographic.com/maps

0 mi. 800
0 km 800
Azimuthal Equidistant projection

Leading Coffee-Producing Countries

Analyzing the Graph

Brazil's highlands have the right soil and climate to grow coffee.

Economics Which leading coffee-producing countries are in South America?

Textbook Update

Visit owt.glencoe.com and click on **Chapter 22— Textbook Updates.**

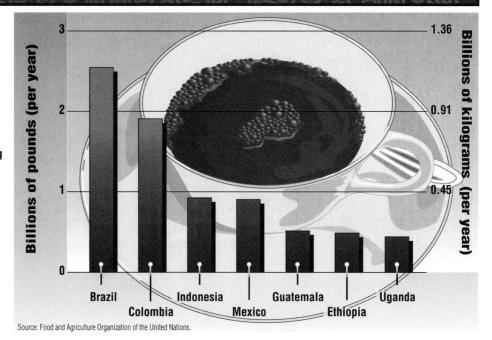

Billions of pounds (per year)

Billions of kilograms (per year)

Brazil   Colombia   Indonesia   Mexico   Guatemala   Ethiopia   Uganda

Source: Food and Agriculture Organization of the United Nations.

living there. This began to change in the mid-1800s. World demand skyrocketed for the rubber harvested from the basin's trees, and new settlers streamed to Brazil's interior. Today, mining companies dig for minerals such as bauxite, tin, and iron ore. Logging companies harvest mahogany and other woods from the rain forest. Farmers use the cleared land to grow soybeans and tobacco and to graze cattle.

Brazil's Economic Challenges Today Brazil's economy is diverse and productive, yet the country still faces serious economic challenges. Brazil's economy has brought wealth to many Brazilians and built a large and strong middle class. Yet as many as one-fifth of Brazil's people live in extreme poverty. Many Brazilian cities are surrounded by favelas, or slum areas. Thousands of poor people move to cities looking for work in the factories. They live in crude shacks with neither running water nor sewage systems. City governments have tried to clean up these areas, but people continue to settle here because they have no money to pay for housing. Many children as young as 10 years old go to work to help earn money for their families.

While Brazil has the largest area of remaining rain forest in the world, it also has the highest rate of deforestation. Deforestation is the cutting down or destroying of large areas of forest. To increase jobs and products for export, the government has encouraged mining, logging, and farming in the rain forest. The main causes of deforestation, however, are the building of roads and the clearing of land for farms. Roads allow both poor farmers and rich landowners to use parts of the rain forest that used to be out of reach. The forest is cut down on each side of the road to create long strips of land for farming.

Deforestation is an issue because tropical forests are of great importance to the regulation of the earth's climate. The rain forests help move heat from the Equator to the more temperate regions. Scientists are still studying the role of the rain forest in global warming. Although the Amazon rain forest belongs to Brazil, the effects of deforestation are worldwide. An important issue, then, is how to balance the needs of a country to develop jobs and revenue for its people and the climate needs of the planet.

In addition, deforestation threatens the Native Americans who live in the rain forest. As more people settle in the Amazon Basin, Native Americans find it difficult to follow traditional ways of life. In the 1990s, the government announced plans to set aside 10 percent of the Amazon forest as parks. Another 10 percent will be set aside for native peoples.

✓ Reading Check **What are the leading causes of deforestation in the Amazon Basin?**

Brazil's History and Culture

With about 172 million people, Brazil has the largest population of all South American nations. Brazil's culture is largely Portuguese. The Portuguese were the first and largest European group to colonize

Literature

BOTOQUE
Kayapo Indian Myth
In this myth of central Brazil, the hero brings fire to his people.

"Botoque and the animals safely returned to their village with Jaguar's possessions. Everyone was delighted to eat grilled meat. They loved being able to warm themselves by the fire when the nights became cool. And they liked having the village fires provide protection from wild animals.

As for Jaguar, when he returned home and found that he had been robbed of his special possessions, his heart flooded with fury. 'So this is how Botoque has repaid me for adopting him as my son and teaching him the secret of the bow and arrow!' he exclaimed. 'Why, he did not even leave me fire. Well, no matter. In memory of this theft, from this time forth and evermore, I will eat my catch raw! This will keep the memory of my adopted son before my eyes and hatred for him—and all who walk the earth as he does—alive in my heart!'"

Botoque, Bringer of Fire, p. 298–299, Folklore, Myths, and Legends, A World Perspective, edited by Donna Rosenberg, NTC Publishing, 1997.

Analyzing Literature
Do you think the Kayapo Indians feared jaguars? Why or why not?

Rio de Janeiro

A huge statue of Christ overlooks Rio de Janeiro (right). Crowds of people in Rio de Janeiro celebrate Carnival wearing brightly colored costumes (above).

Culture What groups make up Brazil's population?

Brazil. Today Brazilians are of European, African, Native American, Asian, or mixed ancestry. Almost all of them speak a Brazilian form of Portuguese, which includes many words from Native American and African languages. Most of the population follow the Roman Catholic religion, but many Brazilians combine Catholicism with beliefs and practices from African and Native American religions.

Influence of History Native Americans were the first people to live in Brazil. In the 1500s, the Portuguese forced Native Americans to work on large plantations that grew tobacco and, later, sugarcane. Many Native Americans died from disease or overwork. To replace them, early Portuguese settlers brought people from Africa and enslaved them. Slavery finally was banned in 1888, but Africans remained in Brazil, most of them living in the northeastern part of the country. Over the years, African traditions have influenced Brazilian religion, music, dance, and food.

Moving to the Cities Much of Brazil is sparsely populated. Millions of people have moved from rural areas to coastal cities in hopes of finding better jobs. Now the government is encouraging people to move back to less populated inland areas. Highways now crisscross the country and reach many once remote regions. In 1960 Brazil moved its capital from coastal **Rio de Janeiro** 600 miles (966 km) inland to the newly built city of **Brasília.** With more than 1.5 million people, Brasília is a modern and rapidly growing city.

The Government Brazil declared independence from Portugal in 1822. At first the new nation was an empire, with emperors ruling from 1822 to 1889. Like some other countries in Middle and South America, Brazil was ruled by military dictators. Today Brazil is a democratic republic, where people elect a president and other leaders. In Brazil, though, citizens cannot choose whether to vote or not vote. People from ages 18 to 70 are required by law to vote. Brazil has more than a dozen political parties—not just two main ones, as in the United States.

The national government of Brazil is much stronger than its 26 state governments. Brazil's president has more power over the country than an American president does in the United States.

Leisure Time Brazilians live for soccer, which they call fútbol. Every village has a soccer field, and the larger cities have stadiums. Maracana Stadium in Rio de Janeiro seats 220,000 fans. Basketball is another important sport.

Brazil is also famous for **Carnival.** This festival is celebrated just before the beginning of Lent, the Christian holy season that comes before Easter. The most spectacular Carnival is held each year in Rio de Janeiro. The celebration includes Brazilian music and showy parades.

Brazil has one of the largest television networks in the world. This network produces prime-time soap operas called *telenovelas*. These programs are wildly popular in Brazil—and viewers in more than 60 other nations enjoy them too.

Reading Check **Why do most Brazilians speak a form of Portuguese?**

Web Activity Visit the *Our World Today: People, Places, and Issues* Web site at owt.glencoe.com and click on **Chapter 22— Student Web Activities** to learn more about the destruction of the rain forest.

Assessment

Defining Terms

1. Define basin, *selva*, escarpment, favela, deforestation, republic.

Recalling Facts

2. **History** Who were the first and largest group of Europeans to colonize Brazil?
3. **Economics** What resources attract companies to the Amazon Basin?
4. **Culture** What is the major religion of Brazil?

Critical Thinking

5. **Drawing Conclusions** In what way is deforestation threatening the Native Americans who live in the rain forest?
6. **Summarizing Information** What economic challenges face Brazilians?

Graphic Organizer

7. **Organizing Information** Create a diagram like this one. Beside the left arrow, write the cause of the government action. On the right, list results of this action.

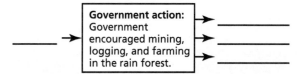

Government action: Government encouraged mining, logging, and farming in the rain forest.

Applying Social Studies Skills

8. **Analyzing Maps** Look at the physical map on page 583. What large landform in Brazil surrounds the Amazon River?

Critical Thinking Skill

Sequencing and Categorizing Information

Sequencing means placing facts in the order in which they occurred. *Categorizing* means organizing information into groups of related facts and ideas. Both actions help you deal with large quantities of information in an understandable way.

Learning the Skill

Follow these steps to learn sequencing and categorizing skills:

- Look for dates or clue words that provide you with a chronological order: *in 2004, the late 1990s, first, then, finally, after the Great Depression,* and so on.
- If the sequence of events is not important, you may want to categorize the information instead. Categories might include economic activities or cultural traits.
- List these characteristics, or categories, as the headings on a chart.
- As you read, fill in details under the proper category on the chart.

Practicing the Skill

Read the paragraphs below, and then answer the questions that follow.

After Brazil's independence from Portugal in 1822, a bill was presented to build a new capital named Brasília. More than 100 years later, in 1955, a planning committee chose the site for the new capital. The first streets were paved in 1958. On April 20, 1960, the festivities to officially "open" the new capital started at 4:00 P.M.

Brasília has both positive and negative aspects. The positive include virtually no air pollution, no threat of natural disasters, many green areas, and a pleasant climate. The negative aspects of the capital include very high housing prices, inefficient public transportation, few parking spaces, and long distances between the various government buildings.

1. What information can be organized sequentially?
2. What categories can you use to organize the information? What facts could be placed under each category?

Brasília

Applying the Skill

Find two newspaper or magazine articles about Brazil or another South American country. Sequence or categorize the information on note cards or in a chart.

GO TO

Practice key skills with **Glencoe Skillbuilder Interactive Workbook, Level 1.**

Brazil's Neighbors

Guide to Reading

Main Idea

Brazil's neighbors give South America a diverse array of landforms, climates, and cultures.

Terms to Know

- pampas
- *estancia*
- gaucho
- hydroelectric power
- llanos
- altitude
- caudillo

Reading Strategy

Create a diagram like this one. Fill in the names of Brazil's neighbors to the south and north, and then write at least one key fact about the people from each country.

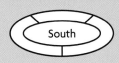

South

North

NATIONAL GEOGRAPHIC Exploring Our World

The traditional music of Paraguay seems to be out of place with the rest of its culture. The harp is the country's national instrument, and Paraguayans are famous for their slow, mournful guitar playing. In contrast, the traditional dances are much livelier. Here, a woman performs the bottle dance—a difficult feat even though the bottles are attached to one another.

South of Brazil lie **Argentina, Uruguay,** and **Paraguay.** To the north of Brazil are countries that border the Caribbean Sea— **Venezuela, Guyana, Suriname,** and **French Guiana.**

Argentina

Argentina is South America's second-largest country, after Brazil. Its southern tip reaches almost to the continent of Antarctica. Argentina is about the size of the United States east of the Mississippi River. Within this vast area, you can find mountains, deserts, plains, and forests.

The Andes tower over the western part of Argentina. Snowcapped peaks and clear blue lakes draw tourists for skiing and hiking. **Aconcagua** (AH•kohn•KAH•gwah) soars to a height of 22,834 feet (6,960 m), making it the highest mountain in the Western Hemisphere.

South and east of the Andes lies a dry, windswept plateau called **Patagonia.** Most of Patagonia gets little rain and has poor soil. As a result, sheep raising is the only major economic activity. Find Patagonia on the physical map on page 583.

The center of Argentina has vast treeless plains known as the pampas. Similar to the Great Plains of the United States, the pampas are home to farmers who grow grains and ranchers who raise livestock. More than two-thirds of Argentina's people live in this region.

Argentina's Economy Argentina's economy depends heavily on farming and ranching. The country's major farm products include beef, sugarcane, wheat, soybeans, and corn. Huge *estancias* (ay•STAHN•see•ahs), or ranches, cover the pampas. Gauchos (GOW•chohs), or cowhands, take care of the livestock on the ranches. Gauchos are the national symbol of Argentina, admired for their independence and horse-riding skills. The livestock that the gauchos herd and tend are a vital part of the country's economy. Beef and food products are Argentina's chief exports. Turn to page 596 to read more about gauchos.

Argentina is one of the most industrialized countries in South America. Most of the country's factories are in or near **Buenos Aires,** Argentina's capital and largest city. The leading manufactured goods are food products, automobiles, chemicals, textiles, books, and magazines.

Petroleum is Argentina's most valuable mineral resource. The country's major oil fields are in Patagonia and the Andes. Other minerals—zinc, iron, copper, tin, and uranium—are mined in the Andes as well. Despite these resources, Argentina's economy has struggled during the early years of the twenty-first century.

Argentina's History In the late 1500s, the Spanish settled in Argentina. In 1816 a general named **José de San Martín** led Argentina in its fight for freedom from Spain. After independence, the country was torn apart by civil war. By the mid-1850s, a strong national government had emerged, and Argentina entered a time of prosperity. During the first half of the 1900s, Argentina's elected leaders governed poorly. The economy suffered, and the military took over. One of these military leaders, **Juan Perón,** became a dictator in the late 1940s. With his popular wife, Eva, at his side, Perón tried to improve the economy and give more help to workers. His crackdown on freedom of speech and the press made people unhappy, however. In 1955 a revolt drove Perón from power, and democracy returned.

Military officers again took control of Argentina in the 1970s. They ruled harshly, and political violence resulted in the deaths of many people. In 1982 Argentina suffered defeat in a war with the United Kingdom for control of the **Falkland Islands,** which lie in the Atlantic Ocean off the coast of Argentina. Argentina's loss forced the military to step down, and elected leaders regained control of the government.

Today Argentina is a democratic republic. As in the United States, the national government is much stronger than the provincial, or state, governments. A powerful elected president leads the nation for a four-year term. A legislature with two houses makes the laws.

Argentina's People About 85 percent of Argentina's people are of European ancestry. During the late 1800s, immigrants in large numbers came to Argentina from Spain and Italy. They drove out or killed many

Exploring Government

Strong Presidency

In Argentina, the president is very strong. Checks and balances present in the American system to control the power of the president do not exist in Argentina. In the United States, the Supreme Court may declare acts or laws unconstitutional. Or, the legislature may pass laws to limit the power of U.S. federal agencies. In Argentina, it is harder to prevent a popular leader from forcing his programs through the legislature.

Native Americans who inhabited Argentina. Their arrival greatly influenced Argentina's society and culture. Many more immigrants arrived from Europe after World War II. European ways of life are stronger in Argentina today than in most other South American countries.

The official language of Argentina is Spanish, although the language includes many Italian words. Most people are Roman Catholic. About 90 percent of Argentina's people live in cities and towns. Buenos Aires and its suburbs hold more than 12 million people. Buenos Aires has wide streets and European-style buildings. Its citizens call themselves *porteños* (pohr•TAY•nyohs), which means "people of the port." Many have a passion for the national dance of Argentina, the tango.

✓Reading Check Why does Argentina have a strong European culture?

About 40,000 workers labored to build Paraguay's Itaipu Dam. Brazil funded its construction. In return, Brazil pays low prices for the electricity it buys from Paraguay.

Human/ Environment Interaction How do both Paraguay and Brazil benefit from this arrangement?

Uruguay and Paraguay

Uruguay and Paraguay differ from each other in environment, population, and development. Uruguay, with its mild climate, low-rolling terrain, and rich grasslands, is a buffer zone between the two large and powerful nations of Brazil and Argentina. Originally settled by the Portuguese, then taken over by Spain, Uruguay revolted against both countries and eventually became independent in 1825.

Immigration from Spain and Italy and the introduction of sheep are keys to Uruguay's modern development. The country's 3.4 million people are mostly of European descent. Its economy is extremely dependent on sheep and cattle raising. In fact, sheep and cattle outnumber people by ten to one in Uruguay, and about 70 percent of the territory of the country is in pasture. These grasslands produce the wool, hides, and meat that Uruguay exports.

In Uruguay, the expected social pattern of a wealthy landowning elite and poverty-stricken working class did not evolve. The large haciendas were complemented by many medium-sized and small farms on which wheat, flax, wine, and vegetable production has doubled in recent years. Enlightened government policies have provided social welfare to the poor. The Uruguayans, half of whom live in the capital city of **Montevideo,** have the highest literacy rate, the lowest population growth rate, the best diet, and one of the highest standards of living of any South American country. Spanish is the official language, and the Roman Catholic faith is the major religion.

Paraguay   In Paraguay, the society and economy followed a quite different course. The eastern third of Paraguay, with its rich soils and fertile grasslands, was settled by the Spanish. The western two-thirds of the country, the great forest area known as the **Chaco,** was brought into the Spanish territory by Roman Catholic missionaries.

In the 1800s and 1900s a series of wars severely hurt Paraguay, destroying the economy of the country. After the worst of these—the five-year War of the Triple Alliance against Brazil, Argentina, and Uruguay in the 1860s—Paraguay's population was cut in half. Experts estimate that only 28,000 adult males were left alive, and Paraguay had lost 55,000 square miles of territory. The country has yet to recover from this catastrophe.

Currently, the only area with success in agriculture in Paraguay is located near the capital city of **Asunción** (ah•SOON•see•OHN). Cotton, tobacco, soybeans, and cassava are among Paraguay's crops.

Paraguay also exports electricity. The country has the world's largest hydroelectric power generator at the **Itaipu** (ee•TY•poo) **Dam,** on the Paraná River. Hydroelectric power is electricity that is generated by flowing water. Paraguay sells nearly 90 percent of the electricity it produces to neighboring countries.

Paraguayans today are mostly of mixed Guaraní—a Native American group—and Spanish ancestry. Both Spanish and Guaraní are official languages, but more people speak Guaraní. Most people practice the Roman Catholic faith. About one-half of the people live in cities.

Paraguayan arts are influenced by Guaraní culture. Guaraní lace is Paraguay's most famous handicraft. Like people in Uruguay, the people of Paraguay enjoy meat dishes and sip yerba maté, a tealike drink.

✓ Reading Check What important export is generated at the Itaipu Dam?

Venezuela

Venezuela (VEH•nuh•ZWAY•luh) is the westernmost country of Caribbean South America. In the northwest lie the lowland coastal areas surrounding **Lake Maracaibo** (MAH•rah•KY•boh), the largest lake in South America. Swamps fill much of this area, and few people live here. The great number of towering oil wells, however, gives you a clue that rich oil fields lie under the lake and along its shores. Venezuela has more oil reserves than any other country in the Americas.

Roping a Capybara

Capybaras are the world's largest rodents. They may grow to be 2 feet tall and 4 feet long, and weigh more than 100 pounds. Found in Central and South America, the capybara (ka•pih•BAR•uh) lives along rivers and lakes, and eats vegetation. Here, a gaucho ropes a dog-sized capybara in Venezuela. Some Venezuelans eat capybara during the Easter season.

The Andean highlands begin south of the lake and are part of the Andes mountain ranges. This area includes most of the nation's cities, including **Caracas** (kah•RAH•kahs), the capital and largest city.

East of the highlands, you see grassy plains known as the **llanos** (LAH•nohs). The llanos have many ranches, farms, and oil fields. Venezuela's most important river—the **Orinoco**—flows across the llanos. This river is a valuable source of hydroelectric power for Venezuela's cities.

South and east of the llanos rise the Guiana Highlands, deeply cut by rivers. **Angel Falls**—the world's highest waterfall—spills over a bluff in this region.

Because it is close to the Equator, Venezuela has a mostly tropical climate. In the Guiana Highlands to the south, you enter a steamy rain forest. As in Mexico, temperatures in Venezuela differ with **altitude,** or height above sea level. Higher altitudes have cooler climates.

NATIONAL GEOGRAPHIC On Location

Angel Falls

Angel Falls—the highest waterfall in the world at 3,212 feet (979 m)—roars over a cliff in Venezuela. It would take 11 football fields stacked end-to-end to reach the top.

Economics **What is one of the rivers that provides Venezuela with hydroelectric power?**

Venezuela's Economy Venezuelans once depended on crops such as coffee and cacao to earn a living. Since the 1920s, petroleum has changed the country's economy. Venezuela is a world leader in oil production and one of the chief suppliers of oil to the United States. Because the government owns the oil industry, oil provides nearly half of the government's income. Iron ore, limestone, bauxite, gold, diamonds, and emeralds also are mined. Factories make steel, chemicals, and food products. About 10 percent of the people farm, growing sugarcane and bananas or raising cattle.

History and Government With its many rivers, the land in South America reminded early Spanish explorers of Venice, Italy, which is full of canals. They named the area *Venezuela,* which means "Little Venice."

In the early 1800s, rebellion swept across the Spanish colonial empire. **Simón Bolívar** (see•MOHN boh•LEE•VAHR), who was born in Venezuela, became one of the leaders of this revolt. He and his soldiers freed Venezuela and neighboring regions from Spanish rule. In 1830 Venezuela became independent.

During most of the 1800s and 1900s, the country was governed by military rulers called **caudillos** (kow•THEE•yohz). Their rule was often harsh. Since 1958, Venezuela has been a democracy led by a president and a two-house legislature.

NATIONAL GEOGRAPHIC On Location

Celebration

Venezuelan dancers in costumes and playing maracas take part in Corpus Christi, a local Roman Catholic celebration.

Religion What is the major religion in Venezuela?

Rising oil prices during the 1970s brought more money to the country. The middle class grew, and many people prospered. When oil prices fell in the 1990s, the country suffered. The government did not have the money to give the poor and unemployed the services they needed. In 1998 Venezuelans showed their impatience with the government. They elected a former military leader, **Hugo Chavez,** as president. Chavez proposed major changes to the country's constitution and economy. In December 1999, the Venezuelan people voted to accept the new constitution.

Venezuela's People   Most of Venezuela's 24.6 million people have a mix of European, African, and Native American backgrounds. Spanish is the major language of the country, and the major religion is Roman Catholicism. About 90 percent of Venezuelans live in cities. Some 3 million people live in Caracas, which holds towering skyscrapers surrounded by mountains.

✓Reading Check **What product changed Venezuela's economy?**

The Guianas

Caribbean South America also includes the countries of Guyana (gy•AH•nuh) and Suriname (SUR•uh•NAH•muh) and the territory of French Guiana (gee•A•nuh). Guyana was a British colony called British Guiana. Suriname, once a colony of the Netherlands, was called Dutch Guiana. As a result, these three lands are called "the Guianas."

The climate in the Guianas is hot and tropical. Most people live on the coastal plains because of the cooling ocean winds. Sugarcane grows in Guyana and French Guiana, while rice and bananas flourish in Suriname. Many people also earn their living mining gold and bauxite.

Guyana   In the early 1600s, the Dutch were the first Europeans to settle in Guyana. They forced Native Americans and Africans to work on tobacco, coffee, and cotton farms and, later, on sugarcane plantations. The United Kingdom won possession of the Dutch colonies in the early 1800s and ended slavery. Still needing workers, the British paid Indians from Asia to move here. Today people from India make up most of Guyana's population. Another one-third are of African ancestry. Small numbers of Native Americans and Europeans also live here. Christianity and Hinduism are the chief religions. Most people speak English. **Georgetown,** the capital, is the major city.

Guyana won its independence from Britain in 1966. Guyana remains a very poor country, however, and depends on aid from the United Kingdom.

Suriname The British were the first Europeans to settle Suriname, but the Dutch gained control in 1667. As in Guyana, the Dutch brought enslaved Africans to work on large sugar plantations. Because of harsh treatment, many Africans fled into the isolated interior of the country. Their descendants still live there today. Later the Dutch hired workers from the Asian lands of India and Indonesia.

Asians form a large part of Suriname's population. About half of Suriname's people practice Christianity. The rest follow Hinduism or Islam. The main language is Dutch. **Paramaribo** (PAH•rah•MAH•ree•boh) is the capital and chief port. In 1975 Suriname won its independence from the Dutch. The country is poor, so it still relies on Dutch aid.

French Guiana French Guiana became a colony of France in the 1600s and remains one today. The country is headed by a French official called a *prefect,* who lives in the capital, **Cayenne** (ky•EHN). The French government provides jobs and aid to many of French Guiana's people.

Most people in French Guiana are of African or mixed African and European ancestry. They speak French and are Roman Catholic. In Cayenne, you see sidewalk cafés, police in French uniforms, and shoppers using francs, the French currency—just as you would in Paris, France. You also see local influences, such as Carnival, Native American woodcarving, and Caribbean music and dance.

✓ Reading Check **What European countries influenced the development of Guyana, Suriname, and French Guiana?**

Assessment

Defining Terms

1. Define pampas, *estancia,* gaucho, hydroelectric power, llanos, altitude, caudillo.

Recalling Facts

2. Region Describe two ways in which the pampas are similar to the Great Plains of the United States.

3. Human/Environment Interaction What is the significance of the Itaipu Dam?

4. History Who was Simón Bolívar?

Critical Thinking

5. Analyzing Cause and Effect Which of Juan Perón's policies led to his removal from office?

6. Drawing Conclusions Why is Hinduism one of the major religions of Guyana?

Graphic Organizer

7. Organizing Information Create a diagram like this one. In the top box, under the heading, list similarities about the Guianas. In the bottom boxes, under the headings, write facts about each country that show their differences.

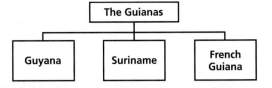

The Guianas

| Guyana | Suriname | French Guiana |

Applying Social Studies Skills

8. Analyzing Maps Look at the physical map on page 583. Which of Brazil's neighbors in this section have mountain elevations above 10,000 feet?

Brazil and Its Neighbors

Making Connections

Poetry on the Pampas

As you learned in Section 2, gauchos herd cattle on the pampas. In 1872 José Hernández wrote the epic poem *El Gaucho Martín Fierro*. The poem tells the story of Martín Fierro, who recalls his life as a gaucho on the pampas. The following lines were translated from the poem.

El Gaucho Martín Fierro
by José Hernández (1834–1886)

A son am I of the rolling plain,
  A gaucho born and bred;
    For me the whole great world is small,
    Believe me, my heart can hold it all;
The snake strikes not at my passing foot,
  The sun burns not my head.

  • • • • • • •

Ah, my mind goes back and I see again
  The gaucho I knew of old;
    He picked his mount, and was ready aye,
    To sing or fight, and for work or play,
And even the poorest one was rich
  In the things not bought with gold.

The neediest gaucho in the land,
  That had least of goods and gear,
    Could show a troop of a single strain,
    And rode with a silver-studded rein,
The plains were brown with the grazing herds,
  And everywhere was cheer.

And when the time of the branding came,
  It did one good to see
    How the hand was quick and the eye
      was true,
    When the steers they threw with the
      long lassoo [lasso],
And the merry band that the years have swept
  Like leaves from the autumn tree.

Excerpt from *The Gaucho Martin Fierro,* adapted from the Spanish and rendered into English verse by Walter Owen. Copyright © 1936 by Farrar & Rinehart. Reprinted by permission of Henry Holt and Company, LLC.

Gauchos on Argentina's pampas ▲

Making the Connection

1. How does the poet describe the land on which the gaucho lives?

2. How can you tell from the poem that a gaucho is often on the move?

3. **Drawing Conclusions** What evidence does the poem give that the gaucho's way of life was a proud and happy one?

Section 1 | Brazil

Terms to Know

basin
selva
escarpment
favela
deforestation
republic

Main Idea

Brazil is a large country with many resources, a lively culture, and serious economic challenges.

✓ History Brazil declared independence in 1822 after centuries of colonial rule by Portugal.

✓ Economics Brazil is trying to reduce its number of poor people and balance the use of resources with the preservation of its rain forests.

✓ Culture Most Brazilians are of mixed Portuguese, African, Native American, and Asian ancestry.

Section 2 | Brazil's Neighbors

Terms to Know

pampas
estancia
gaucho
hydroelectric power
llanos
altitude
caudillo

Main Idea

Brazil's neighbors give South America a diverse array of landforms, climates, and cultures.

✓ History Simón Bolívar led a revolt that freed Venezuela from Spanish rule in 1830.

✓ Culture Argentina's capital, Buenos Aires, is a huge city with European style.

✓ Economics Uruguay and Paraguay have large areas of grass-covered plains that support ranching and industries that depend on raising livestock.

✓ Culture Most Venezuelans are of mixed European, African, and Native American ancestry. Most live in cities in the central highlands.

✓ Culture Guyana and Suriname have large numbers of people descended from workers who were brought from Africa and Asia.

Shepherd and sheep dogs on plateau of Patagonia in southern Argentina ▶

Assessment and Activities

Chapter 22

Using Key Terms

Match the terms in Part A with their definitions in Part B.

A.

1. basin
2. *estancia*
3. escarpment
4. caudillo
5. altitude
6. gaucho
7. *selva*
8. deforestation
9. llanos
10. pampas

B.

a. steep cliff separating two flat land surfaces, one higher than the other
b. cowhand
c. cutting down large areas of forest
d. military ruler
e. large, grassy plains region with many ranches, farms, and oil fields
f. height above sea level
g. broad, flat lowland surrounded by higher land
h. vast treeless plains
i. rain forest in Brazil
j. large ranch in Argentina

Reviewing the Main Ideas

Section 1 Brazil

11. **History** Why are Brazil's inland areas sparsely populated?
12. **Government** What are the voting requirements in Brazil?
13. **History** When and why did Brazil's government move the capital city to Brasília?

Section 2 Brazil's Neighbors

14. **Culture** Why is the European influence so strong in Argentina today?
15. **Culture** What are the major language and religion of Uruguay?
16. **Economics** What are the major economic activities of Paraguay?
17. **Economics** Which of Venezuela's resources is its main source of income?
18. **Culture** Where do most of the people of the Guianas live? Why do they live there?
19. **History** Which of Brazil's neighbors has been a colony of France since the 1600s?

 **Brazil and Its Neighbors**

Place Location Activity

On a separate sheet of paper, match the letters on the map with the numbered places listed below.

1. Brazil
2. Amazon River
3. Argentina
4. Rio de Janeiro
5. Paraguay
6. Orinoco River
7. Río de la Plata
8. Venezuela
9. Brasília
10. Suriname

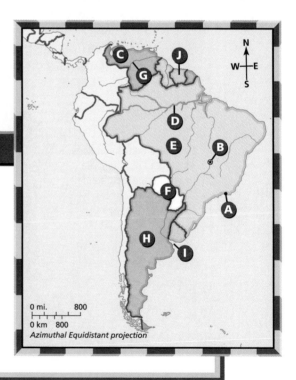

Self-Check Quiz Visit the *Our World Today: People, Places, and Issues* Web site at <u>owt.glencoe.com</u> and click on **Chapter 22—Self-Check Quizzes** to prepare for the Chapter Test.

Critical Thinking

20. **Analyzing Information** What facts support the statement "Argentina is one of the most industrialized countries in South America"?

21. **Identifying Points of View** In a chart like the one below, identify arguments for and against the cutting down of the rain forest.

Cutting Down the Rain Forest	
For	Against

Current Events Journal

22. **Creating a Population Graph** As you have read, the quickly rising populations in many of Brazil's cities have created problems. Many people live in slum areas without adequate sewage or water systems. City governments struggle to cope with the problems. Use the population map on page 576 to create a bar graph of Brazil's five largest cities for the year 2001.

Mental Mapping Activity

23. **Focusing on the Region** Create an outline map of South America. Refer to the map on page 574; then label the following:

- Caribbean Sea
- Brazil
- Atlantic Ocean
- Argentina
- Pacific Ocean
- Brazilian Highlands
- Amazon River
- Río de la Plata
- Venezuela
- Guiana Highlands

Technology Skills Activity

24. **Using the Internet** Conduct a search for information about the Amazon rain forest and create an annotated bibliography of five useful Web sites. Your bibliography should contain the Web address, a brief summary of the information found on the site, and a statement of why you think the site is useful.

Standardized Test Practice

Directions: Read the passage below and answer the question that follows.

The Amazon Basin is a gigantic system of rivers and rain forests, covering half of Brazil and extending into neighboring countries. Much of the Amazon is still unexplored, and the rain forest holds many secrets. Some of the animals found here include the jaguar, tapir, spider monkey, sloth, river dolphin, and boa constrictor. Forest birds include toucans, parrots, hummingbirds, and hawks. More than 1,800 species of butterflies and 200 species of mosquitoes give you an idea about the insect population. In addition, the fish— such as piranha, pirarucu, and electric eel— are very unusual. Biologists cannot identify much of the catch found in markets.

1. **On the basis of this passage, which of the following generalizations is most accurate?**

   F The Amazon rain forest covers about one-third of the South American continent.

   G Native Americans living in the rain forest are losing their old way of life.

   H The Amazon Basin is huge, and its rain forests hold thousands of animal species.

   J The Amazon Basin is located only in Brazil.

Test-Taking Tip: This question asks you to make a generalization about the Amazon Basin. A *generalization* is a broad statement. Look for facts and the main idea *in the passage* to support your answer. Do not rely only on your memory. The main idea can help you eliminate answers that do not fit. Also, look for the statement that is true and that is covered in the paragraph.

VANISHING
Rain Forests

WEST INDIES

CENTRAL
AMERICA

SOUTH
AMERICA

☐ Rain forests

Rain Forest Riches Imagine never tasting chocolate. Think about never eating a banana, chewing gum, or munching cashews. If there were no rain forests, we would have none of these foods. We also would not have many of the drugs used to treat malaria, multiple sclerosis, and leukemia. In fact, rain forest plants provide one-fourth of the world's medicines.

Millions of kinds of plants and animals live in rain forests—more than half of all species on Earth. Scientists have studied only a fraction of these species. So no one really knows what new foods, medicines, or animals are there, just waiting to be discovered.

Rain Forest Destruction Yet we may never know. Why? Because a chunk of rain forest the size of two football fields vanishes every second! The forests are being destroyed for many reasons.

Loggers cut trees and sell the lumber worldwide.

Ranchers and farmers clear land for cattle and crops.

Miners level acres of forest to get at valuable minerals.

People are trying to find ways to use rain forests without destroying them. Changing farming practices and developing different forest industries are possible solutions. However, time is running out. Can we afford to lose rain forests and all their treasures?

Male golden toad

Settlers clear trees for a home in the rain forest.

Making a Difference

New species *Callithrix humilis*, a dwarf marmoset

Discovering New Monkeys How would it feel to discover an animal that no one knew existed? Dutch scientist Marc van Roosmalen knows. He recently discovered a new species of monkey (photo, at right) in Brazil.

Van Roosmalen runs an orphanage for abandoned monkeys. One day, a man showed up with a tiny monkey van Roosmalen had never seen before. He spent about a year tracking down a wild population of the monkeys deep in the Amazon rain forest. Of some 250 kinds of monkeys known worldwide, about 80 live in Brazil. At least 7 new species have been discovered since 1990.

Rain Forest Field Trip With help from the Children's Environmental Trust Foundation, students from Millbrook, New York, traveled to Peru's Yarapa River region, deep in the Amazon rain forest. Students studied the forest from platforms built in the canopy, and they soared among the tall trees using ropes. The students met rain forest creatures at night, went bird-watching at dawn, and swam in the Yarapa River— home to crocodiles called caimans.

Back in Millbrook, the students educate others about saving rain forests. They also raise money to help support a Peruvian zoo that protects rain forest animals.

Millbrook student traps insects for study.

What Can You Do?

Write a Note
Write to your government representatives and encourage them to support plans that help save rain forests.

Check Out Your Community
What environmental problems face your community? What can you do to help solve the problems? For example, does your community have problems with water pollution or water shortages? What steps does your community take to make sure you have clean water to drink?

The Andean Countries

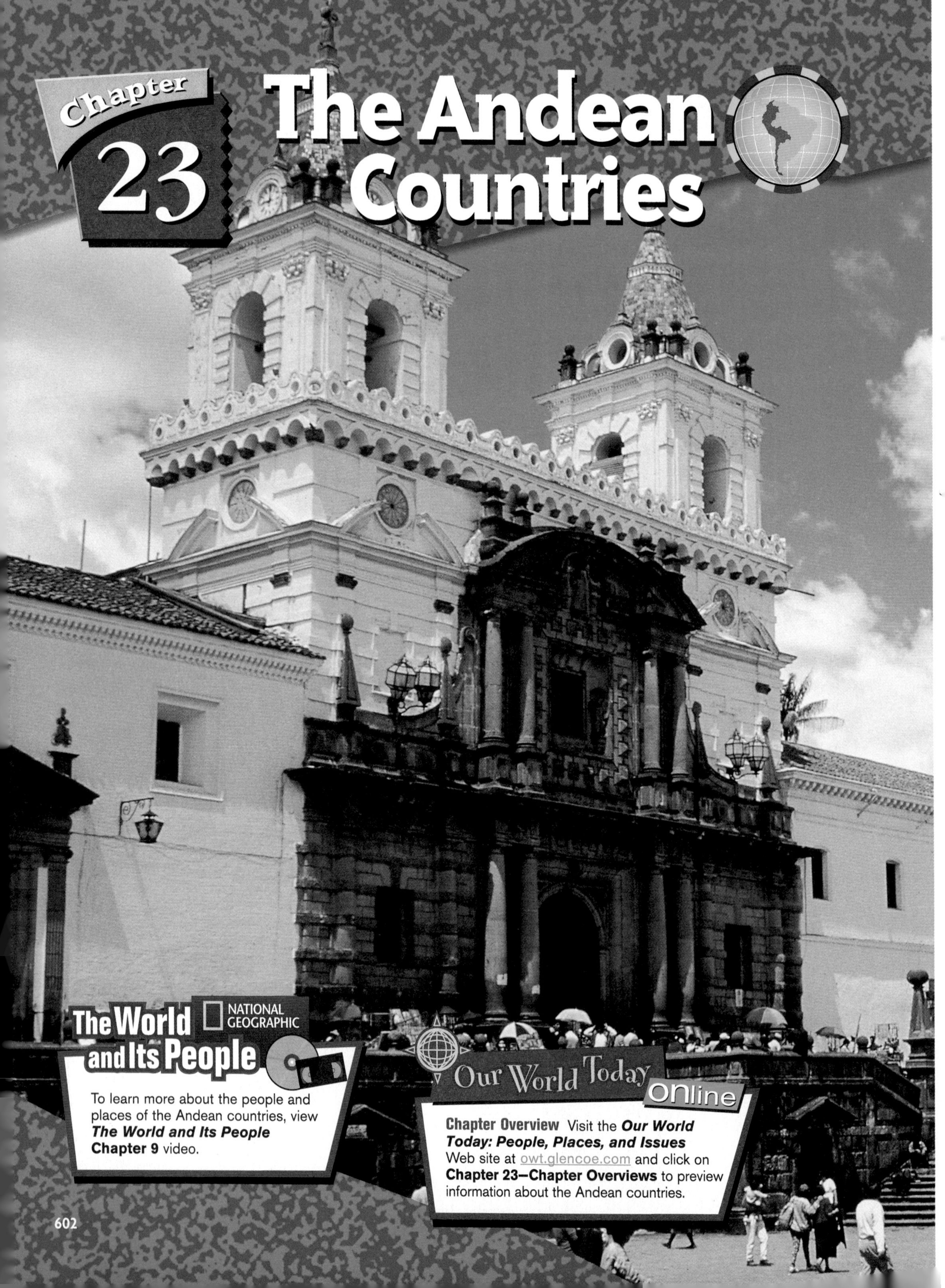

The World and Its People | **NATIONAL GEOGRAPHIC**

To learn more about the people and places of the Andean countries, view *The World and Its People* **Chapter 9** video.

Our World Today online

Chapter Overview Visit the *Our World Today: People, Places, and Issues* Web site at owt.glencoe.com and click on **Chapter 23–Chapter Overviews** to preview information about the Andean countries.

FOLDABLES™
Study Organizer

Summarizing Information Study Foldable Make this foldable and use it to organize note cards with information about the people and places of the Andean countries of South America.

Step 1 Fold a 2-inch tab along the long edge of a sheet of paper.

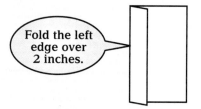
Fold the left edge over 2 inches.

Step 2 Fold the paper in half so the tab is on the inside.

The tab can't be seen when the paper is folded.

Step 3 Open the paper pocket foldable and glue the edges of the pockets together.

Glue here. Glue here.

Step 4 Label the pockets as shown.

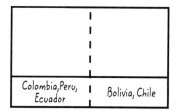
Colombia, Peru, Ecuador | Bolivia, Chile

Reading and Writing As you read the chapter, summarize key facts about the Andean countries on note cards or on quarter sheets of notebook paper. Organize your notes by placing them in your pocket foldable inside the appropriate pockets. (Glue your foldable from Chapter 22 on the front cover of this foldable to form a four-pocket foldable on South America.)

Why It Matters

Wealth in the Andes

The Andes form the spine of South America and are the longest mountain chain on Earth. These high, rocky peaks are the source of some of the world's most highly desired substances, including oil, emeralds, gold and silver, coffee, and "Colombian Gold"—the illegal drug cocaine. Worldwide demand for these products has caused corruption and instability in the countries of this region.

◄ **Monastery of San Francisco, Quito, Ecuador**

Guide to Reading

Main Idea

Although it has many resources, Colombia faces political and economic turmoil.

Terms to Know

- cordillera
- llanos
- cash crop
- mestizo
- republic
- campesino

Reading Strategy

Create a chart like the one below and list advantages that Colombia enjoys in the left column. In the right column, list the challenges that it faces.

Colombia	
Advantages	Challenges

NATIONAL GEOGRAPHIC Exploring Our World

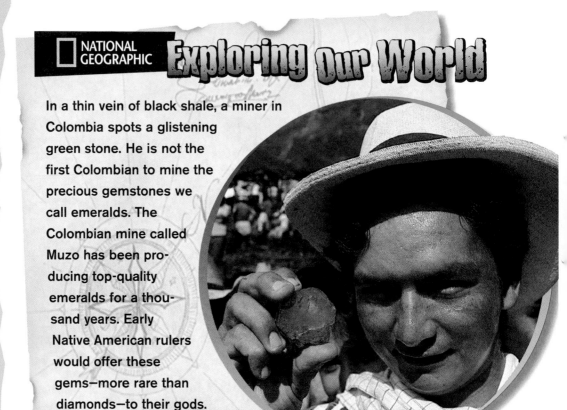

In a thin vein of black shale, a miner in Colombia spots a glistening green stone. He is not the first Colombian to mine the precious gemstones we call emeralds. The Colombian mine called Muzo has been producing top-quality emeralds for a thousand years. Early Native American rulers would offer these gems—more rare than diamonds—to their gods.

Colombia—named after Christopher Columbus—sits astride the lofty Andes mountain ranges at the northwestern edge of South America. These mountains continue south through five other countries—Ecuador, Peru, Bolivia, Chile, and Argentina.

Colombia's Land

Colombia—almost three times larger than Montana—has coasts on both the Caribbean Sea and the Pacific Ocean. The Andes sweep through the western part of Colombia. Here they become a **cordillera** (KAWR•duhl•YEHR•uh), or a group of mountain ranges that run side by side. Nearly 80 percent of Colombia's people live in the valleys and highland plateaus of the Andes. Thick forests spread over lowlands along the Pacific coast. Few people live there.

Only a few Native American groups live in the hot, steamy tropical rain forests of the southeast. In the northeast, ranchers drive cattle across the hot grasslands known as the **llanos.**

Colombia lies within the Tropics. Temperatures are very hot, and heavy rains fall along the coasts and in the interior plains. In the high elevations of the Andes, temperatures are very cool for a tropical area. **Bogotá,** Colombia's capital and largest city, lies on an Andean plateau at 8,355 feet (2,547 m) above sea level. Here high temperatures average only 67°F (19°C).

✓Reading Check **Where do most of Colombia's people live?**

Colombia's Economy

Colombia has many natural resources. The mountains hold valuable minerals and precious stones, and Colombia has more coal than any other country in South America. Second only to Brazil in its potential hydroelectric power, Colombia also has large petroleum reserves in the lowlands. In addition, the country is a major supplier of gold and

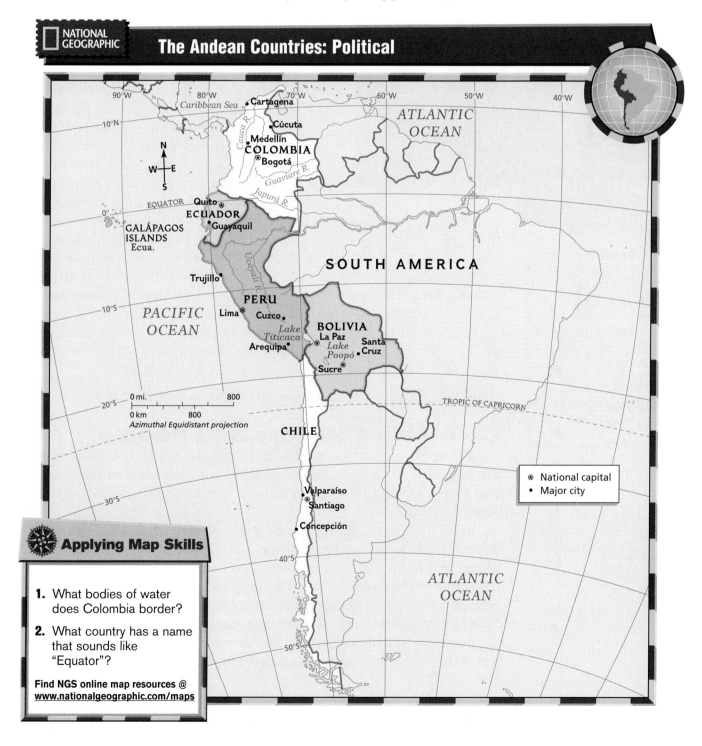

NATIONAL GEOGRAPHIC

The Andean Countries: Political

Applying Map Skills

1. What bodies of water does Colombia border?

2. What country has a name that sounds like "Equator"?

Find NGS online map resources @ www.nationalgeographic.com/maps

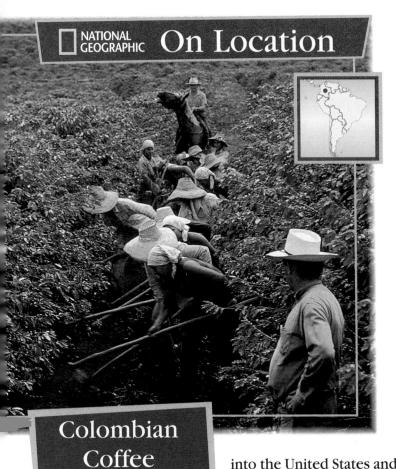

Colombian Coffee

Many historians believe that coffee was "discovered" in Ethiopia, Africa. Eventually Spanish missionaries brought the first coffee plants to Colombia.

Economics What other crops does Colombia export?

the world's number one source of emeralds. Manufacturing produces clothing, leather goods, food products, paper, chemicals, and iron and steel products.

Agriculture The coastal regions and the highlands have good soil for growing a variety of crops. Coffee is the country's major **cash crop**—a product sold for export. Colombian coffee is known all over the world for its rich flavor.

Colombia exports bananas as well as cacao, sugarcane, rice, and cotton. Huge herds of cattle roam large *estancias,* or ranches, in the llanos. The rain forests also supply a valuable resource—lumber.

Economic Challenges Despite many natural resources, Colombia faces economic challenges. In the 1980s, drug dealers became a major force in Colombia. The dealers paid farmers more to grow coca leaves—which are used to make the drug cocaine—than the farmers could earn growing coffee. Much of this cocaine is smuggled into the United States and western Europe. The drug dealers have used their immense profits to build private armies. They have threatened—and even killed—government officials who have tried to stop them.

With U.S. support, the government of Colombia has stepped up its efforts to break the power of the drug dealers. It has had some success, but drug dealers continue to flourish. In addition, the government has tried to persuade thousands of farmers to switch back to growing other crops. See TIME Reports: Focus on World Issues on pages 609–615 for an in-depth study of the drug problem.

✓ Reading Check What crop has been a problem in Colombia? Why?

Colombia's History and People

About 43.1 million people live in Colombia. Nearly all Colombians are mestizos (meh•STEE•zohs), meaning they have mixed European and Native American backgrounds. Most speak Spanish and follow the Roman Catholic faith.

In 1810 Colombia was one of the first Spanish colonies in the Americas to declare independence. Simón Bolívar, whom you read about in Chapter 22, led this struggle for independence. In 1819 Colombia became part of New Granada, an independent country that included Venezuela, Ecuador, and Panama. Later, these other regions broke away and became separate countries.

Colombia today is a republic with an elected president. Political violence has scarred the country's history, though. During the late 1800s alone, Colombia suffered through more than 50 revolts and 8 civil wars. Fighting broke out again in 1948. About 200,000 people died in this conflict, which ended in the late 1960s.

To prevent further turmoil, the two main political parties agreed to govern the country together. Efforts were made to improve the lives of poor farmers by giving them more land. Factories and industrial jobs opened up. Still, a wide gap between rich and poor remained, causing further troubles.

In the 1960s, groups of rebels in the countryside began fighting the government. This latest civil war has lasted more than 35 years and left more than 100,000 people dead. In late 1999, more than 10 million Colombians joined in a massive protest across the country, urging an end to the fighting.

A Diverse Culture Colombia has a rapidly growing urban population. Colombian farmers, or campesinos, and their families have journeyed to cities to look for work or to flee the fighting in the countryside.

You can see Colombia's Spanish, Native American, and African heritages reflected in its culture. Native American skills in weaving and pottery date back before the arrival of Columbus. Caribbean African rhythms blend with Spanish-influenced music.

✓ Reading Check **What is a mestizo?**

Web Activity Visit the *Our World Today: People, Places, and Issues* Web site at owt.glencoe.com and click on **Chapter 23— Student Web Activities** to learn more about Colombia.

Section 1 Assessment

Defining Terms

1. Define cordillera, llanos, cash crop, mestizo, republic, campesino.

Recalling Facts

2. Economics Colombia is the world's number one source of what resource?

3. Culture What language do most Colombians speak? What religion do they practice?

4. History Who led Colombia's struggle for independence from Spain?

Critical Thinking

5. Analyzing Cause and Effect Why does Bogotá, which is located in the Tropics, have an average temperature of only 67°F (19°C)?

6. Drawing Conclusions Why do you think it is so difficult for Colombian farmers to stop growing coca?

Graphic Organizer

7. Organizing Information Create a time line like this one; then put the following events and their dates in the correct order on it: Massive protest held by more than 13 million Colombians, groups of rebels fight the government, Colombia declares independence from Spain, Colombia suffers 50 revolts and 8 civil wars, Colombia becomes part of New Granada.

├─────────┼─────────┼─────────┼─────────┤

Applying Social Studies Skills

8. Analyzing Maps Study the political map on page 605. What rivers run through Colombia? What are Colombia's major cities?

The Andean Countries

Technology Skill

Using a Database

An electronic **database** is a collection of data—names, facts, and statistics—that is stored in a file on the computer. Databases are useful for organizing large amounts of information. The information in a database can be sorted and presented in different ways.

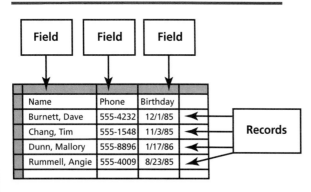

▲ Using a database can help organize statistics, names and addresses, and even baseball card collections.

Learning the Skill

The database organizes information in categories called *fields*. For example, a database of your friends might include the fields **Name, Address, Telephone Number,** and **Birthday.** Each person you enter into the database is called a *record*. After entering the records, you might create a list sorted by birthdays or use the records to create address labels. Together, all the records make up the database.

Scientists use databases for many purposes. They often have large amounts of data that they need to analyze. For example, a sociologist might want to compare and contrast certain information about the people of the Andean countries. A database would be a good place to sort and compare information about the languages, religions, and ethnic groups of these countries.

Practicing the Skill

Follow these steps to build a database about the Andean countries.

1. Determine what facts you want to include in your database and research to collect that information.
2. Follow the instructions in the database that you are using to set up fields. Then enter each item of data in its assigned field.
3. Determine how you want to organize the facts in the database—chronologically by the date, alphabetically, or by some other category.
4. Follow the instructions in your computer program to sort the information.
5. Check that all the information in your database is correct. If necessary, add, delete, or change information or fields.

Applying the Skill

Research and build a database that organizes information about an Andean country of your choice. Explain why the database is organized the way it is.

TIME REPORTS

FOCUS ON WORLD ISSUES

Waging War on Drugs

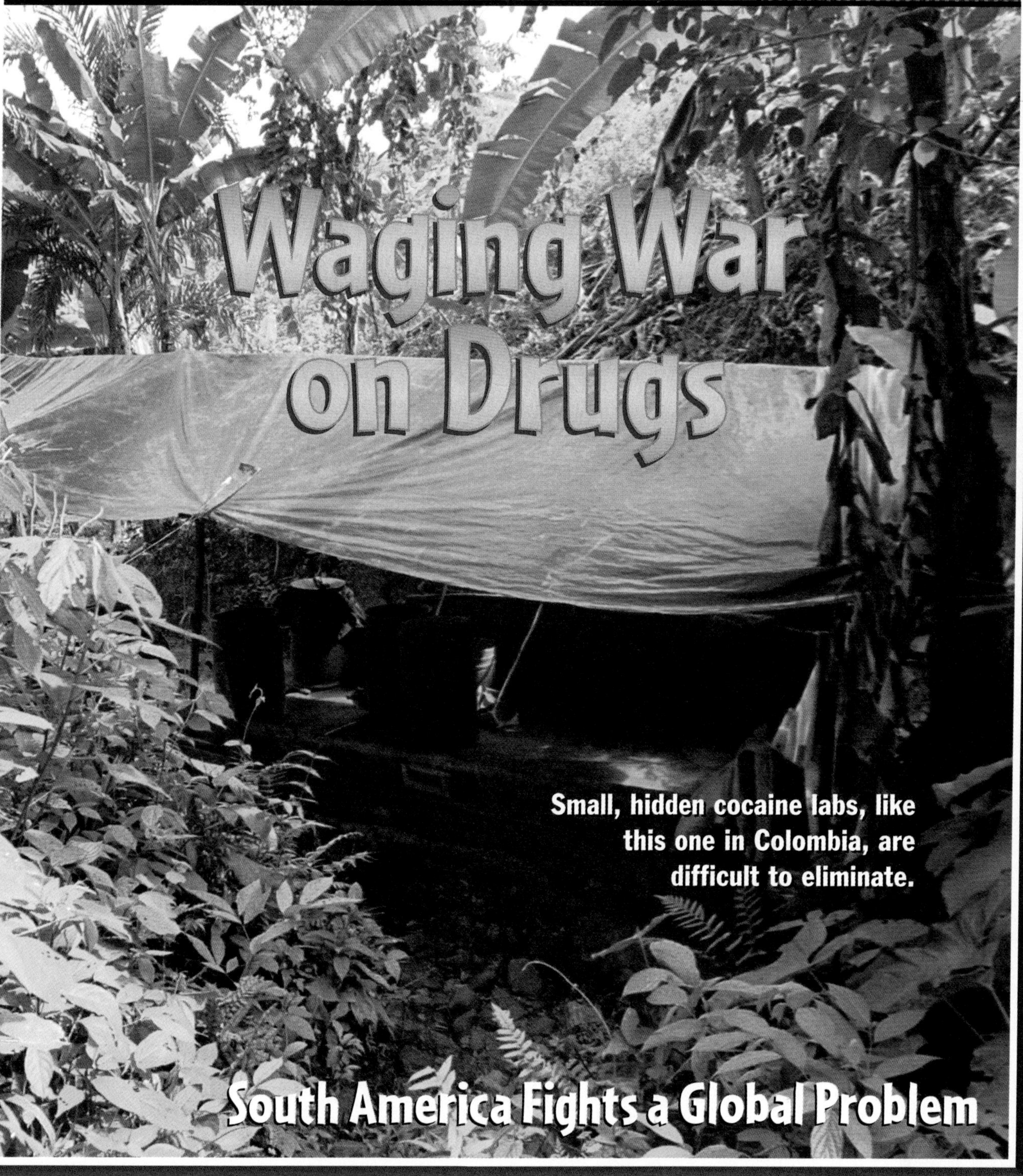

Small, hidden cocaine labs, like this one in Colombia, are difficult to eliminate.

South America Fights a Global Problem

CARLOS VILLALO/GETY/NEWSCOM

A war on many fronts: U.S. Navy Seals on patrol in Brazil. Police in Peru seize cocaine.

LUKE FRAZZA/AFP

MARIANA BAZO/REUTERS

The Drug Trade's Tragic Effect

In June 1986, Len Bias had everything to be happy about. The University of Maryland basketball star had just signed with the Boston Celtics, a pro team. The Celtics promised him $700,000 for his first season. A sneaker company agreed to pay him $325,000 to be their spokesman.

For this instant millionaire, it looked as if things could only get better. "That kid is going to be a superstar," a coach said. "What can't he do? He's got the whole package."

Bias also had a package of cocaine. **Cocaine** is a drug that can cause brain injuries even if taken only once. In 1986, scientists didn't know that for sure. Bias certainly didn't know it. Two days after he signed with the Celtics, he used cocaine and died.

A Deadly Import

The cocaine that killed Len Bias came from South America. And so did the 650 tons of cocaine smuggled into the United States in 2000. Every day Americans died as a result of using this drug.

Cocaine is made from the coca plant, which is grown in only three countries. Colombia is by far the biggest producer followed by Peru and Bolivia.

In all three countries coca is grown high up in the Andes Mountains. Cocaine "factories" there turn coca leaves into a white powder. **Smugglers** use boats and airplanes to slip that powder, cocaine, into countries around the world.

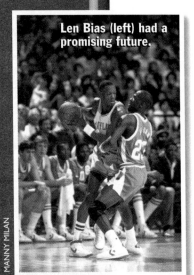

Len Bias (left) had a promising future.

MANNY MILAN

The Drug War in the Andes

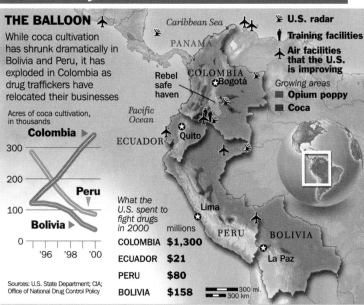

THE BALLOON

While coca cultivation has shrunk dramatically in Bolivia and Peru, it has exploded in Colombia as drug traffickers have relocated their businesses

Acres of coca cultivation, in thousands

Colombia ▶
300
200
100 **Peru** ▶
Bolivia ▶
0
'96 '98 '00

Sources: U.S. State Department; CIA; Office of National Drug Control Policy

Caribbean Sea
PANAMA
Rebel safe haven
COLOMBIA
Bogotá
Pacific Ocean
ECUADOR Quito

⚡ U.S. radar
🏋 Training facilities
✈ Air facilities that the U.S. is improving

Growing areas
■ Opium poppy
■ Coca

What the U.S. spent to fight drugs in 2000   millions
COLOMBIA $1,300
ECUADOR $21
PERU $80
BOLIVIA $158

Lima
PERU   BOLIVIA
La Paz

300 mi.
300 km

INTERPRETING MAPS AND CHARTS

1. **Interpreting Data** What does this map tell you about the U.S. role in South America's drug war?

2. **Making Inferences** Suppose the war against drugs succeeds in Colombia. How might the lines on the graph change?

Len Bias became an instant millionaire. Then he made a choice that cost him his life.

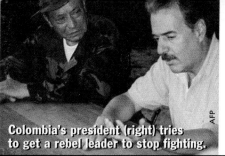

Colombia's president (right) tries to get a rebel leader to stop fighting.

LANE STEWART

AFP

▲ The beautiful poppy is harvested for deadly heroin.

Heroin, another deadly drug, is made from the poppy plant. In South America, poppies are turned into heroin only in Colombia.

Rebels' Businesses

Drugs have nearly brought Colombia to its knees. Colombia is a country about the size of Texas and California combined. Rebel armies based in Colombia's jungles have fought government troops for nearly 40 years. The rebels make and sell cocaine and heroin. Their dirty business brings them more than $1 million a day. They spend a lot of that money on weapons.

Paramilitaries add to Colombia's woes. These are armed men that landowners and businesses hire to protect their workers. In 2000, rebels and paramilitaries kidnapped eight innocent civilians every day and murdered 80 more. The chaos has forced more than 2 million Colombians to flee their homes.

U.S. money had helped Bolivia and Peru tackle drug problems during the 1990s. But many cocaine producers in those countries moved their operations to Colombia. Cocaine production in Colombia doubled between 1995 and 2000.

In 2000, the U.S. decided to help Colombia rid itself of the drug trade. It gave Colombia's government $1.3 billion to equip and train its army to fight drugs.

Fighters for Life

If no one bought drugs, no one would produce them. Len Bias's death in 1986 persuaded millions to avoid cocaine. But millions still use it, so the cocaine business remains strong.

Len Bias's mother, Louise, is happy that some good came out of her son's death. "Len Bias has done more in death," she said, "than he could have done in life. Life is what the fight against deadly drugs is all about." ▪

EXPLORING THE ISSUE

1. **Explaining** What comment does this article's title make about the trade in illegal drugs?

2. **Cause and Effect** Describe how Len Bias's death could have persuaded millions of Americans to avoid cocaine.

Targeting Drug Supplies

Pop works for the U.S. Customs Service in Hidalgo, Texas. He looks for illegal drugs in vehicles that cross into the United States from Mexico.

By any measure, Pop is good at his job. In 1998, he discovered 3,075 pounds of cocaine in a pineapple truck. In 1999, he found 50 pounds of marijuana hidden in an ice chest.

Popsicle's nose earned him a Significant Seizure Medal in 1999.

CHRIS USHER

What makes Pop so successful? His nose. Pop—short for Popsicle—is a pit bull. Like 500 other **customs dogs** in the U.S., he's been trained to sniff out drugs.

Popsicle plays a role in the worldwide effort to stop the flow of drugs. Many thousands of people are also part of that effort. Police officers, for example, arrest people who sell drugs on the street. Members of the U.S. Coast Guard head off smugglers at sea. Soldiers in Colombia destroy coca plants and cocaine factories.

U.S. Help

About 80 percent of the cocaine that reaches the United States comes from Colombia. That's why the U.S. has put more than $1.3 billion behind Colombia's fight against drugs. Colombia's armed forces use most of the money to train soldiers and buy equipment. Planes bought with U.S. dollars drop chemicals that kill growing coca plants. New helicopters rush soldiers to cocaine factories defended by heavily armed rebel troops.

Stopping cocaine at its source isn't just a military job. It's also an effort to change minds. Colombian officials are trying to persuade farmers to stop growing coca and poppies. They pay farmers to grow cocoa, coffee, palm hearts, and other crops instead.

Will those efforts work? Some experts think so. Others aren't so sure. "Those that profit from producing cocaine and heroin are not about to roll over and play dead," said one expert. "There is evidence that drug factories are moving into Brazil and Ecuador."

If she's right, Popsicle has a lot of work ahead of him. ▪

EXPLORING THE ISSUE

1. **Explaining** Why is stopping cocaine at its source important to the war on drugs?

2. **Analyzing Information** Why might it be more effective to wipe out cocaine in Colombia than to stop it in the United States?

Dealing with Demand

I t's tragic but accurate: Someone somewhere is always going to want to buy illegal drugs. And someone else will be willing to **supply** them. Worldwide, about 14 million people use cocaine today. About 5.3 million of them live in the United States. Nine million people in the world use heroin. More than 650,000 of them are Americans.

Suppose those numbers were cut in half. Heroin and cocaine production would plunge. And illegal drugs would cause far less misery than they do today.

Inside Drug Court

Is slashing the **demand** for drugs by 50 percent an impossible dream? Not in Baltimore, Maryland. Baltimore has a Drug Treatment Court. The court's goal is to help people arrested for carrying illegal drugs to stop abusing them. "If you ask for help," a Drug Court judge said in 2001, "you'll get it. If you don't ask, you'll go to jail."

Half the addicts placed in treatment by the Drug Court have stayed away from drugs. Copy that success rate throughout the nation, and the demand in the U.S. for illegal drugs would nose-dive.

Educating Americans

No war on drugs can be won without such a drop in usage, experts say. U.S. President George W. Bush shares their view. "The main reason drugs are shipped . . . to the United States," he said in 2001, "is because United States

PAUL F. GERO/SABA

▲ Phoenix, Arizona, has a drug court like Baltimore's. Here a judge rewards a drug offender's good behavior with tickets to Phoenix's science museum.

citizens use drugs. Our nation must do a better job of educating our citizenry about the dangers and evils of drug use."

Yes, someone somewhere is always going to want to buy illegal drugs. But proper education and treatment will surely reduce the demand for drugs everywhere. ◾

EXPLORING THE ISSUE

1. Analyzing Information Which is more important—reducing the demand for illegal drugs or stopping criminals from producing them? Why?

2. Problem Solving What could schools do to lower the demand for illegal drugs?

Fighting Drug Abuse: What Can One Person Do?

Andy McDonald is a man with a mission—helping kids stay away from drugs. Andy is the national spokesperson for the Partnership for a Drug-Free America. His message: "Kids don't need drugs to succeed."

Andy should know. He's one of the few top-ranked skateboarders in the world. He's so good, he once jumped over three SUVs and one car—all at once. That feat landed him in *The Guinness Book of World Records*. "That right there," he says of skateboarding, "is my idea of getting high."

Speaking Out

You don't have to be a champion athlete to fight drug abuse. Kaelin Weiler proved it. Between 1996 and 1998,

DANJAL BOURQUI

▲ **Andy McDonald: too active to do drugs**

EXPLORING THE ISSUE

1. **Making Inferences** Experts say that the risk for alcohol and other drug use skyrockets in the sixth grade. Assuming that's true, when—and why—should kids begin learning about the dangers of illegal drugs?

2. **Analyzing Information** What makes Andy McDonald an effective promoter of a drug-free life?

---

heroin killed 11 teens in her hometown of Plano, Texas. The youngest victim was a seventh-grader.

Kaelin, 17, persuaded other kids to fight back. They tied white ribbons around traffic lights to remind people

of the problem. They created a "memorial wall"—pictures of kids lost to drugs and the families they left behind. And they made videos about the dangers of drugs and showed them at assemblies.

You can work with school officials to start a similar program in your school and town. Learn as much as you can about the dangers of illegal drugs. Then design a program to teach what you learned to young students and their parents. Launch your program. Afterwards, describe the program and its results in a letter to your local newspaper's editor. E-mail a copy to the "In Your Own Words" website page of the Partnership for a Drug-Free America. (**...www.drugfreeamerica.org**). ▪

REVIEW AND ASSESS

UNDERSTANDING THE ISSUE

1. Defining Key Terms Write definitions for the following terms: *drug abuse, cocaine, heroin, smuggler, paramilitary, customs dog, supply, demand.*

2. Writing to Inform In a 300-word article, explain why the drug trade needs both buyers and suppliers for its survival. Use at least five of the key terms listed above.

3. Writing to Persuade In your view, can the war against drugs ever be won? Support your answer to that question in a brief essay.

INTERNET RESEARCH ACTIVITY

4. Since 1998 the U.S. government has funded ads designed to combat drug abuse among young people. How good are the ads? You be the judge. Browse the Internet or use copies of current magazines and newspapers to find ads designed to combat drug abuse. Choose two you think are effective and two you think are not effective. Either print copies off the Internet, or photocopy ads from magazines and newspapers. Attach a comment to each one explaining why it does or doesn't work well.

5. The No. 1 drug problem in America isn't cocaine or heroin. It's underage drinking. One organization that works to prevent underage drinking is Mothers Against Drunk Driving (MADD). Use Internet resources to learn more about this organization and report your findings to the class.

BEYOND THE CLASSROOM

6. Research the way farmers live in Colombia, Bolivia, or Peru. In a short article, explain why

▲ Venus and Serena Williams: At the top of their game—and far from drugs.

those farmers might see growing poppies or coca as a way to improve their lives.

7. Research the geography and people of Afghanistan and Myanmar. Most of the world's heroin is produced in those two countries. Make a list of conditions—poverty, climate, and location, for example—that each country shares with Colombia. In class, explain what your list suggests about places in which drug production thrives.

WHAT DRUG AND ALCOHOL ABUSE COST

Annual Cost of Alcohol Abuse in the U.S. $148 billion

- Output[1] lost to alcohol-related illness: **$67.6 billion**
- Output lost to early deaths: **$31.4 billion**
- Healthcare: **$18.7 billion**
- Car-crash damage: **$13.6 billion**
- Alcohol-related crime: **$12.7 billion**
- Other costs: **$3.8 billion**

Annual Cost of Drug Abuse in the U.S. $98 billion

- Output[2] lost by crime victims: **$19.9 billion**
- Output lost to lives of crime: **$19.3 billion**
- Property damage, government expenses[3]: **$18 billion**
- Output lost to early deaths: **$14.6 billion**
- Output lost to drug-related illness: **$14.2 billion**
- Healthcare: **$9.9 billion**
- Other costs: **$1.8 billion**

[1]Output is the estimated amount of goods and services that workers would have produced if they had not died or gotten sick, injured, or jailed. [2]Includes output lost by victims of drug-related crimes and criminals jailed for those crimes. [3]Includes the cost of government anti-drug efforts, police, prisons, and other items.
Source: National Institutes of Health

BUILDING GRAPH READING SKILLS

1 Explaining A worker's "output" consists of the goods and services that he or she produces. The abuse of alcohol and other drugs costs the U.S. billions of dollars in lost output. How do these two graphs show that?

2 Making Inferences Suppose the U.S. Congress made cocaine and heroin as legal as alcohol. What might happen to the annual cost of drug abuse?

FOR UPDATES ON WORLD ISSUES GO TO
www.timeclassroom.com/glencoe

Peru and Ecuador

Guide to Reading

Main Idea

Peru and Ecuador share similar landscapes, climates, and history.

Terms to Know

- navigable
- foothills
- subsistence farm
- empire

Reading Strategy

Create two ovals like these. Under each heading list facts about Peru and Ecuador in the outer parts of the ovals. Where the ovals overlap, write facts that apply to both countries.

Peru   Ecuador

NATIONAL GEOGRAPHIC Exploring Our World

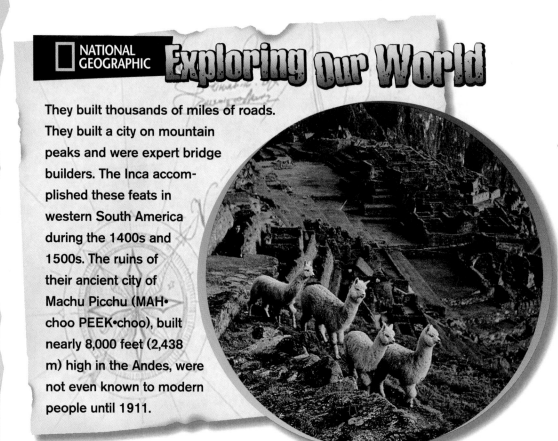

They built thousands of miles of roads. They built a city on mountain peaks and were expert bridge builders. The Inca accomplished these feats in western South America during the 1400s and 1500s. The ruins of their ancient city of Machu Picchu (MAH•choo PEEK•choo), built nearly 8,000 feet (2,438 m) high in the Andes, were not even known to modern people until 1911.

Peru and **Ecuador** lie along the Pacific coast of South America, west of Brazil and south of Colombia. The Andes form the spine of these countries. *Peru*—a Native American word that means "land of abundance"—is rich in mineral resources.

Peru

Dry deserts, snowcapped Andes mountain ranges, and hot, humid rain forests greet you in Peru. Most of Peru's farms and cities lie on a narrow coastal strip of plains and deserts. The cold Peru Current in the Pacific Ocean keeps temperatures here fairly mild even though the area is very near the Equator.

On the border with Bolivia, you can see **Lake Titicaca** (TEE•tee•KAH•kah), the highest navigable lake in the world. Navigable means that a body of water is wide and deep enough to allow ships to travel in it. East of the Andes you descend to the foothills and flat plains of the **Amazon Basin.** Foothills are the low hills at the base of a mountain range. Rainfall is plentiful here, and thick, hot rain forests cover almost all of the plains area.

Peru's Economy Peru's economy relies on a variety of natural resources. The Andes contain many minerals, including copper, silver, gold, and iron ore. Peru's biggest export is copper. The second-largest export—fish—comes from the Peru Current, the cool Pacific Ocean current that parallels the coast.

About one-third of Peru's people farm the land. Some grow sugarcane, cotton, and coffee for export. Like Colombia, Peru grows coca leaves. Most people, however, work on **subsistence farms,** where they grow only enough food to meet their family's needs. The chief crops are rice, plantains (a kind of banana), and corn. Native Americans in the Andes were the first people ever to grow potatoes. Today, potatoes are Peru's main food crop, and farmers grow hundreds of varieties in different colors and shapes.

From Empire to Republic During the 1400s, a Native American people called the **Inca** had a powerful civilization in the area that is now Peru. Their **empire,** or group of lands under one ruler, stretched more than 2,500 miles (4,023 km) along the Andes.

The Incan emperor developed a complex system of tax collection, courts, military posts, trade inspections, and work rules. Work crews built irrigation systems, roads, and suspension bridges that linked the regions of the empire to **Cuzco,** the capital city of the Inca. You can still see the remains of magnificent fortresses and buildings erected centuries ago by skilled Incan builders. The photograph on page 616 shows the ruins of one of the Inca's most famous cities—**Machu Picchu.**

In the early 1500s, Spaniards arrived in Peru, craving the gold and silver found here. They defeated the Inca and made Peru a Spanish territory. Peru gained its freedom from Spain in the 1820s.

Peru is now a republic with an elected president. In recent years, the country's economy has grown very rapidly. Many of Peru's people, however, still live in poverty and cannot find steady jobs.

The Columbian Exchange Besides gold and silver, many other items were found in the Americas. The next time you eat a french fry, you might remember that the potato first grew wild in the mountains of Peru. By the 1400s, the Inca had developed thousands of varieties of potatoes. But no one in the Eastern Hemisphere had ever seen a potato. With Christopher Columbus's second trip, what became known as the "Columbian Exchange" began—an exchange of people, animals, plants, and even diseases between the two hemispheres.

From the Americas, explorers returned home with a wide variety of plants. Nutritious and easy to grow, the potato became one of Europe's most important foods. Corn from the Americas fed European cattle and pigs. Tobacco grown there became as valuable as gold.

The Europeans brought many new things to the Americas, including horses, sheep, cattle, and pigs. Crops that were introduced include oats, wheat, rye, barley, and sugarcane. Some parts of the exchange were disastrous, however. European diseases killed millions of Native Americans.

The Quipu

The Inca did not possess a written language. In order to record events and other aspects of their lives that they wished to remember, they used a system of knotted strings, called the quipu. Men in charge of the quipu recorded on the knots all the taxes brought each year to the Inca, noting details on everything. They recorded the number of men who went to the wars and how many were born and died every year. In short, they may be said to have recorded on their knots everything that could be counted.

EXPLORING CULTURE

Music

Dating back thousands of years, the panpipe is one of the most common musical instruments from the Andes region. Panpipes are made of bamboo in various sizes. Individual bamboo stalks are carefully cut and lashed together with strips of bamboo and woolen string. The notes often alternate from one set of pipes to another. To play a complete melody, then, the two rows of pipes are often stacked one on top of the other.

Looking Closer Which do you think produces a higher note—a short or long stalk? Explain.

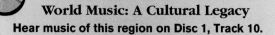

World Music: A Cultural Legacy
Hear music of this region on Disc 1, Track 10.

Peru's Culture Peru's 26.1 million people live mostly along the Pacific coast. Lima (LEE•mah), with more than 5 million people, is the capital and largest city. In recent years, many people from the countryside have moved to Lima in search of work. Because of this sudden rise in population, the city has become overcrowded, noisy, and polluted.

About half of Peru's people are Native American. In fact, Peru has one of the largest Native American populations in the Western Hemisphere. Most of them blend the Catholic faith, Peru's main religion, with traditional beliefs of their ancestors.

Peruvians also include many people of mixed or European ancestry. People of Asian heritage form a small but important part of the population. In the 1990s, **Alberto Fujimori** (FOO•jee•MAW•ree), a Peruvian of Japanese ancestry, was Peru's president for 10 years.

Spanish is Peru's official language, but about 70 Native American languages also are spoken. You can hear the sounds of **Quechua** (KEH•chuh•wuh), the ancient language of the Inca, in many Native American villages.

✓ Reading Check Who built a huge empire centered in Peru?

Ecuador

Ecuador is one of the smallest countries in South America. Can you guess how it got its name? *Ecuador* is the Spanish word for "Equator," which runs right through Ecuador. West of Ecuador and also on the Equator are the **Galápagos Islands.** Owned by Ecuador since 1832, these scattered islands are known for their rich plant and animal life. Turn to page 620 to learn more about the unusual Galápagos Islands.

Ecuador's Economy   Agriculture is Ecuador's most important economic activity. Because of the mild climate, bananas, cacao, coffee, rice, sugarcane, and other export crops grow plentifully in the coastal lowlands. Farther inland, farms in the Andean highlands grow coffee, beans, corn, potatoes, and wheat. The eastern lowlands yield petroleum, Ecuador's major mineral export.

Ecuador's People   About half of Ecuador's nearly 13 million people live along the coast. The important port of **Guayaquil** (GWY•ah•KEEL) is the most populous city. The other half of the population live in the valleys and plateaus of the Andes. **Quito** (KEE•toh), Ecuador's capital, lies more than 9,000 feet (2,743 m) above sea level. The city's historic center has Spanish colonial churches and old whitewashed houses with red-tiled roofs built around central courtyards. You will not find flashing neon signs here because the building of modern structures has been strictly controlled since 1978. In that year, the United Nations Educational, Scientific, and Cultural Organization (UNESCO) declared the "old town" section of Quito a protected world cultural heritage site. Quito does have a "new town" section, though, in the north. This area contains modern offices, embassies, and shopping centers.

 Reading Check   **Why are Ecuador's eastern lowlands important economically?**

Section 2 Assessment

Defining Terms

1. **Define** navigable, foothills, subsistence farm, empire.

Recalling Facts

2. **History** Who were the first people to grow potatoes?

3. **Culture** What has been the result of Lima's sudden population growth?

4. **Economics** What is Ecuador's major mineral export?

Critical Thinking

5. **Analyzing Information** Why is Peru's name, which means "land of abundance," appropriate? Why is it also inappropriate?

6. **Analyzing Cause and Effect** What effect does the Peru Current have on the coastal areas of Peru?

Graphic Organizer

7. **Organizing Information** Create two diagrams like this one, one for Peru and one for Ecuador. Under each heading list facts about the countries.

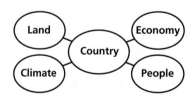

 Applying Social Studies Skills

8. **Analyzing Maps** Turn to the political map on page 605. What Andean capital city lies closest to the Equator?

The Galápagos Islands

The Galápagos Islands are located in the eastern Pacific Ocean about 600 miles (966 km) west of mainland Ecuador. Since 1959 about 95 percent of the islands has been maintained as a national park.

History of Exploration

From the first documented visit to the Galápagos Islands in 1535, people have commented on the islands' unusual wildlife. Sailors, including pirates and whalers, stopped on the islands to collect water and to trap the huge *galápagos*, or tortoises, found on the islands. Sailors valued the tortoises as a source of fresh meat because the giant tortoises could live on ships for months without food or water.

Charles Darwin

The most famous visitor to the Galápagos Islands was Charles Darwin, a scientist from England. He was studying animals all over the world. In 1835 Darwin spent five weeks visiting four of the biggest islands in the Galápagos. He carefully studied the volcanic landscape and the plant and animal life that he saw. He took notes on the differences between animals such as finches, mockingbirds, and iguanas from island to island. Darwin believed that these differences showed how populations of the same species change to fit their environment.

A Fragile Environment

Today the Galápagos Islands are still prized for their amazing variety of animal and plant life. Many of the species found here exist nowhere else on the earth. For instance, the Marine iguana that lives here is the only seagoing lizard in the world.

Unfortunately, years of contact between the islands and humans have had serious effects. Three of the 14 types of tortoises are extinct, and others are seriously threatened. Populations of goats, pigs, dogs, rats, and some types of plants, brought by visitors, have grown so large that they threaten the survival of native plants and animals. Demand for exotic marine life, including sharks and sea cucumbers, has led to overfishing. The government of Ecuador, along with environmentalists worldwide, is now working to protect the islands.

▲ Giant Galápagos tortoise

→ Making the Connection

1. Why did sailors long ago stop at the islands?
2. What did Darwin observe about the islands?
3. **Drawing Conclusions** Why are environmentalists and the government of Ecuador working to protect the Galápagos Islands?

Bolivia and Chile

Guide to Reading

Main Idea

Bolivia and Chile share the Andes, but their economies and people are different.

Terms to Know

- landlocked
- altiplano
- sodium nitrate

Reading Strategy

Create a chart like the one below. In each row, write at least one fact about Bolivia and one about Chile.

	Bolivia	Chile
Land		
Climate		
Economy		
People		

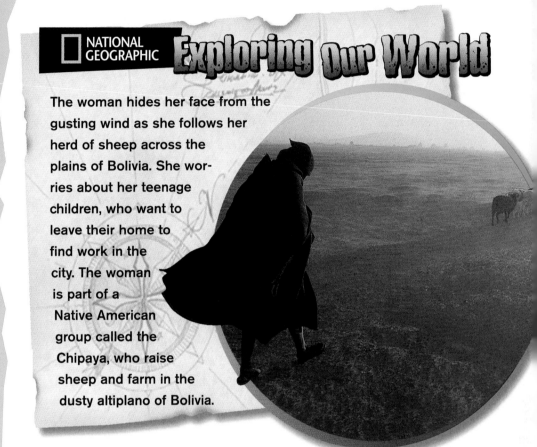

NATIONAL GEOGRAPHIC

Exploring Our World

The woman hides her face from the gusting wind as she follows her herd of sheep across the plains of Bolivia. She worries about her teenage children, who want to leave their home to find work in the city. The woman is part of a Native American group called the Chipaya, who raise sheep and farm in the dusty altiplano of Bolivia.

At first glance, Bolivia and Chile seem very different. Bolivia lacks a seacoast, while Chile has a long coastline on the Pacific Ocean. The Andes, however, affect the climate and cultures of both countries.

Bolivia

Bolivia lies near the center of South America. It is a landlocked country, having no sea or ocean touching its land. Fortunately, in 1993 Peru agreed to give Bolivia a free trade zone in the port city of **Ilo.** In this way Bolivia has had better access for the free flow of people, goods, and ideas. Bolivia also is the highest and most isolated country in South America. Why? The Andes dominate Bolivia's landscape. Look at the physical map on page 574. You see that in western Bolivia, the Andes surround a high plateau called the altiplano. Over one-third of Bolivia is a mile or more high. Unless you were born in this area, you would find that the cold, thin air makes it difficult to breathe. Few trees grow on the altiplano, and most of the land is too dry to farm. Still, the vast majority of Bolivians live on this high plateau. Those areas that have water have been farmed for many centuries.

Chile's Contrasts

Chile has a wide variety of climates and landforms. The moderate capital city of Santiago in central Chile (above) contrasts sharply with the icy southern region (right).

Location What group of islands lies at the southern tip of Chile?

Bolivia also has lowland plains and tropical rain forests. Most of this area has a hot, humid climate. South-central Bolivia, however, has more fertile land, and many farms dot this region.

Bolivia's Economy Bolivia is rich in minerals such as tin, silver, and zinc. Miners remove these minerals from high in the Andes. Workers in the eastern lowlands draw out gold, petroleum, and natural gas.

Still, Bolivia is a poor country. About two-thirds of the people live in poverty. Throughout the highlands, many villagers practice subsistence farming. Subsistence farming is growing just enough food to feed yourself and your family. There is little, if any, left over to sell for cash. They struggle to grow wheat, potatoes, and barley. At higher elevations, herders raise animals such as alpacas and llamas for wool and for carrying goods. In the south, farmers plant soybeans, a growing export. Timber is another important export. Unfortunately, one crop that can be grown for sale is coca, which is made into coca paste from the leaves of the bush. The paste is then made into cocaine. The Time Reports: Focus on World Issues on pages 609–615 looks at the efforts to reduce the flow of drugs from South America.

Bolivia's People What is unusual about Bolivia's capital? There is not just one capital city, but two. The official capital is **Sucre** (SOO•kray). The administrative capital and largest city is **La Paz** (lah•PAHZ). Both capital cities are located in the altiplano. La Paz—at 12,000 feet (3,658 m)—is the highest capital city in the world.

Most of Bolivia's 8.5 million people live in the Andean highlands. About half are of Native American ancestry, and another 30 percent are mestizos. In the cities, most people follow modern ways of living. In the country, you may hear traditional sounds—music played with flutelike instruments.

✓ **Reading Check** **What is the altiplano?**

Chile

Chile is almost twice the size of California. Though its average width is only 110 miles (177 km), Chile stretches 2,652 miles (4,267 km) along the Pacific Ocean.

About 80 percent of Chile's land is mountainous. The high Andes run along Chile's border with Bolivia and Argentina. Except in the altiplano area of Chile's north, very few Chileans live in the Andes.

Also in the north is the **Atacama Desert.** It is one of the driest places on the earth. Why? This area is in the rain shadow of the Andes. Winds from the Atlantic Ocean bring precipitation to regions east of the Andes, but they carry no moisture past them. In addition, the cold Peru Current in the Pacific Ocean does not evaporate as much as a warm current does. As a result, only dry air hits the coast.

A moderate zone lies just north of **Santiago,** Chile's capital. Most of Chile's people live in a central region called the Central Valley. The fertile valleys here have the largest concentration of cities, industries, and farms.

The lake region, also known as "the south," supports thick forests. Chile's far south is a stormy, wind-swept region of snowcapped volcanoes, thick forests, and huge glaciers. The **Strait of Magellan** separates mainland Chile from a group of islands known as **Tierra del Fuego** (FWAY•goh)—or "Land of Fire." This region is shared by both Chile and Argentina. Cold ocean waters batter the rugged coast around **Cape Horn,** the southernmost point of South America.

NATIONAL GEOGRAPHIC

The Andean Countries: Population Density

Cities
■ City with more than 5,000,000 people

● City with 1,000,000 to 5,000,000 people

○ City with 500,000 to 1,000,000 people

Persons per	
Sq. Mi.	**Sq. Km**
Uninhabited	Uninhabited
Under 2	Under 1
2–60	1–25
60–125	25–50
125–250	50–100
Over 250	Over 100

Applying Map Skills

1. How many people per square mile live along the coast of Peru?

2. What cities have more than 5 million people?

Find NGS online map resources @ www.nationalgeographic.com/maps

Chile's Economy In recent years, Chile has enjoyed high economic growth, and the number of people living below the poverty line has fallen by half. Mining forms the backbone of Chile's economy. The Atacama region is rich in minerals. Chile ranks as the world's leading copper producer. The country also mines and exports gold, silver, iron ore, and sodium nitrate—a mineral used as a fertilizer and in explosives.

Agriculture is also a major economic activity. Farmers produce wheat, corn, beans, sugar, and potatoes. The grapes and apples you eat in winter may come from Chile's summer harvest. Many people also raise cattle, sheep, and other livestock.

Chile has factories that process fish and other foods. Other workers manufacture wood products, iron, steel, vehicles, cement, and textiles. Service industries such as banking and tourism also thrive.

Chile's Culture Of the 15.4 million people in Chile, most are mestizos. A large minority are of European descent, and some Native American groups live in the altiplano and "the south." Nearly all the people speak Spanish, and most are Roman Catholic. Some 80 percent of Chile's population live in urban areas. Chile has been a democratic republic since the end of strict military rule in 1989.

✓ Reading Check **What three cultural backgrounds make up Chile's 15.4 million people?**

Assessment

Defining Terms
1. Define landlocked, altiplano, sodium nitrate.

Recalling Facts
2. Economics What part of Bolivia's population lives in poverty?
3. Geography What makes La Paz unusual?
4. Economics Chile is the world's leading producer of what mineral?

Critical Thinking
5. Analyzing Cause and Effect Why is the Atacama Desert one of the world's driest places?
6. Making Comparisons What are differences and similarities between the economies of Bolivia and Chile?

Graphic Organizer
7. Organizing Information Create a diagram like this one. Under each arrow list supporting facts for the main idea given.

Main Idea: Bolivia is rich in minerals but is still a poor country.

↑   ↑   ↑   ↑

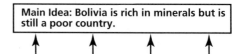

Applying Social Studies Skills

8. Analyzing Maps Study the physical map on page 574. The southernmost tip of South America is part of what country? What is the name of the group of islands at the southern tip of South America? What does its name mean?

Section 1 | Colombia

Terms to Know
cordillera
llanos
cash crop
mestizo
republic
campesino

Main Idea
Although it has many resources, Colombia faces political and economic turmoil.

✓ Culture Most Colombians speak Spanish and follow the Roman Catholic faith.

✓ Economics Colombia is rich in hydroelectric power, gold, and emeralds.

✓ Government The government of Colombia is struggling to combat the power of drug dealers who make huge fortunes from selling cocaine, which comes from the coca plant.

✓ History Civil war in Colombia has lasted for more than 35 years.

Section 2 | Peru and Ecuador

Terms to Know
navigable
foothills
subsistence farm
empire

Main Idea
Peru and Ecuador share similar landscapes, climates, and history.

✓ History The Inca had a powerful civilization in the area that is now Peru. They developed a complex system of record-keeping.

✓ Economics Peru's main exports are copper and fish. Many people farm. Ecuador's economy is focused on agriculture.

✓ Culture Most people in Peru and Ecuador live along the coast.

▲ Dancers in Peru

Section 3 | Bolivia and Chile

Terms to Know
landlocked
altiplano
sodium nitrate

Main Idea
Bolivia and Chile share the Andes, but their economies and people are different.

✓ Human/Environment Interaction Bolivia is a poor country consisting mainly of the towering Andes and a high plateau that is difficult to farm.

✓ Culture Most of Chile's people speak Spanish and follow the Roman Catholic faith.

✓ Economics Chile has a diverse economy that includes mining—especially copper and sodium nitrate—farming, and manufacturing.

Chapter 23 Assessment and Activities

Using Key Terms

Match the terms in Part A with their definitions in Part B.

A.

1. cordillera
2. campesino
3. cash crop
4. llanos
5. navigable
6. foothills
7. empire
8. sodium nitrate
9. landlocked
10. mestizo

B.

a. person of mixed Native American and European ancestry
b. crop grown to be sold, often for export
c. mineral used in making fertilizer
d. group of lands under one ruler
e. group of mountain ranges that run side by side
f. when a body of water is wide and deep enough for ships to pass through
g. land that doesn't have a sea or ocean touching it
h. low hills at the base of a mountain range
i. farmer in Colombia
j. hot grasslands

Reviewing the Main Ideas

Section 1 Colombia

11. **Economics** List four of Colombia's natural resources.
12. **History** What is the heritage of most of Colombia's people?
13. **History** What type of activities have scarred Colombia's history?

Section 2 Peru and Ecuador

14. **Place** What is the highest navigable lake in the world?
15. **History** What ancient Native American civilization of the Andes lived in Peru?
16. **Government** Who owns the Galápagos Islands?

Section 3 Bolivia and Chile

17. **Culture** What is life like for about two-thirds of Bolivia's people?
18. **Government** What type of government does Chile have?
19. **Culture** What is the ethnic background of most Chileans?

 **The Andean Countries**

Place Location Activity

On a separate sheet of paper, match the letters on the map with the numbered places listed below.

1. Colombia
2. Peru
3. Chile
4. Andes
5. Lake Titicaca
6. Quito
7. Bogotá
8. Strait of Magellan
9. Lima
10. Bolivia

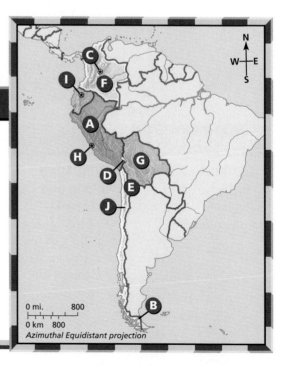

Critical Thinking

20. **Making Inferences** Why are Native Americans who live in the Andean highlands more likely to follow a traditional way of life than those who live in the cities?

21. **Analyzing Cause and Effect** On a diagram like the one below, list factors that have led to political violence during Colombia's history.

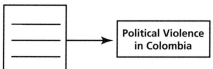

Political Violence in Colombia

Current Events Journal

22. **Writing About Cities** Research one of the major cities of the Andean countries. Prepare a report that includes a map, a fact bank, pictures, and a T-shirt design that shows a famous landmark of the city.

Mental Mapping Activity

23. **Focusing on the Region** Create a simple outline map of South America; then label the following:

- Pacific Ocean
- Peru
- Andes
- Colombia
- Atacama Desert
- Galápagos Islands
- Strait of Magellan
- Lake Titicaca
- Chile
- Ecuador

Technology Skills Activity

24. **Building a Database** Create a fact sheet about the Andean countries by building a database. Create fields for such categories as physical features, natural resources, capital cities, population, and type of government. When you have entered data for each field, print your fact sheet.

The Princeton Review

Standardized Test Practice

Directions: Read the paragraphs below, and then answer the question that follows.

Simón Bolívar, an aristocrat from Venezuela, led many of South America's lands to independence. He believed in equality and saw liberty as "the only object worth a man's life." Called "the Liberator," Bolívar devoted his life to freedom for Latin Americans.

Bolívar was the son of a rich family in New Granada, or what is today Colombia, Venezuela, Panama, and Ecuador. In 1805 he went to Europe. There he learned about the French Revolution and its ideas of democracy. He returned home, vowing to free his people from Spanish rule. In 1810 Bolívar started a revolt against the Spaniards in Venezuela. Spanish officials soon crushed the movement, but Bolívar escaped and trained an army. During the next 20 years, Bolívar and his forces won freedom for the present-day countries of Venezuela, Colombia, Panama, Bolivia, and Ecuador.

1. **What is the main idea of the paragraphs above?**

   A Bolívar was the son of a rich family.

   B Bolívar traveled to Europe and learned about democracy.

   C Simón Bolívar was called "the Liberator."

   D Bolívar devoted his life to freedom for Latin Americans.

Test-Taking Tip: This question asks you to find the main idea, or to make a generalization. Most of the answer choices provide specific details, not a general idea. Which of the answers is more of a general statement?

Unit

9

Fur seal on
the beach,
Antarctica

Boy selling
fish, Samoa

Australia, Oceania, and Antarctica

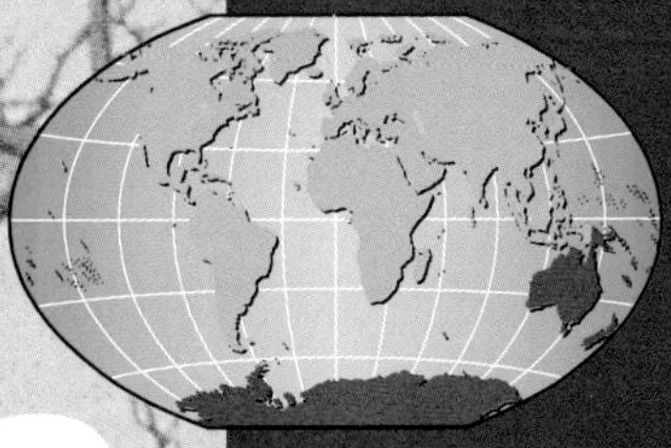

Australia, Oceania, and Antarctica are grouped together more because of their nearness to one another than because of any similarities among their peoples. These lands lie mostly in the Southern Hemisphere. Australia is a dry continent that is home to unusual wildlife. Oceania's 25,000 tropical islands spread out across the Pacific Ocean. Frozen Antarctica covers the earth at the South Pole.

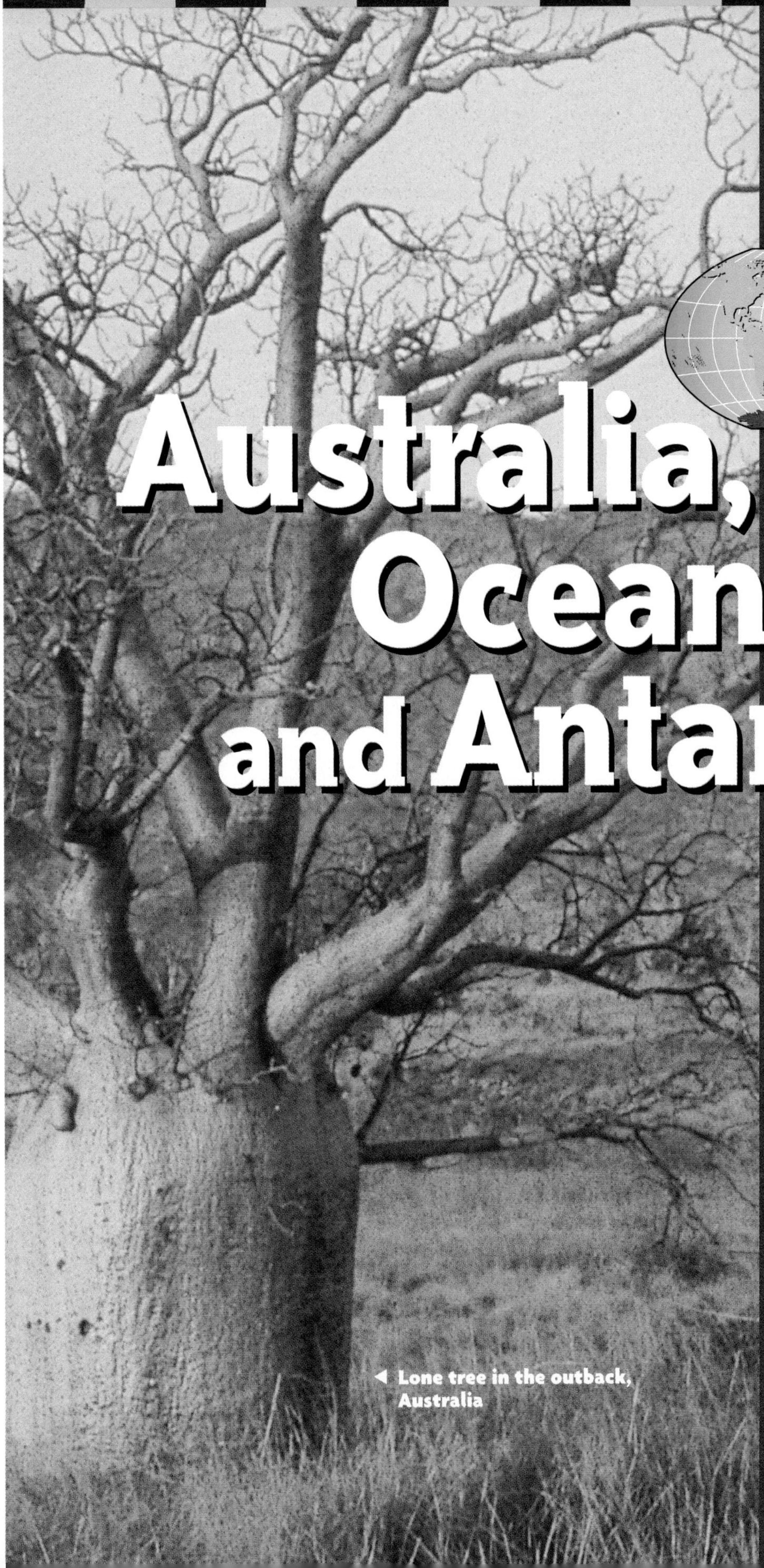

◄ Lone tree in the outback, Australia

NGS ONLINE
www.nationalgeographic.com/education

Australia, Oceania, and Antarctica

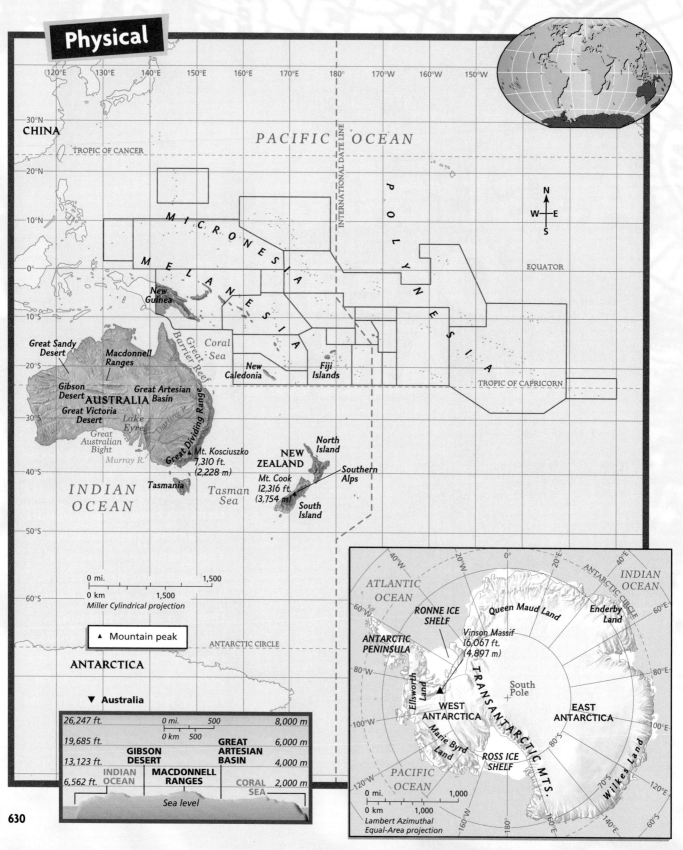

Physical

CHINA

PACIFIC OCEAN

TROPIC OF CANCER

INTERNATIONAL DATE LINE

MICRONESIA

MELANESIA

POLYNESIA

EQUATOR

New Guinea

Great Sandy Desert

Macdonnell Ranges

Great Barrier Reef

Coral Sea

New Caledonia

Fiji Islands

TROPIC OF CAPRICORN

Gibson Desert

Great Artesian Basin

AUSTRALIA

Great Victoria Desert

Lake Eyre

Great Australian Bight

Great Dividing Range

Darling R.

Murray R.

Mt. Kosciuszko 7,310 ft. (2,228 m)

North Island

NEW ZEALAND

Southern Alps

Mt. Cook 12,316 ft. (3,754 m)

South Island

Tasmania

Tasman Sea

INDIAN OCEAN

0 mi. 1,500
0 km 1,500
Miller Cylindrical projection

▲ Mountain peak

ANTARCTIC CIRCLE

ANTARCTICA

▼ Australia

26,247 ft.	0 mi.	500		8,000 m
19,685 ft.	0 km	500		6,000 m
13,123 ft.	GIBSON DESERT	GREAT ARTESIAN BASIN		4,000 m
6,562 ft.	INDIAN OCEAN	MACDONNELL RANGES		2,000 m
			CORAL SEA	
	Sea level			

ATLANTIC OCEAN

ANTARCTIC CIRCLE

INDIAN OCEAN

RONNE ICE SHELF

Queen Maud Land

Enderby Land

ANTARCTIC PENINSULA

Vinson Massif 16,067 ft. (4,897 m)

South Pole

TRANSANTARCTIC MTS.

Ellsworth Land

WEST ANTARCTICA

EAST ANTARCTICA

Marie Byrd Land

ROSS ICE SHELF

Wilkes Land

PACIFIC OCEAN

0 mi. 1,000
0 km 1,000
Lambert Azimuthal Equal-Area projection

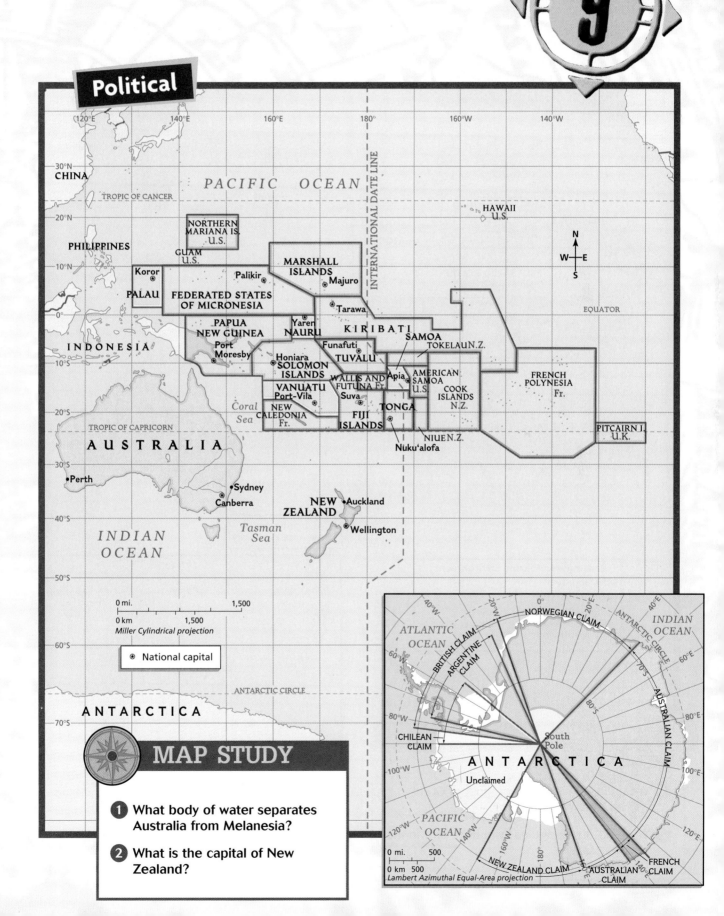

Political

PACIFIC OCEAN

CHINA

TROPIC OF CANCER

PHILIPPINES

INDONESIA

NORTHERN MARIANA IS. U.S.

GUAM U.S.

PALAU
Koror ⊛

FEDERATED STATES OF MICRONESIA
Palikir ⊛

MARSHALL ISLANDS
⊛ Majuro

INTERNATIONAL DATE LINE

HAWAII U.S.

EQUATOR

Tarawa ⊛

KIRIBATI

PAPUA NEW GUINEA
Port Moresby ⊛

Yaren
NAURU

Funafuti ⊛
TUVALU

SAMOA
Apia ⊛

TOKELAU N.Z.

FRENCH POLYNESIA Fr.

Honiara ⊛
SOLOMON ISLANDS

AMERICAN SAMOA U.S.

VANUATU
Port-Vila ⊛

WALLIS AND FUTUNA Fr.

Suva ⊛

COOK ISLANDS N.Z.

Coral Sea

NEW CALEDONIA Fr.

FIJI ISLANDS

TONGA

NIUE N.Z.

PITCAIRN I. U.K.

TROPIC OF CAPRICORN

AUSTRALIA

Nuku'alofa

Perth ●

Sydney ●
Canberra ●

NEW ZEALAND
● Auckland

Tasman Sea

⊛ Wellington

INDIAN OCEAN

0 mi. 1,500
0 km 1,500
Miller Cylindrical projection

⊛ National capital

ANTARCTIC CIRCLE

ANTARCTICA

MAP STUDY

1 What body of water separates Australia from Melanesia?

2 What is the capital of New Zealand?

ATLANTIC OCEAN

NORWEGIAN CLAIM

INDIAN OCEAN

ANTARCTIC CIRCLE

BRITISH CLAIM

ARGENTINE CLAIM

AUSTRALIAN CLAIM

CHILEAN CLAIM

South Pole

ANTARCTICA

Unclaimed

PACIFIC OCEAN

NEW ZEALAND CLAIM

AUSTRALIAN CLAIM

FRENCH CLAIM

0 mi. 500
0 km 500
Lambert Azimuthal Equal-Area projection

Australia, Oceania, and Antarctica

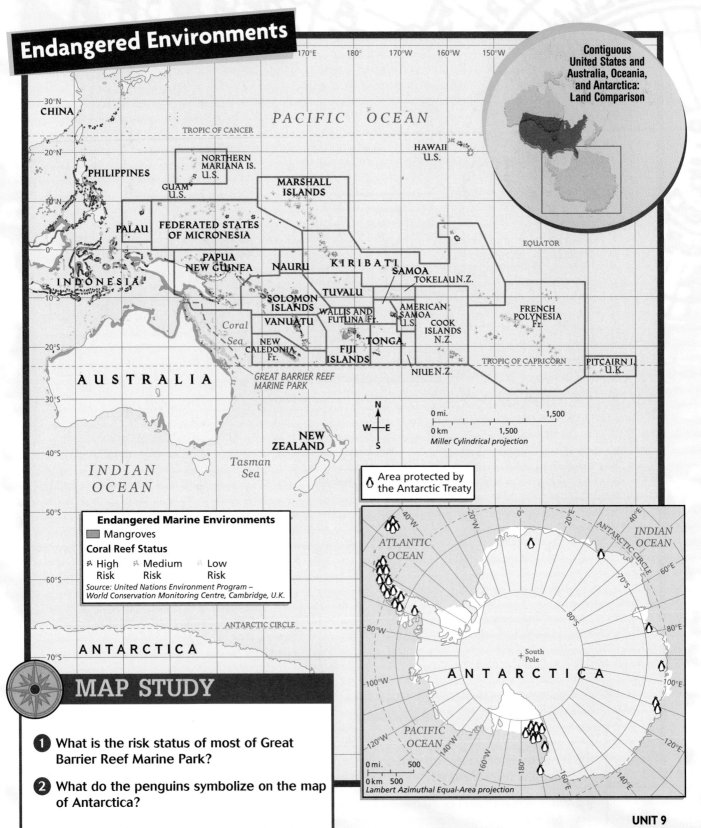

Endangered Environments

Contiguous United States and Australia, Oceania, and Antarctica: Land Comparison

MAP STUDY

1 What is the risk status of most of Great Barrier Reef Marine Park?

2 What do the penguins symbolize on the map of Antarctica?

Endangered Marine Environments
Mangroves
Coral Reef Status
High Risk   Medium Risk   Low Risk
Source: United Nations Environment Program – World Conservation Monitoring Centre, Cambridge, U.K.

Area protected by the Antarctic Treaty

Fast Facts

COMPARING POPULATION:
United States and Selected Countries of Australia, Oceania, and Antarctica

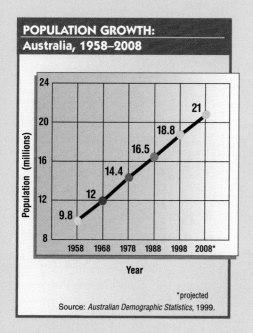

UNITED STATES

AUSTRALIA

PAPUA NEW GUINEA

= 15,000,000

NEW ZEALAND

Source: *Population Reference Bureau*, 2000.

POPULATION GROWTH:
Australia, 1958–2008

24

21

20

18.8

16.5

16

14.4

12

12

9.8

8

1958   1968   1978   1988   1998   2008*

Population (millions)

Year

*projected
Source: *Australian Demographic Statistics*, 1999.

Australia

Data Bits

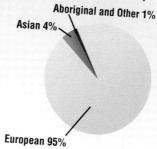	Automobiles per 1,000 people	485
	Telephones per 1,000 people	505
VOTE	Democratic elections	Yes

Ethnic Makeup

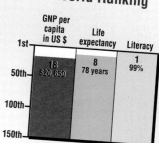

Aboriginal and Other 1%

Asian 4%

European 95%

World Ranking

	GNP per capita in US $	Life expectancy	Literacy
1st			
	18 $20,650	8 78 years	1 99%
50th			
100th			
150th			

Population: Urban vs. Rural

85%   15%

Source: *World Desk Reference*, 2000.

GRAPHIC STUDY

1 What is Australia's world ranking for life expectancy?

2 By how much is Australia's population expected to have grown between 1958 and 2008?

REGIONAL ATLAS

Country Profiles

AUSTRALIA

POPULATION:
19,400,000
7 per sq. mi.
3 per sq. km

LANGUAGE:
English

MAJOR EXPORT:
Coal

MAJOR IMPORT:
Machinery

CAPITAL:
Canberra

LANDMASS:
2,988,888 sq. mi.
7,741,220 sq. km

FEDERATED STATES of MICRONESIA

POPULATION:
100,000
370 per sq. mi.
143 per sq. km

LANGUAGES:
English, Local
Languages

MAJOR EXPORT:
Fish

MAJOR IMPORT:
Foods

CAPITAL:
Palikir

LANDMASS:
270 sq. mi.
699 sq. km

FIJI ISLANDS

POPULATION:
800,000
113 per sq. mi.
44 per sq. km

LANGUAGES:
English, Fijian,
Hindi

MAJOR EXPORT:
Sugar

MAJOR IMPORT:
Machinery

CAPITAL:
Suva

LANDMASS:
7,054 sq. mi.
18,270 sq. km

KIRIBATI

POPULATION:
100,000
355 per sq. mi.
137 per sq. km

LANGUAGES:
English, Gilbertese

MAJOR EXPORT:
Coconut Products

MAJOR IMPORT:
Foods

CAPITAL:
Tarawa

LANDMASS:
282 sq. mi.
730 sq. km

MARSHALL ISLANDS

POPULATION:
100,000
1,449 per sq. mi.
559 per sq. km

LANGUAGES:
English, Local
Languages

MAJOR EXPORT:
Coconut Products

MAJOR IMPORT:
Foods

CAPITAL:
Majuro

LANDMASS:
69 sq. mi.
179 sq. km

NAURU

POPULATION:
10,000
1,111 per sq. mi.
435 per sq. km

LANGUAGES:
Nauruan, English

MAJOR EXPORT:
Phosphates

MAJOR IMPORT:
Foods

CAPITAL:
Yaren

LANDMASS:
9 sq. mi.
23 sq. km

NEW ZEALAND

POPULATION:
3,900,000
37 per sq. mi.
14 per sq. km

LANGUAGE:
English

MAJOR EXPORT:
Wool

MAJOR IMPORT:
Machinery

CAPITAL:
Wellington

LANDMASS:
104,452 sq. mi.
270,531 sq. km

PALAU

POPULATION:
20,000
112 per sq. mi.
43 per sq. km

LANGUAGES:
English, Palauan

MAJOR EXPORT:
Fish

MAJOR IMPORT:
Machinery

CAPITAL:
Koror

LANDMASS:
178 sq. mi.
461 sq. km

PAPUA NEW GUINEA

POPULATION:
5,000,000
28 per sq. mi.
11 per sq. km

LANGUAGES:
English, Local
Languages

MAJOR EXPORT:
Gold

MAJOR IMPORT:
Machinery

CAPITAL:
Port Moresby

LANDMASS:
178,703 sq. mi.
462,841 sq. km

SAMOA

POPULATION:
200,000
182 per sq. mi.
70 per sq. km

LANGUAGES:
Samoan, English

MAJOR EXPORT:
Coconut Products

MAJOR IMPORT:
Foods

CAPITAL:
Apia

LANDMASS:
1,097 sq. mi.
2,841 sq. km

SOLOMON ISLANDS

POPULATION:
500,000
45 per sq. mi.
17 per sq. km

LANGUAGES:
English, Local
Languages

MAJOR EXPORT:
Cocoa

MAJOR IMPORT:
Machinery

CAPITAL:
Honiara

LANDMASS:
11,158 sq. mi.
28,899 sq. km

Countries and flags not drawn to scale

For more information on countries in this region, refer to the Nations of the World Data Bank on pages 690–699.

TONGA
POPULATION:
100,000
345 per sq. mi.
133 per sq. km

LANGUAGES:
Tongan, English

MAJOR EXPORT:
Squash

CAPITAL:
Nuku'alofa

MAJOR IMPORT:
Foods

LANDMASS:
290 sq. mi.
751 sq. km

Nuku'alofa

TUVALU
POPULATION:
10,000
1,000 per sq. mi.
385 per sq. km

LANGUAGES:
Tuvalu, English

MAJOR EXPORT:
Coconut Products

CAPITAL:
Funafuti

MAJOR IMPORT:
Foods

LANDMASS:
10 sq. mi.
26 sq. km

Funafuti

VANUATU
POPULATION:
200,000
43 per sq. mi.
16 per sq. km

LANGUAGES:
Bislama, English,
French

MAJOR EXPORT:
Coconut Products

CAPITAL:
Port-Vila

MAJOR IMPORT:
Machinery

LANDMASS:
4,707 sq. mi.
12,191 sq. km

Port-Vila

BUILDING CITIZENSHIP

Voting Nearly all eligible voters participate in elections in Australia. All citizens over 18 years old are required to vote in all local, state, and national elections. If they don't vote they can be fined up to 50 Australian dollars. To make it easier, elections are held on Saturdays and voting is done at schools, churches, and other convenient locations. In the United States, only about half of eligible people vote in the presidential elections.

Why do so many people in the United States not exercise their right to vote?

WRITE ABOUT IT

Voting and participating in political activities are important parts of belonging to a democratic society. Yet in the United States most people do not vote. Pretend you are the head of elections for your city and it is your responsibility to encourage people to vote in upcoming elections for mayor and the city council. Design a flyer that will be mailed to all households to encourage people to vote.

This woman is exercising her right to vote. ▼

Australia and New Zealand

The World and Its People

NATIONAL GEOGRAPHIC

To learn more about the people and places of Australia and New Zealand, view *The World and Its People* **Chapter 27** video.

Our World Today **online**

Chapter Overview Visit the *Our World Today: People, Places, and Issues* Web site at owt.glencoe.com and click on **Chapter 24—Chapter Overviews** to preview information about Australia and New Zealand.

Why It Matters

An Isolated Region

Australia and New Zealand have been called the last places on Earth because they are so far from other lands. Within Australia, some farmers in the remote outback region often have to drive several hours on unpaved roads to reach a distant rural town. Yet, despite its isolation and distance from other countries, Australia has a prosperous economy that ties it very closely to the rest of the world.

◀ Ayers Rock in central Australia

FOLDABLES™
Study Organizer

Making Predictions Study Foldable Make this foldable to record information about Australia and New Zealand, which you will use to make predictions about the future of the countries.

Step 1 Fold one sheet of paper in half from top to bottom.

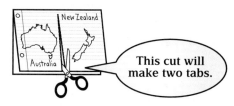

Step 2 Fold it in half again, from side to side.

Step 3 Unfold the paper once. Sketch an outline of Australia and New Zealand across the front tabs and label your foldable as shown.

Step 4 Cut up the fold of the top flap only.

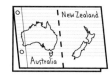

This cut will make two tabs.

Reading and Writing As you read the chapter, write what you learn about these countries under the appropriate tabs of your foldable. Then use that information to make predictions about the future economic growth and development of these countries.

Australia

Guide to Reading

Main Idea

Both a continent and a country, Australia has many natural resources but few people.

Terms to Know

- coral reef
- outback
- station
- marsupial
- immigrant
- boomerang
- bush

Reading Strategy

Create a chart like this one. Then fill in two facts about Australia for each category.

Land	History
Climate	Government
Economy	People

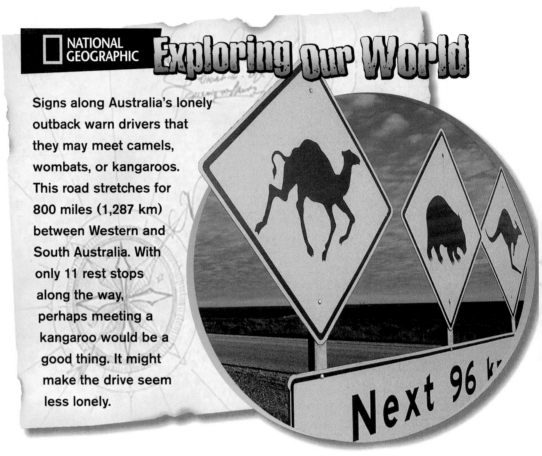

NATIONAL GEOGRAPHIC Exploring Our World

Signs along Australia's lonely outback warn drivers that they may meet camels, wombats, or kangaroos. This road stretches for 800 miles (1,287 km) between Western and South Australia. With only 11 rest stops along the way, perhaps meeting a kangaroo would be a good thing. It might make the drive seem less lonely.

Is Australia a country or a continent? It is both. Australia is the sixth-largest country in the world. Surrounded by water, Australia is too large to be called an island. So geographers call it a continent.

Australia's Land and Climate

Australia is sometimes referred to as the Land Down Under because it is located in the Southern Hemisphere. The island of **Tasmania,** to the south, is part of Australia. The **Great Barrier Reef** lies off Australia's northeastern coast. Coral formations have piled up for millions of years to create a colorful chain that stretches 1,250 miles (2,012 km). A coral reef is a structure formed by the skeletons of small sea animals.

The people of Australia use the name outback for the inland regions of their country. Mining camps and cattle and sheep ranches called stations dot this region. Some stations are huge. One cattle station is almost twice as large as the state of Delaware.

Water is scarce in Australia. In the **Great Artesian Basin,** however, water lies in deep, underground pools. Ranchers drill wells and bring the underground water to the surface for their cattle. Australia's western plateau is even drier. Imagine a carpet of sand twice as large as Alaska, Texas, California, and New Mexico combined. Most people who cross this vast, dry plateau do so by plane. Narrow plains run along the south and southeast of Australia. These fertile flatlands hold Australia's best farmland and most of the country's people.

Unusual Animals   About 200 million years ago, the tectonic plate upon which Australia sits separated from the other continents. As a result, Australia's native plants and animals are not found elsewhere in the world. Two famous Australian animals are kangaroos and koalas. Both are marsupials, or mammals that carry their young in a pouch. Turn to page 642 to read about some of Australia's animals.

√ Reading Check   **Where do most of Australia's people live?**

Australia's Economy

   Australia has a strong, prosperous economy. Australia is a treasure chest overflowing with mineral resources. These riches include iron ore, zinc, bauxite, gold, silver, opals, diamonds, and pearls. Australia also

Literature

GREAT MOTHER SNAKE
Aboriginal Legend
All cultures developed stories to help explain their beginnings. In this Aboriginal legend the Great Mother Snake is credited with creating Australia as well as all of its human and animal inhabitants. In this culture, the snake is a symbol of good rather than evil.

❝ . . . Then finally She awoke and brought from the womb on the Earth itself, man and woman. And they learned from the Mother Snake how to live in peace and harmony with all these creatures who were their spiritual cousins ...And man and woman were now the caretakers of this land. And the Great Snake then entered a large water hole where she guards the fish and other water creatures, so that when the Aboriginal people fish they know to take only as much as they can eat, because if someone should take more than they need through greed or kills for pleasure, they know that one dark night, the Great Mother Snake will come . . . and punish the one who broke this tribal law.❞

Source: Great Mother Snake an Aboriginal legend.

Analyzing Literature

1. What do you think the term "spiritual cousins" means?
2. Why would it be important for people in this culture to take from the earth only as much as they needed?

Music

The *didgeridoo* is the most famous musical instrument from Australia. In its original form, it was made when a eucalyptus branch fell to the ground and was hollowed out by termites. Someone playing a *didgeridoo* creates a variety of sounds by making a combination of lip, tongue, and mouth movements. Once you hear the eerie sounds, it is easy to understand why the Aborigines considered this instrument sacred and made it part of their ceremonies.

Looking Closer What other instruments are similar to the *didgeridoo*?

GO TO

World Music: A Cultural Legacy
Hear music of this region on Disc 2, Track 28.

has energy resources, including coal, oil, and natural gas. Mineral and energy resources make up more than one-third of Australia's exports.

Australia's dry climate limits farming. With irrigation, however, farmers grow grains, sugarcane, cotton, fruits, and vegetables. The main agricultural activity is raising livestock, especially cattle and sheep. Australia is the world's top producer and exporter of wool. Ranchers also ship beef and cattle hides.

Manufacturing, which is growing in importance, includes processed foods, transportation equipment, metals, cloth, and chemicals. High-technology industries, service industries, and tourism also play a large role in the economy.

✓Reading Check What is Australia's main agricultural activity?

Australia's History and People

Despite its huge area, Australia has few people—only 19.4 million. Australia has long needed more skilled workers to develop its resources and build its economy. More than 5 million immigrants, or people who move from one country to live in another, have arrived in recent decades.

Aborigines (A•buh•RIHJ•neez) are a small part of Australia's population. They are the descendants of the first immigrants who came from Asia at least 40,000 years ago. You may have heard of one of their weapons—the boomerang. This wooden tool is shaped like a bent bird's wing. Hunters throw their boomerangs to stun prey. If the boomerang misses, it curves and sails back to the hunter.

Australia was first discovered by the Dutch in the late 1600s. In 1770 **Captain James Cook** reached Australia and claimed it for Great

Britain. At first the British government used Australia as a place to send prisoners. Then other British people set up colonies, especially after gold was discovered in the outback in 1851. Land was taken from the Aborigines, and many of them died of European diseases.

Today nearly 300,000 Aborigines live in Australia. Growing numbers of them are moving to cities to find jobs. In 1967 the Australian government finally recognized the Aborigines as citizens.

The Government   In 1901 the Australian British colonies united to form the independent Commonwealth of Australia. Today Australia has a British-style parliamentary democracy. A prime minister is the head of government.

Like the United States, Australia has a federal system of government. This means that political power is divided between a national government and state governments. The country has six states and two territories, the **Northern Territory** and the **Australian Capital Territory.**

City and Rural Life   About 85 percent of Australia's people live in cities. **Sydney** and **Melbourne** are the largest cities. Sydney gained the world's attention as the host city for the 2000 Summer Olympic Games. About 15 percent of Australians live in rural areas known as the bush. Many rural people also live and work on the stations that dot the outback.

Australians speak English, but "Aussies," as they call themselves, have some different words. For example, Australians say "G'Day," as a form of hello and cook beef on a "barbie," or barbeque grill.

✓ Reading Check   **What kind of government does Australia have?**

Web Activity Visit the *Our World Today: People, Places, and Issues* Web site at owt.glencoe.com and click on **Chapter 24— Student Web Activities** to learn more about the Great Barrier Reef.

Assessment

Defining Terms
1. Define coral reef, outback, station, marsupial, immigrant, boomerang, bush.

Recalling Facts
2. History Why does Australia have animals not found on other continents?
3. Economics What are four mineral resources found in Australia?
4. History Who are the Aborigines?

Critical Thinking
5. Understanding Cause and Effect How does climate affect agriculture in Australia?
6. Drawing Conclusions How does life in Australia show that the country was once a colony of the United Kingdom?

Graphic Organizer
7. Organizing Information Create a time line like this one with at least four dates in Australia's history. Write the dates on one side of the line and the corresponding event on the opposite side.

Applying Social Studies Skills

8. Analyzing Maps Look at the physical map on page 644. What mountain peak represents the highest elevation in Australia? What mountain range is it part of?

Making Connections

Australia's Amazing Animals

Australia is home to some fascinating and unusual animals. In fact, many of Australia's animal species are found nowhere else in the world.

Kangaroos

Ask people what comes to mind when they think of Australian animals, and they will probably say the kangaroo. Kangaroos are marsupials—mammals whose young mature inside a pouch on the mother's belly. The young kangaroo, called a joey, stays there for months, eating and growing. Australia is home to more than 50 species of kangaroo, ranging in size from the 6-foot (2-m) red kangaroo to the 9-inch (23-cm) musky rat-kangaroo. No matter what their size, all kangaroos have one thing in common—big hind feet. Kangaroos bound along at about 20 miles (32 km) per hour. In a single jump, a kangaroo can hop 10 feet (3 m) high and cover a distance of 45 feet (14 m).

Koalas

Because of their round face, big black nose, large fluffy ears, and soft fur, people sometimes call these animals koala bears. Yet they are not bears at all. The koala is a marsupial. The female's pouch opens at the bottom. Strong muscles keep the pouch shut and the young koalas, also called joeys, safe inside. The koala is a fussy eater who feeds only on leaves of eucalyptus trees. Although there are over 600 species of eucalyptus that grow in Australia, koalas eat only a few types. The leaves also provide the animals with all the moisture they need. Quiet, calm, and sleepy, koalas spend most of their time in the trees.

▲ Koala and joey

Platypus and Emu

The odd-looking platypus is one of the world's few egg-laying mammals. Sometimes called a duck-billed platypus, the animal has a soft, sensitive, skin-covered snout. The platypus is a good swimmer who lives in burrows along the streams and riverbanks of southern and eastern Australia. It uses its bill to stir the river bottom in search of food.

After the ostrich, the Australian emu is the world's second-largest bird. Although the emu cannot fly, its long legs enable it to run at speeds of up to 30 miles (48 km) per hour. Another interesting characteristic of the emu is its nesting behavior. Although the female lays the eggs, the male emu sits on them until they are ready to hatch.

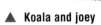

◀ Kangaroo and joey

Emu ▼

Making the Connection

1. What are marsupials?

2. How far can a kangaroo hop in a single jump?

3. **Making Comparisons** Compare two different animals that live in Australia. Tell how they are alike. Tell how they are different.

New Zealand

Guide to Reading

Main Idea

New Zealand is a small country with a growing economy based on trade.

Terms to Know

- geyser
- *manuka*
- fjord
- geothermal energy
- hydroelectric power

Reading Strategy

Create a time line like this one with at least four dates in New Zealand's history. Write the dates on one side of the line and the corresponding event on the opposite side.

NATIONAL GEOGRAPHIC Exploring Our World

Have you ever tasted a ripe green kiwifruit (KEE•wee•FROOT)? If so, it might have been grown on a New Zealand farm like the one shown here. After all, New Zealand is one of the world's leading producers of this tasty fruit. The kiwifruit, once known as the Chinese gooseberry, is now named for the kiwi bird—New Zealand's national symbol.

New Zealand lies in the Pacific Ocean about 1,200 miles (1,931 km) southeast of its nearest neighbor, Australia. In contrast to Australia's flat, dry land, New Zealand is mountainous and very green. Its climate is mild and wet.

New Zealand's Land

New Zealand is about the size of Colorado. It includes two main islands—**North Island** and **South Island**—as well as many smaller islands. The **Cook Strait** separates North Island and South Island.

North Island A large plateau forms the center of North Island. Three active volcanoes and the inactive Mount Egmont are located here. You also find **geysers,** or hot springs that spout steam and water through a crack in the earth.

Small shrubs called *manuka* grow well in the plateau's fertile volcanic soil. Fertile lowlands, forested hills, and sandy beaches surround North Island's central plateau. On the plateau's slopes, sheep and cattle graze. Fruits and vegetables are grown on the coastal lowlands.

South Island   The **Southern Alps** run along South Island's western coast. Snowcapped **Mount Cook,** the highest peak in New Zealand, soars 12,316 feet (3,754 m) here. Glaciers lie on mountain slopes above green forests and sparkling blue lakes. These glaciers once cut deep fjords (fee•AWRDS), or steep-sided valleys, into the mountains. The sea has filled these fjords with crystal-blue waters.

To the east of the Southern Alps stretch the Canterbury Plains. They form New Zealand's largest area of flat or nearly flat land. Farmers grow grains and ranchers raise sheep here.

Plants and Animals   New Zealanders take pride in their unique wildlife. Their national symbol is a flightless bird called the kiwi. Giant kauri (KOWR•ee) trees once dominated all of North Island. About 100 years ago, European settlers cut down many of these trees, using the wood to build homes and ships. Today the government protects kauri trees. One of them is more than 2,000 years old.

✓ Reading Check   **Which island of New Zealand has glaciers and fjords?**

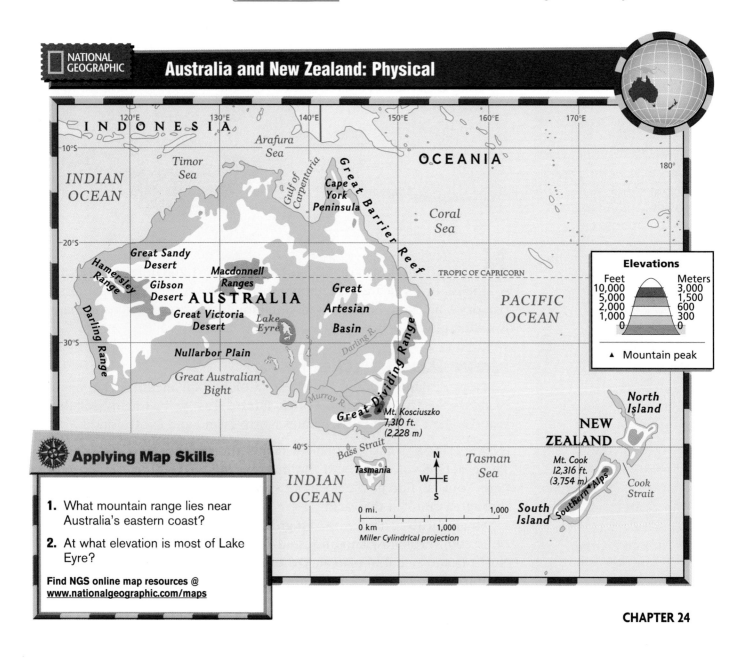

NATIONAL GEOGRAPHIC

Australia and New Zealand: Physical

Applying Map Skills

1. What mountain range lies near Australia's eastern coast?

2. At what elevation is most of Lake Eyre?

Find NGS online map resources @ www.nationalgeographic.com/maps

New Zealand's Economy

New Zealand has a thriving agricultural economy. Sheep are an important agricultural resource. New Zealand is the second-leading wool producer in the world. Lamb meat is another important export. Apples, barley, wheat, and corn are the main crops.

Trade with other countries is an important part of New Zealand's economy. Its main trading partners are Australia, Japan, the United States, and the United Kingdom. Depending on trade brings both benefits and dangers to New Zealand. If the economies of other countries are growing quickly, demand for goods from New Zealand will rise. If the other economies slow, however, they will buy fewer products. This can cause hardship on the islands. In recent years, trade has grown, and New Zealanders enjoy a high standard of living.

Maori

In recent years the Maori culture has experienced a revival in New Zealand. Some Maoris dress in traditional costumes for special celebrations.

History How did the Maoris arrive in New Zealand?

Mining and Manufacturing   New Zealand sits on top of the molten rock that forms volcanoes. As a result, it is rich in geothermal energy, electricity produced from steam. The major source of energy, however, is hydroelectric power—electricity generated by flowing water. New Zealand also has coal, oil, iron ore, silver, and gold.

The country is rapidly industrializing. Service industries and tourism play large roles in the economy. The main manufactured items are wood products, fertilizers, wool products, and shoes.

Reading Check   Why does trade with other countries offer both benefits and dangers to New Zealand?

New Zealand's History and People

People called the **Maoris** (MOWR•eez) are believed to have arrived in New Zealand between A.D. 950 and 1150. They probably crossed the Pacific Ocean in canoes from islands far to the northeast. Undisturbed for hundreds of years, the Maoris developed skills in farming, weaving, fishing, bird hunting, and woodcarving.

The first European explorers came to the islands in the mid-1600s. Almost 200 years passed before settlers—most of them

British—arrived. In 1840 British officials signed a treaty with Maori leaders. In this treaty, the Maoris agreed to accept British rule in return for the right to keep their land. More British settlers eventually moved onto Maori land. War broke out in the 1860s—a war that the Maoris lost.

In 1893 the colony became the first land to give women the right to vote. New Zealand was also among the first places in which the government gave help to people who were old, sick, or out of work.

New Zealand became independent in 1907. The country is a parliamentary democracy in which elected representatives choose a prime minister to head the government. Five seats in the parliament can be held only by Maoris. Today about 10 percent of New Zealand's 3.9 million people are Maoris. Most of the rest are descendants of British settlers. Asians and Pacific islanders, attracted by the growing economy, have increased the diversity of New Zealand's society.

About 85 percent of the people live in urban areas. The largest cities are **Auckland,** an important port, and **Wellington,** the capital. Both are on North Island, where about 75 percent of the people live.

New Zealanders take advantage of the country's mild climate and beautiful landscapes. They enjoy camping, hiking, hunting, boating, and mountain climbing in any season. They also play cricket and rugby, sports that originated in Great Britain.

✓ Reading Check **What group settled New Zealand about 1,000 years ago?**

Section 2 Assessment

Defining Terms
1. **Define** geyser, *manuka,* fjord, geothermal energy, hydroelectric power.

Recalling Facts
2. **Region** How do New Zealand's land and climate compare to Australia's?
3. **Economics** What two animal products are important exports for New Zealand?
4. **History** Most of New Zealand's people are descendants of settlers from what European country?

Critical Thinking
5. **Analyzing Information** Why do you think New Zealand's government guarantees the Maoris a certain number of seats in the parliament?

6. **Making Predictions** With so many different peoples settling in New Zealand, how do you think the country's culture might change?

Graphic Organizer
7. **Organizing Information** Imagine that you are moving to New Zealand. Write a question you would ask for each topic in the chart below.

Physical features	Economy	Recreation
Climate	Government	Culture

Applying Social Studies Skills

8. **Analyzing Maps** Look at the physical map on page 644. Which New Zealand island has higher mountains?

TIME
REPORTS

Closing the Gap

Symbol of Unity: New Zealand's National Rugby Team

TIME REPORTS

FOCUS ON WORLD ISSUES

DAVE E. HOUSER/ CORBIS

In their ancestors' clothes, Maoris do a fierce dance. Land is sacred to Australia's Aborigines (right).

The New World Down Under

When Ngataua Omahuru was five years old, he made a big mistake. Ngataua (en•gah•TOW•ah) was a **Maori**, a native New Zealander. He and his family lived in the forest beneath Mount Taranaki, a volcano on New Zealand's North Island.

One day in 1869, Ngataua made the mistake of wandering away from his parents. A band of British soldiers kidnapped him.

New Zealand was a British colony then. Europeans had been settling there in great numbers for more than 40 years. They had moved onto Maori land, paying nothing or very little for it. Maoris who tried to protect their land were often forced off it at gunpoint.

Ngataua ended up in the home of William Fox, the head of the colony's government. Fox and his wife changed Ngataua's name to William Fox. They sent him to English schools. They cut all his links to the Maori world.

A Rich Culture

Through their religion, the Maoris felt close to their ancestors and to nature. They expressed themselves through song, poetry, weaving, woodcarving, even tattooing. They were brave and clever warriors.

About 200 years ago, New Zealand was home to dozens of iwi, or tribes. This map shows where 10 of them were.

The British, called **Pakehas** (pa•KAY•haws) by the Maoris, did not value the Maori culture. The Pakehas were convinced that no way of life was better

Maori Iwi Lands

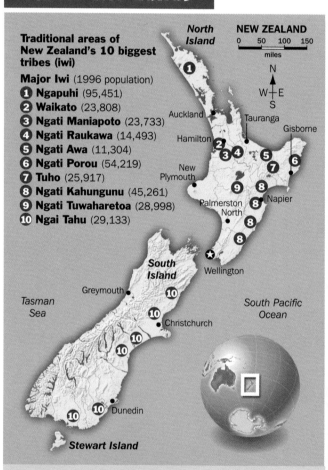

Traditional areas of New Zealand's 10 biggest tribes (iwi)

Major Iwi (1996 population)
1. **Ngapuhi** (95,451)
2. **Waikato** (23,808)
3. **Ngati Maniapoto** (23,733)
4. **Ngati Raukawa** (14,493)
5. **Ngati Awa** (11,304)
6. **Ngati Porou** (54,219)
7. **Tuho** (25,917)
8. **Ngati Kahungunu** (45,261)
9. **Ngati Tuwaharetoa** (28,998)
10. **Ngai Tahu** (29,133)

NEW ZEALAND
0   50   100   150
miles

North Island

Auckland
Tauranga
Gisborne
Hamilton
New Plymouth
Palmerston North
Napier
Wellington
South Island
Greymouth
Christchurch
Dunedin
Stewart Island
Tasman Sea
South Pacific Ocean

INTERPRETING MAPS

Making Inferences Suppose you were a Ngapuhi living 200 years ago. About how far would you have had to travel to reach the Ngai Tahu? What might have made this trip difficult and dangerous?

Maori children. Maoris and Pakehas often marry one another.

Women in traditional dress perform Maori dances.

Aborigine Cathy Freeman lights the Olympic flame in 2000.

than their own. They believed the Maoris would be better off leaving their ways behind.

That **ethnocentric** decision guided Pakeha thinking for a century. The Maoris were taught they had nothing in their culture to be proud of. Cut loose from their traditions but not fully accepted by whites, the Maoris fell on hard times.

They are still trying to recover. Compared with Pakehas, Maoris today learn less and earn less. They die more readily from cancer, diabetes, and heart disease.

New Zealanders are trying to close the gaps between the two groups. They are doing it both to be fair and to keep their nation strong. In 50 years the Maoris will make up almost a quarter of the country's population.

Australia's Ghosts

A similar issue haunts Australia, 1,200 miles (1,931 kilometers) west of New Zealand. Australia's native people, the **Aborigines**, make up 2 percent of the population. For tens of thousands of years, all of Australia was theirs. In 1788 British settlers arrived. They began almost immediately to separate the Aborigines from their culture. They drove them off land that was **sacred** to them. They killed many who resisted.

The Australian settlers repeated the New Zealand settlers' mistakes. They tried to make the first Australians more like them.

Some of their methods were especially harsh. The government decided that Aborigine children would be better off in the hands of white families. So from 1910 to 1971, as many as 100,000 Aborigine children were removed from their parents. White families adopted most of them. Few of the children ever saw their birth mothers again.

Ngataua Omahuru got to see his mother again. As a young lawyer, he returned to his homeland on business. His real family recognized him, and he saw how badly they had been treated. He devoted the rest of his life to helping the Maoris fight for their rights.

It would take the Maoris almost a century to get a fair hearing. By then, Maori foods, words, art, and songs had become part of New Zealand's culture. New Zealanders today realize just how much they would lose if the Maori way of life ever disappeared. ◾

EXPLORING THE ISSUES

1. **Making Inferences** Why do you think British settlers believed their way of life was best?

2. **Problem Solving** If you could, what two things would you change to improve the Maoris' lives?

Broken Promises

Around noon on February 6, 1840, about 75 people stood under a tent in the coastal hamlet of Waitangi, New Zealand. The gathering included Maori chiefs, British settlers, missionaries, and military men.

BETTMANN/CORBIS

▲ Maori children in traditional dress.

They were there to sign a treaty. The treaty gave Britain the right to rule New Zealand. It gave the Maoris Britain's promise to protect them and their land.

The deal made sense to the Maoris. Shady businessmen had begun grabbing Maori land. The chiefs felt that Britain's military muscle was the only thing that could stop the thefts.

Founding Charter

The **Treaty of Waitangi** became New Zealand's founding document. It is as important to New Zealanders as the U.S. Constitution is to Americans. It granted British citizenship to the Maoris. It also described how Maoris and European settlers would share responsibility for New Zealand.

But an agreement is only as strong as the will to enforce it. Greedy settlers took control of New Zealand's government. They used small conflicts as excuses to take over huge pieces of Maori land.

The Maoris tried to embarrass the Pakehas into living up to the treaty. They plowed up the lawns of rich settlers who lived on stolen land. They met Pakeha troops with singing children who offered the soldiers bread.

But in the end nothing, not even the support of many white settlers, could keep the Maoris from losing more land. **Waitangi Day** is a national holiday in New Zealand. Many Maoris refuse to celebrate it, and few people wonder why. ■

EXPLORING THE ISSUES

1. **Explaining** What does the phrase "the will to enforce it" mean? How might the phrase apply to any law?

2. **Making Inferences** Why might it have been hard for Great Britain's government to live up to its side of the agreement?

Closing the Gap

How do you fix a problem that began 200 years ago? New Zealanders have three answers. They hope to keep the Maori culture alive. They want Maoris to have the skills they need to succeed. And they want to pay the **iwi** for land their ancestors lost.

Maoritanga, the Maori way of life, is in trouble. Few people speak the Maori language. To help more people learn it, schools have begun to teach it. They also teach Maori traditions, along with Maori arts and crafts, music, and dance. Maoris now have an "all-Maori" TV channel, too.

Prescription for Success

Equipping Maoris to succeed is another challenge. The government calls its solution "closing the gap"—in skills, wages, housing, and health care. Maoris are being encouraged to stay in school longer, so that they can find and keep good jobs.

The land issue is difficult. The government can't return land to the Maoris that it doesn't own without hurting the people who live on it now. The Maoris will be paid for lost land and other lost "treasures," such as fishing rights.

By 2001, the Waitangi Tribunal had awarded several iwi a total of $300 million. The tribunal, or claims court, won't finish its work until around 2012.

"The process [of sorting through Maori claims] is about more than money," one panel member said. "It is

REUTERS NEW MEDIA INC/CORBIS

▲ New Zealand Prime Minister Helen Clark in 2001. New Zealand was the first nation to let all women vote.

about renewing a relationship that was intended to be based on trust."

That was the spirit of the Treaty of Waitangi. This time, New Zealanders are determined to make it work. ■

EXPLORING THE ISSUE

1. Summarizing What does the title of this article mean? Where is the gap, and why do you think it exists?

2. Making Inferences Why might some Maoris be unhappy with the Waitangi Tribunal's decisions?

Bridging the Gaps at Home: What Can One Person Do?

Ngataua Omahuru, the Maori who was raised in the Pakeha world, did a lot to help his people. He was successful in part because he knew both worlds well.

Americans are fortunate to live in a country blessed with many cultures.

▲ Auckland, with 400,000 people, is New Zealand's largest city.

But how many of us take the time to really understand another culture? Do so, and you will help bridge the gaps that often keep Americans apart.

Here's one way to start. First, choose an immigrant group that you would like to learn more about. You'll have a lot of choices, because all Americans have immigrant roots. And that includes Native Americans, whose ancestors came from Asia thousands of years ago.

Detective Work

Second, get together with a couple of friends who share your interest in this group. As a team, find out all you can about it. When did members of the group come to the United States in large numbers? One of you could find out. Did a particular event prompt them to leave their homeland at that time? One of you could find an answer. How did Americans view the newcomers? How have those views changed? How do members of this group see themselves today—as members of an ethnic group, as Americans, or as both? How has this group changed the way Americans define themselves? By dividing up the work, you could find answers to all these questions fairly quickly.

Share your findings. Publish an article about them in a school newspaper or on a Web page. Illustrate your findings on a poster. Display the poster at school, at your local library, or with a church group. By doing so, you will help others appreciate the glittering mosaic of American life. ■

EXPLORING THE ISSUE

1. **Making Generalizations** In 2001, one of every 10 Americans had been born in another country. Why do you think the United States acts as a magnet for people from other countries?

2. **Cause and Effect** Write a new title for this *TIME Reports* feature. Share it with your classmates. Explain why you think your title fits the story.

REVIEW AND ASSESS

UNDERSTANDING THE ISSUE

1. Defining Key Terms Write definitions for the following terms: *Maori, Aborigine, Pakeha, sacred, Treaty of Waitangi, culture, Maoritanga, iwi, traditions, rights.*

2. Writing to Inform Write a short article describing the history of the Treaty of Waitangi. Use at least five of the terms listed above.

3. Writing to Persuade Why is it important to respect other cultures? Write a short article to support your view, using the experiences of New Zealand and Australia as examples.

INTERNET RESEARCH ACTIVITY

4. With your teacher's help, use Internet resources to learn more about New Zealand. Read about the history of the Maori language. Read about the Maori Language Commission, and what it does. How important is language to a culture's survival? Write a short essay answering that question, using facts picked up on this site.

5. With your teacher's help, use Internet resources to find information on Maori food. Try to find specific sites that list Maori recipes in particular. Browse through the traditional recipes. Then write a 250-word article explaining how those recipes provide clues to where the Maoris live, how they cook, and what foods their great-grandparents ate.

BEYOND THE CLASSROOM

6. Compare the map on the opening page with the physical map of New Zealand in the text on page 644. What does the physical map tell you about the land the iwi occupied? In a short essay, describe in general terms what

BETTINA A. STAMMEN

▲
A banana leaf serves as a plate for traditional Maori food.

one iwi's traditional land may have looked like.

7. Visit your school or local library to find books on the Maoris or Aborigines. (A good but long one is Peter Walker's *The Fox Boy*, which tells Ngataua Omahuru's story.) Prepare an oral book report to deliver in class. Make sure to note the author's point of view.

The Making of a Multicultural Society

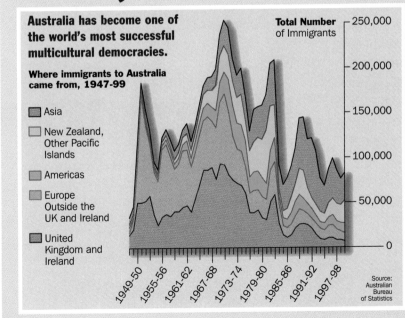

Australia has become one of the world's most successful multicultural democracies.

Where immigrants to Australia came from, 1947-99

- Asia
- New Zealand, Other Pacific Islands
- Americas
- Europe Outside the UK and Ireland
- United Kingdom and Ireland

Total Number of Immigrants

- 250,000
- 200,000
- 150,000
- 100,000
- 50,000
- 0

1949-50 1955-56 1961-62 1967-68 1973-74 1979-80 1985-86 1991-92 1997-98

Source: Australian Bureau of Statistics

BUILDING GRAPH READING SKILLS

1. Analyzing the Data In 1999 there were about 80,000 immigrants. What were the two largest sources?

2. Making Inferences What might make people want to leave their homelands and settle in Australia?

FOR UPDATES ON WORLD ISSUES GO TO www.timeclassroom.com/glencoe

Study and Writing Skill

Outlining

Outlining may be used as a starting point for writing. The writer begins with the rough shape of the material and gradually fills in the details in a logical manner. You may also use outlining as a method of note taking and organizing information as you read.

Learning the Skill

There are two types of outlines—formal and informal. An informal outline is similar to taking notes—you write words and phrases needed to remember main ideas. In contrast, a formal outline has a standard format. Follow these steps to formally outline information:

- Read the text to identify the main ideas. Label these with Roman numerals.
- Write subtopics under each main idea. Label these with capital letters.
- Write supporting details for each subtopic. Label these with Arabic numerals.
- Each level should have at least two entries and should be indented from the level above.
- All entries should use the same grammatical form, whether phrases or complete sentences.

▼ A huge sheep herd pours down a ravine on New Zealand's North Island.

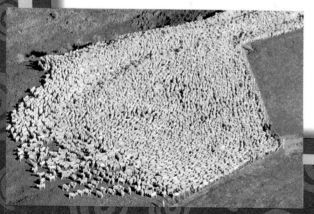

Practicing the Skill

On a separate sheet of paper, copy the following outline for Section 2 of this chapter. Then use your textbook to fill in the missing subtopics and details.

```
I.  New Zealand's Land
    A. North Island
       1. Central plateau surrounded by fertile lowlands
       2. Active volcanoes and geysers
    B. _____
       1. Southern Alps on western coast
       2. _____
    C. Plants and Animals
       1. _____
       2. _____
II. New Zealand's Economy
    A. Agriculture
       1. _____
       2. _____
    B. Trading Partners
       1. _____
       2. _____
       3. _____
       4. _____
    C. _____
       1. _____
       2. Wood products, fertilizers, wool products, and shoes
III. New Zealand's History and People
    A. _____
    B. _____
```

Applying the Skill

Following the guidelines above, prepare an outline for Section 1 of this chapter.

GO TO

Practice key skills with **Glencoe Skillbuilder Interactive Workbook, Level 1.**

Reading Review

Section 1 | Australia

Terms to Know
coral reef
outback
station
marsupial
immigrant
boomerang
bush

Main Idea
Both a continent and a country, Australia has many natural resources but few people.

✓ Place The land of Australia is mostly flat and dry, with little rainfall.

✓ History Because Australia has been separated from other continents for millions of years, unusual plants and animals developed here.

✓ Economics Most of Australia's wealth comes from minerals and the products of its ranches. It is the world's leading producer and exporter of wool.

✓ Culture Australia has relatively few people, most of whom live along the coasts.

▶ Sydney Opera House in Sydney, Australia

Section 2 | New Zealand

Terms to Know
geyser
manuka
fjord
geothermal energy
hydroelectric power

Main Idea
New Zealand is a small country with a growing economy based on trade.

✓ Place New Zealand has volcanic mountains, high glaciers, deep-cut fjords, fertile hills, and coastal plains. The climate is mild and wet.

✓ Economics New Zealand's economy is built on trade. Sheepherding is an important activity, and wool and lamb meat are major exports.

✓ History The people called the Maoris first came to New Zealand about 1,000 years ago.

✓ Culture Most people live on North Island, where the country's two main cities can be found.

✓ History New Zealand was the first land to allow women to vote.

Using Key Terms

Match the terms in Part A with their definitions in Part B.

A.

1. boomerang
2. bush
3. station
4. geothermal energy
5. outback
6. *manuka*
7. marsupial
8. hydroelectric power
9. coral reef
10. geyser

B.

a. electricity produced from steam
b. wooden weapon that returns to the thrower
c. mammal that carries its young in a pouch
d. hot spring that shoots hot water into the air
e. rural area in Australia
f. structure formed by the skeletons of small sea animals
g. name for entire inland region of Australia
h. cattle or sheep ranch in Australia
i. electricity generated by flowing water
j. small shrub found in New Zealand

Reviewing the Main Ideas

Section 1 Australia

11. **Location** Why is Australia called the Land Down Under?
12. **Place** For what is the outback used?
13. **Economics** What does Australia lead the world in producing and exporting?
14. **History** What country colonized Australia?
15. **Culture** What percentage of people live in Australia's cities?

Section 2 New Zealand

16. **Location** On which island do most New Zealanders live?
17. **History** When did New Zealand gain its independence from Britain?
18. **Economics** Why can New Zealand's economy suffer if other countries have economic problems?
19. **Culture** How many New Zealanders have Maori heritage?
20. **Human/Environment Interaction** What leisure activities do New Zealanders enjoy that are made possible by the country's climate?

Australia and New Zealand

Place Location Activity

On a separate sheet of paper, match the letters on the map with the numbered places listed below.

1. Auckland
2. Sydney
3. Tasmania
4. Great Barrier Reef
5. Great Dividing Range
6. Southern Alps
7. Darling River
8. Wellington
9. Canberra
10. Perth

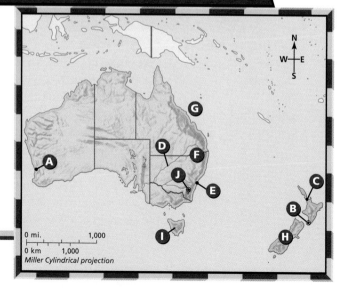

0 mi. 1,000
0 km 1,000
Miller Cylindrical projection

Self-Check Quiz Visit the *Our World Today: People, Places, and Issues* Web site at owt.glencoe.com and click on **Chapter 24–Self-Check Quizzes** to prepare for the Chapter Test.

Critical Thinking

21. **Understanding Cause and Effect** Why do most Australians and New Zealanders live in coastal areas?

22. **Organizing Information** Create two ovals like these. In the outer ovals, write four facts about each country under its heading. Where the ovals overlap, write three facts that are true of both countries.

Australia   New Zealand

Current Events Journal

23. **Designing a Poster** Choose one of the unusual physical features found in Australia or New Zealand. You might choose the Great Barrier Reef, the Great Artesian Basin, or the geysers or glaciers of New Zealand. Research to learn more about this physical feature and how the people of the country relate to it today. Create an illustrated poster that includes a map, four photographs, and four facts about the feature.

Mental Mapping Activity

24. **Focusing on the Region** Create a simple outline map of Australia and New Zealand; then label the following:

- North Island
- South Island
- Sydney
- Auckland
- Tasmania
- Tasman Sea
- Coral Sea
- Darling River
- Great Artesian Basin
- Hamersley Range

Technology Skills Activity

25. **Using the Internet** Use the Internet to find out more about one of Australia's or New Zealand's cities. Prepare a travel brochure aimed at a tourist who might visit the city. Describe the city's main attractions.

Standardized Test Practice

Directions: Study the graph below, and then answer the following question.

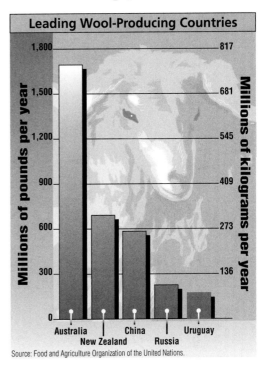

Leading Wool-Producing Countries

Millions of pounds per year: 1,800 / 1,500 / 1,200 / 900 / 600 / 300 / 0

Millions of kilograms per year: 817 / 681 / 545 / 409 / 273 / 136

Australia / New Zealand / China / Russia / Uruguay

Source: Food and Agriculture Organization of the United Nations.

1. **How much wool does Australia produce per year?**

   A 1,800 pounds

   B 1,800,000 pounds

   C about 1,700 pounds

   D about 1,700,000,000 pounds

Test-Taking Tip: Remember to read the information along the sides of the graph to understand what the bars represent. In addition, eliminate answers that you know are wrong.

Chapter 25

Oceania and Antarctica

The World and Its People NATIONAL GEOGRAPHIC

To learn more about the people and places of Oceania and Antarctica, view **The World and Its People Chapter 28** video.

Our World Today **Online**

Chapter Overview Visit the *Our World Today: People, Places, and Issues* Web site at owt.glencoe.com and click on **Chapter 25—Chapter Overviews** to preview information about Oceania and Antarctica.

Summarizing Information Study Foldable Make this foldable and use it to help you summarize what you learn about Oceania and Antarctica.

Step 1 Stack four sheets of paper, one on top of the other. On the top sheet of paper, trace a large circle.

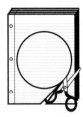

Step 2 With the papers still stacked, cut out all four circles at the same time.

Step 3 Staple the paper circles together at one point around the edge.

Staple here.

This makes a circular booklet.

Step 4 Label the front circle **Oceania** and take notes on the pages that open to the right. Turn the book upside down and label the back **Antarctica.** Take notes on the pages that open to the right.

Oceania

Reading and Writing As you read the chapter, write facts about the people and places of Oceania and Antarctica in the appropriate places of your circular foldable booklet.

▲ Huts on stilts Moorea Lagoon, Tahiti

Why It Matters

A World of Water

The water world of the Pacific Ocean covers one-third of the earth. It is larger than all the world's land areas combined. Tens of thousands of islands lie in this remote part of the world. As technology shrinks the world, many societies of this region are struggling to maintain their cultural identities.

Oceania

Guide to Reading

Main Idea

Oceania is made up of thousands of Pacific Ocean islands organized into countries and territories.

Terms to Know

- cacao
- copra
- pidgin language
- high island
- low island
- atoll
- phosphate
- trust territory

Reading Strategy

Create a chart like this one. In the right column, write two facts about each region.

Region	Facts
Melanesia	
Micronesia	
Polynesia	

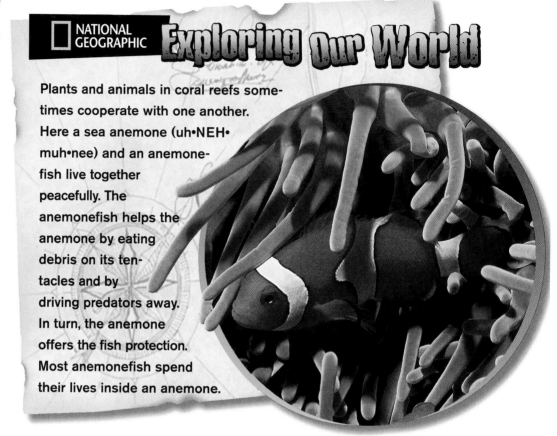

NATIONAL GEOGRAPHIC

Exploring Our World

Plants and animals in coral reefs sometimes cooperate with one another. Here a sea anemone (uh•NEH•muh•nee) and an anemonefish live together peacefully. The anemonefish helps the anemone by eating debris on its tentacles and by driving predators away. In turn, the anemone offers the fish protection. Most anemonefish spend their lives inside an anemone.

Oceania is a culture region that includes about 25,000 islands in the Pacific Ocean. Geographers group Oceania into three main island regions—**Melanesia, Micronesia,** and **Polynesia.**

Melanesia

The islands of Melanesia lie just to the north and east of Australia. The largest country in size is **Papua New Guinea** (PA•pyu•wuh noo GIH•nee). Slightly larger than California, the country's 5 million people also make it Oceania's most populous island. Southeast of Papua New Guinea are three other independent island countries: the **Solomon Islands,** the **Fiji** (FEE•jee) **Islands,** and **Vanuatu** (VAN•WAH•TOO). Near these countries is **New Caledonia,** a group of islands ruled by France.

Rugged mountains and dense rain forests cover Melanesia's islands. Strips of fertile plains hug island coastlines. Most of Melanesia has a tropical climate with nearly constant temperatures between 70°F (21°C) and 80°F (27°C).

Most Melanesians work on subsistence farms. Coffee, palm oil, and cacao are important exports. Cacao is a tropical tree whose seeds are used to make chocolate. Sugarcane is exported as sugar and molasses. Coconut oil from copra, dried coconut meat, is used to make margarine, soap, and other products.

Some Melanesian islands hold rich mineral resources such as gold, oil, copper, and nickel. Several Melanesian islands export timber and fish. Melanesia is also becoming a popular tourist destination.

Melanesia's People   Most Melanesians are originally from the Pacific islands. Exceptions include the people of New Caledonia—about one-third of whom are Europeans—and the Fiji Islands—half of whom are from South Asia. These South Asians are descendants of workers brought from British India in the late 1800s and early 1900s to work on sugarcane plantations. Today South Asians control much of the economy of the Fiji Islands. Fijians of Pacific descent own most of the land. Conflict often arises as the two different groups struggle for control of the government.

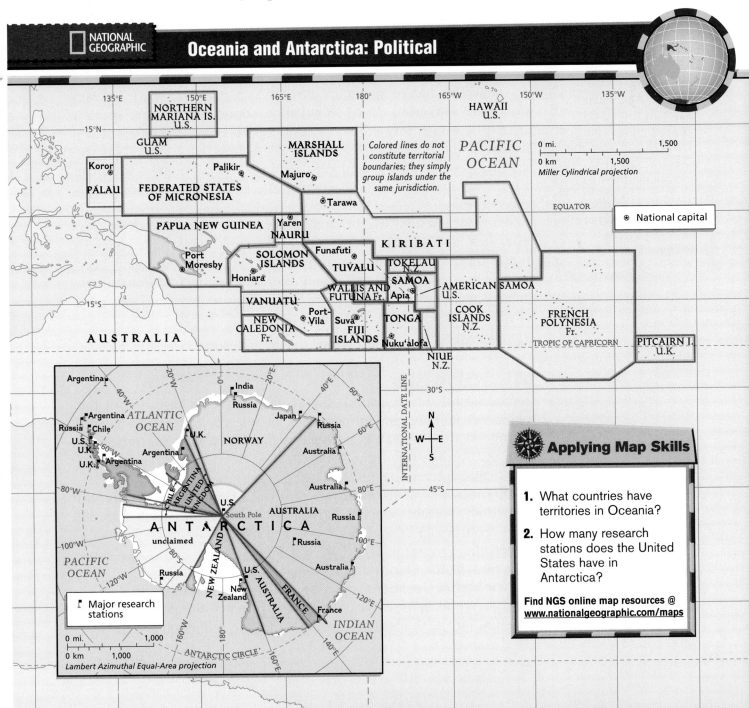

Oceania and Antarctica: Political

Colored lines do not constitute territorial boundaries; they simply group islands under the same jurisdiction.

⊛ National capital

Miller Cylindrical projection

Lambert Azimuthal Equal-Area projection

▮ Major research stations

Applying Map Skills

1. What countries have territories in Oceania?

2. How many research stations does the United States have in Antarctica?

Find NGS online map resources @ www.nationalgeographic.com/maps

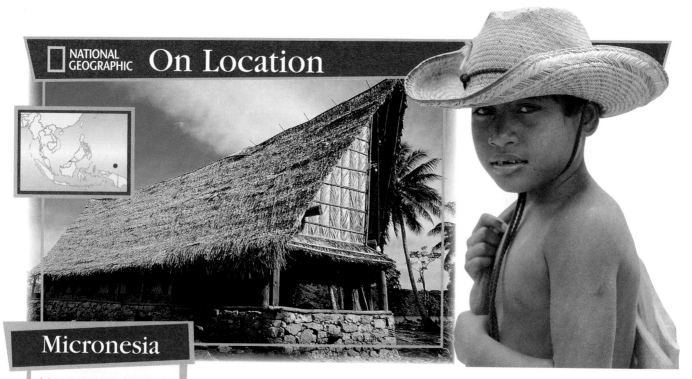

Micronesia

Many of the homes in Micronesia have thatched roofs and no walls (left). This young boy is from the island of Yap in Micronesia (right).

Culture How does the house reflect an adaptation to the environment?

Languages and religions are diverse as well. More than 700 languages are spoken in Papua New Guinea alone. People here speak a pidgin language formed by combining parts of several different languages. People speak English in the Fiji Islands, while French is the main language of New Caledonia. Local traditional religions are practiced, but Christianity is widespread. The South Asian population is mostly Hindu.

Many Melanesians live in small villages in houses made of grass or other natural materials. In recent years, people have built concrete houses to protect themselves from tropical storms. Melanesians keep strong ties to their local group and often hold on to traditional ways. Only a small number live in cities. Many of those living in cities have jobs in business and government.

✓ Reading Check **What is the largest country in Melanesia?**

Micronesia

The islands of Micronesia are scattered over a vast area of the Pacific Ocean. Independent countries include the Federated States of Micronesia, the Marshall Islands, Palau (puh•LOW), Nauru (nah•OO•roo), and Kiribati (KIHR•uh•BAH•tee). The Northern Mariana Islands and Guam are territories of the United States.

Micronesia is made up of two types of islands—high islands and low islands. Volcanic activity formed the mountainous high islands many centuries ago. Coral, or skeletons of millions of tiny sea animals, formed the low islands. Most of the low islands are atolls—low-lying, ring-shaped islands that surround lagoons.

Like Melanesia, Micronesia has a tropical climate. From July to October, typhoons—tropical storms with heavy winds and rains—sometimes strike the islands, causing loss of life and much destruction.

On Micronesia's high islands, soil is rich. Most people here engage in subsistence farming—growing cassava, sweet potatoes, bananas, and coconuts. Some high island farmers also raise livestock. People in the low islands obtain food from the sea. On some low islands, recent population growth has resulted in the need to import food.

Micronesia receives financial aid from the United States, the European Union, and Australia. With this money, the Micronesians have built roads, ports, airfields, and small factories. Clothing is made on the Northern Mariana Islands. Beautiful beaches draw many tourists here.

Several Micronesian islands have phosphate, a mineral salt used to make fertilizer. Phosphate supplies are now gone on Kiribati, and they have almost run out on Nauru. The Federated States of Micronesia and the Marshall Islands have phosphate but lack the money to mine this resource.

Micronesia's People   Southeast Asians first settled Micronesia about 4,000 years ago. Explorers, traders, and missionaries from European countries came in the 1700s and early 1800s. By the early 1900s, many European countries, the United States, and Japan held colonies here.

During World War II, the United States and Japan fought a number of bloody battles on Micronesian islands. After World War II, most of Micronesia was turned over to the United States as trust territories. These territories were under temporary United States control. Some of these islands served as sites for hydrogen bomb testing. Since the 1970s, most have become independent.

Many of Micronesia's people are Pacific islanders. They speak local languages, although English is spoken on Nauru, the Marshall Islands, and throughout the rest of Micronesia. Christianity, brought by Western missionaries, is the most widely practiced religion. Micronesians generally live in villages headed by local chiefs. In recent years, many young people have left the villages to find jobs in towns.

√ Reading Check   In what two ways were Micronesia's islands formed?

Polynesia

Very little is known about the origins of the Polynesians. Historians believe that their ancestors sailed from Asia hundreds of years before the birth of Christ. We know that the first people to settle the islands must have been extremely gifted navigators. Today, the influence of the early Polynesians in language, music, and dance can be seen throughout the South Seas.

When the Polynesian people traveled from island to island, they carried all the supplies they would need with them, including pigs, hens, and dogs. Young banana and breadfruit plants were put in the ground as soon as the Polynesians landed.

Today, Polynesia includes three independent countries—Samoa, Tonga, and Tuvalu. Other island groups are under French rule and are known as French Polynesia. Tahiti, Polynesia's largest island, is part of this French-ruled area. American Samoa, a United States territory, is

The Fate of Nauru
Micronesia's most famous phosphate island is Nauru, an 8-square-mile coral atoll. The name *Nauru* means "nowhere." Over the last 90 years, Nauru's citizens have chosen to "consume" their island by mining the coral as phosphate and selling it as fertilizer. The government of Nauru is now working to develop other industries such as fishing and tourism in preparation for the day when the phosphate is gone.

also part of this region. Most Polynesian islands are high volcanic islands, some with tall, rugged mountains. Some of the islands are of the low atoll type. Because Polynesia lies in the Tropics, the climate is hot and humid.

Polynesians grow crops or fish for their food. Some farmers export coconuts and tropical fruits. The main manufacturing activity is food processing. The tuna you eat for lunch might have come from American Samoa. This island supplies about one-third of the tuna brought into the United States. Tonga exports squash and vanilla. Tourism is one of the fastest-growing industries of Polynesia. Tourists come by air or sea to the emerald green mountains and white palm-lined beaches.

Polynesia's People   Settlers came to Polynesia later than they did to the other island regions. The first to arrive were probably Melanesians or Micronesians who crossed the Pacific Ocean from Asia in canoes.

During the late 1800s, several European nations divided Polynesia among themselves. They built military bases on the islands and later added airfields. The islands served as excellent refueling stops for long voyages across the Pacific. Beginning in the 1960s, several Polynesian territories chose independence, while others remained territories.

About 600,000 people live in Polynesia. Most Polynesians live in rural villages and practice traditional crafts. An increasing number live in towns and cities. **Papeete** (PAH•pay•AY•tay), located on Tahiti, is the capital of French Polynesia and the largest city in the region.

✓Reading Check **What is the largest island in Polynesia?**

Assessment

Defining Terms

1. Define cacao, copra, pidgin language, high island, low island, atoll, phosphate, trust territory.

Recalling Facts

2. Region What three regions make up Oceania?

3. Economics What two kinds of economic activities are most important in these regions?

4. History What groups first settled the lands of Micronesia?

Critical Thinking

5. Summarizing Information What is copra, and what is it used for?

6. Drawing Conclusions Why do many people in Melanesia speak a pidgin language?

Graphic Organizer

7. Organizing Information Create a chart like this one. List all the islands of Oceania under their specific region; then note whether they are independent countries or territories.

Melanesia	Micronesia	Polynesia	Country/Territory of ?

Applying Social Studies Skills

8. Analyzing Maps Look at the political map on page 661. Which two territories are colonies of France?

Study and Writing Skill

Writing a Report

Writing skills allow you to organize your ideas in a logical manner. The writing process involves using skills you have already learned, such as taking notes, outlining, and sequencing information.

Learning the Skill

Use the following guidelines to help you apply the writing process:

- Select an interesting topic. Do preliminary research to determine whether your topic is too broad or too narrow.
- Write a general statement that explains what you want to prove, discover, or illustrate in your writing. This will be the focus of your entire paper.
- Research your topic by coming up with a list of main ideas. Prepare note cards listing facts and source information for each main idea.

An atoll in the Pacific Ocean ▼

- Your report should have an introduction, a body, and a conclusion summarizing and restating your findings.
- Each paragraph should express one main idea in a topic sentence. Additional sentences should support or explain the main idea by using details and facts.

Practicing the Skill

Read the following paragraph, and then answer the questions that follow.

Most of Micronesia's low islands are atolls—low-lying, ring-shaped islands that surround lagoons. An atoll begins as a ring of coral that forms around the edge of a volcanic island. Over time, wind and water erode the volcano, wearing it down to sea level. Eventually only the atoll remains above the surface. The calm, shallow seawater inside the atoll is called a lagoon.

1. What is the main idea of this paragraph?
2. What are the supporting sentences?
3. What might be the topic of an additional paragraph that follows this one?

Applying the Skill

Suppose you are writing a report on Oceania. Answer the following questions about the writing process.

1. How could you narrow this topic?
2. What are three main ideas?
3. Name three possible sources of information.

Antarctica

Guide to Reading

Main Idea

Antarctica is a harsh land of rock and ice, which the world's nations have agreed to leave open to scientific study.

Terms to Know

- crevasse
- ice shelf
- iceberg
- krill
- ozone

Reading Strategy

Create a chart like the one below. Under each heading, fill in at least one fact about Antarctica.

Antarctica	
Land	Climate
Resources	People

NATIONAL GEOGRAPHIC

Exploring Our World

Whee! These Emperor penguins live on the harsh continent of Antarctica. Their shiny "tuxedos" and waddling walk fascinate people. Although they cannot fly, their feathers provide excellent insulation against the ice, snow, and freezing water. Emperor penguins often travel 30 miles (48 km) a day to bring food to their rookeries, or nests. Sometimes walking takes too long, so penguins simply slide on their bellies, which is called tobogganing.

Antarctica sits on the southern end of the earth. Icy ocean water surrounds it. Freezing ice covers it. Cold winds blow over it. The least explored of all the continents, this frigid mysterious land is larger than either Europe or Australia.

A Unique Continent

Picture Antarctica—a rich, green land covered by forests and lush plants. Does this description match your mental image of the continent? Fossils discovered here reveal that millions of years ago, Antarctica's landscape was inhabited by dinosaurs and small mammals.

Today, however, a huge ice cap buries nearly 98 percent of Antarctica's land area. In some spots, this ice cap is 2 miles (3.2 km) thick—about the height of 10 tall skyscrapers stacked upon one another. This massive "sea" of ice holds about 70 percent of all the freshwater in the world.

The Antarctic ice cap is heavy and strong, but it also moves. In some areas, the ice cap forms crevasses, or cracks, that plunge more than 100 feet (30 m). At the Antarctic coast, the ice cap spreads past the land into the ocean. This layer of ice above the water is called an ice shelf. Huge chunks of ice sometimes break off, forming icebergs, which float freely in the icy waters.

Highlands, Mountains, and Valleys   Beneath the ice cap, Antarctica has highlands, valleys, and mountains—the same landforms you find on other continents. A long mountain range called the **Transantarctic Mountains** crosses the continent. The highest peak in Antarctica, the **Vinson Massif,** rises 16,864 feet (5,140 m). The Transantarctic Mountains sweep along the Antarctic Peninsula, which reaches within 600 miles (966 km) of South America's Cape Horn. East of the mountains is a high, flat plateau where you find the **South Pole,** the southernmost point of the earth. On an island off Antarctica's coast rises **Mount Erebus** (EHR•uh•buhs). It is Antarctica's most active volcano.

Climate   Now that you have a mental picture of Antarctica's ice cap, think about this: Antarctica receives so little precipitation that it is the world's largest, coldest desert. Inland Antarctica receives no rain and hardly any new snow each year. Antarctica has a polar ice cap climate. Imagine summer in a place where temperatures may fall as low as −30°F (−35°C) and climb to only 32°F (0°C). Antarctic summers last from December through February. Winter temperatures along the coasts fall to −40°F (−40°C), and in inland areas to a low of −100°F (−73°C).

✓Reading Check   **What landforms are found under Antarctica's ice cap?**

NATIONAL GEOGRAPHIC On Location

Antarctica

Elephant seals lounge on the coast of Elephant Island off the Antarctic Peninsula (below). Mount Erebus, on the opposite side of Antarctica (left), has a lava lake that is often studied by scientists.

Environment How might an eruption of Mount Erebus affect Antarctica?

The *Endurance*

In January 1915, Ernest Shackleton and his crew in the *Endurance* became trapped in Antarctica's freezing seawater. In late October, ice crushed the wooden ship, forcing the explorers to abandon it (below). They spent five more months drifting on the ice until they reached open water and used lifeboats to get away.

Resources of Antarctica

Antarctica has a harsh environment, but it can still support life. Most of the plants and animals that live here are small, however. The largest inland animal is an insect that reaches only one-tenth of an inch in length. Penguins, fish, whales, and many kinds of flying birds live in or near the rich seas surrounding Antarctica. Many eat a tiny, shrimp-like creature called **krill.**

Scientists believe that the ice of Antarctica hides a treasure chest of minerals. They have found major deposits of coal and smaller amounts of copper, gold, iron ore, manganese, and zinc. Petroleum might lie offshore.

These mineral resources have not yet been tapped. To do so would be very difficult and costly. Also, some people feel that removing these resources would damage Antarctica's fragile environment. A third reason is that different nations would disagree over who has the right to these resources. Forty-three nations have signed the **Antarctic Treaty,** which prohibits any nation from taking resources from the continent. It also bans weapons testing in Antarctica.

✓Reading Check **What is the Antarctic Treaty?**

A Vast Scientific Laboratory

The Antarctic Treaty does allow for scientific research in Antarctica. Many countries have scientific research stations here, but no single nation controls the vast continent. In January—summer in Antarctica—about 10,000 scientists come to study the land, plants, animals, and ice of this frozen land. Some 1,000 hardy scientists even stay during the harsh polar winter.

Much of the research focuses on ozone. **Ozone** is a type of oxygen that forms a layer in the atmosphere. The ozone layer protects all living things on the earth from certain harmful rays of the sun. In the 1980s, scientists discovered a weakening, or "hole," in this layer above Antarctica. If such weakening continues, some scientists say, climates around the world will get warmer. By studying this layer further, they hope to learn more about possible changes.

This frozen world attracts more than just scientists, though. Each year, a few thousand tourists come to Antarctica. With such a harsh environment, however, Antarctica is the only continent in the world that has no permanent population.

✓Reading Check **Why are scientists studying the ozone layer?**

Living in Antarctica

Humans can adapt to life under the most difficult of conditions. One example of this is the Villa Las Estrellas, or Village of the Stars. Located in Chile's Antarctic Territory, the "town" has a school, hospital, supermarket, post office, bank, telephone, television, and Internet systems. There is even a gym and a sauna. Village residents are mostly members of Chile's Air Force and their families and scientists from various countries. In all, a total of 88 people live in Villa Las Estrellas, including 24 children under the age of 12.

Like Penguins Daily dress here means wearing thermal under-clothes, warm boots and dark sunglasses to protect against the sun's strong ultraviolet rays. Villagers must survive extreme temperatures—down to −13°F (−25°C) below zero with an even more bone-chilling wind factor—but they don't stay inside all day. Adults walk from house to house to visit their neighbors. The children seem to enjoy the experience more than anyone else. One resident describes outdoor playtime. "The children go crazy over the snow and enjoy sledding or just tobogganing downhill on their stomachs. They look like penguins!"

Global Village The countries of Russia, China, Korea, Brazil, Poland, Argentina, and Uruguay have military or scientific bases close to the village. In Antarctica, normal tensions between countries do not seem to matter. Every Wednesday afternoon the different bases send soccer teams to the Chilean gymnasium for a game of indoor soccer. Once a year a "winter Olympics" is held in volleyball and basketball. Visitors to the different bases mix freely with the people who live in the bases. Villa Las Estrellas may be as close to a real global village as the earth has ever seen.

Our World Today Online

Web Activity Visit the *Our World Today: People, Places, and Issues* Web site at owt.glencoe.com and click on **Chapter 25—Student Web Activities** to learn more about Antarctica.

✓ **Reading Check** What is one way that humans have adapted to the harsh Antarctic environment?

Section 2 Assessment

Defining Terms
1. Define crevasse, ice shelf, iceberg, krill, ozone.

Recalling Facts
2. **Place** What covers nearly 98 percent of Antarctica?
3. **Location** Where in Antarctica would you find the most living things?
4. **Human/Environment Interaction** Why do scientists come to Antarctica?

Critical Thinking
5. **Summarizing Information** Why have countries agreed not to use the resources of Antarctica?
6. **Writing Questions** Imagine that you are planning a trip to Antarctica. What questions would you ask scientists working there?

Graphic Organizer
7. **Organizing Information** Create a chart like the one below, and then look at the political map on page 661. In your chart, list the various national claims made in Antarctica by the world's countries. Then give the number of research stations for each country.

Countries With Claims in Antarctica	Number of Research Stations

Applying Social Studies Skills

8. **Analyzing Maps** Look at the physical map on page 630. What mountain range cuts across Antarctica?

Making Connections

CULTURE   GOVERNMENT   PEOPLE   TECHNOLOGY

Antarctica's Environmental Stations

For nearly 200 years, adventurers, explorers, geographers, and scientists have been drawn to the icy wilderness of Antarctica. Scientific research is the major human activity on this remarkable continent.

Polar Science

In the 1950s, countries began to talk of preserving Antarctica as an international laboratory for scientific research. Today a formal treaty guarantees free access and research rights for scientists of many countries. Antarctica now holds more than 40 research stations.

Types of Research

Geologists, biologists, climatologists, and astronomers are some of the many scientists who come to Antarctica to study. Understanding the earth's environment is a major focus. The Antarctic ice cap contains 90 percent of the world's ice and 70 percent of its freshwater. Changes here can affect the world's oceans and climates.

Scientists in Antarctica were the first to discover the holes in the ozone layer. Such holes can expose life on the earth to too much ultraviolet radiation.

Researchers in Antarctica also study the earth's history. Locked in the continent's ice crystals and air bubbles are clues to the earth's past. Fossils show how landmasses existed before the formation of today's continents.

The harsh living conditions of Antarctica provide another subject for study. The National Aeronautics and Space Administration (NASA) sends engineers and scientists to Antarctica to learn how to survive in extreme conditions, such as those humans might someday encounter on visits to other planets.

Research station at the South Pole ▶

Life at the Edge

Living and working in Antarctica's polar wilderness demands special equipment, well-trained people, and a sizable dose of caution. Freeze-dried food, layers of warm, quick-drying clothes, insulated boots, and specially designed pyramid tents keep researchers well-fed, warm, and dry. Researchers quickly learn the importance of staying inside during whiteout conditions, when snow and fog create a total lack of visibility.

The Antarctic environment is a fragile one, and researchers take care to protect it. All trash and wastes are removed from the continent. Mining of mineral resources is banned, and laws protect native plants and animals. Such care ensures that Antarctica will continue to hold exciting discoveries for years to come.

→ Making the Connection

1. What do researchers study in Antarctica?
2. What discovery did researchers make about the ozone layer?
3. **Summarizing Information** What items do researchers in Antarctica use to stay warm and dry?

Terms to Know

cacao

copra

pidgin language

high island

low island

atoll

phosphate

trust territory

Main Idea

Oceania is made up of thousands of Pacific Ocean islands organized into countries and territories.

✓ Region Oceania is a huge area of vast open ocean and about 25,000 islands.

✓ Region Geographers divide Oceania into three regions: Melanesia, Micronesia, and Polynesia.

✓ Place High islands were formed by volcanoes. Low islands were made from coral.

✓ Place Papua New Guinea, in Melanesia, is the largest and most populous country of Oceania.

✓ Economics The main economic activities are farming and tourism. Some islands have important minerals or other resources.

✓ History Most people of Oceania are descendants of people who left Southeast Asia on canoes thousands of years ago.

Fijian schoolgirls buy ▶
snacks from an Indian
merchant in Suva.

Terms to Know

crevasse

ice shelf

iceberg

krill

ozone

Main Idea

Antarctica is a harsh land of rock and ice, which the world's nations have agreed to leave open to scientific study.

✓ Location Antarctica lies at the southern end of the earth.

✓ Place Most of the continent, which has mountain ranges and a plateau, is covered by a huge, thick ice cap.

✓ Place Most plants and animals that live in Antarctica are small. Larger animals thrive in the waters off the coast.

✓ Economics Antarctica has many minerals, but many nations have signed a treaty agreeing not to remove these resources.

✓ Culture Antarctica is a major center of scientific research, but is the only continent with no permanent human population.

Using Key Terms

Match the terms in Part A with their definitions in Part B.

A.

1. pidgin language
2. copra
3. trust territory
4. ice shelf
5. phosphate
6. ozone
7. low island
8. iceberg
9. high island
10. krill

B.

a. mineral salt used to make fertilizer
b. tiny, shrimp-like animal
c. Pacific island formed by volcanic activity
d. combines elements of several languages
e. chunk of a glacier that floats free
f. dried coconut meat
g. layer of ice above water in Antarctica
h. area temporarily placed under control of another nation
i. Pacific island formed of coral
j. layer in the atmosphere that blocks certain harmful rays of the sun

Reviewing the Main Ideas

Section 1 Oceania

11. **History** Why are there South Asians on the Fiji Islands?
12. **Human/Environment Interaction** Which is likely to have better farmland—a high island or a low island? Why?
13. **Government** New Caledonia is ruled by what European country?
14. **Economics** What attracts tourists to Oceania?
15. **History** From where did the people who first settled Oceania originally come?

Section 2 Antarctica

16. **Place** What is significant about Mount Erebus?
17. **Location** What marine birds feed in the seas near Antarctica?
18. **Economics** What resources have been found in Antarctica?
19. **Government** What agreement bans the use of Antarctica's resources?
20. **Human/Environment Interaction** Why do scientists study the ozone layer in Antarctica?

 **NATIONAL GEOGRAPHIC** **Oceania and Antarctica**

Place Location Activity

On a separate sheet of paper, match the letters on the map with the numbered places listed below.

1. Antarctic Peninsula
2. South Pole
3. Vinson Massif
4. Marshall Islands
5. Papua New Guinea
6. Fiji Islands
7. French Polynesia
8. Coral Sea
9. Solomon Islands

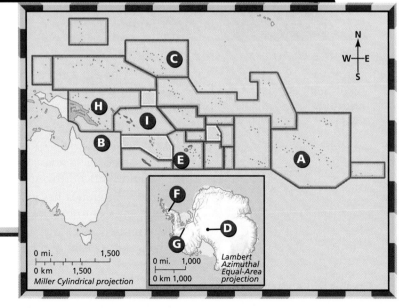

Critical Thinking

21. **Making Generalizations** You have read about two areas with very different climates. Write a generalization about how climate affects the way people live in each area.

22. **Organizing Information** Make a chart like this one. Under each heading, write a fact about each of the four regions you studied in this chapter: Melanesia, Micronesia, Polynesia, and Antarctica.

Region	Landforms	Climate	Economy or Resources	People

Current Events Journal

23. **Writing a Pamphlet** Choose one island country from Oceania. Research current topics such as the land, economy, daily life, or culture. After you have completed your research, create a pamphlet about your island, encouraging tourists to visit there.

Mental Mapping Activity

24. **Focusing on the Region** Create an outline map of Antarctica. Refer to the map on page 630, and then label the following:

- Antarctic Circle
- Vinson Massif
- Antarctic Peninsula
- Pacific Ocean
- Atlantic Ocean
- South Pole

Technology Skills Activity

25. **Building a Database** Research three animals of Oceania. Create a database of the information you find. Include separate fields for the following items: name of species, location, type of habitat, diet, natural predators, and population status. Then use the database information to create a map showing the location of each species.

The Princeton Review

Standardized Test Practice

Directions: Read the paragraph below, and then answer the following question.

Because of the clear Pacific waters, fresh fish is the primary traditional food of Oceania's people. This is especially true in the low coral islands, where there is little land suitable for farming. The rich volcanic soil of the high islands allows pineapples, coconuts, bananas, and sweet potatoes to grow. In Papua New Guinea, pork is a favorite food. Great feasts of pork, greens, and yams are social gatherings for whole villages. At these feasts, pigs are cooked for about eight hours over hot stones set in an "earth oven"—or large hole in the ground.

1. **Which of the following statements best summarizes the paragraph above?**

   F People in the high islands are able to grow and eat pineapples, bananas, and sweet potatoes.

   G In the low islands, the coral prevents much farming.

   H The Pacific Ocean is the source of the fish that most people eat.

   J Physical geography influences the traditional foods of Oceania's people.

Test-Taking Tip: When a question uses the word *best,* it means that more than one answer might be correct. Your job is to pick the *best* answer. This question also asks for a summary of the passage. Read through all of the answer choices before choosing the one that provides a more general restatement of the information.

Alone in a hut in 1934, Byrd fell ill from inhaling stove fumes.

"BYRD'S"-EYE VIEW:
Exploring Antarctica

What sound does your breath make when it freezes? On his first expedition to Antarctica, United States Navy officer Richard E. Byrd found out. When the temperature dropped to –64°F (–53°C), Byrd heard what sounded like Chinese firecrackers as a slight breeze crackled across his frozen breath.

Beyond Imagination

Strange things happen in Antarctica. The sun lurks near the horizon for months and never sets. Then from mid-March to mid-September, the dark winter sets in, and the sun never rises. Antarctica is covered by thick ice and snow. Huge glaciers spill out between mountains that rim the coast, creating shelves of ice that extend out over the sea. Sometimes chunks of ice break off to form huge icebergs.

In the early 1900s, little was known about Antarctica. A few brave explorers had traveled on foot, skis, and sleds across the frozen land. Several made it to the South Pole. Much of the continent remained a mystery, however. Admiral Richard Byrd was determined to change that.

Mapping the Continent

With detailed planning, Byrd launched a major scientific expedition to Antarctica in 1928. He brought 650 tons (590 t) of supplies on his ship—everything from fur clothing and folding bathtubs to 80 sled dogs and a snowmobile! He also brought the latest technology, including radios, cameras, and three airplanes. His team of 53 professionals set up a complete village on the Ross Ice Shelf. They named it Little America.

From this base, Byrd and his crew explored the continent. On short flights from Little America, they used a mapping camera to gather data on vast areas never seen by humans. In 1929 Byrd and three companions were the first to fly over the South Pole. Such feats took great

courage. Twice during flights, the engines failed. Once the pilot was able to restart them; once the plane crashed. Luckily, everyone survived. On a later expedition, huddled alone in a hut for months, Byrd almost died.

Byrd led five Antarctic expeditions and mapped nearly all of the continent. He supervised the completion of five Antarctic stations—Little America I through Little America V. His work paved the way for future researchers.

Today the United States and many other countries have scientific stations on Antarctica. From these stations, scientists record weather data, measure ozone levels, analyze ice and rock samples, and record animal behavior. Such research may answer important questions about future life on the earth.

QUESTIONS

1 What conditions did Byrd and his fellow explorers face in Antarctica?

2 Why is Byrd's work still important today?

Crunching through ice, the U.S.S. *Glacier* brings Admiral Byrd back to Antarctica in 1955. ▶

NATIONAL GEOGRAPHIC

Byrd's First Flight to the South Pole

Polar Plateau

South Pole

A N T A R C T I C A

Byrd's 1st Flight, November 1929

ROSS ICE SHELF

Little America

Bay of Whales

Ross Sea

ANTARCTICA

Ross Ice Shelf

Area enlarged

0 mi. 200

0 km 200

Lambert Azimuthal Equal-Area projection

GB 4

Appendix

Contents

What Is an Appendix and How Do I Use One?

An appendix is the additional material you often find at the end of books. The information below will help you learn how to use the appendix in *Our World Today*.

SKILLS HANDBOOK

The **Standardized Test Skills Handbook** requires you to learn and apply many key skills that you will use throughout the study of geography as well as other subject areas.

NATIONS OF THE WORLD DATA BANK

The **Nations of the World Data Bank** that begins on page 690 lists all of the world's countries and various categories of information for each. For example, each country's type of government, form of currency, and literacy rate—among other topics—are listed in the data bank.

GLOSSARY AND SPANISH GLOSSARY

A **glossary** is a list of important or difficult terms found in a textbook. The glossary gives a definition of each term as it is used in the book. Since words sometimes have other meanings, you may wish to consult a dictionary to find other uses for the term. The glossary also includes page numbers telling you where in the textbook the term is used. The **Spanish glossary** is the English glossary translated into Spanish.

INDEX

An **index** is an alphabetical listing at the end of a book that includes the subjects of that book and the page numbers where those subjects can be found. The index in this book also lets you know that certain pages contain maps, graphs, photos, or paintings about the subject.

ACKNOWLEDGMENTS

This section lists photo credits and/or literary credits for the book. You can look at this section to find out where the publisher obtained the permission to use photographs or to use excerpts from other books.

Test Yourself!

Do you think you can use an appendix quickly and easily? Try it. Find the answers to these questions by using the appendix on the following pages.

1. What does *canopy* mean?
2. Where in the book is the word *canopy* used?
3. What is the Spanish word for *cassava*?
4. What kind of currency does Saudi Arabia use?
5. On what page can you find out about the climate of Norway?
6. On what pages can you find information about Cuba?
7. What is the literacy rate in Nepal?

Appendix User Tip

When using an appendix, be sure to notice and use the guide words at the top of the page. These guide words indicate the alphabetically first and last entries on that page.

Standardized Test

Skills Handbook

Standardized tests are one way educators measure what you have learned. This handbook is designed to help you prepare for standardized tests in social studies. On the pages that follow, you will find a review of the major social studies critical thinking skills that you will need to master to be successful when taking tests.

Standardized Test Skills Practiced in This Handbook:

Interpreting a Map

Interpreting a Political Map

Interpreting Charts

Making Comparisons

Interpreting Primary Sources

Interpreting a Political Cartoon

Interpreting a Circle Graph

Drawing Inferences and Conclusions

Comparing Data

Categorizing and Analyzing Information

Sequencing Events

Author
Tara Musslewhite
Humble I.S.D.
Humble, Texas

Reviewed by

Interpreting a Map

Before 1492 people living in Europe in the Eastern Hemisphere had no idea the continents of North America and South America in the Western Hemisphere existed. That was the year Christopher Columbus first reached the Americas. His voyage of exploration paved the way for other European voyages to the Western Hemisphere. The voyages of the early explorers brought together two worlds. Previously these parts of the globe had no contact with each other. Trade between the hemispheres changed life for people on both sides of the Atlantic Ocean. The trade between the peoples of the Eastern Hemisphere and the Western Hemisphere is referred to as the Columbian Exchange.

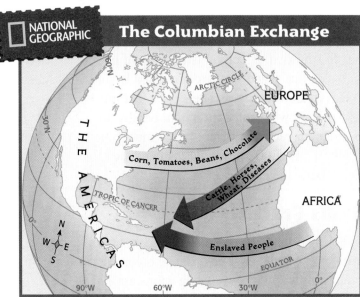

The Columbian Exchange

Skills Practice

Although globes are the best, most accurate way to show places on the round earth, people can more easily use maps to represent places. A map is made by taking data from a round globe and placing it on a flat surface. To read a map, first read the title to determine the subject of the map. Then read the map key or the labels on the map to find out what the colors and symbols on the map mean. Use the compass rose to identify the four cardinal directions of north, south, east, and west. Study the map of the Columbian Exchange and answer the questions that follow on a separate sheet of paper.

1. What is the subject of the map?

2. What do the arrows represent?

3. What continents are shown on the map?

4. What foods did Europeans acquire from the Americas?

5. What did the Americas acquire from Europe?

6. What people were brought from Africa to the Americas?

7. In what direction is Europe from the Americas?

Standardized Test Practice

DIRECTIONS: Use the map and your knowledge of social studies to answer the following question on a separate sheet of paper.

1. Which of the following statements about the Columbian Exchange is true?

   A Food products were traded only between Africa and the Americas.

   B Europeans acquired cattle from the Americas.

   C Europeans introduced corn, tomatoes, and beans to Native Americans.

   D Enslaved Africans were brought to the Americas.

Interpreting a Political Map

By 1750, or the middle of the eighteenth century, there were 13 British colonies in North America. A colony is a group of people living in one place who are governed by rulers in another place. The British colonists in America were ruled by the monarchy and Parliament of Great Britain. That meant that rulers living 3,000 miles away made laws for the American colonists.

Skills Practice

Political maps illustrate divisions between territories such as nations, states, colonies, or other political units. These divisions are called boundaries. Lines represent the boundaries between political areas. To interpret a political map, read the map title to determine what geographic area and time period it covers. Identify the colonies or other political

units on the map. Look at the map key for additional information. Study the map on this page and answer the questions that follow on a separate sheet of paper.

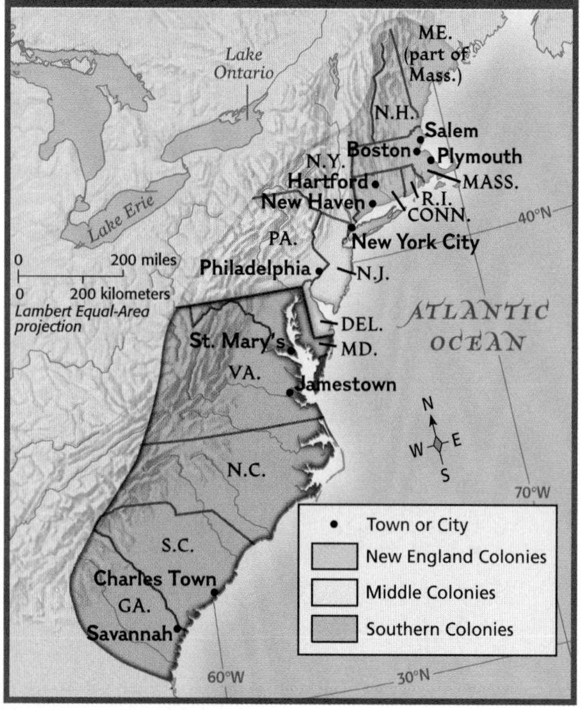

1. List the New England Colonies.

2. Which were the Middle Colonies?

3. Which Middle Colony bordered Pennsylvania to the north?

4. Which was the southernmost early British colony?

5. Name the body of water that formed the eastern border of the colonies.

6. Where was Charles Town located?

The Thirteen Colonies, 1750

NATIONAL GEOGRAPHIC

Map labels: Lake Ontario, ME. (part of Mass.), N.H., Salem, Boston, Plymouth, N.Y., Hartford, MASS., New Haven, R.I., CONN., 40°N, Lake Erie, PA., New York City, 200 miles, Philadelphia, N.J., 200 kilometers, Lambert Equal-Area projection, DEL., ATLANTIC OCEAN, St. Mary's, MD., VA., Jamestown, N, W-E, S, N.C., 70°W, S.C., Charles Town, GA., Savannah, 60°W, 30°N

Map key: Town or City · New England Colonies · Middle Colonies · Southern Colonies

The Princeton Review — Standardized Test Practice

DIRECTIONS: Use the map and your knowledge of social studies to answer the following questions on a separate sheet of paper.

1. The New England Colony that covered the largest land area was

A Virginia.

B Pennsylvania.

C Massachusetts.

D New Hampshire.

2. The northernmost Middle Colony is the present-day state of

F Maryland.

G New York.

H Massachusetts.

J Pennsylvania.

3. The settlement of Plymouth was located

A near Jamestown.

B in Massachusetts.

C in the Southern Colonies.

D in Virginia.

Interpreting Charts

Government is a necessary part of every nation. It gives citizens stability and provides services that many of us take for granted. However, governments can sometimes have too much power.

The United States was founded on the principle of limited government. Limited governments require all people to follow the laws. Even the rulers must obey rules set for the society. A democracy is a form of limited government. Not all forms of government have limits. In unlimited governments, power belongs to the ruler. No laws exist to limit what the ruler may do. A dictatorship is an example of an unlimited government.

Skills Practice

Charts are visual graphics that categorize information. When reading a chart be sure to look at all the headings and labels. Study the charts on this page and answer the questions that follow on a separate sheet of paper.

1. What do the charts compare?

2. Which political systems are forms of limited government?

3. Which form of government often uses military rule?

4. In which political system does the king or queen have complete power?

LIMITED GOVERNMENTS

Representative Democracy	Constitutional Monarchy
People elect leaders to rule	King or queen's power is limited
Individual rights important	Individual rights important
More than one political party	More than one political party
People give consent to be governed	People elect governing body

UNLIMITED GOVERNMENTS

Dictatorship	Absolute Monarchy
One person or small group rules	King or queen inherits power
Little personal freedoms	Usually some freedoms
Rule by force, often military	Officials are appointed by king or queen
Ruler does not have to obey rules	Monarch has complete authority

Standardized Test Practice

DIRECTIONS: Use the charts and your knowledge of social studies to answer the following questions on a separate sheet of paper.

1. Information found in the charts shows that the most restrictive form of government is a
 A dictatorship.
 B representative democracy.
 C absolute monarchy.
 D constitutional monarchy.

2. Under which type of government do citizens have the most power?
 F unlimited government
 G limited government
 H absolute monarchy
 J dictatorship

3. An example of an unlimited government is
 A the United States in the 1960s.
 B Libya in the 1970s.
 C the United Kingdom in the 1980s.
 D Mexico in the 1990s.

Standardized Test Skills Handbook

Making Comparisons

The roots of representative democracy in the United States can be traced back to colonial times. In 1607 English settlers founded the colony of Jamestown in present-day Virginia. As the colony developed, problems arose. Later colonists formed the House of Burgesses to deal with these problems. Citizens of Virginia were chosen as representatives to the House of Burgesses. This became the first legislature, or lawmaking body, in America.

Today citizens of the United States elect representatives to Congress. The major function of Congress is to make laws for the nation. There are two houses, or chambers, of the U.S. Congress. Legislative bodies with two houses are said to be bicameral. The bicameral Congress of the United States includes the Senate and the House of Representatives. Article I of the U.S. Constitution describes how each house will be organized and how its members will be chosen.

Skills Practice

When you make a comparison, you identify and examine two or more groups, situations, events, or documents. Then you identify any similarities and differences between the items. Study the information presented on the chart on this page and answer the questions that follow on a separate sheet of paper.

1. What two things does the chart compare?

2. How are the qualifications for each house of the U.S. Congress similar?

Standardized Test Practice

DIRECTIONS: Use the chart and your knowledge of social studies to answer the following questions on a separate sheet of paper.

1. Which of the following statements best reflects information shown in the chart?

   A The Senate has more members than the House of Representatives.

   B Representatives to the House are elected to two-year terms.

   C House members must be residents of their states for at least 9 years.

   D A state's population determines its number of senators.

2. One inference that can be made from information shown on the chart is that

   F Texas elects more senators than Rhode Island.

   G Texas elects more House members than Rhode Island.

   H Texas elects fewer senators than Rhode Island.

   J Texas elects fewer House members than Rhode Island.

THE U.S. CONGRESS

House of Representatives	Senate
Qualifications: • Must be 25 years old • Must be U.S. citizen for 7+ years • Must live in the state they represent	**Qualifications:** • Must be 30 years old • Must be U.S. citizen for 9+ years • Must live in the state they represent
Number of Representatives: • 435 total representatives; number of representatives per state is based on state population	**Number of Representatives:** • 100 total senators; two senators elected from each state regardless of state population
Terms of Office: • Two-year terms	**Terms of Office:** • Six-year terms

Interpreting Primary Sources

When Thomas Jefferson wrote the Declaration of Independence, he used the term "unalienable rights." Jefferson was referring to the natural rights that belong to humans. He and the other Founders of our nation believed that government could not take away the rights of the people.

Skills Practice

Primary sources are records of events made by the people who witnessed them. A historical document such as the Declaration of Independence is an example of a primary source. Read the passage below and answer the questions that follow on a separate sheet of paper.

> "We hold these truths to be self-evident, that all men are created equal, that they are endowed by their Creator with certain unalienable Rights, that among these are Life, Liberty, and the pursuit of Happiness . . ."
>
> —Declaration of Independence, July 4, 1776

1. What does the document say about the equality of men?

2. List the three natural, or unalienable, rights to which the document refers.

After gaining independence, American leaders wrote the U.S. Constitution in 1787. The Bill of Rights includes the first 10 amendments, or additions, to the Constitution. The First Amendment protects five basic rights of all American citizens. Study the chart on this page and answer the questions that follow.

1. Which right allows Americans to express themselves without fear of punishment by the government?

2. Which right allows people to worship as they please?

3. Which right allows citizens to publish a pamphlet that is critical of the president?

4. What is the Bill of Rights?

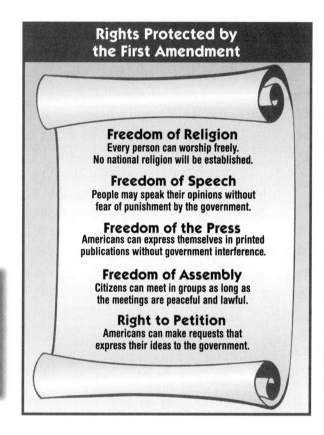

Rights Protected by the First Amendment

Freedom of Religion
Every person can worship freely. No national religion will be established.

Freedom of Speech
People may speak their opinions without fear of punishment by the government.

Freedom of the Press
Americans can express themselves in printed publications without government interference.

Freedom of Assembly
Citizens can meet in groups as long as the meetings are peaceful and lawful.

Right to Petition
Americans can make requests that express their ideas to the government.

Standardized Test Practice

DIRECTIONS: Use the chart and your knowledge of social studies to answer the following question on a separate sheet of paper.

1. Which First Amendment right protects citizens who are staging a protest outside a government building?

A freedom of speech

B freedom of the press

C freedom of assembly

D freedom of religion

Test Practice Lesson 6

Interpreting a Political Cartoon

Just as the government of the United States is limited in its powers, freedoms extended to Americans also have limits. The First Amendment was not intended to allow Americans to do whatever they please without regard to others. Limits on freedoms are necessary to keep order in a society of so many people. The government can establish laws to limit certain rights to protect the health, safety, security, or moral standards of a community. Rights can be restricted to prevent one person's rights from interfering with the rights of another. For example, the freedom of speech does not include allowing a person to make false statements that hurt another's reputation.

Skills Practice

The artists who create political cartoons often use humor to express their opinions on political issues. Sometimes these cartoonists are trying to inform and influence the public about a certain topic. To interpret a political cartoon, look for symbols, labels, and captions that provide clues about the message of the cartoonist. Analyze these elements and draw some conclusions. Study the political cartoon on this page and answer the questions that follow on a separate sheet of paper.

1. What is the subject of the cartoon?

2. What words provide clues as to the meaning of the cartoon?

3. Whom does the person in the cartoon represent?

4. What is the person doing?

5. What do the subject's thoughts suggest about the task faced by those involved in planning the new nation's government?

6. What limits are placed on First Amendment rights? Why are these rights limited?

Standardized Test Practice

DIRECTIONS: Use the political cartoon and your knowledge of social studies to answer the following questions on a separate sheet of paper.

1. The most appropriate title for the cartoon is

   A Limits on Government.

   B Parliament at Work.

   C Limiting Rights.

   D Unlimited Government.

2. The sources of our rights as citizens of the United States come from which of the following?

   F the Declaration of Independence and the U.S. Constitution

   G the will of the president

   H unwritten customs and traditions

   J the United Nations charter

LET'S SEE NOW... WE'LL GIVE THEM FREEDOM, BUT NOT TOO MUCH FREEDOM; LIBERTY, BUT NOT TOO MUCH LIBERTY; RIGHTS, BUT NOT TOO MANY RIGHTS...

P. harris

Interpreting a Circle Graph

"E pluribus unum" is a Latin phrase found on United States coins. It means "Out of many, one." The United States is sometimes called a "nation of immigrants." Unless you are a Native American, your ancestors came to America within the last 500 years.

Groups of people who share a common culture, language, or history are referred to as ethnic groups. American neighborhoods include many different ethnic groups. The circle graph on this page shows the major ethnic groups in the United States.

NATIONAL GEOGRAPHIC — Ethnic Groups in the United States

African American 12.1%
Asian 3.6%
American Indian/Inuit 0.7%
Native Hawaiian/other Pacific Islander 0.1%
Other 1.8%
Hispanic 12.5%
White 69.2%

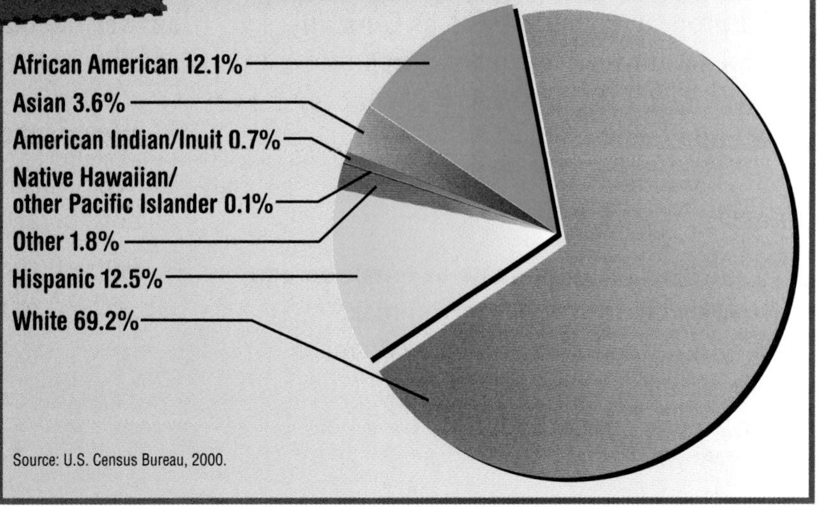

Source: U.S. Census Bureau, 2000.

Skills Practice

A circle graph shows percentages of a total quantity. Each part, or slice, of the graph represents a part of the total quantity. To read a circle graph, first read the title. Then study the labels to find out what each part represents. Compare the sizes of the circle slices. Study the circle graph and answer the questions that follow on a separate sheet of paper.

1. What information does this circle graph present?

2. Which ethnic group includes the largest percentage of Americans?

3. Which groups represent less than 1 percent of the people in the United States?

4. What percentage of the United States population is represented by African Americans?

5. The smallest ethnic groups have lived in the United States the longest. What are these ethnic groups?

The Princeton Review

Standardized Test Practice

DIRECTIONS: Use the graph and your knowledge of social studies to answer the following questions on a separate sheet of paper.

1. Which group's population is about three times greater than the number of Asians and Native Hawaiians/Pacific Islanders?

   A African American

   B White

   C American Indian/Inuit

   D Hispanic

2. How does the Hispanic population compare to the African American population of the United States?

   F It is greater than the African American population.

   G It is the smallest segment of the United States population.

   H It is less than half the size of the African American population.

   J It is slightly less than the African American population.

Drawing Inferences and Conclusions

During the mid-nineteenth century, immigration to the United States increased. People from European countries such as Germany and Ireland traveled to America seeking new opportunities. Life, however, was not easy for these immigrants.

Skills Practice

To infer means to evaluate information and arrive at a conclusion. When you make inferences, you "read between the lines." You must use the available facts and your own knowledge of social studies to form a judgment or opinion about the material.

Line graphs are a way of showing numbers visually. They are often used to compare changes over time. Sometimes a graph has more than one line. The lines show different quantities of a related topic. To analyze a line graph read the title and the information on the horizontal and vertical axes. Use this information to draw conclusions. Study the graph on this page and answer the questions that follow on a separate sheet of paper.

1. What is the subject of the line graph?

2. What information is shown on the horizontal axis?

3. What information is shown on the vertical axis?

4. Why do you think these immigrants came to the United States?

Standardized Test Practice

DIRECTIONS: Use the line graph and your knowledge of social studies to answer the following questions on a separate sheet of paper.

1. The country that provided the most immigrants to the United States between the years 1820 and 1860 was

   A Great Britain.

   B Ireland.

   C Germany.

   D France.

2. In about what year did the number of German immigrants to the United States reach a peak?

   F 1845

   G 1852

   H 1855

   J 1860

3. Irish migration to the United States increased in the mid-1800s because of

   A a terrible potato famine in Ireland.

   B the failure of a German revolution in 1848.

   C the nativist movement.

   D the availability of low-paying factory jobs.

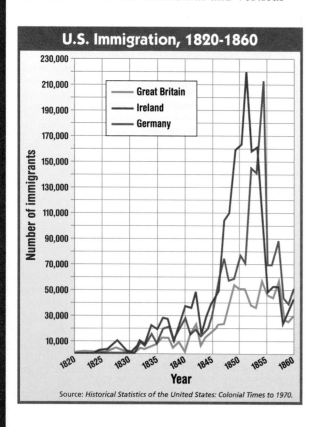

U.S. Immigration, 1820-1860

Legend: Great Britain, Ireland, Germany

Number of immigrants (vertical axis): 10,000 – 230,000

Year (horizontal axis): 1820–1860

Source: *Historical Statistics of the United States: Colonial Times to 1970.*

Comparing Data

The world's earliest civilizations developed more than 6,000 years ago. The discovery of farming led to the rise of ancient cities in Mesopotamia and the Nile River valley. These early cities shared one important characteristic—they each arose near waterways. Since water was the easiest way to transport goods, the settlements became centers of trade.

Since then cities have grown all over the world. Every 10 years, the United States Census Bureau collects data to determine the population of the United States. (A census is an official count of people living in an area.) The first census was conducted in 1790. At that time, there were 3.9 million people in the 13 original states. The most recent census occurred in 2000. The results of that census showed that more than 280 million people reside in the 50 states that make up our nation.

Skills Practice

The charts on this page show populations of the five most populous cities in the United States during different time periods. When comparing information on charts be sure to read the titles and headings to define the data being compared. Study the charts and answer the questions below on a separate sheet of paper.

1. Which U.S. city had the greatest population in 1790?

2. Which U.S. city had the greatest population in 2000?

3. What was the population of Philadelphia in 1790?

4. What was Philadelphia's population in 2000?

5. Which city had the third-largest population in 1790?

6. Which cities are on both lists?

| POPULATION OF FIVE LARGEST U.S. CITIES, 1790 ||
City	Number of People
New York City	33,131
Philadelphia	28,522
Boston	18,320
Charleston	16,359
Baltimore	13,503

| POPULATION OF FIVE LARGEST U.S. CITIES, 2000* ||
City	Number of People
New York City	8,008,278
Los Angeles	3,694,820
Chicago	2,896,016
Houston	1,953,631
Philadelphia	1,517,550

*Numbers do not include metropolitan areas.

Standardized Test Practice

DIRECTIONS: Use the charts and your knowledge of social studies to answer the following questions on a separate sheet of paper.

1. One inference that can be made from the charts is that the most populous cities in the United States

A have good weather.

B were founded early in our nation's history.

C are port cities.

D are in the eastern United States.

2. In 1790 the major cities of the United States were all

F larger than 20,000 people.

G located in the East.

H Northern cities.

J founded for religious reasons.

Test Practice Lesson 10

Categorizing and Analyzing Information

Economic systems describe the ways in which societies produce and distribute goods and services. Early societies, such as Mesopotamia, used bartering as their system of trade. In the seventeenth and eighteenth centuries, European countries practiced mercantilism in which colonies provided wealth to their parent countries. Great Britain used this idea to gain wealth from its North American colonies. The economy of the United States is based on the principle of free enterprise. Americans have the freedom to own businesses with limited interference from the government.

Because Americans are employed in a variety of industries, our economy is the largest and among the most diverse in the world. The U.S. economy includes the following parts:

- Manufacturing makes up 22 percent of the economy.
- Agriculture makes up 2 percent of the economy.
- Mining makes up 1 percent of the economy.
- Service and information industries make up 75 percent of the economy.

Skills Practice

Grouping information into categories is one way of making the information easier to understand. The economic systems of today's world can be classified into four basic groups. Study the chart on this page and answer the questions that follow on a separate sheet of paper.

1. Under which economic system does the government have the most control?

2. Under which system would people be most likely to have the same job as their parents?

3. Use the information about the U.S. economy on this page to create a circle graph. Then answer this question: Industries related to farming represent what percentage of the U.S. economy?

WORLD ECONOMIC SYSTEMS			
Traditional	**Command**	**Market**	**Mixed**
Based on customs	Government controls production, prices, and wages	Individuals control production, prices, and wages	Individuals control some aspects of economy
Trades are passed down through generations	Communism; Government owns businesses	Free enterprise; Individuals own businesses	Government regulates selected industries and restricts others

Standardized Test Practice

DIRECTIONS: Use the chart and your knowledge of social studies to answer the following questions on a separate sheet of paper.

1. Which economic system provides individuals with the most economic freedom?

A traditional

B command

C market

D mixed

2. The United States has this type of economic system.

F traditional

G command

H market

J mixed

Sequencing Events

The free enterprise economic system of the United States has encouraged Americans to invent and produce new technology throughout the history of our nation. Using its rich natural and human resources, Americans are continually advancing the economy through technology. The tremendous economic growth of the United States at certain times in history, such as after the Civil War, resulted from foundations laid early in the nation's history and affects the growth of the United States economy today.

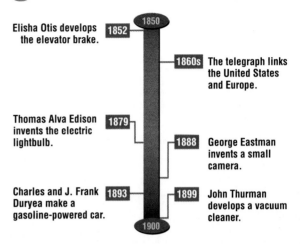

Skills Practice

Sequencing information involves placing facts in the order in which they occurred. Listed below are technological advances that occurred at different times in history, transforming the world economy. Find the date of each invention by studying the time line on page 23 of your textbook. On a separate sheet of paper take notes by writing the date beside a brief description of each invention. Then sequence the events by rewriting them in the correct order.

• Telephone	• Airplane
• Cellular phone	• Automobile
• Steamboat	• Steam locomotive
• Space shuttle	• Internet
• Radio	• Television

In the late 1800s, innovations in technology and new business combinations helped the United States grow into an industrial power. By the year 1900, the United States's industrial production was the greatest around the world.

Read the time line on this page. Determine the subject of the time line and summarize it in a few words. Then write a title for the time line on a separate sheet of paper.

The Princeton Review

Standardized Test Practice

DIRECTIONS: Use the events you have sequenced and your knowledge of social studies to answer the following questions on a separate sheet of paper.

1. Which of the following inventions occurred last?

   A steam locomotive

   B telephone

   C airplane

   D steamboat

2. During the late 1700s, 1800s, and early 1900s, new inventions and developments in the area of transportation, such as the steamboat, steam locomotive, and airplane, resulted in

   F increased poverty among urban Americans.

   G an end to westward migration.

   H the creation of new markets for trade.

   J rural growth.

3. In the early 1800s steamboats dramatically improved the transport of goods and passengers

   A along major roads.

   B along major inland rivers.

   C between the Americas and Africa.

   D in the West.

Nations of the World
DATABANK

Today we are learning to understand the global, connected world in which we live. Technological advances have made global communication and interaction much easier. We must interact with many different countries—more than at any other time in the world. Every country of the world has its own identity, culture, economy and government. This information about each country has been charted here for you. Using the chart will make it easier for you to compare and contrast the information.

	GOVERNMENT		ECONOMICS			SOCIAL & CULTURAL		
COUNTRY	Type of Government	Date Founded	*GNP Ranking	GNP Per Capita	Currency	Literacy	**Infant Mortality	Primary Religion
Afghanistan	Transitional Government	2001	101st	$270	Afghani	32%	147	Muslim
Albania	Republic	1991	131st	$760	Leke	93%	40	Muslim, Catholic
Algeria	Republic	1962	52nd	$1,500	Dinar	62%	41	Muslim
Andorra	Parliamentary Democracy	1993	155th	$15,600	Euro	99%	4	Catholic
Angola	Republic	1975	126th	$260	Kwanzaa	42%	194	Indigenous, Catholic, Protestant
Antarctica	n/a	n/a	n/a	n/a	n/a	100%	n/a	n/a
Antigua & Barbuda	Parliamentary Democracy	1981	166th	$7,380	E.C. Dollar	89%	22	Protestant, Catholic
Argentina	Republic	1816	17th	$8,950	Argentine Peso	96.5%	18	Catholic
Armenia	Republic	1991	137th	$560	Dram	99%	41	Orthodox
Australia	Parliamentary Democracy	1901	14th	$20,650	Australian Dollar	100%	5	Catholic, Protestant

*Gross National Product

**deaths/1,000 live births

COUNTRY	Type of Government	Date Founded	*GNP Ranking	GNP Per Capita	Currency	Literacy	**Infant Mortality	Primary Religion
Austria	Federal Republic	1918	22nd	$27,920	Euro	98%	5	Catholic
Azerbaijan	Republic	1991	118th	$510	Manat	97%	83	Muslim
Bahamas	Parliamentary Democracy	1973	124th	$11,940	Bahamian Dollar	96%	19	Protestant, Catholic
Bahrain	Monarchy	1971	104th	$7,800	Bahrain Dinar	86%	20	Muslim
Bangladesh	Republic	1971	51st	$360	Taka	56%	70	Muslim, Hindu
Barbados	Parliamentary Democracy	1966	145th	$6,560	Barbados Dollar	98%	12	Protestant
Belarus	Republic	1991	61st	$2,150	Belarus Rouble	98%	14	Eastern Orthodox
Belgium	Constitutional Monarchy	1830	19th	$26,730	Euro	98%	5	Catholic, Protestant
Belize	Parliamentary Democracy	1981	162nd	$2,670	Belizean Dollar	70%	25	Catholic, Protestant
Benin	Republic	1960	133rd	$380	CFA Franc	38%	90	Indigenous, Christian, Muslim
Bhutan	Monarchy	1907	172nd	$430	Ngultrum	42%	109	Buddhist, Hindu
Bolivia	Republic	1825	91st	$970	Boliviano	84%	59	Catholic
Bosnia & Herzegovina	Republic	1992	160th	$288	Maraka	93%	25	Muslim, Catholic, Orthodox
Botswana	Republic	1966	105th	$3,310	Pula	70%	64	Indigenous, Christian
Brazil	Federal Republic	1889	8th	$4,790	Real	84%	37	Catholic
Brunei	Monarchy	1984	116th	$14,240	Brunei Dollar	88%	14	Muslim, Buddhist
Bulgaria	Republic	1991	82nd	$1,170	Lev	98%	15	Orthodox, Muslim
Burkina Faso	Republic	1960	130th	$250	CFA Franc	20%	107	Traditional, Muslim
Burundi	Republic	1966	157th	$140	Burundi Franc	35%	71	Indigenous, Catholic
Cambodia	Constitutional Monarchy	1953	125th	$300	Riel	35%	65	Buddhist
Cameroon	Republic	1960	86th	$620	CFA Franc	64%	70	Indigenous, Christian, Muslim

*Gross National Product

**deaths/1,000 live births

 **GOVERNMENT**

 **ECONOMICS**

 **SOCIAL & CULTURAL**

COUNTRY	Type of Government	Date Founded	*GNP Ranking	GNP Per Capita	Currency	Literacy	**Infant Mortality	Primary Religion
Canada	Parliamentary Democracy	1867	9th	$19,640	Canadian Dollar	97%	5	Catholic, Protestant
Cape Verde	Republic	1975	169th	$1,090	C.V. Escudo	72%	53	Catholic, Protestant
Central African Republic	Republic	1960	152nd	$320	CFA Franc	60%	105	Indigenous, Catholic, Protestant, Muslim
Chad	Republic	1960	147th	$230	CFA Franc	48%	95	Muslim, Christian, Indigenous
Chile	Republic	1823	43rd	$4,820	Chilean Peso	95%	9	Catholic, Protestant
China	Communist State	1949	7th	$860	Yuan	82%	28	Atheist, Buddhist, Daoist, Confucian
Colombia	Republic	1819	39th	$2,180	Col. Peso	91%	24	Catholic
Comoros	Republic	1975	181st	$400	Comoros Franc	58%	84	Catholic, Muslim
Congo, Democratic Republic of the	Dictatorship	1960	103rd	$110	Congolese Franc	77%	100	Indigenous, Catholic, Protestant
Congo, Republic of the	Republic	1960	144th	$670	CFA Franc	75%	100	Catholic, Protestant, Indigenous
Costa Rica	Republic	1838	85th	$2,680	Colon	95%	11	Catholic
Côte d'Ivoire	Republic	1960	81st	$710	CFA Franc	49%	94	Indigenous, Muslim, Christian
Croatia	Republic	1991	69th	$4,060	Kuna	97%	7	Catholic, Orthodox
Cuba	Communist State	1959	72nd	$1,650	Cuban Peso	96%	7	Catholic
Cyprus	Republic	1960	92nd	$9,400	Cyprus Pound	94%	8	Orthodox, Muslim
Czech Republic	Republic	1993	48th	$5,240	Koruny	100%	6	Catholic, Protestant
Denmark	Constitutional Monarchy	1849	25th	$38,890	Kroner	100%	5	Protestant
Djibouti	Republic	1977	167th	$750	Djib. Franc	46%	102	Muslim
Dominica	Republic	1978	179th	$3,040	E. Car. Dollar	94%	17	Catholic, Protestant
Dominican Republic	Republic	1865	76th	$1,750	Dom. Rep. Peso	82%	35	Catholic
East Timor	Republic	2002	n/a	n/a	U.S. Dollar	50%	n/a	Catholic

*Gross National Product   **deaths/1,000 live births

Nations of the World Data Bank

COUNTRY	Type of Government	Date Founded	*GNP Ranking	GNP Per Capita	Currency	Literacy	**Infant Mortality	Primary Religion
Ecuador	Republic	1830	71st	$1,570	Sucre	90%	34	Catholic
Egypt	Republic	1953	42nd	$1,200	Egyp. Pound	52%	60	Muslim
El Salvador	Republic	1841	78th	$1,810	Colon	72%	29	Catholic, Protestant
Equatorial Guinea	Republic	1968	168th	$1,060	CFA Franc	79%	93	Catholic
Eritrea	Republic	1993	158th	$230	Nafka	25%	75	Christian, Muslim
Estonia	Republic	1991	107th	$3,360	Kroon	100%	12	Protestant
Ethiopia	Republic	1995	99th	$110	Birr	36%	100	Muslim, Orthodox, Indigenous
Fiji Islands	Republic	1987	139th	$2,460	Fiji Dollar	92%	14	Christian, Hindu
Finland	Republic	1917	31st	$24,790	Markka	100%	4	Protestant
France	Republic	1958	4th	$26,300	Euro	99%	5	Catholic
Gabon	Republic	1960	110th	$4,120	CFA Franc	63%	95	Christian
Gambia	Republic	1970	170th	$340	Dalasi	48%	78	Muslim
Georgia	Republic	1991	111th	$860	Lari	99%	52	Orthodox, Muslim
Germany	Federal Republic	1949	3rd	$28,280	Euro	99%	5	Protestant, Catholic
Ghana	Republic	1960	95th	$390	Cedis	65%	57	Christian, Muslim, Indigenous
Greece	Republic	1975	32nd	$11,640	Euro	95%	6	Orthodox
Grenada	Parliamentary Democracy	1974	174th	$3,140	E. Car. Dollar	98%	15	Catholic, Protestant
Guatemala	Republic	1838	74th	$1,580	Quetzal	64%	46	Catholic, Protestant
Guinea	Republic	1958	119th	$550	Guinea Franc	36%	129	Muslim
Guinea-Bissau	Republic	1974	176th	$230	Guinea Peso	54%	110	Indigenous, Muslim
Guyana	Republic	1970	162nd	$800	Guy. Dollar	98%	39	Christian, Hindu

*Gross National Product   **deaths/1,000 live births

COUNTRY	GOVERNMENT		ECONOMICS			SOCIAL & CULTURAL		
	Type of Government	Date Founded	*GNP Ranking	GNP Per Capita	Currency	Literacy	**Infant Mortality	Primary Religion
Haiti	Republic	1804	128th	$380	Gourde	45%	95	Catholic, Protestant
Honduras	Republic	1838	113th	$740	Lempira	73%	31	Catholic
Hungary	Republic	1989	50th	$4,510	Forint	99%	9	Catholic, Protestant
Iceland	Republic	1944	94th	$26,580	Ic. Krona	100%	4	Protestant
India	Federal Republic	1950	15th	$370	Ind. Rupee	52%	63	Hindu, Muslim
Indonesia	Republic	1949	23rd	$1,110	Rupiah	84%	41	Muslim
Iran	Islamic Republic	1979	34th	$1,780	Ir. Rial	72%	29	Muslim
Iraq	Dictatorship	1958	65th	$950	Ir. Dinar	58%	60	Muslim
Ireland	Republic	1949	44th	$17,790	Punt	98%	6	Catholic
Israel	Republic	1948	37th	$16,180	Shekel	95%	8	Jewish, Muslim
Italy	Republic	1946	6th	$20,170	Euro	98%	6	Catholic
Jamaica	Parliamentary Democracy	1962	117th	$1,550	Jamaican Dollar	85%	14	Protestant
Japan	Constitutional Monarchy	1947	2nd	$38,160	Yen	99%	4	Shinto, Buddhist
Jordan	Constitutional Monarchy	1946	96th	$1,520	Jordanian Dinar	87%	20	Muslim
Kazakhstan	Republic	1991	62nd	$1,350	Tenge	98%	59	Muslim, Orthodox
Kenya	Republic	1964	83rd	$340	Kenya Shilling	78%	68	Protestant, Catholic, Indigenous
Kiribati	Republic	1979	188th	$910	Australian Dollar	98%	54	Catholic, Protestant
Korea, North	Communist State	1948	64th	$1,390	Won	99%	24	Atheist, Buddhist, Confucian
Korea, South	Republic	1948	11th	$10,550	Won	98%	8	Buddhist, Christian
Kuwait	Monarchy	1961	58th	$17,390	Kuwaiti Dollar	79%	12	Muslim
Kyrgyzstan	Republic	1991	134th	$480	Som	97%	77	Muslim, Orthodox

*Gross National Product

**deaths/1,000 live births

 **GOVERNMENT**

 **ECONOMICS**

 **SOCIAL & CULTURAL**

COUNTRY	Type of Government	Date Founded	*GNP Ranking	GNP Per Capita	Currency	Literacy	**Infant Mortality	Primary Religion
Laos	Communist State	1975	142nd	$400	New Kip	57%	93	Buddhist
Latvia	Parliamentary Democracy	1991	100th	$2,430	Lati	100%	15	Protestant, Catholic, Orthodox
Lebanon	Republic	1944	77th	$3,350	Leb. Pound	86%	28	Muslim, Christian
Lesotho	Constitutional Monarchy	1966	150th	$680	Maloti	83%	83	Christian, Indigenous
Liberia	Republic	1847	154th	$330	Liberian Dollar	38%	132	Indigenous, Muslim, Christian
Libya	Dictatorship	1969	57th	$5,220	Libyan Dinar	76%	29	Muslim
Liechtenstein	Constitutional Monarchy	1719	151st	$40,000	Swiss Franc	100%	5	Catholic
Lithuania	Republic	1991	88th	$2,260	Lita	98%	15	Catholic
Luxembourg	Constitutional Monarchy	1868	70th	$45,360	Euro	100%	5	Catholic
Macedonia	Republic	1991	135th	$1,100	Mac. Denar	94%	13	Orthodox, Muslim
Madagascar	Republic	1960	120th	$250	Malagasy Franc	80%	84	Indigenous, Christian
Malawi	Republic	1966	136th	$210	Kwacha	58%	121	Protestant, Catholic, Muslim
Malaysia	Constitutional Monarchy	1963	36th	$4,530	Ringgit	84%	20	Muslim, Buddhist, Confucian, Daoist
Maldives	Republic	1965	173rd	$1,180	Rufiyaa	93%	64	Muslim
Mali	Republic	1960	129th	$260	CFA Franc	31%	121	Muslim
Malta	Republic	1974	122nd	$9,330	Maltese Lira	89%	6	Catholic
Marshall Islands	Republic	1986	185th	$1,610	U.S. Dollar	93%	40	Protestant
Mauritania	Republic	1960	153rd	$440	Ouguiya	47%	77	Muslim
Mauritius	Republic	1992	112th	$3,870	Mauritian Rupee	83%	17	Hindu, Catholic, Muslim
Mexico	Federal Republic	1823	16th	$3,700	Mexican Peso	90%	25	Catholic
Micronesia	Republic	1986	180th	$1,920	U.S. Dollar	89%	30	Catholic, Protestant

*Gross National Product   **deaths/1,000 live births

Nations of the World DATABANK

GOVERNMENT

ECONOMICS

SOCIAL & CULTURAL

COUNTRY	Type of Government	Date Founded	*GNP Ranking	GNP Per Capita	Currency	Literacy	**Infant Mortality	Primary Religion
Moldova	Republic	1991	140th	$460	Leu	96%	43	Orthodox
Monaco	Constitutional Monarchy	1911	106th	$11,000	Euro	99%	6	Catholic
Mongolia	Republic	1992	156th	$390	Tughrik	97%	54	Buddhist
Morocco	Constitutional Monarchy	1956	54th	$1,260	Dirham	44%	48	Muslim
Mozambique	Republic	1975	132nd	$140	Metical	42%	139	Indigenous, Christian, Muslim
Myanmar	Military Regime	1948	41st	$1,500	Kyat	83%	74	Buddhist
Namibia	Republic	1990	123rd	$2,110	Namibian Dollar	38%	72	Christian, Indigenous
Nauru	Republic	1968	187th	$7,270	Aus. Dollar	99%	11	Protestant, Catholic
Nepal	Constitutional Monarchy	1990	108th	$220	Nepalese Rupee	28%	74	Hindu
Netherlands	Constitutional Monarchy	1815	12th	$25,830	Euro	99%	5	Catholic, Protestant
New Zealand	Parliamentary Democracy	1907	47th	$15,830	N. Z. Dollar	99%	6	Protestant, Catholic
Nicaragua	Republic	1838	143rd	$410	Cordoba Gold	66%	34	Catholic
Niger	Republic	1960	141st	$200	CFA Franc	14%	124	Muslim
Nigeria	Republic	1963	55th	$280	Naira	57%	73	Muslim, Christian, Indigenous
Norway	Constitutional Monarchy	1905	27th	$36,100	Norwegian Krone	100%	4	Protestant
Oman	Monarchy	1970	79th	$4,820	Omani Rial	80%	23	Muslim
Pakistan	Republic	1956	45th	$500	Pakistani Repee	43%	80	Muslim
Palau	Republic	1994	186th	$5,000	U.S. Dollar	92%	17	Christian, Indigenous
Panama	Republic	1903	92nd	$2,670	Balboa	91%	20	Catholic, Protestant
Papua New Guinea	Parliamentary Democracy	1975	115th	$930	Kina	72%	58	Indigenous, Catholic, Protestant
Paraguay	Republic	1811	80th	$200	Guarani	92%	30	Catholic, Protestant

*Gross National Product   **deaths/1,000 live births

COUNTRY	Type of Government	Date Founded	*GNP Ranking	GNP Per Capita	Currency	Literacy	**Infant Mortality	Primary Religion
Peru	Republic	1824	46th	$2,610	New Sol	89%	40	Catholic
Philippines	Republic	1946	38th	$1,200	Ph. Peso	95%	29	Catholic
Poland	Republic	1990	29th	$3,590	Zloty	99%	10	Catholic
Portugal	Republic	1910	33rd	$11,010	Euro	88%	6	Catholic
Qatar	Monarchy	1971	93rd	$11,600	Qatar Rial	79%	22	Muslim
Romania	Republic	1991	56th	$1,410	Lei	97%	19	Orthodox
Russia	Federation	1991	13th	$2,680	Rouble	98%	20	Orthodox
Rwanda	Republic	1962	146th	$210	R. Franc	48%	119	Catholic, Protestant, Indigenous
St. Kitts & Nevis	Parliamentary Democracy	1983	177th	$6,260	E. Car. Dollar	97%	16	Protestant, Catholic
St. Lucia	Parliamentary Democracy	1979	163rd	$3,510	E. Car. Dollar	67%	15	Catholic, Protestant
St. Vincent & the Grenadines	Parliamentary Democracy	1979	175th	$2,420	E. Car. Dollar	96%	17	Protestant, Catholic
Samoa	Constitutional Monarchy	1962	182nd	$1,140	Tala	97%	31	Christian
San Marino	Republic	1291	183rd	$7,830	Euro	96%	6	Catholic
São Tomé & Príncipe	Republic	1975	189th	$290	Dobra	73%	49	Catholic, Protestant
Saudi Arabia	Monarchy	1932	187th	$7,150	Saudi Riyal	63%	51	Muslim
Senegal	Republic	1960	109th	$540	CFA Franc	33%	57	Muslim
Serbia and Montenegro	Republic	2002	84th	$900	Dinar, Euro	93%	17	Orthodox, Muslim
Seychelles	Republic	1976	165th	$6,910	S. Rupee	58%	17	Catholic
Sierra Leone	Republic	1971	161st	$160	Leone	32%	146	Muslim, Indigenous, Christian
Singapore	Republic	1965	35th	$32,810	Singapore Dollar	94%	4	Buddhist, Muslim
Slovakia	Republic	1993	66th	$3,680	Koruna	99%	9	Catholic, Protestant

*Gross National Product   **deaths/1,000 live births

GOVERNMENT

ECONOMICS

SOCIAL & CULTURAL

COUNTRY	Type of Government	Date Founded	*GNP Ranking	GNP Per Capita	Currency	Literacy	**Infant Mortality	Primary Religion
Slovenia	Republic	1991	67th	$9,840	Tolar	99%	5	Catholic
Solomon Islands	Parliamentary Democracy	1978	171st	$870	Solomon Is.Dollar	62%	25	Catholic, Protestant
Somalia	Republic	1960	159th	$100	Somali Shilling	24%	124	Muslim
South Africa	Republic	1961	30th	$3,210	Rand	82%	60	Christian, Indigenous
Spain	Constitutional Monarchy	1978	10th	$14,490	Euro	97%	5	Catholic
Sri Lanka	Republic	1972	75th	$800	Sri Lanka Rupee	90%	16	Buddhist, Hindu
Sudan	Republic	1956	90th	$290	Sudanese Dinar	46%	69	Muslim, Indigenous
Suriname	Republic	1975	164th	$1,320	S. Guilder	93%	24	Hindu, Muslim, Catholic, Protestant
Swaziland	Monarchy	1968	149th	$1,520	Lilangeni	77%	109	Christian, Indigenous
Sweden	Constitutional Monarchy	1809	21st	$26,210	Swedish Krona	99%	4	Protestant
Switzerland	Federal Republic	1848	18th	$26,210	Swiss Franc	99%	5	Catholic, Protestant
Syria	Republic	1946	73rd	$1,120	Syrian Pound	71%	34	Muslim, Christian
Taiwan	Republic	1949	20th	$10,320	Taiwanese Dollar	86%	7	Buddhist, Confucian, Daoist
Tajikistan	Republic	1991	138th	$330	Taj. Rouble	98%	116	Muslim
Tanzania	Republic	1964	97th	$210	T. Shilling	68%	79	Muslim, Christian
Thailand	Constitutional Monarchy	1932	26th	$2,740	Baht	94%	31	Buddhist
Togo	Republic	1960	148th	$340	CFA Franc	52%	70	Indigenous, Muslim, Christian
Tonga	Constitutional Monarchy	1970	184th	$1,810	Pa'anga	99%	14	Protestant, Catholic
Trinidad & Tobago	Republic	1976	102nd	$4,250	T&T Dollar	98%	25	Catholic, Protestant, Hindu
Tunisia	Republic	1956	68th	$2,110	T. Dinar	67%	29	Muslim
Turkey	Republic	1923	24th	$3,130	Turkish Lira	85%	47	Muslim

*Gross National Product

**deaths/1,000 live births

GOVERNMENT

ECONOMICS

SOCIAL & CULTURAL

COUNTRY	Type of Government	Date Founded	*GNP Ranking	GNP Per Capita	Currency	Literacy	**Infant Mortality	Primary Religion
Turkmenistan	Republic	1991	127th	$640	Manat	98%	73	Muslim, Orthodox
Tuvalu	Parliamentary Democracy	1978	190th	$330	Australian Dollar	95%	23	Protestant
Uganda	Republic	1963	98th	$330	New Uga. Shilling	62%	91	Catholic, Protestant, Muslim, Indigenous
Ukraine	Republic	1991	49th	$1,335	Hryvna	98%	21	Orthodox
United Arab Emirates	Federation	1971	53rd	$17,400	UA.E. Dirham	79%	17	Muslim
United Kingdom	Constitutional Monarchy	1707	5th	$20,870	Pound Sterling	99%	6	Protestant, Catholic
United States	Federal Republic	1776	1st	$29,080	U.S. Dollar	97%	7	Protestant, Catholic
Uruguay	Republic	1828	63rd	$6,130	U. Peso	97%	15	Catholic
Uzbekistan	Republic	1991	59th	1,020	Som	99%	72	Muslim
Vanuatu	Republic	1980	178th	$1,340	Vatu	53%	61	Protestant, Catholic
Vatican City	Religious State	1929	n/a	n/a	Euro	n/a	n/a	Catholic
Venezuela	Republic	1821	40th	$3,480	Bolivar	92%	25	Catholic
Vietnam	Communist State	1954	60th	$310	Dong	94%	30	Buddhist
Yemen	Republic	1990	114th	$270	Rial, Dinar	38%	69	Muslim
Zambia	Republic	1964	121st	$370	Zambian Kwacha	78%	90	Christian, Muslim, Hindu
Zimbabwe	Republic	1980	89th	$720	Zimbabwe Dollar	85%	63	Christian, Indigenous

*Gross National Product   **deaths/1,000 live births

Honoring America

For Americans, the flag has always had a special meaning. It is a symbol of our nation's freedom and democracy.

Flag Etiquette

Over the years, Americans have developed rules and customs concerning the use and display of the flag. One of the most important things every American should remember is to treat the flag with respect.

- The flag should be raised and lowered by hand and displayed only from sunrise to sunset. On special occasions, the flag may be displayed at night, but it should be illuminated.

- The flag may be displayed on all days, weather permitting, particularly on national and state holidays and on historic and special occasions.

- No flag may be flown above the American flag or to the right of it at the same height.

- The flag should never touch the ground or floor beneath it.

- The flag may be flown at half-staff by order of the president, usually to mourn the death of a public official.

- The flag may be flown upside down only to signal distress.

- When the flag becomes old and tattered, it should be destroyed by burning. According to an approved custom, the Union (stars on blue field) is first cut from the flag; then the two pieces, which no longer form a flag, are burned.

★ ★ ★ ★ ★ ★ ★ ★

The Star-Spangled Banner

O! say can you see, by the dawn's early light,
What so proudly we hail'd at the twilight's last gleaming,
Whose broad stripes and bright stars through the perilous fight,
O'er the ramparts we watched, were so gallantly streaming?
And the Rockets' red glare, the Bombs bursting in air,
Gave proof through the night that our Flag was still there;
O! say, does that star-spangled banner yet wave
O'er the Land of the free and the home of the brave!

The Pledge of Allegiance

I pledge allegiance to the Flag of the United States of America and to the Republic for which it stands, one Nation under God, indivisible, with liberty and justice for all.

GLOSSARY

absolute monarchy   form of government in which the king or queen governs with complete power (p. 31)

acid rain   rain containing high amounts of chemical pollutants (pp. 65, 308)

adobe   sun-dried clay bricks (p. 546)

airlift   system of carrying supplies by aircraft (p. 263)

alliance   political agreement between countries to support each other in disputes with other countries (p. 258)

alluvial plain   area that is built up by rich fertile soil left by river floods (p. 122)

altiplano   large highland plateau (p. 621)

altitude   height above sea level (pp. 537, 593)

amend   to change the basic laws of a government (p. 520)

apartheid   system of laws that separated racial and ethnic groups and limited the rights of blacks in South Africa (p. 456)

apprentice   young worker who learned a trade or skill from a master teacher (p. 244)

aquifer   underground rock layer that stores large amounts of water (pp. 60, 101)

archipelago   group of islands (p. 192)

atoll   low-lying, ring-shaped island that surrounds a lagoon (p. 662)

autobahn   German, Swiss, or Austrian superhighway (p. 289)

autonomy   right of self-government (p. 494)

axis   horizontal (bottom) or vertical (side) line of measurement on a graph (p. 10)

bar graph   graph in which vertical or horizontal bars represent quantities (p. 10)

basin   low area surrounded by higher land (p. 582)

bauxite   mineral used to make aluminum (pp. 309, 444, 564)

Bedouin   nomadic Arab of the deserts of Southwest Asia (p. 111)

bilingual   having or expressed in two languages (p. 493)

bishop   official of the Christian Church (p. 241)

blockade   to forcibly prevent entry to an area (p. 263)

Boer   South African of Dutch ancestry (p. 455)

bog   low swampy land (p. 280)

boomerang   Australian weapon shaped like a bent wing that either strikes a target or curves back toward the hunter (p. 640)

bush   rural areas of Australia (p. 641)

cacao   tropical tree whose seeds are used to make chocolate and cocoa (pp. 436, 660)

calligraphy   art of beautiful writing (p. 172)

campesino   Latin American farmer or farm laborer (p. 607)

canopy   umbrella-like covering formed by the tops of trees in a rain forest (pp. 423, 558)

canyon   deep, narrow valley with steep-sides, which a river has cut through a plateau (p. 59)

capitalism   economic system that allows private ownership and open competition of businesses (p. 265)

cash crop   crop grown to be sold for export (p. 606)

cassava   plant with roots that can be ground to make porridge (p. 405)

caste   social class based on a person's ancestry (p. 147)

caudillo   military ruler (p. 593)

chart   graphic way of presenting information clearly (p. 11)

charter   written agreement guaranteeing privileges and freedoms (p. 244)

circle graph   round or pie-shaped graph showing how a whole is divided (p. 11)

city-state   city and its surrounding countryside (p. 83)

civilization   culture that includes certain elements such as a system of writing, cities, and workers with specialized jobs (p. 82)

civil war   fight between different groups within a country (pp. 212, 438)

clan   family, or group of related people (pp. 195, 412)

classical   relating to the ancient Greek and Roman world (p. 236)

climate   usual, predictable pattern of weather in an area over a long period of time (p. 60)

climograph   combination bar and line graph giving information about temperature and precipitation (p. 12)

coalition government   government in which two or more political parties work together to run a country (p. 298)

Cold War   conflict between the United States and the Soviet Union dating from the late 1940s to the late 1980s, when the two superpowers competed for world influence without declared military action at each other (pp. 261, 370)

colony   overseas territory or settlement tied to a parent country (pp. 492, 543)

Columbian Exchange   process in which people, diseases, ideas, and trade were distributed around the world from the Americas (p. 247)

commercial crop   crop grown for the purpose of selling (p. 564)

common law   unwritten set of laws based on local customs (p. 242)

commonwealth   partly self-governing territory (p. 567)

communism   economic, social, and political system based on the teachings of Karl Marx, which advocated the elimination of private property (pp. 258, 346)

communist state   country whose government allows little or no private ownership of property and has strong control over the economy and society as a whole (pp. 165, 369, 566)

compound   group of houses surrounded by walls (p. 437)

conservation   careful use of resources so they are not wasted (p. 64)

constitution   formal agreement that establishes the basis for a country's laws (p. 248)

constitutional monarchy   government in which a king or queen is the official head of state, but elected officials run the government (pp. 31, 104, 196, 280)

consul   elected chief official of the Roman Republic (p. 238)

consumer goods   goods people buy to use for themselves, such as household products, clothing, and cars (pp. 165, 312, 344)

contiguous   areas that are joined together inside a common boundary (p. 502)

continental divide   mountainous area from which rivers flow in different directions (p. 290)

contour line   line connecting all points at the same elevation on a contour map (p. 8)

convent   place where nuns live, pray, and study (p. 242)

cooperative   farm owned and operated by the government (p. 566)

copper belt   large area of copper mines (p. 460)

copra   dried coconut meat (p. 661)

coral reef   structure formed by the skeletons of small sea animals (pp. 404, 505, 638)

cordillera   group of mountain ranges that run side by side (p. 604)

cottage industry   home- or village-based industry in which family members supply their own equipment to make goods (pp. 145, 257)

crevasse   deep crack in the Antarctic ice cap (p. 666)

crop rotation   varying what is planted in a field to avoid using up all the minerals in the soil (p. 65)

Crusades   holy wars sponsored by the Catholic Church to capture Jerusalem from the Muslims (p. 92)

cultural borrowing   when a group of people adopts another's cultural traits (p. 29)

cultural diffusion   process of spreading new knowledge and skills to other cultures (p. 29)

culture   way of life of a group of people who share similar beliefs and customs (p. 28)

cuneiform   Sumerian form of writing using wooden triangular-shaped sticks in the form of hundreds of different wedge-shaped markings on moist clay tablets (p. 83)

custom   practice handed down from the past (p. 24)

cyclone   intense storm system with heavy rain and high winds (pp. 151, 467)

czar   title of Russian emperor (p. 367)

deforestation   widespread cutting of forests (pp. 65, 211, 426, 466, 584)

delta   area formed from soil deposited by a river at its mouth (pp. 84, 100, 213)

democracy   form of government in which citizens choose the nation's leaders by voting for them (pp. 31, 236)

desalinization   process of removing salt to make seawater drinkable (p. 113)

desertification   process by which grasslands change to desert (p. 441)

deterrence   maintenance of military power for the purpose of discouraging an attack (p. 262)

diagram   drawing that shows steps in a process or parts of an object (p. 13)

Diaspora   collective name for scattered Jewish settlements around the world (p. 89)

dictator   all-powerful government leader (p. 102)

dictatorship   government under the control of one all-powerful leader (p. 31)

dike   high bank of soil built along a body of water to control floods (p. 164)

disciple   follower of a specific teacher (p. 91)

divine right of kings   belief that royalty ruled by the will of God (p. 248)

dominion   self-governing nation that accepts the British monarch as head of state (p. 492)

drought   long period of extreme dryness and water shortages (pp. 411, 441)

dry farming   method in which the land is left unplanted every few years so that it can store moisture (p. 296)

dynasty   line of rulers from the same family (pp. 168, 200)

dzong   Buddhist center of prayer and study (p. 154)

ecosystem   place where the plants and animals are dependent upon one another and their surroundings for survival (p. 64)

eco-tourist   person who travels to another country to view its natural wonders (pp. 408, 559)

elevation   height of land above sea level (pp. 8, 358)

elevation profile   cutaway diagram showing changes in elevation of land (p. 13)

embalm   to treat a dead body to protect it from decay (p. 84)

embargo   order that restricts or prohibits trade with another country (pp. 123, 566)

emperor   absolute ruler of an empire (p. 238)

empire   group of lands under one ruler (pp. 83, 182, 617)

enclave   small territory entirely surrounded by a larger territory (p. 457)

endangered species   plant or animal under the threat of completely dying out (p. 413)

environment   natural surroundings (p. 54)

erosion   process of moving water and wind across the earth's surface, leaving the land less fertile than before (p. 65)

escarpment   steep cliff between higher and lower land (p. 583)

estancia   huge ranch (p. 590)

ethnic cleansing   forcing people from a different ethnic group to leave their homes (p. 313)

ethnic conflict   dispute between two or more ethnic groups (p. 357)

ethnic group   people who share a common culture, language, or history (pp. 24, 523)

ethnocentrism   attitude that one's own ethnic group is superior (p. 29)

Eurasia   region that lies on both the European and Asian continents (p. 342)

Euro   common currency adopted by countries in the European Union (p. 268)

European Union   trade alliance among many European countries, formerly called the Common Market (p. 268)

exclave   small part of a country that is separated from the main part (p. 464)

exile   unable to live in one's own country, usually because of political beliefs (p. 171)

famine   extended, usually widespread, severe lack of food (p. 203)

fault   crack in the earth's crust (pp. 58, 164)

favela   slum area (p. 584)

federal republic   form of government in which the national and state governments share powers (pp. 289, 375, 520, 548)

feudalism   political and social system in which a lord gave land to a noble to work, govern, and defend, in return for the noble's loyalty (p. 242)

five pillars of faith   basic religious obligations of Islam (p. 92)

fjord   steep-sided valley cut into mountains by the action of glaciers (pp. 292, 644)

foothills   hilly region at the base of a mountain range (p. 616)

fossil fuel   fuel formed in the earth, such as coal, oil, or natural gas (p. 508)

free enterprise system   economic system that operates on free competition, in which people start and own businesses with limited government intervention (pp. 165, 405, 507)

free market economy   economy driven by forces including competition, supply, and demand (p. 373)

free port   place where goods can be loaded, stored, and shipped again without import taxes (p. 218)

GLOSSARY

GLOSSARY

free trade   trade without barriers like tariffs or quotas so that goods flow freely between countries (p. 510)

free trade zone   area where people can buy goods from other countries without paying import taxes (p. 567)

gaucho   cowhand (p. 590)

genocide   mass murder of a people because of their race, religion, ethnicity, politics, or culture (pp. 259, 414)

geographic information systems (GIS) special software that helps geographers gather and use different kinds of information about a place (p. 54)

geography   study of the earth in all its variety—land, water, plants, and animals (p. 52)

geothermal energy   electricity produced by natural underground sources of steam (pp. 295, 645)

geyser   spring that shoots hot water through a crack in the earth's crust (pp. 295, 643)

glacier   giant, slow-moving sheet of ice (p. 487)

glasnost   Russian policy of "openness," which permitted Soviet people to criticize the government without punishment (pp. 265, 371)

Global Positioning System (GPS)   group of satellites that travels around the earth, which can be used to pinpoint an exact location on the earth (p. 54)

globalization   development of a world culture and interdependent world economy (p. 37)

greenhouse effect   buildup of certain gases in the atmosphere that, like a greenhouse, prevent the warm air from escaping into the atmosphere, resulting in increase of the earth's overall temperature (p. 61)

green revolution   great increase in food grains production due to the use of improved seeds, pesticides, and efficient farming techniques (p. 145)

guild   medieval workers' organization (p. 244)

habitat   type of environment in which a particular animal species lives (p. 408)

hacienda   large ranch (p. 543)

hajj   religious journey to Makkah that Muslims are expected to make at least once during their lifetimes if they are able to do so (pp. 93,114)

harmattan   dry, dusty wind that blows south from the Sahara (p. 434)

heavy industry   industry that produces goods such as machinery, mining equipment, and steel (pp. 294, 347)

hemisphere   one-half of the globe; the Equator divides the earth into Northern and Southern Hemispheres; the Prime Meridian divides it into Eastern and Western Hemispheres (p. 5)

hieroglyphics   ancient form of writing that used signs and symbols (pp. 85, 542)

high island   Pacific island formed by volcanic activity (p. 662)

high-technology industry   industry that produces computers and other kinds of electronic equipment (p. 180)

Holocaust   systematic murder by Adolf Hitler and his followers of more than 12 million people, including 6 million European Jews, during World War II (p. 259)

human resources   supply of people who can produce goods (p. 256)

human rights   basic freedoms and rights, such as freedom of speech and freedom of assembly, that all people should have (p. 171)

hurricane   tropical storm with winds of more than 74 miles (119 km) per hour and heavy rains (pp. 537, 557)

hydroelectric power   electricity generated by flowing water (pp. 423, 592, 645)

iceberg   chunk of a glacier that has broken away and floats free in the ocean (p. 667)

ice shelf   layer of ice above water in Antarctica (p. 667)

immigrant   person who moves to a new country to make a permanent home (p. 640)

imperialism   system of building foreign empires for military and trade advantages (p. 258)

indigenous   native to or originating from a place (p. 486)

indulgences   pardons for sins, given or sold by the Catholic Church (p. 246)

industrialize   to change an economy to rely more on manufacturing and less on farming (pp. 368, 538)

industrialized country country in which a great deal of manufacturing occurs (p. 454)

intensive cultivation growing crops on every available piece of land (p. 195)

interdependence dependence of countries on one another for goods, raw materials to make goods, and markets in which to sell goods (p. 37)

invest to put money into a business (p. 165)

irrigation system planned series of ditches to control flooding and to better water land (p. 83)

Islamic republic government run by Muslim religious leaders (p. 124)

isthmus narrow piece of land that connects two larger pieces of land (p. 556)

jade shiny, usually green gemstone (p. 541)

jute plant fiber used for making rope, burlap bags, and carpet backing (p. 145)

kibbutz settlement or farm in Israel where the people share property and produce goods (p. 108)

krill tiny, shrimp-like animal that lives in the waters off Antarctica and is food for many other animals (p. 668)

land bridge narrow strip of land that joins two larger landmasses (p. 534)

landfill area where trash companies dump the waste they collect (p. 509)

landform individual feature of the land, such as a mountain or valley (p. 53)

landlocked not bordering a sea or an ocean (pp. 309, 621)

latitude location north or south of the Equator, measured by imaginary lines (parallels) numbered in degrees north or south (pp. 5, 536)

light industry industry that produces goods such as clothing, shoes, furniture, and household products (p. 347)

limited government type of government where, through law, some control is placed on leadership's powers (p. 31)

line graph graph in which one or more lines represent changing quantities over time (p. 10)

literacy rate percentage of people who can read and write (p. 559)

llanos grassy plains (pp. 593, 604)

longitude location east or west of the Prime Meridian, measured by imaginary lines (meridians) numbered in degrees east or west (p. 6)

low island Pacific island formed of coral and having little vegetation (p. 662)

maize Native American name for corn (p. 542)

majority culture dominant or main culture of a region (p. 350)

majority group group in society that controls most of the wealth and power, though not always the largest group in numbers (p. 24)

mangrove tropical tree with roots that extend both above and beneath the water (p. 434)

manor feudal estate made up of a manor house or castle and land (p. 243)

manuka small shrub of New Zealand (p. 643)

maquiladora factory that assembles parts made in other countries (p. 538)

marsupial mammal that carries its young in a pouch (p. 639)

megalopolis huge urban settlement made up of several large cities (pp. 196, 503)

messiah in Judaism and Christianity, a savior sent by God (p. 90)

mestizo person with mixed European and Native American ancestry (p. 606)

migrant worker person who travels from place to place when extra help is needed to plant or harvest crops (p. 549)

migrate to move from one place to another (p. 34)

minister to take care of or tend to another's needs (p. 91)

minority culture culture other than the majority culture in a region (p. 350)

minority group group of people who are different in some characteristic from the group with the most power and wealth in a region (p. 24)

GLOSSARY

missionary   teacher of Christianity (p. 242)

monastery   place where monks live, pray, and study (p. 242)

monotheism   belief that there is only one God (p. 88)

monsoon   seasonal wind that blows for months at a time, often affecting a region's climate (pp. 144, 201, 210)

moshav   cooperative settlement of small individual farms in Israel where people share in farming, but also own some private property (p. 108)

mosque   place of worship for followers of Islam (pp. 106, 314)

multilingual   able to speak several languages (p. 286)

multinational company   firm that does business in several countries (p. 286)

mural   wall painting (p. 542)

NAFTA (North American Free Trade Agreement)   free trade agreement among United States, Mexico, and Canada (p. 539)

national debt   money owed by a nation's government (p. 550)

navigable   describes a body of water wide and deep enough to allow the passage of ships (pp. 283, 616)

neutrality   refusing to take sides in disagreements and wars between countries (p. 290)

nomad   person who moves from place to place with herds of animals (pp. 182, 357)

nuclear energy   power created by a controlled atomic reaction (p. 374)

nuclear weapon   weapon whose destructive power comes from an uncontrolled nuclear reaction (p. 261)

Nunavut   Canadian province created in 1999 from land that was part of the Northwest Territories (p. 486)

oasis   green area in a desert fed by underground water (pp. 101, 358)

obsidian   hard, black glass created by the cooled molten lava of a volcano (p. 542)

"one-country, two-systems"   China's policy of blending capitalism and some Western freedoms with Chinese communism (p. 166)

outback   inland regions of Australia (p. 638)

overfishing   taking fish from the ocean so fast that it threatens their existence in that region (p. 488)

overgraze   to allow animals to strip areas so bare that plants cannot grow back (p. 440)

ozone   type of oxygen that forms a layer in the atmosphere and protects all living things on the earth from certain harmful rays of the sun (p. 668)

pagoda   many-storied Buddhist temple (p. 172)

pampas   vast treeless, grass-covered plains of South America (p. 590)

papyrus   tall plant of the Nile valley, used in making a form of paper (p. 85)

Parliament   supreme legislative body in the United Kingdom and other countries (p. 248)

parliamentary democracy   system of democratic government where the executive power is held by a cabinet composed of members of the legislature; headed by a prime minister (pp. 279, 492, 560)

parliamentary republic   parliamentary government that also has a president, who acts as head of state (p. 297)

peat   wet ground with decaying plants that can be dried and used for fuel (p. 280)

peninsula   piece of land with water on three sides (p. 535)

perestroika   Soviet policy that loosened government controls and permitted its economy to move towards capitalism (p. 265)

pesticide   powerful chemical that kills crop-destroying insects (pp. 64, 146)

pharaoh   ruler of Ancient Egypt (p. 84)

philosophy   term derived from Greek, meaning "love of wisdom" (p. 236)

phosphate   mineral salt used in fertilizers (pp. 444, 663)

pictograph   graph in which small symbols represent quantities (p. 12)

pidgin language   language formed by combining elements of several different languages (p. 662)

plain   low-lying stretch of flat or gently rolling land (p. 59)

plantation   large farm that grows a single crop for sale (pp. 538, 557)

plate   huge slab of rock that makes up the earth's crust (pp. 216, 412)

plateau   flat land with higher elevation than a plain (p. 59)

plate tectonics   theory that the earth's crust is not an unbroken shell but consists of plates, or huge slabs of rock, that move (p. 57)

plaza   public square (p. 546)

poaching   illegal hunting of protected animals (p. 404)

polder   area of land reclaimed from the sea (p. 285)

polis   Greek term for "city-state" (p. 236)

polytheistic   believing in more than one god (p. 84)

pope   head of the Roman Catholic Church (pp. 241, 308)

potash   type of mineral salt that is often used in fertilizers (p. 319)

precious gems   valuable gemstones, such as rubies, sapphires, and jade (p. 211)

prime minister   official who heads the government in a parliamentary democracy (pp. 146, 493)

productivity   measurement of the amount of work accomplished in a given time (p. 256)

Protestant   person who "protested" Catholic practices; today, a member of a non-Catholic Christian church (p. 246)

province   regional political division similar to a state (p. 486)

pyramid   huge stone structure that served as an elaborate tomb or monument (p. 85)

recycling   reusing materials instead of throwing them away (p. 509)

reform   to improve by changing (p. 246)

refugee   person who flees to another country to avoid persecution or disaster (pp. 35, 313, 414)

region   area that shares common characteristics (p. 54)

reincarnation   rebirth of a soul in a new body (p. 148)

relief   differences in height in a landscape; how flat or rugged the surface is (p. 8)

representative democracy   government where the people are represented by elected leaders (p. 520)

republic   nation with a strong national government headed by elected leaders (pp. 238, 284, 560, 587, 606)

responsibilities   duties owed by citizens to their government to make sure it can continue its functions (p. 37)

reunification   bringing together the two parts of Germany under one government (p. 289)

revolution   great and often violent change (p. 247)

rights   benefits and protections that are guaranteed by law (p. 36)

rural   relating to the countryside (pp. 344, 524)

samurai   powerful land-owning warriors in Japan (p. 196)

satellite nation   nation politically and economically dominated or controlled by another, more powerful country (p. 263)

sauna   wooden room heated by water sizzling on hot stones (p. 294)

savanna   broad grassland with few trees, found in the tropics (pp. 422, 434)

scale   relationship between distance on a map and actual distance on the earth (p. 7)

scale bar   on a map, a divided line showing the map scale, usually in miles or kilometers (p. 7)

scapegoat   one who is wrongly blamed by others (p. 89)

scribe   one who records information by writing (p. 85)

secular   nonreligious (p. 103)

selva   tropical rain forest in Brazil (p. 582)

Senate   legislative body of government; the supreme council of the ancient Roman Republic (p. 238)

serf   peasant laborer (pp. 243, 368)

service industry   industry that provides services like banking, education, and tourism to people rather than producing goods (pp. 487, 539)

shah   title given to kings who ruled Iran (p. 124)

shogun   military leader in Japan (p. 195)

silt   small particles of rich soil (p. 100)

sirocco   hot, dry winds from North Africa that blow mainly across Italy (p. 298)

sisal   plant fiber used to make rope and twine (p. 407)

slash-and-burn farming   method of clearing land for planting by cutting and burning forest (p. 466)

Slav person from the ethnic group originating in northeast Europe whose native tongue is a Slavic language (p. 350)

smog thick haze of fog and chemicals (p. 550)

socialism economic system in which many businesses are owned and run by the government (p. 211)

social scientist scientist who studies the interaction of people and society (p. 27)

sodium nitrate chemical used in fertilizer and explosives (p. 624)

sorghum tall grass with seeds that are used as grain (p. 460)

spa resort that has hot mineral springs that people bathe in to regain their health (p. 309)

station cattle or sheep ranch in Australia (p. 638)

steppe partly dry, treeless grassland often found on the edges of a desert (pp. 182, 317, 345)

strait narrow body of water between two pieces of land (p. 218)

strike refusal to work, usually by a labor organization, until demands are met (p. 257)

subcontinent large landmass that is part of a continent but distinct from it (p. 144)

subsistence farm small plot where a farmer grows only enough food to feed his own family (pp. 435, 538, 558, 617)

suburb smaller community located in the area surrounding a city (pp. 344, 524)

taiga huge forests of evergreen trees that grow in subarctic regions (p. 345)

tenant farmer or other who pays rent to another for the use of land or property (p. 243)

terraced field strips of land cut out of a hillside like stair steps so the land can hold water and be used for farming (p. 219)

terrorism systematic use of violence or terror to achieve certain goals (p. 102)

textiles woven cloth (p. 257)

theocracy form of government in which one individual ruled as both religious leader and king (p. 83)

township South African neighborhood created for the housing of people of non-European descent (p. 457)

tropics low-latitude region between the Tropic of Cancer and the Tropic of Capricorn (p. 60)

trust territory area temporarily placed under control of another nation (p. 663)

tsetse fly type of fly found in Africa south of the Sahara, whose bite can transmit a parasite that carries a disease called sleeping sickness (p. 425)

tsunami huge sea wave caused by an earthquake on the ocean floor (p. 192)

tundra vast, rolling, treeless plain in high latitude climates in which only the top few inches of ground thaw in summer (pp. 344, 487)

typhoon hurricane that occurs in Asia (p. 662)

union labor organization that negotiates for improved worker conditions and pay (p. 257)

unlimited government government in which leaders rule without any restrictions (p. 31)

urban relating to the city (pp. 343, 524)

urbanization movement to cities (p. 34)

vaquero cowhand (p. 537)

vassal noble in medieval society who swore loyalty to a lord in return for land (p. 243)

wadi dry riverbed filled by rainwater from rare downpours (p. 113)

welfare state country that uses tax money to provide social services for sick, needy, jobless, or retired people (p. 293)

yurt large, circle-shaped tent made of animal skins that can be packed up and moved from place to place (p. 183)

SPANISH GLOSSARY

absolute monarchy/monarquía absoluta forma de gobierno en la cual el rey o la reina gobierna con absoluto poder (pág. 31)

acid rain/lluvia ácida lluvia que contiene grandes cantidades de contaminantes químicos (págs. 65, 308)

adobe/adobe ladrillos secados al sol (pág. 546)

airlift/puente áereo sistema de transportar suministros por avión (pág. 263)

alliance/alianza acuerdo político entre dos países para apoyo mútuo en disputas contra otros países (pág. 258)

alluvial plain/llanura aluvial área creada por el suelo fértil que se acumula después de las inundaciones causadas por los ríos (pág. 122)

altiplano/altiplano meseta grande y muy elevada; también se llama altiplanicie (pág. 621)

altitude/altitud altura sobre el nivel del mar (págs. 537, 593)

amend/enmendar cambiar las leyes básicas de un gobierno (pág. 520)

apartheid/apartheid sistema de leyes que separaba los grupos raciales y étnicos y limitaba los derechos de la población negra (pág. 456)

apprentice/aprendiz jóven trabajador quien aprende un oficio o destreza bajo tutela de un maestro experto (pág. 244)

aquifer/manto acuífero capa de rocas subterránea en que el agua es tan abundante que corre entre las rocas (págs. 60, 101)

archipelago/archipiélago grupo de islas (pág. 192)

atoll/atolón isla de muy poca elevación que se forma alrededor de una laguna en la forma de anillo (pág. 662)

autobahn/autobahn autopista muy rápida (pág. 289)

autonomy/autonomía gobernarse por sí mismo (pág. 494)

axis/eje terrestre la línea vertical (del lado) u horizontal (de abajo) de una gráfica que se usa para medir (pág. 10)

bar graph/gráfica de barras gráfica en que franjas verticales u horizontales representan cantidades (pág. 10)

basin/cuenca área baja rodeada de tierras más elevadas (pág. 582)

bauxite/bauxita mineral que se usa para hacer aluminio (págs. 309, 444, 564)

Bedouin/beduino persona nomádica del desierto del sudoeste de Asia (pág. 111)

bilingual/bilingüe se refiere a un país que tiene dos idiomas oficiales (pág. 493)

bishop/obispo oficial de la Iglesia Cristiana (pág. 241)

blockade/bloquear impedir por la fuerza la entrada a un área (pág. 263)

Boer/bóer uno de los holandeses que fueron los primeros colonos en Sudáfrica (pág. 455)

bog/ciénaga tierra baja y pantanosa (pág. 280)

boomerang/bumerán arma australiana con forma de ala que se lanza para que golpee un objetivo o de la vuelta y caiga a los pies de la persona que la lanzó (pág. 640)

bush/campo áreas rurales de Australia (pág. 641)

cacao/cacao árbol tropical cuyas semillas se usan para hacer chocolate y cocoa (págs. 436, 660)

calligraphy/caligrafía el arte de escribir con letra muy bella (pág. 172)

campesino/campesino agricultor (pág. 607)

canopy/bóveda techo formado por las copas de los árboles en los bosques húmedos (págs. 423, 558)

canyon/cañón valle profundo y estrecho, con laderas escarpadas, formado por un río que corta una meseta (pág. 59)

capitalism/capitalismo sistema económico que permite propietarios privados y la competencia abierta de negocios (pág. 265)

cash crop/cultivo comercial producto que se cultiva para exportación (pág. 606)

cassava/yuca planta con raíces que se pueden convertir en harina para hacer pan o que se pueden cocinar de otras formas (pág. 405)

caste/casta clase social basada en la ascendencia de una persona (pág. 147)

caudillo/caudillo gobernante militar (pág. 593)

chart/cuadro manera gráfica de presentar información con claridad (pág. 11)

charter/cédula acuerdo escrito garantizando privilegios y libertades (pág. 244)

circle graph/gráfica de círculo gráfica redonda que muestra como un todo es dividido (pág. 11)

city-state/ciudad estado ciudad junto con las tierras que la rodean (pág. 83)

civilizations/civilizaciones culturas altamente desarrolladas (pág. 82)

civil war/guerra civil pelea entre distintos grupos dentro de un país (págs. 212, 438)

clan/clan grupo de personas que están emparentadas (págs. 195, 412)

SPANISH GLOSSARY

classical/clásico   relacionado a la antigua Roma y Grecia (pág. 236)

climate/clima   el patrón que sigue el estado del tiempo en un área durante muchos años (pág. 60)

climograph/gráfica de clima   gráfica que combina barras y líneas para dar información sobre la temperatura y la precipitación (pág. 12)

coalition government/gobierno por coalición   gobierno en que dos o más partidos trabajan juntos para dirigir un país (pág. 298)

Cold War/Guerra Fría   período entre los fines de los 1940 y los fines de los 1980 en que los Estados Unidos y la Unión Soviética compitieron por tener influencia mundial sin pelear uno contra el otro (págs. 261, 370)

colony/colonia   territorio o poblado con lazos a un país extranjero (págs. 492, 543)

Columbian Exchange/Intercambio Colombiano   proceso en el cual personas, enfermedades, ideas y comercio fueron distribuidos alrededor del mundo desde las Américas (pág. 247)

commercial crop/cosecha comercial   cultivo agrícola para propósito de venta (pág. 564)

common law/derecho común   grupo de leyes no escritas basadas en costumbres locales (pág. 242)

commonwealth/estado libre asociado   territorio que en parte se gobierna por sí solo (pág. 567)

communism/comunismo   sistema económico, social y político basado en las enseñanzas de Karl Marx, el cual abogaba por la eliminación de propiedades privadas (págs. 258, 346)

communist state/estado comunista   país cuyo gobierno mantiene mucho control sobre la economía y la sociedad en su totalidad (págs. 165, 369, 566)

compound/complejo residencial   grupo de viviendas rodeadas por una muralla (pág. 437)

conservation/conservación   uso juicioso de los recursos para no malgastarlos (pág. 64)

constitution/constitución   acuerdo formal que establece las bases para las leyes de un país (pág. 248)

constitutional monarchy/monarquía constitucional   gobierno en que un rey o reina es el jefe de estado oficial pero los gobernantes son elegidos (págs. 31, 104, 196, 280)

consul/cónsul   oficial en jefe electo en la república romana (pág. 238)

consumer goods/bienes de consumo   productos para la casa, ropa y otras cosas que la gente compra para su uso personal (págs. 165, 312, 344)

contiguous/contiguas   áreas adyacentes dentro de la misma frontera (pág. 502)

continental divide/línea divisoria continental   área montañosa de la cual los ríos descienden en diferentes direcciones (pág. 290)

contour line/curva de nivel   línea que conecta todos los puntos a la misma elevación en un mapa de relieve (pág. 8)

convent/convento   lugar donde viven, rezan y estudian monjas (pág. 242)

cooperative/cooperativa   granja que es propiedad y es operada por el gobierno (pág. 566)

copper belt/cinturón de cobre   área extensa de minas de cobre en el norte de Zambia (pág. 460)

copra/copra   pulpa seca del coco que se usa para hacer margarina, jabón y otros productos (pág. 661)

coral reef/arrecife coralino   estructura formada al nivel del mar o cerca de éste por los esqueletos de pequeños animales marinos (págs. 404, 505, 638)

cordillera/cordillera   grupo de cadenas paralelas de montañas (pág. 604)

cottage industry/industria familiar   industria basada en una casa o aldea en que los miembros de la familia usan sus propias herramientas para hacer productos (págs. 145, 257)

crevasse/grieta   rajadura profunda en el casquete de hielo de la Antártida (pág. 666)

crop rotation/rotación de cultivos   variar lo que se siembra en un terreno para no agotar todos los minerales que tiene el suelo (pág. 65)

Crusades/Cruzadas   guerras religiosas patrocinadas por la Iglesia Católica para capturar Jerusalén de los musulmanes (pág. 92)

cultural borrowing/adopción cultural   cuando un grupo de personas adoptan las características culturales de otro grupo (pág. 29)

cultural diffusion/difusión cultural   el proceso de esparcir nuevos conocimientos y habilidades a otras culturas (pág. 29)

culture/cultura   modo de vida de un grupo de personas que comparten creencias y costumbres similares (pág. 28)

cuneiform/cuneiforme   forma de escritura sumeria que utiliza trozos triangulares de madera para hacer cientos de diferentes figuras en forma de cuña en tabletas de barro húmedo (pág. 83)

custom/costumbre   práctica aceptada a través de los siglos (pág. 24)

cyclone/ciclón   tormenta violenta con vientos muy fuertes y mucha lluvia (págs. 151, 467)

czar/zar   título de los antiguos emperadores rusos (pág. 367)

deforestation/deforestación   la extensa destrucción de los bosques (págs. 65, 211, 426, 466, 584)

delta/delta   área formada por el suelo que deposita un río en su desembocadura (págs. 84, 100, 213)

democracy/democracia   tipo de gobierno en que los ciudadanos seleccionan los líderes de la nación por medio del voto (págs. 31, 236)

desalinization/desalinización   proceso de hacer el agua de mar potable (pág. 113)

desertification/desertización   proceso por el cual los pastos se convierten en desiertos (pág. 441)

deterrence/disuasión   el mantener el poder militar con el propósito de desalentar un ataque (pág. 262)

diagram/diagrama   dibujo que muestra los pasos en un proceso o las partes de un objeto (pág. 13)

Diaspora/diáspora   nombre dado a los poblados judíos alrededor del mundo (pág. 89)

dictator/dictador individuo que toma control de un país y lo gobierna como quiere (pág. 102)

dictatorship/dictadura gobierno bajo el control de un líder que tiene todo el poder (pág. 31)

dike/dique muros de tierra muy altos construidos a lo largo de los ríos para controlar las inundaciones (pág. 164)

disciple/discípulo partidario de un maestro específico (pág. 91)

divine right of kings/derecho divino de los reyes la creencia de que los reyes gobernaban por la voluntad de Dios (pág. 248)

dominion/dominio naciones que se gobiernan por sí solas que aceptan al monarca británico como jefe de estado (pág. 492)

drought/sequía largos períodos de sequedad y de escasez de agua (págs. 411, 441)

dry farming/agricultura en seco método de cultivar en que la tierra se deja sin sembrar cada varios años para que almacene humedad (pág. 296)

dynasty/dinastía serie de gobernantes de la misma familia (págs. 168, 200)

dzong/dzong centro budista en Bután para rezar y estudiar (pág. 154)

ecosystem/ecosistema lugar en el cual las plantas y animales dependen unos de otros y de sus alrededores para sobrevivir (pág. 64)

eco-tourist/ecoturista persona que viaja a otro país para ver sus bellezas naturales (págs. 408, 559)

elevation/elevación altura por encima del nivel del mar (págs. 8, 358)

elevation profile/perfil de elevaciones diagrama que muestra los cambios en la elevación de la tierra como si se hubiera hecho un corte vertical del área (pág. 13)

embalm/embalsamar tratar un cuerpo sin vida para protegerlo de la descomposición (pág. 84)

embargo/embargo orden que limita o prohibe el comercio con otro país (págs. 123, 566)

emperor/emperador gobernante absoluto de un imperio (pág. 238)

empire/imperio grupo de países bajo un gobernante (págs. 83, 182, 617)

enclave/enclave territorio pequeño totalmente rodeado por un territorio más grande (pág. 457)

endangered species/especie en vías de extinción planta o animal que está en peligro de desaparecer completamente (pág. 413)

environment/medio ambiente alrededores naturales (pág. 54)

erosion/erosión proceso de mover los materiales desgastados en la superficie de la tierra (pág. 65)

escarpment/escarpa acantilado empinado entre un área baja y una alta (pág. 583)

estancia/estancia rancho (pág. 590)

ethnic cleansing/limpieza étnica forzar a personas de un grupo étnico distinto a abandonar el lugar donde viven (pág. 313)

ethnic conflict/conflicto étnico disputa entre dos o más grupos étnicos (pág. 357)

ethnic group/grupo étnico personas que comparten una cultura, idioma o historia en común (págs. 24, 523)

ethnocentrism/etnocentrismo la actitud que el grupo étnico de uno es superior a los demás (pág. 29)

Eurasia/Eurasia región que se ubica en los dos continentes de Europa y Asia (pág. 342)

Euro/eurodólar moneda común adoptada por los países de la Unión Europea (pág. 268)

European Union/Unión Europea alianza comercial entre muchos países europeos formalmente llamado el Mercado Común Europeo (pág. 268)

exclave/territorio externo parte pequeña de un país que está separada de la parte principal (pág. 464)

exile/exilio tener que vivir fuera de su país nativo por causa de sus creencias políticas (pág. 171)

famine/hambruna falta de alimentos (pág. 203)

fault/falla fractura en la corteza de la tierra (págs. 58, 164)

favela/favela barrio pobre y deteriorado (pág. 584)

federal republic/república federal nación en que el poder está dividido entre el gobierno nacional y el de los estados (págs. 289, 375, 520, 548)

feudalism/feudalismo sistema político y social en el cual un lord cedía tierra a un noble para que la trabajara, gobernara y defendiera, obligándose éste rendirle fidelidad (pág. 242)

five pillars of faith/cinco pilares de fé obligaciones religiosas básicas del Islam (pág. 92)

fjord/fiordo valle creado por el movimiento de glaciares en las montañas que deja laderas sumamente empinadas (págs. 292, 644)

foothill/estribaciones colinas bajas al pie de una cadena de montañas (pág. 616)

fossil fuel/combustibles fósiles carbón, petróleo o gas natural (pág. 508)

free enterprise system/sistema de libre empresa sistema económico en que la gente empieza y administra negocios con poca intervención del gobierno (págs. 165, 405, 507)

free market economy/economía del libre comercio economía llevada por fuerzas como la competencia, oferta y demanda (pág. 373)

free port/puerto libre lugar donde las mercancías se pueden cargar, almacenar y embarcar de nuevo sin tener que pagar derechos de importación (pág. 218)

free trade/libre comercio eliminar las barreras al comercio para que se puedan mover productos libremente entre países (pág. 510)

free trade zone/zona de cambio libre área donde la gente puede comprar bienes de otros países sin pagar impuestos adicionales (pág. 567)

SPANISH GLOSSARY

gaucho/gaucho vaquero (pág. 590)

genocide/genocidio asesinato en masa de personas a causa de su raza, religión, etnicidad, política o cultura (págs. 259, 414)

geographic information systems (GIS)/ sistemas de información geográfica (SIG) programas de computadoras especiales que ayudan a los geógrafos a obtener y usar la información geográfica sobre un lugar (pág. 54)

geography/geografía el estudio de la tierra y de toda su variedad (pág. 52)

geothermal energy/energía geotérmica electricidad producida por fuentes de vapor subterráneas naturales (págs. 295, 645)

geyser/géiser manantial de agua calentado por rocas fundidas dentro de la tierra que, de vez en cuando, arroja agua caliente al aire (págs. 295, 643)

glacier/glaciar capa de hielo inmensa que se mueve muy lentamente (pág. 487)

glasnost/glasnost política rusa de "franqueza," la cual permitió que la gente soviética criticara al gobierno sin castigo (págs. 265, 371)

Global Positioning System (GPS)/Sistema de posición global (GPS) grupo de satélites alrededor de la tierra que se usan para localizar lugares exactos en la tierra (pág. 54)

globalization/globalización desarrollo de una cultura y economía interdependiente mundiales (pág. 37)

greenhouse effect/efecto invernadero la acumulación de ciertos gases en la atmósfera que mantienen más del calor del sol, como hace un invernadero (pág. 61)

green revolution/revolución verde gran aumento en la producción de granos debido al uso de semillas, pesticidas, y técnicas agrícolas perfeccionadas (pág. 145)

guild/gremio organización de trabajadores en la época medieval (pág. 244)

habitat/hábitat tipo de ambiente en que vive una especie animal en particular (pág. 408)

hacienda/hacienda un rancho grande (pág. 543)

hajj/hajj viaje religioso a La Meca que todo musulmán debe hacer por lo menos una vez en la vida si puede (págs. 93, 114)

harmattan/harmattan viento seco y lleno de polvo que sopla hacia el sur desde el Sahara (pág. 434)

heavy industry/industria pesada manufactura de productos como maquinaria, equipo de minería y acero (págs. 294, 347)

hemisphere/hemisferio una mitad del globo terráqueo; el ecuador divide la tierra en los hemisferios norte y sur; el primer meridiano la divide en hemisferios este y oeste (pág. 5)

hieroglyphics/jeroglíficos forma de escribir que usa signos y símbolos (págs. 85, 542)

high island/isla oceánica isla del Pacífico formada por actividad volcánica (pág. 622)

high-technology industry/industria de alta tecnología industria que produce computadoras y otras clases de equipo electrónico (pág. 180)

Holocaust/Holocausto la matanza sistemática de más de 6 millones de judíos europeos por Adolfo Hitler y sus seguidores durante la Segunda Guerra Mundial (pág. 259)

human resources/recursos humanos suministro de personas quienes pueden producir bienes de consumo (pág. 256)

human rights/derechos humanos libertades y derechos básicos que todas las personas deben disfrutar (pág. 171)

hurricane/huracán tormenta tropical violenta con vientos y lluvias fuertes (págs. 537, 557)

hydroelectric power/energía hidroeléctrica electricidad generada por una corriente de agua (págs. 423, 592, 645)

iceberg/iceberg pedazo de un glaciar que se ha desprendido y flota libremente en los océanos (pág. 667)

ice shelf/plataforma de hielo capa de hielo sobre el mar en la Antártida (pág. 667)

immigrant/inmigrante persona que se muda permanentemente a un país nuevo (pág. 640)

imperialism/imperialismo el sistema de desarrollar imperios extranjeros para ventaja militar y comercial (pág. 258)

indigenous/indígeno nativo u originario de un lugar (pág. 486)

indulgences/indulgencias perdón por los pecados concedido o vendido por la Iglesia Católica (pág. 246)

industrialize/industrializar cambiar una economía de manera que dependa más de la manufactura que de la agricultura (pág. 368, 538)

industrialized country/país industrializado país en el cual ocurre mucha manufactura (pág. 454)

intensive cultivation/cultivo intensivo labrar toda la tierra posible (pág. 195)

interdependence/interdependencia países que dependen unos de otros para bienes, materia prima para producir bienes, y mercados en los cuales vendan sus productos (pág. 37)

invest/invertir poner dinero en un negocio (pág. 165)

irrigation system/sistema de irrigación sistema planeado de zanjas para prevenir inundaciones y mejorar el riego de tierra (pág. 83)

Islamic republic/república islámica   gobierno dirigido por líderes musulmanes (pág. 124)

isthmus/istmo   extensión de tierra que conecta a dos masas de tierra más grandes (pág. 556)

jade/jade   piedra preciosa reluciente, usualmente de color verde (pág. 541)

jute/yute   fibras de una planta que se usan para hacer soga, sacos y el revés de alfombras (pág. 145)

kibbutz/kibutz   poblado en Israel donde las personas comparten la propiedad y producen bienes (pág. 108)

krill/krill   animales diminutos parecidos a los camarones que viven en las aguas alrededor de la Antártida y sirven de alimento para muchos otros animales (pág. 668)

land bridge/puente terrestre   franja de tierra que une a dos masas de tierra mayores (pág. 534)

landfill/vertedero de basura   lugar donde las compañías que recogen la basura botan los residuos que colectan (pág. 509)

landform/accidente geográfico   característica particular de la tierra (pág. 53)

landlocked/rodeado de tierra   país que no tiene tierras bordeadas por un mar u océano (págs. 309, 621)

latitude/latitud   posición al norte o al sur del ecuador, medida por medio de líneas imaginarias (paralelos) numeradas con grados norte o sur (págs. 5, 536)

light industry/industria ligera   fabricación de productos como muebles, ropa, zapatos y artículos para el hogar (pág. 347)

limited goverment/gobierno limitado   tipo de gobierno donde, a través de la ley, hay algo de control sobre el poder del liderazgo (pág. 31)

line graph/gráfica lineal   gráfica en que una o varias líneas representan cambios de cantidad a través del tiempo (pág. 10)

literacy rate/índice de alfabetización   porcentaje de personas que saben leer y escribir (pág. 559)

llanos/llanos   planicie cubierta de hierba (págs. 593, 604)

longitude/longitud   posición al este o el oeste del primer meridiano, medida por medio de líneas imaginarias (meridianos) numeradas con grados este u oeste (pág. 6)

low island/isla coralina   isla del Pacífico formada por coral que tiene poca vegetación (pág. 662)

maize/maíz   nombre Nativo Americano del elote (pág. 542)

majority culture/cultura mayoritaria   cultura dominante o principal de una región (pág. 24)

majority group/grupo mayoritario   grupo en una sociedad que controla la mayoría de la riqueza y el poder, el cual no siempre es el grupo más numeroso (pág. 24)

mangrove/mangle   árbol tropical con raíces que se extienden por encima y por debajo del agua (pág. 434)

manor/feudo   estado feudal compuesto de una casa solariega o castillo y tierra (pág. 243)

manuka/manuka   pequeño arbusto de Nueva Zelanda (pág. 643)

maquiladora/maquiladora   fábrica donde se ensamblan piezas hechas en otros países (pág. 538)

marsupial/marsupial   mamífero que lleva a sus crías en una bolsa (pág. 639)

megalopolis/megalópolis   área extensa de mucha urbanización (págs. 196, 503)

messiah/Mesías   en judaísmo y cristianismo, el salvador enviado por Dios (pág. 90)

mestizo/mestizo   persona cuya ascendencia incluye indios americanos o africanos y españoles (pág. 606)

migrant worker/trabajador itinerante   persona que viaja a distintos lugares donde hacen falta trabajadores para sembrar y cosechar cultivos (pág. 549)

migrate/migrar   mudarse de un lugar a otro (pág. 34)

minister/atender   cuidar las necesidades de otro (pág. 91)

minority culture/cultura minoritaria   cultura aparte de la cultura mayoritaria en una región (pág. 350)

minority group/grupo minoritario   grupo de gente quien es diferente en alguna característica del grupo con mayor poder y riqueza en una región (pág. 24)

missionary/misionario   maestro del cristianismo (pág. 242)

monastery/monasterio   lugar donde viven, rezan y estudian monjes (pág. 242)

monotheism/monoteísmo   creencia en un solo Dios (pág. 88)

monsoon/monzón   vientos que soplan en un continente por varios meses seguidos en ciertas estaciones del año (págs. 144, 201, 210)

moshav/moshav   poblados en Israel en que la gente comparte alguna propiedad pero también tiene propiedad privada (pág. 108)

mosque/mezquita   edificio de devoción islámico (págs. 106, 314)

multilingual/multilingüe   que puede hablar varios idiomas (pág. 286)

SPANISH GLOSSARY

multinational company/compañía multinacional
compañía que hace negocios en varios países
(pág. 286)

mural/mural   pintura hecha sobre una pared
(pág. 542)

**NAFTA (North American Free Trade
Agreement)/TLC (Tratado de Libre Comercio)**
convenio de libre comercio entre Estados Unidos,
México y Canadá (pág. 539)

national debt/deuda pública   dinero debido por el
gobierno de una nación (pág. 550)

navigable/navegable   describe una masa de agua
ancha y profunda suficiente para que los barcos
puedan viajar por ella (págs. 283, 616)

neutrality/neutralidad   negarse a ponerse a favor
de uno de los adversarios en un desacuerdo o una
guerra entre países (pág. 290)

nomad/nómada   alguien que se muda de un lugar
a otro con sus manadas o rebaños de animales
(págs. 182, 357)

nuclear energy/energía nuclear   energía
producida por medio de una reacción atómica
controlada (pág. 374)

nuclear weapon/arma nuclear   arma cuya fuerza
destructiva viene de una reacción nuclear no
controlada (pág. 261)

Nunavut/Nunavut   provincia canadiense creada en
1999 de terrenos que fueron parte de los Territorios
del Noroeste (pág. 486)

oasis/oasis   área verde en medio de un desierto
adonde llegan aguas subterráneas (págs. 101, 358)

obsidian/obsidiana   piedra vítrea de color negro
formada por el enfriamiento de la lava líquida de un
volcán (pág. 542)

**"one country, two systems"/un país, dos
sistemas**   política de China de combinar el
capitalismo y algunas libertades occidentales con el
comunismo chino (pág. 166)

outback/tierra adentro   el interior de Australia
(pág. 638)

overfishing/sobrepesca   pesca excesiva que pone
en riesgo la existencia de una especie marina en
una región del océano (pág. 488)

overgraze/pastar excesivamente   cuando el
ganado despoja los pastos hasta tal punto que las
plantas no pueden crecer de nuevo (pág. 440)

ozone/ozono   tipo de oxígeno que forma una capa
en la atmósfera que protege a todas las cosas vivas
de ciertos rayos del sol que son peligrosos (pág. 668)

pagoda/pagoda   templo budista de muchos pisos
(pág. 172)

pampas/pampa   llanura de gran extensión en
América del Sur sin árboles y cubierta de hierba
(pág. 590)

papyrus/papiro   planta alta de la valle del Nilo,
usada para hacer un tipo de papel (pág. 85)

Parliament/Parlamento   cuerpo legislativo
supremo en Inglaterra y otros países (pág. 248)

**parliamentary democracy/democracia
parlamentaria**   gobierno en que los votantes
eligen a representantes a un cuerpo que hace las
leyes y que selecciona a un primer ministro para
que sea el jefe del gobierno (págs. 279, 492, 560)

parliamentary republic/república parlamentaria
véase parliamentary democracy/democracia
parlamentaria (pág. 297)

peat/turba   suelo mojado con plantas en
descomposición que se puede secar y usar para
combustible (pág. 280)

peninsula/península   masa de tierra con agua
alrededor de tres lados (pág. 535)

perestroika/perestroika   política soviética que
relajó los controles gubernamentales y permitió
que la economía se moviera hacia el capitalismo
(pág. 265)

pesticide/pesticida   sustancia química poderosa
que mata a los insectos que destruye los cultivos
(págs. 64, 146)

pharaoh/faraón   soberano del antiguo Egipto
(pág. 84)

philosophy/filosofía   término derivado del griego
que significa "amor a la sabiduría" (pág. 236)

phosphate/fosfato   sal mineral que se usa en los
abonos (págs. 444, 663)

pictograph/pictograma   gráfica en que pequeños
símbolos representan cantidades (pág. 12)

pidgin language/lengua franca   lenguaje formado
al combinar elementos de varios idiomas distintos
(pág. 662)

plain/llanura   extensión de tierra plana u
ondulante a elevaciones bajas (pág. 59)

plantation/plantación   granja grande en que se
siembra un solo cultivo para venderse (págs. 538,
557)

plate/placa   plancha de roca inmensa que forma
parte de la corteza de la tierra (págs. 216, 412)

plateau/meseta   planicie a elevaciones más altas
que las llanuras (pág. 59)

plate tectonics/tectónica de placas   teoría que
dice que la corteza de la tierra no es una envoltura
enteriza, sino que está formada por placas, o planchas
de roca inmensas, que se mueven (pág. 57)

plaza/plaza   sitio donde se reúne el público
(pág. 546)

poaching/caza furtiva   cacería ilegal de animales
protegidos (pág. 404)

SPANISH GLOSSARY

polder/pólder   área de tierra ganada del mar
(pág. 285)

polis/polis   término griego para "ciudad estado"
(pág. 236)

polytheistic/politeísta   que cree en más de un
dios (pág. 84)

pope/papa   líder de la Iglesia Católica Apostólica
Romana (págs. 241, 308)

potash/potasa   tipo de sal mineral que a menudo
se usa en los abonos (pág. 319)

precious gems/piedras preciosas   valiosas piedras
preciosas, como el rubí, el zafiro y el jade (pág. 211)

prime minister/primer ministro   líder del
gobierno en una democracia parlamentaria
(págs. 146, 493)

productivity/productividad   la medida de la
cantidad de trabajo ejecutado en un tiempo dado
(pág. 256)

Protestant/protestante   persona que "protestaba"
contra las prácticas católicas; hoy en día, miembro
de la iglesia cristiana pero no católica (pág. 246)

province/provincia   división política regional,
parecida a un estado (pág. 486)

pyramid/pirámide   estructura de piedra gigantesca
que sirvió como tumba o monumento elaborado
(pág. 85)

recycling/reciclaje   usar materiales de nuevo en
vez de botarlos (pág. 509)

reform/reforma   mejorar cambiando (pág. 246)

refugee/refugiado   persona que huye de un país a
otro para evitar la persecución o un desastre (págs.
35, 313, 414)

region/región   área destacada por ciertas
características (pág. 54)

reincarnation/reencarnación   renacimiento del
alma en un cuerpo nuevo (pág. 148)

relief/relieve   las diferencias en altitud de una
zona; lo plana o accidentada que es una superficie
(pág. 8)

**representative democracy/democracia
representativa**   gobierno donde las personas están
representadas por dirigentes elegidos (pág. 520)

republic/república   gobierno nacional fuerte
encabezado por líderes elegidos (págs. 238, 284,
560, 587, 606)

responsibilities/responsabilidades   deberes que
la gente debe a su gobierno para asegurar que éste
puede seguir funcionando (pág. 37)

reunification/reunificación   juntar de nuevo las dos
partes de Alemania bajo un solo gobierno (pág. 289)

revolution/revolución   una órbita completa
alrededor del sol (pág. 247)

rights/derechos   beneficios y protecciones que
están garantizados por ley (pág. 36)

rural/rural   área en el campo (págs. 344, 524)

samurai/samurai   propietarios y guerreros
poderosos del Japón (pág. 196)

satellite nation/nación satélite   nación dominada
o controlada política y económicamente por otro
país más poderoso (pág. 263)

sauna/sauna   cuarto de madera calentado por agua
que hierve sobre piedras calientes (pág. 294)

savanna/sabana   pastos extensos en los trópicos
con pocos árboles (págs. 422, 434)

scale/escala   relación entre las distancias en un
mapa y las distancias verdaderas en la tierra (pág. 7)

scale bar/regla de medida   en un mapa, línea con
divisiones que muestra la escala del mapa,
generalmente en millas o kilómetros (pág. 7)

scapegoat/chivo expiatorio   una persona que es
acusada erróneamente por otras (pág. 89)

scribe/escribiente   persona que registra
información por medio de la escritura (pág. 85)

secular/secular   no religioso (pág. 103)

***selva*/selva**   bosque húmedo tropical, como el de
Brasil (pág. 582)

Senate/Senado   asamblea legislativa
gubernamental; el consejo supremo de la antigua
República Romana (pág. 238)

serf/siervo   labrador que podía ser comprado y
vendido con la tierra (págs. 243, 368)

service industry/industria de servicio   negocio
que proporciona servicios a la gente en vez de
producir productos (págs. 487, 539)

shah/sha   título de los reyes que gobernaban Irán
(pág. 124)

shogun/shogún   líder militar en Japón (pág. 195)

silt/cieno   pequeñas partículas de suelo fértil
(pág. 100)

sirocco/siroco   vientos calurosos y secos que
soplan a través de Italia desde el norte de África
(pág. 298)

sisal/sisal   fibra de una planta que se usa para
hacer soga y cordel (pág. 407)

**slash-and-burn farming/agricultura por tala y
quema**   método de limpiar la tierra para el cultivo
en que se cortan y se queman los bosques (pág. 466)

Slav/eslavo   persona originaria del grupo étnico del
noreste de Europa cuya lengua nativa es un idioma
eslavo (pág. 350)

smog/smog   neblina espesa compuesta de niebla y
sustancias químicas (pág. 550)

socialism/socialismo   sistema económico en que
muchos negocios son propiedad y están dirigidos
por el gobierno (pág. 211)

social scientist/sociólogo   científico que estudia la
interacción de la gente y la sociedad (pág. 27)

sodium nitrate/nitrato de sodio   sustancia
química usada en abonos y explosivos (pág. 624)

sorghum/sorgo   cereal de tallo alto cuyas semillas
sirven de alimento y del cual se hace un jarabe para
endulzar (pág. 460)

spa/termas   balneario con manantiales de agua mineral caliente en que la gente se baña para recobrar su salud (pág. 309)

station/estación   rancho donde se crían ganado vacuno u ovejas en Australia (pág. 638)

steppe/estepa   pastos parcialmente secos que a menudo se encuentran en los bordes de un desierto (págs. 182, 317, 345)

strait/estrecho   masa de agua delgada entre dos masas de tierra (pág. 218)

strike/huelga   una negativa a trabajar, usualmente por una organización de trabajo, hasta que las demandas sean solucionadas (pág. 257)

subcontinent/subcontinente   masa de tierra grande que forma parte de un continente pero se puede diferenciar de él (pág. 144)

subsistence farm/granja de subsistencia   terreno pequeño en el cual un granjero cultiva sólo lo suficiente para alimentar a su propia familia (págs. 435, 538, 558, 617)

suburb/suburbio   comunidad pequeña en los alrededores de una ciudad (págs. 344, 524)

taiga/taiga   bosques enormes de árboles de hoja perenne en regiones subárticas (pág. 345)

tenant/arrendatario   agricultor u otra persona que paga renta a otra persona por el uso de una propiedad o terreno (pág. 243)

terraced field/terrazas   franjas, parecidas a escalones, que se cortan en la ladera de una colina para que el suelo aguante el agua y se pueda usar para la agricultura (pág. 219)

terrorism/terrorismo   uso sistemático de violencia o terror para lograr ciertas metas (pág. 102)

textiles/textiles   tela tejida (pág. 257)

theocracy/teocracia   forma de gobierno en la cual un individuo gobernaba como líder religioso tanto como rey (pág. 83)

townships/municipios   barrios abarrotados de gente en las afueras de las ciudades de Sudáfrica donde viven la mayoría de las personas que no son blancas (pág. 457)

tropics/trópicos   región entre el Trópico de Cáncer y el Trópico de Capricornio (pág. 60)

trust territory/territorio en fideicomiso   área que está bajo el control temporario de otra nación (pág. 663)

tsetse fly/mosca tsetsé   insecto cuya picada puede matar al ganado o a los seres humanos por medio de la enfermedad del sueño (pág. 425)

tsunami/tsunami   ola inmensa causada por un terremoto en el fondo del mar (pág. 192)

tundra/tundra   inmensas planicies ondulantes y sin árboles en latitudes altas con climas en que sólo varias pulgadas del suelo de la superficie se deshielan (págs. 344, 487)

typhoon/tifón   nombre para un huracán en Asia (pág. 662)

union/sindicato   organización laboral que negocia para mejorar las condiciones y pago de los trabajadores (pág. 257)

unlimited government/gobierno ilimitado   gobierno en el cual los líderes gobiernan sin ninguna restricción (pág. 31)

urban/urbano   parte de una ciudad (págs. 343, 524)

urbanization/urbanización   movimiento hacia las ciudades (pág. 34)

vaquero/vaquero   pastor de ganado vacuno (pág. 537)

vassal/vasallo   noble en la sociedad medieval quien juraba lealtad a un lord en cambio de tierra (pág. 243)

wadi/uadi   lecho de un río seco que llenan los aguaceros poco frecuentes (pág. 113)

welfare state/estado de bienestar social   estado que usa el dinero recaudado por los impuestos para mantener a personas que están enfermas, pobres, sin trabajo o jubiladas (pág. 293)

yurt/*yurt*   tienda de campaña grande y circular hecha de pieles de animales que se puede desmantelar y llevar de un lugar a otro (pág. 183)

INDEX

INDEX

INDEX

INDEX

INDEX

269; desalinization plant in, *p131;* oil in, *g129*
Kyrgyzstan, *c338,* 357, 358, *m359*
Kyushu, 192

La Paz, 622
Labrador, 494
lace, 592
lagoons, *p659,* 662, 665
Lagos, 437, 442
lakes, 60
Lalibela, 410
land, 66
land bridge, 534–535
"Land Down Under," 638
"Land of Fire," 623
"Land of Gold," 443
"Land of the Blue Sky," 182
"Land of the Free," 212
"Land of the Morning Calm," *p205*
"Land of the shaking earth," 534
land use, 64–65
landfills, 509, 530, *p530,* 531
landforms, *p53,* 53, 59. *See also* maps, physical
landlocked, 309, 621
LANDSAT images, 54
languages, 53; alphabets of, 84, 238, 312, 313, 320, 353; culture and, 278; of Europe, *m228, g319;* on Internet, 30; knotted strings for (quipu), 617; of North and Middle America, *m478;* pidgin, 662; printing press and, 246; technology and, 30; of world, *g10. See also* bilingual country; multilingual people; *specific countries*
Laos, *c139, m211,* 212; refugees from, *p421*
Laozi, 169
Latin, 299, 312, 313, 353
Latin America, 33, 249, 627; exports by, *g49;* Mexico and, 534; oil in, *g113*
latitudes, *p5,* 5–6, *m7,* 94, *m94;* climate and, 60, 506; GPS and, 54; of U. S. *vs.* Russia, *m347*
Latvia, *c231,* 306, *m307,* 308, *p308;* population of, *m318*
lava lake/flow, *p1,* 667, *p667*
Leaning Tower of Pisa, *p299,* 299
Lebanon, *c77, g77, c78,* 84, 111–112
LEGO® toy building blocks, 294
lemur, *p467*
Lenin, Vladimir, 369, *p384;* statue of, *p364, p379*
León-Castile, *m243*
Lesotho, *c395, c398, m455, m465;* people of, 457
Lesser Antilles, *p554–555*
Liberia, *c398,* 433, *m435, m436,*

442, 444, *m444, m465*
liberties, 515
library resources, 287
Libreville, 426
Libya, *c77, g77, c78,* 101–102, *m103, m465;* Israel and, 117; oil in, 102, *g129*
lichens, 275
Liechtenstein, *c231,* 290
light industry, 347
Lighthouse of Commerce, *p532–533*
Lima, 618
limited government, 31, 405, 507
Limpopo River, 460
line graph, 10, *g11, g12*
Lisbon, 297
literacy rates, 295, *g337, g479,* 559, 592
literature: of ancient Greece, 236–237; of Argentina, 596; of Australian Aborigines, 639; of Brazil, 585; of China, 172; of Denmark, 295; of French Revolution, 248; of Ireland, 281; of Japan, 194, 198, 199; of Kenya, 405; of Native Americans, 526; of Nigeria, 437; Renaissance, 246; of Russia, 352, 353, 355; storytelling as, 30, *p31,* 352; of Uganda, 413; of United States, 524–525, 526
Lithuania, *c231,* 306, *m307;* population of, *m318;* Ukraine and, 318
Little America, 674, *m675*
Little Mermaid, The (statue), *p287,* 294–295
Livingstone, David, 460
llanos, 593, 604, 606
location, 2
Locke, John, 248
Lockerbie, Scotland, 102
logging, 54, 70, *p70,* 600
Loire, 270
London, *p278,* 280, *p280*
London Eye, *p278*
longitudes, *p5,* 5–6, *m7,* 94, *m94;* GPS and, 54; time zones and, 490, *m490*
"Lost Boys of Sudan," 415–417, *p415–417,* 420
Louis XVI, 249
Louvre, *p224–225,* 285
low islands, 662, 663, 665, 673
Loyalists, of Ireland, 281
Ludwig II, 288
lumber, 558, 606
Lusaka, 460
Luther, Martin, 247
Luxembourg, *c231,* 262, 268, 286; European Union and, 324; languages of, 286
Luxor, *p80–81*

macaques, *p132*
Macau, 166
Macedonia, *c232,* 268, 269, *m307,* 313, 314; Albanian refugees in, 314; population of, *m318*
Machu Picchu, 616, *p616,* 617
Madagascar, *c398, m455,* 464, *m465,* 467; plants and animals of, 467
Madeira Islands, *p64*
Madinah, 92, 240
Madrid, 297
Maghreb, 102–104
magma, 534
Magyars, 309
Mahé, 468
mahogany, 558, 584
maize, 542
majority, 24
majority culture, 350
Makkah, 92, 93, 114, 240, 441, 446
Malabo, 427
Malacca, 217, 218
Malawi, *c398,* 459, 460, *m465*
Malawi, Lake, 460
Malay Peninsula, 210, 212, 218
Malay people, 217, 218; in Philippines, 219; in Singapore, 219
Malaysia, *c139, m204,* 209, *m211,* 216, 218; exports/imports of, *g204,* 218; rubber trees in, *p218*
Maldives, *c139,* 144, 150
Mali, *c398, m435, m436,* 440, 441, 442, *m444, m465;* Great Mosque of Djenné in, 446, *p446;* salt trade and, 450, *m451*
Mali Empire, 441, 450
Malta, *c232*
Mandela, Nelson, 400, 456, *p456,* 457
mangroves, 434, 442
Manitoba, *p472,* 489, *g499*
manor, 243
Mansa Musa, 441
manufacturing, 33, 112, 538–539
manuka, 643
Mao Zedong, 169, 170
Maori, 29, 645, *p645,* 646, 648–653
maps, *p3,* 28, 54; basics of, 4–9; colors and symbols in, 8–9; compass rose on, 6, *m6;* contour, 8, 215, *m215;* direction on, 6, *m6;* general purpose, 7–8; GIS software for, 54, 56; keys for, 6, *m6,* 8, 62, *m62;* latitude and longitude on, *p5,* 5–6, *m7 (See also* latitudes; longitudes); mental, 518; physical, 7, 8, 545, *m545;* political, 7, 8; population density, 8–9, 154, *m154;* scale in, 7, *m7,* 431, *m431;* special purpose, 8–9, 46,

INDEX

INDEX

INDEX

Text

37 From **Millennium Report, April 3, 2000** by Kofi Annan, secretary-general of the United Nations. United Nations Press Release SG/SM/7343 GA/9705, April 3, 2000. Copyright 2001 by United Nations; **194** From **Sadako and the Thousand Paper Cranes** by Eleanor Coerr. Copyright 1977. The Putnam Publishing Group; **199** From **Haiku, Volume II.** Copyright 1952 by R.H. Blyth. Reprinted by permission of Hokuseido Press; **248** From **The Scarlet Pimpernel** by Baroness Orzy. Copyright 1961 Doubleday and Company, Inc; **314** From **Zlata's Diary: A Child's Life in Sarajevo** translated with notes by Christina Pribichevich-Zoric. Translation copyright by Fixot et editions Robert Laffont, 1994 Viking, published by the Penguin Group, Penguin Books USA Inc. NY; **353** From **Nobel Lecture, 1972** by Alexander Isayevich Solzhenitsyn in *Bartlett's Familiar Quotations.* Copyright 1992 by Little, Brown and Company Inc, Boston; **355 The Grandfather and His Little Grandson** by Leo Tolstoy in *A Harvest of Russian Children's Literature,* edited by Miriam Morton. Copyright 1967 by Miriam Morton. University of California Press, Berkeley and Los Angeles, CA; **405** From **Where are those Songs** by Micere G. Mugo. Reprinted by permission of the author; **456** From **Long Walk to Freedom: The Autobiography of Nelson Mandela by Nelson Mandela.** Copyright 1994 by Nelson Rolihlahla Mandela. Little, Brown and Company; **493 A Declaration of First Nations.** Copyright 2001 by Assembly of First Nations National Indian Brotherhood. (http://www.afn.ca/About%AFN/a_declaration_of_first_nations.htm) **526 Survival This Way** by Simon J. Ortiz. Reprinted by permission of the author. **I, Too** in *Collected Poems* by Langston Hughes. Copyright 1994 by the Estate of Langston Hughes. Reprinted by permission of Alfred A. Knopf, a Division of Random House, Inc; **585** From **Botoque, Bringer of Fire** in Folklore, *Myths, and Legends, a World Perspective,* edited by Donna Rosenberg. Copyright 1977. NTC Publishing; **596** From **The Gaucho Martín Fierro** adapted from the Spanish and rendered into English verse by Walter Owen. Copyright 1936 by Farrar & Rinehart. Reprinted by permission of Henry Holt and Company, LLC; **639** From **Great Mother Snake** an Aboriginal legend. Glencoe would like to acknowledge the artists and agencies who participated in illustrating this program: Ortelius Design, Inc.

Photographs

Cover (background)3D Spherical Jigsaw™ Puzzle World Globe made in USA by Buffalo Games, Inc., (l)PhotoDisc, (c)William Waterfall/Pacific Stock, (r)Jodi Cobb/National Geographic Image Collection; **iv–v** Ken Stimpson/Panoramic Images; **vi–vii** PhotoDisc; **viii** Art Wolfe; **ix** Jeff Schultz/Alaska Stock Images; **x** Owen Franken/CORBIS; **xi** (l)KS Studios, (r)Thomas E. Franklin/Record/Saba/CORBIS; **1** PhotoDisc; **2** (l)Todd Gipstein/CORBIS, (tr)Craig Aurness/CORBIS, (br)Craig Lovell/CORBIS; **3** (l)Peter Turnley/CORBIS, (tr)Robert Caputo/Aurora, (br)Roger Ressmeyer/CORBIS; **16** Aaron Haupt; **18** (l)Robert Landau/CORBIS, (r)Digital Stock; **18–19** S. Purdy Matthews/Stone; **20–21** Anthony Cassidy/Stone; **22** AP/Wide World Photos; **23** (t)Private Collection/The Bridgeman Art Library, London/New York, (cl)Hulton/Archive/Getty Images, (cr)Glencoe photo, (bl)NASA, (br)Glencoe photo; **24** Richard Hutchings/PhotoEdit; **26** Bettmann/CORBIS; **27** Kenneth Garrett/National Geographic Image Collection **28** (l)Black Star/National Geographic Image Collection, (r)Steven L. Raymer/National Geographic Image Collection **29** (l)Jeff Schultz/Alaska Stock Images, (tr)Will & Deni McIntyre/Stone, (br)Yellow Dog Productions/The Image Bank; **31** Hilarie Kavanagh/Stone; **34** Gerd Ludwig/National Geographic Image Collection; **35–37** AFP/CORBIS; **50–51** Norman Kent Productions/oi2.com; **53** (l)Richard T. Nowitz/National Geographic Image Collection, (r)Yann Arthus-Bertrand/CORBIS; **54** Doug Martin; **56** Courtesy Environmental Systems Research Institute, Inc.; **57** Natalie Fobes/National Geographic Image Collection; **59** (l)Michael K. Nichols/National Geographic Image Collection, (r)Wolfgang Kaehler/CORBIS; **60** Alberto Garcia/Saba; **63** Bryan & Cherry Alexander/National Geographic Society; **64** Robert Harding/CORBIS; **65** Jodi Cobb/National Geographic Image Collection; **67** George Grall/National Geographic Image Collection; **70** (background)Michael Nichols/National Geographic Society, (l)Gerry Ellis/ENP Images, (r)Jose Azel/Aurora/PictureQuest; **71** (bl)Art Wolfe/Stone, (others)Lisa Hoffner/Wildeye Photography; **72** (l)Hugh Sitton/Stone, (r)David Coulson; **72–73** X. Richer/Hoaqui/Photo Researchers; **79** AP/Wide World Photos; **80–81** Wolfgang Kaehler; **82** Kenneth Garrett/National Geographic Image Collection; **85** (l)Gianni Dagli Orti/CORBIS, (c)Wolfgang Kaehler, (r)Charles & Josette Lenars/CORBIS; **87** Steve Vidler/SuperStock; **88** ASAP Ltd./Index Stock; **90** (l)A. Ramey/Woodfin Camp & Associates, (c)Alan Oddie/PhotoEdit, (r)R & S Michaud/Woodfin Camp & Associates; **91** (t)CMCD/PhotoDisc, (c)C Squared Studios/PhotoDisc, (b)Robert Harding/CORBIS; **95** Kenneth Garrett/National Geographic Image Collection; **98–99** Wolfgang Kaehler/CORBIS; **100** George Steinmetz/National Geographic Image Collection; **105** Patrick Ward/CORBIS; **106** James L. Stanfield/National Geographic Image Collection; **107** Richard T. Nowitz/CORBIS; **109** Arthur Thevenart/CORBIS; **110** Dean Conger/National Geographic Image Collection; **111** Brian Haimer/PhotoEdit; **112** Annie Griffiths Belt/National Geographic Image Collection; **122** Charles & Josette Lenars/CORBIS; **123** AP/Wide World Photos; **124** (l)Alexandra Avakian/National Geographic Image Collection, (r)Jon Spaull/CORBIS; **126** National Geographic Society; **130** Ed Kashi; **130–131** James L. Stanfield; **131** (r)courtesy Sandra Postel, (others) Ed Kashi, ; **132** (l)Tim Davis/Stone, (r)David Sutherland/Stone; **132–133** Waranun Chutchawan-Tipakorn; **140** Steve Raymer/CORBIS; **141** Keren Su/Stone; **142–143** Tibor Bognar/CORBIS Stock Market; **144** George F. Mobley/National Geographic Image Collection; **145** Steve McCurry/National Geographic Image Collection; **146** (l)SuperStock, (c)Bettmann/CORBIS, (r)Hulton/Archive/Getty Images; **147** Rudi Von Briel/PhotoEdit; **148** Oriental Museum, Durham University, UK/The Bridgeman Art Library, London/New York; **149** Brian Vikander/CORBIS; **150** Steve McCurry/National Geographic Image Collection; **151** Brian Vikander/CORBIS; **153** Paul Chesley/Stone; **157** Robert Holmes/CORBIS; **160–161** Christopher Arnesen/Stone; **162** Keren Su/CORBIS; **165** (l)Michele Burgess/CORBIS Stock Market, (r)Kevin R. Morris/CORBIS; **167 168** AFP/CORBIS; **171** Joseph Sohm/ChromoSohm/CORBIS; **180** How-Man Wong/CORBIS; **181** Marc Garanger/CORBIS; **182** Nik Wheeler/CORBIS; **184** AFP/CORBIS; **188** Cary Wolinsky/Stock Boston; **189** Keren Su/CORBIS; **190–191** Dallas and John Heaton/CORBIS; **192** Reuters NewMedia/CORBIS; **194** Steve Cole/PhotoDisc; **195** Karen Kasmauski/Matrix; **196** Roger Ressmeyer/CORBIS; **199** Asian Art & Archaeology/CORBIS; **200** Carmen Redondo/CORBIS; **201** (l)Nathan Benn/CORBIS, (r)Wolfgang Kaehler/CORBIS; **202** Chris Lisle/CORBIS; **205** Neil Beer/CORBIS; **208–209** Bob Krist/Stone; **210 212** Paul Chesley/National Geographic Image Collection; **213** Kevin R. Morris/CORBIS; **216** Roger Ressmeyer/CORBIS; **218** Earl & Nazima Kowall/CORBIS; **220** David Hanson/Stone; **221** AP/Wide World Photos; **224** (l)IFA-Bilderteam-Travel/Bruce Coleman, Inc., (r)Robert Everts/Stone; **224–225** SuperStock; **233** Mary Kate Denny/PhotoEdit; **234–235** Christian Sarramon/CORBIS; **236** Ira Block/National Geographic Image Collection; **237** Vanni Archive/CORBIS; **238** Richard T. Nowitz/National Geographic Image Collection; **239** Museo della Civilta Romana, Rome/Art Resource; **241** Richard List/CORBIS; **242** SuperStock; **244** Archivo Iconografico, SA/CORBIS; **245** David Lees/CORBIS; **246** Sistine Chapel, Vatican, Rome/Fratelli Alinari/SuperStock; **247** The Art Archive; **248** Matt Meadows; **250** (l)Bettmann/CORBIS, (r)Gianni Dagli Orti/CORBIS; **254–255** Philippa Lewis/CORBIS; **256** Dagli Orti/Musee National d'Art Moderne, Paris/The Art Archive; **257** North Wind Picture Archives; **258** (l)CORBIS, (r)Craig Aurness/CORBIS; **259** Hulton-Deutsch Collection/CORBIS; **260** V. Yudin/Sovfoto/Eastfoto/PictureQuest; **261** Owen Franken/CORBIS; **264** (l)Culver Pictures/PictureQuest, (r)Bettmann/CORBIS; **267** James Stanfield/National Geographic Image Collection; **268** Shone/Sipa Press; **269** AFP/CORBIS; **274** (l)Robert Winslow, (r)Johan Elzenga/Stone; **274–275** Oliver Strewe/Stone; **275** (l)Martin Bond/Science Photo Library/Photo Researchers, (r)Mike Lewis, Northamptonshire Grammar School, UK; **276–277** Photowood/CORBIS Stock Market; **278** London Aerial Photo Library/CORBIS; **280** (l)Tim Thompson/CORBIS, (r)Stephen Beer/Stone; **282** Adam Woolfit/CORBIS; **283** James L. Stanfield/National Geographic Image Collection; **284** Mandalay/Rolph Konow/Kobal Collection; **285** Michael John Kielty/CORBIS; **287** SuperStock; **288** Ric Ergenbright/CORBIS; **289** Owen Franken/CORBIS; **290** Sisse Brimberg/National Geographic Image Collection; **291** AFP/CORBIS; **292** Tomasz Tomaszewski/National Geographic Image Collection; **293** (l)Buddy May/CORBIS, (r)Richard S. Durrance/National Geographic Image Collection; **294** Sisse Brimberg/National Geographic Image Collection; **296** David Cumming/Eye Ubiquitous/CORBIS; **297** Gerard Degeorge/CORBIS; **298** (t)Glen Allison/PhotoDisc, (b)Vittoriano Rastelli/CORBIS; **299** Louis O. Mazzatenta/National Geographic Image Collection; **300** Gail Mooney/CORBIS;

304–305 Foto World/The Image Bank; 306 Priit J. Vesilind/National Geographic Image Collection; 308 Steven L. Raymer/National Geographic Image Collection; 309 Owen Franken/CORBIS; 311 Steve Raymer/CORBIS; 312 Peter Wilson/CORBIS; 313 Francoise de Mulder/CORBIS; 314 Aaron Haupt; 316 (l)Kelly-Mooney Photography/CORBIS, (r)Craig Aurness/CORBIS; 317 Gerd Ludwig/National Geographic Image Collection; 328 James Stanfield/National Geographic Image Collection; 329 Tomasz Tomaszewski/National Geographic Image Collection; 332 (l)Bruce Dale, (r)Alain Le Garsmeur/Stone; 332–333 Marc Moritsch/National Geographic Society; 338 Paul Harris/Stone; 339 KS Studios; 340–341 David & Peter Turnley/CORBIS; 342 Tom Brakefield/CORBIS; 343 (l)Marc Garanger/CORBIS, (r)David Turnley/CORBIS; 344 (l)Andre Gallant/The Image Bank, (c)Wolfgang Kaehler/CORBIS, (r)A. Solomonov/Sovfoto/Easfoto/PictureQuest; 349 Doug Martin; 350 Gerd Ludwig/National Geographic Image Collection; 352 Bob Krist/CORBIS; 353 Richard Howard/Black Star/PictureQuest; 354 David & Peter Turnley/CORBIS; 355 Roger-Viollet/Musee du Petit Palais, Paris/Bridgeman Art Library, London/New York; 356 Wolfgang Kaehler; 357 Gerd Ludwig/National Geographic Image Collection; 358 Ron Rovtar; 361 Gerd Ludwig/National Geographic Image Collection; 364–365 John Lamb/Stone; 366 Sisse Brimberg/National Geographic Image Collection; 367 Stock Montage; 369 Scala/Art Resource; 370 Kremlin Museums, Moscow, Russia/Bridgeman Art Library; 371 David & Peter Turnley/CORBIS; 373 Dean Conger/CORBIS; 374 Peter Turnley/CORBIS; 375 Reuters NewMedia/CORBIS; 384 Farrell Grehan/CORBIS; 385 Jay Dickman; 388 Giraudon/Art Resource; 389 Sovfoto/Eastfoto/PictureQuest; 390 (l)Hugh Sitton/Stone, (r)Jacques Jangoux/Stone; 390–391 Manoj Shah/Stone; 400 David & Peter Turnley/CORBIS; 401 Nicholas Parfitt/Stone; 402–403 W. Perry Conway/CORBIS; 404 Carol Beckwith & Angela Fisher/Robert Estall Photo Agency; 405 Courtesy of the author; 406 The Purcell Team/CORBIS; 408 Frank Lane Picture Agency/CORBIS; 409 Darrell Gulin/CORBIS; 410 Dave Bartruff/CORBIS; 412 AP/Wide World Photos; 413 Chinch Gryniewicz/Ecoscene/CORBIS; 422 AP/Wide World Photos; 425 Daniel Laine/CORBIS; 426 (l)Ann and Carl Purcell/Words & Pictures/PictureQuest, (r)James A. Sugar/Black Star/PictureQuest; 428 David Turnley/CORBIS; 432–433 Paul Almasy/CORBIS; 434 436–437 AP/Wide World Photos; 437 Robert W. Moore/National Geographic Image Collection; 439 (l)Bowers Museum of Cultural Art/CORBIS, (r)Davis Factor/CORBIS; 440 Steve McCurry/Magnum; 441 Carol Beckwith & Angela Fisher/Robert Estall Photo Agency; 443 Robert W. Moore/National Geographic Image Collection; 446 Nik Wheeler/CORBIS; 447 AP/Wide World Photos; 450 Michael A. Hampshire; 451 James L. Stanfield; 452–453 Pictor; 454 Wolfgang Kaehler; 456 Turnley Collection/CORBIS; 459 Walter Edwards/National Geographic Image Collection; 461 Jack Vartoogian; 463 (l)Chris Johns/National Geographic Image Collection, (r)Aaron Haupt; 464 Des & Jen Bartlett/National Geographic Image Collection; 466 AP/Wide World Photos; 467 Art Wolfe/Stone; 472 (l)Sisse Brimberg, (r)Michael Lewis; 472–473 David R. Stoecklein; 482 Aaron Haupt; 483 Kenneth Garrett/National Geographic Society; 484–485 SuperStock; 486 Raymond K. Gehman/National Geographic Image Collection; 488 David A. Harvey/National Geographic Image Collection; 491 Marie-Louise Brimberg/National Geographic Image Collection; 492 Michael Evan Sewell/Visual Pursuit; 493 Index Stock Imagery/PictureQuest; 494 Marie-Louise Brimberg/National Geographic Image Collection; 496 Reuters NewMedia/CORBIS; 497 Richard T. Nowitz/National Geographic Image Collection; 500–501 Mitchell Funk/The Image Bank; 502 David Hiser/National Geographic Image Collection; 504 (l)Steven L. Raymer/National Geographic Image Collection, (r)Vincent Musl/National Geographic Image Collection; 505 Roy Corral/Stone; 507 Karen Kasmauski/Matrix; 508 Joseph Sohm/ChromoSohm/CORBIS; 509 Joel Satore/National Geographic Image Collection; 519 Mug Shots/CORBIS Stock Market; 521 The White House Historical Association; 523 526 CORBIS; 530 Ray Pfortner/Peter Arnold, Inc.; 531 (background)Michael Mathers/Peter Arnold Inc., (l)David Young-Wolff/Stone, (r)David Schmidt; 532–533 Randy Faris/CORBIS; 534 Bettmann/CORBIS; 536 Nik Wheeler/CORBIS; 538 Joel Satore/National Geographic Image Collection; 541 Tomasz Tomaszewski/National Geographic Image Collection; 543 Vladimir Pcholkin/FPG; 545 James L. Amos/National Geographic Image Collection; 546 David A. Harvey/National Geographic Image Collection; 547 Nik Wheeler/CORBIS; 548 Tomas Tomaszewski/National Geographic Image Collection; 554–555 Sylvain Grandadam/Stone; 556 Art Wolfe; 559 (l)Vincent Musl/National Geographic Image Collection, (r)Jan Butchofsky-Houser/CORBIS; 560 Michael S. Yamashita/CORBIS; 563 Jonathan Blair/National Geographic Image Collection; 565 (l)Tony Arruza/CORBIS, (r)Michael K. Nichols/National Geographic Image Collection; 566 Robert A. Tyrrell; 569 George Mobley/National Geographic Image Collection; 572 (l)David Levy/Stone, (r)William J. Herbert/Stone; 572–573 Ecoscene/CORBIS; 579 Michael K. Nichols/National Geographic Image Collection; 580 Layne Kennedy/CORBIS; 582 Alex Webb/Magnum; 585 Kennan Ward/CORBIS Stock Market; 586 (l)Jim Zuckerman/CORBIS, (r)Yann Arthus-Bertrand/CORBIS; 588 Jeremy Horner/CORBIS; 589 Louis O. Mazzatenta/National Geographic Image Collection; 591 (l)Jack Fields/CORBIS, (r)Louis O. Mazzatenta/National Geographic Image Collection; 592 Robert Caputo/Aurora & Quanta Productions; 593 Pablo Corral V/CORBIS; 594 Jacques Jangoux/Stone; 596 Kit Houghton Photography/CORBIS; 597 Robert van der Hilst/CORBIS; 600 (br)William Albert Allard, (others) Michael & Patricia Fogden; 601 (l)Michael Doolittle, (r)Marc Van Roosmalen Conservation International; 602–603 Paul Harris/Stone; 604 Fred Ward/Black Star; 606 Richard S. Durrance/National Geographic Image Collection; 616 Frank & Helen Schreider/National Geographic Image Collection; 618 Tiziana and Gianni Baldizzone/CORBIS; 620 Art Wolfe; 621 Maria Stenzel/National Geographic Image Collection; 622 (l)Richard T. Nowitz/National Geographic Image Collection, (r)James L. Stanfield/National Geographic Image Collection; 625 William A. Allard/National Geographic Image Collection; 628 (l)David Madison/Stone, (r)David Hiser/Stone; 628–629 Oliver Strewe/Stone; 635 Elaine Shay; 636–637 Larry Williams/CORBIS Stock Market; 638 R. Ian Productions P Lloyd/National Geographic Image Collection; 639 640 Penny Tweedie/CORBIS; 642 (t)Australian Picture Library/CORBIS, (bl)Charles Philip Cangialosi/CORBIS, (br)Earl & Nazima Kowall/CORBIS; 643 Kevin Fleming/CORBIS; 645 Neil Rabinowitz/CORBIS; 654 Kevin Fleming/CORBIS; 655 Australian Picture Library/CORBIS; 658–659 Dana Edmunds/FPG; 660 Hal Beral/CORBIS; 662 (l)Ben Simmons/CORBIS Stock Market, (r)SuperStock; 665 David Doubllet/National Geographic Image Collection; 666 Wolfgang Kaehler/CORBIS; 667 (l)SuperStock, (r)David Madison/Stone; 668 Underwood & Underwood/CORBIS; 670 Galen Rowell/CORBIS; 671 AP/Wide World Photos; 674 Byrd Antarctic Expedition; 675 Andrew H. Brown; 676 PhotoDisc; 684 Sidney Harris.

ACKNOWLEDGMENTS